ISBN 978-0-484-61304-0
PIBN 10630840

Transactions of The Institution of Mining Engineers

Edited by the Secretary

VOL. XXV.

1902–1903.

ARTHUR MARSHALL CHAMBERS,

PRESIDENT OF THE INSTITUTION OF MINING ENGINEERS, 1897-1898.

0

TRANSACTIONS

OF

THE INSTITUTION

OF

MINING ENGINEERS.

VOL. XXV.—1902-1903.

EDITED BY M. WALTON BROWN, SECRETARY.

NEWCASTLE-UPON-TYNE: PUBLISHED BY THE INSTITUTION.

PRINTED BY ANDREW REID & CO., LIMITED, NEWCASTLE-UPON-TYNE.
1904.

ADVERTIZEMENT.

CONTENTS OF VOL. XXV.

GENERAL MEETINGS.

THE INSTITUTION OF MINING ENGINEERS.

89807

CONTENTS.

THE MINING INSTITUTE OF SCOTLAND.

THE NORTH OF ENGLAND INSTITUTE OF MINING AND MECHANICAL ENGINEERS.

APPENDICES.

APPENDICES.—*Continued.*

CONTENTS.

APPENDICES.—*Continued.*

LIST OF PLATES :—

ANY PUBLICATION OF A FEDERATED INSTITUTE MAY BE PLACED AT THE END OF THE VOLUME, *i.e.*, "ANNUAL REPORT," "LIST OF MEMBERS," ETC.

THE INSTITUTION OF MINING ENGINEERS.

OFFICERS, 1902-1903.

Past-Presidents *(ex-officio)*.

Mr. WILLIAM NICHOLAS ATKINSON, H.M. Inspector of Mines, Barlaston, Stoke-upon-Trent.

Mr. JAMES STEDMAN DIXON, Fairleigh, Bothwell, N.B.

Mr. GEORGE LEWIS, Albert Street, Derby.

Sir WILLIAM THOMAS LEWIS, Bart., Mardy, Aberdare.

Mr. JOHN ALFRED LONGDEN, Stanton-by-Dale, near Nottingham.

Mr. GEORGE ARTHUR MITCHELL, 5, West Regent Street, Glasgow.

Mr. H. C. PEAKE, Walsall Wood Colliery, Walsall.

Mr. ARTHUR SOPWITH, Cannock Chase Collieries, near Walsall.

Sir LINDSAY WOOD, Bart., The Hermitage, Chester-le-Street.

President.

SIR LINDSAY WOOD, BART., The Hermitage, Chester-le-Street.

Vice-Presidents.

* Deceased.

*Mr. HENRY AITKEN, Darroch, near Falkirk, N.B.

Sir LOWTHIAN BELL, Bart., Rounton Grange, Northallerton.

Mr. G. ELMSLEY COKE, 65, Station Street, Nottingham.

Mr. JOHN DAGLISH, Rothley Crag, Cambo, R.S.O., Northumberland.

Mr. MAURICE DEACON, Whittington House, near Chesterfield.

Mr. JAMES T. FORGIE, Mosspark, Bothwell, N.B.

Mr. JOHN GERRARD, H.M. Inspector of Mines, Worsley, Manchester.

Mr. A. MAYON HENSHAW, Talk-o'-th'-Hill Colliery, Talke, Stoke-upon-Trent.

Mr. GEORGE MAY, The Harton Collieries, South Shields.

Mr. HORACE BROUGHTON NASH, 23, Victoria Road, Barnsley.

Mr. JOHN BELL SIMPSON, Bradley Hall, Wylam-upon-Tyne.

Mr. JOHN GEORGE WEEKS, Bedlington, R.S.O., Northumberland.

Mr. ROBERT SUMMERSIDE WILLIAMSON, Cannock Wood House, Hednesford Staffordshire.

Councillors.

* Deceased.

Mr. WILLIAM ARMSTRONG, Wingate, Co. Durham.

Mr. FREDERICK ROBERT ATKINSON, Shottle Hall, near Derby.

Mr. JAMES BARROWMAN, Staneacre, Hamilton, N.B.

Mr. JOHN BATEY, St. Edmunds, Coleford, Bath.

Mr. THOMAS W. BENSON, 24, Grey Street, Newcastle-upon-Tyne.

Mr. GEORGE JONATHAN BINNS, Duffield House, Duffield, Derby.

Mr. BENNETT HOOPER BROUGH, Cranleigh House, near Addlestone, Surrey.

Mr. MARTIN WALTON BROWN, 10, Lambton Road, Newcastle-upon-Tyne.

Mr. JAMES COPE CADMAN, Madeley, Newcastle, Staffordshire.

Mr. WILLIAM HENRY CHAMBERS, Conisborough, Rotherham.

Mr. ARTHUR GEORGE CHARLETON, 5, Avonmore Road, Kensington, London, W.

SIR DAVID DALE, Bart., West Lodge, Darlington.

Mr. ROBERT WILSON DRON, 55, West Regent Street, Glasgow.

Mr. THOMAS EMERSON FORSTER, 3, Eldon Square, Newcastle-upon-Tyne.

Mr. WILLIAM GALLOWAY, Cardiff.

Mr. W. E. GARFORTH, Snydale Hall, near Pontefract.
Mr. REGINALD GUTHRIE, Neville Hall, Newcastle-upon-Tyne.
Mr. JAMES HAMILTON, 208, St. Vincent Street, Glasgow.
*Mr. JAMES HASTIE, 4, Athole Gardens, Uddingston, N.B.
*Mr. W. J. HAYWARD, West Bromwich.
Mr. HENRY RICHARDSON HEWITT, H.M. Inspector of Mines, Breedon Hill Road, Derby.
Mr. JAMES A. HOOD, Rosewell, Mid Lothian, N.B.
Mr. W. R. M. JACKSON, Clay Cross Hall, Chesterfield.
Mr. PHILIP KIRKUP, Leafield House, Birtley, R.S.O., Co. Durham.
Mr. CHARLES CATTERALL LEACH, Seghill Colliery, Northumberland.
Mr. GEORGE ALFRED LEWIS, Albert Street, Derby.
Mr. EDWARD LINDLEY, Eastwood, Nottingham.
Mr. HENRY LOUIS, 11, Windsor Terrace, Newcastle-upon-Tyne.
Mr. ROBERT MCLAREN, H.M. Inspector of Mines, 19, Morningside Park, Edinburgh.
Mr. T. W. H. MITCHELL, Mining Offices, Regent Street, Barnsley.
Mr. ROBERT THOMAS MOORE, 142, St. Vincent Street, Glasgow.
Mr. JOHN MORISON, Cramlington House, Northumberland.
Mr. DAVID M. MOWAT, Summerlee Iron Works, Coatbridge, N.B.
Mr. R. D. MUNRO, 111, Union Street, Glasgow.
Mr. JOHN NEVIN, Littlemoor House, Mirfield.
Mr. MATTHEW WILLIAM PARRINGTON, Wearmouth Colliery, Sunderland.
Mr. WILLIAM HENRY PICKERING, H.M. Inspector of Mines, Lawn House, Doncaster.
Mr. RICHARD AUGUSTINE STUDDERT REDMAYNE, The University, Birmingham.
Mr. ARTHUR ROBERT SAWYER, P.O. Box 2202, Johannesburg, Transvaal.
Mr. ALEXANDER SMITH, 3, Newhall Street, Birmingham.
*Mr. JOHN STRICK, Bar Hill, Madeley, Staffordshire.
Mr. EDWARD B. WAIN, Whitfield Collieries, Norton-in-the-Moors, Stoke-upon-Trent.
Mr. GEORGE BLAKE WALKER, Wharncliffe Silkstone Colliery, Barnsley.
Mr. WILLIAM OUTTERSON WOOD, South Hetton, Sunderland.

Auditors.

Messrs. JOHN G. BENSON AND SONS, Newcastle-upon-Tyne.

Treasurers.

Messrs. LAMBTON AND COMPANY, The Bank, Newcastle-upon-Tyne.

Secretary.

Mr. MARTIN WALTON BROWN, Neville Hall, Newcastle-upon-Tyne.

THE INSTITUTION OF MINING ENGINEERS.

FOUNDED JULY 1ST, 1889.

BYE-LAWS

As revised at Council Meeting held on May 29th, 1902.

I.—CONSTITUTION.

1.—The Institution of Mining Engineers shall consist of all or any of the societies interested in the advancement of mining, metallurgy, engineering and their allied industries, who shall from time to time join together and adhere to the Bye-Laws.

2.—The Institution shall have for its objects—

(a) The advancement and encouragement of the sciences of mining, metallurgy, engineering, and their allied industries.

(b) The interchange of opinions, by the reading of communications from members and others, and by discussions at general meetings, upon improvements in mining, metallurgy, engineering, and their allied industries.

(c) The publication of original communications, discussions, and other papers connected with the objects of the Institution.

(d) The purchase and disposal of real and personal property for such objects.

(e) The performance of all things connected with or leading to the purpose of such objects.

3.—The offices of the Institution shall be in Newcastle-upon-Tyne, or such other place as shall be from time to time determined by resolution of the Council.

4.—The year of the Institution shall end on July 31st in every year.

5.—The affairs and business of the Institution shall be managed and controlled by the Council.

II.—MEMBERSHIP.

6.—The original adherents or founders are as follows:—

(a) Chesterfield and Midland Counties Institution of Engineers, Chesterfield.

(b) Midland Institute of Mining, Civil and Mechanical Engineers, Barnsley.

(c) North of England Institute of Mining and Mechanical Engineers, Newcastle-upon-Tyne.

(d) South Staffordshire and East Worcestershire Institute of Mining Engineers, Birmingham.

7.—Written applications from societies to enter the Institution shall be made to the Council, by the President of the applying society, who shall furnish any information that may be desired by the Council.

8.—A.—If desired by the Council, any of the Federated Institutes shall revise their Bye-Laws, in order that their members shall consist of Ordinary Members, Associate Members, and Honorary Members, with Associates and Students, and section B following shall be a model Bye-Law to be adopted by any society when so desired by the Council.

B.—"The members shall consist of Ordinary Members, Associate Members and Honorary Members, with Associates and Students :—

(a) Each Ordinary Member shall be more than twenty-three years of age, have been regularly educated as a mining, metallurgical, or mechanical engineer, or in some other branch of engineering, according to the usual routine of pupilage, and have had subsequent employment for at least two years in some responsible situation as an engineer ; or if he has not undergone the usual routine of pupilage, he must have been employed or have practised as an engineer for at least five years.

(*b*) Each Associate Member shall be a person connected with or interested in mining, metallurgy, or engineering, and not practising as a mining, metallurgical, or mechanical engineer, or some other branch of engineering.

(*c*) Each Honorary Member shall be a person who has distinguished himself by his literary or scientific attainments, or who may have made important communications to any of the Federated Institutes.

(*d*) Associates shall be persons acting as under-viewers, under-managers, or in other subordinate positions in mines or metallurgical works, or employed in analogous positions in other branches of engineering.

(*e*) Students shall be persons who are qualifying themselves for the profession of mining, metallurgical, or mechanical engineering, or other branch of engineering, and such persons may continue Students until they attain the age of twenty-five years."

9.—The Ordinary Members, Associate Members and Honorary Members, Associates and Students shall have notice of, and the privilege of attending, the ordinary and annual general meetings, and shall receive all publications of the Institution. They may also have access to, and take part in, the general meetings of any of the Federated Institutes.

10.—The members of any Federated Institute, whose payments to the Institution are in arrear, shall not receive the publications and other privileges of the Institution.

11.—After explanations have been asked by the President from any Federated Institute, whose payments are in arrear, and have not been paid within one month after written application by the Secretary, the Council may decide upon its suspension or expulsion from the Institution ; but such suspension or expulsion shall only be decided at a meeting attended by at least two-thirds of the members of the Council by a majority of three-fourths of the members present.

III.—SUBSCRIPTIONS.

12.—Each of the Federated Institutes shall pay fifteen shillings per annum for each Ordinary Member, Associate Member, Honorary Member, Associate and Student, or such other sum, and in such instalment or instalments as may be determined from time to time by resolution or resolutions of the Council. Persons joining any of the Federated Institutes during the financial year of the Institution shall be entitled to all publications issued for that year, after his election is notified to the Secretary, and the instalment or instalments due on his behalf have been paid.

IV.—ELECTION OF OFFICERS AND COUNCIL.

13.—The officers of the Institution, other than the Secretary and Treasurer, shall consist of Councillors elected annually prior to August in each year, by and out of the Ordinary Members and Associate Members of each Federated Institute, in the proportion of one Councillor per forty Ordinary Members or Associate Members thereof ; of Vice-Presidents elected by and from the Council at their first meeting in each year on behalf of each Institute, in the proportion of one Vice-President per two hundred Ordinary Members or Associate Members thereof ; and of a President elected by and from the Council at their first meeting in each year ; who, with the Local Secretaries of each Federated Institute and the Secretary and Treasurer, shall form the Council. All Presidents on retiring from that office shall be *ex-officio* Vice-Presidents so long as they continue Ordinary Members or Associate Members of any of the Federated Institutes.

14.—In case of the decease, expulsion, or resignation of any officer or officers, the Council may, if they deem it requisite, fill up the vacant office or offices at their next meeting.

V.—DUTIES OF OFFICERS AND COUNCIL.

15.—The Council shall represent the Institution and shall act in its name, and shall make such calls upon the Federated Institutes as they may deem necessary, and shall transact all business and examine accounts, authorise payments and may invest or use the funds in such manner as they may from time to time think fit, in accordance with the objects and Bye-Laws of the Institution.

16.—The Council shall decide the question of the admission of any society, and may decree the suspension or expulsion of any Federated Institute for non-payment of subscriptions.

17.—The Council shall decide upon the publication of any communications.

18.—There shall be three ordinary meetings of the Council in each year, on the same day as, but prior to, the ordinary or annual general meetings of the members.

19.—A special meeting of the Council shall be called whenever the President may think fit, or upon a requisition to the Secretary signed by ten or more of its members, or by the President of any of the Federated Institutes. The business transacted at a special meeting of the Council shall be confined to that specified in the notice convening it.

20.—The meetings of the Council shall be called by circular letter, issued to all the members at least seven days previously, accompanied by an agenda-paper, stating the nature of the business to be transacted.

21.—The order in which business shall be taken at the ordinary and annual general meetings may be, from time to time, decided by the Council.

22.—The Council may communicate with the Government in cases of contemplated or existing legislation, of a character affecting the interests of mining, metallurgy, engineering, or their allied industries.

23.—The Council may appoint Committees, consisting of members of the Institution, for the purpose of transacting any particular business, or of investigating any specific subject connected with the objects of the Institution.

24.—A Committee shall not have power or control over the funds of the Institution, beyond the amount voted for its use by the Council.

25.—Committees shall report to the Council, who shall act thereon and make use thereof as they may elect.

26.—The President shall take the chair at all meetings of the Institution, the Council, and Committees at which he may be present.

27.—In the absence of the President, it shall be the duty of the senior Vice-President present to preside at the meetings of the Institution. In case of the absence of the President and of all the Vice-Presidents, the meeting may elect any member of Council, or in case of their absence any Ordinary Member or Associate Member to take the chair at the meeting.

28.—At meetings of the Council six shall be a quorum.

29.—Every question shall be decided at the meetings of the Council by the votes of the majority of the members present. In case of equal voting, the President, or other member presiding in his absence, shall have a casting vote. Upon the request of two members, the vote upon any question shall be by ballot.

30.—The Secretary shall be appointed by and shall act under the direction and control of the Council. The duties and salary of the Secretary shall be fixed and varied from time to time at the will of the Council.

31.—The Secretary shall summon and attend all meetings of the Council, and the ordinary and annual general meetings of the Institution, and shall record the proceedings in the minute book. He shall direct the administrative and scientific publications of the Institution. He shall have charge of and conduct all correspondence relative to the business and proceedings of the Institution, and of all committees where necessary, and shall prepare and issue all circulars to the members.

32.—One and the same person may hold the office of Secretary and Treasurer.

33.—The Treasurer shall be appointed annually by the Council at their first meeting in each year. The income of the Institution shall be received by him, and shall be paid into Messrs. Lambton & Co.'s bank at Newcastle-upon-Tyne, or such other bank as may be determined from time to time by the Council.

34.—The Treasurer shall make all payments on behalf of the Institution, by cheques signed by two members of Council, the Treasurer and the Secretary after payments have been sanctioned by Council.

35.—The surplus funds may, after resolution of the Council, be invested in Government securities, in railway and other debenture shares such as are allowed for investment by trustees, in the purchase of land, or in the purchase, erection. alteration, or furnishing of buildings for the use of the Institution. All investments shall be made in the names of Trustees appointed by the Council.

36.—The accounts of the Treasurer and the financial statement of the Council shall be audited and examined by a chartered accountant, appointed by the Council at their first meeting in each year. The accountants' charges shall be paid out of the funds of the Institution.

37.—The minutes of the Council's proceedings shall at all times be open to the inspection of the Ordinary Members and Associate Members.

VI.—General Meetings.

38.—An ordinary general meeting shall be held in February, May and September, unless otherwise determined by the Council; and the ordinary general meeting in the month of September shall be the annual general meeting at which a report of the proceedings, and an abstract of the accounts of the previous year ending July 31st, shall be presented by the Council. The ordinary general meeting in the month of May shall be held in London, at which the President may deliver an address.

39.—Invitations may be sent by the Secretary to any person whose presence at discussions shall be thought desirable by the Council, and persons so invited shall be permitted to read papers and take part in the proceedings and discussions.

40.—Discussion may be invited on any paper published by the Institution, at meetings of any of the Federated Institutes, at which the writer of the paper may be invited to attend. Such discussion, however, shall in all cases be submitted to the writer of the paper before publication, and he may append a reply at the end of the discussion.

VII.—Publications.

41.—The publications may comprise:—
(a) Papers upon the working of mines, metallurgy, engineering, railways and the various allied industries.
(b) Papers on the management of industrial operations.
(c) Abstracts of foreign papers upon similar subjects.
(d) An abstract of the patents relating to mining and metallurgy, etc.
(e) Notes of questions of law concerning mines, manufactures, railways, etc.

42.—Each paper (with complete drawings, if any, to scale), to be read at any meeting of the Institution or of any of the Federated Institutes, shall be placed in the hands of the Secretary at least fourteen days before the date of the meeting at which the paper is to be read, and shall, subject to the approval of the Council, be printed, together with any discussion or remarks thereon.

43.—The Council may accept communications from persons who are not members of the Institution and allow them to be read at the ordinary or annual general meetings.

44.—No paper which has already been published (except as provided for in Bye-Law 41) shall appear in the publications of the Institution.

45.—A paper in course of publication cannot be withdrawn by the writer.

46.—Proofs of all papers and reports of discussions forwarded to any person for revision must be returned to the Secretary within seven days from the date of their receipt, otherwise they will be considered correct and be printed off.

47.—The copyright of all papers accepted for printing by the Council shall become vested in the Institution, and such communications shall not be published for sale or otherwise without the written permission of the Council.

48.—Twenty copies of each paper and the accompanying discussion shall be presented to the writer free of cost. He may also obtain additional copies upon payment of the cost to the Secretary, by an application attached to his paper. These copies must be unaltered copies of the paper as appearing in the publication of the Institution, and the cover shall state that it is an " Excerpt from the Transactions of The Institution of Mining Engineers."

49.—The Federated Institutes may receive copies of their own portion of the publications in respect of such of their members as do not become members of the Institution, and shall pay 10s. per annum in respect of every copy so supplied; and similar copies for exchanges shall be paid for at cost price.

50.—The Local Secretary of each Federated Institute shall prepare and edit all papers and discussions of such Institute, and promptly forward them to the Secretary, who shall submit proofs to the Local Secretary before publication.

51.—A list of the members, with their last known addresses, shall be printed in the publications of the Institution.

52.—The publications of the Institution shall only be supplied to members, and no duplicate copies of any portion of the publications shall be issued to any member or Federated Institute unless by order of the Council.

53.—The annual volume or volumes of the publications may be sold, in the complete form only, at such prices as may be determined from time to time by the Council:—to non-members for not less than £3 ; and to members who are desirous of completing their sets of the publications, for not less than 15s.

54.—The Institution. as a body, is not responsible for the statements and opinions advanced in the papers which may be read or in the discussions which may take place at the meetings of the Institution or of the Federated Institutes.

VIII.—MEDALS AND OTHER REWARDS.

55.—The Council, if they think fit in any year, may award a sum not exceeding sixty pounds, in tne form of medals or other rewards, to the author or authors of papers published in the *Transactions.*

IX.—PROPERTY.

56.—The capital fund shall consist of such amounts as shall from time to time be determined by resolution of the Council.

57.—The Institution may make use of the following receipts for its expenses :—
(a) The interest of its accumulated capital fund ;
(b) The annual subscriptions ; and
(c) Receipts of all other descriptions.

58.—The Institution may form a collection of papers, books and models.

59.—Societies or members who may have ceased their connexion with the Institution shall have no claim to participate in any of its properties.

60.—All donations to the Institution shall be acknowledged in the annual report of the Council.

X.—ALTERATION OF BYE-LAWS.

61.—No alteration shall be made in the Bye-Laws of the Institution, except at a special meeting of the Council called for that purpose, and the particulars of every such alteration shall be announced at their previous meeting and inserted in the minutes, and shall be sent to all members of Council at least fourteen days previous to such special meeting, and such special meeting shall have power to adopt any modification of such proposed alteration of the Bye-Laws, subject to confirmation by the next ensuing Council meeting.

TRANSACTIONS

OF

THE INSTITUTION

OF

MINING ENGINEERS.

THE NORTH OF ENGLAND INSTITUTE OF MINING AND MECHANICAL ENGINEERS.

GENERAL MEETING.

HELD IN THE WOOD MEMORIAL HALL, NEWCASTLE-UPON-TYNE,
FEBRUARY 14TH, 1903.

SIR LINDSAY WOOD, BART., PRESIDENT, IN THE CHAIR.

The SECRETARY read the minutes of the last General Meeting, and reported the proceedings of the Council at their meetings on January 31st and that day.

The following gentlemen were elected, having been previously nominated :—

MEMBERS—

MR. HERBERT AINSWORTH, Mining Engineer, P.O. Box 1553, Johannesburg, Transvaal.

MR. JAMES THOM BEARD, Civil and Mining Engineer, 640, Clay Avenue, Scranton, Pennsylvania, United States of America.

MR. WALTER RICHARD HAIGHTON CHAPPEL, Mining Engineer, Batu Gajah, Perak, Straits Settlements.

MR. RALPH CLOUGH, Mining Engineer, Kilton Mines, Brotton, R.S.O., Yorkshire.

MR. JAMES MILL CRAWFORD, Colliery Manager, Shildon House, Shildon, County Durham.

MR. WILLIAM STEPHEN DAVIES, Colliery Manager, The Poplars, Mountain Ash.

MR. RICHARD PERCIVAL FORSTER, Mechanical Engineer, Mount Pleasant, Spennymoor, R.S.O., County Durham.

DR. CLEMENT LE NEVE FOSTER, Professor of Mining, Royal College of Science, South Kensington, London, S.W.

Mr. GEORGE CHARLES FOX, Consulting Mechanical Engineer, P.O. Box 1961, Johannesburg, Transvaal.

Mr. CHARLES GEORGE HENZELL, Engineer, Catcleugh, Otterburn, R.S.O., Northumberland.

Mr. FREDERIC OCTAVIUS KIRKUP, Colliery Manager, Langley Park, Durham.

Mr. GRAHAM CAMPBELL LATHBURY, Assistant Colliery Manager, East Indian Railway Collieries, Giridih, E.I.R., Bengal, India.

Mr. FRANCIS DOUGLAS OSBORNE, Mine-manager, Gopeng, Perak, Federated Malay States.

Mr. JOHN RIDLEY RITSON, Mining Engineer, Burnhope Colliery, near Lanchester, County Durham.

Mr. THOMAS TROWELL, Mechanical Engineer, 17, Acutt's Arcade, Durban, Natal, South Africa.

Mr. THOMAS WELSH, Colliery Manager, Woodhouse, Whitehaven.

Mr. THOMAS OUTTERSON WOOD, Mining Engineer, Harraton Colliery, Washington, R.S.O., County Durham.

ASSOCIATE MEMBER—

Mr. PERCY COPELAND MORRIS, 79, Elm Park Gardens, London, S.W. ; and 1, Garden Court, Temple, London, E.C.

ASSOCIATES --

Mr. EDWARD STOKOE CLOUGH, Assistant Surveyor, Bomarsund House, Bomarsund, Bedlington, R.S.O., Northumberland.

Mr. JOSEPH HODGSON, Under-manager, West Thornley Colliery, Tow Law, R.S.O., County Durham.

Mr. HENRY MARSHALL IMRIE, Overman, Western Hill, Durham.

Mr. JOHN CHARLTON PEARSON, Back-overman, Montagu Colliery, East Denton, Scotswood, R.S.O., Newcastle-upon-Tyne.

Mr. THOMAS URWIN, Deputy-overman, Dipton Colliery, Lintz Green, R.S.O., County Durham.

STUDENTS—

Mr. ERNEST HUMBLE, Mining Student, Shotton Colliery, Castle Eden, R.S.O., County Durham.

Mr. GEORGE HERON DINSDALE THOMPSON, Mining Student, Dinsdale Vale, Windsor Avenue, Waterloo, Blyth.

SUBSCRIBER—

THE MOST HONOURABLE THE MARQUESS OF BUTE, Bute Estate Office, Aberdare, South Wales.

———

DISCUSSION OF MR. W. C. BLACKETT'S PAPER ON AN "IMPROVED OFFTAKE-SOCKET FOR COUPLING AND UNCOUPLING HAULING-ROPES."[*]

Mr. F. COULSON asked whether there was any chance of the space within the sockets of the coupling being filled with dirt. The principal advantage of the coupling seemed to be quickness of detaching and attaching; but it occurred to him that there

———

[*] *Trans. Inst. M.E.*, 1902, vol. xxiv., page 61.

might be a possibility of the connecting-piece or link coming out in the case of a slack rope, and more especially when the rope was twisted.

Mr. W. C. BLACKETT said that in trials of the socket no trouble had been experienced in the direction indicated. It did not pick up more dirt than any other socketting arrangement; and, in fact, it would pick up less, as it was much smaller. The link could not slip out, unless the two sockets were placed at right angles one to the other, and this was a position which was never assumed in an ordinary way. The first samples were made too weak, and it had been necessary to increase the dimensions and to harden the bulb of the link. The sockets were made of manganese-steel, and it might be of interest to mention that although this metal was extremely hard, it should not be heated to redness. Some men, for instance, were in the habit of removing wheels from their axles by means of heating, but if wheels of manganese-steel were heated to redness, they became short and brittle, and the steel was spoiled.

DISCUSSION OF MR. ROBERT MARTIN'S PAPER ON "SINKING ON THE SEASHORE AT MUSSELBURGH."[*]

Mr. F. COULSON said that the system adopted by Mr. Martin appeared to be an ordinary " sinking-wall," protected by a casing of steel-plates on the outside, and the cost would probably be much greater than to use a " sinking-wall " in the ordinary way —bolting it together and using cast-iron rings. The use of the outer steel frame would prevent the wall from separating while going down. He thought that it was somewhat risky to sink the full size from the surface, and starting with an external diameter of 18 feet and a wall, 2 feet thick, the diameter of the pit was only 14 feet. If the wall stuck, as it did at 48 feet (though it moved again by putting on an increased weight) and stopped there, the pit could not have regained its diameter except by cutting off a certain depth of brickwork, and that under certain circumstances might not have been possible.

He thought that it would have probably been desirable to have used a walling of a less thickness than 2 feet in that

* *Trans. Inst. M.E.*, 1902, vol. xxiv., pages 126 and 209.

particular kind of ground, when the sinking was dry for a depth
of 30 feet.   An important matter had been mentioned in the
previous discussion, as to the relative value of cast-iron or steel
for resisting the action of salt-water.   He thought it would be
generally admitted that cast-iron was better than steel, and he
imagined that if cast-iron was always under water, as it would be
at Musselburgh, there would not be much risk of corrosion.
Experiments had recently been made as to the use of cast-iron
in docks, and it was found that cast-iron, left in the bottom of
the dock and always covered with salt-water, was apparently
uninjured; but cast-iron that was exposed to the air at every tide
was very much corroded, and its strength was almost wholly
destroyed.   He noticed that the cylinder varied 18 inches from
the vertical, so that the pit would be 12½ feet in diameter, unless
some of the brickwork was cut out of the inside walling.   The
method of walling below the pumps was very interesting : short
lengths of walling (9 to 12 feet) were put in.   A system was
largely adopted on the Continent, of hanging cribs and using
concrete, instead of putting in timber to secure the sides, as it
was difficult to wall past the pumps, and it was not always possible
to build short lengths of walling owing to the nature of the
ground.   The barrels of the Evans pumps were brass-lined, but
Mr. Martin did not state definitely what had been found to be the
best kind of bucket for the pumps.

Mr. J. B. ATKINSON (H.M. Inspector of Mines) said that the sinking, at Olive Bank, had been a successful operation, and the only
difficulties experienced had been with regard to the brickwork
and the enclosing cylinder getting slightly out of the vertical.
Also, when the cylinder was nearly on to the rock, they ceased
adding the steel-rings and depended only on the brickwork,
which was caught and held by woodwork at the surface ; the steel-
cylinders and enclosed brickwork continued travelling downward, leaving the unenclosed brickwork hanging at the surface.
In another sinking, near Larbert, where there were many feet
of alluvium, the pit was started by forcing steel-sheaths down
the sides, but was not successful ; cast-iron cylinders were next
tried, and that also was not successful.   Then steel-cylinders
were adopted, following the plan adopted at Olive Bank, but
after they sank down to a certain depth, they became fixed owing

to the occurrence of boulders; and eventually the shaft was enclosed, and sunk with compressed air. He thought in both cases, and certainly in the case near Larbert, that the freezing process would have been more successful. The risk was certainly less, but without details of costs, it would be difficult to decide which would be the best system to adopt.

Mr. A. L. STEAVENSON said that he would like to make a protest against the use of steam-engines and pumps in sinking pits: there was little enough room for the rods and pumps, without introducing steam-pipes. He was not surprised to learn that, owing to the limited area of the shaft, and the number of pipes, it was found impracticable to sink and wall the shaft simultaneously. There had been some very extensive sinkings in South Yorkshire with steam-engines and steam-pipes in the shafts: of course, the pit was sunk in time, but the period would have been reduced if the engines had been placed on the surface and the ordinary pumping-rods had been placed in the pit. Steam-pumps in a shaft were a source of considerable danger, and also of inconvenience. He agreed that cast-iron was more suitable as a lining than steel. The system of forcing down a weighted cylinder was by no means new; it was tried 25 years ago in Cleveland, but it was found that it would deviate on one side, and it had to be pulled out.

Mr. T. E. FORSTER said that he agreed with what Mr. Atkinson had said about the freezing process, but when it came to a question of cost they could not tell in starting the sinking with cylinders what the cost was likely to be, and there were very few sinkings by this method that had not gone to one side and had to be pulled out. On the other hand, it was easy to form an idea of the cost of the freezing process, and there were contractors who would tender a price for carrying out the whole work.

Mr. F. COULSON said that he agreed with Mr. Steavenson, with reference to the use of steam-pumps in shafts; and the best pump for sinking was the old lift-pump, or pumps of that class. There was always much trouble with a hanging steam-pump: if water rose over it from any cause it was necessary to lift it, and if the water rose to a considerable height it was a very slow process to lower it down again.

Mr. W. C. BLACKETT said it appeared to him that there was a proper place for both systems, depending on the depth of the sinking and the quantity of water to be expected. A shallow sinking would usually be put through with a steam-pump, while in a deeper and more important shaft it might be better to use rods worked by an engine on the surface.

The PRESIDENT (Sir Lindsay Wood, Bart.) thought that at the depth referred to, only 100 feet, it would have been much cheaper to have sunk through the strata by the freezing method. With the method adopted there was always a great risk of not being able to press down the cylinder on account of boulders, and when once it stuck, they could not tell what the cost of the sinking would be.

Mr. ROBERT MARTIN, replying to the discussion, wrote that Mr. F. Coulson refers to the use of the steel-cylinder. At Olive Bank, it was used principally to reduce the skin-friction, and no doubt it held the brickwork together until it had time to set. He quite agreed with Mr. Coulson that it would be less risky, when sinking a cylinder through a soft deposit, 108 feet thick, to sink it in two or three divisions in telescopic fashion. This would almost certainly ensure a truly vertical pit. This method was considered and abandoned, because the result of a previous trial afforded a very good idea of the nature of the ground to be penetrated, and it was thought advisable to risk the running of the cylinder in one length, a course which was justified by the result. The cylinders deviated only 18 inches from the vertical, and this can be corrected by cutting and carving the brickwork upwards from the foot.

Sinking and walling, at the same time, may be possible in some circumstances, but risks must always be incurred in removing the cribbing and exposing a portion of soft ground; and it will continue to be dangerous, until the brickwalling becomes set and is strong enough to sustain the pressure against it.

For sinking pumps, indiarubber buckets were found to be the best; in clean water gunmetal cups or angular rings were preferred; and in every case the working-barrel should be lined with brass.

The trouble described by Mr. J. B. Atkinson would have been avoided, if the steel-cylinders had been made the whole depth of the deposit, 108 feet instead of 80 feet. The wooden shaft, in the boulder-clay, was intended to form a sort of guiding frame for the steel-cylinder, but owing to irregular subsidence, the frame became twisted and the upper portion of the cylinder was not built, until the cylinder, 81 feet long, rested on the bed-rock. A cylinder, 110 feet long, would have been more economical than the one used.

He could not agree with Mr. A. L. Steavenson that an engine placed on the surface and working ordinary rods in the pit would be safer and more convenient; and until the cylinder was sunk on to the rock-bed, it was impossible to erect any machinery on the surface. The temporary wooden frame, carrying the winding-pulley, sank and was raised daily, from 4 to 12 inches, until the surface-subsidence ceased. To pump, say, 1,000 gallons per minute from a shaft-bottom (and that is only a moderate quantity with ordinary pump-rods) would require a bucket, at least 30 inches in diameter, with a stroke of 8 feet going over 5 strokes per minute; a good deal of air must be pumped if the bottom is to be kept dry. In an ordinary shaft, there might be room for two such pumps. One pump alone is useless, as it is drowned in a few minutes, after a stoppage for repairs. He presumed that about 200 to 250 feet is the extreme limit that could be dealt with in one lift of such a bucket-pump; and when that point was reached what was to be adopted in a shaft that must reach a depth of 1,000 feet? To travel and secure for safe working two such columns of pumps would be a very difficult task; and it would be interesting to know whether it had ever been attempted. With three Evans steam-pumps (12 inches in diameter) in each shaft, a depth of 300 feet has been attained at Olive Bank; stationary pumps are placed in an offsett to raise the feeders found at that depth; and sinking is again resumed, and in three stages a pit is sunk to a depth of 1,000 feet. The handling of a steam-pump means dealing with a load of only 4 or 5 tons, and, in the use of steam at a pressure of 100 pounds per square inch, so far as his experience had gone, it could not be said to be either dangerous or inconvenient.

He thought that Sir Lindsay Wood seemed to favour the freezing method. It would have been almost impossible to put bore-holes through the large boulders encountered in this sinking. The freezing tubes are passed, he understood, down through a

large number of bore-holes to the necessary depth. The cost of
boring alone, he feared, would be very great. The boulders had
not been very troublesome, and they were removed, in every
instance, before the cutting-edge of the steel-cylinder reached
them. He considered that, where the workmen could stand on the
bottom, even though boards are required for footing, and where
the influx of water could be pumped, the freezing process was
unnecessary and would prove more costly.

Mr. S. J. POLLITZER (Sydney, New South Wales) wrote that
some 16 or 17 years ago, under his direction, a similar shaft was
sunk at Maryville colliery, Newcastle, New South Wales. After
these many years he (Mr. Pollitzer) could not exactly remember
every detail about the sinking; its salient features, however, were
quite fresh in his mind. In this shaft, no iron casing was used,
not even an iron bottom-ring or shoe. This latter was replaced
by a wooden one, made of the hardest timber that could be got,
consisting of four circular rings, each 3 inches thick, well bolted
and secured together; and the inside was bevelled, so as to be of
conical shape, with a sharp, yet solid, cutting edge. The actual
internal diameter had lapsed from the writer's memory, whether
it was 14 feet, or more or less, he could not say. This ring had a
width of 19 inches, allowing for a thickness of 18 inches of brick-
wall laid in cement. Through the centre of this circular wooden
ring or shoe, there went six vertical tie-bolts, 10 or 12 feet long,
and these were embedded in the brickwalling for nearly their
whole height, excepting 6 inches at the top, left projecting to
receive an additional wooden ring of that thickness, and to be
secured to the top of the brickwork by these tie-bolts. Before
this second ring was tied down, another series of 6 tie-bolts was
inserted into this new ring, for the purpose of repeating the operation, which was carried out for the whole depth of the shaft, about
135 feet, the thickness of the drift-sand to be sunk through before
reaching the Carboniferous sandstone (upon which the shaft
rested) and the sandstone had a thickness here of about 45 feet
before the coal-seam was reached. The external surface of this
monolithic brick-cylinder was covered with a layer of cement, for
the double purpose of making the cylinder water-tight and to
obtain for it a perfectly smooth surface, which, perhaps, should
not have a greater frictional resistance than iron. Excepting for

a top-layer of about 15 inches of vegetable soil, all the strata, as stated above, were sand, followed by watery drift-sand. After the removal of the few inches of soil, the bottom shoe was placed upon the sand, and two sections of tie-bolts bricked up; then the removal of the sand began, first by hand, and subsequently as the depth became greater, by an endless bucket-pump driven by steam. As the excavation deepened, the brick-cylinder sank regularly day by day: the bricking and sinking were then a simultaneous process, and there was never any external weight applied to the top of the cylinder to force it down, as its own weight was sufficient for that purpose. The whole operation of sinking to the depth of 135 feet took about 4 months, and when the cylinder was once sitting on the bed-rock it was found to be practically perpendicular. This colliery was worked for 8 or 9 years when, as the writer was informed, a sudden inrush of water from some adjoining abandoned mine flooded it, and it also was abandoned. In his (Mr. Pollitzer's) opinion, if Mr. Martin had not made a square pit through the boulder-clay for a depth of 32 feet from the surface, he would have avoided many of the difficulties that he (Mr. Martin) had to contend with in sinking his shaft at Olive Bank, and he (Mr. Pollitzer) also believed that the shaft would have been perpendicular.

————

Mr. THOMAS ADAMSON's paper on "Working a Thick Coal-seam in Bengal, India," was read as follows:—

· WORKING A THICK COAL-SEAM IN BENGAL, INDIA.

By THOMAS ADAMSON.

System of Working.—The seam worked at the Komaljore mine of the East Indian Railway Company's Serampore colliery, at Giridih, is 21 feet thick, the roof is of hard stone, and the thickness of cover over the seam ranges from 100 to 160 feet. A section of the strata, from the surface downward (Fig. 1, Plate I.), is as follows :—

No.	Description of Strata.				Thickness of Strata. Ft. In.	Depth from Surface. Ft. In.
1	Clay	...	...	...	19 6	19 6
2	Sandstone...	...	...	...	70 0	89 6
3	**COAL**	...	...	...	1 6	91 0
4	Sandstone	...	...	...	24 6	115 6
5	**COAL**	...	...	...	4 4	119 10
6	Sandstone	...	...	...	39 0	158 10
7	**COAL**	...	...	...	21 0	179 10
8	Sandstone	...	...	...	—	—

Fig. 2 (Plate I.) is a plan of the mine, and the area, AB, is that of a goaf-fall.

The system of working the coal, which has been the most successful employed in getting this thick seam, is a modification of the pillar system. Roadways are driven from the shaft to the boundary, and sides of work are then opened out. Large blocks (square or rectangular) are formed next the boundary, these are again split up into pillars, 40 feet square, and these are taken out, working from the boundary towards the shaft (Fig. 3, Plate I.). The roads forming these pillars are driven in the lower part of the seam, and are 6 feet in height by 8 feet in width.

The upper portion of the seam (15 feet thick) is worked as follows:—The pillar (40 feet square) is split, by driving four roads, 6 feet wide by 6 feet high, across it, thus forming nine knobs (Fig. 3, Plate I.) to support the top coal. Small headings, one above the other (called *chatnies* by the natives) are driven (Figs. 4 and 5, Plate I.) so as to facilitate the falling of the top coal, when the knobs are blasted out. In the first dropping operation, these narrow headings (*chatnies*) are driven on the four sides of the block of coal which is to be dropped. But, afterwards, the headings are driven only on two or three sides, the goaf forming the other loose side or sides, as they are called by the Staffordshire overmen.

Fig. 3 (Plate I.) shews a side of work, the pillars (40 feet square) that are formed being taken out in steps. Eight pillars have been removed, and two pillars have been split and are standing on knobs. Fig. 4 is a vertical section shewing the top coal standing on knobs, and the nicking headings (*chatnies*) driven in the top part of the seam.

Two shot-holes, *a* and *b*, are drilled in each knob, the dimensions of the knobs are then reduced to 5 feet or 6 feet square, the size of the knobs depending on the indication of the weight of roof upon them. When the areas of the knobs have been sufficiently reduced, the shot-holes are charged with dynamite, and fuses placed to the charges. All timber is then drawn out, commencing next the goaf. The props are knocked out by means of a bunter, a bamboo rod, 2 inches in diameter, 12 to 20 feet long, and shod with iron (Fig. 7, Plate I.); and then pulled back by the pricker, a bamboo rod, 2 inches in diameter, 15 to 20 feet long, and fitted with a hook at one end (Fig. 8, Plate I.).

When the timber has all been withdrawn, the overman listens to the goaf and for weight on the knobs; and if all be quiet, he, with 2 or 3 trained shot-firers, enters the place, lights the shots, and retires. The knobs are thus blown out, and the top coal falls. The operation of reducing the knobs, until the coal is dropped, is done under the supervision of European overmen at the end of the shift. When the coal has fallen, the place is fenced for the night.

The overman examines the openings (gateways) next morning, and if he finds all safe, orders the workpeople to load the loose coal. As soon as sufficient coal has been cleared away to enable him to get on to the top of the fallen coal, the overman makes a careful examination of the newly-exposed sandstone-roof.

In the lower workings, props are set at distances not exceeding 3 feet apart. When the roof or top coal is being loaded, cogs,

5 feet square, are set at intervals of 15 feet; and props (*long rollas*), 7 to 8 inches in diameter, are set between the cogs, at distances apart not exceeding 5 feet.

When one fall of coal is being loaded, one or more falls are being got ready, so that the loading-gangs are kept constantly at work, and the output maintained.

All the coal is removed in the first lift (when starting a side of work), 40 feet wide by 200 to 600 feet long. In the next lift, knobs from 8 to 10 feet square, the full height of the seam (*chowkidars*, or watchmen, as the natives call them) are left. These are used to indicate when the main-roof begins to weight.

No gas has been found in any of the mines, where explosives are used in working the upper portion of the seams.

By this system of working, about 90 per cent. of the seam is wrought. Six per cent. is left in the tell-tales, D, (*chowkidars*) and about 4 per cent. in the thin ribs, left next the goaf when a fall takes place, and separating the old goaf from the new side of work, which is opened out against it.

Heavy Fall.—The *chowkidars* are left 40 feet apart, one in each fall, after the first lift has been removed. The roof usually gives ample warning, and allows time to the overman to draw the timber and remove the workmen from the mine. As a rule, several days elapse from the first indication of roof-weight until the goaf falls. During this period that part of the mine or district is stopped, owing to the heavy blasts which are caused by falls of main roof. Sometimes, however, the roof quickly follows the first indication of coming weight—as on May 20th, 1902, when the writer was making an inspection of the mine.   :

The writer entered the mine at 8 a.m. and, accompanied by Mr. S. Hancox, overman, he inspected the travelling-roads, haulage-road and working-places; and he had been busy for about 10 minutes, while the roof was sounded, below which loose coal was being loaded, when pieces of coal commenced to roll from the sides of the *chowkidars* (tell-tales) in the goaf. The writer took immediate precautions, and all people who were engaged in loading loose coal close by were removed out of the goaf, some leaving their picks and baskets behind in their haste. Mr. S. Hancox and the writer were the last to retire, and, as we were doing so, pieces of roof began to fall and the main roof began to

FIG.   FIG. 7.—BUNTER. FIG. 8.—PRICKER.

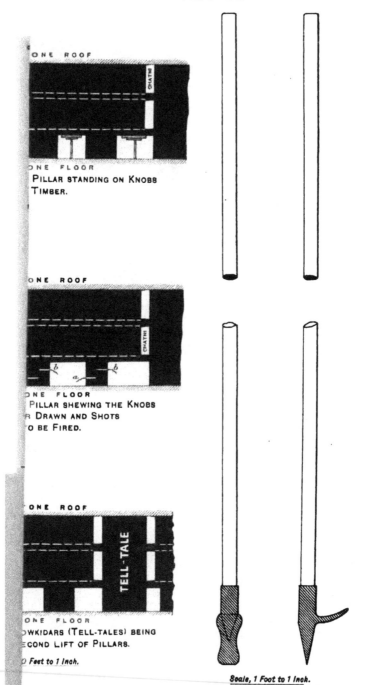

ONE ROOF

CHATRI

ONE FLOOR

PILLAR STANDING ON KNOBS
TIMBER.

ONE ROOF

CHATRI

ONE FLOOR

PILLAR SHEWING THE KNOBS
DRAWN AND SHOTS
TO BE FIRED.

ONE ROOF

TELL-TALE

ONE FLOOR

OWKIDARS (TELL-TALES) BEING
ECOND LIFT OF PILLARS.

Feet to 1 Inch.

Scale, 1 Foot to 1 Inch.

"bump" heavily, to such an extent that we considered it unsafe to send the people out of the mine, as in case of a heavy fall of roof taking place, they would be caught, by the blast, on the travelling-road. The people were ordered to go and sit down in galleries where the force of the blast would least affect them. Mr. Hancox and the writer took up a position about 100 feet from the goaf in a gallery, C, running at right angles to the direction which the blast would take (Fig. 2, Plate I.). While we were there, main-roof bumps occurred at intervals of from 5 to 10 minutes, and in 50 minutes from the time that we arrived at this part of the mine, a bump took place that shook coal off the sides of pillars, 100 feet back from the goaf. We immediately lay down, as a heavy fall of main roof took place, and for 5 or 6 seconds the air rushed past us, taking with it clouds of dust, and moved several loaded tubs that were standing on the rails near the goaf. The place was still "uneasy," and 15 minutes afterwards another fall took place, which caused a blast a little less violent than the first. Then, as all was quiet, we came to the conclusion that all the roof had fallen. We made an inspection around the edge of the goaf, and found that all the openings were filled with fallen roof. We ascertained that all the people were safe, but in darkness, as their lights and ours had been extinguished by the air-blast. The interval between the first indication of weight, and the collapse of the goaf was the shortest that the writer had ever known. The dimensions of the fallen goaf, AB (Fig. 2, Plate I.), were 230 feet by 220 feet, and there were 15 tell-tales (*chowkidars*) in it. The thickness of cover over the seam was about 100 feet, and the fall was visible on the surface, causing it to subside in places to a depth of 12 feet.

Costs.—The coal-getters are paid 4d. (4 annas) per bucket of 10 cwts. for undergoing and driving main galleries; 8d. (8 annas) per bucket for driving chatnies (nicking); and 2d. (2 annas) per bucket for filling the dropped coal, the latter forming 75 per cent. of the whole of the coal worked.

———

Mr. A. L. STEAVENSON said that the method of working described seemed to him very dangerous, and one which would not be allowed by H.M. inspectors of mines in this district. If the coal could not be all taken away with the first passage of longwall, he would advocate taking it out at twice, and not leaving the little knobs to come down at random. It would be interesting to know whether the seam described was in the Coal-measures or in the Mountain Limestone.

Mr. T. E. FORSTER said that no reference was made to the average area, which was wrought out without a very heavy fall. Usually, in working thick seams it is very difficult to get the roof to fall, but in the case described in the paper when a fall came on the trouble seemed to stop it, and the pillars would be all crushed. Certainly, 10 per cent. seemed a very small loss, and it would be interesting to elicit further information on that question. He had, himself, worked a seam in Australia about 12 feet thick, with a strong post roof; and in taking out the pillars a heavy fall took place, which ran over about 60 acres of the workings in a very short time.

Mr. J. B. ATKINSON (H.M. Inspector of Mines) said that in working a seam in Fifeshire, about 18 or 20 feet thick, it was usual to leave 2 or 3 feet of coal at the top and about 3 feet at the bottom —this was not very good coal. The method of working was not that described in the paper, which, he agreed with Mr. Steavenson, seemed to be a somewhat dangerous plan, but it was worked by longwall in two "carries;" up to 6 feet thick in the first carry, and then in the same longwall roads it was brought back in the top coal, 4 or 5 feet thick. The method of working was very successful.

Mr. T. E. FORSTER said that there were several bands in the Fifeshire seam, and it had a very strong roof. On the other hand, it must be remembered in the case described, that they had coolies to do the work, and working longwall there was a very different matter from what it was in this country.

Mr. RICHARD KIRKBY (Leven) wrote that it would be interesting if Mr. Adamson would state the character of the coal, whether it has any partings, or is at all laminated, and whether there is any fire-clay underlying the seam. Is there not even 1 inch of under-clay, or does the coal rest directly upon sandstone? Are there any signs of plant-remains in the beds above and below the seam?

It appeared to be a rather risky mode of working the seam, and yet evidently no other method could be successfully adopted

unless there were material for stowing the waste. If there had
been stone-partings, to the thickness of only 18 inches, in the seam,
or if stone could be taken down the pit and into the working-faces,
then the coal might have been worked by longwall in, say, four
lifts.

In Fifeshire, the Dysart Main seam is worked longwall in two
lifts, where it totals 14 feet of workable coal to 1 foot of stone-
partings. Fires, however, are troublesome, and another method
is now being tried, in which the seam is being worked in four
lifts, with a total thickness of 19 feet of coal and 3 feet of stone.

Where the Dysart Main seam is thinner, it is worked successfully without any waste, in two lifts, the thickness of coal being
9 feet and of stone and fire-clay, $1\frac{1}{2}$ feet. The amount of prop-
wood used is only 3 to 4 feet per ton of coal raised.

Would Mr. Adamson kindly state whether his seam was
troubled with fires?

Mr. J. P. KIRKUP (Burnhope Colliery) wrote that Mr. Adamson's paper upon the method adopted to work the thick seam at
Giridih, afforded the members an insight into the great advance
made in the methods of mining coal in India. The old rabbit-
warren system (Figs. 1 and 2, Plate II.) had, at the collieries
of the East Indian Railway Company, given place to proper
systematic methods under the able control of Dr. Saise, who
introduced the system described by Mr. Adamson about sixteen
years ago. Mr. Adamson's paper indicated some modifications
of the original method adopted and described some years ago.
All galleries were then driven from 10 to 12 feet wide; and the
size of the pillars was proportioned to the depth of the seam from
the surface: thus, the pillars were made 40 feet square, at a
depth of 200 feet; 40 feet by 60 feet, at a depth of 350 feet; and
80 feet square, at a depth of 450 feet.

The method of dropping the coal had been modified, he
supposed, from experience gained in its application. Formerly,
as described by Mr. Ward,* each pillar was split in the lower

* *The Indian Engineer*, 1887, vol. iii., page 146 and plate. Prior to the
introduction of the new system, all galleries were commenced at the top, and the
bottom was dug up until the thill was bared (Fig. 1, Plate II.). This Indian
system necessitated the levering or chipping of every pound of coal by means of
the pick or *sabol* (a crow-bar pointed at both ends). The downward working of
the seam also added to the danger of the process of removing the pillars (Fig. 2,
Plate II.); and, as in the galleries, all the coal was dug out with picks. The coal
at the top of the pillar was cut away, and the natural support of the roof was
removed at an early stage.

coal by a jenkin, A, driven in the middle and towards the goaf or waste; cross-cuts, B, B, were then driven; and at the side of the pillar, next the goaf, a thin piece of coal, C, was sometimes left, to keep back the fallen roof-stone and débris (Figs. 3 and 4, Plate II.). The cross-cut, B, was then widened, until about half of the pillar, D, had been undergone, temporary timber being set to support the upper coal. As soon as signs of weight were evident, they were assisted by drawing as much timber as possible, and shots were put into the fast side. After the fall (Fig. 5, Plate II.) the coal was got from the goaf-edge as far as the roof-stone would allow. To be able to drop and recover 90 per cent. of a pillar of coal, 40 feet square and 14 feet thick would require an unusually good roof-stone, which might exist in opening out a side of work before the first collapse, but hardly after a larger area of goaf had been made. The work would require very careful and skilled attention, and Mr. Adamson was to be congratulated upon his success with Indian labour. The costs for coal-getting were very interesting, when compared with those current in this country: an average hewing price of 5d. per ton for hard coal could hardly be excelled by the much bepraised cheap working in America. Moreover, the percentage of coal actually won from so thick a seam far exceeded the results obtained by the wasteful methods customary in America in working similar seams. He had often observed that the advantage of machinery was not always apparent in competition with Indian labour, and he doubted whether any economy could be gained by machine-cutting against the cheap labour of India. It would have been of interest if Mr. Adamson had told the members how the system of working affected the sample of coal.

The PRESIDENT (Sir Lindsay Wood) moved a vote of thanks to Mr. Adamson for his interesting paper.

Mr. G. MAY seconded the resolution, which was cordially approved.

––––

Mr. S. J. POLLITZER's paper on " A Measuring-tape and its Use in Mine-surveying," was read as follows:—

FIG. 1.—OLD METHOD OF WORKING GALLERIES.

Scale, 30 Feet to 1 Inch.

FIG. 3.—PLAN OF SIDE OF WORK.

Scale, 120 Feet to 1 Inch.

FIG. 2.—OLD METHOD OF REMOVING PILLARS

FIG. 4.—PREPARING PILLAR FOR DROPPING THE COAL.

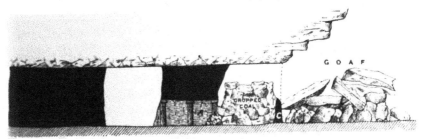

FIG. 5.—COAL DROPPED AND READY FOR LOADING.

Scale, 30 Feet to 1 Inch.

The North of England Institute of Mining & Mechanical Engineers
Transactions 1902-1903.

VOL. LII, PLATE

A MEASURING-TAPE, AND ITS USE IN MINE-SURVEYING.

By S. J. POLLITZER.

The writer's opinion, backed up by experience, is that the connection of a surface and an underground survey can be effected by two plumb-lines, just as accurately as by the use of any optical instrument. Further, to make a comprehensive survey of a mine an horizontal projection alone is not all that is wanted; vertical sections are equally as important, and to be able to prepare these one must have the means of measuring correctly the relative heights of the various points in the mine. These are the reasons that induced the writer to construct a measuring-tape, 3,000 feet long, which he presumes is long enough for use in any mine in Australia.

The wire used for the purpose is a soft pliable steel-wire of No. 24 Birmingham wire-gauge, weighing about 2 ounces per 100 feet, and Prof. Warren, of Sydney University, states that this wire will stand a strain of 84 pounds. In the writer's own experiments, the wire stood a strain of 56 pounds, without showing the least sign of weakness or over-strain. To one end of the wire is attached an excellently-made detachable plummet of brass, weighing exactly 9 pounds, and it is made in such a manner that additional weights can be added to it without greatly disturbing either the wire or the plummet. A weight of 9 pounds was chosen, because, in experiments, (1) it was found that a weight of even less than 9 pounds would keep this wire in a perfectly vertical condition, but there must be no kink in it; and (2) when the wire is played out in a shaft for nearly its whole length, the maximum strain on the upper end of the wire must not exceed one-fourth of that found in above-mentioned experiment.

Based on the 9 pounds' weight, the wire was divided at each length of 10 feet under a 9 pounds' strain and with a noted temperature: the correct strain was obtained by means of a spring-balance that had previously been tested with ordinary standard weights; and for each succeeding length of 100 feet, 2 ounces were added to the original 9 pounds of strain. At the exact 10 feet division-points, small pieces of the same wire were twisted across them, and soldered to them with a drop of lead-solder; at that end of the twisted bit of wire, where an eye was formed, a piece of white calico, measuring 2½ inches by 1½ inches, was fastened, and on the calico was stitched in black wool the number of feet at that spot, and at every 50 and 100 feet mark the calico was replaced by red Turkey cloth, of course, provided with its respective number stitched in black wool. The pieces of calico and Turkey cloth were also used for covering up and enveloping the twisted wires, to which they were attached, so as to prevent them from scratching or otherwise damaging the rest of the wire, when being wound up. The length of wire is wound upon a cedar wheel, 18 inches in diameter and 1⅞ inches thick, with a groove in its periphery, 4 inches deep. The wheel when in use is attached to a strongly-made wooden trestle, which is permanently fixed to a purposely-made staging placed over the mouth of the shaft. The wheel can be stopped at any single inch throughout the depth of the shaft.

Where a joint had to be made in the wire caused by its being accidentally broken, or adding a fresh coil, the two ends were overlapped for a length of about 2 inches and lightly twisted over each other, so as to keep them fairly close together. Afterwards this joint was closely wound with thin brass-wire, which was extended to about ½ inch on either side of the single wire; and the entire joint was covered with a thin coating of lead-solder.

When the tape was finished, it was taken to the Sydney Land Office, and tested on the standard length of 100 feet. It was found that to bring each 100 feet wire within the two standard-marks, the same strain had to be exerted almost in every instance as that under which the particular part was constructed; and in those few instances where the strains did not agree, the differences were never greater than 2 ounces more or less.

Before the wire is brought into use, the surveyor selects two points in the roof of each level that is to be connected, and these two points forming the ends of the base-line in that level are marked by two steel-pins, 5 to 6 inches long and ¾ or 1 inch thick; these pins are permanently fixed in the roof, and have a small hole drilled in their centres. They should as much as possible be placed in such a position, that neither of them shall form an ill-conditioned triangle with each cage-shaft, and the distances between the pins are carefully measured.

After one of the cages has been disengaged and secured over-head, a temporary staging is erected over that shaft-mouth, upon which the trestle for the wire-wheel is nailed down; the staging should be of such a height that there shall be between it and the surface of the shaft a clear height of about 4½ feet. Then the 9 pounds brass plummet is hooked on to the wire, both are lowered 3 or 4 feet, and kept in that position until the plummet is quite still and there is no perceptible movement in it; this is done with the object of preventing any swinging or knocking of the plummet against the sides of the shaft while it is being lowered. The plummet is then slowly and gently lowered by turning the wheel, until it reaches the shaft-bottom or the required level; before the plummet reaches that point, the shaft below must be covered with planks, and a bucket of oil placed upon it to receive the plummet.

After the plummet and wire have been brought to absolute rest, and after having laid down on the surface a similar base-line to those in the various levels, and connected the surface-base to the general surface-survey, the surveyor connects the wire with the base-line by the triangulation-method, by sighting from each point of the base-line on to the wire, and carefully ascertains the angles included between the wire and the base-line at each end. This process he begins on the surface, repeats it in each level in succession, and finishes at the shaft-bottom. The bucket of oil is then removed, and the plummet is left free for the purpose of measuring the heights; if the plummet swings more or less, it does not affect the vertical measurements.

To measure the heights or depths, the surveyor begins, in the reverse order, that is, at the bottom. With the assistance of an ordinary spirit-level, he draws a line across one of the vertical beams of the timbering, level with the lowest point of the plummet; and this line he marks permanently, as the bench-mark for the lowest level. In the second cage, he then goes to the surface and reads the number of feet on the calico-mark next below the surface, and thence to the surface or to a similar bench-

mark, as the one at the bottom, he can measure with a small hand-
tape or any other measuring staff, and in this way the depth to
the bottom is ascertained.   The plummet is then gently wound
up to the next higher level, into a position similar to that in which
it was placed in the one below it, and the identical operation is
repeated.   In fact, this operation is repeated until all the levels
have been treated in the same way, and the survey of the shaft is
finished.

Next, the wire and the second cage change places, and the
operation is repeated through the second shaft, except that the
measuring of the heights is entirely omitted, as the measurements in the one shaft will serve.

In this way, the writer brings down an azimuthal line of
definite co-ordinates in every level, which is a correct way of
connecting the surface- and underground-surveys in vertical
shafts.

It has been stated that the triangulation of the wire from
each end of the base-line begins when the wire is at complete rest.
When is the wire at complete rest, and how is that to be ascertained?   One answer is given in Mr. G. R. Thompson's paper on
the subject; but in addition to that, the writer uses another check.
While he is observing the wire from each end of the base-line,
two assistants stretch a thin strong piece of white cotton diagon-
ally across the shaft at different levels, leaving a few inches in
height between each cotton-strip; one end, the further one, is
fastened to a nail while the assistant holds the other end in his
hands, strains it, and brings it quite close up to the wire, without, however, touching it.   He holds it in this position while
the surveyor is making his observation: the least movement
in the wire will in this way be surely detected, and the instrumental observation is stopped until the wire is again at rest.   Of
course, the surveyor detects the lateral motion of the wire through
his telescope, yet it is desirable to have two assistants, each
stretching a piece of cotton.
    In mines, where there is either an artificial or natural air-
current so strong that the wire will not come to rest, the only way
of stopping the air-current is by placing doors temporarily across

the levels whence the current comes. Of course, this means a cessation of all work below, for the time being, which is justified by the importance of the survey-connection.

The use of a brass plummet prevents all local magnetic attraction, and this latter will hardly ever be strong enough to deflect the steel-wire from the perpendicular, when such a plummet is hanging from it.

Some 15 months ago, the writer's services were requested on one of the largest gold-fields in the western part of New South Wales. On this mine, there are a score of shafts and many miles of levels and drives. At the 800 feet level of one shaft, there was a large quantity of ore in sight ready for treatment at the battery. The battery was close to another shaft, and there was no connection between the 800 feet levels of the two shafts, and between the two shafts there is a fairly deep creek. They are about 1,000 feet apart in a straight line, and are vertical. The question was where and how to make the shortest possible connection between the respective levels, which were by no means at the same horizon. The survey, which was carried out very much in the manner here described, and the requisite calculations, were finished within less than three weeks; and then it was pointed out to the manager that over one of the pins driven in the floor of one level he must make a perpendicular rise in the rock overhead of about 45 feet, and, by driving in the direction indicated by two pins in the roof of the other level for a distance of about 270 feet, the connection would be effected. The country-rock through which this connection was made is very hard diorite, and it took more than seven months to complete the connection. One morning, the writer received a letter from the manager congratulating him on the successful connection, and stating that the survey was perfectly correct in every detail.

———

Mr. ARNOLD LUPTON (Leeds) wrote that Mr. S. J. Pollitzer's paper was interesting, as describing the actual practice of a surveyor in Australia, and the exceedingly small wire used by him for measuring the depth of a shaft. The attachment of the marking tapes to the wire would, however, be apt to cause vibration in cases where there was a strong current of air in the shaft.

Mr. Pollitzer appeared to use two shafts in connecting his surface and underground lines. It was not always easy to secure the use of two shafts, and where one shaft only was available the distance between two wires suspended in the same shaft was not always as large as would be desirable. But, if the distance between the two wires was only 60 inches, a line might be set out from them with great accuracy, provided that the wires were not unequally deflected from their true position by air-currents; and any effect produced by the rotation of the earth, being the same for each wire, would not, he thought, alter the direction of the line connecting the two wires. The accuracy to be obtained in connecting the surface and underground surveys by suspended wires was chiefly dependent on details. For instance, the points of suspension must be perfectly steady, and in order to ensure that, there must be no vibrating machinery at work near these points; there must be no high wind, nor movement of weights; there must be no eddying currents of air in the shaft; also the mean position of the wires at the pit-bottom must be noted with great care, and the observation repeated several times. Some years ago, he (Mr. Lupton) set out an underground line by three methods, namely:— By compass-needle, by marks fixed by a transit-instrument, and by two wires suspended in the same shaft; and the three lines agreed exactly one with the other.

Mr. HENRY JEPSON (Durham) wrote that the title of Mr. Pollitzer's paper seemed somewhat misleading, as the tape, in question, appeared to have been constructed for a specific operation in surveying, and its use was probably limited to that operation. Mr. Pollitzer commenced his paper by giving his opinion "that the connection of a surface and an underground survey can be effected by two plumb-lines, just as accurately as by the use of any optical instrument." As he did not state that his experience extended to the use of an optical instrument, one might be at liberty to doubt it. With 20 years' experience of the use of the transit-instrument for this purpose and some experience of wires, he ventured to assert that it was not possible to make the connection between surface and underground surveys as accurately by the use of suspended wires as by a powerful optical instrument. Throughout Mr. Pollitzer's paper careful and elaborate means were described to avoid movements in the wires suspended in the shaft. Therefore, he concluded that they were free to move, as they ought to be, and they did move; whereas in the use of the transit-instrument, the marks to be observed were absolutely fixed, and could be observed as finely as any wire. Under the circumstances, the particular operation described in the paper might be most convenient. In using a transit-instrument, it would be extremely unlikely that the conditions would permit of more than the surface and one level being connected at one time, as the marks at one level would probably obstruct the view of the marks below. And, in addition, it was not likely that the exclusive use of a shaft could be obtained for so protracted an operation as that described. He suggested that, without materially increasing the strain on the wires, Mr. Pollitzer might devise some means by which the plummet would present an increased surface to the oil in the bucket, and thus still further reduce the tendency to movement.

The PRESIDENT (Sir Lindsay Wood, Bart.) moved that a vote of thanks be accorded to Mr. Pollitzer for his interesting paper.

Mr. T. E. FORSTER seconded the resolution, which was cordially approved.

———

Mr. S. J. POLLITZER's paper on "The Underlay-table" was read as follows:—

THE UNDERLAY-TABLE.

By S. J. POLLITZER.

There has of late been so much written on the methods of connecting the underground workings of a mine with the surface-boundaries, that it might almost appear that nothing new could be written on the subject, and less likely that anything new could be invented; and he would be considered very bold, who would venture to enunciate some new idea in the face of what European and American mining-engineers have accomplished in this branch of science.

The writer wishes to state in all modesty that boldness is not exactly one of his virtues, but he believes that he has invented a useful mechanical appliance, and he desires the members to give an unbiassed opinion on the same. If anything more should be added to this preamble, so as to produce a fair criticism, he wishes to state that he has absolutely no monetary interest whatsoever in his invention, which is not patented, and anybody who desires can have the appliance made without let or hindrance.

The writer has had considerable experience in almost all classes of surveying on land and water, but during the last ten years his attention has principally been occupied with the underground workings of mines, and about five years ago he invented the abovenamed " mechanical appliance." The last two words are written, because, according to accepted notions an appliance that is not an " optical instrument " is simply a " mechanical appliance."

It is quite needless to preface the paper with a disquisition on the necessity of correctly connecting the underground workings of a mine with some fixed data on the surface, as that has been done repeatedly, but particularly so of late by Mr. G. R. Thompson in his excellent paper communicated to the members.[*] Mr. Thompson and his numerous predecessors occupied them-

* *Trans. Inst. M.E.*, 1901, vol. xxii., page 519.

selves with the question in all its various branches, finding the
solution for all cases in the application of the optical instrument,
or the theodolite in one form or another.

With all due deference to these able workers, and in the light
of his practical experience, the writer has come to the conclusion
that in some instances the theodolite is not the most suitable
instrument to use for the connection of the two surveys, belowground and aboveground; and that is, in cases where the shafts
are driven on the "underlay." If an underlay-shaft be driven
in a perfectly straight line longitudinally and transversely from
the top to the bottom, then no doubt, the transit-theodolite or one
fitted with an eccentric telescope, according as the dip may allow
of the use of one or the other, is the best instrument to use;
because the process is not only the quickest, but there it risks the
least possible chances of introducing an error into the survey.

However, let the reader imagine a properly-equipped working shaft, not a prospecting-shaft, that starts from the surface
with a dip of say 75 degrees for, say, a depth of 90 feet; at this
point, the inclination becomes 45 degrees and for a distance of
say 50 feet takes a turn by a few degrees to the right; thence for
100 feet it may deviate several degrees to the left under an inclination of perhaps 85 degrees, and so it will go along at its own
sweet will, to and fro, forwards and backwards, until the bottom
is reached, at a depth of between 600 and 700 feet. This is,
by no means, an hypothetical case, nor is it a solitary one; there
are numerous such shafts in this State, and no doubt, there are
similar cases not only in the other States of Australia, but also
in other countries. Now, the use of any kind of theodolite in
such a shaft may not only be highly objectionable to the accuracy
of the survey, but it will be almost impossible. Take for
instance, the simplest of all possible cases in such a zig-zag shifting shaft, which may hardly ever occur, that none of the various
and continuous altering dips ever exceed an angle of inclination
of 60 degrees with the horizon, when a 6 inches astronomical
transit-theodolite could be used, because at about that angle
the axis of the telescope just passes the tangent to the horizontal
circle of the theodolite. Of course, at every point where an
alteration of either grade or direction occurs, the instrument
will have to be set up and this setting up, by no means an
easy task, will consist in building a strong and rigid platform

in such a way that the new point can be centred to the instrument with facility, and that there is no obstruction by it to the foresight after the telescope has been tied in to the backsight; and, say, if all goes satisfactorily, that there are two points which will vitiate a possible correct result and that is (1) to correctly ascertain the height of the axis of the telescope above the centred point, and (2) vibration in the platform, no matter how rigid it may be. However, to assume that a shaft will be nowhere steeper than 60 degrees, is highly problematic, and it is more than likely that there will be more inclinations of over 60 degrees than under; consequently, under such circumstances, the transit-theodolite becomes useless, and it must be replaced by an instrument fitted with an eccentric telescope. With such an instrument, the case is exactly reversed, it is an excellent appliance for steep grades, but it may be entirely useless for light ones.

When an eccentric theodolite is used, it must be placed on the rigidly-prepared platform in such a way that two legs of the tripod shall be placed on the same side as the telescope, and that their end-points shall be fairly placed in a line parallel to it; and, in this position, the top ends of those legs will afford the greatest opening for the telescope to sight through. This arrangement is convenient so long as only steep sights are to be observed; but so soon as the dip flattens much, the line of sight will be obstructed by the forward one of the two legs, and if all these legs be shifted for this reason, into another position, it (more likely than not) may be found, that the top part of the leg obstructs the sight; in other words, under certain conditions, it may not be possible with this instrument to obtain an unobstructed line of vision.

There is yet another way of overcoming this difficulty, by using a common transit for smooth grades and an eccentric transit for steep ones, or one of American make, as mentioned by Mr. H. D. Hoskold. Such an instrument, certainly would overcome all surveying obstacles, but the greatest care would have to be exercised, so as to prevent one's notes and calculations from becoming confused.

However, to overcome all these obstacles connected with surveying and confusing calculations, the writer has invented and practically tested successfully a new instrument called the "underlay-table," which on account of its simplicity of use in surveying, and still greater simplicity in computing the results, would appear to commend itself better for the purpose of underlay-shaft surveys, than any of the wellknown optical instruments. Moreover, the appliance readily lends itself to surveying every kind of underlay-shaft, of steep or shallow inclination, or twisting in any direction. The only disadvantage which it possesses is, that it is the slowest imaginable process, as will be noticed later on.

The principle, upon which the table is based, is to carry a short horizontal base-line from the surface down to the bottom of the shaft. To carry down this line correctly, a perfect rectangular table must be imagined, fixed level on its supports and level with the surface of the shaft; if the side of the table nearest to the underlay be the initial short base, then the opposite side of the table, that is the one nearest to the hanging-wall, will be the same line, except that it has been shifted parallel to the original line, by a distance equal to the width of the table, which of course is known. If now the second line or the forward side of the table be produced down the shaft by means of two plummets, wherever they may hit the shoot, and marked, and their depth measured, and noted from the surface of the level table; there is thus obtained a new position of the original first line. If now the table be shifted lower down, and be placed so that its back line shall coincide in azimuth and altitude with the two marked points, the front-end can again be produced lower down in the same manner as before, and in this way, the whole depth of the shaft can be passed through until the bottom is reached; by which process, the base-line is brought down, and it is parallel, by so many feet, to the original line and so many feet below it. As the co-ordinates of the original line were previously ascertained on the surface, those of the bottom-line become known also, and all underground surveys can now be safely tied into this bottom-line.

The appliance is not exactly a full table, for such an one would be too heavy and awkward to handle. It consists of a light cedar frame, made up of six pieces, 2 inches by 1 inch, properly screwed together, and shewn in Fig. 1, front-elevation; Fig. 2, side-elevation; and Fig. 3, plan (Plate III.). Before the table is brought into use, two points in the most favourable position,

and as close as possible to the hanging-wall of the shaft, are
dropped down from the surface, until they hit the shoot of the
foot-wall; and, obviously, those two points have, previous to the
plumbing, been tied into the surface-survey or their co-ordinates
have been ascertained. Near the two points plumbed down and
marked in the shaft, a temporary seat for the surveyor is provided: then the table with its three telescopic legs, which can be
extended to a height of nearly 10 feet, is fixed in the shaft in
such a position that its top shall be a few inches above the first
two marks fixed in the shoot. By means of two levels and an
union-screw attaching the table to its stand, the appliance is
levelled in both directions, and permanently fixed. Along the
whole length of the piece a (Figs. 1, 2 and 3, Plate III.), in front
of which the surveyor is sitting, and of the piece b, there are two
swallow-tailed grooves of brass fixed on the top of the frame;
these grooves are for the purpose of receiving four similar
sectioned slides of brass, c, which travel freely in the two grooves,
and to each pair of these slides are permanently attached two
hollow brass tubes, d; by these slides, the two tubes can be shifted
over the frame or table in any direction to either side of the
observer, parallel to themselves and at right angles to the longitudinal direction of the table, namely, the pieces a and b. These
two tubes are of the same length as the width of the table, and
with four mill-headed screws (not shown in the figures), attached
to the slides, these tubes can be permanently fixed at right angles
to the table at any desired point. Through each of these two
tubes, another hollow brass tube, e, is passed, so that in telescopic
manner, they can be made to travel forward or backward, and by
means of another set of mill-headed screws on the outer tube the
inner tubes can be fixed at any desired point in the transverse
direction of the table. In the figures, the four tubes are shown
of square section; however, one that was constructed is of circular
section, and has the appearance of a long telescope (Fig. 4).

The mode of use is as follows:—The table is levelled, as stated
above, the two pairs of tubes are shifted either to the right or to
the left, and the two internal tubes are moved forward or backward, until the centre of the back-ends of the latter will touch
two fine plumb-lines, being the raised points first determined on
the shoot. In Fig. 3, it will be noticed that the two grooves and

four tubes are symmetrically divided into inches, by which correct parallelism and right angularity is obtained as abovementioned. When once the back-ends of the internal tubes have been made to touch the two plumb-lines, the surveyor reads the number of inches and fractions thereof that have been drawn out backwards from each, and if both have been drawn out to exactly the same length, then the instrument stands in perfect adjustment and in condition to the principle it represents; if not, the clamping-screw, f, must be loosened, and the table turned horizontally round its vertical axis by its levels, and the operation must be repeated so often as is required to produce an equal backward extension of the two internal tubes to meet the two plumb-lines, and after two or three trials the adjusted position will be found. When once adjusted, the height of the tubes over the marks in the foot-wall is ascertained and noted, as also is the backward extension of the tubes. Next, both of the movable internal tubes are moved forward, of course both to exactly the same extent, either until near the hanging-wall, or, if the latter will permit, to move them forward until they are nearly played out, and rest in the external tubes by a few inches only, sufficient to clamp down on them the two front clamping-screws. The front central ends of the internal tubes are now plumbed down with heavy plummets of first-class construction, until they hit the foot-walls somewhere below, and a reliable assistant there steadies and marks the two new points. Next the field-notes are entered, consisting of the back travel of the rod, say, for instance, $10\frac{1}{2}$ inches, *plus* the width of the table, which is constant, say 2 feet, *plus* the forward motion of the internal tubes, say 2 feet 6 inches, or a total of 5 feet $4\frac{1}{2}$ inches, the initial base-line has travelled parallel forwards, and this new parallel line has been produced downward, equal to the depth from the tubes to the new points which is measured with a steel-tape, minus the height of the tubes above the first points. The instrument is now removed lower down to the two fresh points, and the operation repeated in this manner, until the whole depth of the shaft has been gone through.

When the above-described operation has been finished, and the information so obtained is plotted on paper, a true trans-verse section of the shaft will be the result. On the other hand, a longitudinal section of the shaft should appear as two vertical

lines, being the perpendicular projection of the two internal
tubes, which are always distant apart from each other by the
measured length of the original base-line.

The length of this base-line is between 24 and 26 inches, being
then 1 or 2 inches less than the length of the table; and this
latter must be at least 1 inch less than the clear opening between
the skids of the shaft which is about 28 inches; hence, it follows
that in metalliferous mines with shafts of this kind, no base-line
can be longer than 26 inches at the most. However, if there are
shafts in mines that are wider, a longer base-line can be obtained
by using a longer table.

The verticality of such shafts, in longitudinal section, in
metalliferous mines is very rarely to be found, and it happens that,
even in the best constructed shafts, their longitudinal axis is
never truly perpendicular, and invariably deviates more or less
from the perpendicular. Suppose that this be the case: the appliance will detect it at once. Assuming that at any one stand, the
table has been fixed and manipulated as described above, and when
the front-ends are being plumbed down, it is found that either one
or the other of the two plummets strikes the shaft-timbering
before reaching the foot-wall: shewing that the shaft twists or
takes a turn to the side opposite to where the plummet touches
the timber: in such a case, the tube from which the obstructed
plummet is hanging is shifted parallel to the other tube and
towards it until the timber is cleared. If the case be reversed,
of course, the reverse side has to be shifted; and in the book it
is noted down by how many inches at each stand the shaft twists
to the right or to the left, as the case may be.

There are a few more details that may be mentioned.
The central longitudinal piece of the frame rests on the base-
board, g; the frame is fixed to it by two sets of clamps and screws,
h; and on this base-board the union-screw works and keeps it
and the table firm and immovable in any desired position. Round
the base of the union-screw, there is a movable collar from which
two arms branch off at right angles one to the other, each of which
carries a vertical thumb-screw, i, intended to counterbalance that
part of the table, which, at times, may, on account of twisting,
happen to want shifting on the base-board more to one side than
on the other. Experience has shewn that, though these two screws

are an extra precaution, they are quite unnecessary, since the union-screw acts efficiently, and will even stand the strain when about half of the table only rests on the base-board and its other half is wholly unsupported, although such exaggerated cases do not occur.

The lower ends of the three legs are provided with three strong iron points, which will find a firm foothold in almost any crevice of the rock in the shaft; the three legs are placed in such a position that two of them lie on either side of the observer and the third leg is placed in front of him, and all three sit in the country-rock in the foot-wall and hanging-wall respectively, where they are free from any vibration. Metalliferous-mine shafts generally consist of three compartments, two for the up- and down-cages, and one for the ladder, air-pipe, and other accessories. For surveying, the middle compartment is usually chosen, while the other cage is always at the disposal of the surveyor to move his instruments and his assistants wherever required; and when moving the appliance, the table is unscrewed and placed upright in the cage, the legs are shut, and the party move themselves in the cage to the next points lower down the shaft where the operation begins afresh. The appliance can withstand a reasonable amount of rough handling, without injuring its efficiency or being materially damaged; there is nothing in it that cannot be repaired or replaced within a few hours in the mine workshop, so as not to stop the progress of surveying. The legs are made of American ash, 1 inch by $\frac{1}{2}$ inch, and braced with strap-iron; the table parts are made of cedar, 2 inches by 1 inch; the base-board is made of iron-bark, 3 inches by 1 inch; and the ball-and-socket of the union-screw is made of jarrah wood. Nearly the whole of the metal used is brass, but the bolt of the union-screw and the counterbalance-collar are made of iron. The weight of the instrument is as follows:—The table, $14\frac{1}{2}$ pounds; two pair of tubes, 5 pounds; and the legs, 16 pounds; or a total of $35\frac{1}{2}$ pounds. When not required for use, the table is completely dismounted, and packed in a strong suitable box; and made singly its cost will amount to about £35.

When surveying, three assistants are required, one on each side of the surveyor to plumb-up the back-points, after which they plumb-down the front-points; and a third, who marks the front-points; besides, there should be one, or preferably two, smart miners to assist in making rough timber-seats, working the travelling cage, and all sorts of odds and ends that may be wanted; and lastly the engine-driver who has to work the winding-engine.

This instrument has been tried practically some three years ago in a large gold-mine in the southern part of this State, where there are three underlay-shafts of the above description. The three

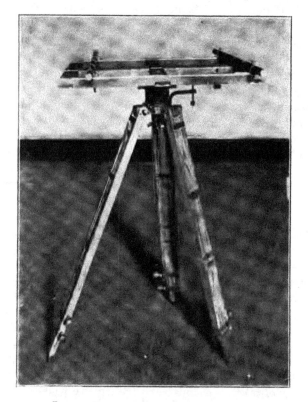

Fig. 4. —THE POLLITZER UNDERLAY-TABLE.

shafts have a depth of from 300 to 700 feet, and in the aggregate they amount to not quite 1,400 feet; the three shafts are in a fairly straight line on the surface, and the distance between each is between 500 and 600 feet. The underlay-table operation was carried out through the three shafts, and as two of the shafts at one of their respective levels were connected by a drive,

.. Mining Engineers
ations 1918 1913

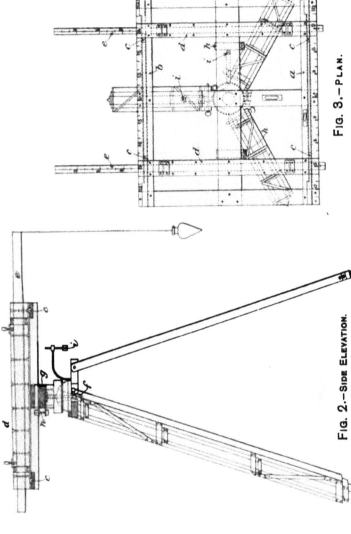

I.—FRONT ELEVATION.

FIG. 2.—SIDE ELEVATION.

Scale, 1 Foot to 1 Inch.

FIG. 3.—PLAN.

though of a very tortuous character, there was an excellent opportunity for testing the efficiency of the table, by having a closed circuit. The result of the close, to be candid, was not satisfactory; the reader must bear in mind that the above-mentioned tortuous course consisted of a good number of traverses, the length of which was a good deal below the focal length of the telescope of the theodolite, and as the latter had frequently to be placed in shifty and unsteady positions, the writer is forced to the conclusion that the mis-close of the survey was not caused by the table, but by the underground traverse. Moreover, the writer is convinced that, given fairly favourable conditions, the underlay-table will give satisfactory results, although the base-line be only 2 feet long.

Engineers and surveyors alike rightly avoid the use of very short base-lines, and so does the writer; but when circumstances arise that short lines must be used, then according to his experience they will assist in obtaining as correct surveying results as with long lines; the difference will only be in the time used, for in short lines much more numerous checks, hence more time, will be required to ensure that no mistakes are made. As an illustration of this, it may be mentioned that from a measured base-line of about 12 miles the earth's quadrant can be calculated, and the proportion between the measured and calculated line is roughly as 1 in 8,300 and such a small fraction might not often, if at all, occur in underground surveying. Certainly, with this underlay-table the difficulties are considerably increased by moving it so many times, but the writer is convinced that it can be used with accuracy.

Fig. 4 shews the underlay-table, and its construction in all its detail is very minutely visible.

———

Mr. ARNOLD LUPTON (Leeds) wrote that Mr. S. J. Pollitzer's paper was interesting, as showing a method of surveying crooked and sloping shafts actually carried out by a practical surveyor. There could be no doubt that an accurate survey might be made by this method and a base-line accurately transferred from the surface to the deepest level. The accuracy depended entirely on the amount of care taken in hanging the plumb-bobs and in marking their exact position, reading the measurements, levelling the table, etc., and as considerable time might be occupied in doing the work it was necessary that the shaft should be at the service of the surveyors for considerable lengths of time, to ensure accuracy in the results. For similar work, he was inclined to think that the Hedley dial could often be used with great efficiency. If there was no attraction in the shaft, and the compass-needle could be used, it would be, by far, the quickest way of surveying the shaft. Even if the compass-needle could not be used, still the Hedley dial was so handy an instrument that the survey could probably be made much more expeditiously with it than by the use of the underlay-table. The Hedley dial, as commonly made, could be used for angles up to 65 degrees from the horizontal, and a slight modification would enable it to be used for any angle. The accuracy of the work would depend on the care with which the instrument was levelled, etc., and where two or more shafts were connected underground, the accuracy of the survey could be readily tested. He (Mr. Lupton) supposed that the Hedley dial would come under Mr. Pollitzer's definition of a "mechanical appliance," except when the magnetic needle was used; but he left it for philosophers to decide as to whether or not it would then become an "optical instrument."

Mr. BENNETT H. BROUGH (London) thought that, with care, accurate results could be obtained with the ingenious underlay-table invented by Mr. Pollitzer. The appliance was, however, costly; and the process of using it would be extremely tedious. It was doubtful, moreover, whether better results would be obtained than those given by the T-square method,* in which the apparatus is made by the mine-carpenter and consists of a straight edge and two T-squares. The former is a planed pine-board, measuring 8 inches by ¾ inch, and 1 foot longer than the distance between the wires. It rests horizontally upon supports fixed across the shaft, and is brought to about ⅛ inch from each wire, and then nailed down to prevent slipping. The T-squares are most serviceable if made with a movable head, clamped by a thumb-screw. Except in cramped quarters, they are set at right angles. The T-squares are slid along the straight edge, until close to the wires, but not touching them, and are there fixed. Another instrument of the same type is that devised by Mr. T. H. B.

* *Mines and Minerals*, 1899, vol. xix., page 242.

Wayne.* This consists of a collapsible triangular frame, fitted with levels and legs so that it can be supported horizontally. A telescope and a vernier are placed at the apex, and the two legs are recessed at their ends so as to fit loosely round the two plumblines. If the telescope is clamped to a sight at one level, the frame may be transferred to another level where the telescope will give the same direction. Mechanical methods of connecting surface and underground surveys are the oldest, for they were used by Hero of Alexandria 2,000 years ago, and they are still most often employed, as the optical methods present serious difficulties. It is doubtful, however, whether, in shafts such as those described by the author, as satisfactory work could be done with his complicated mechanical table as would be possible with a transit-theodolite and an artificial horizon, or with the new form of theodolite invented by Mr. Dunbar D. Scott† in which the auxiliary telescope is interchangeable and may be used either as a side or as a top telescope.

The PRESIDENT (Sir Lindsay Wood, Bart.) moved a vote of thanks to Mr. Pollitzer for his paper.

Mr. J. G. WEEKS seconded the resolution, which was cordially approved.

———

Mr. E. ROUBIER read the following description of "The Max Electric Mining Lamp":—

---

* *Mining Journal*, 1902, vol. lxxii., page 911.
† *Trans. Inst. M.E.*, 1902, vol. xxiii., page 611.

THE MAX ELECTRIC MINING LAMP.

By E. ROUBIER.

The Max electric mining lamp (Fig. 1) has been designed to fulfil the requirements expressed by French mining-engineers, and has already met with considerable favour in the mines of the North of France. It is the outcome of a long series of experiments in which suggestions from practical men have been embodied as fully as possible. The accumulator is of the now wellknown Max type. In this cell, no plates are used, the electrodes are cylindrical, and constituted by rods of anti-moniated lead, round which the active material is squirted, under pressure, and thereafter covered with asbestos braiding, the jackets thus formed effectually preventing disintegration, and all possible chances of the active material falling, and establishing short circuits between electrodes of opposite polarities. Each cell comprises 24 electrodes, arranged in three rows of 8 each, the middle row forming the positive, and the two outer rows the negative poles. The electrodes are arranged so as to give the best possible utilization of the space occupied, and their cylindrical form and relatively small diameter enable the whole of the active material to enter into play, thereby ensuring a very high output per unit of weight and space. The electrodes are made by machinery, by which an absolutely uniform application of the active material is obtained, and all the electrodes are alike. There are two cells in each lamp, and each cell has an electromotive force of about 2 volts. The accumulator is recharged from a continuous-current supply at the end of the day at the rate of 1 ampère, and as the lamp is not generally exhausted at the end of each day, it would be found that 10 hours charging at 1 ampère would be sufficient.

Fig. 1.—Max Lamp.

The maintenance expenses are reduced to a minimum, since no material can drop, and therefore no washings are ever required to keep the cell clean. Moreover, the subdivision of the material in small units allows of a rapid rate of charge, and this consideration is an all-important one at collieries, or in any industry where continuous working must be ensured. The peculiar construction of the accumulator allows of its being regenerated when its capacity has dropped for any reason whatever, by means of reversal of polarities. The spilling of the electrolyte is obviated by a special type of indiarubber plug, prepared in such a manner as to allow of the escape of the gases evolved at the end of charge, whilst preventing any drop of liquid from leaking. It has been found in practice that the immobilization of the electrolyte by admixture of silicates, which renders it gelatinous, very considerably reduces the actual capacity of any cell to which it is applied; moreover, the asbestos coating of the electrodes acts as a very effective absorbent and as a check against rapid movements of the liquid when the cell is accidentally tilted. The type of lamp, lamp-protector and reflector could be easily altered to suit particular requirements. The switch, which is placed under the incandescent lamp, is arranged in such a manner as to prevent the charging-current from going through the lamp instead of charging the cell, as the charging plug pushes back the lighting switch. The Max lamp weighs about $5\frac{1}{2}$ pounds, and it gives a light of $1\frac{1}{2}$ candlepower for 10 to 12 hours, or a smaller light for a considerably longer period. It has not been thought advisable to add a gas-detector, as opinions on this subject appear to be extremely divided.

————

Mr. E. ROUBIER, replying to questions, said that if the charging current had a pressure of 200 volts a 32 candlepower lamp for 200 volts would be placed on the charging-board when charging one or two cells only; but the number of cells that can be charged on a given circuit is equal to the total supply-voltage divided by 2·6.

Mr. A. ANDREWS (Newcastle-upon-Tyne) said that the system of charging the lamp appeared to be a slow one. He supposed that with 100 lamps on a 200 volts circuit the lamps would be joined in series.

Mr. ROUBIER, replying to further questions, stated that there was only a small quantity of liquid in each cell, and the lamp could be turned upside down with safety; but if it were left in that position for 2 or 3 hours, a few drops of the liquid might escape. The lamp used a current of about 0·7 ampère; and the accumulator should be recharged when the pressure fell to 1·8 volts.

Mr. W. C. BLACKETT (Durham) said that he had used electric lamps in mines for many years for purposes of observation, and after provision had been made for the slight danger arising from the fact that electric lamps did not indicate the presence of dangerous gases, they were most useful in enabling one to see objects which could not be seen with the light of an ordinary safety-lamp. He had hitherto used a Bristol lamp, which seemed to be on the same principle as the Max lamp, but he could not remember whether the electrodes possessed asbestos-jackets.

The PRESIDENT (Sir Lindsay Wood, Bart.) asked whether the lamp, when not in use, would last a long time without the charge running down.

Mr. W. WALKER (H.M. Inspector of Mines) said that at Murton colliery some 1,200 lamps of the Sussmann type were in use, and most, if not all, of the parts of these had been manufactured at the colliery; and similar lamps had also been used at South Moor colliery with satisfactory results for some time past.

Mr. W. C. BLACKETT said that it was a great advantage to have the absorbent firmly fastened round the plates, as it prevented the paste from falling out and short-circuiting the lamp.

Mr. J. B. ATKINSON asked whether any case had occurred in France which led to the men working in a place where they ought not to have worked, on account of the presence of gas.

Mr. E. ROUBIER said that no such case had come to his knowledge.

Mr. A. ANDREWS thought that such a difficulty had been overcome by the addition of gas-detectors to electric lamps.

Mr. W. C. BLACKETT said that at Newbottle colliery the electric lamps were taken out, owing to this danger.

Mr. M. WALTON BROWN stated that the gas-tester attached to the Swan lamp, when constantly applied, used up the charge of the lamp in about 30 minutes.

Mr. J. B. ATKINSON said that if the gas-tester depended on the man putting it into operation, he might be suffocated before he thought of using it.

Mr. A. ANDREWS said that there was an automatic gas-tester, in which a porous pot was employed, and the gas raised an india-rubber valve, and agitated a relay which put a switch into contact and lighted a red lamp, and that lamp remained lit until the switch was taken off. Some of the lamps were not fitted with the relay, and, therefore, when the gas agitated the valve the red lamp was only lighted during that time; and it was necessary to watch the red lamp to see if it was lighted. Lamps, fitted with relays, were usually used in the pit over the week-ends.

Mr. M. WALTON BROWN said that the gas-tester referred to by Mr. Andrews depended for its action on the diffusion of gases, and another test for gas could not be made until the appliance had been taken into a purer atmosphere. If this were not done, the pressure of the diffused gas diminished, and the red lamp would give no further indication of the presence of gas.

Prof. H. STROUD (Durham College of Science) wrote that it is important to obtain an electric mining lamp, whose weight is sufficiently small and which at the same time gives enough light for the purpose. From the description, the Max storage-cell seems well constructed and satisfactory.

Mr. G. E. SMITH (Nottingham) wrote that, he trusted, the fact of the plates or electrodes in the Max lamp being circular would give them longer life, and render them better able to resist such rough usage as the miner's lamp frequently received in a mine. He thought that the number of electrodes (24) would also tend to extend the life of the lamp.

Mr. HENRY DAVIS (Derby) wrote that the Max electric lamp, introduced by Mr. Roubier, appeared to be substantially con-structed to stand rough usage in mining work and to give a good light; but he feared that the inventor was far from having pro-duced a lamp which would suit the ideas of the miner or the pocket of the mine-owner. The miner will object to the weight of 5½ pounds, over double that of the lamp it is intended to displace. The mine-owner will hardly be anxious to adopt this or any electric lamp, unless the first cost and upkeep does not seriously exceed that of the oil-lamp. Inventors must bear in mind, notwithstanding, that a portable electric lamp had received an immense amount of thought for years, of encouragement from manufacturers, capitalists and the owners of mines; yet it could not be said to have obtained any foothold, except in cases of exploration after an explosion, when an oil-lamp requiring oxygen could not be supported, and then the electric lamp had rendered excellent and invaluable service.

The PRESIDENT (Sir Lindsay Wood, Bart.) moved a vote of thanks to Mr. Roubier for his description of the Max electric lamp.

Mr. W. C. BLACKETT seconded the resolution, which was approved.

DISCUSSION OF PROF. A. RATEAU'S PAPER ON "THE UTILIZATION OF EXHAUST-STEAM BY THE COMBINED APPLICATION OF STEAM-ACCUMULATORS AND CONDENSING TURBINES."*

Prof. A. RATEAU (Paris) wrote that since his paper was written, the air-pump of the condenser, into which the turbine at Bruay collieries exhausts its steam, has been perfected in such wise that the vacuum now reaches 65 centimetres (25·6 inches) of mercury-column at full load. The writer thereupon requested the mining-engineers at Bruay to be kind enough to repeat some of the experiments upon the turbine-dynamo, by increasing the load as much as possible.

These experiments were accordingly carried out on January 31st, in the presence of Prof. Hubert, and yielded the results summarized in Table II. The turbine, at that date, had been working for five months (with the exception of a six weeks' lapse due to the general strike of miners).

In the course of these experiments the necessary variations of power developed were obtained by transferring progressively to the turbine the load of the reciprocating engines, which in

* *Trans. Inst. M.E.*, 1902, vol. xxiv., page 322.

TABLE II.—EXPERIMENTS WITH THE LOW-PRESSURE TURBINE AND DYNAMO OF 300 HORSEPOWER AT THE BRUAY COLLIERIES.

No. of Experiment.	Revolutions per Minute. n	Electric Horsepower.*	Absolute Pressures.				Calculated flow of Steam per Hour. I		Consumption of Steam per Horsepower-hour.					Efficiency. ρ
			Steam to Turbine. P		Condenser. p				Measured. K′		Theoretical. K		
			Kilogrammes per square centimetre.	Pounds per square inch.	kilos per square centimetre.	Pounds per square inch.	Litres.	Cubic feet.	Kilogrammes.	Pounds.	Kilogrammes.	Pounds.	
1	1,580	0	0·11	1·56	·065	1·21	600	21	—	—	—	—	—
2	1,585	79	0·35	4·98	0·095	1·35	1,920	68	24·30	53·6	13·80	30·4	0·570
3	1,620	115	0·50	7·11	0·115	1·64	2,750	97	23·90	52·7	12·15	28·8	0·515
4	1,610	213	0·75	10·67	0·141	2·01	4,120	146	19·30	42·6	10·55	23·3	0·550
5	1,610	241	0·82	11·66	0·141	2·01	4,500	159	18·70	41·2	10·10	22·3	0·545
6	1,610	282	0·85 to 1·00	12·09 to 14·22	0·154	2·19	?	?	?	?	—	—	—

The barometric pressure was 0·766 metre (30·18 inches) during the experiments

* A French electric horsepower equals 0·735 kilowatt.

ordinary circumstances are conjoined with the turbine in providing the electricity needed for the working of the colliery. At the termination of the series of experiments, this provision was secured by means of the turbine alone, and, in order to increase the load, all the lamps in the pit had been lighted. It was not found possible, however, to increase this load beyond 241 horse-power for any appreciable length of time. For a moment only, on starting a large winch, did the demand on the system attain 282 electric horsepower; but this was merely a matter of a few seconds.

Speaking generally, such experiments made in course of practical working are of only comparative accuracy, because of the continual variations of the load. The figures on the fourth and fifth lines of Table II. are undoubtedly the most exact, being each the mean of three concordant measurements, taken at times when the load was approximately stable. As to those of the sixth line, which correspond as aforesaid to a merely transitory load, they must be considered as unreliable.

If the figures of Table II. be compared with those of Table I.,[*] it will be seen that the results of practical working agree very fairly with those obtained in the course of experiments at the factory. For instance, at loads varying between 213 and 241 electric horsepower, the counter-pressure at the outlet of the turbine being sensibly the same as that obtained experimentally at the works, the consumption of steam per electric-horsepower-hour approximates to 19 kilogrammes (42 pounds) and the efficiency of the turbine and dynamo taken together is 0·54 or 0·55, at a speed of 1,600 revolutions per minute. These results are concordant with those expressed by the curves drawn in Figs. 4 and 5 (Plate IX.).[†]

It may be added that the turbine and the dynamos work silently, without any appreciable vibration.

Mr. M. WALTON BROWN mentioned that after the erection of a high-pressure beam pumping-engine at the Farme collieries, with a cylinder 24 inches in diameter, the exhaust-steam was passed into a boiler, fitted with a safety-valve loaded at 5 pounds on the square inch. A pumping-engine and a winding-engine,

---

[*] *Trans. Inst. M.E.*, 1902, vol. xxiv., page 347.

[†] *Ibid.*, vol. xxiv., page 348.

with cylinders 60 inches and 42 inches respectively in diameter were thus supplied with steam, and any deficiency was supplied from a high-pressure boiler.

———

DISCUSSION OF DR. BROOCKMANN'S PAPER ON "THE GASES ENCLOSED IN COAL"[*] AND DR. P. P. BEDSON'S PAPER ON "THE GASES ENCLOSED IN COAL AND COAL-DUST."[†]

Mr. J. W. THOMAS (London) wrote that the accuracy of Dr. Bedson's and his own experiments on account of using india-rubber connections had been impugned by Dr. Broockmann, who said that "he (Dr. Bedson) used about the same method as that employed by Mr. Thomas . . . . the glass vessel being closed by a perforated indiarubber-stopper," etc.   Dr. Bedson had defended his own method. With regard to his (Mr. Thomas'), Dr. Broockmann had not even made himself acquainted with the facts. He (Mr. Thomas) had never used a stopper—his tubes were always drawn off and bent over in the blowpipe-flame, so as to make one joint with the Sprengel air-pump by an indiarubber tube, which was surrounded by a water-jacket in every experiment with all kinds of coal, a perfectly reliable method. Dr. Broockmann stated that "indiarubber is as permeable to gases as a sieve is to water." Such language was as inaccurate as it was sweeping, and he (Mr. Thomas) did not think that any appreciable error could occur in Dr. Bedson's analyses from the method adopted by him. Dr. Broockmann used the indiarubber chimera to throw discredit on the large volume of gases obtained from the coal of the North of England and of South Wales, but that large volume could be obtained from fresh coal without difficulty.

With regard to the action of oxygen upon coal sealed up in tubes with air, Dr. Broockmann's remarks were incomprehensible.[‡] If acetic acid was formed, why leave it questionable when it was so easy to detect? and further, as to its vapour not being "absorbed by caustic potash?"   He (Mr. Thomas) did not wish to point out absurdities, but only to note that there was good reason for "doubting the accuracy" of Table II.   Lignites and Tertiary coals did absorb oxygen at moderately low temperatures, but the

[*] *Trans. Inst. M.E.*, 1902, vol. xxiv., page 18.
[†] *Ibid.*, page 27.   [‡] *Ibid.*, page 24.

ordinary circumstances are conjoined with the turbin
ing the electricity needed for the working of the colli
termination of the series of experiments, this p
secured by means of the turbine alone, and, in ord
the load, all the lamps in the pit had been lighte
found possible, however, to increase this load bey
power for any appreciable length of time.   For :
on starting a large winch, did the demand on tl
282 electric horsepower; but this was merely a
seconds.

Speaking generally, such experiments m:
practical working are of only comparative ac
the continual variations of the load.   The figur
fifth lines of Table II. are undoubtedly the m
the mean of three concordant measurements,
the load was approximately stable.   As to tl
which correspond as aforesaid to a merely
must be considered as unreliable.

If the figures of Table II. be compared
it will be seen that the results of practi
fairly with those obtained in the cours
factory.   For instance, at loads varyin
electric horsepower, the counter-press
turbine being sensibly the same as tha:
at the works, the consumption of stea
hour approximates to 19 kilogram
efficiency of the turbine and dynamo t
at a speed of 1,600 revolutions per
concordant with those expressed by
and 5 (Plate IX.).†

It may be added that the tu
silently, without any appreciable v

Mr. M. WALTON BROWN men
a high-pressure beam pumping-
with a cylinder 24 inches in
passed into a boiler, fitted with
on the square inch

to
a the
, and
udy of
able to
Walton

indeed,
e electric
responds to
a with the

* Trans. In
† Ibid., vol

E/R, where C is
electromotive
conductors
ors supplied
as it stands,
ole to a single
from two points
. Thus, in any
e is applied, the
pon the law given
of the incandescent
n its filament varies
t its terminals, and
n terminal to terminal,
omplete system is con-
aw applies equally; but
stances, which the electric
been included in the equa-
one likely to be omitted,
tance of the generator itself.
e generator is taken into the
where R is the resistance of the
and r that of the generator itself.
nt, it will be remembered, passes
al to the generator, back to the
generator-coils to the terminal
current passing continuously round
nce of the generator-coils making a
the current through them, just as if
external circuit, and were not engaged
c pressure. Again the electric pressure
ge of the coils, usually of the armature,
field of the dynamo, and the electric
rect conversion from the mechanical energy
. Also the electric pressure created depends
number of conductors assembled on the
n the speed at which it is driven through the
Again, the charge for the passage of the current
ature-coils depends directly upon the square of
the current multiplied by the resistance of the

best steam-coals of the North of England and South Wales were
not appreciably affected. The reason why the latter stood for long
periods practically unweathered and unchanged in the coaling-
depôts of warm countries was because they so strongly resisted
oxidation.

Those acquainted with the analysis of compound and combustible gases would know that all errors of manipulation, where
the real values were less than 100 (and they almost always were less
than 100) appeared as nitrogen, because this was estimated by
difference. A glance at Table I. in Dr. Broockmann's paper*
would reveal figures which any gas-analyst might challenge, and
these, to say the least, were as much open to suspicion as Dr.
Bedson's excess of oxygen in the analysis under dispute. He
(Mr. Thomas) knew of no Carboniferous coal which did not contain
nitrogen among its occluded gases before being subjected to exhaustion by the Sprengel air-pump; and it was always present in
the blower-gases that he had examined, which were free from
oxygen and contained no air.

———

DISCUSSION OF MR. É. GOSSERIES' PAPER ON "THE GUIBAL FAN COMPARED WITH A DYNAMO."†

Mr. SYDNEY F. WALKER (London) wrote that Mr. Gosseries
appeared to have started with a good idea. His reasonings, though
correct in the main, were exceedingly difficult to follow, but in
the end he became greatly involved. He (Mr. Walker) thought,
therefore, that he would render the best service that he could to
the members if he gave simply the idea that he had formed on the
subject of the similarity between the action of air in mines, and
of electric currents in a supply-service, from his previous study of
the question, and from the study which he had now been able to
make of it, from Mr. É. Gosseries' paper and from Mr. M. Walton
Brown's pamphlet.‡

The analogy between the two actions is very close indeed,
especially if a dynamo be used as a generator. The electric
generator corresponds to the fan; and the mine corresponds to
the circuits, the distribution-service in connection with the
electric generator.

* *Trans. Inst. M.E.*, 1902, vol. xxiv., page 23.
† *Ibid.*, 1901, vol. xxi., page 568.
‡ *A Theory of Mine-ventilation*, 1884.

I.—A.—In electricity, there is the law : $C = E/R$, where C is the strength of the electric current passing ; E, the electromotive force, or electric pressure ; and R, the resistance of the conductors of the distribution-system, or any system of conductors supplied by the generator. It will be understood that the law as it stands, in the simple form given above, is really applicable to a single conductor, or system of conductors, all emanating from two points between which the difference of pressure exists. Thus, in any single conductor to which an electric pressure is applied, the current, passing through it, depends simply upon the law given above. This is well illustrated by the case of the incandescent electric lamp. The current passing through its filament varies directly as the pressure which arrives at its terminals, and inversely as the resistance of the lamp from terminal to terminal, while the current is passing. When a complete system is considered, including the generator, the law applies equally ; but it is necessary to be sure that all the resistances, which the electric pressure will have to overcome, have been included in the equation. The most important, and the one likely to be omitted, unless warning is given, is the resistance of the generator itself. The law is usually stated, when the generator is taken into the equation, thus :—$C = E/R + r$, where R is the resistance of the circuits external to the generator ; and r that of the generator itself. In the electric service, the current, it will be remembered, passes through the conductors external to the generator, back to the generator, and through the generator-coils to the terminal whence it first started, the current passing continuously round and round, and the resistance of the generator-coils making a charge for the passage of the current through them, just as if they formed part of the external circuit, and were not engaged in generating the electric pressure. Again the electric pressure is created by the passage of the coils, usually of the armature, through the magnetic field of the dynamo, and the electric energy created is a direct conversion from the mechanical energy of the driving-engine. Also the electric pressure created depends directly upon the number of conductors assembled on the armature, and upon the speed at which it is driven through the magnetic field. Again, the charge for the passage of the current through the armature-coils depends directly upon the square of the strength of the current multiplied by the resistance of the coils, or it depends upon the current-strength simply multiplied by the pressure which is used to force the current through the coils. In any case, as the current which is being taken from the generator increases, so does the charge for its passage through the generator itself, and so does the pressure used in the generator.

B.—Passing on to the distribution-service, the current delivered to the service depends directly upon the pressure delivered to the terminals of the service, and inversely upon the total resistance offered by the distribution-system, the cables, lamps, motors, etc.; and the total current passing through the service, and through the generator, depends upon the pressure created in the generator when the current is passing (the current at any particular instant) and inversely upon the total resistance opposed to it, which is again made up of the resistance of the cables, lamps, etc., and that of the generator itself. The resistance offered by the cables and by any conductors through which the current passes, depends directly upon the length of the conductor, and inversely upon its sectional area. That is to say, the longer that the conductor is the greater is the pressure required to drive a certain current through it; or inversely, with a given pressure, the less will be the current that can pass. On the other hand, the larger the conductor, the greater its sectional area, the more easily will it allow the current to pass; the greater current will pass with a given pressure, and the smaller pressure that will be required for a given current. Further, unless special provision is made to counteract it, the passage of a larger current through the generator, that is to say the demand for a larger current by the external service, the lower will be the pressure delivered at the terminals of the service and the lower will be the pressure delivered at any point in the service. This is well illustrated by the case of a lighting-service at a colliery. The pressure at the pit-top will be, say, 100 volts, and so long as no current is passing it will be the same at the pit-bottom; but so soon as a current passes through the shaft-cables, the pressure is lowered in direct proportion to the strength of the current and the resistance of the cables. In most dynamos used for lighting and power at collieries, provision is made for keeping a constant pressure at the terminals of the generator, no matter what may be the current, and the alternator will, therefore, probably more nearly represent the action of the fan under similar conditions. With the alternator, as the current which is taken from the generator rises, the pressure at its terminals falls, unless provision is made for maintaining it, by increasing the speed, or increasing the strength of the magnetic field in which the armature-coils move. The fall of pressure in the alternator is due to two factors, the increased charge made for the passage of the increased current through the coils, and the increased self-induction of the currents passing in the coils themselves. This is mentioned, as it has a counterpart in the fan.

II.—Now as to the fan and the air-distribution service.

A.—In the fan, the vanes represent the armature-conductors of the electric generator, and the total pressure delivered by the fan depends upon the number of vanes, and upon the speed at which the fan is driven. This, of course, may be taken as referring to any particular type of fan. The same thing applies to electric generators, as all the arguments, which have been enunciated, are to be taken as applying to a particular type of generator. Any type of generator will do, but the laws, for simplicity, must follow one type.

It appears to him (Mr. S. F. Walker) that pressure created by the fan corresponds to pressure created by the dynamo, and depends upon the quantities stated, irrespective of the quantity of air passing. The pressure gives the ability to drive air through air-conductors, mine-galleries, air-passages of various forms, including the passages of the fan itself. The useful pressure, which appears at the fan-outlet, is the total pressure which has been created by the vanes of the fan, less the charge made upon that pressure for the passage of the air-current through the fan, just as the pressure at the terminals of the electric generator is the total pressure created, less the charge for the passage of the current through its own coils. Further, it appears also that there will be additional charges upon the pressure delivered at the outlet of the fan, due to eddy-currents within the fan itself, which will increase in some ratio with the strength of the air-current itself, corresponding to the self-induction in the alternator. What Prof. T. Guibal had named the "temperament of the mine" corresponds to conductivity in an electrical distribution-service, the reciprocal of resistance. In electrical work it had been found more convenient to use resistance in calculation, instead of conductivity, but the latter could be used, and

there is a unit of conductivity, the mho, in contra-distinction to
the unit of resistance, the ohm.   The laws which govern the distribution of air in a mine appear to be the same as those governing the distribution of electricity to lamps, motors, etc., with
certain important modifications.   Thus, the resistance offered by
any airway in a mine depends upon its length, inversely upon its
area, but it also depends directly upon its perimeter, and this
factor has to be included in the equation.   With electricity, the
resistance offered by a conductor depends directly upon the length
of the conductor, and inversely upon its sectional area, or equals
l/a, where l represents the length, and a the sectional area.   With
an airway in a mine, the resistance, it appears, is represented
by lp/a, where l again represents length, a the area, and p the
perimeter; and the two quantities are brought closer into agreement by the fact that the actual resistance of an electric conductor
is found, when its dimensions are known, by multiplying the
above fraction by the specific resistance of the metal of which
the conductor is made, that is to say by the proportion which any
given section of the substance bears to the standard; and by the
fact that, with air-currents, the actual resistance is found by
multiplying the above fraction by the coefficient of friction.
Given, however, the above modifications, the two seem to follow
the same law.   There is, however, another modification, to be
introduced before the two sets of equations can be brought into
line.   With electricity, the current passing simply depends upon
the pressure divided by the resistance; with air-currents, it depends
upon the square of the pressure divided by the resistance.   But,
with this modification, it appears that the two, electrical distribution and air-distribution, follow exactly the same laws.   Thus,
what corresponds to current with air, the quantity of air passing,
is found by dividing the square of the pressure by the resistance,
as stated in the preceding equation.   But the law must be applied
fully, that is to say, the resistance offered by the fan itself must
be taken into account.   Thus, when additional roads are opened
in the mine, offering additional paths for the air, the increased
air-current passing through the fan lowers the pressure available on the outside, while the actual quantity of air passing in
the mine depends upon the square of the available pressure at the
fan-outlet, divided by the resistance offered by the mine itself.
The case is similar to an electrical distribution, where with an

incandescent-lamp service, additional lamps are turned on, and the supply-cables are small. The light given by the individual lamps is less than before, but the total current passing through the cables is larger. The same thing may apply to an electric generator, if it is not properly compounded, or if it is alternating, and provision is not made for raising the pressure as the current taken from it increases. The pressure at its terminals lowers, and with it the light given by the lamps supplied, though the total current being furnished by the generator may be considerably higher than when the lights are brighter. With air-currents, and with electric-currents, the same rule appears to hold good, with the modification mentioned. The actual current passing depends upon the pressure available, or its square, and the resistance offered to it.

The question of splits with air, which Mr. Gosseries had attempted to discuss, appears to follow the same law as with electricity, so far as resistance is concerned, with the above reservations: the addition of the perimeter to one side of the equation, and the square of the pressure instead of the simple pressure. When two or more conductors are bridged between two points, between which a difference of pressure exists, the current which passes in each of them will be inversely as their resistances. That is to say, the total current which passes through the system of conductors connected to these two points will be divided between the conductors in the inverse ratio of their resistances; but the total current itself which passes will depend, again, upon the pressure available at the two points, and inversely upon the combined resistance offered by the system of conductors. In calculating the combined resistance offered by a system of conductors ending in two points, in electrical work, it is customary to make use of their reciprocals. Thus if r, r_1 and r_2 be three conductors connected in parallel, as it is termed in electrical work, to find their combined resistance, their conductivities, $1/r$, $1/r_1$ and $1/r_2$ are added together, giving the fraction $r_1 r_2 + r r_2 + r r_1 \div r r_1 r_2$; and reversing this, the combined resistance is $r r_1 r_2 \div r_1 r_2 + r r_1 + r r_2$, or the combined resistance of the several circuits is equal to the product of their resistances divided by the sum of the quantities shown. With a service, such as that used for incandescent lamps, the matter simplifies itself very much: all the lamps may be taken to have the same resistance, and the combined resistance is equal to the resistance of one lamp divided by their number. It appears that splits of air in mines must follow much the same law. Each will take its share of the air available, in inverse proportion to its own resistance, or in Mr. Guibal's terms, in direct proportion to its temperament. But each will take air, irrespective of all the others, directly in proportion to the square of the pressure available at its ends, and inversely in proportion to its own resistance. The combined resistance of the splits will determine what quantity of air will pass, because the pressure available at the ends of each individual split will be regulated by the quantity of air passing through the fan, and the roads leading up to the split, and away from it. One point, which Mr. M. Walton Brown makes in his pamphlet,[*] appears to correspond directly with electrical conditions:—The return-airway, if not of sufficient area, will throttle the air, and so reduce the available pressure at the ends of any individual branch, just as much as the intake-airways or the branches themselves. The air has to be driven or sucked through the return-airways, before it can pass out of the mine and fresh air pass in, and any resistance offered by it will use up pressure exactly as does a return-cable.

Prof. H. STROUD (Durham College of Science) wrote that the reference to a dynamo occurs in two paragraphs of Mr. Gosseries' paper. In addition to the circuit where the electric energy is utilised, a second (shunt) circuit is arranged, and apparently the idea is to keep the current through the dynamo constant by varying the resistance of the shunt-circuit. It is certainly never desirable in the case of a dynamo (magneto-electric machine) to keep the current through the dynamo constant, even when no electric energy is being utilized outside; and that it is desirable in the case of a fan to introduce air from the outside, by means of a tube communicating directly with the atmosphere, does not seem at all clear. There is also a statement that " the resistance to the movement of the air in each split shall be the same and equal to $P_A - P_B$," in which the author uses the word resistance in a sense different from its usual meaning.

----

* A Theory of Mine-ventilation, 1884, page 7.

NORTH STAFFORDSHIRE INSTITUTE OF MINING AND MECHANICAL ENGINEERS.

GENERAL MEETING,
HELD AT HANLEY, DECEMBER 8TH, 1902.

MR. A. M. HENSHAW, PRESIDENT, IN THE CHAIR.

The minutes of the last General Meeting were read and confirmed.

The following gentlemen, having been previously nominated, were elected:—

MEMBERS—

MR. JOHN CORK, Mechanical Engineer, Midland Coal, Coke and Iron Company, Limited.

MR. W. H. DAVIES, General Manager, Shelton Iron, Steel and Coal Company, Limited.

MR. J. W. HARTLEY, Engineer, Drysdale House, Stone.

MR. EVAN JARVIS, Blast-furnace Manager, Midland Coal, Coke and Iron Company, Limited.

MR. FRANK NEWTON, Mechanical Engineer, Longport.

ASSOCIATES—

MR. FRANCIS W. KNIGHT, Mining Student, Newcastle, Staffordshire.

MR. C. H. CLARK, Assistant Mining Engineer and Surveyor, Newton-le-Willows.

STUDENT—

MR. CHAS. A. WEBBERLEY, Mining Student, Moneta House, Dresden, Longton.

Mr. J. T. STOBBS read the following "Notes on the Map and Sheet-memoir of the North Staffordshire Coal-field, recently published by the Geological Survey of Great Britain":—

NOTES ON THE MAP AND SHEET-MEMOIR OF THE NORTH STAFFORDSHIRE COAL-FIELD, RECENTLY PUBLISHED BY THE GEOLOGICAL SURVEY OF GREAT BRITAIN.* .

By J. T. STOBBS.

1. INTRODUCTORY.

Two years ago, the then President of this Institute (Mr. W. N. Atkinson) said that " the members had long been looking for the publication of the new series of the Geological Survey maps, and they must feel glad that there was a prospect that these would soon be obtainable."† However, in the fulness of time our patience has been rewarded and the maps and the memoir have at last been published. Much has been written about the North Staffordshire coal-field, with and without knowledge, and from some writers one may gather that its true significance has not been apprehended. The following passage may be cited from Mr. A. L. Steavenson's review of the " Report upon the Working and Regulation of Fiery Mines in England by Messrs. Pernolet and Aguillon ":—" ' The basin . . . . of North Staffordshire might be added to this list [that is, of fiery districts], although it is generally admitted that the reports of the inspector [of mines] indicate that the accidents which have occurred in the . . . . basin are more due to a vicious organization of the workings than to the quantity of gas.' So much for North Staffordshire. No wonder the Commissioners go on to state, ' that for this reason we abstained from visiting North Staffordshire.' "‡ The new memoir, one trusts, will make it clear that the French Com.

---

* *Memoirs of the Geological Survey, England and Wales: The Geology of the Country around Stoke-upon-Trent* (Explanation of Sheet 123), by Messrs. Walcot Gibson and C. B. Wedd.

† *Trans. Inst. M.E.,* 1900, vol. xx., page 78.

‡ *Trans. N.E. Inst.,* 1881, vol. xxxi., page 8.

missioners also "abstained from visiting" the most interesting
and (per unit of area) the richest coal-field in Britain, and its
perusal will show that "the vicious organisation of the workings"
is not the cause, but the effect of the natural difficulties due to the
disposition of the seams, which, in variety of faulting and folding, are unequalled in this country.

2. THE MAPS AND THE MEMOIR.

The new maps are beautiful specimens of colour-printing, and
those responsible for their production are to be congratulated on
the result. According to the preface, "they represent the first
attempt to substitute colour-printing for hand-colouring in the
issue of the one-inch Geological Survey maps,"* and it is to be
hoped that the experiment will prove an unqualified success.
Compared with the old map, the new one is very much cheaper,
and is quite within the reach of every miner—even of the poorest.
The sheet also is larger, and contains a greater area of the Pottery
coal-field, although it is a pity that the apical 6 square miles of
the Biddulph area must be sought on a neighbouring sheet, this
being the only missing patch. The map and index-colouring is
absolutely uniform, and the reference-symbols have been most
usefully amplified. The map is, indeed, a model of what a
geological map should be. The publication of the "solid" sheet
is a boon to mining-engineers, and its appreciation will be
evidenced by its sale.

The explanatory memoir has been somewhat carelessly printed,
mistakes of spelling are not uncommon, and cases even occur
where the geological expert has apparently been corrected by the
printer's devil. A serious error, however, occurs in Fig. 2,† where
the "First Grit" is printed "Fourth Grit," the mistake being
aggravated by the fact that the Fourth Grit is found below, and
not above, the Third Grit.

3. HOW TO READ THE MAP.

The present state of knowledge regarding the structure of the
coal-field, the nature, succession, and exposure of its measures is
epitomized in the maps and memoir. As this knowledge advances,
additions, and possibly corrections, will have to be made. The
work embodied in the maps is not placed before us as incapable

---

* *Op. cit.*, page 1.   † *Ibid.*, page 9.

of development; on the contrary, the accompanying memoir frequently indicates subjects of inquiry that demand extended research and investigation. In the course of further exploitation of the coal-mines, some of the details given may require modification—indeed it would be marvellous should such not be the case.

Let us try to form an estimate of the task involved in the production of such a map. The structure of the district is very complicated; the faults, and also the workable coal-seams, are numerous, and their nomenclature is often confusing. Comparing for a moment the work of the Government geologist with that of a private person, the latter is a free-lance, working a set of beds or an area in which he is specially interested, so long as they promise elucidation. The time that he spends on a limited area is his own affair. Should he fail to clear up difficulties in the structure, they are not referred to by him—they are " shelved," for the time being at any rate; but with the official the case is widely different. In a given locality, a marked bed may not be exposed: a characteristic fossil may not have been seen: fundamental observations may be incomplete: the one essential proof that would stamp the entire work with exactness, may be wanting. Whether doubtful or certain, he must fill the sheet, as no blank patches are permissible. It is always possible, therefore, that some obscure areas are marked on the map with the same clearness and precision as those of which the geologist is absolutely certain. So that in a right use of the maps, this aspect should be kept in view. Finality is not claimed for the work; it is only the latest, not the last, interpretation of the geological phenomena of our district.

4. Future Publications.

The statement by the Director (Mr. J. J. H. Teall) that "fuller descriptions of the rocks, and especially of the Coal-measures, will be given in a larger memoir dealing with the coal-field as a whole,"* will be heartily welcomed by mining-engineers. They earnestly desire that it will not be deferred till other localities are surveyed, for North Staffordshire has had more than its share of delays in connection with the Geological Survey maps, and any information concerning this mining district should be published when it is fresh, as nowhere else is information in danger of getting

* *Op. cit.,* page 2.

more rapidly out-of-date. The expression, "the coal-field as a whole," it is hoped may be taken to imply that the adjoining coal-field of Cheadle will be treated as an integral part of this district. The writer also trusts that the very full palæontological lists drawn up by local geologists will be incorporated in this memoir. No coal-field has been more diligently examined, or has yielded richer finds to the fossil-collector. Palæontological material, "in bulk," can only be adequately worked by the residents of a locality, for in many instances the opportunities of discovery are transient, and if not seized at the moment, are lost for ever.

Judging from the sheet-memoir in our hands, it is not too much to expect that the larger treatise will be a type-memoir of a type-coal-field, which should be directly for North Staffordshire and indirectly for all students of Coal-measure geology, what the late Prof. Jukes' memoir has so long been for South Staffordshire, namely, a work, known by heart by the mining-engineers of that district, which has been culled from by every subsequent writer of text-books on coal-mining, and which, moreover, ranks as one of the classics of geological literature.

With reference to the maps, this Institute has already memo-rialized the authorities on the desirability of their publication on the scale of 1 mile to 6 inches. Our district has peculiar claims for exceptional treatment in this matter. In comparatively flat coal-fields (such as those lying on the east of the Pennine chain) with few coal-seams, the lines of outcrop can be distinctly indicated on the 1 inch map; but in North Staffordshire, with its varying dips, its considerable faults, and numerous coal- and ironstone-seams in close proximity to one another, the outcrops cannot be shewn on the 1 inch map (where $\frac{1}{8}$ inch represents 660 feet), and therefore, for the purpose of taking measurements from, it is not of much use. The writer may here be allowed to mention that it was the necessity for reliable maps of the great mineral districts on a working scale that led the illustrious founder of the Geo-logical Survey (Sir H. T. de la Beche) to conceive the public benefit to be derived from such an institution; and the claims of these districts should not be set aside, even for those of pure science, since the well-being of the country, to a large extent, depends on the development of the mining industries, and any assistance rendered them by the Geological Survey promotes the prosperity of the nation.

5. Classification of the Carboniferous Rocks.

The treatment of the Carboniferous rocks, in the memoir, affords much material to stimulate thought in the 66 pages (out of 82 pages) devoted to this subject. The divisions of the Carboniferous system are given as follows :—

(d) Upper Coal Series, Keele Series.
                Newcastle-under-Lyme Series.
                Etruria Marl Series.
                Blackband Series.
(c) Middle Coal Series, Bassey Mine to the Ash Coal-seam.
(b) Lower Coal Series, Ash Coal-seam to the Winpenny Coal-seam.
                Below the Winpenny Coal-seam.
(a) Millstone Grit Series.

There is some discordance between the above classification, and the statement that " This thickness [of Coal-measures] is made up of three groups : a lower portion about 1,200 feet thick in which the seams of coal are of small commercial value; a middle portion, about 4,000 feet thick, containing all the chief seams of coal; and an upper portion, about 2,000 feet thick, with no workable seams of coal, but several bands of valuable ironstone."[*] The " middle portion," referred to in this quotation, embraces part of the (b) Lower Coal series of the above table. The writer lays stress on this confusion of terms, because it is inseparable from any arbitrary division of the Coal-measures, such as that here attempted. No division should be drawn that cannot be mapped, or that may not be of service as a tie-line in connecting the measures of different coal-fields. The " middle " and " lower " measures evidently will not admit of mapping in North Staffordshire, or in any other of our coal-fields, and to make " confusion worse confounded " the same terms are freely used in every coal-mining district; but it must not for a moment be understood that they indicate the same beds in all cases. In this locality, there is no break from the Millstone Grits to the base of the Blackband series that might offer a convenient line of demarcation, and the adoption of certain coal-seams for this purpose, in the face of the statement that " the presence or absence of workable seams of coal or ironstone does not afford a scientific basis for classification "[†] is a retrograde step. The division into " lower " and " middle " is particularly weak, and the

* *Op. cit.*, page 17.   † *Ibid.*, page 17.

adherence to the use of these terms is a remnant of the old system of grouping,* which, in the light of present knowledge of the relationship existing between the measures of different coal-fields, may be seen to have done little or nothing to promote generalization in the past, nor does it promise to be more helpful in the future. From a mining-engineer's point of view, no useful end can be served by their retention. If divisional lines must be drawn, they should be applicable to, and have the same signification in, other coal-fields; they ought, therefore, to be palæontological.

The long continuance of the series of see-saw changes in the marine and freshwater conditions of deposition of the beds, from the Crabtree coal-seam to the Bassey mine, forbids the introduction of compartments into the system other than those of the " zone " type.

In marked contrast to the unsatisfactory state of classification of the beds below the Bassey mine, those above that horizon have been brilliantly worked out by Mr. Walcot Gibson, not only in North Staffordshire but in the other Midland coal-fields. The subdivisions of the Upper Coal series, introduced by him, are natural, and have good palæontological upper and lower limits. Once they have been pointed out to us, they can be readily distinguished. On this sheet they are separately mapped for the first time, and the fact that this has been possible, and that the equivalent beds have been recognized in so many other districts, is evidence enough that the work is thoroughly sound and of permanent value; indeed, it may be taken as an object-lesson on the subdivision of Coal-measures. Unlike the " middles " and " lowers " no confusion can arise with these groups, and, above all, the divisions are of practical use in mining exploration, as proved in the writer's experience.

The chapter on the Millstone Grit, by Mr. G. Barrow, should be read by mining-engineers (especially in those districts where the grits are split up into thinner and more numerous beds) shewing as it does, the rapid and local changes in their character, rendering necessary greater caution in naming single sandstones " the Millstone Grit," simply because of coarseness or low position in the Coal-measures.

* *The Iron-ores of Great Britain*, part iv., page 258.

Dr. Wheelton Hind is to be congratulated on the official adoption of his views, by the deletion of the "Yoredale Sandstones" from the North Staffordshire list of rocks, the beds previously described as such being now included in the "Millstone Grit series."

6. Correlation of Beds.

The large number of comparative sections given in the memoir show that a serious attempt has been made to trace the seams throughout the coal-field. The difficulties of the task have been enumerated by the writer in a previous communication to this Institute,[*] and he sympathizes with anyone undertaking the work. To some of these correlations, however, exception must be taken, namely:—

(1) The Little Row seam of the Talke area has been proved on palæontological evidence to be identical with the Bowling Alley seam of the Potteries,[†] but in Fig. 6 of the memoir, they are represented as being separated by three other coal-seams and measures.

(2) The Brown mine is marked as the equivalent of the Knowles or Winghay seam (Fig. 7), but in the Longton area, the Priorsfield measures are found between them.

(3) The table shewing the characters of the chief seams, whether or not intended as a piece of correlation, is very likely to be interpreted as such.[‡] It may then be pointed out that the Yard coal-seam of the eastern area is not the correlative of the Single Five-feet seam of the western area, nor is the Stoney Eight-feet seam of the eastern area the equivalent of the Hams coal-seam of the western area.

The suggestion that the "light grey grit" below the Winpenny seam is possibly identical with the Woodhead rock of the Cheadle district,[§] may be of importance in fixing the relation between the two coal-fields; but as no supporting evidence is adduced, the correlation probably is purely hypothetical. The question involved is of moment both to science and to industry, and it should, therefore, either have been more thoroughly investigated or not referred to in this way.

* "Recent Work in the Correlation of the Measures of the Pottery Coal-field," _Trans. Inst. M.E._, 1901, vol. xxii., pages 229 to 246.

† _Ibid._, page 242 to 243.　　‡ _Op. cit._, page 24.　　§ _Ibid._, page 18.

7. PALÆONTOLOGY.

The treatment of this subject is much too condensed. It is a matter for wonder how long it will be, ere this science is systematically and seriously utilized in the examination of coal-fields. The reason why we know less of Coal-measure geology than, say, of Ordovician, is because the aid of palæontology has not been enlisted to the same extent in its behalf.

North Staffordshire has been splendidly worked by Dr. Wheelton Hind and Mr. John Ward, and the recognition of their services to the cause of geology is honourable alike to them and to the officers of the Geological Survey.

The only rich bed of plant-remains, mentioned in the memoir, is that lying above the Bowling Alley seam. The writer's experience of Coal-measure plant-beds is, that they vary much in the relative abundance of fossils, from place to place. The material, prior to deposition, would float about on the surface of the water, and according to the prevalence of winds and currents would be drifted together in sheltered spots. It is only to be expected, then, that the beds containing them should be rich in certain limited areas.

Some of the best plant-beds in this district are :—(1) About 60 feet below the Hard mine ; (2) about 36 feet above the Twist coal-seam ; and (3) the roof of the Red mine.

The only fossil mentioned in connection with the Blackband series is *Anthracomya Phillipsii*, which admittedly is the most important, but surely the abundance and the beautiful state of preservation of *Stigmaria ficoides* in the ironstones, is deserving of notice. It is specially interesting to those who rely on its abundance in the underclays of coal-seams as confirmatory proof of the " growth-in-place " theory of the origin of coal, for in these ironstones about 80 per cent. of the ribbon-like rootlets are to be seen attached to the root, testifying that they have not been transported far from their place of growth. In this case, however, they are always found immediately above the coal-seams. Further, of all the underclays examined by the writer, he cannot recollect one that will for an instant bear comparison with these Blackband ironstones, either in point of number, or perfection of preservation, of these roots.

8. ADDITIONS TO OUR KNOWLEDGE OF THE COAL-MEASURES.

The most important advance made during the last 50 years in Coal-measure geology is brought before us in chapter iv. of this memoir,* which is a masterly exposition of all that is known of the rocks termed the Upper Coal Series in this district.   Here we are brought face to face with work of the highest value, which has been carried clean through, from first to last, by Mr. Walcot Gibson. This is the chapter which will be sought out by students of Carboniferous geology : and in every line of it we read the enthusiasm of the discoverer.   By it we are made to feel that we are getting things at first-hand, and we reach the end realizing that a great work has been accomplished in our midst.   The enigma of these Upper Measures has at last been cleared up ; and after being classed as Coal-measures, Permian and New Red Sandstone, they are now shewn by conformity and fossil-contents to be part of the Carboniferous system.   The compartments into which they are divided are distinguished by physical characters easily identified by the unlettered man : moreover, they extend throughout the Midlands, and their recognition has thrown the desired light on the original continuity of these coal-fields.   Such are the features that give to this work the hall-mark of solidity, and it may be relied upon to stand the test of time.   In conclusion, the writer acknowledged the impetus given to the study of the coal-field by these publications, which were now very widely distributed. Students might also consult papers by Dr. Wheelton Hind appearing in *The Geological Magazine* of December, 1902, and *The Staffordshire Sentinel* of November 5th, 1902.

———

The PRESIDENT (Mr. A. M. Henshaw) said that, if he had to make any complaint about the maps, it related to the smallness of the scale ; but, a petition was being forwarded to the proper authorities, praying that the maps might be published on the scale of 6 inches to 1 mile, in order that the details might be more carefully studied.

Dr. WHEELTON HIND observed that Mr. Stobbs had accurately reviewed the mistakes and the good points of the memoir and maps. There was no finality in the matter : they represented the know-

* *Op. cit.*, chapter iv., page 35.

ledge of to-day, and to-morrow they might all be out of date.   He should like to thank the mining-engineers of North Staffordshire, because without the assistance which they had most generously given to the officers of the Geological Survey, the map could not have contained the information or attained the proportions that it had reached.   The survey was more accurate because of machine-printing, and the vagaries of hand-painting were omitted.   All the old geological maps were hand-coloured, and if they compared three or four of the maps the tints were different, and the colours on the margin were different from those on the map itself.   There was no break by which mining-engineers could divide the Coal-measures.   There was no reason, from fossil evidence or stratigraphical evidence, by which to make a sub-division in North Staffordshire of Middle and Lower Coal-measures.   Therefore, he commended those who had made the map in not subdividing the Middle and Lower measures.   The Upper Coal-measures were divided, because of the extension of the Coal-measures towards the west.   Then it was cheering to know that there was underneath Stoke and Wolstanton, and west-ward, a coal-field untouched, at a depth to which they would have no difficulty in sinking.

Mr. J. T. Stobbs said that Dr. Wheelton Hind had been teach-ing for many years that nothing was of such service in mapping the Coal-measures as palæontology.   He (Mr. Stobbs) did not agree with that opinion at first, as it was different from what he had been taught.   It was a new point of view; but the more he had examined the matter, the more certain he was that it was only in that direction that they could effect progress in the geology of the Coal-measures.

Mr. E. B. Wain moved a vote of thanks to Mr. Stobbs for his paper, which was very interesting.

Mr. J. Lockett seconded the resolution, which was cordially approved.

————

Mr. James Ashworth's paper on "The Gray Type of Safety-lamp" was read as follows:—

THE GRAY TYPE OF SAFETY-LAMP

BY JAMES ASHWORTH.

The Gray type of safety-lamp came into prominent notoriety after the publication in 1886 of the *Final Report* of H.M. Commissioners appointed to inquire into Accidents in Mines, because it was one of the four lamps which the Commissioners singled out as being "deserving of special attention."[*] The Commissioners classed it as belonging to the Eloin type (Figs. 1 and 2, Plate IV.), but the writer considers that it has more of the characteristics of the Mueseler-Arnould-Godin lamp (Fig. 3, Plate IV.), which was recommended by the Belgian Commission of 1876 for use in dangerous parts of fiery mines: yet it is so materially different from both these types that he prefers to view the Gray lamp (Fig. 4, Plate IV.) as a type of itself.

When tested by H.M. Commissioners on Accidents in Mines, the Gray lamp passed through all the tests in a most successful manner, and although it was submitted to explosive currents containing from $6\frac{1}{2}$ to $12\frac{1}{2}$ per cent. of gas, moving with velocities varying from 400 to 3,200 feet per minute, it never caused an explosion of the outer atmosphere. H.M. Commissioners also remarked that when the atmosphere became explosive by the addition of fire-damp the entering gas-mixture inflamed at the inlet-gauze, and the oil-flame was extinguished; whereas in the case of the Davy lamp the gauze-cylinder became filled with flame, because the air had access to it from all sides.

The Commissioners further observed that the cap or halo produced by any percentage of fire-damp is considerably affected by the surroundings of the flame, and is most distinct in those lamps which admit air below the flame, no matter what illuminant is used. Also, that as the smallest proportion of gas which a Davy lamp will detect is $2\frac{1}{2}$ to 3 per cent., and as 2 per cent. of gas with suspended coal-dust will produce a violent explosion, this and all other ordinary safety-lamps do not afford the means of ascertain-

[*] *Final Report*, 1886, page 118.

ing whether the workings of a mine are sufficiently free from gas
to ensure safety, and they cannot, therefore, be considered as satisfactory means of testing for gas. H.M. Commissioners also stated
that : —

To be really useful a fire-damp detector must be simple and portable, so that
observations may be readily made in any working place or cavity, in which gas
may accumulate.* . . . . Air which would appear quite free from gas, if examined
by a lamp-flame, may become explosive when laden with fine, dry coal-dust.† . . . .
It is most important that all mines should be carefully examined by means of
indicators capable of detecting as small a proportion as 1 per cent. of gas.‡ . . . .
Four lamps . . . deserving of special attention, as combining a high degree
of security, with fair illuminating power and simplicity of construction : They
are Gray's lamp, Marsaut's lamp, the bonneted Mueseler lamp, and Evan
Thomas' modification of the bonneted Clanny lamp. . . . . In our experiments,
the last lamp has given upon the whole, the best results.§ . . . . It is difficult
to imagine how Gray's lamp can become filled with unconsumed gas, without
extinguishing the flame. . . . . In our trials with this lamp, only the very
slightest internal explosions were produced, but the lamp-flame was extinguished in
nearly every experiment.‖ . . . . As Gray's lamp draws its supply of air nearly
from the top of the lamp, it would seem to be better suited than the Marsaut
or the Mueseler for searching for fire-damp near the roof of a working place.¶

One of H.M. Commissioners (Prof. R. B. Clifton) made an
attempt to produce a modification of the Gray lamp, and he did so
by abandoning the top gauze and altering the form of the chimney
(Fig. 5, Plate IV.), but the only practical result was to call fresh
attention to the fact, ascertained by Mr. Smethurst and the writer
when testing the Gray lamp at Brynn (previous to the experiments
made by H.M. Commissioners), namely, that the Gray lamp will
not explode the outer atmosphere, if the gauze above the chimney
be omitted either purposely or by accident.

The foregoing extracts from the *Final Report* of H.M. Commissioners clearly show that the Gray type of lamp has possessed
from the first very high qualifications to recommend it.

When the writer took the lamp in hand, his main purpose was
to make it applicable to the daily requirements of officials when
testing for gas. He found, of course, that the condition of the
atmosphere close up to the roof could be ascertained with certainty
and despatch, but that after gas was indicated on the flame, the
air-tubes remained filled with gas, and could only be cleared by
passing the whole of the mixture that they contained over the

* *Final Report*, 1886, page 99.   † *Ibid.*, page 116.   ‡ *Ibid.*, page 117.
§ *Ibid.*, page 118.   ‖ *Ibid.*, page 88.   ¶ *Ibid.*, page 85.

wick-flame and through the lamp, consequently in the majority
of instances the wick-flame was extinguished by an excess of gas;
that the gauze above the chimney could not be removed for cleaning; and lastly, that the inlet- and outlet-openings required a
better protection against vertical and angular air-currents.

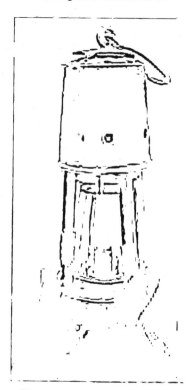

FIG. 7.—ASHWORTH-HEPPLE-
WHITE GRAY SAFETY-LAMP.

FIG. 8.—ASHWORTH-GRAY SAFETY-LAMP.

The first of these difficulties (Fig. 7) was overcome by providing holes at the base of two of the air-tubes, and covering them
with movable shutters, which when testing for gas were pushed
up and the holes closed by a finger and thumb of the hand holding
the lamp. Thus, as soon as gas showed its presence on the flame,
air was admitted from a gas-free level, and the flame was sustained when the forefinger and thumb were lifted (Fig. 8).
Secondly, the chimney was simplified, and a conical outlet substituted for the double truncated chimney of the original lamp.

This change also made room for the insertion of a truncated conical gauze within the chimney and replaced the horizontal gauze originally fixed above the chimney. The cylindrical glass was also replaced by a truncated conical one, so as to lessen the cubic contents of the lamp, increase its lighting power, and to avoid the square shoulder above the cylindrical glass. A different form of hood was also designed to cover the inlet- and outlet-openings, and to preserve the lamp from the possibility of a down-current being created within it.

FIG. 9.—ASHWORTH-CLOWES-REDWOOD HYDROGEN GAS-TESTING ATTACHMENT APPLIED TO AN ASHWORTH-GRAY SAFETY-LAMP. THE SMALL TUBE, EXTENDING TO THE LOWER BAR OF THE SCALE, IS THE HYDROGEN GAS-BURNER.

The greatest fault of this design (Fig. 7) was its sensitiveness to certain air-currents, and the fact that, under some circumstances, the wick-flame became extinguished without the aid of gas, if the air-pressure on the inlet-openings exceeded that on the outlet, or *vice versa*. This difficulty was eventually overcome (Fig. 8), and the more recent forms of this lamp designed by the writer are unaffected by any air-current met with in a mine. The form

most generally adopted is probably that shown in Fig. 8, or the
one fitted to the Ashworth-Clowes-Redwood oil and hydrogen gas-
testing lamp (Fig. 9). But even these lamps will probably give
place to the newer form of ventilation openings shewn in Fig. 10.
The one adopted by Mr. A. H. Stokes for his alcohol flame-testing
arrangement (Fig. 11) will be seen to be similar to that shewn in
Fig. 7.

FIG. 10.—No. 2 GRAY SAFETY-LAMP.

The writer combined separate oil and alcohol-flames in one
lamp for very keen gas-testing (Fig. 12), namely, oil for lighting
purposes and the alcohol for gas-testing, but each flame can be
used separately or both combined. Another modification (Fig.
13), has only one flat air-tube instead of three or four round ones,
and with a shield somewhat like the Fumat lamp, in which all the
air-holes are placed on the front half of the circumference of the

bonnet; also an arrangement of three wick-tubes placed in a triangular form, and regulated by one rack-motion, which produces
a powerful and non-luminous flame when using alcohol or
benzolene, and very perfect combustion when using petroleum.

FIG. 11.—ASHWORTH-HEPPLEWHITE-GRAY SAFETY-LAMP
FITTED WITH THE STOKES ALCOHOL GAS-TESTING
ATTACHMENT, SHEWING ALCOHOL-VESSEL, WITH COVER
REMOVED, SCREWED INTO OIL-VESSEL, AS USED WHEN
MAKING A TEST FOR GAS.

When testing for gas with this lamp, the flame is reduced until
the tip of the flame is just level with the top of the conical deflector
surrounding the flame, and therefore any elongation of the flame
or blue cap caused by gas is at once evident. When using
alcohol or methylated spirit in this lamp, the keenest possible
tests may be made on a shorter scale than with the Pieler lamp
(Fig. 6, Plate IV.) or Chesneau lamp (Fig. 14), and almost equal
to the Ashworth-Clowes-Redwood lamp (Fig. 9).

The latest form, designed by Mr. Thomas Gray, the inventor of this type, is called the No. 2 Gray safety-lamp (Fig. 10), and varies somewhat from all the foregoing patterns, but the principle of the lamp is maintained throughout. Thus, the air to be tested is admitted by one tube only and from the extreme top of the lamp; no slides are absolutely required to admit air to save the

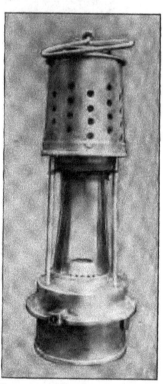

FIG. 12.—ASHWORTH-GRAY COM-
BINED OIL AND ALCOHOL GAS-
TESTING SAFETY-LAMP.

FIG. 13.—ASHWORTH-GRAY TRI-
WICK ALCOHOL OR PETROLEUM
GAS - TESTING SAFETY - LAMP,
WITH SINGLE AIR-TUBE.

flame from extinction by excess of gas, because the air entering below the middle flame is $3\frac{1}{2}$ inches below the top of the gas-testing tube, and therefore supplies air sufficiently clear from fire-damp admixture to save the wick-flame from extinction. The gauze part of the lamp has been removed from the inside of the chimney to the outside, which permits of a considerable reduction in air-friction and provides a larger area of gauze for the passage of the products of combustion. In lamps of this type, the question of safety is not affected, whether the gauze be placed inside or outside the chimney. It might be dispensed with altogether, if there were not a certain amount of nervousness and natural suspicion in the minds of all mining-engineers against leaving any openings between the wick-flame and the outer atmosphere uncovered by wire-gauze, and the gauze has therefore been retained as a necessary part of this safety-lamp.

Both the inlet and the outlet openings in the Gray lamp are so thoroughly protected against all sorts of air-currents that the necessary balance of the inlet and outlet air-forces is maintained in equilibrium ; and therefore the wick-flame is always steady and no reverse current (which is the point of greatest danger in any type of safety-lamp) can be created within the lamp. The shield is provided with double deflector-openings throughout, and completely protects all the vital parts.

FIG. 14.—CHESNEAU ALCOHOL GAS-TESTING SAFETY-LAMP.

Lamps of the Gray type possess many points of excellence, which have been insisted on by scientists and experimenters, as being absolutely necessary qualifications for all lamps which approach as nearly as possible to perfection in safety, and they are so constructed that :—(1) The flow of an explosive mixture into them is always graduated or obstructed ; and that this graduation or obstruction is one of the most important protective points in the construction of a safety-lamp has been amply proved and confirmed by Messrs. Nicholas Wood,[*] Thomas John Taylor,[†] and J. B. Marsaut.[‡] (2) Down-draught within the lamp is totally prevented. (3) The cubical content of that part of the lamp which is liable to become charged with an explosive mixture, is reduced almost to a minimum.

With regard to the second point, all experimenters have observed that if a down-draught can be created within a safety-lamp, the wick-flame is crushed down and the lamp may become filled with an explosive mixture, which on ignition exerts a very high velocity like the firing of a cannon, and in the majority of cases the flame is passed to the outer atmosphere almost instantaneously. As regards the third point, Mr. J. B. Marsaut observed that the Mueseler-Arnould-Godin (Fig. 3, Plate IV.) and the Evan-Thomas lamps,[§] and all others embracing the same principles avoid the phenomenon of internal explosion. A little practical experience with the Gray type of safety-lamp will demonstrate to any one the absolute truth of this assertion, and to others the following figures will convey a large measure of conviction. The total area of the four inlet-air-tubes of the Gray lamp is 0·306 square inch, or equal to one tube $\frac{5}{8}$ inch in diameter; thus, when testing for gas, or if the lamp becomes suddenly surrounded by an explosive mixture, no more gas can explode within the lamp than the very small volume which is contained in the space between the neck of the oil-vessel and the bottom ring-gauze, and the explosion is of so weak a nature that it is described by Mr. J. B. Marsaut, as " an ignition and not an explosion." Weak as it is, however, it has the effect of creating a volume of carbonic acid gas below the wick-flame, which almost immediately extinguishes the wick-flame and the gas also. The avoidance of down-draught and the absolute safety of this type of lamp above the glass part is attained by the gauged outlet for the products of combustion, which is $\frac{3}{4}$ inch in diameter or 0·589 square inch in area.

  [*] *Trans. N.E. Inst.*, 1853, second edition, vol. i., page 301.

  [†] *Ibid.*, 1853, second edition, vol. i., page 316.

  [‡] " Étude sur la Lampe de Sûreté des Mineurs," *Bulletin de la Société de l'Industrie Minérale*, 1883, series 2, vol. xii., page 321.

  [§] *Final Report of H.M. Commissioners appointed to inquire into Accidents in Mines*, 1886, Fig. 18.

The writer considers that these remarks on the Gray type of safety-lamp would not be in any sense complete, if he did not refer to the notable insecurity of the lamps most ordinarily used for the detection of fire-damp. Thus, as regards lamps of the Clanny type, if they are raised into a fire-damp atmosphere when canted on one side, and with a reduced wick-flame, as is usual in practice, such lamps are liable to pass the flame through the gauze, twice out of every 30 tests.[*] Although this fact has often been brought to the notice of mining-engineers, it has not yet been recognized at its real value, but latterly a few of the leading engineers in South Wales have come to the conclusion that it does deserve serious attention. It is only necessary, therefore, for the writer to direct every mining-engineer's attention to the notable failures of safety-lamps, at Whitfield colliery in 1889, Allerton Main colliery in 1894, Shakerley colliery in 1895, and at Wishaw colliery in the same year, to prove, without adding many problematical cases, that the possible failure of a lamp of the Clanny type is not only experimentally possible, but practically certain, if the necessary conditions are present.

Users of the Marsaut double-gauze lamp should observe that all lamps supplied to them have a clear space between the two gauzes, and that the capping of the inner gauze does not fit inside the outer gauze with a metal contact, because if they do so fit, the value of the double gauze is greatly depreciated, and it becomes only equal to the lamps shewn in Figs. 24 and 25 of Mr. J. B. Marsaut's paper, to which the writer had already referred.

H.M. Commissioners of 1886 recommended that every mine should be examined by a lamp or instrument capable of detecting 1 per cent. of fire-damp because they recognized that an atmosphere containing 2 per cent. of fire-damp with suspended coal-dust was an explosive atmosphere; that none of the safety-lamps then in use (and the writer may add "at the present time also") would detect less than 2 per cent. of gas; and that those of the Clanny and Davy types would not indicate less than 2½ to 3 per cent. The writer is able to say, with absolute certainty, that there is no demand for safety-lamps for gas-detection which will indicate 1 per cent. of gas; but the following table shows that the Gray type of lamp may be so constructed that it will indicate anything from ½ per cent. and upwards.

* "Étude sur la Lampe de Sûreté des Mineurs," *Bulletin de la Société de l'Industrie Minérale*, 1883, series 2, vol. xii., page 321.

Methane.	Heights of Cap or Halo above Flames.		
	Colza and petroleum flame: 0·12 inch high.	Tri-wick benzolene flame: 0·12 inch high.	Ashworth-Clowes hydrogen flame: 0·40 inch high.
Per cent.	Inch.	Inch.	Inches.
¼	0·00	0·28	0·72
1	0·00	0·40	0·88
2	0·25	0·56	1·24
3	0·30	0·80	2·08

This table distinctly exemplifies the fact that the hotter and less luminous is the testing flame, the larger and more distinct does the cap become; and, therefore, that the height of the cap depends on the character of the source of heat, as well as on the construction of the safety-lamp in which it is used.

In conclusion, it cannot be too distinctly brought home to all mining-engineers who are interested in fiery mines, and especially where coal-dust is a danger, that the firemen or shot-firers may honestly declare a working-place or roadway clear from fire-damp, and proceed to fire a round of shots in places where the atmosphere is absolutely explosive, and composed of a mixture of 2 per cent. of fire-damp, air and suspended coal-dust (not a cloud of dust, but only the normal quantity of dust in suspension) with the most deplorable results.

———

Mr. JOHN GREGORY said that he had seen the No. 2 Gray lamp. In external appearance, the lamp closely resembled the ordinary type, but it had the addition of a tube, which was movable, and was only placed on the lamp while actual testing was taking place. When used as a safety-lamp, the tube was lifted off, and in that respect it had advantages over the majority of lamps in ordinary use.

Mr. E. B. WAIN observed that no man had given more time and attention to the perfecting of the safety-lamp than had Mr. Ashworth. The Gray lamp was safe, and he did not think it was possible to imagine a better lamp for use in a fiery mine, but it had its disadvantages. The large size of the standards, or air-tubes, made it lose a considerable proportion of the area of light, but the lamp, provided with a flat tube to get the air to the burner, would probably do away with that objection. Mr. Ashworth seemed to

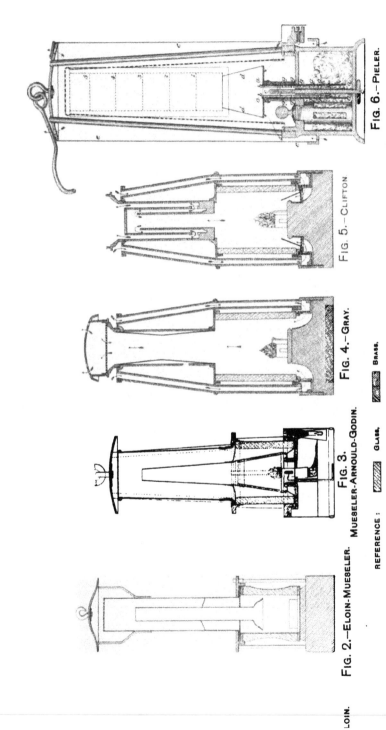

LOIN.

FIG. 2.—ELOIN-MUESELER.

FIG. 3.
MUESELER-ARNOULD-GODIN.

FIG. 4.—GRAY.

FIG. 5.—CLIFTON

FIG. 6.—PIELER.

REFERENCE :

GLASS.

THIN SHEET
METAL

BRASS.

GAUZE.

Scale, 1 Inch to $\frac{3}{10}$ Inch.

VOL. XV. PLATE III.

Andw Reid & Compy Lto Newcastle on Tyne

think that the old type of Gray lamp was not quite what it ought to be, and he had made an improvement by shortening the standards and Mr. Gray had added a loose tube, to be fitted into one of the air-tube standards. But in order to ensure that the flame should not be extinguished, he left the three other standards open at the lower level, so that they had one air-passage drawing down gas and three others open to a moderately pure atmosphere, so that he did not quite see how it was possible to test with the same accuracy the strength of the gas-and-air mixture with the lamp in that form, as with the old type of Gray lamp, with tube-standards, all extending to the top of the lamp.

Reference had been made to certain very notable failures of lamps of the Clanny type, and amongst others one at Whitfield colliery in 1889. He (Mr. Wain) was not aware of any such failure, but probably Mr. Ashworth referred to a case where two men were slightly burned in 1886, and where it is probable that the failure of a lamp was the cause of the explosion. In this case, however, No. 124 lamp was a typical Belgian-made Mueseler lamp, and some few particulars of the case might prove interesting to the members. As he (Mr. Wain) had been for 3 hours previous to the occurrence in company with the injured men, and had only left the place a very short time before the explosion took place, he had accurate details, and the drawing (Plate V.) was prepared on the same day.

Work was in progress for crossing an old level, which had been closed since an explosion and fire in 1881. The road had been subjected to great heat, the débris was very small and covered by very fine dust and soot. The man, who was most seriously burned, had left the place, and had taken his lamp with him, to get his breakfast. A few minutes after he had resumed his work, and was pulling down some loose stuff with a small pick, he felt a puff, and saw a red flame which came, as far as he could judge, from behind him. The flame was not like a gas-flame, as it was dull red and full of sparks. He had examined the place before re-starting work and found it free from gas. The dust was so dense that he had several times to leave the place until it settled.

A man, whose hair was singed and who was working below and behind him, saw the flash, and also saw the flame run a short distance down the dip. The lamp kept burning after the flash, and died out when the flash went out.

The safety-lamp, No. 124, when brought out. was quite bright and clean, whilst the other safety-lamp, No. 330 (an English-made Mueseler lamp), which had been hanging near it, was covered with dust, especially at the bottom of the gauze. There had evidently not been sufficient force in the explosion to shake the dust from the second lamp.

The No. 124 lamp was subsequently tested at Adderley Green colliery by Messrs. A. R. Sawyer and J. R. Haines, in a mixture of coal-gas, air and dust taken from the roadway near the site of explosion. The mixture passed at a velocity of 65 feet per second ; and, although there was an explosion inside the lamp, no flame was passed outside of the lamp in this experiment. Further trials were made, and on admitting a small quantity of dust into the gaseous mixture, sparks were seen spreading sideways and up the chimney of the lamp. When a further supply of dust was admitted, the whole of the space within the glass was immediately filled with flame, and an explosion occurred inside the lamp, which extinguished the wick-flame, but the dust continued to burn under the horizontal gauze.

When a still further supply of dust was admitted, this flame was maintained and described a circuit under the horizontal wire-gauze, about 50 times, indicating a reversal of the direction of the air-current in the lamp.

In his (Mr. Wain's opinion) this explosion may have been due to an accumulation of dust outside the gauze of the lamp being ignited by the heat of the dust burning inside the lamp, as described above, or by a strong internal explosion due to down-draught.* A small proportion of fire-damp, possibly, could have been found in the air, as the road was the only opening into a very large area of workings, which had been closed for five years, and were charged with pure fire-damp. The velocity of the air-current was about 10 feet per second.

Mr. J. T. STOBBS observed that there was one point in Mr. Ashworth's paper, against which he protested, namely, the statement that " in lamps of this type, the question of safety is not affected, whether the gauze be placed inside or outside the chimney. It might be dispensed with altogether, if there were not a certain amount of nervousness and natural suspicion in the minds of all

---

* Such as that described in the *Final Report* of H.M. Commissioners appointed to inquire into Accidents in Mines, 1886, page 87 ; and this (*ibid.*, page 88), is difficult, if not impossible, to obtain in safety-lamps of the Gray type, owing to the small internal volume.

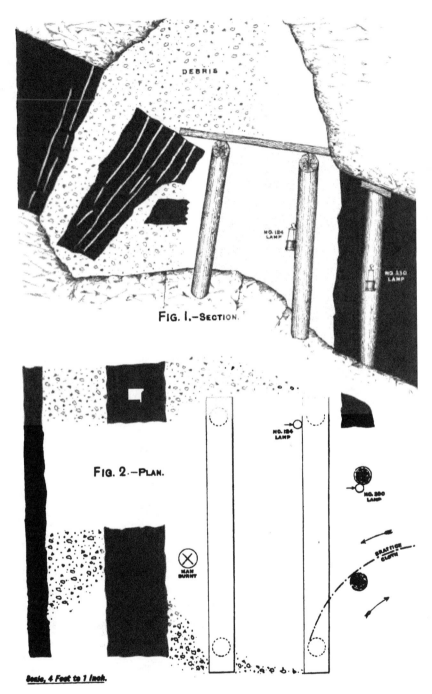

DEBRIS

NO. 184 LAMP

NO. 330 LAMP

FIG. I.—SECTION.

FIG. 2.—PLAN.

NO. 184 LAMP

NO. 330 LAMP

MAN BURNT

BRATTICE CLOTH

Scale, 4 Feet to 1 inch.

The North Staffordshire Institute of Mining & Mechanical Engineers.
Transactions 1902-1903.

mining-engineers."[*] The lamp was not fit to be taken into a mine unless it was fitted with a gauze. He asked whether the three short tubes would be affected if the long testing-tube became heated.

Mr. JOHN GREGORY said that the long tube was only fixed on the lamp during the time when an actual test was being made; and that it would have time to cool before being used again.

The PRESIDENT (Mr. A. M. Henshaw) suggested that the air-current would pass outward and upward through the testing-tube.

Mr. P. BRENNAN said that he had used the Gray lamp for many years. The difficulty of the loss of light, referred to by Mr. Wain, was overcome by using a broader tube. He had used a tube ¾ inch broad, and it afforded a better light than other lamps. The Gray lamp, fitted with a long tube, resembled the trumpet-lamp used in Lancashire a few years ago. The fireman attached the trumpet to his lamp, and the trumpet became very hot before he had finished his round of inspection.

Mr. E. B. WAIN remarked that many of the explosions that had happened in the past had been due to the occurrence of a thin layer of gas close to the roof; and where formerly the examination was made with an ordinary lamp without detecting the gas, if there was gas present, it would be found by the Gray lamp.

The PRESIDENT (Mr. A. M. Henshaw) moved a vote of thanks to Mr. Ashworth for his valuable paper.

Mr. J. LOCKETT, in seconding the resolution, remarked that the light produced by the Mueseler, Marsaut and similar types of lamps was not sufficient to enable the collier to avoid the dangers by which he was surrounded. The Gray lamp gave the best light of any lamp that he had ever used. The Gray lamp could also be used for the detection of small percentages of gas. At one place, in Birchenwood colliery, the Marsaut lamp failed to detect gas, and on the tube being inserted into the Gray lamp the presence of gas was forthwith discovered. Mr. Ashworth, who had given them a safe lamp that afforded a good light, merited the gratitude of the mining community.

The resolution was cordially approved.

----

[*] *Trans. Inst. M.E.*, 1902, vol. xxv., page 69.

NORTH STAFFORDSHIRE INSTITUTE OF MINING AND MECHANICAL ENGINEERS.

GENERAL MEETING,
HELD AT HANLEY, JANUARY 12TH, 1903.

MR. A. M. HENSHAW, PRESIDENT, IN THE CHAIR.

The minutes of the last General Meeting were read and confirmed.

The following gentlemen, having been previously nominated, were elected : —

MEMBER.--
Mr. J. T. DAWES, Mining Engineer, Prestatyn, North Wales.

ASSOCIATE MEMBER—
Mr. E. B. REYNISH, 120, Cliff Vale Terrace, Stoke-upon-Trent.

Mr. J. DUNCAN CRIGHTON read the following paper on " Central Condensing-plants for Collieries " : —

CENTRAL CONDENSING-PLANTS FOR COLLIERIES.

By J. DUNCAN CRIGHTON.

Since the introduction of water-cooling plants, experience has taught engineers that it was possible to achieve the benefits of condensing, without using more water than when non-condensing, inasmuch as the steam evaporated by the boilers did not pass into the atmosphere, but was condensed, and was then either returned direct to the boilers or was used to supply the water evaporated in the cooling of the condensing water, and if the water-cooler were properly constructed this evaporation would not exceed the boiler-feed.

Before considering the question of condensing, the writer will describe four types of water-coolers which are now extensively used. They are as follows :—

(1) *Open-type Cooler.*—The open-type cooler consists of a strong framework, constructed of well-seasoned timber, and braced so as to secure the necessary rigidity. The horizontal cooling frames are made of rough-sawn selected spruce laths. The distributing troughs, at the top of the cooling frames, are designed so that any oil or sediment that may pass over with the water is trapped in the lower part of the trough, and the water to be cooled is distributed in thin films and with perfect uniformity over the upper tier of cooling frames. The tiers of frames are so constructed and arranged with regard to each other, that the flow of the films of water is continually diverted and retarded, thereby ensuring that every particle of water is exposed to the atmosphere for a sufficient time in order that its surplus heat may be dissipated.

This type of cooler occupies more ground-space than the tower or forced-draught cooler, but will reduce the water to a rather lower temperature. The power absorbed in raising the water to the top of the cooling stack is a minimum due to the low lift, which does not exceed 18 feet, and where ground-space is of no moment, it is advisable that this type should be used. It is usual to allow about 3 square feet of ground-space per indicated horse-power.

(2) *Chimney or Natural-draught Cooler.*—Various coolers of this type (with different designs of the cooling stack) have been introduced, most of which, when properly designed to suit the conditions of humidity and temperature prevalent in this country, can be made to maintain an average temperature of about 85° Fahr. The water-cooling apparatus is built into the lower part of the chimney. The efficiency of this type of cooler depends mainly upon the height of chimney, and the temperature of the water delivered to the top of the cooling stack, which regulates the natural draught induced up the chimney. The height that the water has to be lifted to the top of the cooling stack is generally about 25 feet; and it is usual to allow about 1 square foot of ground-space per indicated horsepower.

(3) *Fan or Forced-draught Cooler.*—These coolers are of the closed-chimney type, with one or more fans placed outside and at the bottom. The water-distributing troughs and the cooling frames are placed inside the chimney. The fans are designed to force the quantity of air necessary for cooling the water through the cooling stack. In this type it is of great importance that the water should be evenly distributed, so as to obtain the full benefit of the power absorbed in driving the fans. Owing to the power required to drive the fans and the extra height at which the water is raised to the cooler, which in most cases is about 35 feet, these coolers should only be used when there is very little ground-space available, as they absorb a considerable percentage of the economy derived through condensing. In designing these coolers, it is usual to allow 1 square foot of ground-space per 4 or 5 indicated horsepower.

(4) *Water-sprayers.*—In this system, the water is delivered to nozzles at a pressure of 10 to 20 pounds per square inch. The design is generally such as will impart to the water a rotatory motion on issuing from the nozzles, so that it flies by centrifugal force into fine spray. The writer has not had much experience

with this system, but he may remark that it does not conform to the general principles of water-cooling plant. Although the water is divided into fine particles and offers a large surface, it is not retarded in its fall and, consequently, it is not very long exposed, and further radiation is impeded by the volumes of vapour present. There is also a considerable loss of water, which is carried away in a fine spray, and may become a nuisance to neighbouring buildings. The ground-space occupied by a spraying cooler is generally about twice that required for an open-type cooler.

The foregoing particulars are based on a steam-consumption of 20 pounds per indicated horsepower per hour.

Before the introduction of water-coolers, the evaporative condenser was coming into use, but its great weight and liability to leakage had prevented its general adoption. It is now used only in exceptional circumstances, and, when adopted, it is usual to work, in conjunction with it, a small water-cooler, which cools the surplus water so that it can be used over again.

The writer has had considerable experience among collieries, and has had ample opportunities of forming an opinion as to the conditions under which condensing could be introduced, and he, therefore, proposes to describe one or two arrangements of central condensing plant, which will meet average requirements, and with very little modification will be suitable for almost any case which may occur in practice at collieries.

(I.) The first design of plant consists of the following parts :— (1) An accumulator surface-condenser, in which the steam passes through the inside of the tubes and the water is on the outside ; (2) a water-circulating pump; (3) a dry-air pump; (4) a condensed-water pump; (5) an oil-separator and pump; (6) and a water-cooler for cooling the water in circulation (Figs. 1 and 2, Plate VI.).

The accumulator type of surface-condenser has two special advantages for colliery purposes :—(1) The tubes are immersed in a tank containing a large body of water, sufficient to absorb the surplus heat from a temporary inrush of steam above the average and thus prevent any serious diminution in the vacuum. The heat is diffused through so large a quantity of water that it does not heat the water in contact with the tubes to the same extent; and, therefore, the mean temperature of the condenser does not rise as rapidly, or to as high a point as would occur in an ordinary enclosed condenser, in which there is a limited quantity of water in contact with the tubes. (2) The tubes are easily cleaned when the plant is in operation, and the condenser being built in sections, it is possible to shut off one section for overhauling, while the other sections remain in use.

The water-circulating pump is preferably of centrifugal type, which is cheaper in first cost and working upkeep, and has a fairly good efficiency for the small head against which it has to work.

The dry-air pump is preferably of the Klein or Weiss type, which may be run at a comparatively high speed. The high efficiency of 95 per cent. is due to a valve, which neutralizes the effects of clearance, and, consequently, the air-pump need not have so large a volumetric capacity as that required of an ordinary wet-air pump. The air-pump draws off the air at the highest point, thus preventing the formation of an air-lock, and, consequently, increasing the efficiency of the condenser.

The condensed-water pump takes the water from the condenser at the lowest point, and delivers it into a tank from which it is fed directly into the boilers at a temperature corresponding to the vacuum.

The oil-separator should be of sufficient size to insure the whole of the oil falling out by gravity, and suitable baffle-plates should be fixed for filtering the steam on its way to the condenser. With this apparatus, it is possible to eliminate the whole of the oil used in the steam-engines, and after suitable filtration the oil may be used over again. The oil-pump extracts the oil and water from the oil-separator, and delivers it to the oil-filtering tanks. It will be readily recognized that it is of great moment that an efficient oil-separator should be fitted in the exhaust-steam pipe, as the efficiency of the condenser is considerably enhanced if the tubes are kept clean and free from oil.

(II.) The second arrangement of plant comprizes:—(1) A barometric counter-current jet-condenser; (2) an horizontal steam-driven reciprocating injection-pump; (3) an horizontal steam-driven air-pump or a vertical steam-driven reciprocating injection-pump, with a vertical steam-driven air-pump; (4) an oil-separator and oil-pump; and (5) a water-cooler (Figs. 3 and 4, Plate VI.).

The barometric counter-current jet-condenser consists of a large cylindrical vessel, constructed of mild steel plates, placed with its horizontal axis at a height of 34 feet above the water-level in the hot-water well or tank. The exhaust steam enters at the upper part of the condenser, at one end, and travels in a contrary direction to the water which enters through a dome at the other end, and a thorough mixture of steam and water is thus effected. Suitable baffle-plates are fitted in the interior of the condenser for thoroughly breaking up and spraying the water: for fluctuating loads a considerable body of water is always kept in the condenser, and this must become heated before the vacuum falls. The air liberated from the steam and water is drawn from the top of the dome, at a point adjoining the injection-inlet; the air, therefore, occupies a minimum volume; and the capacity of the air-pump and the driving power is thus reduced to a minimum. The condensing water and condensed steam fall through a barometric tail-pipe into a hot well, placed in a convenient position adjoining the cooling tower, and at such a level that the water will flow by gravity directly into the water-distributing troughs placed on the top of the cooling stack.

The water-circulating pumps lift the water from the cold well under the cooler to the water-inlet on the condenser, and when in operation the head, against which they deliver, is the height from the level of the water in the cold well to the inlet on the condenser, *minus* the vacuum. In this arrangement it is essential that the pumps shall give a positive and steady delivery of water to the condenser, as any fluctuation of the vacuum in the condenser causes a fluctuation of the head against which the pump is working. Consequently, it is not advisable to use a centrifugal pump for this duty, unless a barometric injection is arranged with a fixed water-inlet, which will allow the pump to work against a steady head, and the fluctuation is taken up in the barometric leg.

The air-pump is preferably of the Edwards type, which is well known as an efficient pump and one not very liable to get out of order.

The oil-separator, oil-pump, and water-cooler are the same as described for the first plant.

Most of the members, the writer believes, are interested in the economical working of colliery-plant, and are introducing automatic cut-off valves on winding-engines, electric driving, and other improvements calculated to reduce the consumption of coal at the pit-head for steam-raising. A saving in the steam or coal-consumption of 25 to 50 per cent. may be effected by applying central condensing-plant; and that it is receiving the attention it deserves, is illustrated by the reference made by Mr. James S. Dixon in his presidential address.[*]

Assuming that a saving in coal-consumption of 25 to 30 per cent. may be effected by applying central condensing-plant, irrespective of the saving which in most cases can be made in the water-bill, or in wear-and-tear of boilers due to the use of injurious water, it will be found that a substantial sum is available for condensing-plant equal in most cases to two or three years' purchase.

The accumulator surface-condensing plant is most suitable for collieries where the water-supply to the boilers must be purchased from a public supply, and the available pit-water is unsuitable for boiler-feed. In this plant, the pit-water will be used to make up the loss by evaporation in the cooler, and the whole of the steam condensed is available for boiler-feed and is equal to at least 80 or 90 per cent. of the water evaporated, and therefore a saving of at least that amount of pure water will be obtained. The 10 to 20 per cent. of loss by drains and leakages may be supplied from the pit-water, after it has been passed through a water-softening and purifying plant. The outlay for a water-softening plant may be economical when dealing with 20 per cent. of the water, and yet be unprofitable when dealing with the whole of the water required by the boilers. By this arrangement, the colliery could be made entirely independent of any outside source of water-supply.

The accumulator-plant is not so economical as a barometic counter-current condensing-plant, either in first cost or in the power absorbed in maintaining a vacuum, but in no case should the total power expended in operating the plant exceed 4 per cent. of the horsepower condensed.

Should the pit or surface-water be suitable for the boiler-

* *Trans. Inst. M.E.*, 1902, vol. xxiii., page 374.

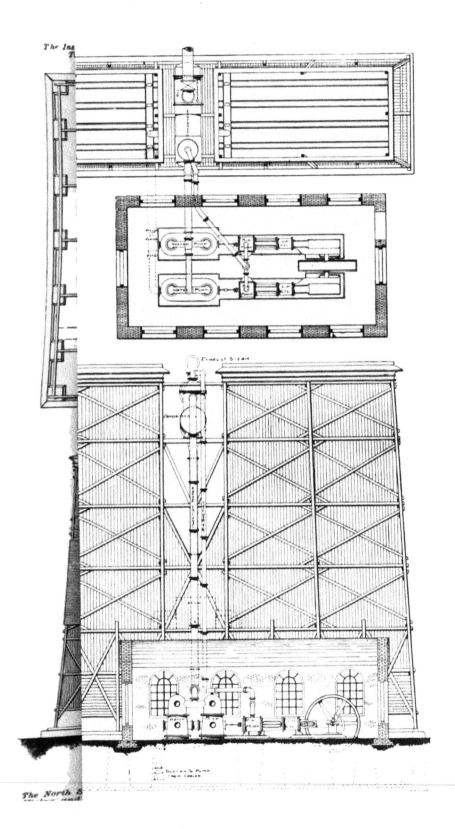

feed, the barometric condenser will be the most suitable plant to adopt, as the vacuum is obtained with the least possible expenditure of power, and the water in circulation is considerably less than is necessary with a surface condensing-plant. The small volume of water in circulation is due to the special system of spraying the water in its passage through the condenser, by which a thorough mixture of steam and water is effected, and the water leaves the condenser tail-pipe at a temperature corresponding to the vacuum; and it may be fed directly into the boiler at the same temperature as the water from the hot well of a surface condensing-plant.

———

The President (Mr. A. M. Henshaw) said that Mr. Crighton in his paper had directed their attention to a very desirable and effective measure of economy and efficiency in colliery steam-plant. Until recent years, the consideration of fuel-consumption at collieries had not received the attention which it deserved, and 10 per cent. of the output had often been stated to be the average fuel-consumption at the collieries of the United Kingdom. There was great room for improvement, and where entire remodelling could not be entertained, a condensing-plant might in many cases be erected with extremely satisfactory results. A case was recorded of a Continental colliery where more boiler-power was required, but instead of spending the money on additional boilers a condensing-plant was erected. The savings repaid the cost in 18 months and the fuel-consumption was reduced permanently by 20 per cent. All engineers probably had some experience of savings of, say, 20 per cent. in isolated engines by condensing, and there was no reason why central condensing-plant dealing with the steam from all the engines should not be even more satisfactory. The barometric jet-condenser plant commended itself to his judgment, as it was cheaper in first cost, less power was required for the air and injection-pumps and probably the cost of maintenance was less. He considered that the power stated as 4 per cent. of the indicated horsepower absorbed by the plant was too low to be realized in regular practice, and he should like to have some figures as to the area of the tube-surface per pound of steam or indicated horsepower, and the weight or the volume of water required for circulation in the case of the surface-condenser plant.

Also, in the case of the jet-condenser, some rule as to the capacity of the condenser and the volume of water used per indicated horse-power, and in each case the approximate cost of, say, a 500 horse-power plant. In designing new plant for a colliery, condensing must be a part of the design in modern steam-practice, and given good boilers working at high pressure, feed-water heaters, steam-superheaters, and each engine designed for its particular duty, the result should approximate to 2½ pounds of coal or 20 pounds of steam per indicated horsepower. This was not too sanguine a calculation even with the ordinary grade of colliery fuel, when they considered that compound condensing engines of the Willans, Corliss, and Sulzer types showed a steam-consumption of 13 to 16 pounds per indicated horsepower, and marine engines even lower, or under 1½ pounds of fuel per indicated horsepower. A compound condensing-engine driving a Walker fan, recently tested by Mr. Higson, showed a consumption of 2 pounds of coal and 20·17 pounds of steam per indicated horsepower, although not working at the time under the most favourable conditions. He should say that one might safely calculate upon a saving of 20 per cent. in fuel by the use of a central condensation-plant, but some rearrangement would often be advisable in the direction of concentration of the engines and minimizing the length of steam-mains; and it was perhaps worth recalling to mind, that every square foot of bare steam-pipe wastes 10 to 12 cwts. of coal per annum by condensation of steam, and even a well-covered pipe loses 2 to 4 cwts. It was not always remembered that ⅛ inch of scale in a boiler would increase the fuel-consumption by 10 per cent., and ¼ inch of scale by nearly 50 per cent. An old plant well tended, with well-considered additions and improvements, effected as opportunity occurred, would give good results, and by only adopting condensing-plant and feed-water heaters, a saving of 30 per cent. in fuel could be realized in most cases. There are plenty of old collieries where the steam-consumption exceeds 100 pounds per indicated horsepower-hour, and many modern collieries where that figure is approached; and where the fuel-consumption is from 2 to 5 times as much as it would be by adopting new, or making the very best of the existing plant.

Mr. J. W. HARTLEY (Stone) asked, supposing steam-engines were placed at considerable distance from each other, whether it had been found necessary to superheat the steam. He had erected a condensing-plant, connecting both steam-engines and steam-hammers to one central condenser. He found that a difficulty frequently occurred, owing to the enginemen at different engines opening the cylinder drain-cocks and, thus, destroying the vacuum. The vacuum varied from 15 to 24 inches, although it frequently fell to nothing when the cylinder drain-cocks were opened. Consideration should be given to this matter, in order that the difficulty might be obviated. His boilers were in a central position, and he had obtained considerable saving by the use of a superheater. He doubted whether any condensing-plant could show a saving of 50 per cent. in the fuel-consumption of any reasonably efficient steam-engine; and he would like to see the results of any experiments substantiating so high a saving as that mentioned by the author of the paper.

Mr. E. B. WAIN remarked that he did not think that the statement that had been made as to a 50 per cent. saving by the use of condensing-plant could be considered extravagant. The average effective pressure, taken from the diagrams of an economical steam-engine working at a boiler-pressure of 80 pounds per square inch, would not be much more than between 30 and 40 pounds, and taking a perfect vacuum at 14. or 15 pounds per square inch, they might expect that the saving of 50 per cent. would be obtained by a perfect condensing arrangement. He had, before him, an indicator-diagram taken from a colliery fan-engine, and it shewed the great economy obtained by the use of a condenser; in this diagram 83 per cent. of the work was done below the atmospheric line. The engine was working at about quarter load, but, in few cases would a more economical engine be found, except perhaps a high-speed triple-expansion engine worked with very high steam-pressure. The value of a condenser was distinctly shown in the case of that engine, which was working with a steam-consumption of only 20 pounds per indicated horsepower-hour, after making all allowances for clearance and cylinder-condensation losses. Another point of the greatest importance was, that condensing should not be the beginning of the economy, but should be the end of it; and before a condensing-plant could be worked economically or to advantage, the engines must be capable of yielding the highest possible degree of economy.   The steam-consumption must be reduced to a minimum, and the condensing-plant would then prove of great value.   If high-pressure steam were turned into a condenser, it would take huge quantities of water for even a small colliery. The size of a cooling-plant as applied to a modern colliery appeared to be a serious matter; the author stated that "it is usual to allow about 3 square feet of ground-space per indicated horsepower" for open-type coolers.   Applying that rule to a colliery using 5,000 horsepower, he found that 1,500 square yards of ground would be required for the erection of the necessary cooling-plant; and he thought that a cooling-plant of the open type would hardly be convenient or suitable in such a case.   The question of water-supply at collieries, in the past, had prevented the application of condensing-plant, for, under ordinary circumstances, a surface-condenser required at least 20 pounds of water (at a temperature of 60° Fahr.) per pound of steam.   An engine of 250 horsepower, with a steam-consumption of 30 pounds per indicated horsepower-hour, would require 250 gallons of water per minute.   There were few collieries where the indicated horsepower was less than 250, and 250 gallons per minute was a large quantity of water for a small colliery.   Consequently, if any cooling arrangement would allow water to be used over and over again after it had been cooled and purified, it must be extremely valuable.   A difficulty was experienced in their own district, because a hard scale formed in the pump, where jet-condensers were used, and in the Edwards condenser, considerable trouble was experienced in keeping the valves clear from scale.

Mr. JOHN GREGORY said that Mr. Crighton's paper dealt with one of a series of subjects requiring consideration in connection with increased efficiency in the burning of fuel and its conversion into mechanical energy.   The President had remarked that, in addition to condensing, higher initial steam-pressure and superheating were important factors in obtaining the maximum efficiency.   These questions were brought forcibly to his notice recently in obtaining tenders for engines for an electrical power-plant.   With steam at a pressure of 90 pounds per square inch and working non-condensing, the best guarantee that they were able to obtain was 27 pounds of steam per brake horsepower-hour.

He could not give the exact figures relating to the highest efficiency, but with superheated steam at a pressure of 150 pounds per square inch, and when working condensing, engines were made capable of working at about half that consumption. Mr. E. B. Wain had very truly remarked that to obtain these results meant the adoption of expensive plants and it was a debatable point how far a colliery-owner was justified in going, to avoid capital charges exceeding the economy in fuel resulting from increased efficiency. A colliery was not to be compared with, say, a large electrical power-station, where the work on the engines was fairly regular, as against the intermittent duty of winding-engines; and the differing conditions demanded special treatment. At Sneyd colliery, a jet-condenser was applied to a compound condensing winding-engine, with an independently driven air-pump; and after an experience extending over several years' working, he could assert that it effected a very considerable economy. It was, however, impossible for a condenser (when attached to a single engine, subject to such rapid and heavy fluctuations) to maintain a steady vacuum; and he could clearly recognize that a central condensing-plant, into which the steam from a number of engines was being exhausted, would, by levelling the load-curve, yield better results. It had been objected that many of the engines about a colliery were small and straggling. In an ideal arrangement, the winding-engine, and possibly the fan-engine, would be connected with a central steam-condenser, and all the auxiliary engines would be replaced by motors deriving their power from conveniently situated generating sets, which should also be connected to the condenser. In his opinion, by this means, the maximum economy would be obtained.

If all that was claimed for power-gas could be obtained in actual practice, more duty would be produced from their fuel than by any system of utilizing steam. It appeared to him, however, that to distribute the power so obtained it would be necessary to convert the whole into electric energy. A difficulty then arose owing to the winding-engines being such disproportionately large units, and although he believed that some such scheme had been worked out on the Continent, he thought that, at present, the better plan was to adopt steam, and practise such economies in its use as had been shown to be possible in the paper communicated by Mr. Crighton.

Mr. Crighton, replying to the discussion, said that in writing his paper he based the horsepower on a consumption of 20 pounds of steam per hour. This was an efficient engine, but he knew that many colliery-engines were using 50 pounds of steam per horsepower-hour owing to the fact that the engines were too large for their work. An engine was bought for an estimated load, which it was thought that it would have to yield at some future time; but it very seldom happened that such a load was ever required. Of course, when dealing with horsepower on a basis of a consumption of 20 pounds of steam per hour, he was dealing with an ideal plant, as at present understood; but considering the tendency towards decreased steam-consumption he thought that it was best to place it on that basis, which was quite consistent with engineering practice and easy of calculation.

The 4½ per cent. of the indicated horsepower mentioned as being absorbed in a condensing-plant, was calculated on this basis, and should the consumption of steam be higher, the percentage of power required would naturally be increased in direct proportion to the weight of steam used per indicated horsepower. With regard to Mr. Hartley's remarks about the cylinder drain-cocks, the difficulty would easily be overcome by connecting them to the main range of exhaust steam-pipes, which, when under vacuum, would easily clear them. The oil and water being removed by the oil-separator, of course, as a safeguard, it was necessary to provide a bye-pass to the atmosphere, to be used when there was no vacuum. Taking a surface-condenser of the submerged type, dealing with a fluctuating load, he found (in practice) that it was necessary to allow no more than 6 or 7 pounds of steam per square foot of tube-surface, taking into account the temperature of the injection-water due to the use of the cooler, and allowing for the average steam evaporated. The amount of water in circulation might be governed under these conditions; for instance, taking a vacuum of 25 inches, providing the condenser was in good condition, the condensed water would be at a temperature of about 133° Fahr. but the circulating-water leaving the condenser would be about 10° to 15° Fahr. below that temperature, owing to the imperfect admixture or the loss in transmission of heat through the tubes. With a good plant, the difference of temperature would be about 10° Fahr.; and, knowing the temperature of the water for the cooler, it was an easy matter to calculate the amount of water necessary to be circulated per hour. It was usual to accept a difference of about 15° Fahr., and its use had been justified in practice. In estimating the cost of a condensing-plant, he must still base the horse-power on a consumption of 20 pounds of steam per hour; and as a rough guess, the complete cost of a plant, such as the submerged condenser, would be about £1 17s. to £1 18s. per horsepower; with the counter-current jet-condenser the cost was greatly reduced, and had been as low as £1 10s. He thought that Mr. Wain's observations would justify the statement that the saving might attain 50 per cent. A condensing-plant had been erected at a colliery, where 3s. 6d. per ton was charged for coal used under the boilers, and a guarantee had been given that the cost of the plant would be repaid in 4 or 5 years.

Mr. W. N. ATKINSON said that anything that tended to reduce the consumption of coal, either at a colliery or elsewhere, was advantageous. There was no doubt that at collieries, where coal was plentiful and cheap, the question of the economical use of coal had not received the attention which it deserved. He had pleasure in moving a vote of thanks to Mr. Crighton for his able paper.

Mr. HODGSON seconded the resolution, which was cordially approved.

MIDLAND INSTITUTE OF MINING, CIVIL AND MECHANICAL ENGINEERS.

GENERAL MEETING,
Held at the Arcade Hall, Barnsley, January 17th, 1903.

Mr. H. B. NASH, President, in the Chair.

The minutes of the previous General Meeting were read and confirmed.

The following gentlemen were elected, having been previously nominated:—

Members—

Mr. Frank Eckersley, Assistant Manager, Sharlston Collieries, Queen's Villas, Crofton, near Wakefield.

Mr. Joseph Middleton Gardner, Mechanical Engineer, Houghton Main Colliery, Barnsley.

Mr. Ernest F. D. Mosby, Surveyor, Houghton Main Colliery, 13, Cliffe Road, Darfield, Barnsley.

Mr. William Edmund Shelley, Civil Engineer, 3, Godstone Road, Rotherham.

Mr. George R. Thompson, Professor of Mining, Yorkshire College, Leeds.

Associate Members—

Mr. Alfred Lucas, Electrical Engineer, Sheffield.

Mr. R. C. Tasker, Electrical Engineer, Brocco Bank, Sheffield.

Associate—

Mr. Herbert Crowther, Enginewright, Elsecar Collieries, near Barnsley.

Students—

Mr. William Dallas, 10, Wharncliffe Road, Broomhall Park, Sheffield.

Mr. Andrew Jackson, Jun., Howley Park Colliery, Batley.

DISCUSSION OF MR. THOMAS MOODIE'S PAPER ON "THE WORKING OF CONTIGUOUS, OR NEARLY CONTIGUOUS, SEAMS OF COAL."*

Mr. M. H. Habershon said that when the Black Mine ironstone was worked in South Yorkshire over twenty-five years ago, a seam of coal about 3 feet thick, lying immediately over the iron-

* *Trans. Inst. M.E.*, 1902, vol. xxiii., page 282; and vol. xxiv., pages 116, 199 and 205.

stone, was also worked. The ironstone was worked on the long-wall system, and the coal was worked after a lapse of from twelve to eighteen months. The gate-roads used for the ironstone were also used for the coal; and it had been found an important matter to provide packing for the lower seam in the first working. Of course, in working an ironstone which lay in the form of nodules and bands, a very large quantity of shale was turned over, probably 40 per cent. of the volume, and it was used to pack the goaf. It was essential in the working of two contiguous seams of coal that the goaf should be packed in the first working, or the upper seam would be afterwards got at great disadvantage.

Mr. J. R. WILKINSON said it was only fair that he should state that he knew nothing whatever about the seams of coal to which the paper under discussion referred. Judging, however, from such experience as he had had, his impressions were that the conditions varied so much, and, at times, within such small areas, that, for anyone to attempt to lay down a hard-and-fast rule as being the best mode of procedure for that kind of work, without first being fully acquainted with all the circumstances, would be presumptuous. There were certain thicknesses of intervening strata, which would cause serious inconvenience in working two contiguous seams, and he would like to know what that thickness was. In his opinion, it was most difficult for some distance prior to that thickness being reached, at which it was necessary to discontinue the single working, and to work the two seams separately; there were many different partings; and difficulties arose at various points, according to their nature and thickness. Another important question was the system of working, and whether longwall, pillar-and-stall, or some modification of these systems should be adopted would depend entirely on the conditions of the seams. In his experience, longwall was the best system, if the roof and other circumstances would allow it. The question as to which seam should be worked first was important, and difficult of decision. There were conditions under which it would be necessary to work either of the seams first. If both the coals were of the same market-value, and there happened to be the same demand for one as for the other, then the manager might be left to his own discretion. But if one seam was more valuable than the other, then that coal must be got whether it be the top or the bottom seam. Another important question was whether the

seams were wet or dry.   If there was any water, especially in dip-
workings, it would seem inadvisable to extract the upper seam
first, because, in that case, the intervening strata would become
wet and soft, and the expense of working the second seam would
be considerably increased.   Under ordinary circumstances, it
would be desirable to extract the bottom seam first, if the mine
were dry and the seam was of such a thickness that it could be
thoroughly packed.   The packing of the lower seam was an
important factor in the successful working of contiguous seams.
The prevailing idea was, that when one seam was taken away,
a certain time should be allowed to elapse before taking out the
second seam, in order to allow the goaf or superincumbent strata
to become thoroughly settled before proceeding with the working
of the other seam.   If the upper seam were of a tender nature, so
long as it was left and the bottom seam was taken out first, in his
opinion a sort of disintegrating process would ensue—not a process
of days or months, but extending over years, and the longer the
top seam was allowed to remain unwrought, the more it would be
damaged when the time came for it to be worked.   If the top seam
was extracted first, then the longer the lower coal was allowed to
remain the worse it would be to get, and in addition to an
enhanced price for getting, the quality of the coal would be
deteriorated.   He thought that there was nothing better for the
protection of the upper seam than solid gobbing, if that could be
carried out; but if the seam was thick, the next best thing he
thought was some system of chequered packing, that is, building
packwalls, 6 or 8 feet thick, and crossing them every two or three
packs.

Mr. I. HODGES said that, when contiguous seams had been
worked in West Yorkshire, it had generally been the seams that
were most marketable that had been worked first, irrespective of
engineering conditions.   Personally, he thought that the lower
seam, unless there was some strong reason for the contrary, should
be worked first, but everything really depended on the commercial department, who gave the mining-engineer instructions
that a certain tonnage had to be worked from certain seams.
The mining-engineer should endeavour to prevent his successor
suffering more than was absolutely necessary, by the seams
being worked out of their proper order; and for this purpose, the
seam should be worked by longwall, and the goafs should be well

packed. For instance, in the West Riding, the Shale coal-seam, 2 to 3 feet thick, overlying the Stanley Main seam at a distance of 40 to 60 feet, had been rendered much more difficult to work by the Stanley Main seam, 7 feet thick, having been previously worked by pillar-and-stall methods. In using coal-cutting machinery in the Shale coal-seam, difficulties were occasioned by the breaks running upward from the pillars in the Stanley Main seam, and also from the unequal gradients in the working-faces produced by these pillars. These occasionally had seriously prejudiced the working of the Shale coal-seam, and in some cases had led to its abandonment.

Mr. RICHARD SUTCLIFFE said that he had come to the conclusion that they must work every seam according to its surroundings and on its own merits. There might, for instance, be a strong roof on the bottom seam and a tender roof on the top seam; and in that case, if the seams were very near together, they would think seriously before extracting the bottom seam first. Sometimes it was of great benefit, when two seams were being worked one over the other, to work the lower seam first, as in the Parkgate seam at High Hoyland colliery, some years ago: there was about 4 feet of bind, and the lower seam was first worked, because the top seam took very little getting, and the roof was able to bear it. The subject was one which would bear ample discussion, but it was undesirable to teach the young members the erroneous idea that there could be any fixed rule for working contiguous seams, as each case must be decided upon its merits and surroundings.

The PRESIDENT (Mr. H. B. Nash) thought that not much would be gained by working the top seam first, unless the roof was so bad that they could not successfully work the top seam after the bottom one had been gotten. If the commercial man could not sell the bottom seam, and could sell the top seam, then the mining-engineer must get the top seam; but in the long run the bottom seam would be seriously depreciated, and the nature would be gone out of it through the exudation of gases, rendering it tough and woody to get and thereby increasing the cost of working.

The discussion was closed.

DISCUSSION OF MR. CLARENCE R. CLAGHORN'S PAPER ON THE "CAMPBELL COAL-WASHING TABLE."[*]

The PRESIDENT (Mr. H. B. Nash) said that the principle of the Campbell washer was almost identical with the Craig table, which had been recently described by Mr. W. Scott.[†]   Both of them, when used singly, were unable to deal with large quantities, but the difficulties of construction were so small that a number of them could be erected for very little money, and a fairly large output could then be treated.   It seemed to him, where they desired simply to wash the small coal, and that not in large quantities, that the Campbell washer was a cheap system and the cost of repairs was low.   Washing was one of the things which was becoming essential to the working of seams that were not of so good a quality as the Barnsley and Silkstone seams—the staple seams of the district—and therefore they would have to devote attention to the washing of small coal.   Large washing-plants were in operation where larger quantities of coal had to be treated.   At many collieries, having an output of 400 or 500 tons per day, they could not afford to erect a washer costing £10,000 or £20,000, but they could spend £2,000 on a Craig or Campbell washing-plant for dealing with 150 to 200 tons of small coal per day.

The further discussion was adjourned.

———

[*] *Trans. Inst. M.E.*, 1902, vol. xxiii., page 435.

[†] *Ibid.*, 1902, vol. xxiii., page 179.

THE MINING INSTITUTE OF SCOTLAND.

GENERAL MEETING,
HELD IN THE CHRISTIAN INSTITUTE, GLASGOW,
FEBRUARY 11TH, 1903.

MR. HENRY AITKEN, PRESIDENT, IN THE CHAIR.

The minutes of the last General Meeting were read and confirmed.

The following gentlemen were elected:—

MEMBERS—

Mr. ALEX. ANDERSON. Farme Colliery, Rutherglen.
Mr. ROBERT B. JARDINE BINNIE, Greenfield Foundry, Hamilton.
Mr. BRYCE FULTON, The Knowe, Bonnybridge.
Mr. JOHN MIDGLEY, 108a, West Regent Street, Glasgow.
Mr. WALTER MUIR, Government Inspector of Mines, Dundee, Natal.
Mr. WILLIAM REID, 41, Garthland Drive, Glasgow.
Mr. DUNDAS SIMPSON, P.O. Box 1,028, Johannesburg.

Office-bearers for the session 1903-1904 were nominated.

It was agreed to celebrate the semi-jubilee of The Mining Institute of Scotland in April next.

The PRESIDENT (Mr. H. Aitken) moved a vote of thanks to the Corporation of Glasgow and their officials for their kindness in allowing the members to visit the electrical works at Pinkston.

The resolution was cordially approved.

DISCUSSION OF MR. ROBERT MARTIN'S PAPER ON "SINKING ON THE SEASHORE AT MUSSELBURGH."*

Mr. JAMES ANDERSON (Rutherglen) said that some years ago he had occasion to widen out a pit which had been sunk in 1810,

* *Trans. Inst. M.E.*, vol. xxiv., pages 126 and 209 ; and vol. xxv., page 3.

through 118 feet of mud near the river Clyde. He could not
tell, except from tradition, how the process of sinking was
accomplished at that time, but it was interesting to note
the fact that cast-iron cylinders had been used successfully at
that period.   The cylinders were cast in Camlachie foundry,
in one piece, each of them 5 feet in depth.   The bottom
one had a cutting edge, and was slightly bellmouthed, the
total length of the cylinder being about 40 feet.   Above the
cylinder was a fine specimen of oaken cribbing, the pieces being
6 inches square, about 3 feet long and cut to the radius of the
pit. Each crib was counterchecked at each end, and an oaken tree-
nail was passed through the joint.   Where the ends joined, when
the circle was completed, the ends of the cribs were left square, and
an oaken wedge was inserted and kept in position by an iron
plate placed in front.   Somewhere between 1840 and 1850, Mr.
Robert Hood sank a pit at Stonelaw through 40 feet of running
sand.   A brick cylinder, 15 feet in internal and 18 feet in external diameter, was built on a cast-iron cylinder, $4\frac{1}{2}$ feet deep,
having a cutting edge.   At every length of 6 feet of brickwork, a wooden ring was built in, and bolts passing from one
wooden ring to another were screwed up at every ring.   There
was no outer metal skin, and the sinking was in every way
successful.

Mr. ROBERT MARTIN wrote that Mr. James Baird admitted
that his cylinder stuck fast before it reached the rock-head
after descending to a depth of 70 feet.[*] He thought that this
was due to the greater friction of the rough brick-and-cement
wall; and that a smooth iron casing would have carried the
wall down to a greater depth. Mr. Cadell's experience[†] was
similar, and the iron outside casing with a loading of 400 tons
reached a depth of nearly 90 feet.[‡] Fortunately for him, as in
the previous case, the ground was hard and dry, and the sinking
could be made in the ordinary fashion to the rock-head. He
thought that if Mr. Cadell had encountered soft ground and a
large feeder of water at that point, the cylinder would probably
not have stuck at a depth of 90 feet.
He agreed with the President's remarks[§] that every case

[*] *Trans. Inst. M.E.*, 1902, vol. xxiv., page 132.

[†] *Ibid.*, vol. xxiv., page 131.   [‡] *Ibid*, vol. xiv., page 243.

[§] *Ibid.*, vol. xxiv., page 209.

of sinking through a bad surface should be considered on its merits; and, it was too early for us, with our limited experience to lay down hard-and-fast rules on the subject. In addition to his experience at Olive Bank, as described in his paper, in 1886, he sank a brick-cylinder to a depth of over 60 feet in running sand with much water. This brick-cylinder was not fitted with a cutting edge or an outside casing of iron. A large quantity of sand was taken out of the cylinder, and the surface-subsidence around the top of the shaft was considerable. The men were in the bottom all the time, except when there was an influx of water and sand, during which the cylinder moved downward. With reference to the President's requirements for work of this nature, he (Mr. Martin) would say as regards (1) that the use of an outside shell of iron is not imperative in every case, but it reduced friction, kept the building in good form, and added strength until the concrete set hard. The use of inside plates and cross-bolts to bind the iron and building together was in his opinion quite unnecessary, unless it was intended to sink to a much greater depth than 110 feet. (2 and 3) If the cutting-edge of the cylinder can be sunk into hard tough till or clay, and the work remains water-tight, then dredging may be the better method of removing the material out of the shaft. If, however, the workman can stand on the bottom until the rock-head is reached, digging, being quicker and surer than dredging, is preferable. It is desirable to allow the cutting edge, where the ground is soft and wet, to penetrate some distance into the rock and this can only be properly done by employing men in the pit-bottom. The open cylinder can be loaded at any time and at any part, and the weight adjusted to keep it in a perpendicular position. (4) He thought that this was a good suggestion. When water was run in at the top, it found its way downward in ruts, and only occasionally, when these became choked or blocked, did the water assist in lowering the cylinder. (5) A guiding-frame erected at the top, unless erected at such a height as to be impracticable, could not have much effect on a cylinder, say, 80 feet below the surface, and with a journey of 27 feet before it. In the case at Olive Bank he thought that he had an excellent guiding-frame in the wooden box or shaft, sunk 30 feet from the surface in stiff boulder-clay—an exact fit for the cylinder. But owing to irregular surface-subsidence, this guiding-frame became twisted out of shape, and proved a hindrance and a source of disturbance. Unless the surface was stable, which was improbable in all circumstances, a guiding-frame was likely to prove a snare.

Mr. ROBT. THOS. MOORE (Glasgow) said that before starting operations at Olive Bank, they considered the method of sinking with brickwork alone; but it was not adopted, as they were afraid that some part of the brick-cylinder would be caught by skin-friction and that the bottom part would fall away, and the cylinder would be completely broken. They also considered the question of cast-iron versus steel. With a large circle (18 feet in diameter), the thickness of the cast-iron would have been considerable. A lining of steel could be made of the same strength and of less thickness, and the cost per running foot was not greater than with cast-iron, while the segments were lighter and more easily handled. The iron lining was only used as a smooth skin, to reduce the friction of the brick-cylinder and hold it together while it was being sunk. The strength of the shaft depended on the 2 feet of brickwork, and it was a matter of indifference whether or not the steel rusted away. Considerable difficulty had been experienced in forcing the cylinder downward, and the skin-friction appeared to be greater than that generally found in the sinking of bridge-foundations. The cylinder repeatedly stuck altogether, and it was only moved by pumping out the water and silt from the bottom until a sufficiently large hole was formed and caused a collapse of the earth right up to the surface and brought down the cylinder along with it. He was inclined to think that the President's suggestion of forcing water through rings of pipes fixed at intervals round the cylinder was a good one, and that it would reduce the skin-friction.

Mr. M. WALTON BROWN (Newcastle-upon-Tyne) wrote that the excellent article on "Mine," written by Mr. Robert Bald in 1818, for the *Edinburgh Encyclopaedia,** contained an interesting description of the methods employed, at that date, in sinking through soft mud and quicksand, including the effectual methods of cast-iron cylinders and drop tubs, used in sections like the tubes of a telescope.†

* Vol. xiv., pages 314 to 378.   † *Ibid.*, pages 337 and 339.

The PRESIDENT (Mr. H. Aitken) said that Mr. Martin took exception to his proposed guiding-frame at the top of the shaft. Undoubtedly, there was no use for a guiding-frame if the adjacent ground was dropped away and into the pit. This dropping away invariably occurred when the inside of the pit was removed by hand-labour. The advantage, however, of using a guiding-frame at the surface was to be found in the fact that the surface remained practically stationary when they kept the water in the pit, and took out the inside with grapples or the like. Mr. Martin, moreover, took exception to his (Mr. Aitken's) remark that the inside of the brickwork should also be lined with plates. His experience, however, had been, when the brickwork was not plated on the inside, that it generally gave way when the cylinder had descended a certain distance. It was in view of that experience that he had recommended that the brickwork should be plated, as described by him, on the inside as well as on the outside. With these exceptions, Mr. Martin and he appeared to agree as to what was the correct way of sinking pits in bad surface. Taking it all in all, he thought it was much better to sink the pit with the water in it, unless in circumstances where they were troubled with large boulders.

The discussion was then closed, and Mr. R. Martin was accorded a hearty vote of thanks for his interesting paper.

―――

DISCUSSION OF MR. HENRY AITKEN'S PAPER ON "FOUR OLD LABOUR-SAVING IDEAS."[*]

Mr. ROBERT CRAWFORD (Loanhead) wrote that Mr. Aitken, when speaking of the idea for filling stored coal, said that when only one trough is used, men are employed with rakes or shovels to draw or fill the adjacent coal on to the band, or in the latter case mentioned the coal could be placed on to the band from the binging-ground by means of a steam-shovel or ram.[†] But the suggested arrangement could be improved by inclining the floor of the binging-ground on both sides of the trough, the inclination of the floor being such that the coal when placed upon

[*] *Trans. Inst. M.E.*, 1902, vol. xxiv., page 211.
[†] *Ibid.*, page 212.

it would be lying at an angle a little over its " angle of repose." By this arrangement, one man only would be required to lift the planks from the trough and to regulate the supply of coals to the band; and the coal would slide quietly down to the band from both sides; and, at any moment he could stop the supply by placing one or more planks across the trough. With this arrangement, coal could be loaded at a moment's notice, the only labour required being one man looking after the distribution of the coal on the band, and one wagon-trimmer; and if screening were necessary more labour would, of course, be required.

Mr. G. A. MITCHELL (Glasgow) asked whether the second method of stocking dirt had ever been put into practical operation?

The PRESIDENT (Mr. H. Aitken) replied that the idea was nearly 30 years old, and at that time Mr. Forgie's firm thought of giving the innovation a trial, at Croy, but it never was carried out. There could be no doubt that it was a cheap way of handling dirt, particularly at oilworks.

Mr. J. T. FORGIE thought that the idea of using a conveyor for removing the coal in the bing was impracticable, except at collieries where only small quantities were handled. For instance, the conveyor would not work at all successfully at Fifeshire collieries.

The PRESIDENT (Mr. H. Aitken) admitted that the idea was only intended for collieries where small quantities were handled: but if the number of conveyors was increased the idea might be feasible at large collieries.

The discussion was then closed, a hearty vote of thanks being given to Mr. H. Aitken for his paper.

———

Mr. ALEXANDER FORBES read the following paper on "The Technical Instruction of Working Miners, with Suggestions as to Mine-managers' Examinations ": —

THE TECHNICAL INSTRUCTION OF WORKING MINERS, WITH SUGGESTIONS AS TO MINE-MANAGERS' EXAMINATIONS.

By ALEXANDER FORBES.

I.—*Introduction.*—Within the past few years, the attention of members has been frequently drawn to the subject of mining education, but more especially to the training of mining engineers. The present paper deals with the instruction of working miners so far as pertains to Scotland; the suggestion as to a preliminary mine-managers' examination applies equally indeed to England, but in the remainder of the subject-matter of the paper it has seemed well to the writer to confine himself to what has come within his own experience. It may be added that the subject of the technical instruction of miners does not appear to have received that attention from mining men which its importance and the amount of money now being spent for the purpose seam to warrant, and the main object of this paper is, therefore, to ascertain the opinions of those who should really be the best judges as to what should constitute such instruction.

II.—*Should Working Miners be Technically Instructed?*—With regard to the utility of educating miners technically, there can hardly, in the writer's opinion, be two views. It is unnecessary here to enter into a discussion of the advantages of technical education in general, but it may be pointed out that mining is not only a skilled but at all times a hazardous occupation, calling for intelligence on the part of the operator in regard to means both for his own safety and that of others. It seems, therefore, that a proper system of instruction, given at an early age, would not only make the recipient a better workman, by increasing his power of adapting himself to circumstances and teaching him some of the economics of his business;

but, by raising the mental standard of miners generally, would reduce the difficulty of supervising the mine, and do much towards the avoidance of dangers which can only be ascribed, in many instances, to ignorance.

III.—*Unsuitableness of the Present Instruction.*—Admitting, then, the necessity for educating miners technically, the questions arise as to what should be the subjects of instruction and how best they can be imparted. At first sight, these problems may appear to have been already solved by the classes, ostensibly for the education of working miners, which are to be found in almost every mining district. But the writer has long been convinced that while the instruction given at these classes may redound greatly to the advantage of the individual, yet it is neither of such a nature nor so general as to result in very material gain to the mining community. At the same time, the impression must not be formed that these classes are a failure: on the contrary, their success, within their own lines, has been most marked, and when the number of miners who yearly take the mine-managers' examinations, and of those who are successful, is considered, nothing but the warmest admiration can be entertained on the one hand for the laudable ambition which impels so many miners to seek to improve their social position, and on the other (the writer hopes he may say so, although a teacher himself) for the patient and painstaking work which in too many instances has to evolve the candidate from the crudest material. But the fact remains that the declared object of these classes is to prepare for the mine-managers' examinations, and for that reason they attract only such as desire to become managers or under-managers, and are, therefore, not likely—at any rate for a very long time—to be attended by numbers sufficiently large to ensure the benefits of the instruction being general. And even if that were not the case, it is very questionable whether after one or two years' attendance a miner would be much the gainer practically, as the syllabuses, framed with a view to these examinations, embrace subjects a knowledge of which is of no value to a working miner, and those which are valuable are as a rule too comprehensive. Further, mere abstract teaching of mining is of little avail in the case of the average mining-class pupil, who, at the com-

mencement of his studies, is, in the majority of cases, either a
very young lad, or, if older, has practically left his mental
faculties unexercised in the interval since leaving school.

IV.—*Responsibility for the Present System.*—That means
should exist for the purpose of enabling working miners to
obtain mine-managers' and under-managers' certificates, while
the general improvement of the mass, with consequent gain to
the country, is neglected, seems so anomalous that it may be of
of interest to enquire how such a state of matters has come into
operation. In the writer's view, the real responsibility seems to lie
with the defective syllabus and method of payment by results
of the now defunct Science and Art Department; and the continuation of the system is due either to custom, or to the mistaken
belief that the possession of a mine-manager's certificate is
a guarantee of practical ability. In the old days, when the
above-named Department was the sole ruling-power, the custom
was to find an isolated class here and there, taught usually by
some local man with an advanced first-class certificate, and who
was either himself preparing for, or had already obtained, his
mine-manager's certificate. The teacher's remuneration consisted of the fees paid by the students (the fee to each was
usually somewhat high) and the grants, if any, earned at the May
examination. The syllabus was, of course, that issued by the
Science and Art Department, but obviously it was to the
teacher's advantage to prepare his pupils at the same time for
the mine-managers' examination, and to have that fact as widely
known as possible. When the County Councils began to give
financial assistance and appoint lecturers, unfortunately the
Science and Art Department syllabus, with its grant-earning
system, still obtained; and notwithstanding that the Scotch
Education Department, into whose hands the control of these
classes passed a few years ago, neither issued any set syllabus (the
writer considers that this showed commendable wisdom on the
part of that Department under the circumstances, but it would
be well to have an uniform syllabus) nor paid grants on the results
of examinations, the old system continued to exist, the only
changes being the increase in the number of classes and the
separation of elementary from advanced instruction, the latter
being to a certain extent the cause of the present tendency to
centralize the advanced work.

V.—*Scheme of Technical Instruction for Working Miners.*—
The writer would not discontinue mining classes as held at
present (though their number would be fewer, unless that of
the students greatly increased). He would have convenient
centres, at which would be taught the higher parts of the
subject, such an arrangement being necessary in view of the
large number of miners desirous of obtaining instruction in
these branches, and whose education would be advanced to the
requisite standard through some such means as afterwards
shown. But, in addition, he would provide for the instruction
of those who have no ambition to obtain a manager's certificate;
and, as this instruction would be designed with the objects
already stated, it would form a necessary preliminary to the
central class. The following (roughly drawn out) suggests
itself as forming a suitable course of the nature required. In
every possible instance, experiment, specimen, or model would
be employed, thus delighting the pupil, as well as at the same
time convincing him, through the evidence of his own eyes; his
examination of the apparatus would be thoroughly supervised,
and the whole capped by a complete system of deductive question-
ing—in this way not only ensuring correct impressions, but tending to make these absolutely permanent.

(1) Ventilation.—Preparation of mine-gases, with experiments showing their properties. Occurrence of these gases in
mines, methods of detection; and, generally, everything which
would tend to make miners acquainted with the nature of the
gases with which they are likely to meet in the course of their
work.

The atmosphere. Composition of air (oxygen and nitrogen
prepared and properties shown). Experiments showing effect
of pressure on a gas. Atmospheric pressure (experiments).
Principle of barometer (experiments), and effect of reduction of
atmospheric pressure on issue of gas.

Dangers of coal-dust (experiments). Explosions in mines.
Entering a mine after an explosion.

Causes of the pollution of air in mines. Principles of ventilation. Conditions under which natural ventilation may occur.
Action of furnace, fan, etc. The circulation of air through the
mine-workings (experiments, models of brattice, etc.).

(2) Lighting.—Principle of the safety-lamp (experiment). Conditions under which safety-lamps are unsafe. Construction of various safety-lamps (lamps taken apart and put together again by pupils). Relighting, etc.

(3) Geology.—The earth's crust. Stratified and unstratified rocks. Occurrence: shale, sandstone, limestone, coal. Varieties of coal. Fire-clay. Ironstone. Dip and strike. Faults and other irregularities. Throw of faults. Rules for searching for part of seam beyond fault. (Models and specimens.)

(4) Blasting.—Action of an explosive. Charging shot-holes. Firing safety-fuse, electric fuse, etc. Miss-fires. Blown-out shots (experiments and specimens).

(5) Shaft-fittings.—A model of a shaft, showing the method of timbering, fixing cage-guides, etc.

(6) Drainage.—The various parts of lift and force-pumps. Action. Construction and action of siphon. (Working models.)

(7) Timbering.—Kinds of timber (specimens). Methods of securing roads with timber (models). Experiments showing how the strength of a bar varies (length, breadth and thickness). Supporting and loading a crown-tree so as to bear the greatest weight. Holing props or sprags, etc. Withdrawing timber.

(8) Modes of Working.—Longwall and stoop-and-room. Extraction of pillars. Jointing. Cleat. Rocks forming roof and floor of coal-seams: tender, etc., roof. (Specimens, models, etc.)

(9) Coal-mines Regulation Act.—Such General and Special Rules as are applicable shall be read with and explained to the pupils.

(10) Accidents in Mines, with special reference to falls from roof and sides. Prevention of accidents.

(11) First Aid to the Injured.

In the writer's opinion, such a course would be best given to lads who have already entered the mine, in conjunction with the evening continuation classes to be found in most mining villages. A night would be set apart for the instruction in mining, and in the case of two villages, not far apart, the classes might be conveniently joined. As a general rule, however, owing to the inclemency of the weather and other conditions, it is better not to ask pupils to walk very far. Managers of mines could do much in assisting the attendance at such classes by indicating that they expected their pit-lads to attend. The course would cover, say, two years, the instruction in first aid being given at the end of that period. Finally, owing to the impossibility of grading such instruction, and in consideration of its importance and the expense of teaching (only one set of apparatus would, however, be needed per teacher), it seems to the writer that grants should be apportioned on a more liberal scale than is customary in the case of work which is merely elementary. But, in any case, the financial responsibility should rest rather with the County Department, the local board acting as managers.

VI.—*Higher Instruction and Mine-managers' Examinations.*— At present, the great obstacle at most mining classes is the imperfect education of the majority of the pupils, and the difficulty experienced in persuading them to study other subjects (the latter following from the former), or indeed to improve themselves in any way except in what directly concerns mine-managers' examinations. Consequently, the period of a pupil's attendance at a class is frequently unduly prolonged, and much of the instruction, especially as concerns figures, is improperly understood. No doubt, if a system of instruction were adopted, such as the writer has already suggested, and a suitable course for further study arranged, improvement might be looked for: as, while keeping in touch with his subject of mining in such a way as to produce little or no mental strain, the would-be manager, by attendance at the local preparatory class, would be gradually rendering himself fit to receive with profit the higher instruction at the central class, which, having a properly organized course, would thus compel him to study the subjects allied to mining. A surer remedy appears, however, to consist rather in the institution of a preliminary or entrance examination to the mine-managers' examination, the former requiring to be passed before the candidate is allowed to sit for his mining certificate. In these days, it is surely not too much to expect that an aspirant to such an important position as that of a colliery-manager should at least be able to spell correctly. While, therefore, the adoption of such an examination is justified by the advancement of education generally, it would place mining on a par with other trades and professions; and there can be no doubt that it would effectually remove the difficulty in the teaching of advanced mining, as the student, knowing that the examination had to be passed, would at once prepare for it. In regard to the subjects of such an examination, they should, of course, be simple—such indeed as would tend to make all acquire the ability to write a description of any article or appliance with due attention to the claims of orthography and syntax. Practical mathematics should also form a prominent feature, but ought not to be more advanced than is required to cover the ordinary ground in mining. In fairness, however, to those who have unsuccessfully attempted the mine-managers' examination, or who are at present preparing for it, a preliminary examination should not be instituted until after reasonable notice.

With the adoption of a preliminary examination, the arithmetic paper as set at some mine-managers' examinations would, no doubt, be dispensed with, and more practical methods rendered compulsory in the performance of calculations generally; at present, however, the use of mathematical tables should be allowed, being supplied by the examining authorities. This in itself would go far towards encouraging the study of practical mathematics (and consequently also of steam and applied mechanics) by mining students; and, although the tables might not be in great demand at first, they would certainly become increasingly so.

In conclusion, it is only proper that the examination questions should be published, and in this respect it is worthy of note that the Scotch Boards for Examinations are behind most of the English Boards.

Mr. FREDK. W. HURD read the following paper on "Electrical Coal-cutting machines":—

ELECTRICAL COAL-CUTTING MACHINES.

By FREDK. W. HURD.

Introduction.—Great Britain stood at the head of the coal-producing countries until 1899, when the United States passed us, producing approximately 230,000,000 against our 223,000,000 tons. Not only was the total output greater, but, what is of much more interest and importance, the output per man was greater, and the American miner will probably now claim to be a better worker than the British miner. The following figures will give the true reason for the apparent supremacy of the American miner. In the United States, out of the total output, some 54,000,000 tons, or 25 per cent., was holed by machines; in this country only 4,000,000 tons, or 1½ per cent., was so holed; and it is, therefore, the miner and his machine which have beaten us, not the miner alone.

The same cause and effect on a smaller scale can be seen in our own country, by taking the case of a single colliery where machines have been successfully introduced, and it will be found that the output per man employed has at least doubled, as compared with the output before the introduction of machines, while at the working-face the tonnage per man may have trebled or quadrupled. In the writer's experience, the output from a certain working-face rose from 2½ tons per man when worked by hand to 7 and 8 tons per man when holed by machine, and in another case from 1¼ to 5 tons. So soon, therefore, as the number of machines in this country approaches that in the United States, the British miner with the machine to help him will, it is hoped, regain the old supremacy as to output per man.

When mining operations began in the United States on an extensive scale, trained miners were of necessity extremely scarce, and in order to get their full value the employers naturally gave them all the assistance possible by providing holing-machines. The men would be glad to have them, as there was no " scarcity-of-work bogey " to alarm them at that time. We are rapidly approaching the same condition as to the scarcity of trained miners, and, from this and other causes, the use of machines is fast becoming a necessity here also.

Causes of Delay in adopting Machines.—The most probable reason for their limited use in this country is the fact that colliery-owners have done well, so far, without them. They also did well, for a time, without the extensive screening arrangements now in general use; and also, for a time, without the elaborate dross-washing plants now dotted all over the coal-counties. It is now time to pay a little attention to the coal before it gets the length of the pit-head, to do away with the hand-pick and to work machinery at the face. A small washery and an electrical coal-cutting equipment, giving a large percentage of round coal, are surely a better investment than a large washery for the treatment of dross that should not be produced. It is more profitable to win solid round coal and to obtain a good price for it, than to break a large percentage of it into dross by hand-labour at the face, then to wash and screen it, and to sell it at a low price.

It is really very difficult to understand why thousands of pounds are spent on surface-machinery, the ultimate effect of which is a gain of say 6d. or 1s. per ton on the sale-price of the total output, while a few hundred pounds are grudged for the purchase of machines, which would save from 1s. to 3s. per ton on the total output and produce a more saleable article.

On referring to old papers on the subject, one repeatedly comes across statements, principally from colliery-owners, that there were no suitable machines on the market: this was not correct. Thirty years ago, there were good and reliable machines, the principal being:—The Gillott-and-Copley, the Rigg-and-Meiklejohn, the Winstanley and the Hurd: and all of these machines, when driven by compressed air, were then as good for ordinary holing in coal to a depth of $2\frac{1}{2}$ or 3 feet as any machine that can be bought to-day.

No doubt some members will be able to speak of good results, obtained many years ago, with coal-cutting machines, and no doubt also some will be able to refer to discouraging failures. From the failures much may be learned. Some were doubtless due to imperfect design or construction of the machine, others to the selection of machines unsuited to the conditions of the particular seams in which they were set to work, but the majority of failures were probably due to the lack of persistent determination, on the part of the management, to achieve success.

Necessity for Close Supervision.—Once it is decided that machines are to be used, the employer must take sufficient interest in the detail of their introduction to convince his manager that he is most anxious that the thing must succeed. If the manager sees his employer fully and actively interested, and the men perceive that the manager does not spare himself to get everything into the necessary swing, there will be no opposition from either.

Machine-cutting demands good organization, and close and persistent supervision. Men will to a certain extent look after themselves, but the machine is helpless if the conditions and facilities for its satisfactory working are not foreseen and provided for.

The manager who is introducing machines must make up his mind to hard work during the inauguration and early stages of machine-cutting. If the machines are to prove successful extra work must be done; and the man who does it, and gets his machines well away, deserves substantial recognition from his employers. He is a better colliery-manager, in every way, after he has successfully introduced coal-cutting by machinery.

The writer ventures to suggest that, as the introduction of machine-cutting entails upon the manager additional work and responsibility, a premium paid to him on the output of the machines would enlist his closer interest and be an incentive and encouragement to achieve the best results. The miner invariably benefits by the introduction of machinery. A machineman expects and is paid an extra wage, and the same may be said of the fillers and drawers; and the case of the official who is chiefly responsible should certainly not be overlooked.

Methods of Working.—Most seams (either thick or thin) which can be worked longwall by hand can be worked much cheaper by machines, and in many cases pillar-and-stall workings could be with advantage altered to longwall machine-faces. There are two principal methods of working a longwall face by machine:—(1) On a comparatively short working-face, say 450 feet long, with one machine cutting backwards and forwards alternately. And (2) on a long working face, 600 feet long and upwards, with one or more machines cutting always in the same direction, and being flitted along the roadways from one end of the face to the other.

In the writer's opinion, the first method is the better; and in Scotland, it particularly suits the rather broken seams which generally obtain. It involves more roads to the face in order to get sufficient men at work to clear the coal during the day-shift. Most of these roads, however, do not require much maintenance, as they are continually being cut off as the face advances, and in most cases goaf-packs and pillars can be dispensed with, the road-buildings being sufficient to support the roof. By this method, the greatest possible output can be obtained from any particular section, the district is quickly exhausted, and the cost of maintaining the roads, owing to the comparatively short time during which they are in use, is greatly reduced. This in itself would represent a considerable saving per ton on the total output, independently of any saving effected by the machine.

This method is illustrated in Fig. 1 (Plate VII.). The main-road, A, leads to about the centre of the working-face, slant-roads branch to the right and left, and from these again temporary roads lead direct to the face, spaced apart as found necessary to clear the coal in a single shift. As the face advances, new slants are set away from the main-road and cut out the longest of the temporary roads. Slant-roads, C, are made, in some cases, to further reduce the length of the drawing along the temporary roads, extra brushing being taken in the road, B, until it is cut off by the new slant-road, D.

The only drawback to this method is that, if for any reason, such as men not coming to work or want of tubs or wagons, the face be not cleared in the day-shift, the machine cannot be set to work that night, and the following day's output is lost; and it is in the avoidance of such delays that good management is shewn. A responsible man should traverse the faces the first thing in the morning and send men from other parts of the pit to any vacant places. The first 2 hours are the best, as it is useless to fill an empty place after the day is half done. Again, from some cause, the machine may not cut the full run: this, however, with a good and well-tended machine should be a rare occurrence.

Although machine-faces can be, and in some cases are, cut up to about 750 feet in length in one night with one machine, the writer holds that in the majority of cases with an ordinary fire-clay or coal-holing, 450 feet is long enough for any face to be cut regularly by one machine.

In the long-face method, the principal advantage is that the effect of a lost cut as just described is partly nullified: and with it there is always face ready for the machine to cut, and coal on sprags for the men to fill. The disadvantages are the cost and trouble of flitting the machine along the roads from one end of the face to the other. In this system, the roads are generally 132 feet apart, and the tubs are taken along the face. This method avoids much of the high-pressure working and constant supervision necessary in the first method. The setting-out of the machine-face is made practically on the same plan as that shewn in Fig. 1, but the face roads are spaced wider apart.

Fig. 2 (Plate VII.) shews one end of an alternative method of laying out the roads on a long machine-face: the purpose being to avoid the flitting of the machine. At the ends of the face, the roads are closely spaced together so as to allow of the coal being cleared in a single shift, and towards the centre of the face the roads are spaced wider apart and two or more days are allowed for their clearance. It will be seen that the machine is not kept standing longer than the single shift required for the clearance of the end places. The face is advanced two or three times a week, according to its special requirements, and the machine may work either during the day or night.

In steep workings, with a good roof, a third method (Fig. 3, Plate VII.) has been adopted by Mr. W. W. Millington, at the Oaks Colliery, Oldham, and it is eminently successful. The face is about 1,000 feet long, and lies at an inclination of 1 in 7. The machine starts from the bottom-end, and rails are laid along the full length of the face: and when the machine reaches the top-end, it is run downward by gravitation on these rails to the bottom-end of the face, where it is ready to cut again. One road only is led into the face, the road-rails being jointed to the face-rails, and an electric haulage-gear, capable of lifting 10 tubs at a time, brings the empty tubs to and takes the full tubs from the spot where they are filled. This method shews that advantage can be taken of the peculiarities of a seam. Thus, the full dip is 1 in 4, but by setting the machine-face at an angle, the inclination is reduced to 1 in 7. There is a dip from the rails towards the coal-face; this throws the weight of the machine in the same direction, and although the holing is hard and full of pyrites, no outside spragging of the rails is required.

The Working-face.—In preparing a machine-face (whichever method is adopted) the first and most vital consideration is straightness. A crooked face is an expensive nuisance to all concerned and means failure from the start. In his admirable paper* read before The Institution of Mining Engineers, Mr. Garforth showed the advantages to be gained in getting down the coal, after it has been holed to the proper depth, by having the face perfectly straight. The advantages to the machine in facilitating cutting are every whit as great; many a failure in introducing machines has been due solely to setting it to work on a crooked face; and for successful working the face must be straight.

If a big shift's work is to be done, the places at each end of the face should be kept at least the depth of a full cut ahead of the machine-face, so as to allow of the machine running clear at both ends.† It is advantageous to do this, as then the straightness of the face is assured, and the machine starts without delay at the beginning of a shift.

These headings are not, however, so necessary with a bar-machine as with a disc-machine. Where a heading is not provided for a disc-machine, a wheel-hole must be cut between the time that the walls are cleared and the time the machine should start, and the holing of it often delays the start of the machine.

---

* *Trans. Inst. M.E.*, 1902, vol. xxiii., page 312.

† The author has invented a small machine, driven either by compressed air or electricity, to hole these end-places and to drive the headings required for opening up the machine-faces. He hopes at some future date to describe it, and give some results of its working.

A bar-machine is brought to the end of the face with the cutter-bar about parallel to the face, and then swung in with a scythe-like motion through an angle of 90 degrees, so cutting itself in and making its own holing. If desired, the bar can extend the cut endwise at the start of each shift, and thus gradually add to the length of the face. As the bar usually works at the back-end of the machine, it is necessary at the end of the face, towards which the machine is cutting, to hole by hand a distance equal to the length of the machine, but this can be done during the cutting-shift and causes no delay.

Opening-up Faces.—Fig. 4 (Plate VII.) illustrates a proposed method of opening a machine-face. A heading, 45 feet wide or upwards, is driven in the direction required, the brushing is packed on each side of the main-road, and a machine-face is formed on each side of the heading. Where necessary or expedient, pillars of coal can take the place of the packs. The work of opening-up the face can be quickly done by the special heading-machine, already referred to, and two straight faces are assured on which machines can start forthwith.

The distance from the coal-face to the pack should be at least 5 feet. The props, if required, may come within $3\frac{1}{2}$ feet, but the pack and buildings should all be left 5 feet back. At the road-heads, as a rule, the coal-face is forward, and the machine crossing them should be able to cut lightly, whereas generally the buildings are kept so close that, unless the machine is put in to its full depth, it cannot get past: thus the defect is exaggerated, and the face thrown out of line.

Before the cutting-shift commences, it should be someone's duty to traverse the face, and see that it is left properly prepared for the passage of the machine.

Where the roof requires support nearer to the face than $3\frac{1}{2}$ feet, straps needled into the coal, should be fixed. An ample supply of timber, of proper lengths, should be laid at each road-head, also a sufficient number of sprags for the coal, and packing for the rails. Many an hour's cutting is lost by the machine-men having to hunt round for timber, etc., where with proper management it should be lying near them; and on a long machine-run, it would probably pay to have one man doing nothing else than looking after the timber. Very little timber, except props, need be sent down from the surface, as, with an axe and saw, he should get all packing that was required from the old timber in the roads and faces.

The sprags for the coal should be heavy wedge-shaped blocks, tapering from 4 to 8 inches and about 15 inches long (Fig. 5, Plate VII.). These would be too large to get lost, and are of too inconvenient a shape and size to be used for any other purpose than that intended. They are easily driven into place, effectually take the weight off the coal, and are readily taken out when the coal is being brought down.

The writer's reason for dealing so fully with this matter of the working-face is that, during 9 years' practical and constant experience of coal-cutting and personally starting about 50 machine-faces, he has not come across half-a-dozen faces which were in a condition even approximately suitable to put on a machine. Fig. 6 (Plate VII.) shews what is usually found. Complaints as to the crookedness of the face are met by the statement that "We will straighten it with the machine." This is almost an impossibility, and the attempt to do so is sure to involve much loss of time, and to result in everybody being at loggerheads before it is half done.

Rules for the Working-face.—For the working-face, the following rules should be laid down, and most strictly enforced:

(1) The working-faces shall be cleared of all coal and dirt and left as sketched, Fig. 7 (Plate VII.), and cross-trees shall be needled into the face, Fig. 8 (Plate VII.), when the roof requires support nearer than 3½ feet from the face.

(2) The places at each end of the face shall be left with the pack not nearer than 6 feet from the face, so as to allow the men passage around the machine to get it into its fresh position.

(3) Pack-dirt and roadhead-buildings must be kept 5 feet back from the face.

(4) All timber required for spragging coal, packing and fixing sleepers, and for roof-props behind the machine must be at the road-heads ready for distribution along the faces before the start of the cutting-shift.

(5) Adequate arrangements must be made for the sure supply of the cutters and oil required.

A man should be specially appointed to attend to, and be responsible for, the above rules being effectually carried out.

Laying Rails.—The writer is of the decided opinion that where it is possible the rails should be laid the full length of

the face. At the start of a machine, the advantages of this
method are so great that they would clear away the bulk of
the initial difficulties of getting the machine properly to work.
It means more capital outlay (about £60 per 300 feet of working-
face), but the cost of cutting will be reduced. Where the tubs
are taken along the face, it is particularly advantageous, as the
same rails can be used for both tubs and machine, and where
it is necessary to blast down the coal or to bore holes for an
hydraulic wedge, a motor-driven drilling-machine can precede
the coal-cutter along the rails.   The advantages gained are :—
(1) A practically straight face, as the rails would not lie along
a crooked one; (2) only two men are actually required at the
machine; (3) a greater length can be cut; (4) the men have
more time to devote to the proper management of the machine;
(5) only half the amount of spragging of rails is required; and
(6) the machine can cut on steadily at an easy cutting speed
instead of intermittently scurrying, probably in an overloaded
condition, over a few pairs of rails.

In the third method of setting out the working-face (Fig. 3),
rail-laying costs 0·54d. per foot of face, and is done by one man who
has laid as much as 450 feet in 10 hours. At another colliery,
2 men regularly laid 300 feet in half a shift and 450 feet in
6 hours, and were paid 1d. per ton in a 30 inches seam. There
were 2 men at the machine, the machineman was paid 1d. per
ton, and the back-tenter, ¾d. per ton. The total cost of cutting
was 2¾d. per ton. The miners used the same rails for drawing,
taking them up and relaying them as required. At first a few
sleepers were lost in the pack, but afterwards the method worked
successfully.

With a bar-machine, there is nothing like the pressure tending to push out the rails from the face, that exists with a disc-
machine; and in many cases where the rails are laid the full
length of the face and cross-jointed (Fig. 9, Plate VII.), no
outside sprags or struts are necessary, but short pieces are
required to prevent the rails from being pressed towards the face
at the front end. Thus a heavy labour-item is substantially
reduced.

Although decidedly of opinion that the above method is
generally the best, the writer is aware that with skilled rail-
layers, and the rigid enforcement of the rule as to the supply of

packing and timber, a machine in a large number of cases need never be standing on account of the rail-layers.

There is difficulty in training a squad of men to be efficient at this class of work. Until they are trained, the machine is frequently standing instead of cutting, and, by the rails not being properly spragged, the machine throws itself away from the face, thereby losing depth of cut and throwing the face off the straight; and it takes some considerable time and labour to get on the right line again. These remarks apply particularly to disc-machines. With a bar-machine, if the rails are not properly spragged, the machine gradually creeps out, and can be stopped as soon as it is noticed and the sprags be re-set.

FIG. 14.—SIX-POLE DYNAMO.

Electrical Plant.—For coal-cutting and general colliery-work the popular opinion is that a comparatively slow-running engine, either vertical or horizontal, with a good governor and a heavy flywheel, turned on the face for belt-driving, is the best prime motor. If properly selected as to size, to ensure its running the greater part of its time at, say, three-quarter load, it will be as economical in steam as any other plant. For efficiency, simplicity of construction and cheapness, the author would recommend a horizontal compound-condensing tandem-engine, designed to give its highest efficiency as to steam-consumption at 85 per cent. of the rated load.

The dynamo (Fig. 14) should be placed in front of the engine, so as to get the tight side of the belt below with the engine running outwards. This arrangement requires rather a long engine-house, which is the only objection. With a vertical engine, space would be saved. A good belt, of ample weight and width, should be used to prevent losses due to slipping. The dynamo should run at a moderate speed, and, like the engine, be designed to give its highest efficiency at 85 per cent. of its rated load.

Fig. 15. - Direct-coupled Slow-running Cross-compound Horizontal Steam-engine and Eight-pole Dynamo.

For installations of 200 indicated horsepower and upwards, a direct-coupled slow-running engine and dynamo (Fig. 15) is to be preferred. It would require less attention than the ordinary high-speed steam-driven dynamo, and would insure a long life of uninterrupted work.

Continuous or Alternating Current.--If the driving of coal-cutting machines is the only or chief use to which electric power is to be applied, continuous current should be adopted, and even where electrical coal-cutters utilize only a small portion of the total electric output, continuous current is preferable under ordinary conditions. There are many cases, however, where other considerations will determine in favour of alternating polyphase current, and it is on this system that power will be provided by the power-supply companies, in various parts of the country.

Where alternating-current must be used, a low periodicity, say, 30 or still better, 25, should, if possible, be chosen. Two of the limiting factors of a coal-cutter motor, namely: small diameter and moderate speed, are essentially opposed to the best practice in the design of polyphase machines. The larger sizes for use in thick seams present little difficulty, but motors to work in medium or thin seams must in their design be a compromise between the best and the attainable.

Polyphase motors have been very successfully applied to reciprocating-bar coal-cutting machines, the pioneer in this direction being Mr. Roslyn Holiday, of Ackton Hall colliery. The difficulty of a similar application to disc-machines is accentuated by the necessity of providing large starting torque and capacity to endure large temporary overload, when the disc is jammed.

Switchboard.—The switchboard, in addition to its usual fittings, should have an automatic overload preventer, ringing an electric bell on falling out, and each separate circuit should be provided with its own ammeter.

Voltage.—As a standard electromotive force for direct current, 500 volts may be taken as the most suitable; and this being the Board of Trade limit for tramway-work and town-lighting, most makers of dynamos and motors stock machines at this voltage. With three-phase current, 400 volts is considered a safe working limit.

Cables in Roadways.—Cables belowground should always be armoured; as no matter what precautions are taken in securing them in the first instance, sooner or later they will come down and be subjected to rough usage. The armouring should be braided over, the object being to keep the armouring properly in place. With unbraided wire, it is common to see parts where the armouring has got a twist and left the insulating material unprotected.

Probably the best cable which can be used in a pit is a concentric armoured cable, braided over, the core of the cable being the positive wire and the outer the return: the return being earthed. The switch and junction-boxes are perfectly water-tight, and the positive wire cannot be touched except by opening the boxes. There is no chance whatever of a man inadvertently coming into contact with the positive wire from which a shock could be got. It is also self-testing, that is to say, if it be not perfectly safe, it cannot be worked. Objections to concentric wiring have been raised by those unfamiliar with it and unacquainted with its advantages, the chief of which is the absolute safety afforded to the men. This method of cabling has been largely and successfully used in mines during the past 12 years. It is a tribute to the safety of concentric cables, that when the question of wiring an explosive-factory of a particularly dangerous character was before the Home Office, this was the only method that the officials would sanction. Few, if any, of the fatalities which have been associated with electric cables could have occurred had concentric cables been used. But whatever type of cable is used in mines, it ought to be substantially armoured; and it is of vital importance that the armouring should be efficiently earthed.

The cables should be carried to the face along as many roads as may be necessary, the cable-roads being spaced about 300 feet apart. Thus on the short-face system, a cable-road in the centre of the face would generally suffice. A length of cable, 330 feet or 660 feet, should be left on a strong drum in order to be paid out as the machine-face advances. The inner end of the cable is passed through the flange of the drum and on the side of the drum a plug and fuse-box can be carried for the attachment of the face-cable. The last junction-box on the main cable before the drum should be fitted with a switch.

Junction-boxes.—Junction-boxes of cast-iron, with closed backs and hinged waterproof fronts, should be used at intervals of 330 or 660 feet, and a spare length of cable should be kept ready for running out and coupling to these boxes, should any working length be damaged and become unworkable by reason of a fall of roof or other accident.

Face-cables.—Face-cables should be from 240 to 300 feet long: this being a sufficient length to be conveniently handled. It should be either a concentric or twin cable, having an extra thickness of insulating material to resist damage, and water-proof-taped (instead of being braided) so as not to absorb moisture, and spirally armoured to prevent kinking, which is the most serious and common cause of damage. The armouring of the face-cable should be electrically connected at the last junction-box with the earthed armouring of the roadway-cable, and also with the framework of the machine. The machine-end of the face-cable should be fitted with such a connection as can be attached and detached almost instantaneously. The other end of the face-cable is attached to a suitable fuse-box.

Fuse-box.—The fuse-box is a small, but most important accessory to the coal-cutting machine. Many an armature has been burned out and other damage done by the use of improper fuses. The fuse is the safeguard of the whole machine, and, if it be too heavy, a breakdown of the machine is almost certain. It should be as small as possible; a motor built to carry 40 ampères should be protected by a fuse blowing at 50 or 55 ampères. In addition to being of the right size, the fuse should be placed in convenient fittings. If it takes 10 minutes to put in a fresh fuse, the machineman will very soon be tempted to put in one that is too strong, and thus avoid the trouble of frequent renewals.

Figs. 16 and 17 shew a fuse-box, in which there are no small contacts or screws to be burned, no tools are required when a new fuse is inserted, there is no danger of coming into contact with the live wires, and it meets all the requirements of mining work.

Testing.—The generating-plant should be thoroughly tested after erection by, say, a 6 hours' run at full load, before being set to its regular work. This can be easily effected by means of a water-resistance.

The Coal-cutting Machine.—In introducing a coal-cutting plant, probably the most troublesome part of the whole scheme is the selection of the particular make of machine to be adopted, and the statements of rival makers all claiming to turn out the best machine must be, to say the least, confusing. So far as actual cutting-speed is concerned, most British machines are on approximately the same footing; Americans are new at this class of machine, most of those in use being heading-machines, and their longwall-machines run very slowly. The other considerations which should therefore decide the question, may be enumerated as follows:—(1) The power required by each class of machine to drive it at a given rate; (2) which machine will make the least small coal; (3) will one type of machine suit all the seams in the mine likely to be worked; (4) which machine is made in the strongest possible form, of the best material, and with the minimum number of working parts open to the dust and dirt of the mine; (5) what type of machine is most easily operated and requires the least labour from the attendants; and (6) which machine will be most suitable under the strenuous conditions of daily work.

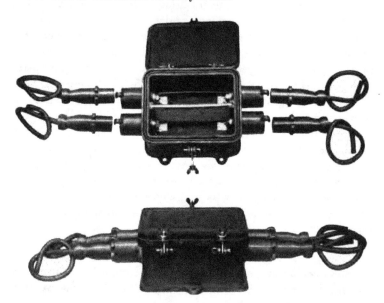

FIGS. 16 AND 17.—KENNEDY-READING FUZE-BOX.

I.--With regard to the first item there can be no doubt that a reciprocating bar-machine requires less power than any other type. In disc-machines there is a great surface exposed to friction, but with the reciprocating bar-machine there is none, and the whole of the power put into the bar is absorbed in the proper place at the pick-points.

Fig. 10 (Plate VII.) shews the power taken by a disc-machine and a reciprocating bar-machine respectively, working in the same seam, cutting to an equal depth and travelling at the same rate:—For the disc-machine the average was 50 ampères at 400 volts, and for the bar-machine 30 ampères at 400 volts. In this case the coal did not fall down as soon as cut. In cases where the coal or dirt does come down as soon as it is cut, the friction is enormous, and the difference in power taken by the two machines is even more marked.

Fig. 11 (Plate VII.) also shews a case in point, which came under the writer's observation. The machine was driven from a plant which was to be capable of driving four disc-machines. The dynamo generated about 200 ampères at 400 volts. The diagram shews one disc-machine at work, taking an average current of 110 ampères. On the same figure, the bar-machine is also shewn taking about 55 ampères. Both machines holed to the same depth, and cut at the same rate in an extremely hard stony fire-clay. The cutting-speed, when under-cutting to a depth of $4\frac{1}{2}$ feet, was 8 inches per minute.

Fig. 12 (Plate VII.) shews $5\frac{1}{2}$ hours' work with a small bar-machine. The holing was made in a hard fire-clay below the coal in a seam, 20 inches high, and the total length cut was 180 feet to an average depth of 3 feet. The motor is built for 25 ampères at 400 volts, and the current never exceeds 23 ampères during the whole shift—the average being 16 ampères. There is one run of 57 minutes, one of 53 minutes, one of 25 minutes, and six others of shorter duration. The diagram was taken during the first fortnight that the machine was at work. The miner would not hole in the fire-clay in which the holing was made by the machine.

A convenient comparison of the power absorbed in various coal-cutting machines is obtained by stating the electric horse-power required per square foot of surface cut per minute; and by the same method, exceedingly interesting and useful comparisons may be made of the powers absorbed by a given machine when cutting in materials of different hardness (Table I.).

It follows, therefore, that the capital-outlay in generating-plant, including boiler, steam-engine, dynamo, cables, etc., for a reciprocating bar-machine plant need only be half or at most two-thirds of that for a disc-machine plant, and as the machines cannot be overloaded by the coal weighing down on them, very little margin of power is required.

Table I.—Showing the Powers Required when Cutting in Various Materials by Disc- and Bar-Machines.

No. of Experiment.	Type of Coal-cutting Machine.	Material §		Thickness of Seam. Feet, Inches.	Height of Under-cut. Inches.	Depth of Under-cut. Feet.	Rate of Feed per Minute. Inches.	Surface Cut per Minute. Square Feet.	Dynamo. Volts.	Dynamo. Ampères.	Dynamo. Horsepower.	Per Square Foot of Surface. Horsepower.	Per Square Foot of Surface. Watts.
1	Disc* ...	Blackband	(6)	4 6	5	4½	18·0	6·75	400	50	27·0	4·00	3,000
2	Bar* ...	Do.	(6)	4 6	5	4½	18·0	6·75	400	30	16·0	2·36	1,750
3	Disc† ...	Stony fire-clay	(1)	3 0	5	4½	8·0	3·00	400	110	59·0	19·86	14,666
4	Bar† ...	Do.	(1)	3 0	5½	4½	8·0	3·00	400	55	29·5	9·83	7,333
5	Bar‡ ...	Hard fire-clay	(5)	1 8	3½	3	12·0	3·00	400	16	8·6	2·87	2,130
6	Disc ...	Do.	(7)	3 3	5½	3¾	18·0	5·65	350	60	28·0	5·00	3,700
7	Disc ...	Fire-clay	(8)	3 3	5½	3½	18·0	5·25	300	70	28·0	5·30	3,950
8	Bar ...	Do.	(9)	4 6	5	5	9·6	4·00	350	32	15·0	3·75	2,800
9	Bar .	Hard fire-clay	(4)	7 0	5½	4½	12·0	4·50	400	35	19·0	4·20	3,120
10	Bar ...	Do.	(3)	7 6	5	4½	12·0	4·50	380	40	19·0	4·20	3,120
11	Bar ...	Do.	(2)	1 10	4	3	10·0	2·50	450	20	12·0	4·80	3,560
12	Bar ...	Hard blackband	(5)	4 0	6	5	15·0	6·25	400	40	21·5	3·35	2,500

* Load-diagram, Fig. 10, Plate VII.   † Load-diagram, Fig. 11, Plate VII.   ‡ Load-diagram, Fig. 12, Plate VII.

§ The numbers in parentheses denote the relative hardness of the various bolings.

Apart from the proof afforded by Figs. 10 and 11 (Plate VII.) and Table I., as to the small relative power used, there is the fact that a break-down of armature, field-coils or starting resistances is almost unknown; and for the past three years, not a single complaint has been received nor has an armature been returned for repairs. This record is a substantial and practical endorsement of the claim.

II.—There is really no comparison between a disc and a bar-machine as to the quantity of dross made. In coal of a friable nature or in seams where partings occur some inches above the holing, the disc-machine tears away a large quantity

Fig. 18.—Hurd Bar Coal-cutting Machine, arranged as an Undertype, and Cutting in the Lowest Position.

of coal, brings it down on the machine where it has to be broken up, and mixes it amongst the holing-dirt. With a bar-machine, there being no surface under the coal except the cutting surface, the width of the cut made is practically the exact dimension over the cutters, sprags can be put in within a foot from the solid coal (Fig. 13, Plate VII.), and the weight being taken instantly on the sprags the coal is kept up. In seams with many partings or slips, this gives the bar-machine a great advantage.

III.—In a paper* recently read before The Institution of Mining Engineers it was stated that it was impossible to design

* "Coal-cutting by Machinery," by Mr. R. W. Clarke, *Trans. Inst. M.E.*, 1902, vol. xxiii., page 96.

a coal-cutting machine to suit the varying conditions of different seams, and that each colliery should be treated on its merits and a special machine built to suit it. The writer considers that the reciprocating bar-machine, in its three standard sizes, fulfils all the varying conditions likely to be met with in longwall working. This result is attained as follows:—Fig. 18 shews a standard machine arranged as an undertype, for making its own pavement. Fig. 19 shews the same machine, arranged as an over-type, at the lowest point of its cut. Fig. 20 shews the machine, arranged as an undertype, and mounted on high wheel-brackets for a mid-holing, where a large quantity of dirt falls on the rails. Fig. 21 is a standard machine, arranged as an overtype, mounted on extra-high wheel-brackets, for a seam with a top holing. All the machines have an up-and-down travel of 8 or 9 inches.

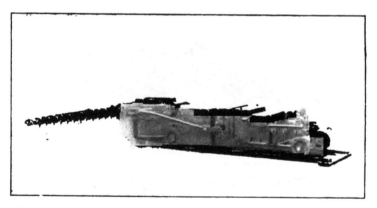

FIG. 19.—HURD BAR COAL-CUTTING MACHINE, ARRANGED AS AN OVERTYPE, AND CUTTING IN THE LOWEST POSITION.

It is obvious, therefore, that, for collieries with a variety of seams, this machine possesses every advantage, and as a standard type can be used in all seams; and the spare gearing, with the exception probably of wheel-brackets, will suit any of them.

IV.—So far as strength of gearing is concerned, most coal-cutting machines could probably be made equal, but in the pro-tection of the gear the bar-machine has every advantage, as a glance at the machine will show. There are no bearings under the coal, as in the centre and periphery of a disc; and the lubrication of the gear is assured by its running in oil.

V.—Most disc-machinemen, after they have once used a bar-machine, never wish to go back to the disc-machine.

With a reciprocating bar-machine there is a minimum amount of noise or vibration, and less spragging of the coal is required. The bar being placed at the back-end of the machine, it is rarely that the machine has to be cleared of a fall of coal or dirt. Should a small hitch or a large ironstone-ball be encountered, no pick-work is required; the bar is tilted up or down, while cutting, and the obstacle is cut round instead of an attempt being made to cut through it, with the risk of the machine being damaged, while doing so.

Fig. 20.—Hurd Bar Coal-cutting Machine, arranged as an Undertype, and Mounted on High Wheel-brackets.

The men following a bar-machine have easier work, in fact, in most instances tne daughing, or clearing-out of the cuttings from below the holing, can be safely left to be done by the filling shift on the following morning. The worm-thread on the bar is a conveyor, the front of the holing forms one side of the conveyor-trough and the trailer-bar the other, consequently so much of the cutting dirt is brought out that the back of the holing can be easily seen, a clear space of from 2 to 3 inches being left. Another advantage is that the whole of the top part of the bar can be inspected, and by stopping the machine for 2 or 3 minutes and turning the bar by hand (which can easily be done), the machineman can at once know whether or not the cutters are in good order.

VI.—From the foregoing description, it will be seen that the reciprocating bar-machine amply fulfils all the requirements o a good machine, and therefore is the most likely one to endure the strenuous conditions of daily work at the coal-face.

The machine being decided on, it is of importance in giving the order that a condition be made to the effect that the machine must be tested at full load for 4 hours' continuous run at the makers' works before being despatched. The makers of the Hurd bar-machine invariably make this test. A test shaft fitted with a belt-pulley is placed in the cutter-bar bearings, and from this pulley a belt is taken to drive a dynamo working against a liquid resistance. The coal-cutter motor is then connected with the current of the testing dynamo and driven at full power. Thus the whole gearing is fully tested running from 4 to 6 hours at full load. During such tests the motor is entirely enclosed, the doors being screwed tightly up. Table II. is a note of a test recently taken.

TABLE II.—TEST OF 16 INCHES BY 8 INCHES COAL-CUTTING MACHINE.

Time.	Speed of Motor per Minute.	Motor.		Parts of Motor.	Temperature after running for 2 Hours.	
a.m.	Revolutions.	Volts.	Ampères.		Degrees.	Rise of Temp. Degrees.
7·15	Start.	—	—	Air	61	—
7·30	741	448	45	Commutator	126	65
8·30	784	456	45	Core	125	64
9·45*	794	450	45	Coils	124	63
					After running for 4 Hours.	
10·15	Start.	—	—	Air	63	—
10·45	848	460	45	Commutator	148	85
11·45	840	460	45	Core	145	82
12·45†	820	450	45	Coils	153	90

* Stop to take temperatures after running for 2 hours.
† Stop to take temperatures after running for 4 hours. All gearing working well, and no heating.

Before closing his remarks, the writer wishes to confute some statements respecting the bar-machine, which have been mentioned on various occasions:—(1) The cutter-bars will not stand the overhang and are continually breaking. The reply to such a statement is that during the last 3 years, he has known of only one bar having broken and this had been at continuous work for about 2½ years. (2) Although this is a good machine for soft or medium holing, it is not suitable for hard holing. On this point, the samples of hard holing exhibited to the members, most of them showing cutter-marks, constitute the best answer. (3) The bar does not clear itself. This, as already

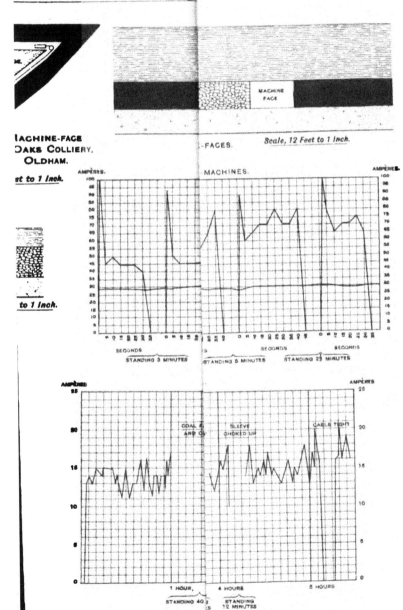

MACHINE-FACE
OAKS COLLIERY,
OLDHAM.

Scale, 12 Feet to 1 Inch.

explained, is incorrect, and it is impossible for the bar not to clear itself.   (4) The bar is continually rising.   The writer never saw a bar rising or dipping, except when all the cutters were absolutely blunt.   Then, it might either rise or dip, depending upon whether it was cutting under or over, independently of the nature of the holing and of the adjacent parts of the seam. As soon as fresh cutters were inserted, it would cut back to its proper place almost instantaneously.

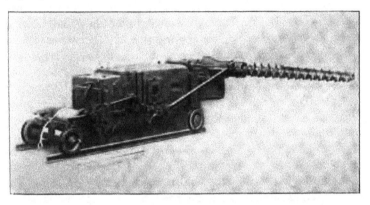

FIG. 21.—HURD BAR COAL-CUTTING MACHINE, ARRANGED AS AN OVERTYPE, AND CUTTING IN THE HIGHEST POSITION.

Mr. DANIEL BURNS (Carluke), referring to Fig. 10 (Plate VII.), said that the diagram showed that the disc-machine had five stoppages and five startings, while the bar-machine was shown to have run continuously for 70 minutes.   It was evident that the disc-machine had been encountering difficulties, and the diagram was not fairly representative of the work done.   Personally he had seen a disc-machine running for an hour, on a length of 150 feet, and he could furnish a diagram which would compare more favourably with the bar-machine than the one given by Mr. Hurd.   He also wished to mention that such diagrams were largely dependent upon the person driving the machine, and if the disc were kept clear of cuttings and in its proper cutting position the nature of the diagram would be completely altered.

Mr. HURD replied that the diagram (Fig. 10) was not intended to represent the various stops and starts of the respective coal-cutting machines, but to record the power taken to drive them.

Mr. GEORGE A. MITCHELL (Glasgow) said that he had seen the bar-machine, a few days ago : it had just started, and was cutting fairly well.   He had always been afraid that the bar-machine would not clear the cuttings out of the holing, but in this instance, it seemed to be doing this work efficiently.   The conditions, however, were very favourable, because the holing was quite dry, and he was doubtful whether the cuttings would come out so readily if the holing were damp.   Regarding the question of the percentage of dross, Mr. Hurd had stated that in all cases there was a decrease ; but he (Mr. Mitchell) did not think that this statement could be substantiated by experience.   Mr. Hurd was of the opinion that any seam could be worked with advantage by a coal-cutter, and that in all cases there would be a saving in cost.   He did not agree with that statement, and he thought that there were certain seams where the cost would be cheaper with the pick than with the coal-cutter.   For example, there were cases where there was dirt to turn over on the top of the seam and when the holing was done by a machine the dirt came away in large blocks which were most difficult to handle and in consequence the cost of ridding the dirt was considerably enhanced. Other circumstances, which he need not enumerate, must be taken into account : for instance, the question of timbering must be considered, and if they added 2d. or 3d. per ton to the cost of the coal for extra timbering, this would seriously reduce the saving effected by the machine in other respects.   He was pleased with Mr. Hurd's paper, but he felt that it would have been wiser if he had omitted the reference to the comparison between this country and the United States of America.   Personally he thought that there was a fallacy in the comparison and that the extra output in the United States was due more to the conditions under which the men wrought than to the more general adoption of coal-cutting machines.   The seams in the United States were thicker and easier to work, and the output per man was necessarily greater than in this country with or without the use of coal-cutting machines.

The further discussion was adjourned.

THE MINING INSTITUTE OF SCOTLAND.

EXCURSION MEETING,
GLASGOW, FEBRUARY 11TH, 1903.

GLASGOW CORPORATION TRAMWAYS : PINKSTON GENERATING-STATION.

The site of the generating-station extends to 18,997 square yards, and is bounded on one side by the Forth and Clyde canal, from which abundance of water can be obtained for condensing purposes. Both the Caledonian and North British Railway Companies have sidings connected with the coal-bunkers which have a capacity of about 4,000 tons, and are built on the outside of the end-wall of the boiler-house. Electric-locomotive cranes are used for handling and tipping the coal-wagons in the sidings.

The main building, 244 feet long and 200 feet wide, consists of a steel framework, comprising about 1,300 tons of steel, and supporting the roof independently of the walls. Stanchions are built into the walls, which are of plastic-clay bricks, faced with terra-cotta bricks on the outside, and cream and French-grey enamelled bricks on the inside. It is divided into three bays :— The eastern bay, forming the boiler-room, is 244 feet long by 84 feet wide ; the central bay, forming the engine-room, is 244 feet long by 75 feet wide ; and the western bay, containing the auxiliary plant, is 244 feet long by 40 feet wide. The two chimney-stacks are 263 feet in height from the ground-level.

Boiler-room.—In the boiler-room there are 16 Babcock & Wilcox water-tube boilers, each capable of producing 20,000 pounds of steam per hour at a working pressure of 160 pounds per square inch. The furnaces are fed by mechanical stokers, of the chain-grate type. The coal-bunkers on the top of the boilers can hold 2,400 tons. The coal-conveyors can deal with 50 tons per hour. The coal is automatically weighed as it passes through the shoots into the furnaces. The mechanical stokers are driven by two electric motors of 30 horsepower. Two water-storage tanks for the boilers, each holding 1,800 gallons, are placed on girders between the two chimneys. The two fuel-economizers are capable of dealing with 12,000 gallons of feed-water per hour.

Engine-room.—In the engine-room are 4 main engines of the vertical inverted compound-condensing three-cylinder type, Nos. 1 and 2 being of American, and Nos. 3 and 4 of British make. Each of these engines is designed to work at 4,000 indicated horsepower at its normal rate of working, but is capable of developing 5,000 indicated horsepower at its normal speed of 75 revolutions per minute. The steam-consumption is about 14 pounds per brake-horsepower per hour. There are one high-pressure and two low-pressure cylinders in each engine. The diameter of the high-pressure cylinder is 42 inches, and of each of the low-pressure cylinders 62 inches. The diameter of the shaft is 36 inches; and the fly-wheel is 24 feet in diameter and weighs about 120 tons. The weight of each engine is about 700 tons. Measured over all, each engine is 43 feet long (or 52 feet including the fly-wheel and generator), 35 feet high, and 24 feet deep from back to front.

Each engine is directly connected to a three-phase generator, designed for an output of 2,500 kilowatts at a pressure of 6,500 volts. The weight of the revolving fields is, in each machine, nearly 38 tons, while that of the armature and its ring-frame is 44 tons. Each generator weighs about 90 tons.

In addition to the 4 main engines, there are 2 auxiliary engines of the vertical cross-compound type, each of 800 indicated horsepower, but capable of developing 1,000 indicated horse-power. The weight of each of these engines is 122 tons. Each of the engines is directly coupled to a 500 volts direct-current generator of 600 kilowatts capacity. These two units are used for driving the auxiliary plant, and for supplying power and light to the car-depots during the night.

Between the auxiliary engines and the switchboard, there are 6 exciter-engines and dynamos. Each of these engines has a capacity of 85 indicated horsepower, when running at 300 revolutions per minute. These machines are used for exciting the fields of the main generators.

The 2 electric cranes in the engine-room are each capable of
lifting 50 tons.

The marble switchboard consists of 4 generator-panels, 6 panels
for the 6 exciter-dynamos, 4 inter-connector panels, and 20 feeder-
panels, with all the instruments and appliances.

Auxiliary-room.—There are 5 surface-condensers—one for
each of the 4 main engines, and 1 for the 2 auxiliary and 6
exciter-engines.   Each of the larger condensers can condense
60,000 pounds of exhaust-steam per hour, and the smaller condenser 24,000 pounds per hour.

There are 4 centrifugal circulating pumps, having a total
capacity of 240,000 gallons per hour, and one with a capacity of
96,000 gallons per hour.

There are 4 air-pumps, each with a total capacity of 60,000
pounds of exhaust-steam per hour, and one with a capacity of
24,000 pounds.

There are 4 boiler-feed pumps capable of delivering 8,000
gallons per hour; and the boiler-feed steam-pump is capable of
delivering 16,000 gallons of water per hour.

The electric crane in this room can lift 30 tons.

Sub-stations.—There are 5 sub-stations, namely :—Coplawhill,
Kinning Park, Partick, Dalhousie and Whitevale. In these
stations, there are in all 24 units, each consisting of three
transformers of 200 kilowatts each and 1 rotary converter
of 500 kilowatts.   At each station, there are two switch-
boards—one for the alternating current, and one for the direct
current.   The static transformers transform the current from
6,500 volts down to 330 volts, and the rotary converters convert
the current from 330 volts alternating to 500 volts direct current.
In this form it is sent out underground by the feeder-cables to
the overhead wires, and thence to the cars.

THE MIDLAND COUNTIES INSTITUTION OF ENGINEERS.

GENERAL MEETING,
Held in the Town Hall, Mansfield, March 7th, 1903.

Mr. G. ELMSLEY COKE, President, in the Chair.

SHERWOOD COLLIERY.

By J. W. FRYAR.

Shafts.—On July 12th, 1901, the position of the pits, 20 feet in internal diameter, was fixed on the western side of the Midland railway, to the south of Mansfield Woodhouse station.

On March 10th, 1902, the sinking of the upcast or No. 2 pit was commenced. The foundations had been put through 15 feet of sand which lies at the top of the limestone. The limestone, which is 141 feet thick, was sunk through at an average rate of 22½ feet per week. Compressed-air percussive rock-drills were worked from the Walker shaft-sinking frame. A small quantity of water was encountered in this limestone, and sealed off with coffering and tubbing: the maximum quantity being 52 gallons per minute, and another feeder of 80 gallons per minute was found at a depth of 270 feet. Sinking was then resumed, and the pit tubbed down to a depth of 303 feet from the surface. Considerable delay was experienced in the delivery of the tubbing, and it was not until October 28th, 1902, that the tubbing was completed and the water sealed off. At that date, the pit was 444 feet deep and lined to a depth of 303 feet. The pit has been sunk since then to a further depth of 771 feet (an average of 41·4 feet being sunk and bricked per week), and the total depth of the shaft is 1,215 feet.

The downcast or No. 1 pit has been sunk to a depth of 435 feet, and is tubbed to a depth of 351 feet. The water is not yet sealed off, as the closing rings of the last length of tubbing have not yet been received.

A steam-winch, with double drums, is used at each pit to work the sinking-scaffolds.

Shops.—The temporary workshops are fitted with 2 forges, drilling-machine, lathe, pipe-and-bolt screwing-machine, planing-machine, circular and band-saws, and wood-planing and thicknessing machine. There are also a stone-crusher and 2 mortar-mills. These machines were at first worked by a large portable engine and boiler, but they are now driven by tri-phase motors.

Electric Plant.—The power-house contains two Robey compound engines of 100 horsepower with cylinders 11½ inches and 18½ inches in diameter by 2 feet stroke, each driving a tri-phase dynamo with an output of 50 kilowatts. One engine also drives a Riedler air-compressor of 50 horsepower, and the other drives a ventilating-fan, 3 feet in diameter, used to ventilate the sinking pits. A duplicate fan is also provided, and is driven by a tri-phase motor.

Boilers.—Three Lancashire boilers, 30 feet long and 9 feet in diameter, are at work, fitted with Bennis stokers, Green economizers, and Dixon superheaters. A fourth boiler is ready for coupling to the main range. The boiler-pressure is 150 pounds per square inch.

Winding-engines.—The permanent winding-engine at the No. 2 or upcast pit has two cylinders, each 26 inches in diameter by 4½ feet stroke, fitted with Corliss valves, Seymour automatic cut-off gear, steam reversing-gear, Whitmore steam-brake and overwinding-prevention gear. The drum is 11 feet in diameter. The Seymour cut-off gear has been working during the sinking.

The sinking-engine at the No. 1 or downcast pit has two cylinders, each 20 inches in diameter by 3 feet stroke, and is geared 8 to 3 to a drum, 8 feet in diameter.

Pumping-plant.—As a somewhat large quantity of water was expected in the sinking, two Evans Cornish sinking-pumps (steam-cylinder, 21 inches in diameter; and a working barrel, 16 inches in diameter, by 2 feet stroke) were provided, but they

have never been used.   Two small Evans pumps (steam-cylinder,
10 inches in diameter; working barrel, 6 inches in diameter,
by 1 foot stroke) were slung on ropes (one in each pit) and proved
sufficient for all the water encountered.

———

Mr. J. A. LONGDEN (Stanton-by-Dale) understood that it was
hoped that the colliery would be worked with a coal-consumption of 1 per cent. of the output: at present the whole of the
work of sinking, driving electric plant, etc., was being carried on
with practically one boiler.   Although a study had been made
of economical working, yet he (Mr. Longden) was somewhat
sceptical of a coal-consumption as low as 1 per cent.   At
Pleasley colliery, the coal-consumption had been reduced from
7 per cent. (which it was formerly) to 3½ per cent. (which it is
now) at a depth of 1,560 feet.   The plant, at Sherwood colliery,
comprized economizers, superheaters, ample boiler-power, engines
with Corliss valves, and every economical appliance except a
central condensation-plant, in order that economical results
should be obtained.   The sinking had been very fortunate, and
the large sinking-pumps had not been required, as the water
was much less in quantity than it was expected that it would be.

Mr. S. A. EVERETT (Gedling) remarked that the result of the
work so far accomplished at Sherwood colliery indicated an
advance upon anything done previously, in two respects.   The
shafts were 20 feet in diameter; the speed of sinking (41·4 feet
per week) was exceptional for such large pits; and, further, the
successful use of cast-iron tubbing, 20 feet in diameter, to a
depth of over 350 feet was a step in advance of previous practice.
He was pleased to hear that the automatic expansion-gear on
the winding-engine had been used throughout the sinking; not
so much on account of any economy effected by its means—this
was a minor consideration during sinking—but because com-
pliance with the exacting demands made on a sinking-engine
proved that the gear was of unusually good design.   He had not
heard of another case in which an automatic cut-off gear had
been continuously used for such a purpose.

Mr. A. B. MARKHAM, M.P. (Stuffynwood) said that he had
recently erected a colliery-plant, entirely worked by electricity,

with the exception of a Corliss compound-condensing winding-engine. With high-pressure steam, economizers and every possible contrivance for reducing the steam-consumption, the coal-consumption at Oxcroft colliery did not fall below $1\frac{3}{4}$ per cent. He could not imagine how Mr. Fryar, who had not erected Corliss compound-condensing engines, but had two high-pressure winding-engines—though he believed that ultimately compound winding-engines would be erected—could possibly reduce the coal-consumption to 1 per cent., especially as he (Mr. Markham) employed a central-condensing plant, into which the whole of the exhaust-steam of those two engines was passed. He thought that the members of the Institution would be interested to learn how it would be possible, when the coal was reached, to be able to wind successfully with small drums; because many engineers had failed to accomplish this in the past. The first Corliss compound-condensing winding-engine was erected in this country at the Oxcroft colliery, and he (Mr. Markham) subsequently ordered a second winding-engine for one of the collieries of the Tredegar Coal and Iron Company, and both of these engines had worked without giving the slightest trouble. The success of these winding-engines depended upon placing a large receiver between the high- and low-pressure cylinders, and the addition of a condensing-plant enabled the engines to start quickly and effected a large economy in steam-consumption.

Mr. J. W. FRYAR thought that Mr. Longden had hardly treated him fairly by repeating information given to him in confidence that day, that he hoped to reduce the coal-consumption to 1 per cent. of the drawings. He had mentioned what he hoped to attain, and Mr. Markham's statement that he had reduced it to $1\frac{3}{4}$ per cent. at Oxcroft colliery, made him (Mr. Fryar) feel more confident of reducing the coal-consumption to 1 per cent., because a large proportion of the steam used at a colliery was not taken up by the engines, but was used in keeping the steam-pipes warm. The engines were intermittent in their working, and there was a considerable amount of condensation, which was fairly constant, with a well-equipped plant, whether the colliery had a large or a small output. The output at Oxcroft colliery was about 800 tons per day, and the output at

Sherwood colliery would probably reach 3,000 tons per day. The large proportion of steam, used by the pipes, being constant in each case, he (Mr. Fryar) thought that more work could be performed with 30 tons at Sherwood colliery than with 14 tons at Oxcroft colliery, and therefore a lower percentage of coal-consumption would result at Sherwood colliery.

He would like to mention that all the steam which could be saved at a colliery was not to be saved by expansion in the steam-engines. The bulk of the waste was due to negligence in taking care of the steam: the steam-pipes, steam-pipe joints, and cylinders should be covered, and every possible loss of heat should be prevented. At a large colliery, with an output of 1,600 tons per day, and no coke-ovens, the coal-consumption was 1 per cent., so that he (Mr. Fryar) would not be the first engineer to attain so low a result. He might mention that, at that particular colliery, the amount of steam used per indicated horsepower-hour averaged from 70 to 80 pounds with economical expansion-engines: these engines probably used only 35 pounds per indicated horsepower-hour, but the total worked out to 70 or 80 pounds per horsepower; and under such conditions the coal-consumption was as low as 1 per cent.[*]

He preferred the use of small drums on the winding-engines at Sherwood colliery, for several reasons:—(1) The winding could be effected with smaller cylinders, and less steam was used; and (2) the small drums gave a much better turning-moment, enabling the cages to run more smoothly and with less vibration of the ropes than with large drums. It remained to be proved whether the winding-drum was too small when heavier weights were wound and larger ropes were used.

The PRESIDENT (Mr. Elmsley Coke) moved a vote of thanks to Mr. Frank Ellis and Mr. J. W. Fryar for kindly allowing the members to visit the Sherwood colliery, and to Mr. J. W. Fryar for reading his instructive and useful paper.

Mr. J. A. LONGDEN (Stanton-by-Dale) seconded the votes of thanks, which were unanimously adopted.

———

Mr. W. A. MOLLER's paper on "Mining in Manchuria" was read as follows:—

[*] *Trans. Inst. M.E.*, 1902, vol. xxiv., page 547.

MINING IN MANCHURIA.

By W. A. MOLLER.

Coal-fields.—The coal-fields of Manchuria form part of a large belt, which runs from the south of Pekin in a north-easterly direction to the northern parts of Manchuria. It comprises, generally, a multitude of synclines of varying size, the anti-clines having all the coal-beds denuded off, leaving, usually, the blue Mountain Limestone cropping out at the surface. The coal-basins do not usually cover a very large area, but they contain a great number of seams, separated in many cases one from the other by only a few feet of intervening strata.

Chinese Mines.—Nearly all these basins have been worked for many years by the Chinese, and, in most cases, the outcrops are visible. Water only collects in the workings after the summer rains: the native plan being to commence work in September and to leave off in June; and during the months of July and August no work is done owing to the lack of ventilation in the hot months, and to the percolation of water through the strata during the rainy period. In many places, where the coal is of a superior quality, work is continued in the old dips, and the water is taken out in water-baskets: a series of ladder-like basins being formed, about 3 feet apart in vertical height. The writer has seen 150 men working on one ladder, baling water to a height of 450 feet; and, in some cases, the amount of water averaged 75 gallons a minute from one set of ladders.

Owing to the entire absence of plans of the mines worked by the natives, many accidents occur in the deeper mines by holing into old workings and running large quantities of water into their faces.

The haulage of coal in the native mines is done by coal-carriers, who use a specially-constructed basket (holding from 130 to 150 pounds) which places the weight directly on the shoulders.

Earthenware-lamps, with an iron hook attached, are used, burning an oil, costing 1½d. a pound, made from a native bean.

Many accidents occur through the presence of gas, and such occurrences lead to the abandonment of the mine, as the superstitious natives believe that these accidents are caused by the wrath of the " Fire Dragon."

Mining by foreign methods has only been carried on to any extent in two areas, namely:—(1) The Kai-ping coal-field, which has been controlled by foreign engineers for about 25 years, and whence coal is supplied to the Imperial Chinese Railway; and (2) the Liao-yang coal-field, worked by the writer since 1899 for the supply of coal to the Chinese Eastern Railway. The latter coal-field is principally described in this paper.

The mines in the Liao-yang coal-field were entirely destroyed by the Chinese in the war of 1900, and it was impossible to resume work until the end of October in that year.

Liao-yang Coal-field.—This coal-field is an oval basin, some 4 miles long and ½ mile to 1½ miles broad. There are fourteen seams, of which five are of sufficient thickness and of good enough quality to be worth working. The seams dip, on the surface, at angles of from 60 to 75 degrees, gradually flattening until they attain a depth of about 400 to 600 feet below the level of the valley.

The overlying and intervening strata consist of white and red sandstones, shales and hard clays; and immediately below the lowest seam (separated only by a few feet of clays and shales) is blue Mountain Limestone, which is used largely over the whole of Southern Manchuria for lime-burning and building houses, forming a very excellent material for the latter purpose.

The strata are not water-bearing, the principal source of trouble arising from leakage of water from the old Chinese workings: this water passes for long distances through the sandstones.

Along the longitudinal axis of the field, there is a small range of hills rising about 300 feet above the valley, and cut, in one or two places, by gulches. In one of these gulches, a shaft was commenced, and two inclines were run under the hill-range.

Inclines.—The inclines, dipping 28 degrees from the horizontal, were commenced so that they would pass through the seams at a depth measured on the incline of from 300 to 500 feet, and enter them again at 800 to 1,200 feet near the bottom of the basin. The larger incline is being driven, 11 feet wide by 6¼ feet high, inside the timbers, a length of 100 feet from the entrance being lined with limestone-arches and sandstone side-walls, and will accommodate two lines of railway of 2½ feet gauge. The haulage-engine is a horizontal direct-acting wind-ing-engine, with a steam-cylinder, 12 inches in diameter, by 2 feet stroke, with a drum, 5 feet in diameter, and is capable of hauling 2 tons of coal at a rate of 10 miles an hour. This incline will be continued to an inclined depth of 1,500 feet.

The smaller incline, used as a pump-way, a return air-way and a travelling-road for the men, was commenced 7 feet high by 6¼ feet wide inside the timbers, but at a depth of 250 feet, so much water had to be pumped from the old workings that more room was necessary, and the drift was continued of the same dimensions as the larger incline. The haulage-engine, a winch, with steam-cylinders, 5 inches in diameter, is capable of hauling 1 ton of coal.

Evans bucket sinking-pumps were used for draining the inclines, the exhaust-steam being passed into the suction-pipes: these pumps were lashed to a flat trolley and followed the incline as it extended downward. At times, three pumps were working in the two inclines.

In the course of driving the inclines, old Chinese workings were continually being tapped, the accompanying outbursts of water being dangerous, and the cause of much delay in driving the inclines. The exploring bore-holes were frequently stopped with mud and shale, preventing the wastes from emptying themselves, and it was eventually found that the best plan was to withdraw the pumps from any danger of drowning, fire a charge of dynamite in the bore-hole, and allow the water to flow to its fullest content: and then the water was gradually drawn off with the available pumping-plant.

The most serious outburst of water occurred after one of the old workings had been passed through, and drained more or less dry: suddenly there was a terrific rush of water, rising in the incline for a length of about 200 feet on the first day, and a further length of 200 feet on the two following days. It is probable that the water had been banked up against some old Chinese coal-pillars, left either as water-barriers or as a support to the workings, and that these pillars had suddenly given way, when the water had been withdrawn from the other side.

One incline has now reached a depth of 1,200 feet, and the other one is stopped at the point where it recrossed a seam at a depth of 800 feet.

Shaft.—The shaft, commenced some time after the inclines, is at present 250 feet deep, measures 12 feet 4 inches by 10 feet 8 inches, inside dimensions, and temporary timber-sets, 8 inches square, are used.   The shaft, from the surface to a depth of 225 feet, during the rainy season, gives off large quantities of water, and this portion will be lined with walling, 14 feet in internal diameter, of limestone and cement, while below this depth, permanent wooden sets of square timber will be used.

The permanent horizontal winding-engine for the shaft has a steam-cylinder 20 inches in diameter and a drum 10 feet in diameter, but sinking is being effected with a steam-winch. Pulsometers and Evans sinking-pumps are used in the sinking, with a permanent water-lodge containing a Worthington pump. The volume of water depends largely upon the rain : about 450 gallons a minute have been pumped.   Trouble has also been experienced from carbonic acid gas flowing in large quantities into the shaft, through the crevices in the sandstone above the water, from the old Chinese workings in the upper seams, and time after time it has driven the pump-men to the surface.

Plant.—There are four locomotive-type boilers at the inclines, and although not very economical in repairs, they are handy to transport by horses across country.   They supply steam at a pressure of 95 pounds per square inch to the winding-engine, the steam-winch, the pumps, a small engine driving the machinery in the shops, and a small direct-acting fan.   The feed-water is slightly warmed, in a tank containing 20,000 gallons, by the exhaust-steam being led through it in pipes.

The shops at the incline are fitted with two lathes, two drilling-machines, and a small planing-machine.   The black-smith-shops contain four forges and a fan-blower, driven from the main shafting of the machine-shop.

Owing to the mine being gassy, all men except those working on the stone-incline use safety-lamps of the bonnetted Marsaut type, and they are cleaned by a rope-driven lamp-cleaning machine.

The boilers at the shaft are of small Cornish type, with Galloway tubes. They were made 20 feet long and 6 feet in diameter, as they had to be transported for a distance of 10 miles by horses. The working-pressure is 100 pounds per square inch.

There is a similar machine-shop at the shaft to that at the inclines, with the exception of a larger planing-machine.

Water is supplied to both of the inclines and to the shaft from a well some 2,000 feet distant, and a tower 40 feet high; and by blowing off the sludge-water twice daily, and not running a boiler longer than three weeks to a month, there has not been much difficulty with scale.

Labour.—The labour employed is entirely Chinese, and they work well under efficient supervision. All work, where possible, is let by contract, the only class of men paid day-wages being machinists and fitters, and a small number of carpenters engaged on tub-repairs. At first, it was difficult to get men who would contract at a reasonable figure, but the Chinese have now attained a certain amount of confidence in the management, and they take the contract-prices that are offered to them.

The Chinese have many superstitious dislikes, and a certain lack of ambition, and when work is required to be done quickly, these qualities are a great hindrance. They dislike to work in water or rain, and in labour-contracts they often prefer to earn just a living-wage. An increase of price simply means less work done, or an increase in the number of day-men. Most labour-contracts are let upon a speed- or quantity-basis, and they invariably gauge their earnings to make a wage that will just keep them. This may be partly due to the insecurity of personal property, arising from the depredations of robbers or the squeezes of native officials, or to the wonderful manner in which the poorer and lazier members of families attach themselves to their richer relations, so that they prefer to make the work last longer on a living wage to saving any of the extra money that they could easily earn.

The better qualities of Chinese workmen are their contentment, power of imitation and their phlegmatic temperament. They make good engine-drivers, machinists and blacksmiths, bad pump-men, and good roadmen and timbermen.

They have an almost frightful indifference to life and danger. In one case, where six men were burnt, through one of them smoking on the rise side of an end-stall, the men working on the dip-side hardly turned round, and had to be forced to leave their work to help in carrying out the burnt men. The man's comrades saw him smoking, and did nothing to prevent it, although they were fully aware of the danger.

The miners are fair pickmen and drill-men, are entirely ignorant of the use of wedges and bars, and have no idea of the advantages resulting from deep holing under the coal. Possibly the latter arises from the fact that there is a demand for slack-coal by the natives. The Chinese are exceedingly careless loaders of the coal, and great care is required in the screening in order that the shale and other impurities may be eliminated. The mine-officials are, however, getting a certain amount of improvement in holing and in timber-setting.

The confidence of the Chinese can only be attained if they are treated with perfect justice, and without mercy. If a man be punished (a fine is the best form of punishment) unjustly, or by mistake, he will leave immediately. If a penalty be not imposed upon a man for an offence, even from a merciful point of view, it is put down to want of sense, and advantage will be taken of it. A limited knowledge of colloquial Chinese is also necessary, so as to avoid the services of that most detested man —the interpreter, who is invariably at the root of all troubles. If any man, however insignificant his position, is given an opportunity of approaching and communicating directly with you, if even only partly by signs, it is greatly appreciated by him, and when you have received their confidence and acquired a limited colloquial knowledge of the language, there is no easier man to manage than the Chinese labourer.

Wages.—The following daily wages are paid by the writer:— Surface-labourers, 5d.; miners, 6d. to 9d.; machinists, 8d. to 1s. 8d.; blacksmiths, 7d. to 1s. 6d.; stokers, 6d. to 9d.; and carpenters, 6d. to 8d. The labour for rock-work, in an incline, 11 feet wide and 6¼ feet high, costs 6s. to 10s. per linear foot. The labour for sinking a shaft, 12 feet 4 inches by 10 feet 8 inches inside the timbers, costs 30s. to 40s. a foot, including all surface-labour employed in tipping the rock. Coal-getters are paid 1s. 2d. to 1s. 4d. per ton, including loading and tramming to a distance varying from 50 to 200 feet. Gate-roads, ripping and packing walls, cost 1s 3d. per linear foot of road.

Materials.—Mine-timber, 6 to 12 inches in diameter, costs 1s. to 2d. per linear foot delivered; and mine-sets for roadways, 11 feet wide by 6¼ feet high, cut and delivered, cost 2s. 6d. each —if 7 feet wide by 6¼ feet high, they cost 2s. each.

Markets.—The coals of the Liao-yang coal-field have different qualities, ranging from semi-anthracite to bituminous caking-coal, and contain from 8 to 12 per cent. of ash. The coals are friable, a large percentage of small coal being produced in work-ing, and the writer proposes to use this in the manufacture of briquettes. The coal is practically smokeless, and a very economical fuel, as compared with most of the other coals of Northern China. An average percentage analysis is as follows:

Fixed Carbon	62
Volatile Hydrocarbons	26
Sulphur	...traces.
Water	2
Ash	10

There is a good local sale for small coal, lump coal is sent away to different stations along the railway, and when the new screening and briquetting-plants are erected, the output will be increased as quickly as possible, until 1,200 tons per day is reached.

———

The PRESIDENT (Mr. G. Elmsley Coke) proposed a vote of thanks to Mr. Moller for his paper.

Mr. J. A. LONGDEN, in seconding the vote of thanks, said that there were only four Europeans at the pits, and the work was carried on entirely by natives, with very little skilled supervision.

The resolution was adopted, and the further discussion of Mr. Moller's paper was adjourned.

———

DISCUSSION OF MR. JAMES KEEN'S "DESCRIPTION OF THE SINKING OF TWO SHAFTS THROUGH HEAVILY-WATERED STRATA AT MAYPOLE COLLIERY, ABRAM, NEAR WIGAN."*

The PRESIDENT (Mr. G. Elmsley Coke) said that the paper described in a straightforward and simple way the actual operations at Maypole colliery; and he always thought that the manager who described his difficulties and how he had overcome them was conferring a great favour upon the members. Mr. Keen had told him that the shafts had not been tubbed, and that he had hopes—in consequence of the number of faults which practically enclosed the colliery—of removing the whole of the water within that area.

Mr. S. A. EVERETT (Gedling) said that Mr. Keen appeared to have used all classes of appliances to pump the water—pulsometers, direct-acting, slung and fixed pumps, and ordinary lifting and forcing sets—and to have obtained good results with each of them. In his opinion it was a mistake to use so many kinds of appliances. In fact, the secret of sinking through heavily-watered strata quickly was to decide upon what appeared to be the most suitable class and size of pump to meet the conditions, and then to adhere to it throughout. This course had two advantages :—(1) Only one set of spare parts and duplicates for renewals, and repairs when breakages occured, was needed; and they would fit any pump: and (2) the, by no means unimportant, advantage was secured of accustoming the men to one kind of pump, and getting them to understand it and its working so thoroughly that the most efficient result was obtained from it.

Mr. W. P. ABELL (Duffield) said he was afraid that sinking was often started with expectations of few and small difficulties. Perhaps the first pump which was tried gave trouble, and then it was thought that they ought to try another, with the result, in some cases, that worse difficulties still were encountered. The best way was to find out the pump best suited to the conditions

* *Trans. Inst. M.E.*, 1900, vol. xix., page 462 ; and 1901, vol. xxi., page 258.

before starting, and to recognize from the beginning that difficulties would have to be overcome, and then there would be less disappointment.

Mr. G. A. Lewis (Derby) said that it was interesting to note that Mr. James Keen stated that " if he had to commence another sinking of this character . . . he would use pulsometer-pumps alone for the pumping during the sinking."* He (Mr. Lewis) had had no experience in the use of pulsometer-pumps, and as there were such conflicting accounts with regard to the use of pulsometer-pumps, with regard to the large amount of steam that they required, etc., perhaps some member would give his experience in their use.

Mr. Walker (Sherwood Colliery) said that, at a colliery with which he was acquainted, there was a bucket-lift, 18 inches in diameter, in one shaft, and a pulsometer-pump in the other shaft. They sank 210 feet with the bucket-lift, in the first shaft, while they were sinking about 30 feet with pulsometer-pump in the other shaft.

Mr. Markham, M.P., said that a centrifugal pump was at present working at Oxcroft colliery, lifting from a depth of 300 feet. It was working satisfactorily, and it was the first centrifugal pump which had been made for a high lift. The cost was about one-third of that of a three-throw pump, and it was intended that these pumps should be tried in a sinking in South Wales.

Mr. Walker (Sherwood Colliery) enquired whether the centrifugal pump would be damaged by the firing of shots in the pit-bottom.

Mr. Markham replied that he did not think that the pump would be strong enough to be shot at in the sinking-pit, and it would have to be lifted out of the way before shots were fired.

The discussion was then closed.

————

* *Trans. Inst. M.E.*, 1900, vol. xix., page 473.

SINKING OF THE RHEIN-PREUSSEN COLLIERY, NOS. IV. AND V. SHAFTS.

By W. H. HEPPLEWHITE, H.M. INSPECTOR OF MINES.

The Rhein-Preussen Colliery Company possess mining rights over concessions which have a mineral area exceeding 23,000 acres, and three coal-drawing shafts, about 4,000 workmen being employed. The territory lies chiefly in the flat valley of the Rhine, and the pits have to be sunk through a considerable thickness of Tertiary and Quaternary strata. The Tertiary strata had been found to be about 420 feet thick in the south, and 1,140 feet in the north.

The surface-measures consist of beds of alluvium possessing very little coherency, very open and full of water. These deposits are so open that, when sinking was in operation, the water always rose to the level of the Rhine. It will be readily recognized that, under the most favourable conditions, shafts could only be sunk through such strata by means of boring-machinery at enormous cost, and under extraordinary difficulties.

The No. I. shaft occupied 20 years in reaching the Coal-measures from the commencement of the sinking. The original diameter was 24 feet 9 inches, and this was reduced to 8 feet 10 inches when coal was reached; the No. II. shaft, after 9 years' sinking, with an original diameter of 32 feet 5 inches, and with linings of masonry and cast-iron, was reduced to a diameter of 14 feet; and the No. III. shaft was sunk to the Coal-measures in 3 years, with a finished diameter of 14 feet 9 inches.

The sinking of the Nos. IV. and V. shafts was commenced with the Pattberg percussive drill or chisel, A. The shaft was commenced with a diameter of 30 feet. The débris was scooped out by means of an elevator to a depth of 55 feet; and then the bottom for a depth of 10 feet was filled with concrete, which was allowed to stand for a few months so as to set hard. When sinking was resumed, the percussive drill was used to penetrate through the concrete, the rate of progress being about 4 feet per day (Fig. 1, Plate VIII.).

The lining of the shaft consisted of an outer ring of cast-iron tubbing, and an inner wall of masonry, so as to increase the resisting power of the shaft and to prevent collapse. The cast-iron tubbing is about 2 inches thick, and is strengthened at intervals by box-shaped rings, 26 inches in width, supporting the inner lining, which is two bricks thick. The joints of the tubbing are packed with asbestos, in lieu of ordinary wooden sheeting. This compound cylinder of brick and cast-iron tubbing, B, did not realize anticipations, as it could not be forced downward to a greater depth than 200 feet.

A new start was made with a cylinder of a diameter of 19½ feet, with 3 inches of cast-iron tubbing, C, exclusively, in order to prevent any undue reduction of the diameter of the shaft. It is probable, however, before sufficiently suitable ground is found whereon to build a curb, and to commence the ordinary method of sinking, that the diameter of the shaft will be reduced to 18½ feet.

The first drill that was used had a cutting-face, 21 feet in diameter, a vertical height of 27 feet, and weighed 9 tons. The next was 19 feet in diameter, with a weight of 7½ tons (Fig. 2, Plate VIII.). The wrought-iron block, r, which carries the teeth or cutters, is V shaped, and pierced with holes, a, a, on both sides; and a number of small tubes, b, b, are connected to channels in the steel-drills or cutters, z, z. The tubes, a, are connected to the hollow boring rod, G, and deliver the flushing water upon the bottom of the shaft. The vertical guides, u, the horizontal guide, v, and the various struts and stays are made of wood.

It was feared that the wooden struts and horizontal guides of the first drill would not stand the strain imposed on them; and a square tool of the same dimensions, made entirely of iron, weighing 12 tons, as being less likely to get out of order, was in use at the time of the writer's visit (Figs. 3 and 4, Plate VIII.). Furthermore, a continuous cutting-edge was substituted for separate drill-teeth and a central drill. The drill is hollow, V shaped, and, at intervals along the cutting-edge, small holes, b, reach into the hollow part, a, to allow of the passage of the water through them (Fig. 4, Plate VIII.).

On each side of the hollow rods, G, 6 inches in diameter, that carry the drill is fixed a mammoth or air-lift pump, R, R, each consisting of a pipe, 5½ inches in diameter, surrounding a pipe, 4 inches in diameter.* The annular spaces between the pipes, R, is placed in connexion with the compressed-air pipes, and the hollow rods, G, with the hydraulic pump stationed on the surface. The air-compressor works at a pressure of 750 atmospheres or 1,100 pounds per square inch, and the hydraulic pump at 1,000 pounds per square inch.

As the drill chops the ground into small pieces, the water sent down the pipe issues at enormous pressure from the drill-point, and assists in churning the débris into mud and minute particles. The compressed air issues into the inner pipe of the air-lift pump, R, near the shaft-bottom, and gathers the whole of the débris from the deepest part of the shaft-bottom and forces it up the inner tube mingled with the water; and it passes from the shaft-mouth along spouting to a settling-reservoir, the water being again used when it is sufficiently cleared.

The progressive lowering of the rods is accomplished by means of the small engine and drum. The drum is mounted loosely on the shaft, and the outer part, which is separate from the drum proper, is keyed on the shaft. Above the boss of each, for a short distance between the arms, it is solid, with a number of perforated holes 1 inch in diameter. Bolts are fitted through these holes, and serve to keep the drum rigid on the shaft; and when any alteration is to be made by letting out, or taking up, rope, the bolts are removed, and the drum moved as required. The holes are so arranged that at any position two-thirds of the bolts will fit. For sinking operations, a quadrant-rack is fitted on the outer part of the drum, and gears into cog-wheels; and then by means of belting and worm-gearing, the rope which supports the drilling-tool is run out.

On the landing-stage at the shaft-mouth, there is mounted on the rods a large round tree, cut in two and clamped. The tree fits tightly into guides, so as to keep the rods perpendicular. A few poles, D, are attached to the tree, and the workmen on the landing-stage slowly turn the drill round by means of the poles. A turn of the drill is made in about 15 minutes.

* *Trans. Inst. M.E.*, 1899, vol. xvii , page 584.

To illustrate Mr. W. H. Hepplewhite's Paper on the "Sinking of the
Rhein-Preussen Colliery" etc.

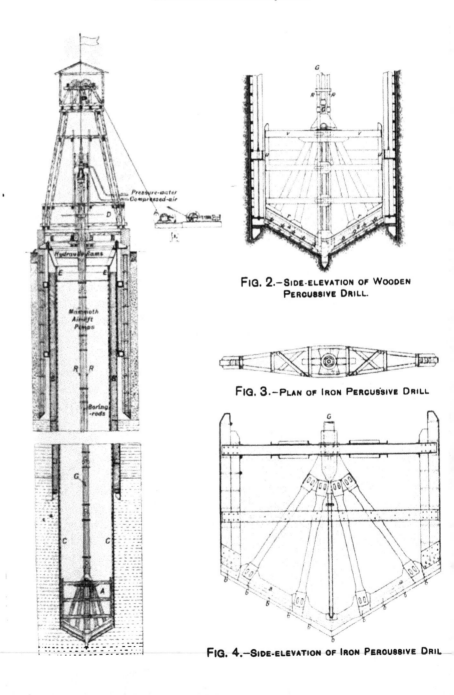

FIG. 2.—SIDE-ELEVATION OF WOODEN
PERCUSSIVE DRILL.

FIG. 3.—PLAN OF IRON PERCUSSIVE DRILL

FIG. 4.—SIDE-ELEVATION OF IRON PERCUSSIVE DRILL

The drill makes from 60 to 70 strokes per minute, and the length of stroke varies from 8 to 12 inches. The rate of progress made in sinking was at the time of the writer's visit about 4 feet per day.

The rings of tubbing at the shaft-bottom are fitted with a sharp-pointed ring, intended to force itself into position after the completion of the sinking. The downward forcing is accomplished by means of 14 hydraulic-rams, E, one placed on each segment, at the top of the tubbing, and the same pressure is exerted on each as that sent down to the drill-points. When the depth of a segment has been forced down, the hydraulic rams are removed, another ring of tubbing is added, and the process is repeated.

The drill is changed about once a fortnight, and the process can be accomplished in 12 hours. It must be understood that two shafts are being sunk, and that all tools are interchangeable.

The advantages to be derived from this method of sinking seem to prove it infinitely superior to that of dealing with a large body of water by means of pumps, and it might profitably be employed in sinking through the Bunter and Permian strata in the Midland districts.

———

DISCUSSION OF MR. W. E. GARFORTH'S PAPER ON "THE APPLICATION OF COAL-CUTTING MACHINES TO DEEP MINING."[*]

Mr. J. A. LONGDEN (Stanton-by-Dale) observed that, with the Electrical Committee sitting, it was extremely difficult to know what was to be the future of electricity as a motive power in mines. If the result be that restrictions are placed on the use of electricity on account of sparking, it would limit the use of coal-cutters. So far as the two kinds of current (altternating and direct) were concerned, engineers who had adopted the alternating-current system approved of it, and those who had a direct-current system said that it was the best. Although the alternating current was supposed to be more safe, it seemed to him that in the matter of cables, switch-boards and resistance there was little to choose between them. Professor Crapper

[*] *Trans. Inst. M.E.*, 1902, vol. xxiii., pages 312 and 346 ; and 1902, vol. xxiv., pages 201 and 260.

recently distinctly stated that the alternating current was not suitable for driving coal-cutting machines; and if that were the case, they were practically confined to the use of the direct current. There was no doubt that when coal-cutting machines were worked by electricity it was much easier to detect any weight on the roof, than when they were being driven by compressed air. The latter made such a loud noise that it was impossible to gauge the safety of a working-place by sound; and they all knew that colliers were led more by sound as to safety than by anything else. It had been suggested to the Electrical Committee that the coal-cutting machine should be stopped every 5 minutes, so that the men might listen for any sound in the roof; but if that were enacted as a bye-law, they had better be without the machines. He had worked a Garforth machine, and it ran satisfactorily, but their roof, unfortunately, was not suitable for coal-getting by machinery. A face, 240 feet long, had been holed in one night, and left at 4 a.m. timbered with two rows of props, and at 6 a.m., the same morning, the entire length was " in." They were compelled to stop the machine, which, however, had since been working satisfactorily elsewhere. Coal-cutters had been introduced at Creswell colliery, at a depth of 1,500 feet, and taken out; and he had a similar experience at Pleasley colliery, with a " nesh " roof on a seam 5 feet thick, at a depth of 1,500 feet. He was introducing a machine in a seam 3 feet thick with a good roof, and he hoped that it would work satisfactorily.

He (Mr. Longden) was told that, at two collieries where machines were largely used in this district, the saving was 3d. per ton, and at another colliery, the saving was slightly more than that amount. This saving was due to labour, that is, in holing by machines instead of by hand; but with machines they should get about 10 per cent. more round coal than previously. Probably that would represent a saving of another 3d. per ton, so that a coal-cutting machine, if working satisfactorily and producing a proper proportion of round coal, should save about 6d. per ton. For working in thin seams, on which we may have to rely in future, coal-cutting machines will prove invaluable.

Mr. M. H. MILLS said there was no doubt that there were collieries where coal-cutters became an absolute necessity, while there were collieries which were not fit for the use of coal-cutting machines of any description. In deep mines, he was under the impression that the best way would be to carry the power a considerable distance in-bye by electric cable, and there install motors for generating compressed air, and use the compressed air for working the coal-cutters. He did not, however, think that it was impossible to make a non-sparking electric coal-cutter, because he believed that that difficulty would be easily overcome. There was, however, the danger of sparking along the cables. but manufacturers of electric cables were now working to overcome this difficulty. His experience with coal-cutters had not, so far, been altogether a happy one, but sometimes it was equally advantageous to learn what to avoid as to learn what was successful.

Mr. J. R. JAMESON (Blackwell) remarked, with regard to the suggestion of using electricity to compress air, that by so doing they lost one of the great advantages of electric motors. The waste of mechanical work in compressed-air machines was excessive, and certainly much greater than with electric motors. There was not only more friction in the engine, but the necessary use of connecting-rods and cranks resulted in considerable loss of power. It was essential in coal-cutting machines to have plenty of power. In many cases, coal-cutters were designed with a cylinder of a certain size to do a certain amount of work, but they failed to do it satisfactorily owing to the loss of pressure between the plant and the face. With a pressure of air of 40 pounds per square inch at bank, they could only anticipate obtaining 20 pounds when it reached the coal-face, and the machine was running. He was inclined to think that the loss was often due to the length and small diameter of the pipes in the gate-roads, and to the inferior quality of the hose-pipe. The latter was roughly handled, doubled and cracked, and as it was usually of great length the leakage was considerable. An important point, too, was that they should be able to prove to the collier that he was going to make more money by the adoption of machines, a point usually capable of proof; as no class of men cared to make changes resulting in diminished earnings, and colliers, perhaps, least of all.

DISCUSSION OF MR. DANIEL BURNS' PAPER ON THE "WEIGHT OF WINDING-DRUMS FOR DEEP SHAFTS."[*]

Mr. J. W. FRYAR, in answer to a question, said that one of the drums at Sherwood colliery would be a partly-conical drum. There would be five revolutions on a diameter of 12 feet, then 2½ from 12 to 16 feet, and then on the parallel 16 feet in diameter. The great advantage of that arrangement was the help given to the winding-engine at the lift, by the leverage due to the different diameters; and he believed that it would probably help their winding-engine to the extent of over 3 tons at the beginning of the wind.

Mr. J. R. JAMESON asked whether Mr. Fryar anticipated that there would be any extra wear of the rope, owing to the wide movement on the drum, or additional side-wear of the headgear-pulleys.

Mr. FRYAR replied that he did not anticipate any serious amount of wear, as he already knew collieries at which there was a similar amount of rope-travel on the drum. There need be no side-wear of the pulley-wheels, if they were properly made for the wide amount of travel on the drum.

The discussion was then closed.

––––

NORTH STAFFORDSHIRE INSTITUTE OF MINING AND MECHANICAL ENGINEERS.

___

GENERAL MEETING,
HELD AT HANLEY, MARCH 9TH, 1903.

___

MR. A. M. HENSHAW, PRESIDENT, IN THE CHAIR.

___

THE LATE COL. STRICK, C.B.

The PRESIDENT (Mr. A. M. Henshaw) said that the death of Col. John Strick would be a great loss to the Institute and to mining in North Staffordshire generally. He was one of the founders of the Institute, and during the 30 years of its existence he had always taken a deep interest in building up the high position which it held at the present time. He proposed that a vote of condolence be sent to Mrs. Strick expressing their deep sense of the loss sustained, and of sympathy with her and her family in their sad bereavement.

The resolution was agreed to.

___

The following gentleman, having been previously nominated, was elected :—

ASSOCIATE MEMBER—
MR. DONALD CRAIG, Kilhendre, Gresford.

___

Mr. B. WOODWORTH's paper on " Condensing-plant for Winding-engines " was read as follows :—

CONDENSING-PLANT FOR WINDING-ENGINES.

By BENJAMIN WOODWORTH.

Winding-engines being regularly subject to intermittent working, often under extreme variations of power, the problem of arranging the condensing-plant in a form to suit the work becomes a complicated matter. It is generally overcome by providing the maximum effort necessary, as a continuous supply, so that a considerable waste of power takes place when little or no work is being done by the winding-engine; and the following arrangements are suggested, with the view of meeting the variable requirements of such work in a practical and economical manner.

The system proposed is a combination of an independent central-condensation plant with a direct plant driven from the winding-engine, and coupled to what may be termed an intermittent accumulative injection-arrangement, all so arranged as to supply the variable requirements of the engine in the way of injection-water supply and air-pump power without the need of special attendance, and to suit the variation of the work or undue waste of power during the time when the winding-engine is not running.

For good working (either compound or non-compound), it will be assumed that the engine runs 30 revolutions per wind, with equal unbalanced load throughout: running 25 revolutions under steam in 30 seconds, then finishing the wind without steam, and simultaneously changing the loads in 20 seconds more or a total of 50 seconds per wind, and working under steam for 60 per cent. of the actual time occupied in the whole wind: this being a maximum obtainable in such work. The writer would advise full-gear steam-admission during the first and second revolutions only, together with the most suitable arrangement for expansion-working during the remainder of the wind that will ensure economy in working.

In the general arrangement proposed, A is the central condensing vessel, B and B_1 are the air-pumps worked from the winding-engine, C is the cylinder of the continuous air-pump used for maintaining the vacuum, while the winding-engine is at rest, and it has coupled to it in addition to the ordinary air-pump, H, a cold-water pump, D, for supplying water to the intermittent accumulative injection-arrangement contained in the vessel, E (Figs. 1 and 2, Plate IX.).

A continuous injection and steam-supply is maintained for the continuous air-pump engine, C, but the main injection water-supply is controlled by a valve, G, which is operated simultaneously with the steam stop-valve, so that the water is shut off and turned on to the winding-engine simultaneously with the steam. The cold-water pump, D, being in constant work, gives a regular supply to the accumulative injection-vessel, E, and while the winding-engine is under steam, this water-supply passes, practically, direct to the condenser, through the auxiliary injection-pipe and valve, F, but as the latter is also controlled by lever-gear to the main steam-valve, this water-supply to the condenser is closed at the same time as the main steam stop-valve is closed ; and, during the interval, when the winding-engine is not under steam, the water-supply continues to be forced into the vessel, E, and the principle of the accumulative injection comes into operation in the following manner : —

In the lower part of the vessel, E (Fig. 2, Plate IX.), there is a self-acting valve (operated by a floating ball) which is kept open so long as the supply of water is above the level at which the valve is designed to act.   Immediately the main steam-valve and the accumulative injection-valve, F, are closed, the water-supply commences to accumulate in the vessel, E, and the air, in the upper part of the chamber, becomes gradually compressed to a smaller volume, and should the interval of rest of the winding-engine be sufficiently long to permit of the maximum pressure (desired) in the vessel, E, being accumulated, then a suitable escape-valve allows the surplus water to escape either direct to the boiler-feed main or to any other supply-purpose that suits the maximum pressure desired in the accumulator-vessel. At the end of the usual rest-interval, it may be assumed that the pressure in the vessel, E, has reached 60 pounds per square inch above that of the atmosphere, and this pressure would represent practically 75 per cent. in volume of water to 25 per cent. in volume of air over the working range of the accumulator, E.

Consequently, the moment that steam is turned on to the winding-engine, the injection-valve, F, is also opened and allows a strong forced current of injection-water to act immediately on the condenser, and prevents any risk of spoiling the vacuum before the main injection-supply gets fully into motion (delay might occur through the intermittent action of the main water-supply), and this forced current gradually reduces its velocity as the air expands in the vessel, E, until, at the working line of level, the self-acting internal valve is closed by the sinking of the floating ball and connections, thus preventing the supply of atmospheric air enclosed in the vessel, E, from escaping to the condenser, and vitiating the action of the accumulative property that would occur, if a vacuum was created in the vessel, E.

Figs. 1 and 2 (Plate IX.) show the principle of action of this proposed arrangement, and the details can be varied to suit the general constructional arrangements, so long as the working proportions are maintained.

_____

The PRESIDENT said that the winding-engine is, as a rule, the principal engine at a colliery, and Mr. W. N. Atkinson stated that, in the North Staffordshire mines-inspection district, out of 70,000 horsepower employed at the collieries, 40,000 horse-power was applied to winding coal, so that it might be realized how important a matter was being dealt with in the paper before the members. Generally speaking, a winding-engine did not attain so high an efficiency as was obtained in engines running continuously at regular loads, but there was every reason to believe that a larger measure of economy might be achieved by following the suggestions given in Mr. Woodworth's paper. A winding-engine is now a simple engine taking steam nearly the whole length of the stroke, to enable a start to be made at full load from any point. The first few seconds called for the utmost power of the engine to start the load, and to overcome the inertia of the heavy drum, while a few seconds later reverse strains were required to check the momentum and to bring the ponderous machine to rest, before reversing for the next wind. This inter-mittent working required that all the mechanism should be as free from complications as possible. Great responsibilities rested upon the attendant, and the mechanical operations concentrated in the levers in his hands must be perfect, and as little liable to

Institution of Mining Engineers.
Transactions 1902-1903.

VOL. XXV. PLATE

To illustrate Mr. B. Woodworth's Paper on "Condensing-plant for Winding-engines."

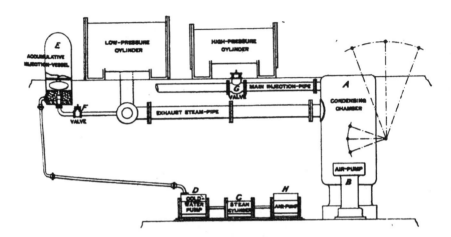

FIG. 2.—SIDE-ELEVATION.

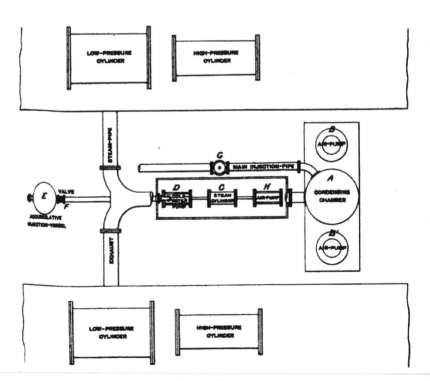

FIG. 1.—PLAN.

derangement as human ingenuity could devise.   One of the difficulties to be overcome in a winding-engine, especially for a deep pit, was the equalizing of the load and the reduction of the weight of the drum, and he would recommend the use of a balance-rope and small drum rather than a cumbrous spiral drum of 100 tons weight.   He would be inclined to adopt the Whiting modification of the Koepe system, and to use two small pulleys in place of the drum; and with a perfect balance of the load, and steam at a pressure of 150 pounds per square inch or more, a small high-speed compound, or even triple or quadruple condensing engine, would form as nearly as possible a perfect winding-engine, securing results as economical as any engine running intermittently could attain.   By adopting some of the excellent modern devices for automatically shutting off steam and applying the brake when speed was exceeded, or the end of the run was approached, safety detaching-hooks, and possibly one of the Continental safety-arrangements for suspending the cages in case of a broken rope, one would have a safe and an efficient winding-plant.   The application of condensation to a winding-engine offered the following advantages:—(1) Considerable economy in fuel-cost; (2) placing in reserve one or more of the present boilers; (3) increasing the power of the engine without forcing the boilers; and (4) providing a reserve of power for winding in boilers and engines in times of pressure.

Mr. JOHN HEATH said that condensing-plant had been in use at the Sneyd collieries for three or four years, and it had worked economically.   There had been occasional trouble with the condensing-engines, but it had arisen through the negligence of the attendant, and not through any fault of the machine.

Mr. E. B. WAIN said that Mr. Woodworth was one of the pioneers of steam-saving at North Staffordshire collieries, and his papers were always read with special interest.   He would be pleased to hear what particular advantage this form of condenser would have over an accumulator surface-condenser, such as had been recently described in Mr. Crighton's paper.*   The extreme variation in duty of a colliery winding-engine made it difficult to apply condensation in such a manner as to obtain the highest efficiency or economy.   He had, in his mind, a case where, in

* *Trans. Inst. M.E.*, 1903, vol. xxv., page 79.

the first five strokes, 1,000 horsepower was developed, the average horsepower whilst the engine was in steam being 680, and, including the five strokes run after steam was cut off, the mean throughout the run of 40 seconds was 517 horsepower; and it was quite evident that the apparatus suitable for a continuous-running engine would be useless in such a case; hence, the necessity for some such special form of condenser as that proposed by Mr. Woodworth in his paper. To provide suitable condensing appliances for winding-engines, heavy capital outlay would be required, and care must be taken that the charges for interest on the expenditure did not outbalance the economy due to fuel-saving. He moved a vote of thanks to Mr. Woodworth for his paper.

Mr. MILWARD said that there could be no question as to the economy that might be obtained by a scheme of condensation. More than 100 years ago, there were condensing-engines of 80 to 100 horsepower, and he had seen one working with steam at a pressure of 5 pounds per square inch. It had been a problem for many years in connection with winding-engines, and the question was surrounded with difficulties. But all these could be overcome, and he believed that no one would in the near future erect a winding-engine, unless it were compounded and condensing. He seconded the vote of thanks to Mr. Woodworth for his paper.

The resolution was approved.

———

DISCUSSION OF MR. R. W. CLARKE'S PAPER ON COAL-CUTTING BY MACHINERY."*

The PRESIDENT (Mr. A. M. Henshaw) observed that there was difficulty in finding a machine to overcome some of the obstacles which were perhaps met with in North Staffordshire to a greater extent than in any other district in the world. A machine might work well under some conditions, but it would find a premature grave when it came to their faults, soft coal and bad roofs. Next, they got a 10 feet seam, worked by the pillar-and-stall system, with a roof that would not allow of stalls being more than a few

* *Trans. Inst. M.E.*, 1902, vol. xxiii., page 96.

feet wide—say from 6 to 9 feet—and he did not know of any machine that was capable of going into such a seam. He then referred to the fact that, in some instances, they had seams dipping up to 45 degrees, and even steeper than that. Some time ago, he was trying to work longwall in a seam 3 feet thick : the seam was lying at an angle of 45 degrees, with a band of inferior coal near the top, and it was found that for the machine to work properly the seam should be horizontal. It was almost as if the machine were trying to run along the roof of a house. At the same time, there were ideal seams for machines, and in these times, when wages were high and labour was not always to be obtained, it behoved mining-engineers to turn their attention to coal-cutting machinery, which undoubtedly had a great future.

Mr. E. B. WAIN said that any maker of a coal-cutting machine would tell them that in a pit arranged as he wanted it, his machine would cut the coal; but they would have to arrange the work to suit the particular machine that they wanted to use to get the best results.

Mr. JAMES LINDAY remarked that the conditions of the coal-mines in North Staffordshire rendered it very difficult to determine what kind of machine was most suitable for use in this district. He did not experience difficulty from the thickness of the seams—but, when the inclination varied from 45 to 90 degrees, it was no easy matter to find a machine suitable for working in such a seam. When they obtained a machine, which they thought would be suitable for cutting in a seam at a given inclination, it would be tried, and they would probably find that the pit was not laid out to correspond with the way in which the machine would cut most efficiently, and this could only be attained by mining-engineers and colliery-managers giving their whole personal attention whilst the coal-cutter was in action, and noting all particulars in regard to the working, causes of delay and timbering. The subject was occupying universal attention, and ultimately more coal would be cut by machinery than at the present time. He was convinced that, at Great Fenton collieries, the whole of the coal would be cut by machinery, but sufficient money would be required to carry it out. And, further, more coal would be wrought at less expense and of better quality. However, there were many difficulties, and whoever started a coal-cutting machine must not think that he was going to buy his experience without cost and much trouble; and, even under the best conditions, great difficulties would be encountered.

Mr. W. G. COWLISHAW thought that some of the chief difficulties that had occurred in the introduction of coal-cutting machines, until recently, had arisen from inferior workmanship and the use of inferior materials. A coal-cutting machine, to work efficiently, should be made of as good materials and as good workmanship as a first-class locomotive.

The PRESIDENT (Mr. A. M. Henshaw) remarked that the Clarke machine was constructed for use in a colliery in South Yorkshire. The thin seams of coal could not be profitably worked with hand-labour; but since the introduction of coal-cutting machines, the colliery had been worked successfully and profitably.

The discussion was closed.

———

DISCUSSION OF MR. J. ASHWORTH'S PAPER ON "THE GRAY TYPE OF SAFETY-LAMP."[*]

Mr. JAMES ASHWORTH, replying to the discussion, wrote that the failure of a safety-lamp at Whitfield colliery in 1889, to which he had referred, occurred with a Mueseler, and with regard to the tests made on this lamp by Messrs. J. R. Haines and A. R. Sawyer, he might observe that the most dangerous condition to which any safety-lamp might be exposed was not necessarily that of a high velocity.[†] So unsafe had this type of lamp (Mueseler) proved at various times that it had been declared to be no safer than a naked Davy lamp. Many of the conditions, which have been present, and which have brought about the failure of a safety-lamp in a pit, are doubtless difficult to reproduce, but the Whitfield, Allerton and Shakerley examples all occurred in a low velocity of current: and, probably in the Shakerley case, the only influence that current-velocity had on the lamp was caused by a man moving the lamp. He (Mr. Ashworth) was inclined to conclude that the conditions most dangerous to a safety-lamp

———

[*] *Trans. Inst. M.E.*, 1903, vol. xxv., page 62.

[†] *Ibid.*, vol. xxv., page 73.

occur when the air surrounding the lamp contains a low percentage of gas, and when the lamp, having become highly heated, is suddenly moved by the man using it, into, possibly, almost pure gas and fine dust.   Under such conditions, a reverse-current may be created within many safety-lamps, and if such a reverse-current be created, the failure of the lamp is practically certain. The strong point of safety in the Gray type of lamp is that it is not possible to obtain a reverse-current in the lamp.   No one has yet succeeded in forming a reverse-current in the lamp, but if anyone could accomplish this feat, the Gray lamp would pass the flame through the protective coverings without hindrance.   It would therefore be seen that it is not necessary for the gauze of a safety-lamp to become heated to any degree of redness to allow the flame to pass, as the cannon-like force exerted by the explosion of a lamp full of gas is quite sufficient to force the flame through instantaneously and without any preliminary heating.   Thus it occurs that a lamp, which has failed to resist an explosion within it, is found, as at Whitfield colliery, with the gauze quite bright, and this is practically the only indication of its failure.

He would direct Mr. Stobbs' attention to the fact that the gauze above the glass part is the one referred to in his (Mr. Ashworth's) paper which might be omitted, but the one below the flame is as necessary as in any other safety-lamp.   The great point of safety in the Gray type of lamp lies in the chimney, and not in the gauze which is placed either within or outside of it. He might add, however, that he did not advocate the abandonment of the gauze, for the simple reason that too great care can not be taken in dealing with fire-damp, and an extra factor of safety is desirable rather than one less.

The long gas-testing tube of the No. 2 Gray safety-lamp cannot be heated sufficiently to become an upcast outlet under any known conditions, as the whole volume of air, passing down the tubes, is required to supply the wick-flame, and as the interior of the chimney is always hotter than the inlet-tubes, a reversal of the current, as suggested, is impossible.   The gas-tube of the No. 2 Gray lamp brings the air to be tested direct to the flame, whereas the trumpet, referred to by Mr. Brennan, did not do so, and it was consequently not a success and is now out of use.   He might add that the best lamp of the Gray type for a miner was one with a single tube and a cylindrical glass, 3 inches high: this gave a magnificent light from a wick ⅝ inch broad; and although the tube barred out some light, it also formed a reflector and a most useful protection to the miners' eyes, and was a real comfort where men were following one another in single file. It would be easily understood that this lamp lent itself naturally to the use of pure petroleum, paraffin, or mineral colza as the illuminant, because the most perfect combustion could always be obtained where the air-feed was from below the flame. The lamp was more costly than the ordinary types in use, but this was compensated for by the use of one of the oils named, and users would find, on experiment, that 50 per cent. additional light might be obtained without any increased cost for oil.

He (Mr. Ashworth) had pleasure in thanking the members for their remarks on the unrivalled power of the Gray type of lamp to discover gas close up to the roof or in cavities, and also for their kind appreciation of his efforts to improve the lighting and safety of mines.

THE MIDLAND COUNTIES INSTITUTION OF ENGINEERS AND THE MIDLAND INSTITUTE OF MINING, CIVIL AND MECHANICAL ENGINEERS.

JOINT GENERAL MEETING,
HELD AT THE ROYAL VICTORIA STATION HOTEL, SHEFFIELD, APRIL 25TH, 1903.

MR. H. B. NASH, IN THE CHAIR.

The following gentlemen were elected to the Midland Institute of Mining, Civil and Mechanical Engineers, having been previously nominated:—

MEMBERS—

Mr. JOHN SHAW BARNES, Engineer, 427, Warrington Road, Abram, Wigan.

Mr. HARRY W. MACRONE, Mining Engineer, 16, Bank Street, Sheffield.

Mr. ALFRED JAMES ROUTLEDGE, Mechanical Engineer, Regnald Cottage, Denby Station, near Derby.

Mr. JOHN WHITTON, Mechanical Engineer, Linden Villa, West field, Wakefield.

STUDENTS—

Mr. GERALD BAGLEY, Mining Student, 8, Alexandra Crescent, Ilkley.

Mr. RONALD LEWIS, Mining Apprentice, Glass Houghton Collieries, Castleford.

Mr. WILLIAM AINSWORTH WOOD, Mining Apprentice, Purston Hall, near Pontefract.

DISCUSSION OF MR. W. E. GARFORTH'S PAPER ON "THE APPLICATION OF COAL-CUTTING MACHINES TO DEEP MINING."[*]

Mr. W. E. GARFORTH said that Mr. J. T. Todd had drawn attention to several points.[†] The first was about the thickness of the holing, a difficulty being found in breaking up the coal where the excavation made by the machine was only 5 inches high. He wished to explain that in one seam, where they had a difficulty

[*] *Trans. Inst. M.E.*, 1902, vol. xxiii., pages 312 and 346; and 1902, vol. xxiv., pages 201 and 260.

[†] *Ibid.*, 1902, vol. xxiv., page 260.

with the thickness of the cut, they had increased it to 8 or 9 inches, by cutting under the dirt and leaving a kind of pancake. There was a lower seam of coal which formed a kind of bench, so that, although there was an extra quantity of dirt for the machine to cut, or grind up, this fell away upon the bench and was not again carried in by the wheel. They had had some difficulty, during the last few months, in working the Silkstone seam by advancing longwall or longwall pack-gate. The information that he (Mr. Garforth) had given in his paper was principally based on retreating longwall-faces, which they had to partly abandon in favour of advancing longwall. Mr. Todd also referred to the extra amount of shot-firing required, and he agreed to a certain extent with him. When the men commenced to get the coal by retreating longwall, they were forced by the officials to carry out many little details that were now neglected. At first, 36 or 37 shots were fired over a face 1,800 feet long, and these were reduced, when the depth of holing was increased to 5 feet, to about 17 or 18 shots; and when the depth of holing was increased to $5\frac{1}{2}$ feet, the shots were reduced to 5 or 6 per day. Now, with advancing longwall, breaking-in shots were required, and the men were using a very large number of " pop-shots," which contained about 2 ounces of explosive. The ventilation on the coal-cutting faces had been largely increased.

In reply to the question which had often been asked as to the proportion of small coal, he might explain that on a particular day a certain quantity, afterwards reduced to 1,000 tons, was put over the screens, and sent out to customers. The best coal, cobbles, nuts, down to the smallest coal mined by machine, realized £31 more than the same quantity got by hand, or 7.44d. per ton of enhanced selling-price.

With further reference to Mr. Todd's remarks, it would be remembered that his (Mr. Garforth's) paper was on the application of coal-cutting machinery to deep mining, and he tried to show that in the future, when they were working seams at the same depth as the New Moss or Pendleton mines, say, 3,000 or 3,500 feet deep, the coal-face would have to advance much more quickly, so as to save the coal from the crushing effect of the superincumbent weight. He had, therefore, assumed that a face 1,100 feet long, multiplied by 16 feet, would produce the same weight of coal as a face 3,000 feet long multiplied by the usual advance of 6 feet.* The members might assume that the gateways were 60 feet apart, and that a man and his filler would work 30 feet on either side of the gateway, and produce 10 tons per shift; for 1,100 feet of face, that mine could produce in the morning shift 1,800 tons per week, and 700 tons during the afternoon and night shift. At another colliery working coal-cutters in a face 645 feet long, 1,500 tons were produced per week, and this amply justified his statement that 2,500 tons per week could be got from 1,080 feet of face.

With respect to the cost of plant, coal-cutters, pipes, repairs, depreciation, redemption of capital, etc., everything had been paid out of revenue; that is, at the end of each year, they had been able, at the Altofts collieries, to pay for everything that had been bought, out of the profits made from machine-cutting. The recent actual underground costs had fully confirmed the statements made in his paper.

With regard to motive power, Mr. J. T. Todd had expressed the opinion that electricity was more economical than compressed air, but as that question was the subject of an enquiry by a Departmental committee appointed by the Home Secretary, and as he had been asked to give evidence, he thought it better to say nothing on that occasion. About $4\frac{1}{2}$ years ago, he erected an installation on the polyphase system; the machine made a cut 7 feet deep on a coal-face about 1,500 feet long, and the height of the cut varied between 9 and 10 inches. They had no trouble with the generator, etc. A three-cored cable insulated with bitumen was used on account of the damp roadways.

Mr. W. PRICE ABELL (Duffield) said that, about three years ago, when considering Mr. Garforth's scheme to reduce the loss accruing from transmitting compressed air over long distances to the coal-cutter, he, with Mr. Garforth, went into the question of driving the air-compressor by an electric motor near the coal-face, thus reducing the loss between it and the coal-cutter. Then, mainly owing to the slow speed at which recognized makes of air-compressors ran, the installation worked out too clumsy and large. However, to-day, owing to the adoption of mechanically-moved valves there are recognized makes of air-compressors giving the required quantity of air at 150 or even more revolutions per

* *Trans. Inst. M.E.*, 1902, vol. xxiii., page 343, and Fig. 21, Plate XXI.

minute, with the result that on reconsidering the scheme more than one large undertaking considers the financial advantages so favourable as to warrant the adoption of an electrically-driven air-compressor placed near the working-face, to supply compressed air to drive the drills and cutters. In making this statement, he (Mr. Abell) particularly desired to draw attention to the important fact bearing on this conclusion, that, at the present time, there is no reliable electric drill. With regard to the relative danger of polyphase and direct currents, this is still an open question, arising mainly from the fact that the voltage of the peak of an alternating-current, though higher than the voltage of a direct current, possesses the advantage of repelling personal contact; and it is this fact, of not holding and retaining its victim within its grip, that still leaves the intermittent, higher voltage, alternating or polyphase current of possibly no greater danger than the direct current, of voltage equal to the polyphase average.

Mr. A. H. STOKES (H.M. Inspector of Mines, Derby) said that one difficulty with regard to coal-cutters was that, when the holing-sprags were taken out, the coal fell on to a pillow of dirt (which the cutters did not remove) and in one large lump, so that a great amount of labour was expended in breaking it up and making it into saleable coal. He advised that the cutters of every machine should clear out the holing-dirt, and leave the holing clean for the coal.

Mr. ISAAC HODGES (Whitwood Collieries) said he found that increasing the height of the holing from 5 to 7 inches, although it did break the coal a little more, did not give a sufficient amount of breakage to enable it to be loaded into the coal-tub. At the Whitwood collieries, in a seam similar to that mentioned by Mr. W. E. Garforth, he was quite satisfied with what had been done: the cost had been less than anticipated, they had less trouble and less labour, and had not had a breakdown. They obtained on the average 1,200 tons per week with a single cutter working in this seam, 4 feet 3 inches thick. But he had the startling fact to disclose that shot-firing had increased in a most phenomenal manner: in a seam where they had no shot-firing when hand-holing they now had a shot fired for every 30 tons, and although the thickness of the holing had been increased from 5 to 7 inches, it had not greatly affected the question, and, as Mr. Stokes had stated, the dirt in a way cushioned the fall. In hand-holing, it was not unusual to take out 2 feet 6 inches of holing-dirt to a depth of 9 or 10 feet : the coal then had a considerable fall, and it explained the non-necessity for shot-firing. He disliked shot-firing in mines, and did all that he could to diminish it, but he now found that by introducing coal-cutters instead of diminishing shot-firing he had very largely added to it. The shots were greater in number than they were serious in point of the quantity of explosive used. They were " pop-shots," put in for the purpose of reducing labour. Unlike Mr. Garforth, he had not found the coal-cutter able to pay all costs out of revenue, nor had he found it able to make large profits; but he believed that the profits were not made so much out of the reduction of labour in the pit, as out of the increased value of the coal produced on the surface.

Mr. M. H. HABERSHON (Sheffield) said that he had found some difficulty when cutting at the floor-level in getting the dirt out of the holing, and unless it was removed, it prevented the coal from breaking down properly : and he thought it was a point in which manufacturers of coal-cutting machines should try to effect some improvement.

Mr. W. H. PICKERING (H.M. Inspector of Mines, Doncaster) said it was somewhat disquieting to hear that more shot-firing was necessary, for there was quite as much danger in firing " pop-shots " as there was with large ones. They were put into the coal after it had fallen and was probably cracked to a certain extent. Sometimes a shot would spread into a crack, and with the modern high explosive, this was a danger. It had been hoped that coal-cutting machinery would have lessened the dangers of colliery-working. He thought that it would undoubtedly do so in the case of falls of roof, and it would be a great pity if they had simply transferred the danger to another cause. There was the danger of electricity, but at many collieries in Yorkshire compressed air was used, so there that danger did not obtain : but shot-firing was most important, because the shots would have to be fired when the men were in the pit, while the coal was being got.

Mr. ISAAC HODGES said that he had expected to see some development in wedging, which would minimize or prevent shot-firing, but in his communications with manufacturers they frankly admitted that they had no possible chance of competing against the cost of explosives.

Mr. W. E. GARFORTH said that he agreed with Mr. Hodges, but wished the members to understand that the statement which he made, some time ago, as to the number of shots being reduced from 37 down to 5 or 6 per day was quite true, on the retreating face. Unfortunately, the men were not carrying out the system as well as they did some years ago, and were taking advantage in little things. Efforts were being made, however, to reduce the number of shots, and he quite appreciated what Mr Pickering had said.

————

DISCUSSION OF MR. J. H. WHITTAKER'S PAPER ON "SPARKLESS ELECTRIC PLANT FOR USE IN MINES AND IRONWORKS,"[*] AND OF MR. SYDNEY F. WALKER'S PAPER ON "ALTERNATING CURRENTS," ETC.[†]

The CHAIRMAN (Mr. H. B. Nash) said the great danger one feared or had feared was that, unless the cables were thoroughly well insulated, a pony-lad, trying what he could do, would get a shock which would end his existence. If they could carry high-pressure currents without that danger, and without the danger of sparking at the motors, they would be taking a step in the right direction and bringing the application of electricity more into the position in which they wanted to get it.

Mr. A. LUCAS said that there was a great tendency prevalent, to avoid having anything approaching earth-connections about electric machinery, and by neglecting this safeguard they were doing the very thing to cause shocks to workmen, and harbouring what might become a dangerous plant. In his opinion, electric plant that could not be run safely with frames and bases earthed was unsuitable for mine-work.

Mr. WILLIAM MAURICE (Tibshelf) said that the question of the relative safety of earthed and other transmission-systems appeared to him to be somewhat beside the mark. Systems were condemned

————

[*] *Trans. Inst. M.E.*, 1902, vol. xxiii., page 170; and 1902, vol. xxiv., page 233.

[†] *Ibid.*, 1901, vol. xxi., page 451 : 1901, vol. xxii., page 566 ; and 1902, vol. xxiv., page 489.

on account of faulty applications of them. He did not think that it was possible, at the present day, for any final statement to be made as to the merits or demerits of any system, because a very large proportion of the failures were traceable to neglect of the essential conditions of safety. Confusion appeared to exist at times as to what constituted an earthed system. Sometimes, for example, a single copper-cable insulated and sheathed with metallic armouring was used as a conductor, the sheathing serving as the return wire. This would be properly referred to as an earthed return-system. If, on the other hand, an entirely insulated line were laid, and that line for the purpose of mechanical protection was covered with armouring, the latter metallic shielding would require to be earthed. The circuits, however, were completely insulated.

The attractive title of Mr. Whittaker's paper—"Sparkless Electric Plant"—might lead one to expect that at last something really safe had been discovered. There was no electric mining machinery which was sparkless under all conditions. Mr. Whittaker in stating that "it is scarcely advisable to fix a continuous-current motor of any kind, in a fiery mine,"[*] would appear to be holding a brief for alternating machinery, although it was well-known that polyphase motors could and did frequently burn out. References to electricity "in a fiery mine," were becoming somewhat frequent and at first sight were apt to scare people, but since "a fiery mine" was not defined the condemnation of electrical plant was not so sweeping after all. He (Mr. Maurice) believed that electrical apparatus was safe to use in any mine, where it was otherwise safe to work, but he would take special precautions according to the conditions obtaining and would restrict its application in the case of faces liable to sudden outbursts of gas. Mr. Whittaker touched on an important point in his section on coal-cutting. When referring to safety-fuses, he gave an instance of the difficulty of efficiently proportioning the size of fuse-wires, as a coal-cutter would frequently take four or more times its normal current for a short period. A fuse should blow on a 50 per cent. overload, and would thus bring the machine to a standstill. The tendency, in the case of men cutting by the yard, was to stop such interruptions by the use of thick wire-fuses,

[*] *Trans. Inst. M.E.*, 1902, vol. xxiii., page 172.

that was to say, by using no fuses at all. His (Mr. Maurice's) remedy for this difficulty was to have at the motor-end of the trailing cable a magnetic cut-out which could not be tampered with, and could yet at any moment be reset by the motor-man in charge.

Mr. W. PRICE ABELL (Duffield), speaking upon the necessity for meeting the varying demands on a generating-station, particularly when using squirrel-caged motors, drew attention to the effective Highfield arrangement, now adopted for the varying work of docks, etc. In this system of high-tension rotary transformers in connexion with batteries used as a reservoir, the batteries are charged with the minimum rise of voltage in the main current; in fact a rise of $2\frac{1}{2}$ volts sends the current to the batteries; and a fall of $2\frac{1}{2}$ volts causes the batteries to assist the main current: thus in a practical way avoiding the often prohibitive variation of voltage hitherto necessary in the main current before the reservoir or battery comes into play. In other words, it gives to an installation a reservoir of power in the form of the battery and booster, without materially interfering with the strength of the main current. It thus increases the capacity and economy of a generating-station that has to deal with intermittent loads, such as starting a pump without draining the discharge-pipe; and this without exhausting the current, or dropping the voltage to such an extent as to interfere with the regular working of the other motors or lights. This system has apparent advantages worth considering for colliery generating-stations where the demands on the main current vary, and will vary still more as electric haulage, winding, and intermittent pumping are adopted.

Mr. J. GERRARD (H.M. Inspector of Mines, Manchester) said he was sorry to hear that there would always be deaths from electricity in mines, in the same manner that, if a man went before a locomotive engine, and that engine ran over him, he would be killed. That was obvious, and electricians should make the dangers of electricity equally obvious: probably some of them could be made so, or at any rate precautions could be taken so that workers might know that it was dangerous to put their hands on certain parts, or those parts could be so secured that hands could not be put on them. With regard to the difficulty of defining a fiery mine: there are many mines ventilated by mechanical ventilators, lighted by the best form of safety-lamp, in which only safety explosives are used—or none at all, from which matches, etc., are rigorously excluded; it is only reasonable that the owners of these mines should seek to be satisfied that no danger could arise from the introduction of electricity.

Mr. J. S. BARNES said that he would not advocate the adoption of any earthing system.

The CHAIRMAN (Mr. H. B. Nash) said that, after having heard such varied expressions of opinion from experts, they were pretty much in the same position as before, so far as knowing whether any safer method had been introduced of applying electricity to mining. Sparkless electric plant was desirable so far as the plant itself was concerned, but the great danger was in carrying the power to the point where the machine was worked. That was where they wanted to be assured of safety; if they at any moment were to have falls of roof, with broken cables in roads heavily timbered and fires or any other danger from short circuits, he was afraid that electricity underground would not make that headway which one felt it ought to make when its great economy in working was considered.

THE NORTH OF ENGLAND INSTITUTE OF MINING AND MECHANICAL ENGINEERS.

GENERAL MEETING,

HELD IN THE WOOD MEMORIAL HALL, NEWCASTLE-UPON-TYNE,

APRIL 4TH, 1903.

SIR LINDSAY WOOD, BART., PRESIDENT, IN THE CHAIR.

The SECRETARY read the minutes of the last General Meeting, and reported the proceedings of the Council at their meetings on March 21st and that day.

The following gentlemen were elected, having been previously nominated : —

MEMBERS—

MR. JOSEPH MACLEOD CAREY, Mining Engineer, Coatham, Redcar, Yorkshire.

MR. CHARLES CHANDLEY, Mining Engineer, 120, Musters Road, West Bridgford, Nottingham.

MR. FRED T. GREENE, Mining Engineer, Butte, Montana, United States of America.

MR. RALPH LIDSTER, Engineer, Langley Park Colliery, Durham.

MR. TOM PATTINSON MARTIN, Mining Engineer, Cumberland Coal-owners' Association, Workington, Cumberland.

MR. JOHN HENRY MILLER, Colliery Manager, South Hetton, Sunderland.

MR. WILLIAM OUGHTON, Engineer, 33, Westgate Road, Newcastle-upon-Tyne

MR. FRANK HOLMAN PROBERT, Mining Engineer, 230 and 231, Bradbury Building, Los Angeles, California, United States of America.

MR. ITHEL TREHARNE REES, Mining and Civil Engineer, Guildhall Chambers, Cardiff.

MR. JOHN CAVERLEY WASLEY, Mining Engineer, 184, Saltwell Road, Bensham, Gateshead-upon-Tyne.

MR. STEPHEN WATERS, Mechanical Engineer, Apartado No. 96, Pachuca, Mexico.

MR. JAMES WILSON WILLIAMS, Mining Engineer, 15, Valley Drive, Harrogate.

ASSOCIATE MEMBERS—

MISS ROSALIND WATSON, Victoria, British Columbia.

MR. GEORGE HENRY WRAITH, Moor House, Spennymoor, R.S.O., County Durham.

DISCUSSION OF MESSRS. A. H. MEYSEY-THOMPSON AND H. LUPTON'S PAPER ON " SOME OF THE CONSIDERATIONS AFFECTING THE CHOICE OF PUMPING MACHINERY."*

Mr. A. H. MEYSEY-THOMPSON said that at the previous meeting questions were asked to which a complete answer could not be given at the time, and the authors wished to supplement their replies with the following remarks:—In the first place, some members spoke as if they thought that the writers were opposed to the electrical system of pumping. If they gave any such impression, the writers wished to correct it, as this was quite contrary to their belief. Indeed their view, as stated in the paper, was that for conveying power for pumping purposes to a distance from the shaft-bottom, electricity was, in the majority of cases, the most convenient medium, and they anticipated that, for such conditions, the electrical system of pumping would be largely adopted in the near future. On the other hand, the figures at their disposal (which were quoted in the paper) led inevitably to the conclusion that if pumping could be done direct by a steam-engine placed upon the surface, the conversion of mechanical into electrical energy and its subsequent re-con-

* *Trans. Inst. M.E.*, 1903, vol. xxiv., page 276.

version for use in driving a pump was a mistake when economy
of fuel was of importance. And this contention was emphatically supported by the very great economy obtained over a long
period of years by Cornish pumping-engines and by the Bradley
and Moat engines of the South Staffordshire Mines Drainage
Commissioners.

Messrs. T. Y. Greener, H. Louis and others, questioned the
liability of an electric spark to fire gas, and the former enquired
whether any case could be given where such had occurred. The
authors referred him to the inquest on the victims of the fatal
explosion at Edge Green colliery last year, where the verdict
of the jury was that " the explosion was caused by the ignition
of gas, which ignition was brought about by some damage to
the electric cable." Mr. A. M. Henshaw stated that the
spark of an ordinary signal-bell would ignite fire-damp, and
he further stated that it would be most unfortunate if any indiscretion in its use should result in disaster, thereby tending to
check the employment of electricity.* With these words the
authors thoroughly agreed, and they were of opinion that the
system adopted in South Wales and Derbyshire of burying the
cable in the ground and adopting three-phase current was a
wise precaution, well worthy of more general application.
Regarding the use of three-phase induction-motors, the authors
agreed with Prof. Louis that they presented some disadvantages
when used for driving pumps, on account of the difficulty in
starting and the impossibility of greatly varying the speed. The
first difficulty could, in practice, be surmounted by fitting the
pump with a large bye-pass, thus enabling it to run practically idle until the motor got into step with the generator on
the surface. Where there were great differences in the make
of water at different seasons, the second difficulty remained, but
if there was sufficient lodge-room it could be met by varying
the number of hours of work, or, if pumping must be constant,
by duplicating the plant. On the other hand, the alternating
current had certain advantages, and there was no difficulty in
using it for pumping purposes, if desired.

Prof. Louis also drew attention to the Riedler express-pump,
with mechanically-governed valves, as suitable to be driven electrically, but the authors had been unable to hear of any important

* *Trans. Inst. M.E.*, 1902. vol. xxiv., page 149.

example of the use of such a pump, or of any case of such a pump being placed underground. They had seen a small pump of this class which was running at about 150 revolutions per minute, it was in use on the surface for pumping water for boiler-feeding; the suction-valves only were mechanically-governed, being forced down by the ram on its return stroke; and the pump was working apparently satisfactorily with but one train of gearing, but the noise and jar were considerable. As to Riedler steam-driven pumps with mechanically-worked valves, the authors doubted whether the additional speed was worth the increased complication. The difficulty with all pumping-plant was to get the water rapidly enough into the pump, the only available force being the difference between the atmospheric pressure and the more or less perfect vacuum in the pump. In slow-running engines, this pressure would keep the pump solid, that is, the water would follow up the ram when it was moving at a moderate speed, but with high-speed pumps, special precautions, such as raising the water to or above the pump-level, with an auxiliary pump, had to be adopted. Mr. C. W. Martin expressed the opinion that where a large single valve failed to act, the consequent shock could be averted by the use of relief-valves, but the authors' experience did not bear this out. To be really effective in large pumps, the relief-valve must be placed on the ram-case, that is, on the pump-side of the delivery-valve. If then this relief-valve by any means got out of order, the pump was liable to draw air, and under such circumstances the pump might suffer a shock worse than the one which the valve was intended to obviate.

Mr. J. J. Prest expressed the opinion that for unwatering sinking shafts, rod-pumps were not suitable. Whether this be so or not, it was certain that some of the most heavily-watered sinkings had been effected by their use at a very low cost. Where special appliances were adopted for sinking and then pulled out and replaced by permanent plant, the cost of the double installation must necessarily be high. The authors believed that the system adopted at Clara Vale colliery of sinking with an engine, which at one time during the sinking raised 2,000 gallons a minute and was still working keeping the pit dry, was the most economical both in time and money. There was no difficulty in instantly varying the speed of pump-

12

ing or the length of stroke, and any duration of pause desired could be made between the strokes.

Allusions had been made by speakers to the first cost of the different systems of pumping, and the authors regretted that they were unable in a few words to comply with the request for fuller information: in fact, to reply satisfactorily would require a paper on this subject alone. They might remark, in brief, that from an examination of the costs of the different systems as supplied by themselves, they found that the cheapest in first cost per horsepower was the steam-driven underground engine, next, the overground-engine driving pumps below by means of rods, while both hydraulic and electrically-driven pumps were more expensive in first cost than either of the two former.

Mr. F. R. SIMPSON (Blaydon-upon-Tyne) said that he had read the paper with great interest, but at the same time with some disappointment at the results of the working of the engines given by the authors. These fell considerably below the duty of the old Cornish engines, which, although cumbrous, were certainly economical. The writers quoted Mr. J. S. Dixon's statement that 10 pounds of fuel per indicated horsepower at collieries was very near the truth.[*] He (Mr. F. R. Simpson) had taken out particulars from six collieries, three of which were heavily watered, and found that this figure was about 8·5 pounds where high-class engines were adopted for pumping, and probably 10 pounds would be used where pumping did not bear a large proportion to the total horsepower. With regard to the percentage of output used for engines, their experience did not agree with Mr. J. S. Dixon's 7·39 per cent, as the following instance would shew:—At Clara Vale colliery, the average water-feeder, last year, was 1,578 gallons pumped to a height of 336 feet, and 215,810 tons of coal were raised (17 tons of water to 1 ton of coal); and the consumption of coal was only 4·9 per cent.: in this case a great proportion of the steam was used by a large compound condensing pumping-engine, actuating pumps by means of quadrants and developing 246 indicated horsepower at a consumption of 4·18 pounds of coal per horse-

_____
* _Trans. Inst. M.E._, 1902, vol. xxiii., page 373.

power-hour. He calculated from a trial which was made when the engine was running at half speed, and allowing for a little more economy when the engine was working at a larger load, that this engine would consume 4,021 tons of coal per annum out of a total of 10,566 tons, leaving only 6,545 tons, or 3 per cent., for two winding engines, two hauling-engines, fan-engine, electric lighting, screening and other auxiliary engines necessary for the production of 215,810 tons of coal. There were 5 Lancashire boilers 30 feet long and $7\frac{1}{2}$ feet in diameter, working at a pressure of 80 pounds per square inch, and if the pumping-engine be omitted the colliery would be worked by 3 boilers. So far as economy of fuel was concerned, engines must necessarily be placed on the surface, but it must be remembered that convenience in many cases counteracts economy, and where large pump-spears are used there are considerable expenses in maintenance, which are almost entirely saved in the case of ram-pumps placed underground.

The remarks as to cutting-off steam at an early part of the stroke in the Cornish engine, applied also, in his opinion, to large engines of other types working pumps by means of quadrants; and cutting off earlier than three-quarter stroke was probably confined to 24 hours' trials.

If his (Mr. Simpson's) calculations were correct, the Bradley engine was using 8·5 pounds per pump-horsepower, or 6·4 per indicated horsepower, and gave the low duty of about 26,000,000 foot-pounds, whereas the Moat engine, though slightly better, only shewed 7·2 pounds and 5·4 pounds, equal to about 31,000,000 foot-pounds; while the average duty of the old Cornish engine was said to be 60,000,000 foot-pounds. He presumed that there must be some explanation of this low duty, even after making allowance for the low evaporating-power of the Staffordshire slack; if not, it need not be a matter for surprise that colliery-engines, which had to work under notoriously unfavourable conditions, except in pumping and ventilating, should get dangerously near the consumption of 10 pounds per horsepower-hour.

Smaller pump-valves had proved satisfactory, so far as their experience was concerned. In the Worthington pump, at Blaydon Main colliery, that idea had been carried still further, and there were no less than 48 small valves, 4 inches in diameter in each pump, and these had worked, without trouble, for many years.

With respect to the length of stroke, in the Worthington engine this was only 2 feet, and the flow of water at the delivery was of remarkable continuity when 1,680 gallons per minute were being pumped (24 double strokes).   The consumption of steam at underground pumps must necessarily be high, and their experience pointed to 7 pounds per indicated horsepower when the engine was placed at a depth of 270 feet from the surface.

He had found the use of hydraulic engines for pumping from dip-workings very convenient, though he did not favour taking the power-water from the rising main, but from a pond on the surface, so that in the event of any stoppage of the main engine the hydraulic motor could still be worked.

With regard to the low efficiency of electricity, he would draw the writers' attention to the fact that, in the instance given, the results were quite equal to those of the Moat engine and considerably better than those of the Bradley engine.   The generating-engine only used half the coal consumed by the Moat engine, and the comparison was manifestly unfair to the electric installation.

As to the inconvenience caused by the movement of the floor, where engines were placed underground, they had never had the slightest trouble at any of the four engines which they had so situated at the Towneley collieries.   In his opinion, it was of the greatest importance to have sufficient standage for water; and, where pumping-engines could be worked at night at their most economical speed and allowed to stand during the day, economy would ensue, and fewer boilers would be required, as the load-line throughout the 24 hours would be kept level, and there would be better opportunities of carrying out repairs necessary for the economical working of the engines.   With a view to carrying out this policy, and having regard to the possibility of feeders increasing, engines were frequently erected of a capacity very much exceeding the volume of water to be pumped, and while this might cause a slightly lower duty, the wisdom of providing for contingencies was desirable.   Mining-engineers were fully alive to the importance of the use of economical engines, but the difficulty in obtaining a pumping-engine which could give a duty of 80,000,000 foot-pounds over a length of time was, as shewn by the paper under discussion, not an imaginary one.

Mr. F. Coulson (Durham) said that for pumping from the bottom of a shaft, probably a pumping-engine on the surface, with spears, and everything well balanced, except the quantity of water to be pumped, was the most economical, having valves so arranged that the water would pass readily through them, but so that they would close quickly, and, if possible, mechanically. With regard to in-bye pumping at a distance from the shaft, it seemed to him that there would be a great saving in using electricity, where such could be employed, and in other cases in using compressed air. The question of cables sparking was an important one, as they knew that gas could be ignited by sparks. He thought that there was a danger from the wires being taken in-bye, even when buried in the ground, for if there was any leakage it was not easily detected. He knew of one case where from such leakage a shot was fired, by one of the fuse-wires being dropped on the ground, when the other was connected with the shot. Where there was a danger from taking electricity in-bye, owing to the presence of gas, compressed air was the proper means to employ for pumping, and it appeared to him that in a colliery they could generate electricity and compress the air by a much higher class of engine, and save probably 40 to 50 per cent. in the quantity of coal used. In that case they would find that the efficiency both of electricity and compressed air would compare very favourably with other methods of pumping.

Mr. Henry Lawrence (Newcastle-upon-Tyne) said that the paper opened up a large field for discussion, though, as the principal methods were so well known to the members, he would not enter into any details of design or manufacture, but state generally that the best and most effective pump was that which would bring the water from the bottom of the shaft to the surface with the least amount of power; and in order to arrive at that satisfactory state he would say never, if it could possibly be helped, stop or divert the flow of the water; avoid also the use of bends and T pipes and badly-designed valves, with contracted areas, as they created much friction, and the pressure taken by a gauge at the bottom of the shaft, at the entrance to the air-vessel, was very much, and in many cases surprisingly, greater than what was due to the statical load. Pumping water from dip-workings to the main pumping-machinery at the bottom of the shaft necessitated consideration, where the power required was much less than could be got from the hydraulic pressure due to the rising main. Hydraulic pumps had been in the past very much used: the power-water was taken underground to the site of the pump at the dip, and as the water subsided it could be easily moved nearer and nearer to the dip, but of course they had the expense of forcing the power-water back to the main pumping-engine; this method, however, was not frequently used at the present day, and then only under very favourable conditions. The most frequent method used at the present time was to work the in-bye pumps with compressed air. Compressed air, as a motive power, was considered to be a very expensive one, but this had been very much exaggerated. Many colliery-owners, who had been using compressed air very largely for pumping, coal-cutting, and other purposes for the last 35 years and were still using it, were quite satisfied with it. It was very useful and convenient, and the exhaust-air aided the ventilation of the mine. The most important motive power was electricity, but as he was not an electrical engineer he would leave that for discussion by electrical and mining engineers, and he hoped, as the matter was most important, that the discussion would be an exhaustive one.

Mr. W. C. BLACKETT (Durham) said it was not every engineer who would allow the statement as to the danger of an electric spark firing underground gases to go unchallenged; and it must not be admitted that there was a parallel between ordinary coal-gas and that found in mines. The conditions under which fire-damp could be fired by a spark underground varied and did not always exist. He hardly thought that Mr. Meysey-Thompson should put forward that a coroner's jury ought to be accepted as the best judges of what would fire gas in the pit. His own experience of juries did not lead him to accept their decision as that of scientific experts. As to the power to be employed for pumping in-bye, he had had some little experience in that direction, and could not altogether give a very good character to compressed air. He agreed with Mr. Lawrence, who probably had in his mind one of their own collieries, when he said that compressed air had been advantageously used for many years, but it might be interesting to know that at a colliery with which that gentleman was familiar compressed air had been replaced by hydraulic power and a Lancashire boiler, in consequence, saved. Compressed air was convenient and easily handled, but it was expensive to instal and it was generally wasteful in use. Mining-engineers spent a large amount of thought and care in erecting air-compressing engines, fitted with efficient valves, and yet when they got the air down the pit, it was far too often applied to defective engines and machinery. American engineers had adopted air-compressing machinery to a greater extent than engineers in this country: they employed displacement-pumps driven by the air, like a kind of steam-trap, and by applying these in stages they got a most economical form of expansive working.

Mr. H. LAWRENCE said he did not wonder that Mr. Blackett found the hydraulic engine more economical than compressed air. In compressed air they had to create the power, but hydraulic power was taken from the rising main or a pond, which was already made.

Mr. T. E. FORSTER (Newcastle-upon-Tyne) said that compressed air was very often compared with electricity, by taking out an old air-compressing plant and putting in a new electric plant, which naturally gave better results. A compressed-air plant really lasted a great deal longer than the average electric plant, and if interest on capital and depreciation at anything like a fair figure was taken, there was very little difference in the cost.

Mr. G. P. LISHMAN, with regard to the question of igniting pit-gases by means of an electric spark, said that he had frequently done this with a spark from a 5 volts accumulator at atmospheric pressure.

Mr. C. C. LEACH wrote that he had a geared pumping-engine, on the surface, with a cylinder 16¾ inches in diameter by 3 feet stroke, geared 6 to 1, supplied with steam at a pressure of 100 pounds per square inch. It worked satisfactorily, and economically, and was pumping from a depth of 500 feet in two lifts: it had lifted 585 gallons per minute, and frequently pumped more than 500 gallons per minute.

Mr. J. Kenneth Guthrie (North Shields) wrote that the most important feature about a pumping-plant is that everything should be of liberal design. From experience with these plants he had found that where the plant has been designed to do just the stated service, and has no margin in hand for an emergency, in most of the cases trouble has ensued and breakdowns have occurred. If an electrically-driven pump is to be installed, it is necessary to see that the motor, which is to drive the pump, has a considerable output beyond its normal rating, so that there will be no danger of "burning-out" if the pump be called upon to do somewhat more than its rated capacity. With regard to motors it is well to understand that everything in a mine tends to reduce their efficiency: the temperature is higher, there are particles of dust flying about, and there is also a considerable amount of damp in the atmosphere. At the power-end of the pump, the bearings should be very ample: as when one erects a mine-pump of any capacity, it is with the idea that it will run continuously for ever and a day. The gearing, between the pump and motor, should be of the highest class, and the motor-pinion should be constructed of raw hide or fibre in preference to steel. Although a steel-pinion will outlast several of the raw-hide type, yet a great gain is made in the reduction of the vibration and the quieter working of the pump. A raw-hide pinion will run without attention for 18 months, and if a steel-pinion were adopted it would probably last from 6 to 10 years.

The pump-end of an installation is probably the most important item. In the case of a steam-driven one, there is a certain amount of elasticity in the working, and if anything should happen to be wrong in the design of the water-end, or an obstruction takes place from any cause, in most cases the steam-end will adapt itself, at least to some extent, to the changed conditions. If, however, in the case of an electrically-driven pump, something were to go wrong with the valves, or if these were of too small an area, nothing could save the pump from working noisily, and perhaps it would damage itself. It is well, therefore, to be careful as to the pump-end, especially as regards the arrangement of the valves and the valve-area. If the valve-area is too meagre, it cannot be expected that the pump will work quietly, as an excessive speed of water will take place in the passages, causing the valves to hammer.

In the planning of pumps it is generally advisable to adopt a design in which the water is flowing, as a general rule, in one direction, that is to say, coming in from the suction-side and flowing up towards the delivery. This is not the case in many pumps, and any strictures of the passages through which the water is led means a restricted output, or at the best noisy working. The suction-end of a pump is by far the most important, and it is advisable to see that the pump is properly supplied with suitable pipe-connexions. On the suction-side, the level of the water to be pumped should not be more than 15 feet below the level of the pump-valves. The suction-pipe should be of somewhat larger diameter than the delivery and provided with a foot-valve, which should have an area considerably in excess of that of the pipe. A strainer should be placed next to the pump so as to keep out all destructive matter, and an easy examination for cleaning can be made by removing the upper cover. The area through the strainer should not be less than four times that of the pipe. They are sometimes put in having the total area of the small holes drilled into the strainer, just the same as the pipe. As a matter of fact only about one-third of this area is effective, a great loss being caused by the friction of the water through the small openings. As the metal continues to corrode, one can understand how small a quantity will in due time reach the pump. This strainer dispenses with the dang·· of gags, mentioned by the authors, and in the Riedler pump, which he mentioned in the former discussion, even should a g··g of a moderate size, say of the thickness of a lead-pencil, get into the valve, no damage is done to the gear, as there is a special buffer provided for the contingency.

A check-valve should be placed on the discharge-pipe, so as to relieve the pump of the weight of water held in the rising main. Beyond this check-valve, a charging-pipe should be led into the suction-pipe so that the pump may start with a load. There should also be placed in the delivery-pipe, or on the delivery-valve covers of the pump, suitable starting-valves to enable the air to escape from the pump while the water is flowing from the rising main. The operation of starting the pump should, therefore, be to open the charging- and starting-pipe valves some minutes before starting the pump. After starting the pump, the starting-valves should be closed so soon as the air has been discharged.

In his experience of pumps, where there has been trouble, almost in every case it has been found to occur on the suction-side of the pump, and it is due either to an obstruction in the pipe, or to the insufficient area of the foot-valve or strainer. In every case where the head against the pump is considerable, a relief-valve should be fixed as near as possible to the pump on the delivery-side, and set at a pressure slightly in excess of the working-pressure.   This accessory prevents the bursting of the pump, due to a valve being closed, or to a sudden stoppage in the discharge-pipe.

In the case of the water failing to reach the main pump, or to a valve being held up, the governor should immediately come into action, thus preventing the racing of the engine and the bursting of flywheels.   In a paper read in 1891, he mentioned in describing the Napier cut-governor on the Eltringham pumping-engine, that "on one occasion, with 14 inches setts, the ball of the governor was seen to fall, the engine stopped and then went quietly on: a man being sent down the shaft found that a spear of one of the setts was broken."[*]

As to the consumption of steam in underground pumps, of course there will always be a loss with a steam-pipe, but at two collieries in Scotland, the owners are erecting small super-heaters in the flues of the boilers, sufficient to send the steam down the shaft in a dry state; this is more like effectual steam-drying than superheating.

He (Mr. Guthrie) could not agree with the authors as to the great advantage of using surface-engines; facts spoke for themselves, the old system had of late been going out, under-ground force-pumps were being substituted for those worked from bank, and now that electrical power could be so economically transmitted they were coming still more to the front.   He had had considerable experience with pumping since serving his apprenticeship with Mr. J. B. Simpson, who was then erecting a Hathorn-Davey horizontal underground engine, which he understood had since given capital results; and while he had known of several serious stoppages of engines through breakages of spears, quadrants, teethed wheels, etc., he had seldom encountered serious cases with underground pumps.   These in his experience had not been hidden away in holes and corners, but were,

* *Trans. Inst. M.E.*, 1891, vol. ii., page 462.

as a rule, placed in roomy chambers near the pit-bottom, well lighted, and with all parts perfectly accessible, and he considered that the cost compared very favourably with that of plants placed on the surface.

Mr. A. H. MEYSEY-THOMPSON, replying to the discussion, said that the ordinary engineer wanted to know how many gallons of water could be raised so many feet high for each pound of coal, and in the paper they had tried to reduce everything to that one standard. In the Moat and Bradley engines, which had been running for nearly 20 years, the pounds of coal and the number of gallons were given, and it was easy to ascertain the foot-pounds of work, and the table at the end of the paper (Plate VI.)* reduced all the comparisons to one standard. Mr. J. S. Dixon practically agreed with Mr. A. M. Henshaw as to the consumption of fuel, and the reason for these figures being adopted as . the standard was merely that they might be independent. It did not matter how they fixed a standard for the steam-engine, as they had to compare hydraulic with electric power, and if a more economical engine was used to produce electricity it should also be applied to the hydraulic motor. Mr. F. R. Simpson said that he had not experienced any difficulty from the shifting of the foundations of underground engines; and if some of their friends were in a similar position they would not have had so many difficulties to overcome. In some cases, it had been necessary to fix girders to support the engines, so that they did not rest upon the floor. As to overground engines cutting off at three-quarter stroke, the object of the second cylinder was to enable the high-pressure cylinder to cut off later in the stroke. The difficulty with the old Cornish engine was with an early cut-off: they required a high initial velocity to enable the engine to finish its stroke, and the object of the second cylinder was to secure the high expansion without this initial velocity. Mr. F. R. Simpson had found small valves satisfactory, but it was their later practice to enclose each valve in its own box, with the advantage that it could be easily taken away and replaced. The figures quoted by Mr. F. R. Simpson as to the consumption of fuel by underground engines agreed very nearly with those given in the paper. As to the desirability

* *Trans. Inst. M.E.*, 1903, vol. xxiv., page 286.

for a plant working only for a few hours in the day, attention
had been drawn to this in the paper, more especially as regards
electric pumping. If the power was purchased from a power-
station, it was wellknown that the power-companies had a difficulty in providing power for only a few hours, and if the pumping could be done at a time when the station had a small demand,
the current would be supplied at a cheaper rate.

Mr. H. Lupton said that the duty of the Bradley engine was
25,500,000 foot-pounds, averaged over 13 years, including losses
for banking fires, etc., and the average speed of the engine was
rather under half the speed for which it was erected. The
average duty for 10 years at the Moat engine was 31,000,000
foot-pounds, and that also included all losses for banking fires,
etc., so that these were not, in any respect, trial-duties. The
coal used had an evaporative efficiency of 4½ pounds of steam
per pound of coal; and the engines ought not to be compared
with a modern engine working with high-pressure steam, as they
were working at a pressure of only 60 pounds per square inch.
If they took more comparable duties, that was, where the coals
were of better quality, it would be found that at Yarlside, taken
over a 5 days' trial, one surface-engine gave 53,000,000 foot-
pounds, and a second engine, 51,000,000 foot-pounds, while the
engine belonging to the Cleator Iron-ore Company gave 50,750,000
foot-pounds. These engines had a steam-pressure in each case of
80 pounds per square inch. An electric plant, at South
Durham colliery, had a duty of 29,000,000 foot-pounds, very
near the duty of the Staffordshire engines, but the coal was of
a much better quality. It was a day's trial, and not a 10 or
13 years' trial, so that it was hardly comparable, and the
29,000,000 foot-pounds should be compared with the engines
yielding 51,000,000 or 53,000,000 foot-pounds. He (Mr. Lupton)
found that the cost in South Staffordshire over 10 and 13 years
was £11 2s. 5d. per actual continuous day-and-night horsepower
in the water lifted per annum, which was equivalent to a charge
of 0·22d. per electrical unit; and he did not know that electrical
units were sold anywhere at so low a rate. At Chamber colliery,
where the engine was underground, and the steam taken down the
shaft, a trial gave a duty of 37,000,000 foot-pounds, and the consumption of coal was 6 pounds per pump-horsepower. So far as

the underground engine was concerned, the duty of the electric plant was very near the duty of the underground engine, but the cost of the electric plant was at least double the cost of the underground steam-engine.

Mr. T. E. FORSTER pointed out that the electric plant at South Durham collieries was not working at its full power.

The PRESIDENT (Sir Lindsay Wood, Bart.) said it was quite possible to take electricity into mines where it would be absolutely safe, at any rate, in the main ways. The three-phase system, with properly constructed covered cables, would be quite safe. Of course in a gaseous mine they would not adopt it.

———

DISCUSSION OF MR. M. FORD'S PAPER ON "SINKING BY THE FREEZING METHOD AT WASHINGTON COLLIERY, COUNTY DURHAM."*

Mr. W. B. WILSON (Easington) asked whether in sinking through the 41 feet of wet sand any difficulty was experienced in supporting the sides of the shaft.

Mr. F. COULSON (Durham) thought that it would be interesting to know something as to the ice-forming capacity of the plant at Washington, or more correctly speaking, of the heat-extracting power, and although Mr. Gobert said it was most satisfactory that the capacity was taken in Centigrade degrees, it would be better if it were given in Fahrenheit degrees, with which they were more familiar, and he thought that the capacity of the plant might be given in British thermal units. It would also be interesting to know to how much greater depth the plant at Washington was capable of sinking. He believed himself that it was a question of time, and that the lighter plant required more time. Mr. Ford's paper also referred to the heat-conductivity of the clay, and this appeared to him to have an important bearing on the freezing of the pits. Taking the conductivity of the slate at 1,000, flag-stone would be 1,110, brick 730, clay 564, and cement as low as 200. In freezing the alluvial deposits of clay and sand and gravel, this seemed to have an important bearing, because having formed the ice-walls with a layer of slow conductivity

* Trans. Inst. M.E., 1903, vol. xxiv., page 293.

between two layers of a higher conductivity, if the ice formed
around, above, and below the layer of slow conductivity, then,
unless some means were taken to provide against it, this layer
would freeze, and in freezing, if it contained much water, it
would expand to the extent of about one-tenth of its bulk, and
the expansion being irresistible, it would probably break through
the ice-walls. In some cases, or even where there was a thick
bed of clay over a bed of sand resting on the stone-head, the
force of the ice would be so great that it would break the ice-
walls. It would certainly find out the weakest place, and it
might destroy the freezing-tubes. This can be provided against
in some cases by putting bore-holes in the middle of the pit, and
it occurred to him that they might, in some cases, with advantage
freeze the bore-holes before freezing the side-walls. At Washington, all the tubes except the freezing-tubes were withdrawn, but,
in some cases, light tubes were inserted to protect the freezing-
tubes. At Washington, there was an accident from the explosion
of a gelignite-cartridge; and it might or might not be caused by
the freezing. For some unknown reason, odd cartridges of gelignite were occasionally found not to have exploded. He considered
that compressed powder was the best and safest explosive, and it
would give the best results in a sinking pit. There was no necessity to use gelignite in frozen ground, as its great advantage was
found in wet pits, where there was difficulty in drying the holes.
It seemed to him that it was absolutely necessary to bore to the
stone-head, because the form of the frozen ground at the bottom of
the shaft was like the bottom of a bottle and left a weak place.
And there was also uncertainty as to how the sand-beds lay
in alluvial deposits. A short distance from the pit, the sand-
beds might occur, and in all cases where there was no knowledge of the thickness of the sand, or the direction in which
it lay, it would be desirable to bore and freeze down to the
stone-head. There was another system, which might be applicable, not so much in quick-sands as in shales and in the Coal-
measures where water was encountered, if it could be relied on
that the partings were all horizontal and would be reached
by the bore-holes: this was by forcing in what was called
" cement-broth," which would flow almost anywhere where water
would flow. If this method were successfully carried out, it
would replace and render tubbing unnecessary. A vertical

gullet might be met with, however, which was not touched by one of the bore-holes and they could not force cement into it.

Mr. MARK FORD, replying to the discussion, said that no difficulty had been experienced in sinking through the wet sand, the sides were quite solid, and with the exception of the 6 feet length inserted near the surface, nothing was used to support the sides of the shaft. With regard to the ice-forming capacity, the quantity of water supplied to the condenser was 4,000 gallons per hour and was heated 18° Fahr., equal to 720,000 British thermal units per hour. It was noted that the clay was more or less frozen into the centre of the pit, while the wet sand was in the ordinary condition in the centre of the pit. Further, the frost seemed to endure longer in the clay than it did in the sand, that portion of the shaft-walling backed with clay was wet for several weeks, while the walling which kept back the sand was dry during the same period of time; and the water that came from the freestone into the pit, and filled it to a height of 20 feet from the bottom, was, for 5 or 6 weeks after freezing ceased, only 5° or 6° Fahr. above the freezing-point. The clay sunk through was of a very tenacious character; bellite was tried, but with rather poor results, and better results were got from gelignite. Diversity of opinion still existed as to the necessity for freezing the clay, and even the German operator confessed that the clay at Washington was not thoroughly understood by his firm; they did not realize that it was of so tough a nature, and evidently thought that it resembled the clay referred to by Prof. Louis in the previous discussion.

Mr. W. C. BLACKETT (Durham) thought that it was not a question of the conductivity of the clay and sand, so much as the conductivity of the water in small bulk and water in large bulk. They had water in clay in small bulk, but the water in the sand was of greater bulk than that in the clay, and possibly capable of circulation, and as water was such a well-known non-conductor of heat, the water in the clay would freeze more quickly because it was of lesser bulk and less free to circulate. Then there was an anomaly in Mr. Ford's statement that whereas the clay showed that it conducted heat quickly while freezing, yet while thawing it conducted it more slowly; this was difficult to understand.

Mr. M. FORD said that the behaviour of the clay as mentioned by Mr. Blackett would be accounted for by the specific heat of the clay and the fact that the ice-wall had penetrated further into the clay than into the sand.

____

DISCUSSION OF MR. T. ADAMSON'S PAPER ON "WORKING A THICK COAL-SEAM IN BENGAL, INDIA."[*]

Mr. R. R. SIMPSON (Calcutta) wrote that, as one who had seen and studied the methods of working the thick coal-seam at the East Indian Railway Company's collieries at Giridih, he was unable to agree with those critics who had expressed the opinion that the method of working was unsafe; on the contrary, the very opposite was the truth. At the present moment, he was unable to refer to the statistics of accidents to workmen employed in extracting coal by this method, but he was convinced that they would compare most favourably with those for other methods of working in India, and still more favourably with those for those districts in Great Britain in which thick coal-seams were worked. The inspectors of mines in India thoroughly approved of the system and were accustomed to hold up the methods in vogue at Giridih as an example to the rest of Bengal. The secret of the success of the system was two-fold: (1) The sandstone-roof is an excellent one, and an area, 500 feet square, has been known to remain suspended for some considerable time after the coal has been extracted. The roof, when it falls, breaks off sharp, and very little coal is lost by crushing. Working is safely resumed alongside a fallen area on the day after the fall has occurred. (2) Excellent discipline is observed at the Giridih collieries, and the officials and workmen are thoroughly trained to appreciate and guard against the dangers incidental to mining operations. The Special Rules of the collieries, which are strictly enforced, far exceed the requirements of the Indian Coal-mines Act, 1901; and an elaborate and carefully-checked system of written reports is also in operation. The testing of the roof and sides was carried out in a manner which he had not seen surpassed in any British colliery.

The *chowkidars*, or "watchmen," left in the goaf, vary from

[*] *Trans. Inst. M.E.*, 1903, vol. xxv., page 10.

8 to 10 feet square, the original pillars are 100 feet square with roads 8 feet wide ; and four *chowkidars* are thus left in an area 108 feet square. The loss in working from *chowkidars*, therefore, varies from 2·2 to 3·4 per cent.; adding to this the 4 per cent. loss in thin ribs, the total loss in working is seen to average from 6·2 to 7·4 per cent. He thought that this small loss compared most favourably with the loss incurred in working thick seams by longwall, in two or three carries, and certainly a much larger percentage of round coal was obtained.

Mr. THOMAS A. WARD (Giridih, India) wrote that the impression evidently left by the reading of Mr. Adamson's paper was that the system described was dangerous. This impression was, he thought, due to the omission of the writer to make it clear that the heavy blasts of air, caused by roof-falls, are not inherent in the particular system described, but are common to all the systems which had been tried—even to the "rabbit-warren system," to quote Mr. J. P. Kirkup's apt expression—and are due to the massive character of the roof. The sandstone is not laminated and hardly any, and no regular, bedding-planes can be distinguished. The intercalated seams of coal are the only regular dividing-planes up to which it is usual for large areas of roof to break. An accident in which several lives were lost occurred last year, in the Raneeganj coal-field, from this cause. The system of working was the "rabbit-warren," and the pillars were being brought back. The air-blast killed men who were a long way back from the face.

In 1885, in this coal-field, the pillars in a seam only 8 feet thick, which had been opened out on the pillar-and-stall system, were being brought back. The mine was worked from the outcrop by adits. The miners had been withdrawn because the roof was uneasy, and three had seated themselves at the entrance to the mine. They were not under cover, but they were killed, one body being thrown about 100 feet. A small party, just before engaged in drawing timber, were actually in the mine at the moment that the roof fell, but as they were not in the direct current, they escaped uninjured.

These instances—which could be multiplied—shew that there is no increased danger from this source when working the system described. The substantial pillars, or *chowkidars*, which are left, help, in fact, to cause the roof to break up in falling and diminish the force with which the air is displaced. These *chowkidars* are spaced 30 feet apart, and not 40 feet, as stated in Mr. Adamson's paper.

The system is really a modification of the method by which the Thick coal-seam of South Staffordshire is won. The sides of work, there, are laid out, nearly always, with 30 feet openings. The main difference in working has arisen from the fact that the coal in India has no partings, so that it cannot be " cut in " in layers, as in South Staffordshire, but has to be brought away *en bloc* as described. The openings are not so dangerous as in South Staffordshire, as the ground is in every respect much stronger.

Mr. T. ADAMSON (Giridih, India), replying to Mr. A. L. Steavenson, wrote that the result of over 12 years working at the collieries of the East Indian Railway Company was that 70 per cent. of the total output (which in 1902 amounted to 614,000 tons) was wrought by the system described in his paper, without a fatal accident having occurred directly from the system, and this proved that it was not a dangerous one. The conditions were altogether different at Giridih, to what existed in the county of Durham. The Giridih seam usually comprizes 21 feet thick of good coal, without a single stone-band in it. The floor and roof are hard sandstone and a roof better adapted for the working of the system is not, perhaps, to be found. The knobs are blasted out, and are not left to fall out " at random," as stated by Mr. Steavenson. The Giridih seam occurs in the Coal-measures and not in the Mountain Limestone.

The area wrought before a fall of roof takes place, varies from 150 feet by 150 feet to a much greater area. He (Mr. Adamson) knew of a goaf, 800 feet by 200 feet, where all the coal had been removed, without even a tell-tale being left in it, which stood for 3 or 4 years before it came on. As the blasts caused by large goaf-falls are great, it was found advisable to leave the tell-tales or indicators as stated in the paper.

With reference to the system worked by Mr. T. E. Forster in Australia, he (Mr. Adamson) had seen mines, which were worked 15 years or more ago, in Bengal, in the same way. The whole of the area from the shaft to the boundaries was cut into pillars (square or rectangular). The pillars were made small, and the roads were driven wide. In recent years, attempts had been made to remove these pillars, with the result that, when a few pillars had been removed, the roof instead of breaking down (on account of its great strength, like the roof over the Giridih seam), settled upon the remaining pillars and crushed out several acres of them.

The coal is of a somewhat hard nature, with thin dirt-partings (2 to 4 in number in the seam 21 feet thick), and the undergoing is made up to one of these partings. The coal rests directly upon hard sandstone, and there is no under-clay. Traces of plant-remains have been found in the beds above the seam, but they are very rare.

An area took fire over 20 years ago, which had been worked on the old Bengal pillar-and-room system, the pillars being made small and the galleries wide. As the pillars were too small to maintain the roof, the latter collapsed, and buried a large area of crushed coal, which after a time took fire, and the fire-area was enclosed by masonry-dams.

No fires had taken place since the system described in his (Mr. Adamson's) paper was introduced. A fire on a large scale could not occur, as there was so small a quantity of coal left, the tell-tales being spaced 40 feet apart, and the ribs 200 feet or more apart. If there was a fire, it would only be necessary to build 2 stoppings in the gate-ways through the ribs, and shut off the fire from the rest of the mine-workings.

The old system, referred to by Mr. J. P. Kirkup, was introduced for dropping the top portion of pillars that had been formed before the dropping system was introduced. As the galleries had been driven, in most cases, to the full height of the seam, *chatnies* were not required to be driven, and a less percentage of the seam was dropped than when the roads were driven in the bottom portion of the seam, and narrow *chatnies* driven in the upper portion of the seam. Therefore, the advantages were less than those accompanying the present system, and it was not so safe, as a large portion of the roof was exposed at an early stage, and needed constant attention. The roof is an unusually good one, and enabled 90 per cent. of the coal in the seam to be obtained.

The 15 feet of dropped coal was mostly in large pieces, and

had sometimes to be broken with picks, before it could be loaded
into tubs. The coal may be sampled as follows:—Steam-coal
and rubble (rubble is 1½ inches cubes) 85 per cent. and less than
rubble (smithy and dust) 15 per cent. The smithy is consumed
under the colliery-boilers and the dust-coal is all made into coke.

DISCUSSION OF MR. S. J. POLLITZER'S PAPER ON "THE UNDERLAY-TABLE."[*]

Mr. JAMES HENDERSON (Truro) wrote that Mr. Pollitzer's
apparent object is to show that by means of the underlay table,
a base-line, run at the surface, can be accurately transmitted
down an underlying shaft; and that the system he recommends
is preferable to that adopted by mine-surveyors generally.
Having read the description of the above instrument and its
applicability to underground surveying, the writer is desirous
of making the following comments thereon. It appears that
as the length of a base-line afforded by the underlay-table cannot
exceed 2 feet or thereabouts, the survey of a diagonal or underlie-
shaft, would be a very hazardous business as to accuracy, and
even if this base-line could be extended to a greater length, the
time occupied in the repeated fixing of the instrument and the
suspension of the necessary plumb-lines, would be so great as
to render its adoption practically useless. While, on the other
hand, by means of a miner's theodolite or the Henderson rapid
traverser,[†] in both of which the vertical limb so overhangs the
horizontal circle or plate as to render the reading of the vertical
angle from 0 to 90 degrees perfectly easy, the various changes
in the vertical and horizontal deflections in a shaft can be taken
and recorded with accuracy and rapidity. If it be assumed that
the measured draft-line down the shaft is, say, 100 feet in length,
and the underlie is 60 degrees, which is of frequent occurrence
in metalliferous mines, it would seem that the underlay-table
would require to be placed in position no less than 25 times,
with the probable result that the operator's mind or temper would
not be in a very enviable condition when in the drawing-office
calculating or plotting his work afterwards. By the same rule,
an underlying shaft of 70 degrees would require 17 fixings of

[*] *Trans. Inst. M.E.*, 1903, vol. xxv., page 24.
[†] *Ibid.*, 1893, vol. v., page 199.

his instrument; one of 80 degrees 8 fixings, and so on; the only angle suitable for one fixing would be 88 degrees 51 minutes, or 1 degree 9 minutes from the perpendicular. An unnecessary number of draft-lines is what every surveyor should avoid, whether at the surface or underground and he (Mr. Henderson) failed to see what advantage would be gained by the use of the underlay-table, when the traverse of a shaft of any angle of underlie could be so readily and accurately ascertained by means of the instruments before referred to.

DISCUSSION OF DR. E. D. PETERS' PAPER ON THE "TREATMENT OF LOW-GRADE COPPER-ORES."*

Mr. J. J. MUIR (Tasmania) wrote that Dr. Peters, in referring to low-grade disseminated sulphide-ores of copper, in a siliceous matrix, and of a character similar to that of the Australian ores investigated by himself (Mr. Muir) emphasized his dissent from the necessity of a preliminary concentration. He fully agreed with Dr. Peters on this point, but his contention for this operation as a preliminary to lixiviation, was especially provided to meet an objection peculiar to Australian mining investors, namely:—That to raise capital, for the development of copper-mines, it is essential to show the immediate return of a sufficient amount of metal to pay the working expenses of the mine; and the profits that are obtained from the deferred process of lixiviation are then acceptable. In his (Mr. Muir's) paper† on the subject, he made a special effort to provide this condition, and no doubt Dr. Peters would agree that, under the local circumstances, his advocacy was legitimate. In following out his investigations on these ores to a metallurgical conclusion (which he had finally completed since writing his paper on the subject) he had fully established the great advantage of preliminary dry crushing and roasting of the ore. He was gratified to hear that the same result had been independently arrived at by Dr. Peters. The operation of roasting, after the ore had been crushed fairly fine, imparted to it a physical change, and fully bore out what Dr. Peters claimed for it. The ore assumed a granulated form, and was amenable to quick percolation and

. * Trans. Inst. M.E., 1902, vol. xxiv., page 315.

† Ibid., 1902, vol. xxiii., page 517.

perfect leaching.   As the final outcome of his (Mr. Muir's)
extended researches on Australian ores, as described in his
paper, he had devised an original forced process, whereby the
copper was extracted, and precipitated in 12 hours, on any tonnage
up to the capacity of the plant that he had designed.   In the
circumstances of the ores under discussion, it appeared that the
only possible solution of the problem, was a method based on the
lines that he had indicated—utilizing the chemical conditions
of the ore itself in the process, and finally melting the precipitates into coarse copper-bars at the works.   In the method
that he had finally adopted for his personal operation, the costs,
under general Australian conditions, were 15s. per ton of crude
ore for the complete operation; and from Dr. Peters' description
of the conditions prevailing in the United States, he saw no
reason why similar results should not be easily arrived at in
that country.

Mr. Horace J. Stevens (Houghton, Michigan, U.S.A.) wrote,
in explanation of the existence of Lake Superior copper-mines
successfully extracting the metal from ores averaging less than
1 per cent. of copper, that it might be said that the Lake Superior
mines were all very low grade under ordinary circumstances, the
richest mine returning less than 3 per cent.; there were no gold-
values, and the silver-values were so small as to be scarcely worth
consideration.   The Lake mines, however, were unique because
they mined native copper exclusively.   In most copper-districts,
more or less native copper was found in the alteration-zone, especially in association with cuprite, but outside the Lake Superior
mines native copper was mined upon a scale of importance only
in the Coro Coro district of Bolivia, and from one or two
mines in Eastern Canada, although extensive and possibly workable deposits of native copper, disseminated in igneous and
sedimentary rocks, were known to occur in Australia and elsewhere.   The Lake Superior mines possessed a great advantage, by
reason of the remarkably simple and inexpensive process of reduction required.   The metal was won by crushing the rock and concentrating the copper-values, while smelting was simple, requiring
merely the addition of limestone for flux and the necessary fuel-
charge.   The complex smelting conditions noted in many
copper-districts were entirely lacking.   In addition to the advan-

tages already enumerated, the Lake Superior mines possessed the great advantage of being able to draw upon almost unlimited capital, and the equipment of the larger mines was upon a scale of magnitude parallelled only by mining operations on the Rand in South Africa.   For the reasons already set forth it was unsafe to use the figures of values from Lake Superior copper-mines in estimates of the possibilities of other fields.

The conditions in Australia were much the same as in certain important cupriferous fields of the United States and Mexico, although perhaps more aggravated.   The drawbacks to the operation of low-grade ore-deposits included aridity, lack of adequate transportation-facilities, and scarcity and high prices of fuel.   Without good water or rail-transportation, and without adequate supplies of coal or coke of fair grade at a reasonable price, smelting was too costly with low-grade ores in Australia as elsewhere.   Many people had thought otherwise, but the results had proved them to be mistaken.

The tendency of copper-production seemed against the suggestion of Mr. Muir, that ores should be transported to the seaboard for smelting.   Under exceptional circumstances this plan was feasible, but ordinarily the process of reduction should be carried on as far as possible at the mine.   Fifty years ago, the bulk of the world's copper-production was shipped to Swansea as ore, for reduction.   Since that time, the tendency had been toward smelting at the mines.   It was true that a large part of the American copper-production was smelted in the western half of the United States, and that the blister-copper, produced by the local smelters, was sent to the Atlantic seaboard for electrolytic refining ; but in this case the impurities carried in the blister-copper averaged only 1 or 2 per cent., and the excess freight on such impurities was so trivial a matter that it was much more than offset by the superior advantages offered for electrolytic plants along the sea-coast.

In the way of direct smelting of low-grade ores, some remarkable results had been secured in the United States and Canada during the past few years.   In the Deadwood district of British Columbia, the cost of smelting had been reduced to 1·35 dollars (5s. 7½d.) per ton of ore.   This location, however, afforded advantages in the way of transportation-facilities and fuel that were entirely lacking in the more remote districts of Australia.

It would seem, from the examination of available data, that direct smelting was out of the question in the case of many or most of the low-grade deposits of Australia. A combined reduction-process of mechanical concentration, followed by smelting of the concentrates was a highly satisfactory process where conditions were favourable. This system, however, called for a large and steady water-supply, which was often lacking in the interior of Australia, and it should also be borne in mind that not all low-grade ores were adapted to economical concentration. While certain ores were being successfully concentrated 20 into 1 at the present time, 3 or 4 into 1 was nearer the average of what was considered successful practice, and even in such cases the losses in fines was very considerable, although much reduced latterly by the very general use of concentrating tables of the well-known Overstrom, Wilfley, or Bartlett types.

It seemed to him (Mr. Stevens) that lixiviation held forth the greatest promise for the low-grade ore-deposits of Australia, and of many other cupriferous fields suffering under somewhat similar disadvantages. The highly successful results secured by comparatively inexpensive processes of lixiviation in Spain and Portugal, and more especially at the Rio Tinto mines, were worthy of careful study. Mr. Eissler's comprehensive little work on the *Hydrometallurgy of Copper* contained a great fund of information and might be studied to advantage by those interested in the treatment of low-grade ores.

The Neill leaching process was now undergoing its initial test at the Coconino mine in Arizona, and the results secured, while not final, owing to the short time during which the method had been employed, were highly satisfactory and very promising. This process was based on the solvent action of sulphurous acid upon copper oxides, carbonates and silicates, in the presence of water and an excess of the acid. The copper in the oxidized ores went into solution as a sulphite, which, though insoluble in water, was soluble in an excess of sulphurous-acid solution. When the excess of sulphurous acid was driven out by heat the cuprous sulphite was precipitated as cupro-cupric sulphite, a finely crystalline, dark red salt, carrying 49·1 per cent. of copper. The precipitate was smelted to blister-copper, practically without fluxing. The sulphurous acid was made from roast-gases, and would dissolve more copper than sulphuric acid, while being weaker, it did not dissolve other metals which fouled the solution and absorbed sulphur. The process, as applied to the oxidized ores of the Coconino mine, was briefly, as follows:—The ore is reduced by crushers and rolls to pass through a 20 meshes screen. The crushed ore is charged into a covered wooden tank containing wash-water, carrying a little sulphurous acid, 2 tons of this watery solution being used to 1 ton of ore. The washed sulphurous-acid gas is forced by an air-compressor through the charge in the leaching-tank, entering near the bottom. The process can be completed in one tank, but tanks in series are preferable. The pulp from the last leaching-tank is drawn into another tank and forced into a filter-press under air-pressure. The cuprous solution from the filter-press is conveyed by pipes to steam-tanks and the pulp remaining in the press is rewashed. The cuprous solution is heated by exhaust-steam to boiling-point, thus expelling the free sulphurous acid. Part of the copper is precipitated by gravity, under heat, the balance of the solution being drawn off into other tanks and the copper precipitated therefrom by scrap-iron or lime. The process is adapted to low-grade oxidized ores carrying neither gold- nor silver-values. It is also applicable to sulphide ores, previously roasted to drive off sulphur, and oxidized in the process. If gold- and silver-values are carried in the ore, these can be secured by chlorination after leaching.

Mr. H. LIPSON HANCOCK (Moonta Mines, South Australia) wrote that, with the low price now ruling for copper, the treatment of low-grade sulphide-ores in a profitable manner was a perplexing and difficult question, and required all the skill that could be centred on it in order to bring it to a successful issue. The friendly discussion on the subject should, therefore, be helpful in possibly throwing some further light on an important matter. As far as his experience in the matter of low-grade copper-ores was concerned, he endorsed Dr. Peters' comments relative to concentration preceding the wet treatment, where mining costs are very low. Mr. Muir's paper,[*] however, relative to this part of the subject seemed to present a rather extreme case, as his tailings appeared to be nearly as rich as the crude ore.

At the Wallaroo and Moonta mines concentration is chiefly accomplished by Hancock jigs, and the quantity passed over each

[*] *Trans. Inst. M.E.*, 1902, vol. xxiii., page 517.

machine per week (working one shift of 8 hours per day) totals from about 500 tons at the Moonta mines to a little over 1,000 tons at the Wallaroo mines, or equal to the rate of 1,500 to 3,000 tons each machine per week for full time. At the Moonta mines, the crude vein-stuff contains from 2·5 to 3 per cent. of copper, the tailings being reduced to under 1 per cent. At the Wallaroo mines, where the crude vein-stuff—the matrix of which is of a higher specific gravity—represents an average copper-content of a little over 4 per cent., the tailings contain between 1 and 1·5 per cent. of copper. Both these cases indicate a fair recovery of copper by concentration.

The tailings from the dressing machinery at both mines are passed on for cementation. At the Wallaroo mines, the latter work is not sufficiently advanced to describe the results, but at the Moonta mines, where over 1,000,000 tons are in process of treatment, copper is being extracted for £25 or £30 per ton. The concentration has paid all costs of these heaps being placed in position. In this way it has, in addition to profit, saved huge expenditure as capital-cost which would otherwise have been necessary, had the vein-stuff been placed to cementation direct, which capital-cost most likely would have prohibited the adoption of the hydro-metallurgical process.

It is, of course, well known that the sulphides of copper are scarcely dissolved with weak acid, and oxidation is required to make the work a success. Mr. Muir appears not to have expressed himself on this point; and experience here has shown that, without oxidation, the application of weak acid under the circumstances is of very little avail. The action of wind and weather opens the way for the use of acid, but the judicious and systematic application of ferrous salts, which gradually become converted more or less into the ferric state, extraordinarily hastens the process and promotes good oxidation. The addition to the liquors at this stage of a very small percentage of sulphuric acid dissolves the oxides of copper which have been formed, and tends to make a very satisfactory leaching proposition under the conditions described. If a larger proportion of iron-pyrites be associated with the sulphides of copper, the proposition should be an easier one for leaching, although it would militate more or less against the success of the concentration problem according to the proportion of the iron-pyrites present.

Of course, it is impossible to set out a process equally applicable to all cases, even should the copper-ores and matrix be somewhat similar, as even the minerals themselves appear to assume very different characteristics in different localities. He (Mr. Hancock) would suggest, however, that unless the grinding or roasting could be done very cheaply, it might be a heavy tax on the cost of extracting the copper, and as the grinding of the ore produces a slimy portion, this latter would be likely to interfere with a percolation-treatment of any depth, even though the roasting should improve the porosity of the mass. In the case where concentration precedes cementation, the finer slime-products are, of course, kept separate for purposes which are manifest.

Mr. C. FERNAU (Alston) wrote that, in discussing in general terms the treatment of low-grade copper-ores, a distinction must be drawn between copper-ores, where the copper is the main or only valuable metal present, and copper-ores, the value of which is enhanced by the presence of gold and silver in appreciable quantities. Where copper alone is the object of treatment, the problem is one that, only in exceptional cases, admits of more than one solution. Low-grade ores are those containing a maximum of 3 per cent. of copper.

Leaching requires either exceptional ores, decomposing in the presence of air and moisture, as in the case of the Rio Tinto ores; or exceptional circumstances in the way of the cheap supply of salt and acids. The capital required is large, and in the case of the Rio Tinto ores, the time required for the completion of the process is very great—10 years or more. After the first 10 years, the process no doubt becomes continuous, and about as much copper is obtained in any one year, as the amount of copper laid down. The percentage of extraction is in the long run very high, higher probably than by any other method, but the ores suitable for such treatment do not occur very generally. Nor does it often occur that mines are so situated as to permit of the cheap handling by means of acids that other leaching processes require.

Smelting is a method more generally applicable, and gives a fairly high extraction. Still, the loss is appreciable, and rarely is less than 0·50 per cent. of copper left in the slag. He (Mr. Fernau) had obtained slags from pyritous material, smelting with charcoal with a low blast, as low as 0·25 per cent., and had seen many million tons of slags in Spain, smelted by the Romans, containing only 0·12 per cent.   How the smelting was done, however, he had no accurate knowledge—probably in hearths not much larger than blacksmiths' forges, and with a blast produced by a bellows.   He considered that 0·50 per cent. in modern practice was not bad, and it would be unsafe to reckon on less.   No doubt a highly siliceous slag can be allowed in a blast-furnace, but, as a general rule, at the expense of fluidity in the slag.   In one case, he smelted with 45 per cent. of silica in his charge, and obtained slags containing 1·20 per cent. of copper.   This was reduced to about 0·70 per cent. on reducing the silica-contents to 28 per cent.   The higher loss, however, paid better, as the flux was very costly. Assuming a loss of 0·50 per cent., it requires cheap fuel and labour to make a 3 per cent. ore profitable, as well as cheap ore.   The cost of smelting, including fuel, flux and labour only will rarely be below the value of 1 unit of copper and the loss ½ unit, leaving only 1½ units for the cost of mining and reducing the matte to copper or sending to market.   As a general rule, the writer considers that a 3 per cent. ore is too low for smelting, that is to say, that for a low-grade ore, smelting is rarely possible.

There remains the process of concentration followed by smelting.   This process is more universally applicable, and is the one most generally in use.

Whether it will pay or not will depend on the nature of the ore, and the percentage of recovery.   With a good recovery, most 3 per cent. ores ought to be profitable, if the ore be regularly supplied, and the cost of mining be not exceptional.   Concentration can be accomplished in a well-designed plant at an exceedingly low cost per ton.   In one instance in his (Mr. Fernau's) experience, it was as low as 1s. 1d., and he is now concentrating lead-ore in the North of England, at about 1s. 6d. per ton.   The question to be determined is, therefore, how much the ore costs to mine, and what proportion of recovery can be obtained.   The recovery in the case of copper-ores is not so good as in the case of lead, tin, or even auriferous iron-pyrites.   When the ore is chalcopyrite, its flakiness occasions loss, and when bornite, its liability to slime.   When in the metallic state, as in the Lake Superior district, the recovery is very high indeed, but that kind of ore is limited in distribution.   On the whole, in his (Mr.

Fernau's) experience, concentration is the process most generally adapted for low-grade copper-ores.

If copper be not the only valuable metal present, the problem alters completely. Smelting and lixiviation then become much more serious competitors, as the losses in concentration, especially of silver-bearing ores, are sometimes very heavy. The field of enquiry also widens very much, and embraces considerations, which are out of place in a discussion of the treatment of low-grade copper-ores.

DISCUSSION OF MR. S. J. POLLITZER'S PAPER ON "A MEASURING-TAPE, AND ITS USE IN MINE-SURVEYING."*

Mr. JAMES HENDERSON (Truro) wrote that great credit was due to Mr. Pollitzer for his explanation of the use of a steel measuring-tape of an unusual length, and for the description which he gave of a very successful application of the same in a complicated survey.

Mr. D. G. KERR's paper on "Air-compression by Water-power: The Installation at the Belmont Gold-mine," was read as follows :—

* *Trans. Inst. M.E.*, 1903, vol. xxv., page 17.

AIR-COMPRESSION BY WATER-POWER: THE INSTALLATION AT THE BELMONT GOLD-MINE.

By D. G. KERR.

The water-power is situated in the township of Belmont, county of Peterborough, Ontario, about 3 miles in a north-westerly direction from the Belmont gold-mine. At the outlet of Deer Lake, there are falls and rapids which give a head of 75 feet in a distance of 1,600 feet; and still farther down the river, there is another drop of 25 feet. Deer Lake is about 4 miles long by 1 mile wide, and holds a splendid reserve of water for a dry season. The lake is fed by a chain of smaller ones, which extend northward about 100 miles, and it makes an ideal situation for a power-plant.

After the water-power was acquired, the question of electricity or compressed air was considered. The generation and transmission of electricity would have cost less at the power-house and to the mine; but it would have been necessary to have erected a motor-driven air-compressor at the mine to supply the rock-drills with compressed air and to apply motors to the hoists and engines. This would have brought the first cost of the electric installation to a higher figure than that of a large air-compressor plant; and besides, the attendance, etc., at the motor-driven air-compressor at the mine would swell the working costs. By installing a large air-compressor at the water-power and carrying the compressed air in pipes to the mine, branching it off in all directions to the shafts and mill, without having to make any alterations to any of the engines or hoists, all that was then required to be done was to shut off steam and turn on air to the engines, hoists and pumps without any loss of time while air was turned on at the power-plant. This left the steam-plant, boilers, etc., with all their connections as a reserve of power, in the event of anything going wrong with the air-power. And further, this arrangement permitted of the utilization of all machinery that was only two or three years old.

An important point was the erection of an air-compressor plant large enough to do the mining work for many years to come. As the underground requirements for air increased, more power could be developed at the falls by electricity to work the surface-machinery. There would also be ample time for considering the proper size of motor-driven machinery for handling the quantity of ore, as it would then be better understood.

The outlet of Deer Lake consisted formerly of two channels, 300 feet apart, through a fine-grained diorite-rock. The southern channel was closed with a concrete-and-cement-masonry dam, 85 feet long, 9 feet wide at the top, 16 feet wide at the base, and 15 feet high at the deepest part. On the top of the dam, small piers, 18 feet apart, are erected for bridging with timbers for a passage across. Underneath this bridge and over the top of the dam, the surplus water passes when stop-logs are placed in the slide-way on the northern dam. The northern dam is 75 feet long, with a slide, 25 feet wide, for the passage of logs. In front of the northern slide, is a forebay with a 30 feet rack; and this is the point where the water is taken out of the lake for the power, through an opening, 7 feet square, in the dam, with a gate on the side next to the lake. This gate is worked by means of rack-and-pinion wheels and a worm-shaft and wheel. The water-intake to the flume is reduced from the square to a cylindrical shape by means of steel-work, with flanges and fasteners for the wooden staves of the flume-pipe. On the top of the dam, behind the gate and going down into the water-entrance of the flume-pipe, is a man- or air-hole; and without this, the shutting-down of the gate at the dam, allowing water to pass through the wheel, would create a vacuum in the flume, causing a tendency to collapse or disturbance of the staves, resulting in much trouble and annoyance through leaks when the water was again turned on.

The flume-pipe, 1,550 feet long and 6 feet in internal diameter, is made of pine-staves, $2\frac{1}{2}$ inches thick and $6\frac{1}{4}$ inches wide, with radial edges and butt-joints, with saw-drafts cut 2 inches into both ends, in which is placed a steel-plate, $\frac{1}{4}$ inch wider than the stave, so as to embed it in the staves on both sides. No two joints come together, as they are made at irregular intervals, the staves being cut in 12, 14, 16 and 18 feet lengths, clamped with 2,000 steel-bands, $\frac{3}{16}$ inch by 2 inches, and fastened with grip-fasteners. The pipe is carried on timbers, 12 inches square, shaped to take the outside circle of the flume, and these bearer-timbers are spaced 8 feet apart, centre to centre. The steel-bands are spaced 3 inches apart at the lower end and 24 inches at the upper. There are two curves in the flume, each of 20 degrees. The staves, 6¼ inches wide, were too wide and rigid to be sprung into place on the top of the flume, and, in consequence, one-third of the top staves going round curves are made 3½ inches wide.

The bed for the flume is cut through ridges of rock for 900 feet from the dam, 3,960 cubic yards of rock-excavation being made by a steam-drill in the winter season. At the lower end, there is 217 cubic yards of stone-piers to carry the flume over a low piece of ground before arriving at the power-house, and inside the power-house a steel-tube takes the place of the wooden flume.

The cost of the wooden flume, made of pine, is 3 dollars (12s. 6d.) per foot, while the estimated price of a steel-flume, was 15 dollars (£3 2s. 6d.) per foot. The power-house building lies north and south, and the part which contains the air-compressor is 50 feet long and 40 feet wide. Southward of this, is the cooler-room, 43 feet long and 16 feet wide, and northward of the main part is the water-wheel section, 64 feet long by 35 feet wide. The water-wheel is a double bronze Leffell wheel, 50 inches in diameter, fitted with a double discharge ; and when running at 210 revolutions per minute, it has a capacity of 800 horse-power, and takes 7,500 cubic feet of water per minute. The water-gates of the wheel are made of cast-steel, and the casing of ½ inch steel-plates, with cast-iron heads. The water-wheel is carried on a steel-shaft, which is extended at one end for the transmission of the power, by means of a rope-pulley, 5 feet in diameter and 6 feet 4 inches across the face, and fitted with 30 grooves for 30 cotton ropes, 1¾ inches in diameter.

On the top of the wheel-casing is a dome, 2 feet in diameter and 10 feet high, with a valve, and just above this valve are two pipes, 12 inches in diameter, fitted with spring-valves, leading into the draught-tubes. This is an arrangement for the relief of any undue pressure from the water-ram, such as might be caused by the water-wheel gates on a long flume, through which water is travelling at a certain rate, being shut quickly. This arrangement takes the place of a stand-pipe; it costs less, and there is no danger from freezing, as it is placed under cover. On the wheel-case is a gauge shewing the water-pressure and head in feet, and on the draught-tube is a gauge giving the vacuum in inches.

The water-wheel, wheel-casing, etc., were furnished by the William Hamilton Manufacturing Company, Peterborough.

Underneath the wheel is the tail-sump, and from that the tail-race passing into the river. This was excavated out of solid rock to a depth of 20 feet, and it has cement-masonry walls with steel-beams and bolts, by which the wheel-casing is held in place. This tail-sump is carried westward underneath the wheel, in order that it may take the water from another wheel of 350 horsepower, and provision is made for its water-supply by means of a T piece on the steel part of the flume. When this second wheel is at work the water-velocity through the 6 feet flume will be increased to about 10 feet per second. It is intended to apply this 350 horse-power to the generation of alternating current by means of a direct-driven dynamo.

The horizontal compound air-compressor, made by Messrs. Walker Brothers, Wigan, has a high-pressure cylinder, 30 inches in diameter, a low-pressure cylinder, 48 inches in diameter, and a stroke of 4 feet. The cylinders are water-jacketted, with accessible inlet-valves, and fitted with metallic packing on the piston-rods. It is rope-driven by means of a pulley, 20 feet in diameter, and 6 feet 4 inches across the face, weighing 60,500 pounds, and built in sections on massive concrete-and-cement foundations, 14 feet high. When running at 65 revolutions, or a piston-speed of 520 feet per minute, it will have a capacity of 6,500 cubic feet of free air per minute. The low-pressure air-cylinder intake is connected by a branch pipe from the 3 feet pipe to the external atmosphere. This 3 feet pipe lies horizontally on the top of the low-pressure air-cylinder, one end going to the southern and the other to the western end of the building. The air is compressed in the low-pressure cylinder to a pressure of 30 pounds, and is then discharged through a pipe, 14 inches in diameter, to the inter-cooler; thence, after being cooled, to the high-pressure cylinder, from which after being compressed to a pressure of 100 pounds per square inch, it passes into the after-cooler. The inter-cooler and after-cooler are filled with brass tubes, through which flows cold water, and the compressed air passes and repasses over the outside of the tubes, and is cooled down to within 10 degrees of the temperature of the water used. In this cooling process, considerable moisture is deposited, as it is only by cooling the air to the lowest temperature attainable that a high extraction of moisture can be obtained.

The air leaves the after-cooler through a 12 inches pipe or ordinary oil-well casing, having fine screwed couplings and tested to a pressure of 600 pounds per square inch. An air-receiver is placed ½ mile away from the air-compressor, to collect any moisture, which may have passed the after-cooler; and this moisture is drawn off daily.

The 12 inches pipe-line, from the compressor to the mine, is 15,000 feet long. At the end of this pipe-line, at the mine, another air-receiver is placed to collect any moisture, which may have been carried into the pipe-line. The only time of the year when any moisture is expected to be carried this distance is when spring sets in, and the heat of the sun drives off any moisture from the inside of the pipe. This will be very little, as the air-receiver near the air-compressor lies in a low swamp, and the air-line leaving it for the mine has a gradual rise of 50 feet in 2,000 feet, thus draining the moisture back into the receiver. The pipe-line has 18 expansion-joints, and is mostly buried in sand, so as to maintain an equable temperature, and thus prevent expansion or contraction.

The plant started running in August, 1902, but has not yet been driven up to its full capacity, and the loss of pressure in transmission has not been determined. The loss of pressure, at present, is less than 1 pound, but using the full quantity of air, it is expected to reach 3½ pounds. The writer has found it impossible in the short time that the plant has been running to make a complete comparison, between summer and winter, of the temperature at which the air is taken into the compressor, the amount of moisture extracted, and the temperature of the water used in the coolers.

During the past winter, there has only been one shut-down owing to freezing, and it was caused by the moisture in the air-receiver, ½ mile from the compressor, being allowed to freeze, as it had not been drained off every day. The ice formed in a honey-comb form, until it interfered with the air-pressure at the mine, reducing it to 65 pounds, while there was 105 pounds at the air-compressor. After taking out the honey-comb ice, the air-receiver was covered with a shed, and banked so as to protect it from the intense frosts, and thus permitting of the moisture being drained off.

There has been no difficulty at the mine from freezing, except with a Corliss engine and a Duplex pump; but engines fitted with slide-valves gave no trouble, as the air is not expanded in the cylinders.

————

DISCUSSION OF MR. É. GOSSERIES' PAPER ON "THE GUIBAL FAN COMPARED WITH A DYNAMO."*

Prof. A. S. HERSCHEL (Slough) wrote that he had read Mr. Gosseries' paper, and appreciated its value very highly; but seeing that it turned on two things which were to him entirely problematical :—(1) That for a given rate of revolution of a Guibal fan there is a particular supply of air at which it shows a maximum water-gauge; with lower water-gauges for any greater or less air-flow; and (2) that the laws of air-flow under a driving-head are just the same as those of an electric current under a potential difference or electromotive force—he would have replied immediately that want of familiarity with fans and fan-theories for the last 10 or 12 years quite disqualified him from holding any opinion at all, either *pro* or *con*, upon the merits of the paper. But before writing, he read, with considerable scruples of conscience for having disregarded them before, two papers by Mr. H. W. Halbaum on "The Equivalent-orifice Theory treated Graphically"† and "An Extension of the Equivalent-orifice Theory;"‡ he also read again Prof. A. Rateau's "Remarks on Mr. M. Walton Brown's Report on Mechanical Ventilators,"§ and this now stirred him up to read Mr. M. Walton Brown's admirable exposé of theory and practice,‖ with more care and leisure than he had been able to give to it before, since it had evidently roused mining-engineers in every country to express their views most skilfully, and to discuss the fan-question in a series of most able papers. He was very sensible now, with the little reflection that was needed to understand these recent papers, how immensely far ahead of all

* *Trans. Inst. M.E.*, 1901, vol. xxi., page 568; and 1903, vol. xxv., page 44.

† *Ibid.*, 1900, vol. xx., page 404. ‡ *Ibid.*, 1901, vol. xxi., page 355.

§ *Ibid.*, 1902, vol. xxiii., page 472. ‖ *Ibid.*, 1899, vol. xvii., page 482.

his previous notions of it the art of mechanical ventilation had advanced.

In regard especially to Mr. Halbaum's two papers on the equivalent-orifice theory, the latter had done a great deal more than merely translate Mr. Murgue's views into graphic language. He had shown their meaning by a feat of composition (fairly rivalling, he considered, in manner and in matter, Fourier's work on heat-conduction), and his first paper, at least, was rhetorically as well as geometrically, a most brilliant performance. The rhetoric was indispensable to the establishment that he effects of Ohm's law in air-flow and in all streams of fluids, because rather large concessions and revolutions of one's natural ideas have to be made to take in the truth of the conception:—Air-head, square of discharge, and square of clear-voidance (rather than thin orifice) of an air-channel, answer in just the manner of Ohm's law to potential-difference, current-quantity and electric conductivity of an electric lead or cable. As regards the discharge or volume-output of the flow, no notice whatever need be taken of any velocity that it may possess at the point of measurement of its water-gauge and volume, since any such terminal velocity is just exactly one individual part of the whole series of similar velocities, which the head essentially produces all along the channel to be killed by friction; and it belongs, where the water-gauge was measured, to the friction of the mine. The air's faster inrush into the mouth of the fan is similarly one individual part of the fan's internal friction or resistance, since it is produced by a later and lower water-gauge than that measured at the drift-mouth, and would also be killed if the fan cannot transform some or much of it into back-pressure. The velocity of the discharge's escape into the open, in leaving the fan, is again just similarly a part of the fan's internal frictional resistance; because, although blades can produce pressure, only fluid-pressure can produce fluid-motion, and some head of pressure must therefore have been spent uselessly in the fan, on the discharge, to produce a velocity in it, which is bound to be killed in the open air. There may perhaps be tracts, at times along a mine, that act like embouchures or the expanding chimney of a Guibal fan, to recuperate, here and there, some of the flow's velocity and transform it into back-pressure to the relief of the final water-gauge, and they are, so far, little fans; but in so far as they are permanent and persistent, they may be lumped together into the mine's whole conductivity or temperament for the time being, and do not, like the mechanical fan, need a separate consideration.

These were the modern through-and-through views of a mechanical air-draught that he gleaned from the above-named papers; but besides skilful rhetoric and geometry Mr. Halbaum had also used examples (as when, in his second paper, he differed from Mr. Murgue on a rather fundamental point), when argument failed to prove his case, with great practical experience; so that his share in the thorough clearing up of the orifice-aspect, of the subject struck him (Prof. Herschel) as a remarkably original and substantial one. Considering that a thin orifice has a very uncertainly known *vena contracta*, and does not therefore flow full any more than a mine does, he (Prof. Herschel) could never imagine why Mr. Murgue chose it as a proper representation of a mine's discharge, or as an equivalent description, with its own peculiar obstruction, of the mine's impeding action. It seemed to be rather a simple practical form of scale by which to measure and describe the air's flow than anything theoretical, as it was hardly probable that the *vena contracta* of a thin orifice and the airway of a mine would constantly match each other well at all water-gauges. If a theoretically equivalent, and not merely a conveniently definite form of aperture, was really meant, it should be a full-flowing one, giving the discharge freely and fully through it with the theoretical gauge-speed, not through a fraction of it only of hardly fixed amount for all depressions. The choice of a true equivalent to the mine being a purely theoretical one, it seemed absurd to take a practical one, especially a not very perfectly explored practical one, to represent it; therefore, he (Prof. Herschel) never credited Mr. Murgue's equivalent orifice with any claim to theoretical significance. But once it was allowed that a mine's limitations to its air-discharge's free and full response to an aspirating-pressure was simply a limitation of aperture, it then seemed quite natural to suppose that the aperture which defines and denotes the limitation should flow as full as it could to give the resulting limited discharge through it at the velocity due to the water-gauge. He (Prof. Herschel) would call it the " free outlet," or voidance, as the square root of the conductivity, and its reciprocal, the " barrage " of the mine,

as the square root of its resistance.* This is just what Mr.
Halbaum had done; for though he termed his papers expositions
and applications of the orifice-theory of mechanical ventilation,
he nowhere used the name "orifice" for his constants, M and
F, that denote them for the mine and the fan throughout his
papers, and the consequence was that one saw at once through the
very perfect theory that he was trying to expound and illustrate:
while he (Prof. Herschel) had read through Mr. Murgue's book,†
at different times, treating of the substitution of an airway by
the technical device of a thin-plate orifice, without being ever
led to suspect that with its use he was clearing the way (as
had thereby come to pass) for a recognition of Ohm's law in
ventilation-currents. He (Prof. Herschel) had often tried to
find a solution of that problem, but always unsuccessfully, as
without the useful hint of an equivalent free orifice in an air-
flow, no one would ever dream of taking the square of the discharge as the quantity of flow in a current produced by a driving
head through the square of a limiting free aperture as a conductivity, or against the reciprocal of that square as a resistance. If Mr. Murgue had set forth that consequence of his
proposition, divesting it for that end of its technical shape, and
expressing it more simply and abstractly, he (Prof. Herschel)
thought he could hardly have overlooked the appearance of so
notable a discovery in Mr. Murgue's book as the Ohm's law
one, nor, he thought, would Mr. Murgue have failed to mention it, if that all-important and invaluable consequence and
conclusion from his proposition had been obvious to him;
but he would read Mr. Murgue's work on mechanical ventilation again more carefully, and see whether it promulgated
the law distinctly.‡ Mr. Halbaum, himself, nowhere stated it

---

* A note of some rather more suitably adapted terms for conveniently
denoting and dealing with the several new quantities that require to be considered
in the conduction-like view of an air-flow's maintenance, will be furnished hereafter, to more fully elucidate and illustrate the principles of that somewhat novel
and surprising aspect of the subject of aërial and other fluid currents.

† "Essai sur les Machines d'Aérage," *Bulletin de la Société de l'Industrie
Minérale*, second series, 1873, vol. ii., page 445; 1875, vol. iv., page 747 (1877,
vol. vi., page 210); and 1880, vol. ix., page 5.

‡ The writer must here correct a mistake, which was indeed a pardonable
one, in the pronunciations which he was led to make originally in the above
remarks (since they were not, in fact, written originally with a view to their being
published): that the work above said to have been read through at different
times, which his memory imperfectly recalled as having, many years ago, afforded
him most valuable instruction in the theory of mine ventilation, was not actually

generally, but simply accepted it, and used and applied it unreservedly throughout his papers.

In a supplementary note to Mr. Gosseries' paper on a case of resemblance between a Guibal fan and a dynamo-machine, Prof. Macquet stated that if u be the velocity of discharge produced by a water-gauge depression through a hole in a thin plate of area, S, Su is the volume of the discharge.[*] Now in his *Applied Mechanics*,[†] on flow of gas from an orifice, Prof. W. J. M. Rankine wrote:—"Let the pressure of a gas within a receiver be p_1, and without, p_2; let A be the effective area of an orifice with thin edges; that is, the product of the actual area by a co-efficient of contraction, whose value is 0.6, nearly . . . ;" and further on:—"For small differences of pressure, such that p_2/p_1 is nearly equal to 1, the following approximate formula may be used . . . $v^2/2g = p_0\,\tau_1\,(p_1 - p_2)/\zeta_0\,\tau_0\,p_1$" for the velocity v, "and the flow of volume is $Q = vA$ at the contracted vein." This showed that engineers may disagree about the properties of flow through a thin orifice, when they write thus about it; and be the contracted vein or equivalent free aperture of a thin-plate orifice what it may, at different depressions (for it is hardly likely to be constant) no one who has seen smoke-rings produced in that way round the issuing jet, will be ready

Mr. D. Murgue's "Essai sur les Machines d'Aérage," in the *Bulletin de la Société d'Industrie Minérale*, 1873-1880, at all, nor the translation into English by Mr. A. L. Steavenson (which he had not seen yet) of Mr. Murgue's work on *The Theory and Practice of Centrifugal Ventilating-machines*, but the very useful handbook, *Ventilation des Mines* (Mons, 1875), by Mr. A. Devillez. He must therefore retract the quite mistakenly presented estimate expressed above of Mr. Murgue's really most complete indication of the law of resistance and conduction found to be always governing aërial currents, since it is most definitely and distinctly laid down as he found it represented and propounded in the first division of Mr. Murgue's "Essai," published in 1873; for, in a note there given, on the fitting capabilities and mode of calculation of the equivalent orifice of an air-flow, the two different descriptions of resistance, by surface-friction and by constriction of a current-channel, are shown to act quite similarly in impairing in a constant manner in a given channel, the driving head's power of producing squared discharge of volume through it; and the law thus shown to hold in two, at least, of the most important forms of current-friction (though sudden deflections also form another serious source of obstruction to a flow), in a fluid at least (like air in ventilating currents), of practically constant density, provides at once the fullest and a perfect means of solving problems of air-flows' distributions through ramified and varied channels by exactly the same rules as those which Ohm's law supplies to answer similar questions of distribution of galvanic currents. The same law is recognised and used, but in a less general way, as it is only proved and affirmed exactly with regard to the effects of surface-friction, without numerical estimates of other forms of friction, along any series of branching and successive airways, in Mr. Devillez' rather later book.

[*] *Trans. Inst. M.E.*, 1901, vol. xxi., page 574.

[†] Article 637, page 581.

to believe that the outflow of a gas from a thin-plate orifice is as free from eddies that affect water-gauge depression, as its ideal outflow is; and hence not the whole but only an unconsumed part of the working pressure can act over the area of the orifice to produce velocity through it, or the full speed through the orifice is less than the theoretical speed for the depression, or the discharge is necessarily less to some (not well-known) extent than the free-flow discharge with the full theoretical air-speed through the whole open area. It is a pity that mining engineers have introduced this idea of a thin-plate orifice into the question of fluid-flow resistance, as a right conception of such a resisted flow's natural discharging power may easily be obscured and made difficult to apprehend in consequence.

Prof. Macquet spoke also of a Pitot tube.[*] He (Prof. Herschel) had seen it in use, and the process of summing up its readings for a mean resulting water-gauge; and he was much inclined to think that considerably more errors were likely to occur in its use than might arise by simpler means without it; especially when a sufficient ball of cotton-waste or wool was tied over the open end of a water-gauge tube, which performed exactly the part of a Pitot tube, without demanding a multiplicity of water-gauge readings and averaging of figures.

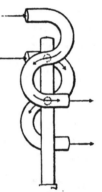

FIG. 1.

Possibly, two brass tubes might be bent like S hooks, and fixed as shewn in Fig. 1, on either side of a straight brass mouth-tube (closed at its upper end) attached to the water-gauge tube; both the straight and curved tubes having pretty wide-open side-holes for communication, where they meet and cross, and are there soldered to each other. An air-draught from the left, would then enter the two S tubes at their top-ends, and would leave them at the bottom; and as their two currents would oppose each other where they cross and communicate with the straight water-gauge mouth-tube, they would baffle each other there in their action of aspiration, or otherwise, on the air in the straight tube, and would obviate the necessity of pairs of readings and averages with a Pitot tube; but the device would be a poor

* *Trans. Inst. M.E.*, 1901, vol. xxi., page 574.

make-shift, he suspected, for the well wool-padded end of an ordinary water-gauge tube.

The radial velocity, which Prof. Rateau considered,[*] in his comments on the report on " Mechanical Ventilators," should be superadded to the blade-tip speed in some fans (that is, open-running ones) in calculating their theoretical effects, can hardly be great and surely must scarcely exist at all in a cased-in Guibal fan, so that there it must be negligible. In open-running fans, it might more depend, perhaps, as a correction to the blade-tip speed, on the volume passed by the fan than on slight forward curvatures of the blade-tips in some fans. He (Prof. Herschel) did not follow Prof. Rateau's mathematics there, and in the rest of his comments, from having failed to read his other papers.

Dr. HENRY STROUD (Durham College of Science) wrote that it might not be superfluous to draw the attention of members to the article on " Warming and Ventilation " by Dr. W. N. Shaw, in Messrs. Stevenson and Murphy's *Hygiene and Public Health*, in which was given a method of treatment suggested by Mr. Murgue's work on *The Theory and Practice of Centrifugal Ventilating-machines*, translated by Mr. A. L. Steavenson. In that the resistance of a mine in a given condition was constant, whatever the flow, there was an analogy to the electrical case where the resistance of a conductor in a given condition was constant, whatever the current; but since the square of the air-flow through a mine was proportional to the head, it seemed doubtful whether any further assistance could be obtained by attempting to pursue the analogy.

Prof. A. MACQUET (École des Mines du Hainaut, Belgium) wrote that Prof. A. S. Herschel began by avowing himself absolutely unqualified to express an opinion for or against Mr. Gosseries' paper, the great merit of which he is, however, pleased to recognize. This does not prevent him from embarking upon the most curious and fantastical definitions, of the equivalent orifice (as he conceives it); of the pressure and motion of a fluid; of the recuperation of energy due to the Guibal expanding chimney; of the orifice in a thin plate; and of the contraction-vein, etc. Next, Prof. Herschel took him (Prof. Macquet) personally to task, because he wrote, $Q = Su$, terming Q the volume of flow through the section, S, at a velocity, u, this volume being reckoned at

[*] *Trans. Inst. M.E.*, 1902, vol. xxiii., page 475.

the pressure proper to the fluid in the section, S.   He agrees,
however, with Prof. Rankine, who writes $Q = vA$, for the volume
of flow under the same hypothetical conditions, that is, through
the contracted section of the effective surface, A, at a velocity, v.
In what do these formulæ differ? He (Prof. Macquet) failed
to see it.   Wherefore, he took the liberty of attaching no
significance to the Professor's remark that "this showed that
engineers may disagree about the properties of flow through a thin
orifice, when they write thus about it." He preferred to leave
Prof. Herschel on the watch for " smoke-rings produced . . .
round the issuing jet " of a fluid through " a thin-plate orifice,"
and to leave him to the pity with which he is filled for the
engineers, adepts in the theory of orifices in thin plates.

Nor did he (Prof. Macquet) propose to insist upon the remarkable instance cited by our esteemed controversialist, who has
had the opportunity of seeing the Pitot tube in use, as he
thought that such opportunities were common among mining-
engineers. Engineers, at all events, would hardly forgive him
for lecturing to them on matters of which he did not understand a
syllable, even were it on a certain " Pitot tube."

DISCUSSION OF MR. W. C. BLACKETT'S NOTE ON AN "IMPROVED OFFTAKE-SOCKET FOR COUPLING AND UNCOUPLING HAULING-ROPES."[*]

Mr. J. H. RONALDSON (Johannesburg) wrote that he was
interested on seeing that the coupling for haulage-ropes was to
all intents and purposes the same as one that he designed and
used in 1893 at Trabboch colliery, Ayrshire, for tail-ropes: it
worked well and expeditiously.   It was designed in the first
instance to allow the couplings to pass through some particularly
narrow sheaves, that gave trouble with the ordinary coupling.

Mr. W. C. BLACKETT said it would have been more interesting
had Mr. Ronaldson favoured the members with a sketch of his
alleged similar offtake socket.

---

* *Trans. Inst. M.E.*, 1902, vol. xxiv.. page 61.

DISCUSSION OF MR. E. ROUBIER'S PAPER ON "THE MAX ELECTRIC MINING LAMP."*

Mr. W. O. WOOD (South Hetton) wrote that the Max electric mining lamp did not appear to be any improvement on the Sussmann lamp, 1,200 of which had been in daily and successful operation at Murton colliery for some years. The Max lamp was 1½ pounds heavier than the Sussmann lamp, while the light was no better, nor was the period of burning longer, as the Sussmann lamp easily burnt for 10 hours, if required.

Mr. E. ROUBIER (London) wrote that he could only congratulate the makers of the Sussmann lamp, and perhaps Mr. Wood would be able to state the cost of maintenance and the life of this type of lamp, which is worth consideration, but as apparently Mr. Wood has neither seen the Max accumulator, nor the lamp in operation, he could not see that much weight reposed on his (Mr. Wood's) opinion. He (Mr. Roubier) thought that the Sussmann lamp weighed 4½ pounds, which is only one pound less than the weight of the Max lamp: anyhow, as the Max lamp gave a light of 1½ candlepower for 12 hours, while, he believed, the Sussmann gave 1 candlepower for 10 hours only, the extra 40 per cent. of light was well worth 20 per cent. of enhanced weight.

———

* *Trans. Inst. M.E.*, 1903, vol. xxv., page 36.

NORTH STAFFORDSHIRE INSTITUTE OF MINING AND MECHANICAL ENGINEERS.

GENERAL MEETING,
HELD AT HANLEY, MAY 11TH, 1903.

MR. A. M. HENSHAW, PRESIDENT, IN THE CHAIR.

The minutes of the last General Meeting were read and confirmed.

The following gentlemen, having been previously nominated, were elected:—

MEMBER—

Mr. WILLIAM TELLWRIGHT, Sneyd Colliery, Burslem.

ASSOCIATE MEMBER—

Mr. R. S. SHOWAN, Albert Terrace, Wolstanton.

ASSOCIATES

Mr. T. H. EDWARDS, Knypersley Villas, Biddulph.
Mr. JOHN GREEN, 270, High Street, Alsager's Bank.
Mr. J. W. MILNER, Biddulph Valley Ironworks, Stoke-upon-Trent.

DISCUSSION OF MR. JAMES ASHWORTH'S PAPER ON "THE GRAY TYPE OF SAFETY-LAMP."[*]

Mr. JAMES ASHWORTH said that he had had a few lamps made with one air-tube and two standards, in accordance with suggestions that had been made when his paper was read. One of these lamps had been sent to Mr. E. B. Wain, who had informed him that his firemen were satisfied with the lamp, which gave a good light. The other lamp, which had been sent to Mr. A. M. Henshaw, was a collier's lamp.

The PRESIDENT (Mr. A. M. Henshaw) said that the lamp which had been sent to him by Mr. Ashworth gave a good light,

---

[*] *Trans. Inst. M.E.*, 1902, vol. xxv., pages 62 and 162.

probably due to the breadth of the flame and, to some extent, to the reflecting power of the broad tube at the back. The lamp became very hot, and the conical glass was smoked in rather a short time, but with these exceptions he formed a high opinion of the lamp.

Mr. E. B. WAIN said that he had had the particular lamp, which had been mentioned, in his possession for about two months. He had had it tested by over 30 firemen. Each man spoke highly of it, and stated that it gave the best light and was a most sensitive lamp for detecting gas. Only one of the men made any complaint, and that was as to the heat of the lamp, which had been referred to. He (Mr. Wain) believed that it was a most admirable lamp, particularly as a fireman's lamp for gas-testing.

The PRESIDENT (Mr. A. M. Henshaw) asked Mr. Wain whether the glass became smoky.

Mr. E. B. WAIN replied in the negative, and thought that it depended on the oil that was burnt and the method of trimming the wick. If the wick was properly trimmed, the flame would not spire up in the lamp.

Mr. JAMES ASHWORTH said that " mineral colza " oil should be used, flashing at about 300° Fahr., and it would not smoke the glass. The lamp would not show less than 2 per cent. of gas. In Dr. Clowes' test-box at Nottingham, the lamp showed a cap with 2 per cent; but practically the caps at 2 and 2½ per cent. were alike. The light was of about 1 standard sperm candlepower.

Mr. J. T. STOBBS asked whether, in making a gas-test, there was any difference between the flat-top flame and the ordinary flame. The statement that the flame-cap did not show any difference between 2 and 2½ per cent., was of a negative character and not quite satisfactory. In quantitative testing for gas, the flame of the safety-lamp could not accurately determine whether the gaseous mixture contained 1 or 2 per cent. of gas, because of the unequal diffusion of the gas and the air. Could Mr. Ashworth give further information as to the behaviour of the flame in the presence of gas?

Mr. James Ashworth remarked that the gas-cap on all flames was of conical shape. In testing for gas, they could not take two lamps and state, if the caps were $\frac{1}{2}$ and $\frac{1}{4}$ inch high respectively, that different percentages of gas were indicated; because no two lamps of different construction gave the same cap with the same percentage of gas. The size and the density of the cap varied with the heat of the flame; and he did not think that it was possible to see a cap indicating less than 2 per cent. of gas, with a lamp of this class.

Mr. W. N. Atkinson (H.M. Inspector of Mines) stated that the Gray lamp was a very excellent one. The removal of the thick air-tubes would improve the lighting quality of the lamp; and, if there was no mechanical difficulty, the fitting of a broad tube at the back would be an improvement. He was surprised to find that the lamp was fitted with a screw-lock, as it was not a safe lock, and it could be easily opened and closed without detection. He thought that lamp-manufacturers should refuse to make a lamp with a screw-lock.

Mr. Ashworth said that the lamp produced was a composite one, which had not come out of a lampmaker's workshop. The lamp that would be put on the market would have a proper lock on it.

Mr. W. N. Atkinson said that some lamps were manufactured with very thin glass, about $\frac{1}{8}$ inch thick.

Mr. J. Lockett said that some time ago Mr. Ashworth had sent him a Gray lamp with four fixed standards, and he was satisfied with it. The only defect that he found in that lamp, compared with others, was that in going down the shaft the light was extinguished by the current. The light of the No. 2 Gray lamp was satisfactory.

Mr. J. T. Stobbs thought that it was possible for thin glass to be better annealed than thick glass, and thus to be more able to withstand the impact of a blow. In testing the illuminating-power, the lamp ought to be revolved, so that the average candle-power might be determined.

Mr. J. Lockett thought that there was something in what had been said, as to a flat wick giving more light on the flat side than on the edge.

Mr. E. B. WAIN said that a man who used a lamp in a mine did not rotate it, but he fixed it where he wanted to use it. As to the luminosity or otherwise of the flame, there was a large percentage of non-luminous flame, looking at the flat side of the wick, but looking at the edge, there was practically a luminous flame down to the wick-tube. He thought that there was the same proportion of luminous flame produced from the edge as from the flat side of the wick, especially as the edge of the flame was surmounted by a considerable depth of luminous flame.

Mr. A. HASSAM asked whether the No. 2 Gray lamp could be lighted electrically.

Mr. JAMES ASHWORTH said that was a detail which could be attached whenever the purchaser required it.

Mr. G. A. MITCHESON asked whether the glass should be tightly or slackly screwed up, and whether a lamp was safe if it could be " blown " at the bottom or the top of the glass.

Mr. JAMES ASHWORTH replied that the glass ought to be perfectly tight with ordinary screwing up. He was opposed to the savage way in which glasses were sometimes screwed up, and, to prevent overscrewing, the rings were now formed with coarse threads. It should not be possible to rotate the glass. If asbestos was used, about $\frac{1}{16}$ inch thick, it should prevent the lamp from being " blown."

Mr. E. B. WAIN stated that it would be possible for a lamp glass to be 0·01 inch slack, and it would still be incapable of passing the flame.

Mr. A. HASSAM said that the glass ought not to be slack.

Mr. J. LOCKETT considered that it was a mistake to screw a lamp together, so that the glass could not be moved by the thumb and finger.

The PRESIDENT (Mr. A. M. Henshaw) thought that the lamp-glass ought to be fixed, and it ought not to be possible to move it with the thumb and finger. If there was a ring of a proper thickness of asbestos at the bottom of the glass, and the brass-ring was screwed sufficiently so that the glass could not be moved by the finger and thumb, the lamp was as near as possible to perfection in that respect.

Mr. G. A. MITCHESON said that he did not think it right to screw up the glass too tightly. A thick asbestos-ring should be used; the glass should be just free to move; and there would be less numerous breakages of glasses.

Mr. E. B. WAIN said that glasses which could be turned with the finger and thumb when cool would be rigid when heated. It was sufficient if the glass would turn round gently when cold. Asbestos-rings were not very elastic, and more glasses were broken by expansion than from any other cause. Some of the speakers appeared to have overlooked the fact that the longer the opening, the larger the area permissible without any danger of passing flame.

Mr. J. T. STOBBS said that the question depended upon the different rates of expansion of brass and glass. The brass standards would expand more than the glass for a given rise of temperature. It was possible for a glass to be tight in a cold intake, and it was possible for a previously tight glass to become slack by heat.

Mr. JOHN GREGORY desired to emphasize the danger of having loose standards in a lamp, as they might cause a true glass to fit badly.

The PRESIDENT (Mr. A. M. Henshaw) said that he was surprised that after so many years of study they had not got a better safety-lamp than they had to-day; and that the best lamp only gave about 1 candlepower of light. If they wished to enhance the safety of life in mines, and if they wanted to work mines economically, it was necessary to give the miners a better light. If a safety-lamp could be devised giving 3 or 4 instead of 1 candlepower, the mines would be worked more safely and more economically than they were to-day.

Mr. ASHWORTH observed that the amount of light had its limit in a safety-lamp. If they increased the light, they increased the heat, and the outlet-temperature must be so low that it would not inflame coal-dust. He could give the members a lamp of almost any candlepower, but they must not limit the size of such a lamp. If they got increased light, it would be accompanied with increased heat.

THE MINING INSTITUTE OF SCOTLAND.

ANNUAL GENERAL MEETING,
HELD IN THE HALL OF THE INSTITUTE, HAMILTON, APRIL 9TH, 1903.

MR. DAVID M. MOWAT, VICE-PRESIDENT, IN THE CHAIR.

The minutes of the last General Meeting were read and confirmed.

The report of the Council was read as follows:—

ANNUAL REPORT OF THE COUNCIL, 1902-1903.

In presenting the semi-jubilee report of the Mining Institute of Scotland, the Council have to congratulate the members on the completion of the twenty-fifth year of the Institute, and to state that its affairs are in a flourishing condition.

The Institute had its origin at a meeting held, in Hamilton, on January 24th, 1878, when ten colliery-managers of the district met, and adopted a scheme which had been suggested by one of their number, Mr. James T. Forgie, for the establishment of a mining institute. Of these ten, five are still members. Rules were formulated and office-bearers appointed, and the first general meeting of the West of Scotland Mining Institute was held, at Hamilton, on April 4th, 1878. In March, 1880, the name was altered to that which the Institute now bears.

At the end of the first year, 46 members had been enrolled, 208 at the end of the second year, 295 at the end of the third year, and there was a steady increase until 1892, when the members numbered 546. From that time until 1901 there was a slight yearly falling-off to 449; and since then there has been an increase, until, at present, the number is 482.

The Presidents during the 25 years have been as follows:—

Mr. Gilbert B. Begg, one year	1878-1879
Mr. Ralph Moore, four years	1879-1883
Mr. James McCreath, two years	1883-1885
Mr. James S. Dixon, three years	1885-1888
Mr. John M. Ronaldson, three years	1888-1891
Mr. J. B. Atkinson, three years	1891-1894
Mr. George A. Mitchell, four years	1894-1898
Mr. James T. Forgie, three years	1898-1901
Mr. James S. Dixon, one year	1901-1902
Mr. Henry Aitken, one year and continuing	1902-1903

While the members have been much indebted to all the Presidents for their kindness in devoting themselves to the interests of the Institute during their respective terms of office, it is only fitting that special mention should be made of the late Mr. Ralph Moore, who, in the midst of his onerous duties as H.M. inspector of mines, took great pains to advance the cause of the Institute in its early years and to see it established on a firm basis. In this he was ably seconded by Mr. J. T. Robson, then his assistant, and now H.M. inspector of mines for the Swansea inspection-district, whose labours on behalf of the Institute in its early stages merit special reference.

With four vice-presidents and twelve councillors, elected annually, the business of the Institute has been conducted in a manner which has met with the approval of the members; and it may not be amiss to mention that the work entailed on the members of Council is considerable.

The office of Secretary was filled by Mr. James T. Forgie for fully a year, by Mr. James Gilchrist for nearly 4 years, and by Mr. James Barrowman for 20 years since then; while the Treasurer was Mr. Forgie for fully a year, followed for a short time by Mr. Gilchrist, the late Mr. Michael Ross for 6 years, and Mr. Archibald Blyth for the 17 years since then. Mr. James Blackwood has been Librarian for 16 years.

The Institute has existed coincidently with a progressive period of the coal-trade. Since 1878, the output of the United Kingdom had increased from 132,000,000 to nearly 230,000,000 tons, that of Scotland from 17,000,000 to 34,000,000 tons and that of Lanarkshire from 10,000,000 to 16,500,000 tons.

The opening-up of numerous collieries in Scotland gave scope for the exercise of that talent of inventiveness and adaptation of means to ends, with a view to the best practical and financial results, that the Institute is meant to foster; and there can be no doubt that the quickening of interest in the best modes of working in all departments of mining operations has been an accompaniment of the Institute's progress, and the Institute can fairly place itself among the influences that have brought the newer undertakings in Scotland well to the front as examples of colliery-enterprise.

The *Transactions* of the Institute remain a record of the work accomplished during these twenty-five years, and have been a great benefit to all concerned in the furtherance of the objects of the Institute.

The federation of the Institute with The Institution of Mining Engineers, which took place on August 1st, 1893, widened the scope of its members, bringing under their notice examples of mining in all parts of the world, and providing, in the *Transactions* of the Institution, a valuable record of the most varied mining practice.

With reference to the proceedings of the past year, the Council have to report that the number of members on the roll at this date is as follows:—

Honorary Members	4
Life Members	9
Life Associate Member	1
Members (subscription £2 2s.) ...	159
Members (subscription £1 5s.) ...	235
Associate Members	23
Associates	10
Students	18
Non-federated Life Member ..	1
Non-federated Members (subscription £1 1s.)	14
Non-federated Members (subscription 10s.6d.)	8
Total ...	482

These figures compare with those of last year as follows:—

On the roll at April, 1902	461	
Added during the year	52	
Total ...		513
Died	7	
Retired	10	
Cut-off through non-payment of subscriptions	14	31
At present on the roll ...		482

This brings out the substantial net increase of 21 members during the year; and this rate of increase can be easily maintained, or even quickened, if members will recommend the Institute to those who, while eligible, are not cognizant of its advantages.

The following papers have been read during the year:—

"Four Old Labour-saving Ideas." By Mr. Henry Aitken.

"The Working of Contiguous, or nearly Contiguous, Seams of Coal." By Mr. Thomas Arnott.

"Description of Underground Haulage at Mossblown Colliery, Ayrshire." By Mr. James Baird.

"Slips in a Sandbank." By Mr. James Barrowman.

"The Technical Instruction of Working Miners." By Mr. A. Forbes.

"The Working of Contiguous, or nearly Contiguous, Seams of Coal." By Mr. John Hogg.

"Electrical Coal-cutting Machines." By Mr. Fredk. W. Hurd.

"Douglas Colliery." By Mr. Douglas Jackson.

"The Dysart, Wemyss and Leven Coal-field, Fifeshire." By Mr. R. Kirkby.

"Sinking on the Seashore at Musselburgh." By Mr. R. Martin.

"Apparatus for Controlling Railway-wagons while Loading at Colliery-screens." By Mr. J. D. Miller.

"The Working of Contiguous, or nearly Contiguous, Seams of Coal." By Mr. Thomas Moodie.

These papers are of the greatest practical use, and, with the discussions thereon, form a valuable addition to the *Transactions* of the Institute. The value of the papers read would be further enhanced if more members would take part in the discussions, either in person or by writing.

The annual excursion was the occasion of a most enjoyable visit to Douglas colliery, which exhibits some special features of modern mining-plant and fittings. A description of the colliery appears in the *Transactions*.

On the date of one of the General Meetings in Glasgow an opportunity was afforded the members of visiting the electric-power station at Pinkston of the Glasgow Corporation Tramways, and a large number availed themselves of it. A description of this station is printed in the *Transactions*.

At the meeting of the British Association for the Advancement of Science held in Belfast in September last, your delegate attended the Conference of Delegates of Corresponding Societies, and has to report to the same effect as he did last year.

No additions have been made to the list of Societies and technical periodicals with which the Institute exchanges *Transactions*, the number still being 64.

The donations to the library for the past year, in addition to those received by exchange, are as follows:—

DONORS.	DONATIONS.
Executive Committee of the International Engineering Congress, Glasgow, 1901.	Proceedings of the Congress:— Section I.—Railways. Section II.—Waterways and Maritime Works. Handbook of Glasgow Industries. Report of the Proceedings and Abstracts of Papers.
Mr. John M. Ronaldson.	Mines Reports and Statistics for 1901, 12 volumes.
Mr. J. T. Robson.	Mines Report for Swansea District for 1901.
Mr. John D. Miller.	Four Views of the Miller-Yates Wagon-controller.
Proprietors of the *Colliery Guardian*.	Coal-cutting by Machinery in America. Coal-cutting by Machinery in the United Kingdom.
Mr. Robert W. Dron.	The Coal-fields of Scotland.

A number of valuable books have also been added to the Library by purchase.

The Treasurer's accounts shew that the Institute is in a good financial state.

There have been seven meetings of the Council during the year.

————

The report was unanimously adopted.

————

The SECRETARY read the annexed abstract of the Treasurer's accounts for the year, which was adopted.

————

Messrs. Thomas Jameson and William Howat were thanked for their services as auditors. ————

The following gentlemen were elected:—

HONORARY MEMBER—

Mr. JAMES HASTIE, 4, Athole Gardens, Uddingston.

MEMBERS—

Mr. ALEXANDER BAILLIE, Bankend, Coalburn.

Mr. ROBERT BARROWMAN, Equitable Coal Company, Barakar, East Indian Railway, Bengal, India.

Mr. ROBERT BONAR, Home Farm Colliery, Hamilton.

Mr. JAMES CONSTABLE, Woodside, Broxburn.

Mr. WILLIAM H. MURRAY, JUN., Ellenbank, Broxburn.

Mr. CHARLES CARLOW REID, Fife Coal Company, Limited, Cowdenbeath.

Mr. JAMES STRACHAN, Westburn Colliery, Cambuslang.

Mr. JOHN WHITE, Banknock, Hollandbush.

ASSOCIATE—

Mr. DAVID CUNNINGHAM, Prestongrange Colliery, Prestonpans.

THE TREASURER IN ACCOUNT WITH THE MINING INSTITUTE OF SCOTLAND.
FOR THE SESSION 1902-1903.

RECEIPTS.

		£	s.	d.
To Balance brought forward	...	333	16	3½
" Subscriptions—				
1 Member, Session 1900-1901	at 42s. 0d.	2	2	0
10 Members, Session 1901-1902	" 42s. 0d.	21	0	0
1 Associate Member, "	" 42s. 0d.	2	2	0
5 Members, "	" 25s. 0d.	6	5	0
142 Members, Session 1902-1903	" 42s. 0d.	298	4	0
1 Member, "	" 40s. 3d.	2	0	3
1 " , "	" 12s. 6d.	0	12	6
2 Members, "	" 21s. 0d.	2	2	0
223 " , "	" 25s. 0d.	278	15	0
23 Associate Members, "	" 42s. 0d.	48	6	0
10 Associates, "	" 25s. 0d.	12	10	0
17 Students, "	" 25s. 0d.	21	5	0
17 Non-federated Members, "	" 21s. 0d.	17	17	0
8 " , "	" 10s. 6d.	4	4	0
3 Members, Session 1903-1904	" 42s. 0d.	6	6	0
1 Member, "	" 29s. 6d.	1	9	6
2 Members, "	" 25s. 0d.	2	10	0
1 Member, "	" 15s. 0d.	0	15	0
1 Non-federated Member, "	" 21s. 0d.	1	1	0
1 " , "	" 10s. 6d.	0	10	6
" Donations to Library	...	3	18	6
" *Transactions*, &c., sold	...	4	16	4
" Rents of Halls, &c.	...	6	5	0
" Interest on Deposit-receipts	...	7	17	9
		£1,086	10	7½

PAYMENTS.

	£	s.	d	£	s.	d.
By The Institution of Mining Engineers				469	3	6
" Printing and Stationery				32	12	7
" Books and Book-binding				13	11	4
" Rent of Halls				29	13	0
" Gas and Assessments				5	19	0
" Stamps, Carriages, and Telegrams				15	19	7½
" Sundry Payments				2	13	7½
" Cleaning Halls				8	4	0
" Expenses of Council				23	9	6
" Salaries				96	8	0
" Cash in Bank	385	17	0			
" " Treasurer's Hands	2	19	5¼	388	16	5¼
				£1,086	10	7½

March 23rd, 1903.—Examined, compared with vouchers, and found correct.

THOS. JAMIESON, AUDITORS.

ELECTION OF OFFICE-BEARERS, 1903-1904.

The CHAIRMAN (Mr. David M. Mowat) declared the following office-bearers elected for the session 1903-1904:—

PRESIDENT.

Mr. HENRY AITKEN.

VICE-PRESIDENTS.

Mr. JAMES HAMILTON.	Mr. ROBERT THOMAS MOORE.
Mr. ROBERT McLAREN.	Mr. DAVID M. MOWAT.

COUNCILLORS.

Mr. THOMAS ARNOTT.	Mr. JOHN MENZIES.
Mr. ADAM BROWN.	Mr. PETER MILLIGAN.
Mr. ROBERT W. DRON.	Mr. THOMAS STEVENSON.
Mr. DOUGLAS JACKSON.	Mr. THOMAS THOMSON.
Mr. RICHARD KIRKBY.	Mr. THOMAS TURNER.
Mr. JAMES M'PHAIL.	Mr. WILLIAM WILLIAMSON.

DISCUSSION OF MR. F. W. HURD'S PAPER ON "ELECTRICAL COAL-CUTTING MACHINES."[*]

Mr. GEORGE A. MITCHELL (Glasgow) wrote that the Hurd electric coal-cutting machine, referred to at the last meeting of the members as having been started at Gateside colliery, had been working more or less constantly since that time.

The machine was started in the Virgin seam, for which it was intended, and cut in a dirt-parting about 6 inches from the pavement, the coal being in two plies, with about 10 inches of dirt between them. The length of the wall-face available was only 180 feet, and this length was cut six different times, the shortest time occupied in cutting being $3\frac{1}{2}$ hours, this time not taking into account the times when the machine was standing.

On March 11th, the machine was removed to the Upper coal-seam, where a Clarke-and-Steavenson coal-cutter had been holing in the fire-clay above the coal; but, as some difficulty had been found in keeping the roof, it was thought advisable to try the effect of holing in the coal at the pavement, and the most convenient way of trying this experiment was with the Hurd machine. It was not intended to continue this machine with the coal-holing, as, owing to the size of the bar, the thickness of the cut is nearly 7 inches at the front of the seam and this meant too much waste of coal. The machine cut fairly well in this

[*] *Trans. Inst. M.E.*, 1903, vol. xxv., page 108.

seam, the best cut being made on April 6th, when 420 feet were cut, between 4 p.m., when the men attending the machine started their shift, and 4·32 a.m. on the following day. The actual cutting began at 6 p.m.; there were no stoppages of any consequence, except for change of rails, etc.; and the machine cut fairly steadily throughout the shift.

The current used varied from 30 to 40 ampères at 450 volts on the surface. When the machine was started, there was some difficulty owing to the gun-metal bevelled-wheel in the gear-case having been made of bad material, but this was changed to steel at the end of March and so far had not given further trouble.

After what he (Mr. Mitchell) had seen of the machine, he believed that it would be suitable for a fairly dry fire-clay holing without ironstone-balls, and that it possessed advantages in some respects over the disc-machine. There is no danger of the bar being jammed by anything coming down on the top of it, as is the case with a disc-machine. The machine is well enclosed, and there is not the same difficulty as in the case of the disc-machine (especially with a top-holing) with the parts becoming worn through the dust getting among the wheels. He understood that, where there is a coal-holing, the diameter of the cutter-bar and of the picks can be reduced so that less coal is wasted, but there is always the objection that the coal is ground into very small particles, and he did not think, under ordinary circumstances, that the machine is as suitable as a disc-machine for a coal-holing. The same machine can be made to cut from the pave-ment-level to a height of 2 feet 9 inches above the pavement, by the use of wheel-brackets supplied with it, and this makes the same machine available for many different conditions.

The machine is to be altered to cut in the fire-clay above the Upper coal-seam, at Gateside colliery, and for this kind of work it appears to be very suitable and he had little doubt that the results would be satisfactory.

Regarding the question of coal-cutting generally, he was of opinion that the saving is in many cases greatly exaggerated. Save in very exceptional circumstances, even when machines are a success, the economy does not amount to more than a few pence per ton. There are many items of cost, in addition to the actual wages paid for cutting and filling:—The maintenance of machines, in some cases, may amount to as much as 4d. per ton.

The cost of stores used may, including coal, amount to fully 1d. per ton. It is often overlooked that it takes nearly 2 hours of 4 men to run the machine into the face to begin its work, and this has to be recouped during the shift by the saving of machine-cutting over hand-cutting. Then jib-holes must be cut at the end of the longwall-face, to enable the machine to get into its work; and this may cost as much as 2d. per ton. The Hurd machine has some advantage in this respect. Extra timber is used in some cases, as roofs that may not require straps when miners only are employed, may require straps when the coal is worked by machines, and these straps cannot always be recovered. In one case the estimate of the extra cost of timber was about 2d. per ton. When everything was taken into consideration, he was satisfied that, in many cases, the use of coal-cutting machinery would result in an actual loss. There are several advantages on the other hand to be gained in suitable circumstances by the use of machinery, such as the increase and the concentration of output, and, in some cases, a reduced percentage of dross is obtained.

When machines are adopted, they should be used in considerable numbers, and cutting should be reduced to a regular system. Where a number of coal-cutters are in use, spare machines could be kept for emergencies, and spare parts could be kept on hand without serious expense. Labour-cost could be minimized, and a good man should be employed to have an oversight over all details. So far as he (Mr. Mitchell) could tell from a short experience of a Hurd machine, he believed that in many cases, where coal-cutters can be used, it is as suitable as any other.

Mr. JOHN CADMAN (H.M. Inspector of Mines), with regard to the question of selecting the most suitable coal-cutting machine, said that much depended upon the particular circumstances, and he was of opinion that no type of machine could be adapted to all conditions: one design of machine might give admirable results in one seam under conditions favourable to its type, while another machine of different make and principle would prove an absolute failure, and *vice versa*. He (Mr. Cadman) had seen the machine described by Mr. Hurd working at the Fenton colliery, in Staffordshire, where it appeared to be doing excellent work. He was informed that the average rate of cutting was 1 foot per minute; a speed of 3 feet per minute had been obtained, but owing to the high cost of repairs, it was found more economical to work at the low speed. The maximum length cut in 9 hours was 324 feet. The machine is cutting in the Littlemine seam, lying at an inclination of 1 in 7, the undercut being made in the coal to a depth of 5½ feet. The machine works along a face of 420 feet in length. Mr. Hurd had suggested that a premium paid to the manager on the output would be a means of enlisting his closer attention to the working of the machine; but he (Mr. Cadman) could not concur with this suggestion.

The CHAIRMAN (Mr. David M. Mowat) remarked that Mr. Hurd had shewn a comparison between the United States and this country, in regard to the amount of output by coal-cutting machinery. In that comparison, Great Britain did not appear to stand in a very good position; it should be noted, however, that Mr. Hurd did not mention that the majority of the machines used in America were stoop-and-room machines, while, in Great Britain, longwall machines were almost entirely used. He believed that the reason for the increased output in America was due to the special adaptability of the seams to the use of coal-cutting machines. He was satisfied that we were on the eve of a revolution in the coal-trade of Great Britain, in respect of the wholesale introduction of coal-cutting machinery. In the past, in this country, we had wrought mostly in thick seams, which did not require the skill of a practical miner; we had men who were good workers in seams 5 feet thick where "stooping" was going on, but who were useless in seams, 2 feet thick, where holing was required. These men, however, might make good "strippers," behind the machines. He (Mr. Mowat) agreed with what Mr. Hurd had said, as to the necessity for close supervision of the machines; and in his opinion, many failures in coal-cutting were attributable to a lack of proper supervision. He could assure the members that the supervision of a coal-cutting machine was not an agreeable task; if a miner lay off, the output was reduced 2 or 3 tons, but if the coal-cutting machine was stopped there was a loss of 100 or more tons—a very serious difference. He (Mr. Mowat) thought that the continuous method of working a coal-cutting machine, afterwards flitting it back, was superior to working it backward and forward. When cutting backward and forward, if anything occurred to prevent the machine from finishing the cut, a day's work would be lost; if there was a shortage of wagons, or if, for any other reason, the coal was not filled, a day's work would be lost; and for these reasons the length of face laid off for cutting backward and forward must be easily attainable, or else the output would often be curtailed.

He (Mr. Mowat) had seen the Hurd coal-cutting machine doing excellent work at the Oaks colliery, Oldham; in other collieries, the machine did not work so effectively; but generally he considered that it was a good type of machine for holing in the top of a seam. It was almost impossible, on the other hand, to hole in the top of a seam with a disc-machine. He agreed with Mr. Hurd's suggestion that rails should be laid along the whole length of the wall; and he had tried it, in order to lessen the labour of pushing the rails over the machine, which was hard work. Unless the wall is straight, however, the rails begin to lead on one side; and, although this can be corrected when working with three pairs of rails, it cannot be adjusted with a continuous road. Eventually there are bad joints, the machine gets off the road, and delay ensues.

Mr. DANIEL BURNS (Carluke) stated that the official returns for the United States showed that there were 48 longwall machines: the remainder, and considerably the greater number, being stoop-and-room machines.

Mr. S. MAVOR (Glasgow) said that in explanation of the extraordinary development of machine-mining in America the statement was often made that the seams there were specially adapted for machine-working, but, in his opinion, there was no doubt that in this country the advantages of machine-mining were relatively much greater. In America, the seams are thick and are largely worked by machines, whereas in this country it is often objected that it is not worth while to use machines in thick seams. It is worth while, and there are many examples to prove it, because the output can be more than doubled and the percentage of dross coal reduced. If it pays in America to work thick seams by pillar-and-room, with machines, it should pay much better here to work thick seams by longwall, with machines. It cannot be said that the longwall-machine in this country will not cut as many square feet per shift, as the best that can be done by American stoop-and-room machines. In short, it has been demonstrated that machine-holing pays in thick seams where the coal is easily got; and it should pay better in thin seams where the coal is hard to get, and the percentage of dross with hand-holing is greater. The advantages to be derived from machine-holing are really much greater in this country than in America: this is now being realized and a rapid development of the system is in progress. The seams in which, owing to faults or bad roof, machine-holing is impracticable are comparatively few. Mr. Cadman had referred to a machine, which was capable of cutting at the rate of 3 feet per minute, being eased down to 1 foot per minute in order to reduce wear-and-tear, and this, if accurate, seemed a shortsighted policy. Speaking generally, the most economical method of working machine-tools of any class is to work them hard, and to get the largest output. Assume that the total cost of operating a coal-cutter, including labour at the machine, engineman at the surface, and all on-cost charges, amounts to £1 10s. a shift:—If, at the rate of 3 feet a minute, 540 feet be cut in a shift, the cost is £1 10s. for this distance; but if the speed is reduced to 1 foot per minute, only 180 feet is cut in a shift, and three shifts each costing £1 10s. are required to cut 540 feet, the cost being £4 10s., or a difference of £3. If the machine, working at the higher rate, cost £3 a shift for repairs, it must be defective.

The CHAIRMAN (Mr. David M. Mowat) said that the area cut by a longwall-machine could not be compared with the result obtained by a stoop-and-room machine, as the machines were wholly different. The stoop-and-room machine worked intermittently and, in fact, it could not work continuously. In this country, however, the longwall-machine worked under the very best conditions: he had seen a longwall-machine cut a length of 660 feet, continuously; while a stoop-and-room machine would only make three or four cuts, before it was removed to another room.

Mr. FREDK. W. HURD thought that the bar-machine was suited to almost any seam where coal-cutters could be introduced. Mr. G. A. Mitchell had told the members what the Hurd machine was doing at Gateside colliery, Cambuslang. At first, when the machine was introduced into the Virgin seam, holing in dirt, the cut measured $7\frac{1}{2}$ inches high at the front of the holing and 5 inches at the back. For coal-holing, the bar had been altered, and the cut was now $5\frac{1}{2}$ inches high at the front and $3\frac{1}{2}$ inches at the back of the holing, as low a holing as could be made by any other machine. The bar-machine had the especial advantage that no small was made except that removed from the actual height of the cut, as the coal was not pulled down and broken up, as with a disc-machine.

The further discussion was adjourned.

———

Mr. H. D. D. BARMAN's paper on "The Riedler Pump" was read as follows : —

THE RIEDLER PUMP.

By H. D. D. BARMAN.

Since the early days of the Cornish pump little or no improvement has been made, until recently, in pumping-plants. A highly satisfactory duty is obtained with this type of pump, but it labours under the disadvantages of occupying a large space, both on the surface and in the shaft, and entailing a heavy first cost for the slow-moving engine and the pit-work.

The chief difficulty in designing pumps to overcome these disadvantages is the limited speed at which the ordinary pump-valve can be worked; it was while investigating the general action of pump-valves that Prof. Riedler, of Berlin, found the benefit arising from the operating of these valves by mechanical means, with the result that he obtained much higher plunger-speeds; and, in consequence was able to reduce the size of the plant for a given duty, as well as the space occupied, and also to diminish the loads and shocks on the working parts. In many cases, the engine driving the pit-work is geared to the crank-shaft of the pump, and by this means the engine can get away at a higher speed, but then the gearing has to be taken into account both for friction, and also for first cost. A plunger-speed of 100 feet per minute used to be looked upon as a fair speed, but, in Riedler pumps, speeds of 500 feet per minute are common, and some plants even run, when required, at 640 feet per minute.

A comparison of the two styles of pumps may be of interest, taking in both cases pumps directly connected to the piston tail-rod with a 4 feet stroke, and each pump delivering about 500 gallons per minute against a head of 500 feet :—

	Cornish Pump.	Riedler Pump.
Speed, revolutions per minute ....·.	12	75
Plunger and piston-speed, feet per minute	96	600
Diameter of plunger, inches	12¼	5
Area of plunger, square inches	120	19·6
Load on plunger, pounds	26,000	4,250
Mean effective pressure in steam-cylinder, say, pounds per square inch	36	36
Area of steam-cylinder, square inches	866	141·6
Diameter of steam-cylinder, inches	33¼	13½

Another obvious gain would be the reduced steam-consumption of a flywheel engine running at normal speed (according to stroke) thereby having far less cylinder-condensation, and more important still, an early cut-off with a high rate of expansion of the steam (Fig. 7, Plate X.).

In the design of the valve, the most important detail of a pump, the first care is to have sufficient area: in this respect, the Riedler valve has great superiority over the ordinary single-beat or double-beat valve. The lift of an ordinary valve is limited, owing to the shock and slamming due to the velocity with which it seats itself, because of its being dragged back to its seat by the returning water or the slip, which becomes worse the greater the head. As an example of the limited lift permissible, with a main-line locomotive having its pump-plungers driven by the crossheads, the noise and crushing of the valve and the seat were only cured by adopting a valve, 4 inches in diameter with a lift of $\frac{1}{8}$ inch. This locomotive, at a speed of 50 miles an hour, would be making 216 revolutions per minute, with a piston and plunger-speed of 935 feet per minute. Fig. 8 (Plate X.) illustrates the gentle action of the Riedler valve.

Some pumps are designed with a number of small valves, usually with springs behind them, but then there is a heavy loss due to friction, set up by forcing water at a high speed through a number of small openings, and the resistance of the springs has to be overcome. On the Continent, recently, several high-speed pumps have been designed with a single valve of the circular-grid type, and having the small lift of, say $\frac{1}{16}$ inch; in this case, again, there is the same loss due to the small openings in the grid-seat. The Riedler valve on the contrary has the very

considerable lift of, say, 2 inches, the lift only being limited by
the necessity of keeping within the range of the tappet.   The
average lift given for small pumps being about 1½ inches, it is
therefore interesting to compare the areas of the valve-seats with
the free opening of the valve.   Fig. 1 (Plate X.) represents
such a valve, with an area of 4·07 square inches through the seat,
an area of 3·56 square inches past the valve, when it is at its
maximum lift, and a seating-area of 3·347 square inches.   It
would be necessary to use 11 or 12 such valves to be equal to
the Riedler valve (Fig. 2, Plate X.), which has an area of
42·608 square inches through the seat, an area of 52·45 square
inches past the valve, when at its maximum lift, and a seating-
area of 24·15 square inches, so that per 100 square inches of free-
opening, there is a seating-area of 94 square inches for the
small valve and 56·5 square inches for the Riedler valve: thus
with the latter there is very much less surface on which grit
could rest, and less chance of being gagged.

The Riedler valve, usually of the annular type, works on a
central spindle, which has a very large wearing surface. It opens
automatically to the full extent of the required lift (a special
arrangement of water-cushion being introduced in the stop-nut,
which prevents the metals from coming together). The maximum
area past this valve is obtained, and owing to this large and
uncontracted area, the water can be driven through at a very
high velocity: many pumps delivering the water through the
valves at speeds of 530 feet per minute. The mechanical gear
(Fig. 2, Plate X.) only comes into operation towards the end
of the stroke of the plunger: that is, when the plunger has
performed about 90 per cent. of its travel, the tappet begins to
descend, and the crank is about to pass over the dead-centre
when the speed of the plunger is decreasing to *zero*, its displacement then is rapidly diminishing while the valve-area is being
gradually restricted until it is either entirely closed, or closed to
within a fraction of an inch of its seat, when a slight fall would
not cause either shock nor noise.

Owing to the valves being closed absolutely at the moment
that the plunger commences its return stroke, there is little or
no slip. In fixing the dimensions of a Riedler pump-plunger,
it is not necessary to allow more than 277 cubic inches dis-

placement per imperial gallon of water pumped: in one case there was good reason for suspecting that the pump was delivering more than the actual swept-volume, but this was difficult to certify unless special measuring-tanks were erected.   It should be pointed out that the design favours a negative slip, as the water has a very direct flow through the barrels. Special arrangements are made to avoid any deformation in the tappet-gear should any gag or grit be caught between the valve and the seat, by the introduction of a rubber-buffer between the tappet and the valve. The success of this arrangement may be gauged by the fact that while testing a pump a handful of gravel was thrown into the sump, or rather, right into the mouth of the suction-pipe.

One detail in the design of the Riedler pump to which special attention is always given, is the dimensions and positions of the air-vessels, and this has been so successfully arranged that, with a 600 gallons pump, running at 90 revolutions per minute at $2\frac{1}{2}$ feet stroke (Fig. 3, Plate X.), the water-level, as shown in the gauge-glass (Fig. 10), does not vary more than $\frac{1}{8}$ inch while running, the discharge at the surface is absolutely steady, and the pressure-gauge on the rising-main only shows a variation of 3 or 4 pounds. It is, of course, occasionally necessary to charge the air-vessels. This may be done by a small air-compressor, driven by some reciprocating part of the engine, or by a hydraulic compressing-cylinder charged from the rising-main.

With a small double-acting pump of 15 inches stroke, at about 50 revolutions per minute, the knocking and hydraulic shocks will become annoying and dangerous, even when fairly large air-vessels are fitted, unless some special means of charging be adopted. As a result of properly charged air-vessels, one is able to get not only an absolutely steady flow of water through the rising-main, but a steady flow at the maximum velocity, and thus the rising-main can be of smaller dimensions. In one case, water to the extent of 600 gallons per minute is being delivered through a rising-main, 6 inches in diameter, at a speed of 489 feet per minute. Of course, the friction in this case somewhat increases the head by about 70 feet (30 pounds pressure).

The knock of the valves is cured sometimes, it is said, by snifting air into the suction-pipe just below the suction-valve, but this must have a very detrimental effect upon the displacement efficiency of the pump. After careful watching of this system with a view of finding its effect as an air-charger for the air-vessels, it was found to be very inefficient; for this purpose, the snifting valve was placed upon the vacuum-vessel, or as it is often called the air-vessel, of the suction-pipe; the level of the water in this vessel would then be lowered until some of the free air admitted would be drawn into the pump-barrel, and from there discharged through the delivery-valve, and trapped in the delivery air-vessels. It was concluded that the pressure of the air in the suction-vessel, being considerably below atmospheric pressure (say, 5 pounds absolute), and the pressure in the delivery air-vessels being, say, 385 pounds for a head of 850 feet, some 25 or 26 cubic inches that would be drawn into the suction-pot would be represented by but 1 cubic inch in the delivery air-vessel. Even this small quantity is overstated, owing to the greater part of the admitted air being absorbed by the water during the period of its passing through, and being thoroughly churned in the pump-barrel. Thus on all points it will be seen that it is better to pass any additional air, required in the air-vessels, direct to these receptacles; and the cylinder-arrangement with its hydraulic piston appears to be best, as it has no working-parts beyond two or three cocks, ½ inch in diameter. Further, when the pump is started slowly with an empty rising-main, the engineman can leave his throttle-valve and put charge after charge into the air-vessels, till the water-level is lowered in them to the correct height.

The vacuum-vessel on the suction-pipe, already referred to, known in Riedler pumps as the "suction-pot," is invaluable, and usually forms the base or foundation of the main pump. It consists (Figs. 3 and 5, Plate X.) of a rectangular or cylindrical box on the side or end of which is formed the inlet of the suction-pipe, and the suction-valves, A, are placed above the bellmouthed pipes. These bellmouthed pipes dip into the suction-pot to within a few inches of the bottom, the water-level stands about B, so that the plunger has only to lift its water a few inches. Fig. 9, (Plate X.) shows the steady flow of the water through the suction-pot. At the lower end of the bellmouthed pipe, a few holes, $\frac{1}{2}$ inch in diameter, are bored, so that, should the water-level in the suction-pot tend to be lowered too much, the plungers would draw a small volume of air at each stroke, and such a small amount would be drawn in at each stroke that no damage could occur, as would certainly happen if the small holes were not there, and the plungers missed their water entirely. With a suction-pot, the flow of the water in the pipes from the sump is kept constant, the pot, acting as a balancer, taking in any excess and giving out the deficiency of water between the intermittent strokes of the plungers.

The Riedler valves and seats are so arranged that, by raising the air-vessel $\frac{3}{8}$ inch and pushing it aside, and on removing three tapered stoppers from the delivery-valve, it is immediately free from its seat. The shell of the suction-valve being slightly smaller can be freed in a similar manner, and drawn out after the removal of the delivery-valve: the whole operation not taking more than 10 minutes. This compares favourably with the time required to remove the ordinary "conical clack-shell," driven into a tapered seat. The plungers in a double-acting Riedler pump are made single-ended, so that it is only necessary to place the pump on its front or back centre, and the plungers will then be clear of the front or back valves.

Figs. 3 and 4 (Plate X.) and Fig. 10 show a single-side, double-acting Riedler pump erected at the Niddrie colliery, to deliver 600 gallons per minute against a head of 750 feet, running at 90 revolutions per minute on a $2\frac{1}{2}$ feet stroke. This pump is steam-driven, having one cylinder, it not being considered desirable to use a compound engine, as the steam, under certain circumstances is apt to drop to a low pressure. This was the first Riedler pump erected in Scotland. The condenser is of the jet-type, and the air-pump is arranged so that all the water to be pumped is drawn by it from the sump, and then delivered to a tank at the back of the engine-room. This tank stands a few feet above the level of the main pump suction-pot: so that the water from this tank will flow with a slight pressure to the pump. One benefit from this arrangement was shown recently, when the sump had been pumped dry, the engine lost its vacuum

and stopped dead, the attendant then foun
empty. The engine, under other circumstan
working on air, and have run at a dangerou
engines driving Riedler pumps are always
governors and on two occasions these have
tion on emergencies, both from burst mains.
of these occasions, the writer noticed that t

FIG. 10.—PUMP-BODY OF RIEDLER PUMP FOR N

sure-gauge had fallen to 5 pounds; it was th
had broken, as the engine had but a ve
speed.

Indicator-diagrams have been taken fro
ing-engine at Niddrie colliery, when runni
per minute; pump-pressure, 370 pounds per
pressure, 40 pounds per square inch; and

by 27½ inches of mercury-column.   Fig. 7 (Plate X.) is a
diagram taken on the steam-cylinder; Fig. 8, a diagram taken
from the pump; and Fig. 9, a diagram taken on the suction-
pipe of the pump.

Figs. 5 and 6 (Plate X.) and Fig. 11, show a smaller
Riedler pump with a capacity of 250 gallons per minute at a
speed of 720 feet and 150 revolutions per minute on a stroke of
15 inches, also driven by steam at a low pressure. This plant,
erected at Dudley colliery in Northumberland, has but one

FIG. 11.--RIEDLER PUMP FOR DUDLEY COLLIERY.

suction and one delivery-valve, the plunger being of the differential, sometimes called the Armstrong, type (the late Lord
Armstrong being the inventor), which acts on the same principle
as a bucket and plunger. All the suction-water, being drawn in
on the out-stroke, half of the delivery-water goes straight up
the rising-main on the in-stroke, the other half falling over to the
differential chamber to be sent to the rising-main on the next
out-stroke.   This pump is fitted with a surface-condenser, or
rather a steam-killer arranged in the suction-pipes.

The Riedler pump is designed to take but little breadth, to avoid roof-troubles, and even then could be made narrower by introducing a bent crank with a main bearing on each side, and one or even two flywheels, one on each side of the main bearings. Fig. 12 is a Riedler pump erected at the Elliot pit of the Powell Duffryn Coal Company, South Wales. Its capacity is 1,000 gallons against a head of 1,600 feet, it is driven by a cross-compound, jet-condensing Corliss engine, and it has a stroke of 4 feet. It is so arranged that when both sides are working it does its work at 40 revolutions per minute, but should it be necessary to overhaul one side, that side can be disconnected in 15 minutes; and still the other side will do its full duty at 80 revolutions per minute, running as a non-compound

FIG. 12.—RIEDLER PUMP FOR POWELL DUFFRYN COLLIERY.

engine. This plant has been tested in this way, and when run for some weeks, it was then quite apparent that 80 revolutions (that is 640 feet of piston and plunger-speed) was not its limit. On another occasion, one of the four suction-valves was taken out, and the pump restarted, no trouble being experienced when running, one might say, three-legged.

It can safely be said, that the engineman has quite as easy a time with a Riedler as with any other pump. Many people may imagine that the two or four valve-tappets make all the difference, but then these additional working-parts are well worth putting into a pump, when it is borne in mind that they reduce the necessary multiplication of valves, and also prevent the wear-and-tear that usually comes on them. The Riedler, of course, has to receive fair attention, like other pumps; the most neglected parts being the plunger stuffing-boxes. These are often packed too tightly, and the rams have been found to be hotter than would be desirable for a crank-shaft journal, sometimes even stopping the pump; often they are leaking on the delivery-stroke, then on the suction-stroke the air drawn in is not noticed, although very detrimental to the displacement.

FIG. 13. – ELECTRICALLY-DRIVEN RIEDLER PUMP.

Another point, often thrown up, is the so-called high speed, but the speeds given are not so high as with the average well-built land engine: and further, any competent designer would allow proper bearing-surfaces and scantlings. It is not more difficult to keep machinery at a colliery in good order than it would be in a cotton-factory. As a proof of this, dozens of winding-engine rooms in this district could be favourably compared to the engine-rooms of the mills in Lancashire.

Steam-driven pumps have only been referred to, so far; but for electric driving, the Riedler pump lends itself even more

readily than to steam. By its higher speed much less gearing
required, one pair of gear-wheels with a ratio of 1 to 4 be
about the usual form: this gives a motor-speed of 600 revo
tions per minute to 150 revolutions of the pump (Fig. 13).
geared type is most favoured, owing to its compactness, althou
rope-and-belt connections are also installed. The direct co
nection of the motor to the crank-shaft is sometimes introduc
but this involves a somewhat slow speed for the motor, th
running up the first cost, but making a gain in the brake-hor
power by the saving of the friction and cost of gearing.

With the ordinary three-throw pump, single-acting plung
type, when one has a three-throw shaft, with its intermedia
bearings, its three connecting rods and slippers, three suctio
valves and three discharge-valves running at, say, 45 revo
tions per minute, that is 135 pulsations, the cost of upkeep m
be more than with the Riedler pump, which would have but o
plunger of the differential type, running at 150 revolutions
minute, that is 300 pulsations, and both the suction-valve a
delivery-valve would be placed quietly on their seats.

The pumps are all on the lines of those built by Mess
Fraser & Chalmers, Limited, who have built some hundre
and who kindly gave to the writer the benefit of their experien

In conclusion, it is only right to say that the Riedler syst
has now been thoroughly tested, some 1,500 pumps having be
built for the Continent, America, South Africa, in fact for
parts of the world.

———

Mr. ROBERT THOS. MOORE (Glasgow) said that he remember
seeing a Riedler pump at Kimberley in 1893, and he was su
prised that more of them were not at work in this country. T
members knew that in a rotary engine applied to pumping the
was no pause at the end of the stroke, and there was no time f
the valve to close before the pressure was brought upon it. Th
difficulty was overcome in the Riedler pump as the valve wa
almost shut at the slow part of the stroke, near the end; an
immediately the motion of the pump was reversed the valve wa
shut. The chief advantage of the Riedler pump was that, owin
to the high speed, a reduced size of pump was required to deal wit
a given quantity of water. Personally, he (Mr. Moore) though

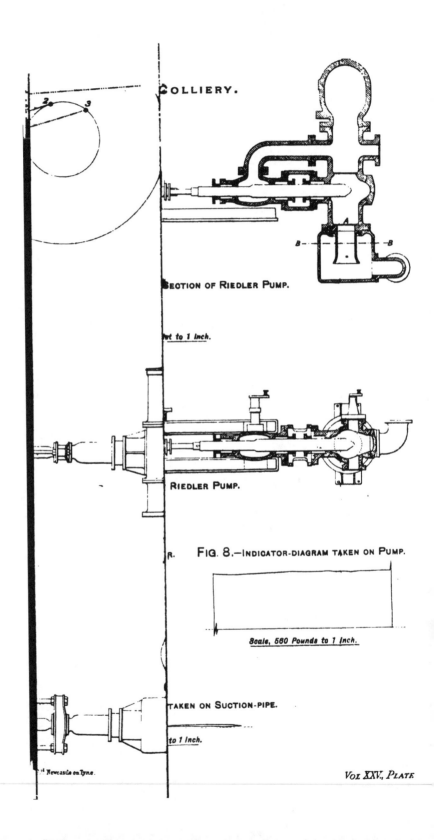

COLLIERY.

SECTION OF RIEDLER PUMP.

...rt to 1 Inch.

RIEDLER PUMP.

FIG. 8.—INDICATOR-DIAGRAM TAKEN ON PUMP.

Scale, 560 Pounds to 1 Inch.

...TAKEN ON SUCTION-PIPE.

...to 1 Inch.

that a Riedler pump driven at 600 feet per minute was running too fast: he did not care to have a colliery-engine running at this speed; and, in his opinion, a pump running at 300 feet per minute was likely to run longer, and was less liable to breakdown, than one running at 600 feet per minute. He would prefer, in applying a Riedler pump, to have it large enough to pump the required quantity of water when running at a speed of 300 feet per minute. The Riedler pump, at the Niddrie collieries, had been running satisfactorily for some months.

Mr. JAMES GILCHRIST (Workington) asked whether the Riedler pump would last as long as many of the slow-running pumps; and whether any additional expense was incurred in maintaining high-speed pumps.

The further discussion was adjourned.

___

SEMI-JUBILEE DINNER.

In celebration of the semi-jubilee of The Mining Institute of Scotland, the members and guests, to the number of about 120, met at dinner in the Windsor Hotel, Glasgow, on April 8th, 1903.

The usual loyal and patriotic toasts were proposed.

PROVOST KEITH (Hamilton) proposed the Mining Institute of Scotland, referring to the progress that had been made by the Institute during the 25 years of its existence, from the original 10 to the present number of 482; the catholic character of the Institute, embracing as it did among its members, colliery-owners, colliery-managers and H.M. inspectors of mines, and the quickening of interest in all departments of mining that had been an accompaniment of the progress of the Institute.

The PRESIDENT (Mr. Henry Aitken), in his reply, hazarded the opinion that not much more than 10 per cent. of the available coal of Scotland had been worked up to the present time. He further stated that it would be the ambition of the Institute to be considered first in everything that promoted the sciences of engineering and mining.

Other toasts were:—Kindred institutions, the coal, iron and steel trades, the guests, etc.

___

THE MINING INSTITUTE OF SCOTLAND.

GENERAL MEETING,
HELD IN THE ODDFELLOWS' HALL, KILMARNOCK, JUNE 13TH, 1903.

MR. HENRY AITKEN, PRESIDENT, IN THE CHAIR.

The minutes of the last General Meeting were read and confirmed.

The following gentlemen were elected:—

MEMBERS—
Mr. WILLIAM BARR, Rosebank, Bargeddie.
Mr. C. AUGUSTUS CARLOW, Leven Collieries, Leven.
Mr. WILLIAM CLARK, Broxburn.
Mr. HENRY CONDIE, Calderbank, Airdrie.
Mr. WALTER DAVIDSON, Shawfield Colliery, Law.
Mr. FREDERICK DUNCAN, Mount Vernon.
Mr. HENRY CHARLES HARRIS, Technical School, Coatbridge.
Mr. JOHN KYLE, 18 Brewland Street, Galston.
Mr. HUGH SCOTT MACGREGOR, Cadzow Colliery, Hamilton.
Mr. ROBERT OGILVIE, Buen Retiro Mines, Coronel, Chile.
Mr. JOHN POLLOCK, Burnfoothill, Dunaskin.
Mr. ARCHIBALD M'KERROW RUSSELL, Blairholm, Newmains.
Mr. THOMAS RUSSELL, Plevna, Newmains.
Mr. WILLIAM H. TELFER, Glencraig Colliery, Lochgelly.
Mr. J. B. THOMSON, Fairview, Hamilton.
Mr. ROBERT WALLACE, Greenfield Colliery, Hamilton.
Mr. ARCHIBALD WILSON, Leith Electric Works, Prince Regent Street, Leith.
Mr. JAMES WILSON, Avonhead Colliery, Longriggend.

ASSOCIATE MEMBER—
Mr. THOMAS THORNTON, Hermand House, West Calder.

ASSOCIATES—
Mr. JAMES S. COMRIE, Bog Colliery, Larkhall.
Mr. THOMAS NISBET, Cadzow Colliery, Hamilton.
Mr. JOHN WILSON, Glencraig Colliery, Lochgelly.

STUDENT—
Mr. ANGUS BREMNER CROLE, Durie Street, Leven.

DISCUSSION OF MR. FRED. W. HURD'S PAPER ON "ELECTRICAL COAL-CUTTING MACHINES."*

Mr. PERCY J. E. KENNEDY (Wyken Colliery) wrote that, having had several opportunities at different collieries during the past six years of comparing disc and bar-machines, working on the same face, and under precisely the same conditions, he could endorse the statements of Mr. Hurd as to the relative power required to work these respective types of coal-cutting machines. In the cases which had come under his notice the power absorbed by the bar-machine was never more than two-thirds of that required to drive the disc-machine, and generally only about one-half. In a bar-machine, which commenced work in June, 1901, and which worked continuously for 14 months before being taken out of the pit to be repaired, the only renewals found necessary were the two small wheels, working the reciprocating gear. The main gear-wheels were not half worn, they were replaced, and have been working ever since. This machine has now been working night-and-day (two shifts) for 22 months, without any breakage due to wear-and-tear. There have been two or three small breakages due to heavy falls of roof or coal; but they were insignificant, and did not cause any serious stoppage of the machine. During the past 9 years he had been in personal contact with nearly all the most successful types of coal-cutters at work in this country, more especially of the disc-type, and from his experience, the average life of the main-gearing of a disc-machine did not exceed 6 to 8 months, under the most favourable circumstances. About a month ago, the writer had occasion to start a bar-machine, on a new working-face, where it was necessary that the cutting should be done in the interval of the colliers changing their shifts, that is from about 1·30 to 2·30 p.m., the cables being used for other purposes during the shifts. Every arrangement having been made beforehand, the bar-machine started on full feed as soon as the colliers had ceased work, and cut at the rate of slightly over 3 feet per minute: the depth of the holing being 4½ feet and the material a fairly soft but rather tough fire-clay. About 5 years ago, at a colliery, in Derbyshire, the writer cut in hard batts with a bar-machine

* *Trans. Inst. M.E.*, 1903, vol. xxv., pages 108 and 231.

82 yards in 2¼ hours, including one stop to change some of the cutters. This high speed of cutting was, however, made under exceptional circumstances, and was of very little practical use. What is wanted is a machine that cuts steadily on at the rate of about 15 inches per minute, and runs continuously for 2 or 3 hours, without excessive heating. The makers of most disc-machines commit the error of employing motors that are too small for the work, and it is often impossible to make a cut of longer duration than ½ hour, without damaging the motor through overheating. With a machine of this description it is a waste of money to lay rails along the whole length of the face, as the machine will have to be constantly stopped to allow of cooling down, whereas with only a few pairs of rails, these can be cut over and the machine can cool down while they are being taken up and relaid.

The further discussion was adjourned.

————

Mr. M. Brand's paper on the "Calcination of Blackband Ironstone at Dumbreck" was read as follows:—

CALCINATION OF BLACKBAND IRONSTONE AT DUMBRECK.

By MARK BRAND, B.Sc.

Seams.—In the Kilsyth and Gartshore mineral-fields, leased by Messrs. William Baird & Company, Limited, the following seams of stratified blackband ironstone have been worked at different times. A stratified seam of clayband ironstone also exists, but it is not of workable thickness. The general section (Fig. 1, Plate XI.) shows the approximate relative positions of the various seams and of the strata associated therewith.

The No. 1 or Upper blackband ironstone-seam does not exist in the Gartshore field, and it is not continuous in the Kilsyth field. It has been worked in the eastern and western portions of the latter estate, when found of workable thickness.

The No. 2 or Haugh blackband ironstone-seam only occurs in the eastern portion of the Kilsyth field, over a comparatively small area, and it was practically all worked out some years ago.

The No. 3 or Neilston blackband ironstone-seam is continuous and of workable thickness over a large portion of the Kilsyth and Gartshore mineral-fields, from which, until about 10 years ago it was raised in large quantities from various pits. The only pit working this seam at present is Dumbreck No. 3, and this paper describes the methods there employed for raising and preparing the ironstone for use in the blast-furnace.

The No. 4 or Banton blackband ironstone-seam is found all over the Kilsyth and Gartshore fields, but it only occurs in workable condition in small portions of those fields.

The Garibaldi or clayband ironstone-seam has not been worked in this district.

Generally speaking, the above seams occur in basin-shaped deposits, the ironstone becoming coaly, or else thinning out altogether at the edges of the deposits.

Method of Working.—At Dumbreck No. 3 pit, the Neilston ironstone is found at a depth of 804 feet, and it is worked on the longwall system. The holing is made in the shale, and the ironstone is brought down by wedging. The roads are spaced about 60 feet apart, and are brushed by the miners. The waste is stowed solid between the roadside-buildings, and the rubbish, which cannot be stowed, is sent to the surface. About 1·4 tons of rubbish are sent out of the pit to every ton of ironstone. The hutches, which hold about 15 cwts. of ironstone, are delivered by the miners at sidings on the main roads, from which points the material is conveyed to the pit-bottom by means of horse-haulage.

Surface-Arrangements.—The surface-arrangements (Figs. 2 and 3, Plate XI.) for burning the raw ironstone, and loading the *char* into trucks are as follows:—The hutches containing the ironstone are run from the pit-head along a gangway, about 9 feet above the level of the hearths, on to which the ironstone is tipped to form an open heap. The hearths are formed on open level ground, each of them measures about 200 feet in length by 68 feet in width, and is causewayed with brick set on edge. Each bin is laid out with a falling gradient, towards the tip-end, of about 1 in 100, making the average height of the bin 8 feet. The object of the gradient is to enable the loaded hutches to be more easily handled. Railways run between the hearths, the rails being placed about 3 feet below the level of the hearths, so as to facilitate the loading of the char. The railway retaining-walls are built of rubble-masonry, 2 feet thick. To enable the work to be carried on continuously, there are three hearths at the pit, each capable of holding 3,000 tons of raw ironstone. While one of these hearths is being filled with raw ironstone from the pit, the ironstone on a second hearth is in process of calcination; and from the third hearth, the calcined ironstone is being filled into trucks for transmission to the blast-furnaces. As the hearth is being filled, the ironstone on the sides and top of the bin is all broken by a labourer to about the size of road-metal.

Calcination.—When one of the hearths has been filled with raw ironstone in the manner described, a coal-fire is lighted at the base of the bin, right across one end, and the bituminous matter in the stone itself provides the necessary combustible material for carrying on the calcination : this forming the essential difference between blackband and clayband ironstone. The process is also assisted by the further oxidation of the iron protoxide (FeO) to the more highly oxidized peroxide of iron (Fe_2O_3). This action is similar to the burning of carbonic oxide (CO) to carbon dioxide (CO_2): heat being generated in both cases.

As the calcination advances, care must be taken that no blow-holes are allowed to form in the mass, as this is an indication that combustion is proceeding too rapidly at local points, and if this be not arrested, the local temperature becomes too high, causing the char to fuse and run together into a solid mass ; and it is not only difficult for the char-filler to break up and load, but it is not in so suitable a condition for reduction in the blast-furnace.

The breaking of the ironstone, at the sides and top of the bin, materially assists in securing a uniform draught throughout the mass ; and prevents the formation of blow-holes, which are to be avoided for the reasons already stated. The further regulation of the draught is secured by the burner placing mine-dust, obtained from a previous fire, upon the surface of the bin at the points where the fire is advancing too rapidly, thus maintaining the uniform progress of the calcination. Another object served by the breaking of the ironstone on the surface of the bin to a small uniform size, is to enable the pieces to be thoroughly calcined. If the ironstone were left unbroken, the lower temperature prevailing at the outer surfaces would not be sufficient to penetrate the larger pieces, and would result in imperfectly-calcined stone, which would require to be laid aside when the bin was lifted, and burned over again in the next fire, causing unnecessary labour and expense. The burner assists the calcination of the ironstone on the surface of the bin, by beating it down into the fire as far as possible.

A bin takes from 5 to 6 weeks to burn, and during the process shrinks to about one-half of its original height. The char at the end of the bin, where the fire was lighted, is cool enough for filling by the time that the calcination of the whole bin is completed. It is found that the fire advances along the bin at a quicker rate with the wind against it, than with the wind in its favour. This is due, in the first case, to the volatile products of combustion being blown clear of the fire, allowing of the admission of fresh air, which promotes combustion. In the other case, the products of combustion are blown into the fire, preventing the admission of fresh air, and thus retarding combustion.

The height of the bin is an important element in the successful calcination of ironstone, and is determined by the percentage of bituminous matter in the stone. In the case of ironstone with a large percentage of bituminous matter, the best results are obtained when the height of the bin does not exceed 6 or 8 feet. When the percentage of bituminous matter falls, the height of the bin is increased. If the stone does not contain sufficient bituminous matter to calcine itself, it is sometimes necessary to mix a proportion of coal with the stone, when the bin is being laid down. The bin is kept low, when calcining ironstone with a high percentage of bituminous matter, so as to prevent too high a temperature being generated. If the bin be too high, the amount of heat, generated in the mass, is sufficient to run the char, making it unsuitable for the blast-furnace.

In the case of ironstone with little bituminous matter, the higher the bin (provided sufficient air can penetrate it for combustion) the better is the heat retained in the mass, as less is lost by conduction and radiation, and consequently less bituminous matter or added coal is required for the calcination of the ironstone.

Analyses of the raw and calcined blackband ironstone yield the following results: —

	Raw ironstone. Per cent.		Calcined ironstone or char. Per cent.
Carbonate of iron	70·00	Iron protoxide (FeO) ...	3·60
Sulphide of iron	0·65	Iron peroxide (Fe_2O_3) ...	75·20
Oxide of manganese ...	0·60	Manganese oxide	1·20
Carbonate of lime	4·10	Lime	3·80
Carbonate of magnesia ...	0·46	Magnesia	0·70
Clay (silicate of alumina)	8·04	Clay (silicates)	13·00
Phosphoric acid	1·02	Phosphoric acid	1·70
Bituminous matter ...	14·95	Sulphuric acid	0·90
	99·62		100·10
Metallic iron	34·10	Metallic iron	55·45

It will be seen from the preceding analyses that very considerable changes have taken place in the chemical composition of the ironstone during the process of calcination.

The physical condition of the ironstone has also altered in this process, from being originally a comparatively impervious material to a laminated porous substance, easily penetrated by the blast-furnace gases. The most important change that takes place is in the condition of the iron itself, which existing in the raw stone as a carbonate ($FeCO_3$) loses by the action of heat all the carbonic acid (CO_2) leaving an oxide, which at the same time absorbs a further quantity of oxygen, resulting in the formation of iron peroxide (Fe_2O_3). Simultaneously, the lime which exists in the stone as limestone ($CaCO_3$) loses its carbonic acid (CO_2), becoming quicklime (CaO). The clay is not altered in the burning. The calcination of the ironstone eliminates a proportion of the sulphur, which exists in the raw ironstone as pyrites or sulphide of iron (FeS_2). The portion eliminated passes off as sulphur dioxide (SO_2), while that remaining is mainly changed into sulphur trioxide (SO_3), and combines with a portion of the lime to form sulphate of lime.

Benefit of Calcination.—The advantages of calcining the ironstone, before sending it to the blast-furnace, are as follows:—

(a) The chief advantage lies in the loss of weight that the raw ironstone sustains during calcination, and the concentration of the iron in the residual material or char. Every 100 tons of raw ironstone put on the bin yields an equivalent weight of metallic iron in $61\frac{1}{2}$ tons of char, giving a saving of $38\frac{1}{2}$ tons in the weight of material to be handled. and a corresponding saving in railway-carriage, and at the blast-furnace. The proportion of metallic iron is raised from 34·1 per cent. in the raw ironstone to 55·45 per cent. in the char, solely due to the elimination of the carbonic acid, bituminous matter, etc., from the raw ironstone.

(b) As already stated, the condition of the ironstone after calcination is more favourable for reduction in the blast-furnace, than in the raw state, as it has been altered from being of an impervious, stony nature to a porous substance easily acted upon by the reducing gases in the blast-furnace.

(c) The oxidation of the iron from iron protoxide (FeO) to iron peroxide (Fe_2O_3) greatly improves its value in the blast-furnace, preventing the formation of scouring slags, with the accompanying loss of iron. The iron peroxide (Fe_2O_3) will require rather more fuel for its reduction to the metallic state, but it will lose the disadvantageous property of the protoxide (FeO), of forming readily fusible silicates with the clayey material, derived from the stone itself, and also from the ash in the fuel, which silicates pass readily into the slag, carrying iron (which is lost) with them. There is a little iron protoxide (FeO) in nearly all blackband chars, but this should be eliminated as far as possible, during calcination, for the reasons already stated.

———

The discussion was adjourned.

———

Mr. Andrew Watson's paper on "Stuffing-Boxes dispensed with on Engines and Pumps" was read as follows:—

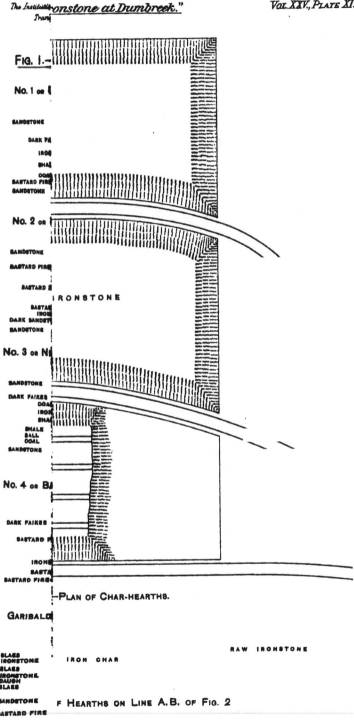

Fig. 1.-

No. 1 or

SANDSTONE

DARK FA

IRO

SHA

COA
BASTARD FIRE
SANDSTONE

No. 2 or

SANDSTONE

BASTARD FIRE

BASTARD

IRONSTONE

BASTA
IRO
DARK SANDST
SANDSTONE

No. 3 or Ni

SANDSTONE

DARK FAIKES

COA
IRO
SHA

SHALE
BALL
COAL

SANDSTONE

No. 4 or BA

DARK FAIKES

BASTARD F

IRON
BASTA
BASTARD FIRE

PLAN OF CHAR-HEARTHS.

GARIBALD

RAW IRONSTONE

BLAES
IRONSTONE   IRON CHAR
BLAES
IRONSTONE
DAUGH
BLAES

SANDSTONE   F HEARTHS ON LINE A.B. OF FIG. 2

BASTARD FIRE

STUFFING-BOXES DISPENSED WITH ON ENGINES AND PUMPS.

By ANDREW WATSON.

Forty years ago, Mr. Henry Aitken, feeling that an improvement could be made on the old arrangement of boiler-floats which had a copper-wire passing through a hole in the shell of the boiler, the steam being in part prevented coming out by a "shangey," on which rested an iron ball, designed the arrangement shewn in Fig. 1 (Plate XII.). This arrangement proved most satisfactory, as no steam escaped, and there was great ease in working owing to the almost entire absence of friction.

Seeing that the results were so satisfactory, Mr. Aitken applied the same principle to the stuffing-boxes of the piston-rods of engines and force-pumps: the only difference being that grooves were cut on the cast-iron "horn" instead of on the piston-rod.

The horn, A (Figs. 2, 3 and 4, Plate XII.) is a solid casting of iron, the inside of the horn being bored out to a certain diameter, so as just to allow the piston-rod to move, in and out, with freedom. The horn has grooves cut out, 1 inch apart along its whole length, each being $\frac{1}{4}$ inch wide and deep. At the end of the horn, which enters the stuffing-box of the cylinder, a ring of white-metal, the internal diameter of which is the same as that of the piston-rod, with a thickness of $\frac{3}{8}$ inch, and a breadth of $1\frac{1}{4}$ inches and "saw-cut" through the same, fits into the check turned out of the end of the horn and the brass-bush.

This arrangement has been a complete success. The first engine, fitted in this way, had the horn cast upon the cylinder-cover, and bored out as before described. As a precaution, a stuffing-box was placed upon the end of the horn of the first engine so made, but no packing was used. This engine was in use for a period of fifteen years, after which it was sold, and all trace of it has been lost.

The three coupled winding-engines at Cadzow colliery, with cylinders 24 inches in diameter and 5 feet stroke, have been altered by simply fitting the horn upon the stuffing-boxes as before described. The horns might be made longer than they are at present, but they work most satisfactorily, and allow no steam to pass out.

An iron casing, B, (Figs. 2 and 4, Plate XII.), with an inside diameter $\frac{5}{8}$ inch greater than that of piston-rod, is fixed to the outer flange of the horn, enclosing the piston-rod; and into the space between the casing and the rod a small quantity of exhaust-steam is allowed to pass: this maintains the piston-rod at a temperature of about 212° Fahr.

If large slow pumping-engines were so fitted, it is certain that a considerable saving in steam would ensue, as the loss caused by using cold piston-rods must be great.

At Cadzow colliery, working under full load, it has been found that since the horns have been fitted to the winding-engines, full steam is admitted for $8\frac{1}{2}$ revolutions, whereas when ordinary packing was in use steam was admitted for $9\frac{1}{2}$ revolutions. It should, however, be noted that before the horns were fitted, the piston-rods were not in good condition.

Experiments, made with one of the pistons of these winding-engines, shewed that when disconnected from the cross-head and in a position ready for work, the piston and rod, weighing fully 9 cwts., could be easily pulled by one man along the cylinder. A Salter spring-balance was fixed to the outer end of the piston-rod and registered 100 pounds, the power necessary to start and move the piston and piston-rod.

If a marking of white lead be put on the rod, it remains on it for hours. These horns have been in use at Cadzow colliery for six years, and the repair and upkeep of the same are practically *nil*.

The cost of fitting the winding-engine with a set of 4 castings or horns, amounted to £6 12s. 6d., comprizing 4 castings weighing 5 cwts. 1 qr. 7 lbs. at 8s. per cwt., or £2 2s. 6d.; and the wages of an engineer, turning, boring out, fitting to engine and use of lathe, £4 10s. The cost of packing the piston-rods of both cylinders amounted to £6 16s., comprizing 12 lbs. of $\frac{1}{2}$ inch

Fig. I.—The Aitken Boiler-Float.

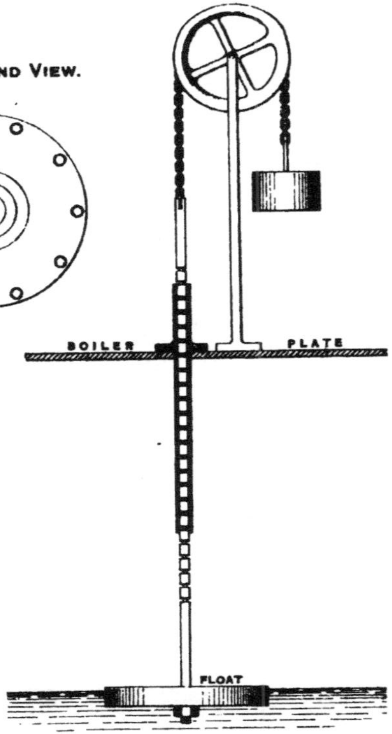

Fig. 3.—End View.

Fig. 2.—Side-Elevation.

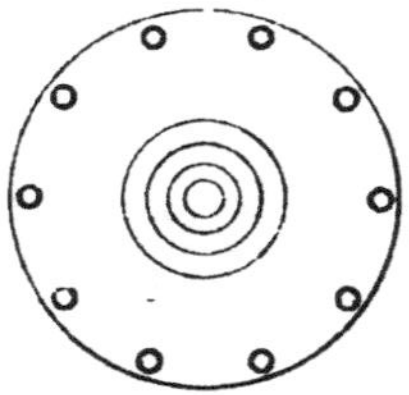

Fig. 4.—Section.

Scale, 20 Inches to 1 Inch.

Vol. XXV. Plate VI.

And⁰ Reid & Comp⁰ Lᵗᵈ Newcastle on Tyne.

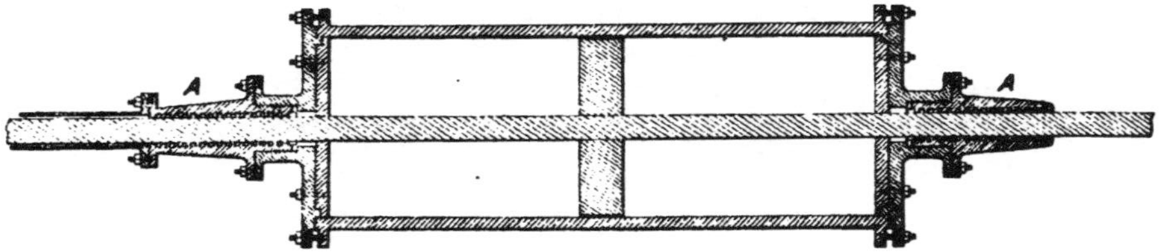

...ning Institute of Scotland.
Transactions 1902-1903.

indiarubber sheet at 2s., £1 4s.; 6 lbs. of asbestos, ¾ inch in diameter at 1s. 6d., 9s.; 4 pounds of fine rope at 6d., 2s; and wages of men packing the rods, 3s. The indiarubber was renewed once each year, and the asbestos and fine rope eight times. The cost per annum including labour was £6 16s.; comprizing indiarubber, £1 4s.; asbestos and fine rope, £4 8s.; and wages, £1 4s.

It seems strange that so simple and effective an arrangement should have been in use for over forty years without being generally adopted, seeing that independently of the saving of steam, the saving in packing alone in a year amounts to fully £6 for each winding-engine.

The advantages of having engines fitted with these castings or horns are as follows:—(1) There is a minimum of friction and consequent saving in the consumption of steam; (2) the piston-rods last very much longer, owing to the diminished friction and wear; (3) the rods carry the piston, so that the piston may be said to float, and does not rest on the bottom of the cylinder; and (4) there is a saving in packing and labour, and it is no longer necessary to stop the engine at possibly inconvenient times in order to pack the stuffing-boxes.

———

Mr. D. G. DUNN (Cambuslang) said that much of the success of the arrangement described by Mr. Watson must be due to the absence of grit in the water used, otherwise the piston-rod was certain to become "corded," especially in the event of the boiler priming, and this often occurred at collieries, through unsteady firing, and inattention to the boiler-feed; and if on the other hand priming of the boilers be absolutely prevented, this system of metallic packing (if he might call it such) seemed worthy of adoption, where the steam-pressure was not high.

The PRESIDENT (Mr. Henry Aitken) thought that all members of the Institute were agreed that cording was the result of bad packing or good packing badly put in, or the use of too soft metal for the piston-rod; there should, of course, be no priming, but it did take place occasionally and it did not affect the rods in question. The ordinary pit-water of the district was used, containing a good deal of lime and a very considerable quantity of magnesia. The steam-pressure was 80 pounds per square inch.

Mr. THOMAS TURNER (Kilmarnock) remarked that the white metal might be the element that made the packing so efficient.

The PRESIDENT remarked that, in the case described, the white metal was used as a make-shift to partly fill up the old stuffing-box, and it was not used when the horn was cast as part of the cylinder-cover, when new engines were so fitted.

Mr. T. TURNER said it seemed surprising that such an idea had not been adopted more generally, as it proved so beneficial in its results in the case described. He knew, of course, of occasional cases where it was used between the cylinders of tandem compound-engines.

The PRESIDENT (Mr. H. Aitken) observed that some American firms were now advertising the same idea in connection with small cheap engines.

Mr. T. TURNER said that his firm in all their locomotives only used white-metal packing-rings to the piston-rods, and these were so satisfactory and permanent that they very seldom had orders for their renewal. It was perhaps rather odd that the steam did not manage to find its way along the piston-rods past the grooves, as one might expect it to do after some wear had taken place; and he presumed it was the uncertainty about wear that prevented others from adopting the plan. In a compound-engine the effect, on condensing in the low-pressure cylinder, would be to draw back to the cylinder any steam that was seeking to get away. The arrangement certainly appeared to be a very sensible one, and he would be glad to get the opportunity for his firm to make some engines fitted with these boxes, so as to give the system a trial.

Mr. HARRY D. D. BARMAN (Airdrie) asked for information as to the lengths of the sleeves required for different pressures; and whether any leakage was observed at different speeds of the engine. The arrangement of grooves had been in use for a long time for air-pump and other pump-buckets, many engineers preferring that air-pump buckets should be made with grooves, owing to the number of accidents occurring through the bucket-rings breaking and becoming jammed against the side of the pump. The engine to which Mr. Turner referred as having the grooves cut in the piston-rod, between the high-pressure and low-pressure cylinders, instead of in the sleeve, was the Kingdon engine, and he believed that the arrangement had worked with a steam-pressure of 200 pounds per square inch when running at a speed of 200 revolutions per minute.

The PRESIDENT (Mr. H. Aitken) said that no experiments had been made to determine what length of horn was necessary for a pressure of 80 pounds per square inch. There was no escape of steam whatever at that pressure, and steam was kept in the cylinders at that pressure, without leak, even when the pistons were stationary.

Mr. T. TURNER thought that the drawback to the arrangement perhaps lay in the fact that sooner or later the piston-rods would be damaged, either from corrosive causes or from wear-and-tear.

The PRESIDENT (Mr. H. Aitken) replied that the piston-rods did not wear, and one had run for fifteen years without any outlay being incurred. The arrangement on the Cadzow winding engine had run six or seven years, and the tool-marks were not yet worn off the piston-rods.

Mr. F. W. HURD said that the principle set forth by Mr. Watson in his paper had been adopted by a firm of steam-carriage and motor-car makers. The piston and valve-spindles were made in a similar manner, and no packing whatever was used on the engine, although high-pressure steam was used. He thought that the satisfactory character of the packing was due to the fact that the condensed water filled up the grooves in the horn. Other engineers seemed to be adopting the idea, and quite recently he had noticed it in use at a place in Shropshire.

Mr. CUTHBERTSON (Kilmarnock) agreed with Mr. Hurd's idea that this was a clever arrangement of water-packing.

Mr. THOMAS B. DUNN (Glasgow) said that he had had an opportunity of inspecting the engine at the Cadzow colliery, and he had no doubt as to the effectiveness of the arrangement.

The further discussion was adjourned.

THE SOUTH STAFFORDSHIRE AND
SHIRE INSTITUTE OF MININ

GENERAL MEETING
HELD AT THE UNIVERSITY, BIRMINGHAM,

MR. T. J. DAVIES, PRESIDENT, I

The minutes of the last General M Meetings were read and confirmed.

The following gentlemen were electe

MEMBERS—

Mr. W. J. BATES, Lyttleton Colliery, Hunt
Mr. C. E. COLE, Hamstead Colliery, Great
Mr. D. ROGERS, jun., Lyttleton Colliery, H

ASSOCIATE MEMBER—

Mr. THOMAS HILL, Old Hill.

STUDENT—

Mr. R. A. PASSMORE, Erdington, Birmingh

A MAGNETIC UNLOCKING AND ELECTRICALLY-IGNITED SAFETY-LAMP.

By W. BEST.

The bolt is arranged so that a suitably-shaped magnet can be directly applied to it, thus obtaining greater withdrawing power. The bolt cannot be tampered with, and the lamp cannot be opened without the aid of the electromagnetic unlocking apparatus.

The lamp is so constructed that it cannot be locked together unless the gauze be in its place. The glass and the gauze are fixed in the lamp, independent of each other, resulting in better joints and obviating the need for making new joints each time that the lamp is cleaned.

The lamp can be electrically ignited when burning an oil with a flash-point as low as 300° Fahr., or if need be as high as 500° Fahr. The lamps are arranged in the lamp-stand in series of twenty, and are lighted at one "sparking." Each lamp is placed in an earthenware-dish, through which one pole protrudes: an arrangement which avoids sparking at the poles, as in the case where the two-pole system is adopted. Ignition is made by the igniter, which on being brought into contact with an insulated copper-ring, fixed at the bottom of the glass, induces sparking at the copper wick-tube.

THE SOUTH STAFFORDSHIRE AND EAST WORCESTERSHIRE INSTITUTE OF MINING ENGINEERS.

GENERAL MEETING,
HELD AT THE UNIVERSITY, BIRMINGHAM, APRIL 6TH, 1903.

MR. T. J. DAVIES, PRESIDENT, IN THE CHAIR.

The minutes of the last General Meeting and of Council Meetings were read and confirmed.

The following gentlemen were elected:—

MEMBER—

Mr. ROCHFORD UNDERHILL, Aldridge Collieries, Walsall.

STUDENT—

Mr. FRANK D. PEACOCK, Aldridge Collieries, Walsall.

Mr. F. G. MEACHEM read the following paper upon "Underground Temperatures":—

UNDERGROUND TEMPERATURES.

By F. G. MEACHEM.

Earth-temperatures.—Many and various speculations have been made regarding the origin and early condition of the earth. The history of the science of geology provides examples of hosts of theories, novel in themselves, but which in the light of subsequent research have vanished and taken themselves off to the dim world of the undemonstrable, and become " as the baseless fabric of a vision." For the purpose of this paper, it would be safe to assume that the nebular hypothesis is in the main well proved, so far at least, as that the interior of the earth is in a highly heated condition with an enclosing crust through which the internal heat is escaping by various vents and even through the solid crust. This being so, we may naturally expect to find, in the strata composing the crust of the earth, evidence of heat other than that derived from the sun, and this solar supply is acting conservatively by slowing down the rate of earth-cooling, for if no such modifying causes as solar heat and chemical and mechanical forces are operating in the crust itself, the earth-heat would long ago have radiated, and like its satellite, the moon, the earth would wander through space a lifeless void, a world of barrenness.

Earth-temperature presents many difficulties to the observer, yet much interesting material has been collected upon the subject. As long ago as 1664, observations were made, and these have been continued, more or less, until the present time: the most elaborate and trustworthy being that of the Committee of the British Association for the Advancement of Science formed for that purpose in 1867.

It is found that the influence of seasonal temperature extends downward to a limited extent: in tropical countries, such as Java, not more than a few feet, but in Germany surface beds at a depth of 82 feet have an average temperature of 48° Fahr. In Siberia, frozen soil is met with at a depth of 260 feet, there being but little variation of the annual temperature at that depth. The writer is of opinion from this fact that it may be assumed, where frozen soil is met with at such depths, that a greater thickness of crust must intervene between it and the earth's heated interior, or that the rocks themselves are at the lowest point of conductivity.

The only means that we have of ascertaining rock-temperatures is by placing slow-action thermometers in bore-holes made during mining or sinking operations: care being exercised also in taking the mean annual temperature of the place where the observations are being made.

Many causes modify rock-temperature, and this subject has afforded and still affords a wide field for observation. It is generally considered, that, below a depth of about 65 feet, the temperature increases at the rate of 1° Fahr. for every 80 feet of further descent. It is, however, very interesting to note the variations recorded in the reports of the Committee of the British Association for the Advancement of Science, whose observations were made with slow-action and maximum thermo-meters.* In the case of mines and shafts, the thermometer was placed in a bore-hole driven to a depth of a few feet, the end being plugged with wood and clay, and left for lengths of time varying from days to months. Table I. contains a summary of the results taken from the fifteenth annual report† of the Committee; and the mean increase deduced from these results is 1° Fahr. in 64 feet.

Since 1882, other important observations have been made, which are recorded in Table II. The highest rock-temperature obtained at a depth of 4,580 feet was 79° Fahr., while the rock-temperature at a depth of 105 feet was 59° Fahr. The difference of temperature in the column of 4,475 feet of rock was 20° Fahr., averaging 1° Fahr. for every 224 feet. The average annual temperature of the air at the surface of the Calumet and Hecla mine is 48° Fahr., and that of the air at the bottom of the shaft is 72° Fahr.

* Reports of the British Association for the Advancement of Science, 1868, I., page 510; 1869, II., page 176; 1870, III., page 29; 1871, IV., page 14; 1872, V., page 128; 1873, VI., page 252; 1875, VII., page 14; 1875, VIII., page 156; 1876, IX., page 204; 1877, X., page 194; 1878, XI., page 178; 1879, XII., page 40; 1880, XIII., page 26; 1881, XIV., page 90; 1882, XV., page 72; 1883, XVI., page 45; 1885, XVII., page 93; 1889, XVIII., page 35; 1892, XIX. page 129; 1894, XX., page 107; 1895, XXI., page 75; and 1901, XXII., page 64.

† Ibid., 1882, XV., page 88.

Table III. records the temperature-observations made by Mr. H. A. Wheeler at other mines in the Lake Superior copper-mining district in 1886, and they differ considerably from those obtained at the Calumet and Hecla copper-mines. The mean increase is 1° Fahr. in 100·8 feet. The variations at the different mines are very striking: those nearest the shore of Lake Superior show the lowest, and those farthest away show the highest increase.

TABLE I.—COMPARATIVE OBSERVATIONS OF UNDERGROUND TEMPERATURES.

Localities.	Temperature Observed. Degs. Fahr.	Depth of Observations. Feet.	Rate per 1° Fahr. Feet.
Bootle, Liverpool, Well, Waterworks	59	1,302	130
Przibram Silver-mines, Bohemia	61½	1,900	126
St. Gothard Tunnel	87	5,578	82
Talargoch Lead-mine, Flintshire	60¾	1,041	80
Mont Cenis Tunnel	85	5,280	79
Nook Colliery, Manchester	62¼	1,050	79
Bredbury Colliery, Manchester	62	1,020	78½
Ashton Moss Colliery, Manchester	85¼	2,790	77
Denton Colliery, Manchester	66	1,317	77
Fowler's Colliery, Pontypridd, South Wales	—	855	76
Schemnitz Mines, Hungary	—	1,368	74
Astley Colliery, Dukinfield	86½	2,700	72
Wearmouth Colliery, Durham	71¼	1,584	70
Scarle Bore-hole, Lincolnshire	79	2,000	69
Kingswood Colliery, Bristol	74¾	1,769	68
Manegaon Bore-hole, India	84¾	310	68
Radstock Collieries, Bath	63	1,000	62
South Hetton Colliery, Bore-hole, Durham	77	1,929	57½
Grenelle Well, Paris	74½	1,312	57
St. André Well, Paris	64½	830	56½
Military School Well, Paris	61¾	568	56
Kentish-town Well, London	70	1,100	55
Rosebridge Colliery, Wigan	94	2,445	54
Kirklandneuk Bore-hole, Glasgow	53½	354	53
Yakutsk, Siberia	28¾	540	52
Sperenberg Bore-hole, Berlin	116	3,492	51½
Seraing Collieries, Belgium	87	1,657	50
Blythswood Bore-hole, Glasgow	53¾	347	50
Boldon Colliery, Durham	79	1,514	49
Anzin Collieries, France	67¾	658	47
Whitehaven Collieries, Cumberland	73	1,250	45
St. Petersburg Well, Russia	54	656	44
Carrickfergus Salt-mine, Ireland	66	770	43
South Balgray Bore-hole, Glasgow	59½	525	41
Carrickfergus Salt-mine, Ireland	62½	570	40
Slitt Lead-mine, Weardale	65	660	34

TABLE II.—OBSERVATIONS OF UNDERGROUND TEMPERATURES.

Localities.	Depth of Observations. Feet.	Rate per 1° Fahr. Feet.
Calumet and Hecla Copper-mines, Lake Superior	4,580	224
Rand Victoria Bore-hole, Transvaal	2,500	82
Port Jackson Bore-hole, New South Wales ...	2,929	80
Wheeling Wells, Virginia	4,462	72
Dolcoath Mine, Cornwall	2,124	70
Schladebach Bore-hole, Prussia	5,734	65
Baruschowitz Bore-hole, Upper Silesia	6,573	62
Comstock Lode, Nevada	2,230	33

TABLE III.—OBSERVATIONS OF UNDERGROUND TEMPERATURES.

Localities.	Depth of Observations. Feet.	Rate per 1° Fahr. Feet.
Quincy Copper-mine	1,931	123
Tamarack Copper-mine	2,240	111
Central Copper-mine	1,950	101
Atlantic Copper-mine	907	99½
Conglomerate Copper-mine	617	95
Osceola Copper-mine	996	76½

The deepest bore-hole, up to 1896, was made by the Prussian Government in Silesia, where a depth of 6,572 feet was attained, and the increase of temperature was found to be 1° Fahr. for every 62·1 feet. In the bore-hole at Schladebach, 387 temperature-determinations were made, taken at 58 points at equal distances down to the depth of 5,734 feet, and giving greater value to the results obtained.

With these facts in mind, the writer made various temperature-tests at Hamstead colliery, extending over a number of years. Taking a series of observations on the surface extending over a number of years, it is found that the mean annual temperature is about 49° Fahr. The temperature of the air of the intake air-current at the pit-bottom is more constant than at the surface. During the long frost of January and February, 1895, the average temperature of the air at the pit-bottom was 50° Fahr., while on the surface it was 25½° Fahr. The lowest temperature of the air observed at the pit-bottom was 43° Fahr. on February 8th, 1895, with a surface-temperature of 4° Fahr. In hot weather, the air at the pit-bottom is cooler than at the surface. The average temperature of the air at the pit-bottom is about 60° Fahr. or 11° higher than at the surface; and in ordinary weather the range of variation is small. At its ordinary velocity, about 3½ minutes is required for the passage of air down the Hamstead shaft, and in that time the air has reached the approximate temperature of the shaft-walls. Their temperature, at any given point in the shaft, will depend chiefly on the average temperature of the air passing them, and partly on the natural temperature of the strata. The air is compressed as it descends the shaft, the heating effect due to this cause being about 5½° Fahr. for every 1,000 feet of descent, and the mean temperature of the air will be about 10° Fahr. higher at the pit-bottom than at the surface. Thus, the increase of temperature at the pit-bottom is mainly accounted for, without assuming any heat to be extracted from the strata round the shaft; but considering that a certain amount of moisture is taken up in the shaft, and that this must cool the air somewhat, some heat must certainly be acquired from the strata, or the air would be decidedly cooler than it actually is.

The temperature of the undisturbed strata at the pit-bottom, 1,950 feet below the surface, is 66° Fahr. This was ascertained by inserting a maximum and minimum thermometer protected by a metal case, into a bore-hole driven 10 feet into freshly-cut coal. The hole was closed with clay, and left for various periods from 1 to 14 days. Repeated observations gave the same result, all observations showing an increase of temperature in undisturbed strata of 1° Fahr. for every 110 feet of descent beyond 65 feet from the surface.* It may be assumed that the mean average temperature of stratified rocks lies somewhere between the two extremes shown by the results of temperature-tests. The great variations of temperature obtained require some effort in the direction of understanding them; and perhaps they may be more readily understood, if they are taken in regular order, and the various modifying causes carefully considered.

Chemical Causes.—This head comprizes the generation of heat by local chemical action in such rocks as contain copper-pyrites or iron-pyrites, by percolation of water or even oxygen alone, which would be derived from air reaching the strata through fissures or lines of fault. It is well known that pyrites

* *Trans. Inst. M.E.*, 1895, vol. viii., page 411.

readily decomposes when combined with oxygen, evolving heat, thus resulting in a higher temperature-record than any non-pyritous rock would give. The pyrites in the veins, which fill old fractures in the Thick coal-seam, when decomposing, gives off a great amount of heat which results actually in fires, some-times of serious extent.[*]

Mechanical Causes.—Along lines of fault, where any move-ment has taken place, heat would be produced, sometimes on a grand scale: this would be confined mostly to regions of earth-quake phenomena, or districts where mining operations affect the stability of the rocks. Experience derived from the mining operations at Hamstead colliery has shown the extent to which a coal-seam may be heated by pressure and mechanical action, although it is scarcely possible to isolate completely the results of chemical activity from the production of heat in strata by mechanical force. At two places in a main road, where the temperature in a bore-hole, 10 feet long, was 66° Fahr. in 1891, it had risen to 83° in 1898, and ultimately reached 90° Fahr.

On reopening the mine after the fire of 1898, fresh roads were driven at a height of 120 feet above the Thick coal-seam, and the old workings were reached by an incline. About 3 months back, an old bore-hole was opened which had been closed since November, 1898, and on the thermometer being withdrawn, it was found that the mercury had been boiled, the column broken and the mercury diffused through the length of the tube. This bore-hole was situated 4,500 feet away from the seat of the fire, and the heat which had boiled the mercury could not have been derived from the fire; because food left by the workmen on leaving the pit on November 11th, 1898, was found in a comparatively fresh condition, the bacon was quite sound and sweet, a bottle of ginger-beer was found to be good when uncorked, and tubs of water left for the horses showed no signs of evaporation. This would not have been the case if the fire had heated the mine sufficiently to affect the thermometer sealed up in the bore-hole. The boiling of the mercury was entirely due to the heat of the coal-seam itself, and possibly to chemical and mechanical forces combined.

[*] *Trans. Inst. M.E.*, 1898, vol. xvi., page 478.

Moisture.—The proximity of springs of cool or heated water must modify considerably a temperature-record. The nature of a rock, as regards its capacity for holding moisture, such as a soft sandstone or a bed of marl, which is the reverse, will of necessity give different answers to our theories. The physiographical character of a district will give many variations, namely, inclined strata, curved and foliated beds, beds with cleavage-planes, in addition to stratification, jointed strata such as limestone, metamorphic rocks, rocks whose base is siliceous or of a basic character, all contribute to afford a chapter of varieties in the matter of thermal phenomena. The writer cannot enter into the matter of degrees of conductivity of rocks in this paper; but much useful information may be found on this question in the late Sir Joseph Prestwich's paper on "Underground Temperatures."[*]

Effects in Coal-mines.—In a paper by Dr. Haldane and the writer,[†] it is shown that the temperature of the air of the mine increases with the diminution of oxygen; and experience has shown that when men have to work in a temperature of 80° or 90° Fahr., and an atmosphere deficient in oxygen to the extent of 8 per cent. to which is added a large amount of moisture, it then becomes a physical impossibility for a man to do a good day's work. Perspiration is excessive, breathing is difficult, and the lack of oxygen causes sluggishness: this applies both to men and horses, but with men the loss of energy is serious, and is, in the writer's opinion, the primary cause of many accidents, as a man working under such conditions is not so alert to danger as one working under more favourable conditions. The sense of hearing is also affected.

Mining.—In mines working the Thick coal-seam, this increased heat is one of the causes of fires, and when worked on the longwall system, where slack is left in the goaf, very few fires would happen if the heat could be removed as quickly as it is generated, so that the ignition-temperature could not be reached.

From the foregoing remarks and particulars, it is evident

* *Proceedings of the Royal Society*, 1885, vol. xli., page 1.

† *Trans. Inst. M.E.*, 1898, vol. xvi., page 465.

that the increase of temperature is less in stratified than in
unstratified rocks, and as our coal-seams lie in the former, it is
greatly to the advantage of the miner.

The writer will now consider the effects of the heat of the
mine, in so far as it effects the mining of the coal.

It is evident to every one from the facts now available, that
in nearly all cases the temperature of rocks and coal is lower
than that of the air of the mine, and, as mining engineers, our
endeavours should be to introduce cool air into the mines in
larger volumes.

This brings us to the initial stage of the laying out of our
future large collieries at great depths, namely, the area of the
shafts. It was thought 20 years ago that a shaft with a
diameter of 15 feet was large enough for all purposes, but the
writer thinks that, where the coal is more than 1,500 feet deep,
the shafts should be not less than 20 feet in diameter, and that
the ventilating apparatus should be made of correspondingly
larger dimensions: this would enable greater volumes of air to
pass through the mine, and would produce cooler mines.

These improvements would entail a large increase in
the capital-account for plant, and a corresponding increase in
depreciation-charges. Very large areas would be required, containing enough coal to cover this enhanced expenditure. It may
be urged that the supply of more air, and consequently more
oxygen, would result in more fires in coal-seams so favourable
to the absorption of oxygen as is the Thick coal-seam. It is
a puzzle to the writer, as many cases of actual fires recur to
his mind, where 80,000 to 90,000 cubic feet of air per minute
were passing; but such cases have been few as compared with
those that occur where not so much air is passing, as in a
return-airway having a temperature of 70° to 80° Fahr.

Summary.—A brief summary of the foregoing results as
regards deep mining will help to fix the points in our minds:—

(1) The extra depth is easily surmounted, so far as mechanical
power is concerned. Electricity has made such rapid strides
that the conveyance of power to any distance is easily
accomplished. The cables and motors can be conveniently and
cheaply placed at any part of the mine, where it was formerly
difficult to use ropes for the transmission of energy to the main
distributing station.

(2) The heat of the rocks will not be so great as was anticipated, and will probably not exceed 1° Fahr. for each 120 feet of increase in depth.

(3) The temperature of the air can be reduced by sinking larger shafts, and the adoption of more efficient ventilating machinery.

(4) If the mines are supplied with larger volumes of fresh air the men will be able to do as much work at the depth of 3,000 feet as is now done at the depth of 1,000 feet. Thus at a depth of 2,100 feet the temperature of the rock is about 66°, and at 3,500 feet it may be about 80° Fahr. But with ample ventilation a man will be able to do all the superintendance that will be required from him, in a few years hence, when further improvements have taken place in coal-cutting machines, which have even now removed much of the hardest work from the pikeman.

And, further, there can be no doubt but that mining engineers will be able to reach any depth at which workable coals are likely to be found in this country, and work the same. It will then be a question of capital outlay and the competition of the deeper with the shallower mines. The difficulties of deep mining can be overcome by large outputs, and also by working the mine continuously, day-and-night. Often the expenditure of £200,000 to £500,000 is only being employed 8 or 9 hours out of the 24, with the result that the standing charges eat up the profits. Whereas, if the mine could be worked, say, for 16 hours, the standing charges would be spread over a much larger output, and this reduction would pay interest and form a redemption-fund for the extra capital that will be required for deeper mining, without increasing the costs of the coal to the consumer.

—— ——

Mr. S. L. THACKER asked whether Mr. Meachem had considered the possibility of the use of liquid air in deep mines.

Mr. W. N. ATKINSON questioned Mr. Meachem's generalization from observations at Hamstead colliery that the temperature of the air of the mine was usually higher than that of the strata; no doubt in some instances that was the case, but in others, and in the majority, he did not think that it was so.

He thought that the temperature of the air was chiefly derived from the heat of the strata, and that as a rule the temperature of the air did not become higher than that of the strata. No doubt much of the heat at Hamstead colliery was due to oxidation of the coal, and if in any particular instance the air of the mine attained a higher temperature, a new element had been introduced which probably only applied to that particular mine.

With regard to the rate of increase of the temperature of the strata, it appeared to him (Mr. Atkinson) that the ratio found by Mr. Meachem at Hamstead colliery was a slow one. In a recent sinking in North Staffordshire, down to a depth of 2,625 feet, the rate of increase was 1° Fahr. in 68 feet.

Prof. R. A. S. REDMAYNE said that he could not concur with Mr. Meachem in his opinion that the air acquired a higher temperature than that of the strata, except in extreme cases. At Rosebridge colliery, when first mined, the strata reached a high temperature which fell considerably afterwards, and it was found that the temperature of the air fell with the temperature of the rocks until it reached a normal result. It might be said that the heat of the air was derived from two sources, physical and chemical, and one might add locality. The temperature varied considerably with locality, and this was true of the deep American mines to a surprising extent, the temperature in some mines in the western states being almost unbearable, while in others in the Lake Superior districts the temperature was by no means excessive. He believed that the primary cause of the failure to work the great Comstock lode was the excessive temperature generated by chemical action, due to the action of water on the ores, the sulphides being converted partly into sulphates, a considerable amount of heat being evolved in the process. He thought that the old ratio of an increase of 1° Fahr. in temperature for every 60 feet after the first 100 feet was probably incorrect, but before he could accept the high ratio of increase advanced by Mr. Meachem, he thought that many more observations were required.

Mr. W. CHARLTON said that, no doubt, in some mines temperatures were observed higher than the natural temperatures of the strata, due in some measure to mechanical but principally to chemical causes. The absorption of oxygen and the amount of oxidation taking place would vary very much with .the character of the coals and the system of working. In some methods a very large coal-surface is constantly exposed, whereas in others, such as the longwall system, a minimum amount of coal is subject to oxidation.

Mr. F. G. MEACHEM said that he had considered the possible utility of applying liquid air, but he did not think that it came within the range of practical mining at the present time.

The amount of oxygen would be greatly influenced by the system of working, the amount of face exposed, and the amount of carbonaceous material left behind in the mine: a system of working that would seal up refuse and prevent its oxidation would be of great benefit. With reference to refrigeration, he was informed that to reduce the temperature of 10,000 cubic feet of air per minute to the extent of 20° Fahr. would require a capital-expenditure of £10,000.

The PRESIDENT (Mr. T. J. Davies), in proposing a hearty vote of thanks to Mr. Meachem for his interesting paper, said that one of the chief difficulties to be overcome in deep mining was certainly that of high temperatures ; and he had little doubt that means would be found to cope with them satisfactorily. The problems of deep mining were really those of finance : when capitalists could foresee a return for their money, there was little doubt that the difficulties would be surmounted.

Prof. REDMAYNE seconded the resolution, which was cordially approved.

____

DISCUSSION OF PROF. R. A. S. REDMAYNE'S PAPER ON "THE TRAINING OF A MINING ENGINEER."*

The PRESIDENT (Mr. T. J. Davies) remarked that Prof. Redmayne's retrospective account of mine-managers and viewers of old enlightened the members as to the quaint ideas and methods of primitive mining. Practical mining was first acquired, then followed the philosophy of mining, built upon the experience acquired in mining practice; and as difficulties, owing to deep mining and other causes, increase, the mining philosopher is led to the adoption of mechanical aids, which

* *Trans. Inst. M.E.*, 1902, vol. xxiv., page 243.

science invents or reveals. Prof. Redmayne did not appear to be satisfied with the efforts being made to introduce coal-cutting machines into mines. The difference between this country and America is striking. While 20 per cent. of the output in America is worked with the aid of coal-cutting machines, in this country, it is only 1½ per cent.; but the fact that the employment of such machines is on the wane in this country is proof that the conditions of our mines are unfavourable to their use.

The course sketched out for the training of mining-engineers is an admirable one, and proves the author's keen insight into the scientific requirements of the mine-manager of the future.

Mr. E. H. ROBERTON, referring to the British system of apprenticeship, said that it was too conducive to habits of conservatism. A man is apprenticed in his youth to a mining-engineer at a colliery in a certain district, and in due course obtains his certificate as a colliery-manager and becomes the manager of a mine, without ever going far out of the district or even out of the immediate neighbourhood. He is only acquainted with the methods of mining in vogue in the district, and the method that he practises may perhaps be a generation old. In some favoured districts, there is a mining-school, but, after a day's work in the pit, he has not usually much energy left for concentration of thought and brain at a night-school. In a few, and the more favoured instances, the apprentice mining-engineer is enabled to visit other districts or countries, and learn other methods. This course is perhaps the soundest way of learning the business of a mining-engineer, but it is open only to the few. He thought it was desirable that the student mining-engineer should spend some years at a mining-school, where he would mix with students from other collieries and, probably, other districts, and he would be able both to hear their views and give them his own. Further, he would have the advantage of having nothing to distract his attention from his studies, and, in addition, he would obtain practical experience by visiting mines in different localities, and worked under different conditions.

Mr. W. CHARLTON said that Prof. Redmayne's paper was a strong plea for systematic and scientific training of mining-

engineers, colliery-managers and mine-officials; it was difficult, however, to arouse interest in the subject.   The present
system of apprenticeship was not altogether satisfactory, as a
young man on being articled was usually turned loose about
a colliery, and only a few made the best of their opportunities.
Some mining-engineers prescribed a routine of duties so as to
give their pupils a knowledge of all the practical details of
their profession; but, in the majority of cases, technical education or training is unsought for until the near arrival of the
time for examination, and then there is a period of excessive
cramming that cannot conduce to efficiency.   After an experience of fourteen years as a teacher, with pupils of all ages and
character, including those training to become mine-managers
and workmen anxious to improve their theoretical knowledge
and to qualify for official positions in mines, he had come to the
regrettable conclusion that much of the teaching had been
wasted, owing to the educational deficiences of the students.
The Boards for Examinations might improve matters by requiring evidence from candidates of their having undergone, over
a period of years prior to their participation in the examination
for certificates of competency, systematic training and education
in the sciences pertaining to the profession of a mining-engineer.

Mr. S. L. THACKER said that if full advantage was to be
taken of the educational facilities afforded by the Universities,
and if the equipment of our mining-engineers was to attain the
level desired by Prof. Redmayne, the whole scheme of the
training and qualification of mining-engineers would have to
be remodelled.   It had been urged that the five years of
"practical experience in a mine," required by the Coal-mines
Regulation Act, was a stumbling block in the way of a collegiate
training, but he thought that the point had been overlooked that
the examinations for certificates of competency held by the
Boards for Examinations were only intended as qualifications
for colliery-managers under that Act.   There was very little
likelihood of the period of practical experience being reduced,
and in his opinion it would be a very unwise thing to do so,
as a mine-manager could not have too much practical knowledge
in the handling of men and in the working of the mine.
What it seemed to him was needed was a higher and more

comprehensive qualification for general managers and consulting mining-engineers. He would like to see a Central Board for Examinations for the whole of Great Britain (and possibly also for the Colonies), leaving the existing boards to deal with colliery-managers precisely as at present constituted: for, owing to the variations in local conditions, their work must of necessity be decentralized. The Central Board would establish an examination, the passing of which would at once be recognized as indicative of proficiency and intellectual equipment, equal to any position or branch of mining-engineering, and equal to any qualification obtained abroad. A certain length of time spent at a mine would be compulsory, but not necessarily to the extent of five years; and, that being so, he believed that the mining-schools of the technical universities would soon be better filled. There would be a preliminary or students' examination for which matriculation at one of the universities would be considered an equivalent, and the final examination would be in part general and compulsory and in part alternative, that is, coal-mining or metal-mining. Generally, the course would broadly follow that of the German State qualifications for chief engineers, described in a presidential address, by Mr. G. Blake Walker in September, 1896.*

The next question was how was this Central Board to be established; there was, he thought, very little probability of getting the Government to do it and he very much questioned any such necessity. What came more within the province of The Institution of Mining Engineers, or what would be more in keeping with the original purposes of its inception than the establishment of such a Board for Examinations? And it would, he submitted, raise the status and give a prestige to mining-engineers equal to that of any other profession.

Prof. R. A. S. Redmayne, in replying to the discussion, said that in view of the Coal-mines Certificates Bill about to be brought before Parliament, the subject of the training of mining-engineers was assuming some importance. Mr. W. Charlton, whose opinions were backed by many years' experience as a teacher of mining, had pointed out the inefficient elementary education of the average mining-pupil. The pre-

* *Trans. Inst. M.E.*, 1896, vol. xii., page 132.

liminary scholarship was as a general rule poor and indifferent, and the student was seldom in a condition to utilize to advantage the training mapped out for him.

With regard to Mr. Davies' remarks on coal-cutting machinery, he was of opinion that the conditions at a large proportion of the coal-mines of this country were favourable to the adoption of such machinery, and certainly a much larger proportion than had coal-cutting machinery at work at present.

—

THE SOUTH STAFFORDSHIRE AN
SHIRE INSTITUTE OF MINII

GENERAL MEETIN
HELD AT THE UNIVERSITY, BIRMINGHAM

MR. J. T. DAVIES, PRESIDENT, I

The minutes of the last General
Meetings were read and confirmed.

The PRESIDENT (Mr. T. J. Davies), i
dolence with the families of the late
Hayward and John Aston, who had die
Meeting, remarked that both gentleme
the Institute: Mr. Hayward had se
Treasurer, and was a mining engineer
district and of great ability.

The motion was carried in silence.

The following gentlemen were electe
MEMBERS—
Mr. WILLIAM STANDLEY, Mining Engineer, Ne
Mr. G. R. WARDLE, Mining Engineer, Poleswo
Mr. T. WARTH, Mining Engineer, The Univers

Messrs. R. S. Williamson and W. '
representatives of the Institute on the Co
of Mining Engineers.

Mr. W. F. Clark was appointed Tre

DISCUSSION OF MR. F. G. MEACHEM'S PAPER ON " UNDERGROUND TEMPERATURES."*

Mr. F. G. MEACHEM, in opening the discussion, laid stress on the following points:—(1) It had until recently been thought that the limit of practical mining would be reached at a depth of 3,000 feet, thus greatly restricting the working of the lower or deeper mines: the main reason being the excessive heat of the strata, alleged to be probably 98° Fahr. at a depth of 3,000 feet, making manual labour impossible; and (2) the weight of the superincumbent strata would be so great; but this latter point had not been considered in his paper. The result of his own research was that the probable ratio of increase of rock-temperature would be 1° Fahr. for every 110 feet of increase in depth, and not 1° in every 65 feet, as per the generally accepted rule; and, further, the probability was that, as the depth increased, the ratio of temperature-increase would probably not increase so rapidly until very considerable depths had been attained, so that the limit of depth in mining, which had been set, might reasonably be much increased. Further, much would depend upon the nature of the rock forming the base of the Coal-measures, and in proportion to their conductivity they would affect the temperature of the mine.

The South Staffordshire coal-field rests upon Silurian rocks. He only knew of one place where there was a thin layer of Devonian or Old Red Sandstone-measures intervening, but when the eastern boundary fault of the South Staffordshire coal-field was passed, it was very likely that instead of Silurian rocks forming the base, it would be found that older rocks—quartzites and kindred formations—would form the base of the new coal-field, as at Barnt Green and other places in Warwickshire and Leicestershire. This unstratified mass, by its greater conductivity, would exert a great influence upon the temperature of the mines. Stratified rocks would, he considered, be absent beyond a depth of 7,000 feet.

Mr. W. F. CLARK remarked that the question of the conductivity of rocks, stratified or unstratified, was a most important one, as affecting the temperature of the mine, and if older rocks, probably unstratified, formed the base of the future Mid-

* *Trans. Inst. M.E.*, 1903, vol. xxv., page 267.

land coal-field, they would no doubt influence the temperature of the mine. He had lately been conducting some boring operations on the western side of the coal-field, and instead of either the Old Red Sandstone or Silurian rocks forming the base, Millstone Grit had been found. This was at first doubted, even by expert geologists, but finally it was pronounced to be Millstone Grit. He assumed that the writer—supposing the Coal-measures overlay the Millstone Grit—would consider that the Millstone Grit was a more favourable base for mining operations than the older rocks.

Mr. F. G. MEACHEM thought that the Millstone Grit, as in the Dean Forest coal-field, would form the most favourable base.

Mr. ALEXANDER SMITH remarked that the Millstone Grit formed the base of the Warwickshire coal-field.

Mr. T. H. BAILEY pointed out that very much depended upon the conductivity of the rocks, and the great variety of sources of heat in the mines. Many mines were subject to spontaneous combustion, and in such the heat would necessarily be greater; while in others, the mechanical causes of heat were great, and a large amount of heat was produced by mechanical action alone. He thought that Mr. Meachem was right in advocating more ventilation to carry off the heat, irrespective of its source.

Mr. W. N. ATKINSON (H.M. Inspector of Mines) said that he had conducted some observations in South Staffordshire, and had arrived at somewhat similar results to those of Mr. Meachem with regard to the temperature of the rocks. Chemical action was one of the principal sources of heat in the mine; and he thought that the heat due to the pressure of the overlying strata was slight—sudden pressure no doubt might produce a temporary increase of heat, but, as a general rule, it had no appreciable effect. Frictional pressure was not the cause of spontaneous combustion, as it was rather due to oxidation and other chemical causes. He had, some time ago, made a few observations which tended to prove that coal, liable to spontaneous combustion, rapidly gave off heat, thus tending to make the temperature of the air higher than that of the strata, and this was not surprising when oxidation was so rapid; but it could not be accepted as a general rule.

Mr. E. H. Roberton, referring to the temperature-records given by Mr. F. G. Meachem, said that at Cremorne, Sydney Harbour, the temperature recorded in a bore-hole, put down by a diamond-drill, shewed an increase of 1° Fahr, for every 80 feet of increase in depth. In the Birthday shaft, distant about 3 miles from the Cremorne bore-hole, the increase was 1° Fahr. in every 94 feet of increase in depth. The methods adopted to ensure accuracy were interesting. It was necessary to protect the thermometers against water-pressure in the bore-hole, and those thermometers which were not protected by glass were enclosed in an iron-tube, sealed at both ends by caps screwed on hot, with molten lead in the threads of the screws. In the temperatures taken in the sinking shaft, readings were commenced at a depth of 600 feet: the methods employed being those recommended by the Underground Temperatures Committee of the British Association for the Advancement of Science. From 600 feet to a depth of 950 feet, bore-holes were drilled 3 feet into the rock; while from 950 to 1,150 feet, they were drilled 4 feet; and below that depth, 5 feet. Two thermometers, a slow-action and a maximum, were used in each hole. After drilling, the holes were allowed to remain open for a period of from 34 to 84 hours, so that the heat generated by the drilling could escape. The thermometers, enclosed in copper-cases, were inserted, and the hole plugged with greasy cotton-waste and clay. The instruments were left in the holes for 37 hours and upwards. The slow-action thermometer had a layer of paraffin-wax surrounding the bulb, to enable readings to be taken before any appreciable alteration in the height of the mercury-column had taken place.

Mr. F. G. Meachem mentioned that reliable readings could not be taken if the thermometers were not inserted into holes drilled for a depth of 10 or 15 feet. In his investigations at Hamstead colliery, the holes were bored to the latter depth.

The President (Mr. T. J. Davies) agreed that the rocks forming the base of a coal-field would have great influence on the temperature of deep mines, and if they adopted the nebular hypothesis, the conductivity of the rocks would be one of the greatest factors. He thought that liquid air and refrigeration as a means of cooling the air traversing a mine would play a most important part in the deep mines of the future.

NORTH STAFFORDSHIRE INSTITUTE OF MINING AND MECHANICAL ENGINEERS.

EXCURSION MEETING,
HELD AT BOLTON, JUNE 22ND, 1903.

HULTON COLLIERY.
DEEP ARLEY PIT.

The turbo-fan is of the screw type, made of manganese-bronze, in one casting. There are eight blades, and the diameter of the casing in which the fan runs is 3½ feet. The fan is direct-driven by a steam-turbine of 250 horsepower. In addition to driving the fan, the turbine also drives direct an electric generator of 70 horsepower.

The generator supplies electricity for driving coal-cutters, the condenser-pump, and for surface-lighting. The turbine, the fan, and the armature of the generator run at 3,000 revolutions per minute. The fan exhausts 90,000 cubic feet of air per minute, under a water-gauge of 3½ inches.

The Evence-Coppée washer is capable of washing 15 to 20 tons of slack per hour.

CHEQUERBENT, No. 2 PIT.

The electric generating-plant comprizes two horizontal engines, each driving by belts a 100 horsepower dynamo of the continuous-current type. The day-work of these generators consists in supplying current for:—2 coal-cutters, 1 hauling-engine of 70 horsepower, 1 hauling-engine of 10 horsepower, 1 hauling-engine of 15 horsepower, with an attached rope-driven pump; and a three-throw pump of 10 horsepower. The night-work consists in supplying current for 6 coal-cutters and 1 pump.

The hauling-engine, close to the pit-bottom, in the Arley mine, comprizes a 70 horsepower motor, driving, by means of cotton-ropes, through suitable gearing, the main endless haulage-rope, travelling about 6,000 feet per hour. The secondary ropes are taken into various districts, and actuated by clutch-pulleys.

Morgan-Gardner chain and Diamond disc coal-cutters are working in the Arley mine on longwall faces, taking out 4 feet of holing.

Seven hydraulic mining-cartridges are used in place of explosives, and no shot-firing whatever takes place in any seam worked at these collieries.

Nos. 3 and 4 Pits.

The No. 3 pit winding-engine has two cylinders each 32 inches in diameter by 6 feet stroke, supplied with steam at a pressure of 100 pounds per square inch. The drum is 15 feet in diameter, and the depth of winding is 900 feet.

The No. 4 pit winding-engine has two cylinders, each 36 inches in diameter by 6 feet stroke, supplied with steam at a pressure of 100 pounds per square inch. The drum is 18 feet in diameter, and the depths of winding are 450 and 1,350 feet.

There are six boilers, four working at a pressure of 100 pounds per square inch and two at 150 pounds, fitted with forced-draught.

The Green economizers are driven by an electric motor.

The electric generating-plant comprises two turbine-driven sets, each of 400 horsepower at 470 volts. Electricity is supplied to the screens, an underground pump, a boiler feed-pump, forced-draught fan for boilers, economizers, small fan, hauling-engine and condenser-pump.

The screens are electrically driven, and are arranged for dealing with 600 tons or more per day. They are fitted with creepers, machine-tippler, lowering arms for passing coal into wagons, etc.

The electric three-throw pump, fixed in the Trencherbone mouthing, is capable of raising 10,000 gallons of water per hour against a head of 600 feet.

The electric hauling-engine, of 8 horsepower, drives an endless-rope in the Trencherbone mine.

The setting-out of the mouthings at the pit-bottom has been arranged with a view to economy in labour, and convenience in passing the tubs quickly to and from the cages.

There are eight hydraulic mining-cartridges regularly in use at these mines, so as to discard the use of explosives, and thereby obtain a better quality of coal. The weight of the apparatus is about 56 pounds, and one man carries it about from place to place, breaking down coal from 25 to 30 places per day.

THE NORTH OF ENGLAND INSTITUTE OF MINING AND MECHANICAL ENGINEERS.

GENERAL MEETING,

HELD AT THE KESWICK HOTEL, KESWICK, JUNE 10TH, 1903.

SIR LINDSAY WOOD, BART., PRESIDENT, IN THE CHAIR.

The SECRETARY read the minutes of the last General Meeting and reported the proceedings of the Council at their meetings on May 30th and that day.

DEATH OF MR. A. L. COLLINS.

The following letters were read :—

NEWCASTLE-UPON-TYNE,
7th February, 1903.

SIR,

The Council desire me to call your attention to the death of Mr. Arthur Launcelot Collins, a member of this Institute, who was murdered at Telluride, Colorado, United States of America, in November last.

The Council trust that the British Ambassador to the United States may be instructed to use every endeavour in order that the perpetrators of this murder may be brought to an early trial.

I enclose herewith, for your information, an account of " The Tragedy at Telluride," taken from the *Engineering and Mining Journal* (published in New York) of November 29th, 1902. *

I have the honour to be, Sir,
Your obedient Servant,
M. WALTON BROWN,
Secretary.

The Right Honourable the Secretary of State,
Foreign Office, London, S.W.

---

* THE TRAGEDY AT TELLURIDE.—Just after going to press last week. our correspondent in Colorado telegraphed that Mr. A. L. Collins, the manager of the Smuggler Union mine, at Telluride, had been shot in his house, with results that were fatal within 36 hours. The act was that of an assassin who had fired buck-shot through the window of the library, in which Mr. Collins sat with two or three friends.

Throughout the membership of that profession which Arthur Collins adorned by his high intelligence and earnest work, this sad news will provoke a feeling of great pity that a useful and honourable career should be terminated in such a tragic manner, and to this pity will be added the bitterness of resentment that an

FOREIGN OFFICE,
April 21st, 1£03.

SIR,

With reference to the letter from this office of the 16th of February last, I am directed by the Marquess of Landsdowne to inform you that a despatch has been received from Sir Michael Herbert, His Majesty's Ambassador at Washington, on the subject of the murder of Mr. A. L. Collins, at Telluride, Colorado, in. November last.

unoffending man of high character should be sacrificed to the spirit of lawlessness which has prevailed at Telluride for the past two years, and of which this tragedy is the logical outcome. Our readers will remember that in July, 1901, there was a strike among the workmen at the Smuggler Union mine, consequent upon the introduction of the contract system; they will also recall the fact of a murderous attack made by a body of strikers, in the course of which the latter shot indiscriminately into the dwellings and office buildings at the mine, killing eight men and severely wounding many more. Nothing whatever was done by the county or the State to punish this outrage, and not a single individual has ever suffered any punishment for this act of cowardly ruffianism. In July of the present year, a marble monument was erected to the memory of the single member of the attacking party who was killed in that assault, and at the un-- veiling of this monument to the murderer a number of flowery orations were spouted by local politicians. Amid these happenings, Mr. Collins, as manager of the mine, stood out fearlessly for the maintenance of law and order, and when the Sixteen-to-one Miners' Union, of Telluride, sent him a list of scabs, or non-union men, warning him not to give them employment, he immediately inserted a paid advertisement in each of the local newspapers promising work to any man on that list so long as he was manager of the property then under his control. In speaking of this incident he would exclaim, with indignation, that the names on. that list "could be pronounced!" That is, they were men of American and English extraction, in contrast to the bulk of the miners in the district who are Austrians, Slavs, Italians, etc., with enough hot-headed ag'in-the-government Irish to lead these foreigners into devilry.

Throughout these troubles, the local and State authorities have been shamelessly negligent of their duty to the community in the enforcement of the law. The present Governor of Colorado is an ordinary political accident, and has permitted the exigencies of corrupt politics to tie his hands. It may be necessary to explain that the Miners' Union is influential in politics, not only locally, but also through its affiliations with similar organisations throughout the State, and the members as a body belong to the political party which of late has been dominant in Colorado on account of the socialistic and anarchistic tendencies. developed by the agitation following the fall in silver. The Smuggler Union mine is owned by Boston capitalists, who are likely to cease operations at the mine and keep it shut down until there is better evidence of protection to life and property in this particular district; they will doubtless be glad to meet the heavy expenses of a thorough investigation, and will spare no effort toward the arrest of the criminal, even though the county and State prove as supine as heretofore. Whether this culminating tragedy will arouse the State authorities to action we do not know, but if it does not it appears to us that it is high time for the civic spirit of Colorado to awaken and to see to it that punishment awaits the perpetrators of this and other crimes committed at Telluride.

From considerations such as these one turns again sadly enough to the realisation of the cost of all this frontier lawlessness and political expediency. Arthur Collins was a member of an old Cornish mining family. He was barely 40 years of age, with a young wife and two lovely children. He was well-travelled and extremely well-informed in technical matters. A few years ago he had been adviser in mining to the Ameer of Afghanistan, and had been placed amid surroundings requiring a high type of courage. As manager of mines first at Central City and then at Telluride, he had evinced great energy and capacity. He seemed destined to occupy a very honourable place in the profession, and he could look forward to a career of domestic happiness and professional distinction. This—citizenship of the very best—is suddenly and without warning wantonly destroyed because local politics have made it inexpedient to enforce the administration of the law.

Sir M. Herbert has received a report from the British Vice-consul at Denver, who was instructed to make enquiry into the steps being taken to bring the murderers of Mr. Collins to justice, and has been in communication with the Governor of Colorado with that object.

The latter has furnished a full statement of the measures taken by the United States authorities for the purpose of investigating the murder, from which it appears that after an investigation lasting nearly three weeks, during which time a grand jury, specially empanelled, examined a large number of witnesses, no clue has been obtained as to the identity of the assassin.

The Governor states that every effort, without regard to cost, has been made to clear up the mystery, and that every resource at his command will be devoted to discovering the perpetrators of the crime and securing their conviction.

I am, Sir,

Your most obedient humble Servant,

F. H. VILLIERS.

AWARDS FOR PAPERS.

The SECRETARY read the following list of papers, communicated during the year 1901-1902, for which prizes of books had been awarded by the Council to the authors:—

"A Method of Socketing a Winding-rope, and its Attachment to a Cage without the Use of Ordinary Chains." By Mr. W. C. Blackett, M.I.M.E.

"Mechanical Undercutting in Cape Colony." By Mr. John Colley, M.I.M.E.

"Electric Pumping-plant at South Durham Collieries." By Mr. Fenwick Darling, M.I.M.E.

"Some Silver-bearing Veins of Mexico." By Mr. Edward Halse, M.I.M.E.

"Apparatus for Closing the Top of the Upcast-shaft at Woodhorn Colliery." By Mr. C. Liddell, Stud.I.M.E.

"Standardization of Surveyors' Chains." By Prof. Henry Louis, M.I.M.E.

"A Visit to the Simplon Tunnel: the Works and Workmen." By Dr. Thomas Oliver.

"The Carboniferous Limestone Quarries of Weardale." By Mr. A. L. Steavenson, M.I.M.E.

"Auriferous Gravels and Hydraulic Mining." By Mr. W. S. Welton, M.I.M.E.

"Tapping Drowned Workings at Wheatley Hill Colliery." By Mr. W. B. Wilson, jun., M.I.M.E.

The SECRETARY read the balloting-list for the election of officers for the year 1903-1904.

The following gentlemen were elected, having been previously nominated:—

———

Mr. W. E. WALKER's paper on " Hæmatite-deposits and Hæmatite-mining in West Cumberland " was read as follows :—

HÆMATITE-DEPOSITS AND HÆMATITE-MINING IN WEST CUMBERLAND.

By W. E. WALKER.

Introduction.—The writer believes that it is a long time since a paper relating to hæmatite-deposits and hæmatite-mining in West Cumberland has been placed before the members, and, as discoveries of great importance have been made in recent years, in West Cumberland, it occurred to him that the subject might be one of interest to members of the Institute.

When the writer commenced hæmatite-mining over thirty years ago, the ore being worked was, with very few exceptions, at a, comparatively speaking, shallow depth. Horse-gins and small combined winding-and-pumping-engines were employed at the shafts, and horses were, in several cases, used underground for the purpose of drawing the ore along inclined roads.

In those primitive days, many shafts were sunk to the rise of the ore-deposit for the purpose of reaching it as quickly as possible, regardless of future cost in working. As this ore, after a time, was much reduced in quantity, it became necessary to prospect for other and deeper deposits by means of boring, some of the bore-holes going down to the slate-rock. Several of these bore-holes were unsuccessful, while others found ore in such quantities as to revolutionize iron-ore mining in West Cumberland. From 200 to 300 feet was formerly looked upon as a fairly deep shaft, whereas the deep bore-holes, above referred to, have necessitated the sinking of pits from 700 to 900 feet in depth. Even that depth is now being far exceeded, south of Egremont, by the Wyndham Mining Company, who are sinking a shaft to a depth of over 1,200 feet; and by the Beckermet Mining Company, who are sinking a shaft near to Winscales, about 1 mile south of Egremont.

The former system, of sinking the shaft to the rise of the deposit, naturally led to great expense in pumping water and drawing the ore from the dip workings. Now, all this has been changed, for, after the discovery of the dip of the "sole," that is, the floor of the deposit, by means of bore-holes, the shaft is sunk to the dip, and thus gravitation aids the conveyance of both ore and water to the foot of the shaft. There is no comparison in the cost of boring, sinking and providing plant with the cost as it stood formerly: for instance, where £2,000 would cover the whole expense in olden days, £10,000, £20,000, and even £30,000, do not always cover it now. Many firms spend thousands of pounds in boring alone, before they feel justified in sinking a shaft, and it has sometimes happened that the boring-operations have been entirely unsuccessful.

Accidents.—In opening out, developing, working and robbing an iron-ore mine, the first consideration of the owner and management is the safety of the men employed, and the efficient ventilation of the mine. During the whole time that the writer has been connected with hæmatite-mining, he has been brought into contact with accidents, to workmen, of all kinds, but he has never known of one where any blame whatever could be justly charged to default on the part of either the owner or the management. Every accident, so far as he is aware, has been caused either by the workman not obeying the instructions of the manager, or by utter carelessness; or, in such a way, that it was entirely impossible for anyone to foresee such dangers as "slips" from smooth joints, which, when sounded by the inspector and miners, showed no indication that movement was likely to take place.

After the safety of the miners, the next consideration is to obtain as much ore as possible from the mine, clean, and at as low a cost as possible, seeing, at the same time, that the workmen employed receive proper remuneration for their labour. It is the universal custom in the district for the miners to take contracts, which are regulated according to the selling-price of the ore.

Machinery.—At present, the large winding-engines and pumping-engines, employed on the surface, are worked by steam, but, in the not far-distant future, the writer hopes to see electricity (undoubtedly the power of the future) used for this purpose. He also anticipates that the hand-drills used for boring shot-holes, both in ore-workings and limestone-drifts and sinkings, will be supplanted by more scientific and efficient methods, which will be of the greatest advantage, both to mine-owners and miners, by producing an increased output at practically the same all-round cost and with much less manual labour.

Dip-workings are, of course, occasionally necessary even now. The ore from them has to be drawn by haulage, and the water removed by steam or hydraulic engines in a way, arranged in detail, according to circumstances.

Spain as a Competitor.—As the members are no doubt aware, Spain is the chief competitor of the iron-ore mines in West Cumberland and North Lancashire. In Spain, the ore has simply to be quarried from the hill-side, and it varies in thickness in the Bilbao district from about 3 feet to about 40 feet. Some years ago, when the writer was visiting one of the largest mines in that neighbourhood, about 20,000 tons of ore were sent away weekly, the total cost being less than 4s. per ton free-on-board ship in the Bilbao river. At that time, about 1,500 work-people were employed: six Englishmen directed the drilling of shot-holes, about 200 women carried the ore in small baskets on their heads from the bank to the railway-wagons; and the remainder were Spanish labourers employed in breaking ore, filling baskets, and lifting them on to the heads of the women, as they passed by, all working with the regularity of a machine.

The men were paid, at that time, from 2s. to 2s. 4d. per day, and they lived chiefly in wooden sheds on the mountains, 20, 30 and even 40 men frequently sleeping in one of these, their beds each consisting of a sack laid on the ground. Needless to say that such wages and mode of life would be inapplicable to the English miner. During the summer, these men work from 5 a.m. to 8 p.m., with 1 hour for breakfast and 2 hours for dinner and siesta. In the autumn, winter and spring, they work from daylight until dark, and, although they live on the poorest of fare, the amount of work that they perform is astonishing.

The main part of the Bilbao hæmatite-deposit has now been worked out, and mine-owners have to go farther into the country for their supplies of ore. This implies increased cost and an increase of capital that will gradually bring the Spanish mines more on to a level with those of this district. There is no concealing the fact, however, that the iron-ore resources of Spain are enormous, though the quality of the ore, taken as a whole, is not so good as formerly.

The writer simply mentions these matters regarding Spanish iron-ore to show what we, in this district, have to compete with. If labour in Spain were put on anything approaching the same footing as it is here (and the writer believes that it is gradually moving in that direction) it would bring the cost of Spanish iron-ores nearer to a level with those of West Cumberland and North Lancashire. We must, however, not lose sight of the fact that if Spain had not been sending iron-ore to West Cumberland and North Lancashire for some years past, neither West Cumberland nor North Lancashire could have kept the number of furnaces in blast that they have done.

Ore-deposits.—In some parts of the hæmatite-district of West Cumberland there are four deposits of ore underlying each other. The No. 1 bed of ore is found in the first limestone; the No. 2 bed in the second limestone, about 30 feet beneath the first bed; the No. 3 bed occurs in the third and fourth limestones, about 250 to 300 feet below the second bed; and the No. 4 bed is found lying on the slate-rock or on calcareous limestone, which is occasionally found immediately above the slate-rock. If the ore be lying on the calcareous limestone it is much more easily worked, because the calcareous limestone, being of a hard and strong character, allows pillars of ore to stand for the purpose of supporting the workings without much expense in timber, whereas, if the iron-ore pillars are immediately on the shale, they are liable to shrink, thus necessitating extra labour and the use of artificial support to the workings.

These four deposits extend over a considerable area, though the main part of the first and second deposits, so far as is known, has been worked out. The quality of the ore is good, particularly in the top or No. 1 bed, which generally reaches 60 per cent. and over of metallic iron, and is low in silica.

In searching for hæmatite in a new field, the first bore-holes may be termed " sporting "; they, however, as a rule, provide information which can be used profitably by the mining engineer, such as the location of faults in the Carboniferous Limestone, where the ore is nearly always found. The writer may mention, however, that indications as to the neighbourhood where faults

may occur in the Carboniferous Limestone may frequently be
traced from the mountainous district, or in quarries or outcrops
in the district where you are desirous of boring. It may be
added that if there were no faults, there would be no hæmatite-
ore, because there would have been no spaces in which iron-ore
could have been deposited.

When, by boring, a sufficient body of ore has been proved, a
shaft is sunk, if possible clear of the deposit, thus leaving the shaft
stable and free from injury, when all the iron-ore is worked out.
There are, however, exceptions to this rule, for instance, in some
mines the deposit is so extensive and the dip of the sole so great
that a second shaft is sunk for the purpose of working the ore
lying to the dip of the first shaft; and also for the purpose of
robbing the first mine of the last remnants of ore; and, in this
case, the position of the first shaft is not of such vital importance
as if it had to be the only one.

Furness.—In the Furness district of North Lancashire, at
Ronhead, Mr. Myles Kennedy has recently proved a very large
deposit of hæmatite-ore, near the shores of Duddon Bay. The
depth of the ore is not yet known, but, so far, it has been proved
up to 60 feet, and a level has been driven through over 300 feet
of iron-ore.

West Cumberland.—North of the Duddon estuary, there is the
great deposit of iron-ore worked by the Hodbarrow Mining Company, who have in hand the building of a huge sea-wall. This,
when completed, will enclose a large area of the deposit, which
they would not otherwise have been able to work.*

To the north of Frizington, the ore occurs in vein-like deposits,
lying at a steep angle, with a general trend a few degrees to the
west of north, and many of these have been traced from the outcrop to nearly the great fault, which runs between the Coal-
measures and the iron-ore measures, through the Frizington and
Arlecdon districts. The writer believes that these deposits will
be found to run up to and abut against the fault; and it is
possible, and more than probable, that they may widen considerably before approaching the fault.

* *Trans. Inst. M.E.*, 1899, vol. xvii., page 313.

To the north of Lamplugh, and extending towards Eaglesfield, about 3 miles from Cockermouth, is a large area of Carboniferous Limestone still unproved. About 24 or 25 years ago, a little "scratching" for surface-ore was done near Eaglesfield, and a few hundred tons of hæmatite were procured; but when it was found that the lease had been inadvertently taken from the wrong party, the undertaking was abandoned. Since then, nothing has been done, although iron-ore can be seen cropping out through the Carboniferous Limestone, and, no doubt, this part of the district will be again prospected.

The borings made in the neighbourhood of Lamplugh Cross were not successful, and it is mainly for this reason that the above-named district has not been further explored. Narrow veins exist at Lamplugh Cross: there is no reason, however, why they should not widen out to the north, and bear a considerable amount of payable ore.

About 25 years ago, some engineers thought that, long ere now, all the West Cumberland hæmatite-deposits would be exhausted. Those years have dropped into the past, but hæmatite still remains.

New hæmatite-fields are being opened out to the south of Egremont, two shafts have been sunk, one by the Ullcoats Mining Company, Limited, and the other by the Millom and Askam Iron Company, Limited, and large quantities of hæmatite are being raised. The Ullcoats Mining Company, Limited, are sinking a second shaft; the Beckermet Mining Company, Limited, are sinking a deep shaft; and the Wyndham Mining Company, Limited, are sinking a shaft to about 1,300 feet in depth near to the town of Egremont. Then, farther northward, the Winder Pit, belonging to Messrs. Ainsworth, is producing a considerable quantity of iron-ore; but, as this ore, like a few other bottom-deposits in West Cumberland, lies on the slate-rock, the writer believes that it is difficult to work, necessitating extra labour and the use of a very large quantity of timber.

Bigrigg and Moor Row districts, the writer believes, have a bright future in store, and he has no hesitation in saying that, so far as it is possible for any human being to judge, the mining of hæmatite in West Cumberland is only in its infancy. He has no doubt that, by employing economical methods of working, with proper railway-facilities and fair royalty-rents, West Cumberland will be able to hold its own against all competitors for many years to come.

The writer believes that large deposits of hæmatite will be found extending through Haile, Calderbridge and Gosforth, because the same geological conditions exist there as to the south of Egremont; and that they will connect, by the narrow veins of the Eskdale and Irton Hills, with the great Hodbarrow deposit. At Haile, Calderbridge and Gosforth, in the near future, the writer hopes to see still another hæmatite-district prospected and developed, and it will be of the greatest importance to the future welfare and prosperity of West Cumberland.

———

Mr. CEDRIC VAUGHAN (Hodbarrow Mines) wrote that Mr. Walker, in his interesting paper on hæmatite-deposits, stated that "if there were no faults, there would be no hæmatite-ore." He agreed with him so far, but not in the reason which he gave for this conclusion, namely, "Because there would have been no spaces in which iron-ore could have been deposited." This would imply that the ore had been deposited in cavities or hollows previously existing in the Carboniferous Limestone, a view which he (Mr. Vaughan) frankly admitted had at one time attractions for him. However, during the working of the Hodbarrow deposit certain phenomena had presented themselves, and to his mind they were quite inexplicable except by a theory of substitution; and having again read that portion of Mr. J. D. Kendall's book, in which he deals with the origin of hæmatite,[*] he (Mr. Vaughan) had come to the conclusion that his theory of the replacement of limestone by hæmatite is the only one by which these phenomena can be explained.

With regard to the presence of fossils in hæmatite-ores, he (Mr. Vaughan) had three fossils which were found in the Hodbarrow mine, one especially being a perfect pseudomorph of pure hæmatite, which could not possibly have been produced by any other means than that of replacement or substitution. Again, detached pieces of limestone are occasionally found in the body of the ore-deposit, that have apparently resisted the action of the salt of iron, and so remained unchanged. If, as Mr. Kendall suggests, this salt of iron came from below in a gaseous condition

_____

[*] *Trans. N.E. Inst.*, 1878, vol. xxviii., page 148.

through the then recently formed faults in the Carboniferous Limestone, and coming into contact with water formed a solution of perchloride of iron, then, as he has shown, the whole of the facts become simple and easily understood by the chemical formula which he gives.

Moreover, in some parts of the Hodbarrow deposit there are traces of the action of heat, whereas in other parts of the same deposit there are none, which might be accounted for by the fact that considerable heat is evolved when perchloride of iron is dissolved in water, but the solution thus formed would cool down while acting upon the limestone. In other words, the parts affected by heat must have been near the vent, where the gaseous emanations from below came in contact with water and formed the solution which acting upon the limestone caused peroxide of iron (hæmatite) to be deposited in its place.

The difficulty, however, still presents itself that a given volume of limestone so acted upon would not produce an equal volume of hæmatite, and as the Hodbarrow ore is very compact and comparatively free from "loughs" and foreign matter, he (Mr. Vaughan) could not quite fall in with Mr. Kendall's calculations on this point.

Mr. J. D. KENDALL (London) wrote that, in discussing Mr. Walker's paper, the first observation that one felt called upon to make was with regard to such tautological expressions as "hæmatite-ore" and "calcareous limestone." Hæmatite is a red ore of iron, therefore it is unnecessary to speak of "hæmatite-ore," while limestone is of necessity calcareous, even when most impure.

When Mr. Walker speaks of hæmatite-mining in West Cumberland as being "only in its infancy" one feels that there ought to be an *arrière-pensée* which would associate that infancy with second childhood. Compare the district as it is to-day with what it was 25 to 35 years ago. Then, new and important deposits were being found in all directions; now, it is the rarest thing to hear of the discovery of a new deposit. The last ironworks erected in the district are now 24 years old. Then glance at the production. In 1882, it was 1,725,478 tons; in 1892, 1,355,007 tons; and in 1901, 1,009,869 tons. To-day, the output is probably not more than 50 per cent. of what it was 20 years ago.

Beneath the large area of Permian rocks south of Egremont, there may be an extensive tract of Carboniferous Limestone, which may contain many unknown deposits of hæmatite, but exploring for them must be very costly and uncertain work: and when deposits are found, they will, from the great thickness of water-bearing strata, by which they are overlain, require large sums of money to develop and equip them. Whilst, afterwards, they must be more costly to work than the shallow and comparatively dry deposits that occur to the north of Egremont. The area north of Rowrah has been much more fully explored than Mr. Walker appears to be aware of. As was pointed out to the Institute many years ago* that area is very little intersected by faults, and therefore deposits of hæmatite are less likely to be found in it than in the area between Rowrah and Egremont, which has been very severely faulted.

Mr. Walker's idea of the deposits in the Carboniferous Limestone south of Egremont being connected with the immense deposit in the same rocks at Hodbarrow, by the narrow veins in the granite at Eskdale, is something like suggesting that the Cretaceous Coal-measures at Crows Nest Pass in British Columbia and seams on the same geological horizon in New Mexico are connected by the Carboniferous Coal-measures of Pennsylvania.

Mr. JOSEPH ADAIR (Egremont) wrote that, being only an amateur in mining matters, it might be considered out of place for him to make any remarks beyond saying that so far as Mr. Walker's sketch extended, it seemed to be in accordance with the generally received opinions on the subject. It was over 30 years since a party of adventurers, of which he (Mr. Adair) was the active member, endeavoured to take royalties in the Pardshaw and Dean district, but after a year's endeavour to take a royalty the attempt was abandoned. He (Mr. Adair) had the idea that there was ore there. It was after their fruitless endeavour to take a royalty that the scratching to which Mr. Walker referred took place near Eaglesfield. It had been held as an opinion by some mining engineers for many years, that the deposit of ore extended from Egremont to Hodbarrow. The members of the

* "The Hæmatite-deposits of Furness," by Mr. J. D. Kendall, *Trans. N.E. Inst.*, 1882, vol. xxxi., page 211.

Institute would see on their visit to Egremont that the trend of exploration was in a southerly direction, but whether ore would be found all the way through south-west Cumberland was problematical.

Mr. W. E. WALKER (Whitehaven), replying to the discussion, remarked that Mr. Cedric Vaughan wrote on a subject very much discussed in past days, that is, as to whether or not there were spaces where the hæmatite-ore could have been deposited. To his mind, there were such spaces and, occasionally, spaces had been and still were met with in many of the mines. Some of the spaces, when holed into, were found to contain water, others shewing plainly, by their water-worn state, that, at one time, water passed through them. Heat might have acted and he (Mr. Walker) believed that it did act, to a certain extent, on hæmatite-ore, but this happened after the deposition took place. He (Mr. Walker) had found several fossils in the mines, but always at the junction of ore and limestone; and, although these fossils were mostly coated with hæmatite-ore, they, when broken, were found to contain limestone, and the coating of hæmatite-ore was very thin. He had never found a fossil consisting of hæmatite-ore only, nor did he ever find one in the body of the ore.

Before he wrote his paper, he had perused the journals of all the bore-holes put down north of Lamplugh as far as Eaglesfield, and they were, comparatively speaking, very few in number. As to the water in the Haile district, to which Mr. Kendall referred, if they were successful in finding hæmatite-ore there they would erect plant to cope with what water might be encountered. The water-question was an insignificant matter, and he had never known nor heard of any mining-engineer considering a few thousand gallons of water per hour to be a stumbling-block. The chief point was to ascertain whether hæmatite-ore was there in quantity; if so, the rest was a simple matter for those experienced in such work as might be required to be done.

The PRESIDENT (Sir Lindsay Wood, Bart.) moved a vote of thanks to Mr. Walker for his interesting paper.

Mr. T. E. FORSTER seconded the vote of thanks, which was cordially approved.

____

Mr. J. POSTLETHWAITE's paper on "The Geology of the Lake District" was read as follows:—

THE GEOLOGY OF THE ENGLISH LAKE DISTRICT.

By J. POSTLETHWAITE, F.G.S.

INTRODUCTION.

Introduction.—The central portion of the English Lake District is composed of Lower Palæozoic rocks, which all geologists agree in dividing into three separate groups. These Lower Palæozoic rocks are almost entirely surrounded by a girdle of Carboniferous strata, and these again by an outer girdle of deposits of Permian and Triassic ages. The tripartite grouping of the rocks, which form the central portion of the area, was first recognized by Mr. Jonathan Otley, of Keswick, and his views were recorded in a paper, contributed to the *Lonsdale Magazine* in 1820, in which the groups were named, in descending order:—Greywacke, Greenstone and Clayslate. These divisions were afterwards described in fuller detail by Profs. Sedgwick, Phillips and Harkness, and still later by Prof. Nicholson, by whom they were classified as Skiddaw Slates, Green Slates and Porphyries, and Coniston Limestone, Flags and Grits. In 1876, Mr. J. Clifton Ward's *Memoir of the Geology of the Northern Part of the English Lake District* was published, and in 1897 *The Physical History of the English Lake District*, by the same author. In these, and in papers contributed to learned Societies, the various formations in the Lake District were described and illustrated in a very full and able manner, and all later workers are more or less indebted to him for the result of his labour during the six years that he was employed on the Geological Survey. Mr. J. C. Ward renamed the Green Slates and the Porphyries the " Volcanic Series of Borrowdale ; " and he also adopted the Falcon Crag section as being typical of the whole volcanic area.

The most recent contribution to the literature bearing on the geology of the Lake country, which comprises the result of many years of patient, laborious and painstaking work, accomplished by Mr. John E. Marr and Mr. Alfred Harker, including microscopic examination and chemical analysis of the rocks, was written by Mr. J. E. Marr, and read by him to the members of the Geologists' Association on July 6th, 1900, in preparation for their excursion to the district in the latter part of the following month, in which he occupied the position of director.[*]

In some important points, Mr. Marr dissented from Mr. Ward's mapping, and to these points further reference will be made later. In this paper the old name, "Green Slates and Porphyries," is again given to the Volcanic Series, and the three great divisions of strata are described under the appellations of Lower Slates, Middle Slates and Upper Slates.

The Skiddaw Slates.—The Skiddaw Slates are exposed in the north-western portion of the Lake District, where they cover an area of about 200 square miles, but the presence of isolated patches on the south-western and eastern margins are indications . that there is very probably a more extensive area buried beneath strata of later date, and it was no doubt once continuous with the sedimentary rocks of the Isle of Man and those of North Wales.

The base of the Skiddaw Slates is nowhere visible, being surrounded or overlain on every side by rocks of later age, namely, on the western, north-western and eastern boundaries by Carboniferous Limestone; on the north and north-east by the volcanic rocks of the Caldbeck mountains; and on its south-eastern margin, which extends in a somewhat irregular course from Swinscale Beck, near Penruddock, by Mellfell and Threlkeld Common to the foot of Walla Crag, thence along the eastern shore of Derwentwater, and over Maiden Moor, to the northern end of Honister Pass; it is succeeded, as far as the western side of Mellfell, by conglomerate, and from that point onward to Honister Pass by rocks of the Volcanic Series; and on the south from the northern end of Honister Pass to Dent Hill, near Egremont, it is bounded partly by the Buttermere and Ennerdale Granophyre and partly by the Volcanic Series.

The Skiddaw Slate-rocks consist of a series of beds of soft shale, or slaty mudstone interstratified with occasional bands of grit. The slaty mudstone, which forms the most important

[*] *Proceedings of the Geologists' Association*, 1900, vol. xvi., page 449.

part of the series, is a soft dark-grey rock, sometimes almost black, and frequently containing oval-shaped ferruginous nodules of various sizes. This rock has been cleaved, crumpled and contorted by lateral pressure, and most of it breaks up into long splintery fragments; but, in localities, where the cleavage coincides with the bedding, it splits readily into thin leaves.

At one time, attempts were made in these localities to quarry the rock for roofing-slate, and portions of old farm-buildings may still be found in the neighbourhood of Keswick, and near the Skiddaw group of mountains, that are wholly or partly roofed with Skiddaw Slate. The ruins of the Earl of Derwentwater's mansion on Lord's Island, of which the foundations only, and some mounds of rubbish remain, afford proof that it, too, was either wholly or in part roofed with Skiddaw Slate. This slate cannot, however, compete with any possibility of success with slate of volcanic origin, being defective in hardness and durability and inferior in colour; nor with the sedimentary rocks of North Wales, in the quality of lightness, or capability of being split into very thin leaves.

The beds of grit are in some cases of considerable, but irregular, thickness, generally more or less micaceous, and of a hard and durable character. Occasionally, the grit is coarse and sandy, breaking up into square or oblong blocks, and in some localities it passes into a true conglomerate; in other cases it is formed of finer materials, and is of a flaggy nature. The flags are frequently ripple-marked, or show a structure similar to ripple-marking, which may have been caused by earth-movements subsequent to the deposition of the rock. These beds have also suffered from the lateral pressure which acted on the softer strata, but in a much less degree.

These alternating beds of fine and coarse mud and sand, hardened by age and pressure, which probably attain a thickness of from 10,000 to 12,000 feet, represent old marine-deposits of river and current-borne materials. There are no indications of deep-sea deposits, but rather those of shore-conditions. To allow of such an accumulation of deltaic deposit, there must have been a continual depression of the area of deposition during a very long period of time. The thickening of the beds towards the west, or a little to the north of west, would indicate that the continent, which supplied the river-borne materials, lay in that direction, and that the river, which brought down, and deposited the materials, must have been a large one to form a delta of such width and thickness.

During the deposition of the Skiddaw Slates, the Ordovician sea obeyed the Divine command and "brought forth abundantly the moving creature that hath life," and the moving creatures thus produced were mostly of a very low type, but some were a little more highly organized. There was vegetable life, too, in the shape of sea-weed and marsh-plants, but the species recorded are few in number, and their preservation often imperfect. Considering their frail and perishable nature, and the vast lapse of time since they were buried in those ancient deposits of sand and mud, we need not be surprised at their imperfection, but rather that any trace of them still remains. The forms of animal life, which were fossilized in those ancient mudstones, were of a much more enduring character, as they were enclosed in chitinous or calcareous cases, which would resist the ordinary decay of nature much longer, and thus afford greater facilities for fossilization. The animal-remains thus preserved in the Skiddaw Slates were chiefly graptolites, the characteristic fossils of the Ordovician age. About 120 species of graptolites have been found in the Lower and Middle Ordovician rocks of Great Britain, and of these upwards of 60 have been obtained from the Skiddaw Slates. Numerous impressions of worm-tracks also occur in the Skiddaw Slates, both in the flaggy beds, and also in the softer and finer mudstones.

Next to the graptolites, the most numerous and persistent of the Skiddaw Slate fossils is the bivalved phylopod, *Caryocaris Wrightii*. Its remains occur in almost every locality where graptolites are found.

A small number of trilobites (probably not more than 30) have been found in the Skiddaw Slates, and most of these differ from all known species; indeed the trilobite-fauna of the Skiddaw Slates may almost be said to stand alone.

The fossil mollusca also stand alone. The remains of molluscs are scarcer than those of trilobites, indeed it is probable that the Skiddaw Slates have not yielded 20 specimens, and of these there is not one that agrees with any known species.

In addition to the alteration caused by lateral pressure, extensive areas of Skiddaw Slate have been metamorphosed by proximity to, or contact with, instrusive masses of igneous rock. The amount of alteration around the larger masses of granite, microgranite, granophyre, picrite, diabase, etc., is very variable, while the slate around the smaller bosses and dykes is often very little altered. In many places, the slate is traversed by mineral veins, and by strings of white opaque quartz, and it is generally much hardened along the sides of the veins and where the strings are numerous.

The most interesting example of metamorphism in the Skiddaw Slate area occurs round the Skiddaw Granite. It measures about 6 miles from east to west, and 5 miles or more from north to south. On approaching the metamorphosed area, the first indication of change observable in the slate is the appearance of small undeveloped crystals of chiastolite. Proceeding in the direction of the granite, the white crystals of chiastolite become larger and more numerous, measuring where they are best developed upwards of 1 inch in length. The chiastolite-slate passes into a hard schistose, foliated, and sonorous rock of a darker colour, in which dark oblong spots composed of andalusite mixed with flakes of mica, take the place of crystals of chiastolite in the former. Proceeding onward, the spotted schist gradually passes into mica-schist, a brownish granular, foliated rock, containing a considerable quantity of mica. This rock is in contact with the granite.

The Volcanic Series.—The Volcanic Series consists of a vast accumulation of bedded volcanic ashes and breccias, interstratified in some parts with sheets of lava. It attains a thickness of from 10,000 to 12,000 feet, and occupies an area of about 400 square miles. This area is bounded on the north and north-west by Skiddaw Slate, except where it abuts against the Buttermere and Ennerdale Granophyre. The south-western boundary is formed partly of Permian Sandstone, partly of Eskdale Granite, and partly of the Skiddaw Slate of Black Combe. On the north-east, it is succeeded partly by Carboniferous Limestone, partly by conglomerate, and partly by Skiddaw Slate, and the south-western margin is formed of Coniston Limestone. The ashes and breccias compose the largest portion, probably nine-tenths of the rocks of the Volcanic Series; they are alike in being formed of materials that have been ejected from volcanoes, but differ with regard to the size of the fragments which they contain. The ash, properly so-called, varies from a rock containing fragments of the size of walnuts, which is considered a very coarse ash, to a rock formed of fine impalpable powder. Breccias are of all degrees of coarseness, from rocks made up chiefly of fragments about the size of walnuts, to those in which they measure 5 or 6 inches across, and in some cases they contain huge blocks several feet in diameter. In both coarse ashes and breccias, the fragments are generally angular and unworn at the edges, but in some rare instances they are water-worn like a conglomerate. In colour, the ashes and breccias vary from dark green to light grey, or sometimes purple. The green tints, which are very prevalent, are due to the presence of chlorite diffused through the rock, in a greater or less degree, while the red and purple tints are due to hæmatite. There is generally more or less stratification discernible in the ashes, shown in some cases by clearly defined bands of lighter and darker colour, and in others by finer and coarser materials. These alternating beds or layers of finer and coarser ash are often very irregular, some layers thinning out and others thickening; and occasionally very clear examples of false-bedding are seen among the finer varieties. Sometimes, a fragment of an older ash of considerable size, showing all its original structure and stratification, may be seen embedded in a newer ash, the stratification of the fragment being at a different angle to that of the rock in which it is enclosed. The best section of lavas, ashes and breccias exposed in the Volcanic Series occurs on the eastern shore of Derwentwater, extending from the margin of the Lake to the summit of Bleaberry Fell, a height of 1,694 feet. The lavas, of which there are 12 distinct sheets, are mostly very hard and compact, of a grey or bluish-green colour, and often contain crystals of altered augite, garnet and other minerals. The sheets differ greatly in thickness, varying from 15 to 150 feet, the upper and under surface in each case being more or less vesicular and cinder-like, while the remainder is massive and compact. Messrs. Marr and Harker have, however, shown that this section is typical of a very small portion only of the Volcanic Series, which they classify as follows, in descending order:—(1) Shap rhyolites; (2) Shap andesites; (3) Scawfell banded ashes and breccias (Kentmere-Coniston slate-band); (4) Ullswater basic-lava group (Eycott group); and (5) Falcon Crag and Bleaberry Fell andesites.

The Ullswater basic group, or Eycott group, is well developed on the western side of Borrowdale, on the north and south of Lake Ullswater, and on the north and north-eastern side of the Caldbeck mountains, from Berrier, or Eycott Hill, to the neighbourhood of Cockermouth. Also on the west and south-west of the volcanic area, from the head of the Vale of Ennerdale to the Duddon Estuary.* At Eycott Hill, a series of lava-flows of this group occur, interstratified with thin beds of ash, one of the lavas being, with the exception of Shap Granite, the handsomest rock of its kind in the Lake District; it consists of a general porphyritic base, containing dark green spots and large crystals of felspar, many of them an inch or more in length.

The Scawfell banded ashes and breccias, or Kentmere-Coniston slate-band, cover a much smaller area than the Eycott group, but form some of the highest mountain-masses, extending from Scawfell to Dunmail Raise; from Helvellyn summit to near Haweswater, and thence westward to the Duddon Valley. This group consists principally of ashes, lavas being few in number.

The Shap andesites and Shap rhyolites occur as rather narrow bands, extending from Shap to a point a few miles north-west of Coniston, where they abut against the Coniston Limestone.† The andesites consist of more or less vesicular lavas, probably made up of a succession of comparatively thin flows, interstratified with ashes and fine agglomerates.

The rhyolites consist of lavas, compact, and often of a flinty appearance; some are laminated, and others coarsely nodular; and associated with the lavas are a series of rhyolitic ashes and breccias.

The Coniston Limestone Series.—The Coniston Limestone consists of a dark-greyish blue, hard and compact limestone, in which the calcareous element is very poorly developed; it is also less jointed than most rocks of its kind. The limestone is a good deal mixed up with cleaved shales, some of which are calcareous. On the eastern side of Kentmere, there is a bed of

---

* *Proceedings of the Geologists' Association*, 1900, vol. xvi., page 455.

† "The Shap Granite, and the Associated Igneous and Metamorphic Rocks," by Messrs. A. Harker and J. E. Marr, *Quarterly Journal of the Geological Society of London*, 1891, vol. xlvii., page 266.

contemporaneous felspathic rock of a reddish colour, interstratified with the beds of limestone and shale; some of the latter are also mixed with volcanic ashes. The whole of the beds, which are somewhat irregular, make up an average thickness of about 300 feet.

The Ordovician river-delta, which afterwards hardened into Skiddaw Slate, sank at the close of the volcanic period, and the Coniston Limestone was deposited in a moderately deep sea, but the large river still brought down great quantities of mud, which were carried far out to sea, and became mixed with the calcareous ooze, rendering it very impure, while fitful outbursts from the expiring volcano scattered showers of fine ashes, which also mingled with the calcareous ooze and river-borne mud on the bed of the Ordovician sea.

The Coniston Limestone forms a band stretching across the Lake country, from Millom, by Broughton, Coniston and Low Wood (near Ambleside), across the valleys of Kentmere and Long Sleddale, to Wasdale Crag (near Shap), where it is partly cut off by the granite. It is again seen on the opposite side of the Shap Granite, then passes beneath the conglomerate and Carboniferous rocks to re-appear once more at Keisley, where it forms part of the Cross Fell inlier. There is also a small exposure of Coniston Limestone in the Furness district, near Dalton, where it is brought up by an anticlinal fold. In the Lake District, it overlies the rocks of the Volcanic Series with perfect conformity, dipping to the south-south-east at high angles. In some places, it is much disturbed and shifted by faults. Although there is no want of conformity between the Ashgill Shale, the uppermost bed of the Coniston Limestone Series, and the superincumbent strata, there is complete discordance between the fossil-remains found in them; therefore, it is at this point that the line of division between the Ordovician and Silurian systems is placed.

The Coniston Limestone is no doubt the equivalent of the Bala Limestone in North Wales, the same characteristic fossils being found in each of them. Deposits of much the same age, as shown by their included fossils, also occur at one point, amongst the rocks of the Volcanic Series, north of the main area of the Skiddaw Slates. This section is exposed in Drygill, between Carrock Fell and High Pike.

Stockdale Shales (Lower Silurian).—At the commencement
of the Silurian age, the sea-bottom was again elevated, and the
formation of the river-delta was resumed.   The sea again
swarmed with graptolites, but they differed greatly from those
which were entombed in the Skiddaw Slates.   During the long
ages that elapsed, while the rocks of the Volcanic Series and the
Coniston Limestone were being formed, changes had occurred
in the graptolite-fauna, but whether of progression, or retro-
gression, it is not easy to determine.

The Stockdale Shales succeed the Coniston Limestone with
perfect conformity; they consist of a belt of highly fossiliferous
shales or mudstones, measuring from 250 to 400 feet in thickness,
and extending from the Duddon Estuary to Shap Fells, where
they disapper beneath the basement-bed of the Carboniferous
series.   They re-appear again, however, near Knock, in Westmorland, as part of the Cross Fell inlier.   There is also another
small exposure of Stockdale Shales, on the eastern side of the
Duddon Estuary, near Dalton-in-Furness.

Coniston Flags and Grits (Lower Silurian).—The Stockdale
Shales are succeeded on the south-east by the Coniston Flags
and Grits, which attain a total thickness of about 6,800 feet of
flags and lower grit-beds, and 4,000 feet of upper grits, the
beds being somewhat thinner on the eastern side of the district.
The flags consist of dark-blue laminated sandy mudstones, which
split into moderately thin flags, where the cleavage is well
developed.   The lower part of the flags has yielded a considerable number of graptolites, trilobites and encrinites, but in the
upper beds fossils are somewhat scarce.

The grits consist of thick beds of tough grit or sandstone,
interstratified with flaggy beds.   They, too, have yielded a considerable number of fossils, one bed near the base being remarkably prolific.

The flags and grits occupy two separate areas—one where
they succeed the Stockdale Shales and older rocks, in regular
order, between the Duddon Estuary on the south-west, and the
Shap Fells on the north-east; and the other where they occupy
the high ground around Howgill and Langdale Fells on the
east side of the Lune, and on Grayrigg Common and Whinfell
Beacon on the west side of the Lune.   There is also a considerable area of flags and grits in the northern part of Furness.

Bannisdale Slates and Kirkby Moor Flags (Upper Silurian, Ludlow Series).—The Coniston Flags and Grits are succeeded on the south and south-east by rocks of the Ludlow formation, comprising beds belonging to both the lower and upper divisions of that series, namely, the Bannisdale Slates and the Kirkby Moor Flags. They attain their greatest development between Kendal and Kirkby Lonsdale, where they measure about 11 miles across the outcrop of the beds. This area is bounded on the north by the Coniston Grits, on the east by the Lune-valley fault, and on the south-west by Carboniferous Limestone. The Upper Ludlow rocks re-appear from beneath the limestone in a narrow strip on the eastern side of Kendal Fell, and there is an outlier of the same rock, extending in a narrow band from about ¾ mile north-west of Staveley to Borrow Beck, near Hollowgate. The Bannisdale Slates (Lower Ludlow) consist chiefly of sandy mudstones, interstratified with beds of grit and thin bands of hard sandstone. Rough and inferior slates are sometimes obtained from these beds, but they rarely yield even a tolerable quality. The mudstones are, however, extensively quarried for paving and building-stone, and for rough slabs. The Kirkby Moor Flags (Upper Ludlow) consist chiefly of thick beds of hard concretionary sandstone, interstratified with thinner beds of fine-grained sandstone of a flaggy nature.

Fossils are scarce in the Bannisdale Slates, but are very plentiful in the Kirkby Moor Flags, consisting chiefly of brachiopods, lamellibranchs and gasteropods.

Deltaic conditions prevailed, and a gradual sinking of the land and ocean-bed continued during the deposition of the Silurian rocks; then the downward movement became more rapid, and the Lake District suffered extensive denudation; while at no great distance, on the west and north-west, still more drastic changes occurred, resulting in the total and final submergence of the continent from which the great river had, during so many ages, worn away and emptied into the Ordovician and Silurian seas, its burden of mud and sand. We have seen that, during the deposition of the Coniston Limestone, the mud far exceeded the calcareous element; but the Carboniferous Limestone was formed almost entirely of calcareous matter, and the rocks of later age, which surround the Lake country, bear no record of deltaic deposits, such as those which prevailed during the deposition of the Lower Palæozoic rocks.

Carboniferous Rocks.—The Lake District is almost surrounded by Carboniferous strata, extending in a continuous belt from Egremont on the west around the northern and eastern margin, to Ingleton on the south-east, thence in detached patches to Silecroft, on the western side of the Duddon Estuary; the remaining 20 miles being the only important break in the circle.

The Carboniferous strata, in descending order, include:— (1) Coal-measures; (2) Millstone Grit; (3) Yoredale Series; (4) Carboniferous Limestone Series; (5) Lower Limestone-shale Series; and (6) the Basement Conglomerate.

These divisions are not continuous, and they vary considerably in their development at different points.* One of the most important of the divisions, from an economic point of view, is the Coal-measures; indeed, it contains in its fossil fuel one of the most valuable mineral assets associated with the Lake country. The West Cumberland coal-field covers an area of about 100 square miles, and it has been yielding its wealth, and providing employment for a large number of men, since the early part of the seventeenth century.

On the eastern margin of the Lake District, at Argill, near Brough, there is a small area of workable coal, discovered by Mr. J. G. Goodchild, and on the south-eastern margin, a more extensive area, about 20 square miles in extent, namely, the Ingleton and Burton coal-field, which has yielded a considerable quantity of coal.

A narrow strip of Carboniferous Limestone, in West Cumberland, extending northwards from Egremont, about 8 or 9 miles, and measuring from 1 mile to 1½ miles in width, has been, and is still, a repository of vast mineral wealth. The south-eastern edge of this strip of limestone reposes on Skiddaw Slate, and its north-western edge is overlain partly by Permian breccia, partly by Millstone Grit, and partly by Coal-measures. The deposits of hæmatite in the area here described are of immense value, and the members of this Institute will have an opportunity during their stay in Cumberland of visiting some of the mines. Similarly rich deposits also occur in the Carboniferous Limestone in South Cumberland, near the mouth of the Duddon Estuary, at Hodbarrow and Whicham mines; indeed, the output

---

* The Carboniferous rocks have been very ably described by Mr. J. D. Kendall, *Trans. N.E. Inst.*, 1883, vol. xxxii., page 319; and 1884, vol. xxxiv., page 125.

at the former probably exceeds that of any mine in Cumberland. Equally rich deposits occur in connection with the Carboniferous rocks of Low Furness. The deposits here are very irregular, but the largest masses of ore are found at or near the junction of the slate-rocks and the Carboniferous Limestone.

The Carboniferous rocks, as their name indicates, are remarkable, above all other geological formations, for the profusion of contained fossil-vegetation, but they are also rich in fossilized remains of animal-life. The graptolites, which were so numerous in the Ordovician and Lower Silurian rocks, had disappeared, and trilobites were dying off, but the remains of shell-fish had largely increased, while ganoid fishes, corals, encrinites, and a higher class of amphibian mammalia had appeared, together with insects of various kinds.

Permian and Triassic Rocks (New Red Series).—The outer girdle of Permian strata which succeeds the Carboniferous rocks, covers a much larger area, but in all probability its thickness is not far in excess of the latter. On the north and north-east the New Red rocks extend from Maryport to a point about 3 miles from Netherby, and from Bowness-on-Solway to Kirkby Stephen, being cut off on the east by the Pennine fault. On the south of the Lake District, there are detached patches in Cartmel and Low Furness, and on the west a strip extending from the Duddon Estuary to St. Bees Head; but they attain their greatest development between Penrith and Carlisle, and in the valley of the river Eden. The lowest bed of the series is the Penrith Sandstone, coarse-grained, and of a dark red colour, with its underlying and overlying breccias (locally called "brockram"). Above the upper brockram is a series of alternating beds of pale yellow sandstone, with thin films of clay between, of reddish dolomitic sandstones, marl-slate bands and shale. The yellow sandstones, clays and shales are crowded with carbonized plant-remains, mostly in a very fragmentary condition. These are overlain by gypseous marls, red shales and micaceous flags; and these again by a great thickness of red sandstones with white and mottled bands, which are known as the St. Bees and Kirklinton Sandstones.

The Permian and Triassic rocks, which girdle the Lake country, are not so prolific in the remains of animal-life as those

of Carboniferous age.   The Magnesian Limestone yields its
ganoids, shell-fish and reptilian remains, and some of the sandstones and shales give evidence of an abundance of vegetable
life, but the thick masses of New Red Sandstone are singularly
barren. The conditions under which they were formed were
probably not favourable for fossilization, as the footprints only
of certain animals are left to show that animal life did exist in
these localities and at that time.

Surface-deposits.—Many of the valleys beneath the highest
mountain-masses contain numerous irregular mounds of moraine-
material, some of which have been cut through by streams. The
rocks on the mountains, and on the sides of the valleys, are
in many places rounded and grooved by glacial action. In the
valleys, there are also numerous patches of Boulder-clay, containing occasionally striated and grooved boulders. On many
of the Skiddaw Slate mountains, notably Latrigg, the eastern
end of Skiddaw, Blencathra, Newlands Hause, etc., are perched
blocks of ash and trap from the Volcanic Series, at all elevations
up to 1,500 feet, and the river-courses in the Skiddaw Slate
valleys are filled with boulders that have been transported from
the mountains of the Volcanic Series.

Numbers of peat-mosses and bogs on the mountains, and in
some of the valleys, mark the sites of old tarns and lakes;
numerous masses of drift-material also occur in the valleys,
consisting of sub-angular gravel and beds of sand, often stratified and false-bedded.

Igneous Rocks: Eskdale Granite.—This is the largest exposure of granite in the Lake District, covering an area of about
35 square miles. It lies at the south-western margin of the
area occupied by the Volcanic Series, and is succeeded on the
western side by New Red strata. Generally the rock is rather
coarse, but in some parts there are bands of fine-grained granite.
It consists of quartz, orthoclase and triclinic felspar, and dark
brown mica; in some portions of the fine-grained granite the
mica is absent. The felspar is more or less impregnated with
hæmatite, which gives a reddish tint to the rock, and that tint
is more apparent on a weathered surface than where it has been
recently fractured; as the hæmatite, when liberated from the

decomposed felspar, overspreads the whole surface of the rock. There are also three small exposures of granite, which are no doubt connected with the Eskdale mass, namely, at Burnmoor Tarn, at the foot of Wastwater, and the foot of Scawfell. The first-named is rather coarse, and the two latter fine-grained, but all have the reddish tint which characterizes the Eskdale Granite.

Shap Granite.—On the opposite margin of the Volcanic Series, near Shap, is a much smaller, but in some respects, a more interesting exposure of granite. The rock consists of a base made up of grains of white felspar, crystalline quartz and black mica; in this base are embedded large oblong crystals of pink felspar (orthoclase), which are often of gigantic size, and form the distinguishing feature of the granite. It is much altered where it is in contact with the surrounding rocks, the latter also being greatly metamorphosed. The mass, as it now exists, measures about 2 miles in length by 1¼ miles in width, but large quantities of it have been removed by denudation, multitudes of boulders of all sizes being scattered over the country to the south and east of the parent mass, some of them having been carried to a distance of 60 miles from their original home.

Skiddaw Granite.—A small mass of Skiddaw Granite is exposed in Syning Gill, near the base of Blencathra, and a larger mass at the upper end of the Caldew valley; there is also a third exposure a little lower down the course of that river, which is connected, on its northern margin, with the igneous rocks of Carrock Fell. The granite is composed of white opaque felspar (both orthoclase and triclinic), dark-brown or black mica, and quartz containing numerous liquid cavities; also some grains of magnetite. The three exposures in the localities named doubtless form portions of a continuous mass, which extends over a large area at a very short distance beneath the surface. The metamorphosed Skiddaw Slate surrounding and overlying it covers an area of about 36 square miles.

Buttermere and Ennerdale Granophyre.—This mass of granophyre occupies a large area, extending from Buttermere on the north to Wastwater on the south, a distance of 9 miles, and for a great part of the distance forms the boundary between the Skiddaw. Slate and the Volcanic Series. The rock is generally of a pale-red colour, sometimes changing to dark-grey, and is fairly uniform in appearance. It consists of pink felspar (plagioclase and orthoclase), transparent quartz, dark hornblende, chlorite, and a little mica.

St. John's Microgranite.—This rock occurs in two masses, occupying both sides of the lower end of the Vale of St. John. Each of these masses measures about 1 mile from north to south, and from ½ to ¾ mile from east to west; they are each bounded on the east, west and north by Skiddaw Slate, and on the south by the Volcanic Series. It is probable that the fault which separates the two formations may have been the channel through which the mineral matter of these masses, as well as the Buttermere and Ennerdale Granophyre, welled up from beneath in a molten state.

The microgranite consists of orthoclase and microcline-felspar, quartz, calcite and schorl, also some epidote and serpentine: part of the last-named has probably replaced mica. The colour is generally grey, but in some places it assumes a reddish tinge. The rock is very much jointed, and frequently contains masses of considerable size that are greatly altered; in some of the joints crystals of galena and blende have been found. Close to the junction with the Skiddaw Slate, the microgranite is slightly altered in appearance and becomes more brittle: the Skiddaw Slate is also hardened, and contains irony spots.

Igneous Rocks of Carrock Fell.—On Carrock and the adjoining mountain, Great Lingy, there is a series of igneous rocks, consisting of (*a*) granophyre, (*b*) gabbro, and (*c*) diabase. The granophyre is a coarsely grained rock, very hard and tough, and of a pale red or brownish-grey colour, with a number of greenish crystals, which show a spherulitic structure, scattered through it. The gabbro is generally coarsely crystalline, but sometimes fine-grained. The base of this rock consists of large crystals of opaque white felspar (plagioclase) and interstitial quartz; and through it are scattered crystals of dark olive-green diallage, which is very nearly allied to hypersthene; also occasional grains of magnetite and apatite. Large blocks of lava of the

Eycott Hill series are here and there enclosed in the gabbro. The diabase is also coarsely crystalline, of a dark-green colour, and in some parts shows distinct lines of bedding.

Hornblende-picrite of Little Knot.—This is a coarsely crystalline rock of a dark olive-green colour; it is largely composed of hornblende, but there are also present in small quantities, quartz, felspar (plagioclase), iron peroxide, epidote, apatite, olivine and calcite. The rock occurs in an oblong mass, about 1,800 feet in length by 120 feet in width, and extends from Little Knot, at the northern end of Longside, down nearly to the bottom of Barkbeth Dale.

Dioritic Picrite of White Hause and Great Cockup.—This rock is very nearly allied to the hornblende-picrite of Little Knot; it is exposed on both sides of Hause Gill, which separates White Hause from Great Cockup. The exposure on the former measures from 36 to 45 feet square, that on the latter mountain being somewhat smaller. The two masses are about ⅓ mile apart, and are doubtless connected beneath the till which forms the floor of the little valley.

The dioritic picrite is a coarsely crystalline rock of a dark olive-green colour, consisting of several varieties of hornblende, also quartz, felspar, calcite, serpentine, iron peroxide, and probably a little apatite, ilmenite and viridite. Olivine is also probably represented by some of the serpentinous mineral, but its presence cannot be clearly determined.

Sale Fell Minette.—At the summit of Sale Fell, is a minute exposure of a very beautiful rock. It consists of a pink crystalline felspathic base, in which are numerous crystals of dark-green mica. The base is chiefly composed of orthoclase, but some triclinic felspar is also present. There is no quartz visible to the naked eye, but small crystals may be detected under the microscope; there is also a little hornblende present. The rock is very hard and tough, and in lithological character is unlike any other rock in the neighbourhood.

Igneous Rocks of Seatoller Fell.—On Seatoller Fell, Borrowdale, there is a dyke of diorite lying between two masses of in-

trusive diabase. The masses of diabase together measure about
¾ mile in width. The rock is very compact, and of a dark-blue
colour; it consists of a felsitic base, in which there are crystals
of triclinic felspar, quartz, augite, and some magnetite. The
diorite, which is much altered, is made up of numerous small
crystals of felspar, hornblende, magnetite and chlorite. The
dyke is about ⅛ mile in length and 40 or 50 feet in width. It
is in connection with these masses of blue diabase, the dyke of
diorite, and the adjoining ash-rocks, that the rich deposits of
graphite have been found. The ash all round the igneous
rocks is much altered.

Garnet-bearing Rocks, below the Banded Ashes of Scawfell.—
These rocks attain their greatest development beneath Scawfell
and the adjoining mountains, and extend in a narrow band,
from 1½ miles to 2 miles in width, from near the head of
Wastwater to Dunmail Raise, and in a narrower band from
Scawfell to the upper part of Langdale valley. They also occur
in another area around and to the west and north-west of Haweswater. The rock is generally of a greyish or greenish colour,
with a compact base in which crystals of felspar and augite
occur: in some places it appears to pass into ash, and occasionally into breccia. The garnets are sometimes small, and very
sparsely scattered, and in others they are numerous and of considerable size. Formerly these rocks were regarded as contemporaneous lavas and ashes, but Messrs. Marr and Harker
consider that the evidence, on the whole, points to the intrusive
character of most of them.

Dolerite of Castlehead, Keswick.—This mass of dolerite,
although not remarkable for beauty, is in some respects one
of the most interesting of the igneous rocks of the neighbourhood, as it probably represents the last expiring effort of the
great volcano from which flowed, or were ejected, all the lavas,
breccias and ashes of the Falcon Crag and Bleaberry Fell series.
On this last occasion it succeeded only in raising the lava part
of the way up the vent, where it solidified, and at some period,
long afterwards, became exposed through the removal of the cone
by denudation. The rock is highly crystalline, and is composed
or augite and pale felspar, with crystals of a soft dark-green

pseudomorphic mineral, in which both serpentine and chlorite are present. Small veins of quartz and calcite occur in it: also flakes of brown mica, and some magnetite.

Armboth Dyke.—This dyke is a quartz-felsite or microgranite, being precisely the same in chemical composition as the St. John's microgranite. It is a very beautiful rock, consisting of a dull red felsitic base, studded with numerous crystals of pink felspar and transparent quartz, also a little serpentine, and occasional grains of green mica. The dyke is from 20 to 30 feet in width and extends in a north-north-westerly and south-south-easterly direction across Armboth Fell, where it may be traced for a distance of about 1½ miles. It may be seen in the wood above Armboth House, and appears again on the opposite side of the valley, near the seventh milestone on the road from Keswick to Ambleside, and extends in a south-easterly direction beyond the crest of Helvellyn.

In addition to those already described, there are small bosses and dykes of igneous rock on Skiddaw Dodd, Seathwaite Howe (near Embleton), Wythop Fell, Robin Hood, Burtness Combe (near Buttermere), Hindscarth, Swirrel Edge, Helvellyn, Langstrath (in Borrowdale), Troutbeck, and on Matterdale Common.

Faults and Mineral Veins.—The faults and veins of the Lake District belong to two systems, one having a prevailing east-north-easterly and west-south-westerly, and the other a north-north-westerly and south-south-easterly direction. The former coincide with the strike of the older rocks, and are much older than the latter, as they do not affect any strata of later date than the Upper Ludlow formation. No date can be assigned to the newer, or north-north-westerly and south-south-easterly system. The best known example of this system is the Pennine fault, which dislocates the strata to the extent of not less than from 6,000 to 7,000 feet, and forms that important geographical feature known as the Pennine Chain. In the Lake District, the most important fault of the newer system is one that extends from Skiddaw Forest, through the Glenderaterra and Thirlmere valleys, thence through Dunmail Pass along the depression occupied by Grasmere, Rydal and Windermere lakes. Its presence can be detected near Ambleside by the shifting of the Coniston Limestone to the south. Another parallel fault probably extends through the depression occupied by Bassenthwaite and Derwentwater lakes, through Borrowdale and the valley in which Coniston Lake lies. This has also shifted the Coniston Limestone at Coniston Waterhead. A third fault probably passes along the valley of Lorton, through Crummock and Buttermere Lakes, terminating in the Duddon Estuary. It is also probable that there are faults in the Shoulthwaite and Watendlath valleys, and many more of minor importance.

Of the older or east-north-easterly and west-south-westerly system, it may be noted that one fault probably extends from the coast of West Cumberland, up the valley of Eskdale, and is prolonged through the Grisedale Valley and the depression of Ullswater Lake; but the most important of this older system is no doubt the great boundary fault between the Skiddaw Slates and the Volcanic Series, which extends in a zigzag course from near Egremont to the eastern side of Mellfell. The nature of this boundary has given rise to some divergence of opinion amongst geologists. Mr. Clifton Ward mapped it as a faulted boundary or junction, and with the dip or hade of the fault approaching the vertical, corresponding in amount with that of the mineral veins of the district. He shows, moreover, that the main east-north-easterly and west-south-westerly fault is complicated by the intersection of a number of north-north-westerly and south-south-easterly faults, the two frequently meeting at right angles with each other, and letting down the newer rocks between them. On the other hand, Sir Archibald Geikie, in *The Ancient Volcanoes of Great Britain*,[*] records his entire disbelief in the existence of such a fault, and furthermore states that he went over most of the ground with Mr. Ward, heard his arguments in favour of it, and yet remained unconvinced. He does not, however, offer any alternative theory to account for the present position of the rocks. Messrs. Marr and Harker hold that there is a fault, which they describe as a " lag-fault," at the junction, not almost vertical, as Mr. Ward mapped it, but nearly horizontal; and that a similar lag-fault forms the boundary between the Ordovician and the Silurian rocks. They have stated as their opinion, " that the folding and faulting which

[*] Vol. i., page 229.

have affected the Lower Palæozoic rocks of the district are primarily due to the pushing forward of the rocks in a general northerly direction by a force acting from the south. Further, that the rocks moved forward at unequal rates, and that, so far as the main mass of rocks now exposed is concerned, the Skiddaw Slates moved farthest forward, causing the Green Slates and Porphyries [Volcanic Series] to lag behind, and the Upper Slates [Silurian] in turn to lag behind the Green Slates and Porphyries. As the result of the lagging, . . . a fault, whose fissure approaches the horizontal, was formed between the Skiddaw Slates and the Green Slates and Porphyries, and a similar fissure between these volcanic rocks and the Upper Slates."[*] The few sections, however, where the junction can be inspected seem to support the theory of a nearly vertical fault rather than of one that is nearly horizontal.

If an observer takes his stand near the junction, between Castle Head, Keswick and Walla Crag, and faces the south, he looks toward the high mountains of the Volcanic Series; in the foreground he has Bleaberry Fell, High Seat and Armboth Fell; in the distance Langdale Pikes, with the Scawfell group a little to the west, and Helvellyn, Fairfield and others a little to the east. Then if he turns round and faces the north he has in front of him Skiddaw, with Saddleback on the east, and Lord's Seat and other Skiddaw Slate mountains on the west. It will be seen that the older mountains of sedimentary origin have their summits practically on a level with the newer mountains formed of volcanic materials; and on looking back into the past history of our Lake country, bearing in mind the fact that the rocks of sedimentary origin are and have been wearing away at a speed at least ten times more rapid than the volcanic rocks, he will see that there was a time in the past when the Skiddaw Slate mountains must have been several thousands of feet higher than those of volcanic origin; whereas, according to their age and the general sequence of the strata, they should have been buried to the extent of some thousands of feet beneath them. This then is the condition of things that is so hard to understand, and which has caused such divergence of opinion.

* Proceedings of the Geologists' Association, 1900, vol. xvi., page 461.

All geologists, from Mr. Jonathan Otley downwards, seem agreed that there is an anticlinal axis, or an axis of elevation, of which the Skiddaw Slates are the centre. Mr. Otley regarded the Skiddaw Granite as the original nucleus and uplifting agent, and believed that the newer rocks were wrapped around it like a mantle.

It is more probable, however, that at a time when the Skiddaw Slates were entirely covered by volcanic, and possibly also by Silurian strata, the granite welled up as a molten magma, into a portion of a huge cavern formed by the upward bending of the superincumbent strata into an arch of vast dimensions, having its ridge in an east-north-easterly and west-south-westerly direction; and that the upward bending, caused by lateral pressure, resulting from the contraction of the earth's surface while cooling, continued until the arch broke open along the ridge. The Skiddaw Slate, which was then in a plastic condition, yielding to the immense pressure of the superincumbent strata, began to flow toward the opening thus formed, and upwards through it, pushing the walls of the fracture more widely asunder, the upward flow continuing until a vast mass accumulated far above the then existing surface. This theory is to some extent supported by the fact that the bedding-planes of the Skiddaw Slates are almost everywhere in a more or less vertical position. Then at the close of this movement the irregular fissure was formed, as mapped by Mr. Ward (which let down the strata on the south-eastern side), and it probably penetrated downwards until it tapped reservoirs of molten rock, which welled upwards and crystallized into what are now the Threlkeld microgranite and the Buttermere and Ennerdale granophyre. An immense thickness of rock has since that time been removed from the district by denudation.

Mineral veins are very numerous in the Skiddaw Slate rocks; many of them are large, and they are filled with a variety of metallic and non-metallic minerals; but the deposits of metallic minerals or metalliferous ores are irregular, occurring in detached sops more frequently than in continuous pipes. Veins are not so numerous in the Volcanic Series as they are in the Skiddaw Slates, although some of them have been more productive, and the deposits of ore are more persistent and extensive. But the most valuable and extensive deposits of metalliferous ore occur in connection with the veins in the Carboniferous Limestone—of hæmatite in West Cumberland, South Cumberland and Furness, and of galena in the Cross Fell Range.

————

Mr. J. L. Shaw (Whitehaven) wrote that he had had some experience of the occurrence of iron-ores in the Skiddaw Slates of Kelton Fell and Knockmurton. In other districts than the above in West Cumberland, many bore-holes had been put through the Carboniferous Limestone rocks into the Skiddaw Slates below, but when the latter are reached the almost universal practice is to stop the boring under the belief of its being useless to proceed further. Shafts, also, are often sunk into the Skiddaw Slates, but more for convenience of working the mine than for any other purpose. There was, no doubt, commonsense in the above idea, but in his (Mr. Shaw's) experience of upwards of 30 years he had never been able to see why a vein of iron-ore in the deeper Carboniferous Limestones should necessarily cease on reaching the Skiddaw Slate. He knew cases of both: for instance, in the Egremont district, in the No. 4 Pit of Gillfoot Park, the vein-ore swept out on the flat at a depth of about 600 feet, on the top of a layer of shale covering the upturned edges of the Skiddaw Slate, but did not penetrate into the slate below this depth; while in the Falcon Pit of the Wyndham mines, which was not quite 1,800 feet to the south-east of the No. 4 Pit already named, a vein of ore (but not on the same fault as above) did descend from the Carboniferous Limestone through a bed of grit (7½ to 9 feet thick and even thicker in places) the stratification of which was parallel with the limestone, into the Skiddaw Slate-rock below, ore having been proved in the last named rock, on the 624 feet level, to be 3 feet wide at one side of this level, and 18 inches wide at the other side; and it probably descended deeper though not yet proved to do so. The sole of the 624 feet level was some distance below the top of the Skiddaw Slate.

The Skiddaw Slates are frequently ripple-marked. An example was and probably is still to be seen, near Foxfield station, but he (Mr. Shaw) had no new facts to give in reference to Silurian fossils. The strike of the bedding of these slates is north-east and south-west in the Kelton mines, the dip being to

the north-west, at an angle varying f
taken in the cross-cuts.

Another point might be mentioned w
surface of Silurian strata having been
rocks crop up at the Moors farm, and
barrow mines were proved to lie at a d

In the Volcanic Ash-rocks, veins c
their general inaccessibility prevents t
A mineral vein is shewn upon the map
between Scaw Fell and Bow Fell. It w
he (Mr. Shaw) visited it, but there w
if not tons of good blast-ore lying abou
bourhood, although no attempt had bee
in capacity or even in width.

He (Mr. Shaw) had also had some e
Limestone at Waterblean, near Mill
thickness as deduced from actual mea
lying in one band. Mr. Postlethwait
important break in the band near Gra
described by Prof. Sedgwick in his " St
Mountains,"* as an " enormous fault."
surprise that this excellent position fo
his (Mr. Shaw's) knowledge been teste
his method would be to test both broke
band, where these abut upon the fault.

At Waterblean, where the colour-
some years, the ore lay in pockets, not
in diameter, these pockets being joined
generally carrying a little ore, but in
consisted of the mere joints in the r
for colour-making was found lying cl
there freest from silica; and at a g
not exceed 40 or 50 feet, the ore becam
with silica as to be unfit for pigment
blast-ore. It had not been proved th
pockets of ore at Waterblean, and the gr
further trials by boring.

* " Introduction to the General Structure of t
Transactions of the Geological Society of London,
page 50.

With reference to the Carboniferous rocks it was scarcely correct for Mr. Postlethwaite to state that the break in the continuity of these rocks was 20 miles wide, because no one knew how far south from Egremont either the coal from nearer to the sea at St. Bees or the limestones extended. Further proof of this was very much wanted, but Carboniferous Limestone was known to lie below the Permian rocks to the south of Egremont. It had not been proved to extend within his (Mr. Shaw's) experience, to the north-west of the Whicham mines, or rather, farther in that direction than the Arrow Hill, but this was by no means the same thing as saying that it extended no farther.

Mr. D. BURNS (Carlisle) wrote that the members were to be congratulated on having added to their *Transactions* such an able epitome of the geological features of Lakeland, by one who had not only mastered all that had been done by others in the district, but who had been a patient and successful searcher for its many hidden treasures, resulting in important additions to the life-records of its oldest sedimentary rocks. Mr. Postlethwaite traced a steady succession of beds from the base of the Ordovician to the top of the Silurian, without any break in the deposition, and hence without any unconformity. The Ordovician strata he estimated at 24,300 feet, and the two lower members of the Silurian at 11,000 feet. He did not give the thickness of the uppermost section of the Silurian, namely, the Bannisdale Slates; but as their outcrop measured 11 miles across, they probably represented a greater thickness. He considered them identical with the Ludlow Series, or Upper Silurian, and Sir Archibald Geikie gave the Ludlow group as 1,950 feet thick, and the whole Upper Silurian group at 5,050 feet.* There was thus a total thickness of say 35,000 feet or approximately of 7 miles of steady deposition, beginning and ending in an estuary within a few feet doubtless of the sea-level. In the middle of this tranquil subsidence, but not extending throughout it, beginning with a burst and ending with spasmodic ash-showers over the calcareous sea; but not till 12,000 feet of pure volcanic products had been spread over an area extending an indefinite distance in excess of 400 square miles, did the Castlehead volcano subside into æsthetic quietude.

Mr. Postlethwaite states that at the close of the Silurian period, the downward movement became more rapid, resulting in

* *Text-book of Geology*, 1885, page 665.

the denudation of the Lake District. It is difficult to see how an increased downward movement could have this effect. A slow uprising exposed to the full beat of the ocean-waves would be more likely to have the denuding effect now noted.

Thus, long after all traces of volcanic action had left the district, the whole of the rocks round the Lake mountains were tossed up here and down there, followed by denudation slicing obliquely across the formations, and giving rise at the beginning and towards the close of the Carboniferous period to the most violent unconformities. One could easily understand that there would be a steady and great depression of surface over an area that recognised Castlehead as a centre, while 1,000 cubic miles at least, and probably several times this amount of volcanic material was being ejected. But this seems not to have perceptibly affected the depression that had continued for long ages before, and was destined to continue long ages after. There was clearly much to learn in such matters, and the evidence of Mr. Postlethwaite's paper was to the effect that, contrary to received opinion, volcanic action, as evidenced by the outpouring of volcanic rocks, was quite an insignificant factor in producing the changes of level of the earth's surface required to explain the stratigraphical evidence presented.

With regard to the disputed fault between the Skiddaw Slates and the Volcanic Series, Mr. Postlethwaite some considerable time ago took him (Mr. Burns) to see the junction in Borrowdale. He could not see at that spot conclusive evidence of a fault, although in newer rocks the evidence would have been most conclusive, as the line of junction approaches the vertical. Certainly it had no "lag" about it. Mr. Postlethwaite's argument for a fault, based on the relative perishability of the rocks on the Skiddaw range, would not carry wide conviction. The many and large faults that upset the calculations and plans of the colliery-manager, and which gave no indication of their presence at the surface, were the greatest trouble of his (Mr. Burns') professional work. Unless a fault had been large enough to determine the course of a river or glacier it had little effect on the relative levels of the rocks on its two sides. The Pennine fault and Stublick dyke were in places apparent exceptions, but they were extreme cases in several ways, and glacial action had been in each case at work. Indeed, denudation was extremely slow on the mountain-tops, and when the mountains were covered with forests, it would be nearly infinitesimal.   At Walltown Crags, north-west of Haltwhistle, there was a good instance of the differentiation of denudation.   At this point the Whin Sill altered its horizon and traversed a limestone.   On the slope, above the influence of running water, there was at one point a dip-slope of whin and in line with it to the east a dip-slope of limestone. Assuming that at the close of the Glacial period these two slopes were reduced to one plane (and certainly the limestone-plane would not project beyond the other), it was evident by the striæ on the whin that it had not been denuded $\frac{1}{64}$ inch since Glacial times, and that the limestone had only suffered about 1 foot and that almost entirely by chemical action.

Mr. J. D. KENDALL (London) wrote that Mr. Postlethwaite had described the rocks of the Lake District with considerable fulness of detail, especially those in the neighbourhood of Keswick, but he had not noticed, except incidentally, a most interesting piece of ground in Low Furness.   Within 1 square mile, just south of Ireleth, occurred the following formations and series :—(1) Carboniferous Limestone ; (2) Basement Conglomerate ;   (3) Stockdale Shale ;   (4) Coniston Limestone ; (5) Borrowdale Volcanic Series ; and (6) Skiddaw Slate.   These formations and series were all shewn on the very imperfect geological map published by the Geological Survey, but so inaccurately and incompletely, that it was impossible to make out the geological structure of the ground therefrom.   He had tramped over the area many times, and was not satisfied that he had yet unravelled its structure, but of one thing he felt quite certain, namely, that the Geological Survey map gave no clue to it.   Two separate and distinct bands of Coniston Limestone were shewn on that map, separated by Skiddaw Slate, but there was no evidence of more than one band.   From the structure of this piece of ground, it was quite clear that the Skiddaw Slate and the Volcanic Series had been subjected to a considerable amount of faulting and denudation before the deposition of the Coniston Limestone, for that formation was found resting on the Skiddaw Slate at one point, and at another, quite close to it, on the Volcanic Series.   It also shewed that the Volcanic Series was thinning out rapidly in that direction, the thickness of these rocks there not being more than about 700 feet.

Under the head of " Faults and Mineral Veins," Mr. Postle-thwaite had described three main north-west and south-east faults as running respectively through Windermere, Coniston Lake, and the estuary of the Duddon; and a north-east and south-west fault as running from the coast of West Cumberland through Grisedale valley to Ullswater Lake. The only evidence referred to as a basis for these statements was the shifting of the Coniston Limestone near Ambleside and Coniston. At both these places, however, the direction of the shifting faults was about north 10 degrees east, and therefore they did not come within the system of north-westerly faults referred to by Mr. Postlethwaite. But apart from that, the suggestions were at variance with all the results obtained in careful mapping; and the fault which shifted the Coniston Limestone near Ambleside did not pass down Windermere, but about 2 miles to the west of that lake.

The boundary between the Skiddaw Slate and the Volcanic Series from Honister to Great Mell Fell was probably only a faulted boundary in part. The facts could be explained without assuming either the complicated system of faults adopted by Mr. Ward, or the drag-fault of Messrs. Marr and Harker, by looking upon the boundary as the outcrop edge of the Volcanic Series, shifted here and there out of its regular course by a number of intersecting faults.

Mr. ALFRED HARKER (London) wrote that of especial interest from the point of view of mining were the intrusive igneous rocks of the district, since it was probably with these that the metalliferous deposits (excepting perhaps those in the Carboniferous formation) were directly or indirectly connected. It was clear that the ore-deposits belonged to more than one epoch, though they had not, he believed, as yet, been classified from this point of view. Mr. Marr had recently shewn that there were metalliferous veins older than the Shap Granite, and therefore of Lower Palæozoic age; while many of the veins in the district were doubtless younger.

Two, and probably three, distinct suites of igneous intrusions were to be recognized in the Lake District:—The first included rocks which belonged to the same period of igneous activity as the Borrowdale Volcanic Series, though they were somewhat later than the lavas and tuffs with which they were associated. Here the Buttermere and Ennerdale granophyre might be placed, with some other acid rocks, such as those of St. John's, Armboth, etc.; also most of the basic intrusions of the district, including the Castlehead and other dolerites, and the rocks which have been termed picrites. To what extent the garnetiferous rocks inter-calated in the Volcanic Series were intrusive was a somewhat doubtful question. The second suite of intrusions, connected with the principal crust-movements of the district, and therefore of Mid-Palæozoic age, embraced the granites of Skiddaw, Eskdale and Shap, with a number of acid dykes and others of minette and allied types. Other igneous intrusions in the district, and in particular those of Carrock Fell and of the Caldbeck Fells, were perhaps to be referred to a distinct later epoch, being younger than the great crust-movements. Indeed, a comparison of the rocks themselves with known rocks in the Western Islands of Scotland suggested that the Carrock Fell intrusions, and conse-quently the associated ore-deposits, might be of Tertiary age.

It seemed doubtful whether the different directions of the lodes could be used with any confidence for classifying them accord-ing to age. The movements with which faults and lodes were connected were apt to recur on the same lines at widely separated intervals of time; and the displacement along a given fault-line—probably, too, the infilling of a given lode—must often have been accomplished at more than one epoch.

Mr. J. Postlethwaite wrote that his (Mr. Postlethwaite's) statement that there is a break in the girdle of Carboniferous rocks in South-west Cumberland, 20 miles in width, is made because of the absence of any direct evidence that these rocks exist in that area. They may extend beneath the Permian strata, over a considerable part of, or possibly the whole, distance between the points named; but, so far as he is aware, there is no evidence to prove it. With regard to the omissions in his paper referred to in the discussion, he (Mr. Postlethwaite) might explain that his aim was to give a description of the geology of the Lake Country, which would be concise, and at the same time as com-prehensive as the space allotted to such a paper would allow; consequently many interesting features had necessarily to be passed over unnoticed. while in other cases, the descriptions had to

be materially curtailed. He (Mr. Postlethwaite) thought that
Mr. Kendall would confer a favour upon all future students of
the geology of the Lake District by embodying in a paper the
results of his examination of the structure of the interesting piece
of ground in Low Furness.

The PRESIDENT (Sir Lindsay Wood, Bart.) moved a vote
of thanks to Mr. Postlethwaite for his valuable paper.

Mr. W. LECK (H.M. Inspector of Mines) seconded the resolution, which was cordially approved.

Mr. WM. H. BORLASE's "Description of the Lead-ore Washing-plant at the Greenside Mines, Patterdale," was read as
follows : —

DESCRIPTION OF THE LEAD-ORE WASHING-PLANT AT THE GREENSIDE MINES, PATTERDALE.

By WM. H. BORLASE.

Adit-level.—The vein-stuff, comprizing quartz, baryta, copper-pyrites, iron-pyrites and galena, is brought from the mine through a long adit-level by an electric locomotive, and the wagons are discharged into stone-made hoppers or kilns, sloping gradually toward the fronts, at the bottom of which are fixed two steel-barred grids or grates, one over the other. The roughs on the top grate are hand-picked for conveyance to the stone-breaker; the stuff on the second grate is small enough for the roll-crushers; the stuff passing through is sized sufficiently for treatment at the Green plunger-jiggers; while the fines, slimes, etc., are caught in settling-tanks (Figs. 1 and 2, Plate XIII.).

First Floor.—This point (the picking grates) being 150 feet lower than the site of the crushing-and-dressing plant, the several sizes are taken, in turn, as required, up the incline in self-tipping wagons, which the writer claims to have erected, the first of its class made in this country, in 1873, at the Ruthers iron-mine, Cornwall, from a design of his deceased father. The power employed in working the incline is derived from the No. 1 vortex-turbine of 20 horsepower. The tip is arranged so that two hoppers are served at one point (Fig. 2, Plate XIII.): the first with rough stuff for the stone-breaker, and the second with the smalls for the crushing-rolls. The stone-breaker is driven by the No. 2 turbine of 15 horsepower, and is fixed, so as to allow the stuff passing through it to join the smaller stuff, being conveyed to the crushing-rolls, and the whole can be crushed together or separately as desired.

The crushers, driven by an overshot water-wheel, 30 feet in diameter and 4½ feet wide, taking the exhaust-water from the No. 1 winding-turbine, comprize three sets of rolls (Fig. 2, Plate XIII.), each 16 inches in diameter and 17 inches long.

The top set of fluted rolls, made of s
crushes the stuff before dropping it
rollers below.

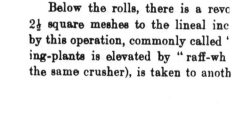

Below the rolls, there is a revo
2½ square meshes to the lineal inc
by this operation, commonly called '
ing-plants is elevated by " raff-wh
the same crusher), is taken to anoth

ish rolls and crushed alone. This method, in the writer's opinion, is much to be preferred, the stuff, being composed of smalls, varying from ½ inch to 1 inch cube, when mixed again with larger stuff, escapes the rolls, and consequently may be several times so returned to the crushing-rolls, thus seriously reducing the work of the crusher.

The stuff passing through this screen, A, is conveyed by water and shoots to the screen, B, with 3 holes to the lineal inch. The stuff, passing to the end of this sizer, feeds into the No. 1 plunger-jigger, with four compartments: the perforated

FIG. 4.—ELECTRIC LOCOMOTIVE.

plates being made of steel sheets of No. 10 wire-gauge with ⅜ inch round, punched holes. The plunger runs about 140 strokes of ½ inch per minute, and makes four grades of quality. The first grade consists of about 75 per cent. of galena; the second, of lead, baryta and blende-ores; the third and fourth is chats, composed of stone, quartz, etc., with particles of the before-mentioned ores attached.

The next screen or sizer, C, meshed to 4 holes to the inch, receives the stuff passing through the meshes of screen, B, the rougher size going to No. 2 jigger, also of four compartments, with steel plates and $\frac{1}{4}$ inch punched holes, and the plunger runs 140 strokes of $\frac{1}{2}$ inch per minute. The results are similar to those of the No. 1 jigger, excepting that the first compartment produces almost pure galena.

The next screen, D, contains 6 holes to the lineal inch and its rough stuff supplies the No. 3 jigger of 3 compartments, fitted with copper-plates and conical punched holes of No. 7 Cornish gauge. The plunger makes 180 strokes of $\frac{3}{8}$ inch per minute. The results comprize three qualities of stuff: the first, equal to about 80 per cent. of metallic lead; the second, of lead-ore, baryta, pyrites and blende-ore; the third, portions of the same ores mixed with quartz, etc.

From this point, the classification is continued through four spitzkasten boxes, E, (Fig. 2, Plate XIII.), each supplied with clean water from the pressure-pipe, P (this pipe also supplies the extra water required for the jiggers), and each supplying separate jiggers of three compartment-capacity, according to the grade of the stuff.

The No. 4 jigger is fitted with copper-plates having conical punched holes of No. 14 Cornish gauge, and produces results equal to those of the No. 3 jig. The plunger runs at a speed of 200 strokes of $\frac{1}{16}$ inch per minute. The No. 5 jigger is fitted with copper-plates with conical punched holes of No. 18 gauge, and the plunger runs at 220 strokes of $\frac{1}{4}$ inch per minute. The No. 6 jigger, with holes of No. 23 gauge, runs at a speed of 230 strokes of $\frac{1}{8}$ inch per minute. The No. 7 jigger, with holes of No. 29 gauge, runs at a speed of 250 strokes of $\frac{1}{8}$ inch per minute. The results at each jigger are as follows:—The first compartments contain 80 per cent. of metallic lead; the second, a mixture of lead-ore, baryta, blende and pyrites; the third, small particles of ores attached to fine quartz, sand, etc.

Second Floor.—From this point, the overflow from the last classifier is taken by way of the downfall-launder, F, to a large spitzkasten, G, where the heavy portions, dropping to the bottom, supply alternately two automatic buddles, H, of the convex type. The overflow from the spitzkasten, G, goes to the classifier, I, which supplies two other automatic convex buddles with finer grade stuff (Fig. 1, Plate XIII.). The overflow from the classifier, I, is then caught in settling-tanks for further treatment in other buddles on the slime-floor, below, when sufficient stuff has accumulated.

The whole of the above plant is being supplied with stuff, which has been only once handled at the crusher.

FIG. 5.—ELECTRIC LOCOMOTIVE ENTERING THE ADIT-LEVEL.

The automatic buddles, H, of simple construction, are capable of dealing with a large quantity of stuff of low grade. Their diameter is 24 feet, with a head or centre 8 feet in diameter. The only bearing is an upright iron bar, with a top turned to a pivot-point, J, on which the cover, K, rests, acting as a coupling to the pipe, 1½ inches in diameter, encasing the upright bar, and carrying the whole of the moving parts (Fig. 2, Plate XIII.). Outside the iron pipe is fixed the bell-mouthed

receiver, L, conveying clear water for the sprays, fitted with
a flange at the bottom and four waterways, M. The flange is
secured to the bottom of the larger receiver, N, into which the
slimes and water to supply the buddles are conveyed by feed-
launders, O. Into this receiver, N, four short arms are screwed,
having T pieces screwed to their ends, through which the stuff
flows unto the head of the buddle. The resistance, or check of
the flow at the T pieces and at the sprays in the long arms or
pipes for the clear water, the outer ends being plugged, supplies
sufficient power to give a rotary motion to the moving parts of
the machine, which, as before mentioned, simply rests on the
pivot-point of the upright bar. Each buddle is capable of treating the fines from the crushers escaping the jiggers for a run
of 200 tons of stuff, requiring little or no attention beyond the
blocking or "stepping" up of the outlet or tail as the head of
the buddle is gradually filled. The contents of the buddle are
divided into four qualities, the tails being run direct to a classifier
and another buddle on the waste-floor. The second and third
partings are hand-fed to a small buddle, while the heads are
treated in a small, mechanically-driven buddle, where the ore
is brought up to the standard.

Second Floor.—The writer must now return to the treatment
of the stuff caught in the hopper at the end of the first screen, A.
The stuff is conveyed by a tram-wagon to a Cornish crusher for
further crushing, and is afterwards classified for four jiggers in
a similar manner to that already described in the first-floor
jigging-house. Both sets of jiggers on the first and second
floors are driven by the No. 3 vortex-turbine of 12 horsepower.

The ore caught in the first compartments of the jiggers,
Nos. 1, 1A, 2, 2A and 3, is separately conveyed to the cleaning-jig,
No. 12, of another type, and thence to the ore-bings. The ore of
the first compartment of the jigs, Nos. 3A, 4, 4A, 5, 6 and 7, is
severally tipped into a trunking-box or rectangular buddle,
through which a clean stream of water is passed to run through
the stuff, as it is being thrown against the sloping head of the
trunk. The water washes back any small impurities which may
have happened to pass through the perforated plates of the
jiggers. The produce is lead-ore, containing from 82 to 84 per
cent. of metallic lead.

The stuff from the second compartments of the jiggers, Nos. 1, 2 and 3, on the first floor, consisting of lead-ore, blende, baryta, copper-pyrites and iron-pyrites is conveyed by tram-wagon to the No. 3 crusher or chat-mill, especially set apart for this class of stuff. It is placed in the hopper, Q, automatically fed by an eccentric arrangement to a belt, and conveyed to the rolls. The speed of the feed can be altered to suit each class of stuff. From this crusher, the stuff first passes into the octagonal, cone-shaped revolving classifier, R, made of wood, containing a spiral screw made of hoop-iron strips, conveying the heavy and coarser grains to the top-end of the classifier, where a feed of clean water is introduced, that washes the fine lead and slimes back over the ribs of the screw, to a spitzkasten-classifier, supplying a jig suitable for treating such fine stuff; the finer stuff going on to the settling-tanks. The coarse stuff having passed upward through the revolving classifier, R (the breaks or angles, causing the stuff to turn over many times in transit, greatly assisting the separation and classifying), falls into a jig, S, fitted with copper-plates, with conical holes of No. 25 gauge and a plunger-speed of 200 strokes of $\frac{1}{2}$ inch per minute. The fines pass into a jig, T, with holes of No. 30 gauge and a plunger-speed of 280 strokes of $\frac{1}{4}$ inch per minute. The results are practically standard galena in the first compartment of each jig. The second compartments contain baryta and blende, with a fair amount of lead-ore, and requiring further jigging. The third compartments contain a very complex mixture, which must be further reduced in the stamps, while the overthrow is exceedingly poor in any mineral. This plant is driven by a Pelton wheel, 12 inches in diameter, under a fall of 400 feet, and capable of running at a speed of 1,400 revolutions per minute.

Fourth Floor.—The third compartments of the whole of the jiggers of fines, Nos. 3, 4, 5, 2A and 3A, are further reduced (as in the case above-named) in the small stamps battery, and classified; the rough portions are jigged, the fines passing on to Luhrig horizontal jigging-tables (Fig. 1, Plate XIII.). These tables also treat the stuff from the third compartments of jiggers, Nos. 6, 7 and 4A.

The contents of the fourth compartment from jiggers, Nos. 1

and 2, and the third compartment of jiggers, Nos. 1A and 2A
taken to No. 4 crusher, and crushed, classified and jigged:
fines being sent to the Luhrig tables.

The table-house is equipped with a very fine jig, runnin
a speed of 280 strokes of $\frac{1}{16}$ inch per minute, fitted with cor
punched holes of No. 36 gauge. The overflow of the class
supplying this jig passes successive spitzkasten-classifiers,
supplying a table. These tables are doing excellent work,
it is intended to erect more of them, so soon as practicabl
treat the slimes, etc., now caught in the zig-zag settling-pits,
now treated by mechanically driven buddles of the old type
the slime and waste-washings.

The writer need not point out the advisability of arran
any washing-plant to avoid as much as possible the hand
of the stuff, so as to dispense as far as practicable with lal
and to reduce the charges to a minimum. The writer does
claim to have, as yet, accomplished this object in the plant u
review, having met with too many drawbacks by way of
appliances and fixtures, which (to prevent a stoppage of
whole of the work) have had to be considered, and conseque
stood in the way of the desired reforms. It would have
better for the installing of an improved plant, had the gro
been cleared of all such obstacles. As far as practicable u
existing circumstances, the plant is on the principle of
tinuous ore-dressing, to the extent that at least 70 per cent
the produce is being delivered to the ore-bing ready fo
smelt-mill, with only two handlings on the shovel after
ing through the stonebreaker.

The settling of the fine-slimes in the pits for clarifyin
water in some instances entails more cost than the value of
contents.

The total labour-costs, including the picking of the
stuff at the grates (containing about 7 per cent. of lead-ore
the delivery of the ore (containing 82 per cent. of metallic le
to the smelt-mill, is 10s. 6d. per ton of ore.

A dynamo, for supplying the washings, smelt-mills, w
shops, offices, etc., with electric light, is fixed in the N
turbine-house, and is connected to the turbine (belt-driven)
clutch-gearing.

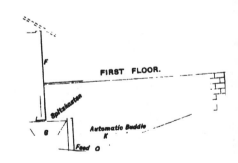

Mouth of Adit-level

TIP

Inclined

Roughs

No. 1 Grids

Tramway

No. 2 Grids

Smalls

Water

Door

Fines

Winding-drum

Fines

Fines

Slimes

Water
Tank

FIRST FLOOR.

F

Spitzkasten

G

Automatic Buddle
K

Feed O

Plunger-jiggers

SECOND FLOOR.

ITEM. I.,

ATIC BUDDLES,

Fig. 1.

The PRESIDENT (Sir Lindsay Wood, Bart.) moved a vote of thanks to Mr. Borlase for his valuable paper.

Mr. A. D. NICHOLSON (H.M. Inspector of Mines) seconded the resolution, which was cordially approved.

———

Mr. GEO. H. BRAGG's paper on "Granite-quarrying, Sett-making and Crushing; and the Manufacture of Concrete-flags and Granitic Tiles," was read as follows:—

GRANITE-QUARRYING, SETT-MAKING AND CRUSH-ING; AND THE MANUFACTURE OF CONCRETE-FLAGS AND GRANITIC TILES.

By GEO. H. BRAGG.

I.—GRANITE-QUARRYING, ETC.

Granite.—Granite is perhaps a painful subject to the inhabitants of large towns. As a material for road-making, and electric-tramway paving it is very familiar, and the frequent upheavals of our thoroughfares makes its presence somewhat disagreeably felt. The Lake District abounds in granite, slates, and many other minerals, but this paper will chiefly deal with the working of the Threlkeld quarries, St. John's-in-the-Vale, where over 300 men and boys find employment.

Mr. Frank Rutley stated that the "Threlkeld rock is closely allied to granite in composition, although it differs from it in structure; its fineness of grain and poorness in mica rendering it really a quartz-felsite, the mineral constituents being felspar, quartz, calcite, a little epidote and serpentine, some of which has apparently replaced mica. A general idea of the hardness of the rock may be realized from the fact that felspar is almost as hard as steel, while quartz is harder." The specific gravity is 2·63, the weight per cubic foot is 164 pounds, and the crushing-strain, as certified by Mr. Kirkcaldy, is 31,912 pounds on the square inch, or 2,052 tons on the square foot.

Site of Quarry.—In opening a quarry, much expense is saved if particular attention be paid to the position of the joints, which should be the first consideration in selecting a site on the side of a hill, instead of going in an haphazard fashion. Very often, a quarry has been opened in such a position that the rock has to be lifted against solid beds, or straight into it. In such cases, the amount of powder or other explosive used is excessive: as the rock is held so tightly, and hemmed in on all sides. The large amount of drilling rendered necessary is also costly, as·

the tonnage of stone blasted per hole drilled is small. The author has noted the difference of cost between drilling and quarrying in a hill-side, and the same labour employed in driving straight into it on the wrong side, and found that the latter cost was as much as 30 per cent. in excess of the former, for reasons which are no doubt obvious. Fig. 1 (Plate XIV.). illustrates this feature:—If the quarry be opened at the side, E, a block blasted out at the base of the pillar will practically dislodge the whole of the pillar, but if opened "straight in" at S, every shot will be solid against the bed.

Joints.—The joints in the Threlkeld quarries are irregular, and may be found at every possible angle. As a matter of observation, it is found that joints are closer together in the finer-grained granites than in the coarser-grained varieties, and it is perhaps owing to this that large blocks for engineering purposes are generally obtained from the coarser-grained granites of the Cornish quarries, which fulfil such requirements better than the finer-grained rock found at Threlkeld, where blocks of large dimensions cannot be secured. But, on the other hand, the finer qualities being very tough, and compact in structure— especially when mica is absent—are eminently adapted for road-metals, for which purpose the rougher varieties are practically useless.

Blasting.—The granite (quartz-felsite), at Threlkeld quarries, is almost invariably dislodged by means of blasting with gunpowder of coarse grain (C.B. quality), excepting in the case of wet holes, when dynamite or gelignite is used. Very little, however, of the latter explosive is allowed, as it shatters the rock too much for manufacture into square blocks. In fact, it has been demonstrated that a given weight of gunpowder in a hole, 20 feet in depth and 3 inches in diameter, will dislodge more rock than the same weight of either dynamite, gelatine, roburite or ammonite. Rules are frequently suggested for determining the quantity of powder to be used in blasting, and the position of the holes. These are excellent in theory, but almost useless in practice. As a matter of fact, before a hole is bored, a good foreman looks at the pillar or block, and if he is a practical man, he soon ascertains whether the mass has to be blown up or against the bed, and he judges also how it may be wedged between surrounding blocks. He then instructs the workmen as to the position of the hole, the necessary depth, and the probable quantity of explosive requisite. In the Threlkeld quarries, the average result of rock from blasting is 458 tons per 100 pounds of gunpowder used.

Rock-drills.—For heavy blasting, holes are bored 3 inches in diameter to any desired depth, up to 20 or 25 feet. Some few years ago, this work was accomplished by hand, but circumstances led to the adoption of machine-drills, which have proved an unqualified success, not only in increasing the output, but also in very materially reducing the cost.

A drill with a $2\frac{1}{4}$ inches cylinder is used for holes 1 inch in diameter up to 4 feet in depth and plug-holes; a drill with a $2\frac{3}{4}$ inches cylinder for holes 2 inches in diameter up to 10 feet in depth; a drill with a $3\frac{1}{2}$ inches cylinder for holes $3\frac{1}{2}$ inches in diameter up to 20 feet or more in depth; and a drill with a 5 inches cylinder, the largest made, will bore holes 6 inches in diameter up to 50 feet in depth. A drill with a $3\frac{1}{2}$ inches cylinder will bore holes $3\frac{1}{2}$ inches in diameter at the rate of 4 feet per hour; a drill with a $3\frac{1}{2}$ inches cylinder will bore holes $2\frac{1}{4}$ inches in diameter at the rate of 8 feet per hour; a drill with a $2\frac{3}{4}$ inches cylinder will bore holes 2 inches in diameter at the rate of 10 feet per hour; and a drill with a $2\frac{1}{4}$ inches cylinder will bore holes 1 inch in diameter at the rate of 12 feet per hour. With a working-pressure of 80 pounds per square inch, the speed is 300 to 360 blows per minute. These rates fluctuate according to the number of times that the drills have to be re-set, and the quantity and depths of the holes. Much also depends upon the length of the stroke of the piston of the drill, which is almost entirely controlled by the feed or the turning of the handle. This is an important point, as, with a short stroke, the blow is too weak to be effective, and a long one is liable to stop the drill. A little experience, however, soon enables a handy man to regulate the speed of working.

Where a hole has to be started in rock, at an angle to the drill, it is advisable to prepare the point of entering with a punch (Fig. 2, Plate XIV.), as an immediate start results in a quicker speed of drilling.

The shape of the drill-bits is important. In jointy rocks, as at Threlkeld, X bits (Fig. 4, Plate XIV.) have proved the best, particularly for commencing holes, but sometimes a chisel-bit can be used, after entering with the X bit, providing that the block of stone is solid. The + bit (Fig. 3, Plate XIV.) produces an oval hole, but the X bit makes an almost perfectly circular one. If chisel-bits can be used at all, they are much preferred, as they are easier to sharpen than any other form: it is of the utmost importance to give them the right shape, as in harder rock, the angle of the bit should be the more obtuse.

There is scarcely any plant that will sooner repay any reasonable expenditure than air-compressors and machine-drills at large quarries and similar work. The time wasted by hand-drilling is enormous, and the output is considerably retarded, as it will take two men $1\frac{1}{2}$ days to drill a hole $2\frac{1}{2}$ inches in diameter to a depth of 10 feet, while it can be bored by a machine in $1\frac{1}{2}$ hours. By the latter means, more rock can thus be dislodged in a given time in the quarry where the working-face or front is limited in extent. Hand-drilled holes also lose much in gauge or diameter: the hole will be commenced with a tool of $2\frac{1}{2}$ inches gauge, and finish at $1\frac{1}{2}$ inches gauge at a depth of 8 or 10 feet, whereas machine-drills start with a tool of $3\frac{1}{2}$ inches gauge, and finish with 3 inches at a depth of 20 feet with a round hole. It will, therefore, be readily understood that the explosive is used with marked advantage in larger holes, and will dislodge considerably more rock. Generally speaking, it has been found that the saving effected by the use of machine-drills amounts to at least £18 per 1,000 tons of rock quarried.

Power.—The most suitable power for working rock-drills is compressed air: it possesses many advantages over steam, especially in conveying power to a distance without serious loss of efficiency. It can be delivered to the machines in a dry condition, and at a good working-pressure; whereas there is great loss in the transmission of steam owing to the loss of power by condensation, the drills are unduly heated, and the constant hot drizzle is disagreeable to all men working in the vicinity. It is stated that 1 cubic foot of water wasted by condensation represents 1 horsepower.

Electrically-driven percussive drills have been introduced, but the results obtained have not borne out the representations of the makers. Their construction is faulty, they are heavier than compressed-air drills of the same capacity, and their use involves the purchase of expensive plant. Electric drills may be successfully applied in metalliferous mines, but the makers hesitate to say whether they will work equally well in hard granite. It is stated that the cost of repairs and renewals of electric drills used in driving a short tunnel was about one third the cost of driving the tunnel.[*]

The writer is responsible for the installation of a compressed-air plant, consisting of a Schram compressor with steam- and air-cylinders 14 inches in diameter and 24 inches stroke : and it is now about to be compounded, so as to ensure ample power to drive 5 rock-drills, and 2 engines with cylinders 8 inches and 9 inches respectively in diameter. Three of the drills will be worked at a distance of 1¾ miles from the air-compressor and the air will be carried in Albion-joint tubes, 4 inches in diameter. The pressure at the air-receiver in the compressor-house is generally maintained at 80 pounds on the square inch, and the pressure at the supplementary air-receiver, 1¾ miles distant, is expected to be 75 pounds per square inch, so that the loss arising from friction, and perhaps some slight leakage in the pipes, will not exceed 5 pounds: From experiments conducted for the Italian Government at the Mont Cenis tunnel, the losses of pressure due to the friction of the air in traversing a length of 3,000 feet of pipe, 4 inches in diameter, was as follows : —

Volume of Free Air Per Minute. Cubic Feet.				Loss of Pressure. Pounds.
200	...	...	...	0·14
400	...	...	...	0·54
800	...	...	...	2·12
1,000	.	...	...	3·27

At Coniston lead-mines, compressed air has recently been conveyed a distance of 7,800 feet over the hills—7,000 feet of pipes being 2 inches in diameter and 800 feet of pipes 3 inches in diameter. Only one rock-drill, however, is being used at the extreme end of the pipe-line, and there is no means of testing

[*] *Transactions of the Institution of Mining and Metallurgy*, 1902, vol. x. page 222.

the quantity of air passed through, as when the drill is working the safety-valve is blowing off, a little, at the air-receiver. The pressure at the compressor-end is 60 pounds, and the air-receiver at the other end of the pipe-line blows off at 50 pounds. This does not, of course, necessarily mean a loss by friction of 10 pounds per square inch ; but the loss will be heavy, as pipes, 2 inches in diameter, are much too small for so long a distance.

Quarrying and Blasting.—After the holes are bored, they are cleaned out with hay and fine granite-dust to dry the sides and bottom. Then the holes are charged with gunpowder in the usual way, granite-dust from the stone-breakers is used as stemming, and it is tamped with a rod of wood or of iron tipped with copper in order to prevent the quartz from striking sparks. Tamping is, of course, a dangerous operation, and as " familiarity breeds contempt " quarrymen are liable to become very careless. It is, therefore, essential that strict rules in reference to blasting should be enforced. All persons are warned by horn or otherwise of any explosion about to take place, and they are compelled to take shelter outside of the quarry before the fuses are ignited. Should a hole miss-fire, no one is allowed to go near—according to the Special Rules of the Home Office—for 30 minutes. Miss-fires are generally caused by the use of defective fuse, or to its being damaged or cut while stemming the hole. In such cases, No. 12 of the Special Rules states that it is "illegal and dangerous to unram, bore or pick out a shot that has missed fire ;" and it is recommended that a hole should be bored parallel to the charged one, with such care as not to interfere with it. This rule is difficult to enforce, as quarrymen will rather unram a shot than drill a new hole, especially if they are on piece-work, and the rock is dislodged at a price per ton. Electric firing is probably the safest, but it cannot be applied economically in all quarries.

The height of a quarry-face should never exceed 60 feet, or thereabouts ; and, when that height is exceeded, another level should be started. The work is then much safer, the drilling is easier, and the men are not required to hang from the long ropes employed in quarrying down the loose rock. The higher level for some time, of course, interrupts the work in the lower level until a width of 30 feet is obtained, in order that men in the

bottom may work with safety from falling pieces; and when this
width is attained both faces can then be worked simultaneously
with very little interruption.

Setts.—After the masses of rock are detached from the face
of the working, the large blocks are reduced to a workable size
by plug-holes, or short holes blasted with a few ounces of powder.
The stones are then transported outside of the quarries by travelling-cranes to the sheds provided for the sett-makers. These
men, using large steel hammers, weighing 28 pounds each, called
" bursters " or " mells," square on one end and obtusely pointed
on the other (Fig. 5, Plate XIV.), break and shape the blocks
into various sizes. The other hammers used are rectangular
" mashes," 24 pounds and 14 pounds each in weight (Fig. 6, Plate
XIV.). When the sett-maker has " blocked " his stones into
suitable sizes, he sits in his shed on a wooden seat about 18 inches
high, and by means of steel " tifflers " (Fig. 7, Plate XIV.)
weighing 5 pounds each with a short handle, he dresses the
blocks into any of the following dimensions that he can obtain:—
4 inches by 5 inches, 4 inches by 6 inches, 3 inches by 6 inches,
3 inches by 5 inches, and 6 inches by 6 inches, all up to lengths
of 8 inches or 9 inches, and cubes of 4 inches.

An experiment in sett-making, interesting to quarry-owners
in particular, was made under the supervision of the writer,
which consisted in erecting a saw-frame, capable of admitting
blocks, 9 feet by 4 feet by 4 feet; and six blades, 6 inches wide by
$\frac{3}{16}$ inch thick, were used, driven by an engine of 10 horsepower,
to saw the block into seven slabs of various thicknesses suitable
for the usual sizes of paving setts. Chilled shot and a copious
supply of water fed continuously into the slits, assisted the blades
in cutting the slabs. The rate of cutting was about 5 inches in
depth per hour—good work for six blades—but, owing to the
hardness of the stone, the cost of chilled shot and the renewal of
the saw-blades was high for the quantity of setts produced and
the total cost was as follows:—

	s.	d.	
Labour : wages paid	7	9	per ton.
Chilled shot ...	6	9	,,
Saw-blades used	4	7	,,
Total cost at saw of setts produced	19	1	,,

To this cost must be added the cost of splitting and dressing the slabs into various sizes of setts.

The stone selected is generally that which will cut the straightest in both directions, and the remainder is broken into rough lumps not exceeding 7 inches in thickness in any part, and these again are broken into macadam.

Two locomotives, with 8 inches cylinders and 2 feet 4 inches gauge, transport about 500 tons per day of lump-granite to a distance of 1 mile to $1\frac{3}{4}$ miles from the quarries to the crushing-works. The heaviest gradient is 1 in 45.

Crushing-plant.—At the Threlkeld works, the crushing-plant (Fig. 8, Plate XIV.) consists of four large Blake-Marsden stone-breakers, 20 inches by 12 inches, one of which is of the latest form of lever-action. This type can be recommended for efficiency and capacity: it runs smoothly and takes less power than the other machines, and it will crush 150 tons per working day of 9 hours. The safety-coupling on the fly-wheel is an admirable device, as in case of any hammer or similar object getting into the jaw, a bolt shears in two and allows the fly-wheel to revolve loose on the shaft.

After the material has passed through the jaws of the crushers, A, it is conducted between a pair of knobbed rolls, B, $2\frac{1}{2}$ feet in length by 2 feet in diameter, which treat the whole of the stone passed through two of the breakers, and very effectively cube the broken granite into macadam. All the material then falls into a revolving screen, C, 30 feet in length by $4\frac{1}{2}$ feet in diameter, which sorts out the various sizes as follows:—$\frac{1}{4}$ inch mesh: with proportion of fine sand used for mortar and plastering; $\frac{1}{2}$ inch mesh: with proportion of fine sand used for concreting; $\frac{3}{4}$ inch mesh: used for topping asphalt and tar-macadam roads; $\frac{3}{4}$ inch to $1\frac{1}{2}$ inches mesh: used for tar-macadam and railway-ballast; and $1\frac{3}{4}$ to $2\frac{1}{4}$ inches mesh: used for general macadamizing.

All stones passing along the screen, which are too large for the holes $2\frac{1}{4}$ inches in diameter, and called "rejectors" or "tailings," are elevated by buckets, D, on to a conveyor, E, 2 feet in width, and carried back again to the cubing-rolls. By using crushing-rolls, the percentage of tailings is very small. Some quarries adopt additional stone-breakers to deal with the

" rejectors ; " but they are far from being efficient, and still leave a large amount of tailings unbroken to the required size. It may be interesting to know that the amount of dust and fine material made by plain surface-rolls is about 6 per cent. in excess of the amount made by knobbed rolls. The proportion of fine material up to ½ inch mesh is generally about 18 per cent. of the total quantity of stone crushed. In order to secure a successful working of a crushing-plant, the speed of the stone-breakers should not be less than 250 revolutions per minute for granite, and about 200 revolutions per minute for softer stone.

Manganese-steel jaws are now largely used for breakers. They cost about three times the price per cwt. of chilled-iron jaws, and one serious objection to the former is that the ribs of the recessed-backs groove themselves into the cast-iron body and swing-jaw stock of the breaker, making very uneven beds. The writer found that one set of manganese-jaws crushed 21,360 tons of stone before being worn out. The working life of a set of chilled-iron jaws is about 14,000 tons, but one exceptional set actually exceeded manganese-steel by crushing 25,200 tons. The teeth of the jaws are of 2¼ inches pitch, and 1¼ inches deep. For the side-cheeks, in the mouths of the stone-breakers, manganese-steel is, however, almost a necessity. The cast-iron cheeks, usually supplied with the machine, will in crushing granite be worn out in a few days; and, unless carefully watched, pieces are apt to chip off and pass into the cubing-rolls, thereby endangering a heavy smash. Manganese-steel cheeks have worn as long as 18 months of regular working.

The eccentric or crank-shafts of the machines should be of the highest quality of Siemens-Martin steel, capable of withstanding a tensile strain of 30 tons on the square inch, with 20 per cent. of elongation on a 6 inches length before fracture : as ordinary qualities will not stand the strain of working for many days. Delta-metal used for the bearings—on account of the dust—has proved superior to brass, as it never runs hot.

Each stone-breaker is driven by one cotton-rope, 1¼ or 1½ inches in diameter. Formerly as many as three ropes were used on each machine, but several years ago it was found that the single rope was ample, and caused much less trouble.

It may be mentioned that the Threlkeld quarries and works have supplied nearly 250,000 tons of ballast for main-line permanent way during the past 10 years, and the total output of the works is about 110,000 tons per annum. All this material is run down inclined tramways, of 2 feet 4 inches gauge and 1 in 12 gradient, in tubs of 1 ton capacity to the sidings, where it is tipped by tumblers into railway-wagons.

The works are provided with siding- and wharf-accommodation for 200 private 10 tons wagons, which are all rebuilt or repaired on the premises.

II.—CONCRETE-FLAGS.

Concrete-flags for footpaths are made by three methods, namely :—(1) Vibratory machine, (2) hydraulic pressing-plant, and (3) hand-labour in wooden moulds. The latter method may be dismissed as inefficient and obsolete.

The manufacture at Threlkeld is carried on in two buildings, each 300 feet in length by 33 feet in width, comprising a cement-store capable of holding 400 tons, and large mixing-platforms. The aggregate, consisting of $\frac{3}{8}$ inch crushed granite with a small proportion of fine material, is run in tubs direct from the crushing-machines into these buildings, where it is elevated in buckets into a hopper, placed above the mixing-platform, to facilitate the measuring in boxes of proper dimensions. The requisite proportion of Portland cement is then added, and all is thoroughly turned over and mixed in a dry state with a 3 pronged fork. This material is then introduced into a square mixer, with tubular projections, where it receives the necessary quantity of water (about 22 gallons per cubic yard of concrete). On emerging from the mixer, the concrete is lifted by hand into moulds placed on a vibratory table, which not only consolidates the material efficiently and quickly, but thoroughly expels any air-bubbles, which rise to the top and disperse. The moulds, made of timber and lined with zinc, give a perfectly smooth face to the flags; and before being filled, the bottoms of the moulds are rubbed with an oily concoction to prevent the cement from adhering to them. After filling, the moulds are transported on bogies along the floor of the building where they are stacked in tiers, and accurately levelled and trowelled on the surface, before the concrete has had time to set. The flags are allowed to remain, at least, 4 to 5 days, before they are discharged from the moulds, and are placed in the open air to

mature for at least 12 weeks.   Under the conditions just
described, the material retains every article of cement originally
introduced into it, and owing to the wellknown property of
cement all superfluous moisture is thrown off after it has taken
up what it requires.

When flags are made under hydraulic pressure, the aggregate
consists almost entirely of very fine material, and it is a well-
known fact that too large a proportion of fine sand reduces the
strength of the concrete.   It is then more like a coarse spongy
mortar and less like an artificial stone; and it is more pervious
to damp and, of course, frost.   The only thing in its favour is
that it is much easier to obtain an apparently-better result.   A
certain amount of fine material is required to fill the interstices
between the larger portions, or the concrete would have a honey-
combed appearance.   The best aggregate is that which has the
largest number of angles, say, from $\frac{3}{8}$ inch mesh graduating down
to pieces not larger than coarse sand; and it enables perfect
bonding to take place.   If the material used is all fine, the
matrix or cement has too many particles to join together, and
the absence of larger portions prevents the bonding that they
should impart to the mixture.   The writer had experience with
one of the first hydraulic presses introduced, and found that
under pressure the water expelled from the flag carried with it
the best and finest portion of the cement, which was consequently
wasted.

In some works, the concrete-flags are immersed in a solution
of silicate of soda, but it is very doubtful whether this is of the
slightest benefit, as after a month's soaking the solution only
penetrates to a depth of the thickness of a thin veneer.   The
idea is to form calcium silicate, but granite is not sufficiently
porous to absorb any of the liquid, and, in Portland cement, the
calcium is mostly in the form of calcium silicate already.

The flags are generally made of 2 feet gauge, and in lengths
increasing every 3 inches from $1\frac{1}{2}$ up to 4 feet.   Architectural
concrete in the form of window-sills, heads, jambs, steps, etc.,
is made in a similar manner and to any design.   The material
weighs about 140 pounds per cubic foot, and the crushing-strain
of a well-seasoned flag, 2 inches thick, is 389 tons on a square
foot.   To give an idea of its weight-carrying strength, a flag,
$2\frac{1}{8}$ inches in thickness and 2 feet square, was placed with 1 inch

of bearing on two sides only; and, weighted with iron blocks distributed over its surface, the flag sustained a total weight of 26 cwts. before it broke.

Granite concrete-flags have the following advantages over York and other stone, namely:—They are more durable; the material is of the same hardness throughout, thus ensuring uniformity of wear; they are of even thickness; no dressing of joints is necessary, so that there is no waste in cutting; and there are no laminations or scalings.

The plant at Threlkeld can produce over 3,000 superficial yards, $2\frac{3}{16}$ and $2\frac{1}{2}$ inches thick, weekly.

III.—Coloured Granitic Tiles.

After exhaustive experiments extending over many years, coloured cement-tiles were, at first, perfected on the Continent; and the owners of the Threlkeld granite-quarries secured the right to use the special machinery and adopt means to ensure the successful manufacture and fixing of the various colours.

A ball-mill is used to incorporate the colours thoroughly with Portland cement, and a small machine will mix 1 hundred-weight of tinted cement in 1 hour in a more satisfactory manner than could be done by hand in 10 hours. Unless the colour is perfectly intermixed, the surface of the tile will be tinted in light and dark patches.

The colours employed, red, blue, black, yellow, chocolate and zinc-white, are carefully sieved into steel-moulds, 8 inches square by 2 inches deep, through stencil-plates of the desired pattern—a sheet-zinc cover being required for every colour that is placed in a design. After the cement-colours are placed in position to a depth of $\frac{1}{4}$ inch, a mixture of pure granite-sand and cement is filled into the remaining depth of the steel-mould, and it is subjected to a pressure of 1 ton on the square inch in a double-geared press, worked by a pulley from a main shaft. The thickness of the tile produced is $\frac{7}{8}$ inch. All the materials, up to this point, are used in a semi-dry condition. On emerging from the press, the tiles are ejected by a mechanical contrivance from the mould, and placed by hand on a shelf or rack for a few hours, when the colours are treated with a solution, which prevents any efflorescence appearing in the colours after the tiles are dried. Without the use of the solution, the colours would be covered with an unsightly white powder on the surface. After this process, the tiles are moistened at intervals until the following morning, when they are totally immersed in tanks of water for some time, so as to render the material thoroughly sound, in the usual way adopted for cement-concrete.

The tiles are made with a plain surface, or with the design shewn in relief, in imitation of mosaic. All the colours are permanent, and the tiles are matured for several weeks.

———

Mr. J. A. G. Ross (Newcastle-upon-Tyne), referring to the crushing of the granite at the quarries, asked whether there were any recorded experiments on the resistance to crushing and fracture, and the tensile strength of the granite-concrete.

Mr. G. H. BRAGG said that the only tests which had been made was the crushing-strain of a 2 inches cube of concrete—389 tons on the square foot.

Mr. GEORGE HEELEY said that Mr. Bragg had drawn attention to the trouble which he had had with the manganese-steel jaw-faces, owing to the ribs at the back wearing the front or resting part of the swing-jaw stock. This trouble had recently been overcome by an improved type of manganese-steel jaw-face, which had been cast with the same outside dimensions as the ordinary chilled-iron face, but cored through instead of being recessed, as formerly. By this means, a perfectly level resting portion on the swing-jaw stock was obtained. The cost of manganese-steel was referred to in the paper as being three times that of chilled-iron; he presumed that this meant the price per cwt.; but the weight of the jaw-faces, of course, was not the same.

Prof. HENRY LOUIS (Newcastle-upon-Tyne) said that it was mentioned in Mr. Bragg's paper that the Blake-Marsden crusher required less driving power than other machines. Did this mean that it took less power than any other machine, or that there were worse machines than this in existence? His own predilection in the case of large quantities of materials to be crushed was for a machine of the Gates rotary type, which did not absorb an undue amount of power. The difficulty with regard to the jaws was overcome in a works with which he was connected by casting in wrought-iron chipping-strips at the back

ECTION OF CRUSHING-PLANT
AT
GRANITE QUARRIES.

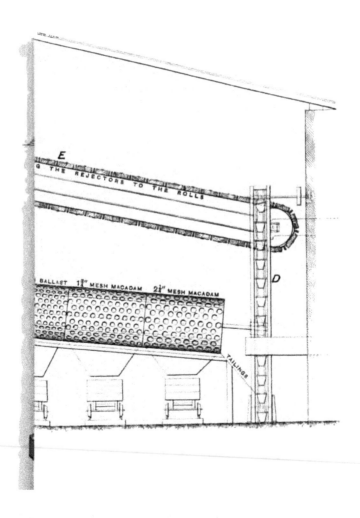

of the chilled-iron casting, and so obtaining an absolutely perfect fit.

Mr. G. H. BRAGG said that the Blake-Marsden lever-type of crusher took less power than the eccentric type. If the jaw was full of stone or granite, and the belt was thrown off, or broken, the machine would finish its work until the jaw was emptied; and he found that this was not the case in crushers of the eccentric type.

Mr. J. A. G. Ross remarked that this, of course, would depend on the weight and power of the fly-wheel.

Mr. W. WATKYN-THOMAS asked whether Mr. Bragg was referring to the same conditions as to power between the different types of crushers; or was it a question of momentum, and not of efficiency? Were the conditions similar under which the different results had been obtained, or was there always the same quantity of stone to the same steam-power?

Mr. G. H. BRAGG said that four machines were working together, and two were working on a separate shaft. It was quite easy to watch the result, as the same material was being fed at the same moment into each; and the speeds of rotation and the weights of the fly-wheels were the same in each case.

It was anticipated on taking the compressed air for a distance of $1\frac{3}{4}$ miles in tubes 4 inches in diameter, that there would be a loss of 5 pounds in that distance arising from friction. Since writing his paper he (Mr. Bragg) had completed the work, and the loss is actually only 3 pounds on the square inch, that is, 85 pounds on the square inch at the air-receiver, in the compressor-house, and 82 pounds on the square inch at the supplementary air-receiver $1\frac{3}{4}$ miles distant.

The PRESIDENT (Sir Lindsay Wood, Bart.) moved a vote of thanks to Mr. Bragg for his interesting paper, and for the interesting excursion which the members had made, that day, to the Threlkeld quarries.

Mr. M. WALTON BROWN seconded the resolution, which was cordially approved.

————

Mr. WILLIAM LECK's paper on "Ambulance-instruction at Mines" was read as follows:—

AMBULANCE-INSTRUCTION AT MINES.

By WILLIAM LECK, H.M. Inspector of Mines.

It is quite possible that the subject of ambulance-instruction at mines may, at first sight, appear to many persons as having but a remote connection with the specific business, which a body of experts like the members of the North of England Institute of Mining and Mechanical Engineers might reasonably be expected to consider. Ambulance-enthusiasts, on the other hand, are more likely to claim that the promotion of this particular knowledge is of paramount and vital importance. Without attempting at the outset to differentiate between the two extremes, the writer thinks that the catholic character of the Institute is not inaptly illustrated by its readiness to consider a subject, which may be described as humanitarian rather than commercial.

Orthographically, an ambulance is, of course, simply a vehicle for the conveyance of sick or wounded persons, but ambulance-instruction, as a generic term, indicates the system of "first aid to the injured," as formulated by the St. John Ambulance Association, whose headquarters are at St. John's Gate, Clerkenwell, London. The objects of the Association include:—(a) The instruction of persons in rendering first aid in case of accidents; (b) the manufacture and distribution of ambulance-material, and (c) the formation of ambulance-depôts at mines and other centres of industry.

The organization, which the Association has popularized, undertakes the formation of "centres" and "detached classes" throughout the country. The centres assume the functions of district-committees; they usually embrace large towns and are formed on application to the St. John Ambulance Association. A president, treasurer, secretary and committee are appointed, and the centre is responsible for the formation and carrying-on of classes within the sphere of its operations.

In recent years, much good work has been done by means of detached classes, each of which is conducted by a local committee; but, instead of working under the supervision of a centre, the local committee is in direct communication with the headquarters in London. This method has been found to produce excellent results in smaller places where a centre would not be practicable.

At each class, a course of not less than five lectures is given by a qualified medical man. An examination is held at the end of the course, and successful pupils receive a certificate signed by the lecturer, examiner and local officers. Certificated pupils who pass two subsequent re-examinations with an interval of one year between each of them are entitled to receive a medallion, and are exempt from further examination.

Some idea of the magnitude of the work carried on by the St. John Ambulance Association may be obtained from the fact that since its institution 26 years ago, upwards of 500,000 certificates of proficiency, and 72,000 medallions have been distributed throughout the British Isles and the British dominions beyond the seas.

Two recent publications of the Association may be mentioned in passing:—The new official text-book, by Mr. James Cantlie, which supersedes the well-known work (Shepherd's *First Aid to the Injured*) familiar to all ambulance-students. The other publication, described as an *Emergency Book*, is an enlarged *aide-mémoire*, containing brief but pregnant instructions on first aid, for almost every imaginable injury. It is constructed to hang on the wall, and the pages are made of strong cardboard, so that they may be readily turned. An excellent index, on the front page, forms a ready reference, and the whole of its graphic information is readily accessible. This valuable work requires only to be known to be appreciated.

The general character of ambulance-work has so far been referred to by way of preface to what, from our point of view, is its more important application to the mining-industry.

In every walk of life the acquisition of that special form of knowledge which teaches how to render first aid to the injured is of great advantage, but to the miner, employed as he perforce must be in remote places underground, it becomes a prime necessity.

The late Mr. J. L. Hedley stated in 1899, that "There is probably no class in the community to whom a knowledge of ambulance-work is of so much value as to miners. Their daily occupation is of necessity carried on in places and under conditions which obviously preclude the advantage of immediate professional attendance. It therefore becomes of the utmost importance that, in every mine, there should be, amongst the workmen themselves, men who are capable of affording skilled assistance at the critical moment."* This has long been recognized, and in the early history of the St. John Ambulance Association it is recorded that in 1878, the second year of its existence, " a great advance was made, especially amongst the Derbyshire and Nottingham collieries." No less than ten centres have been formed at various times in direct connection with mines and quarries, but only two or three appear to be now in active existence.

The Home Office, as the Department of State responsible for the enforcement of mining enactments, nearly a decade ago, issued an important circular to H.M. inspectors of mines, in which they were explicitly desired to take every opportunity which might present itself for the purpose of promoting the formation of ambulance-classes among all persons employed in mines in the United Kingdom, and also to encourage mine-managers and foremen to acquire a knowledge of how to render first aid to the injured. The issue of such a precise recommendation has naturally given a great impetus to ambulance-work among mining employees, and many of H.M. inspectors of mines have frequently referred to the subject in their annual reports.

In 1896, Mr. J. T. Robson (South Wales mines-inspection district), while mentioning one colliery in terms of praise as having a large proportion of trained ambulance-workmen, regretted that at some fairly large collieries in his district there was not a single person qualified to render first aid to the injured.† Mr. Robson's emphatic pronouncement has no doubt had the desired effect of stimulating interest in this important subject.

Mr. A. H. Stokes (Midland mines-inspection district), in 1900, refers at considerable length to ambulance-work in its

* *Annual Report*, 1899, page 63.   † *Annual Report*, 1896, page 19.

special relation to accidents by electricity. He dwells on the necessity of promptly dealing on the spot with cases of electrocution, and inserts a paragraph from the *Official Handbook* of the St. John Ambulance Association describing, with two illustrations, Dr. Sylvester's method of artificial respiration.[*]

The late Mr. F. N. Wardell (Yorkshire and Lincolnshire mines-inspection district) in 1899 remarks that " classes continue to be formed throughout the district in connection with the St. John Ambulance Association, with the most useful and beneficial results. Medical men assert that when injured persons are brought to them now, the requisite first aid which has been rendered is such as hardly to require any alteration at the time."[†]

The testimony of doctors on similar lines to the foregoing has been one of the writer's most pleasant experiences in ambulance-work. Medical men in West Cumberland have repeatedly assured him that during the last few years the improvement in the administration of first aid in cases of accidents has been most marked and beneficial.

Mr. R. D. Bain (Durham mines-inspection district) in 1901 states that "The value of ' first aid ' it is impossible to overestimate. A great amount of suffering and the loss of a limb, or even loss of life, may be prevented by skilful handling and manipulation immediately an accident has occurred, and before a surgeon can possibly be in attendance. I commend this to the consideration of all classes in the district—owners, managers, officials and workmen."[‡]

The late Mr. J. L. Hedley (Newcastle-upon-Tyne minesinspection district), in his last report, wrote that "A period of six years has now elapsed since the inauguration of the Cumberland Mines and Quarries Centre of the St. John Ambulance Association, and it continues to make very satisfactory progress. As its name implies, the centre was formed for the purpose of promoting ambulance-instruction amongst the employées of mines and quarries in Cumberland, and it was the direct outcome of the Secretary of State's circular-letter on the subject of ambulance. . . . . During the past six years, the centre has been the medium of conveying a complete course of instruction in

---

[*] *Annual Report*, 1900, pages 19 and 20.   [†] *Annual Report*, 1899, page 30.
[‡] *Annual Report*, 1901, page 34.

first aid' to upwards of 1,400 workmen, representing every mine and important quarry in West Cumberland. The actual amount of good accomplished can never with certainty be known, but that much needless suffering has been averted by the centre's successful operations is incontrovertible. The opinion has also been freely expressed by competent authorities that these classes are unmistakably improving the *morale* of the workmen attending : . . . which gives additional encouragement to all who are spending time and energy in the prosecution of this humanitarian work."*

If we consider the evidence of statistics, the necessity for ambulance-knowledge amongst miners becomes even more apparent. In 1901, reports were received by H.M. inspectors of mines, of 1,229 deaths at mines and quarries and of 5,326 persons injured. It is wellknown that, under the provisions of the Mines and Quarries Acts, the immense majority of non-fatal accidents are not reported to H.M. inspectors of mines. The acts provide that, except in the case of accidents from three specific causes, only those which have occasioned serious personal injury need to be reported, and "serious personal injury" is an exceedingly vague expression, which allows of much latitude in its interpretation.

A few details of the Cumberland Mines and Quarries Centre will be of interest, as so far as the writer knows, there is no other branch of the organization working on exactly similar lines. The centre was formed in 1895 by the late Mr. J. L. Hedley, as the result of a conference with the mine-owners of Cumberland. Lord Lonsdale undertook to act as president, and on many occasions he has exhibited his warm sympathy with the movement. Mr. Hedley accepted the position of chairman, and the late Mr. Oswald and the writer were appointed respectively vice-chairman and secretary. Mr. A. D. Nicholson, as successor to Mr. Oswald, was appointed vice-chairman, and Mr. J. B. Atkinson, H.M. inspector of mines for the Newcastle-upon-Tyne district, has consented to act as chairman of committee, the position held by Mr. Hedley up to the time of his decease. There is, therefore, in Cumberland, an organization for the promotion of ambulance-instruction at mines directly officered by H.M. inspectors of mines, who, with the mine-managers, constitute the executive committee. The expenses

* *Annual Report*, 1901, page 55.

incurred in carrying on the work are voluntarily subscribed by
the mine- and quarry-owners on a *pro-rata* basis in accordance
with the number of persons employed by each firm.   The Cumberland Mines and Quarries Centre has been fortunate in
securing the services of Mr. George Scoular as treasurer, and he
has materialy encouraged the good work by presenting a silver
challenge-shield to be competed for yearly by workmen attending the ambulance-classes.

The effect of this combination of H.M. inspectors of mines
and mine-managers has been to induce the workmen to attend
and take an interest in ambulance-classes.   In the writer's
opinion, the executive committee only needs to be strengthened
by the accession of direct representatives of the workmen to make
it perfect.

It has been found from experience that members who have
passed through the classes, and obtained the medallion, are apt
from want of practice to allow their knowledge to lapse.   The
inception of the organization known as the St. John Ambulance
Brigade was intended to prevent or remedy this state of affairs,
and the striking success which has attended its efforts is at
once its justification and a proof of its necessity.   The Brigade
is an offshoot of the Association, and was established independently in 1880.   During the last few years it has been brought
prominently before the public, owing to its magnificent record
during the recent war in South Africa, when upwards of 2,000 of
its members volunteered for service as hospital-orderlies or as
auxiliaries to the Royal Army Medical Corps.   Every member
of the Brigade must hold the certificate of the Association, and
the Brigade is, in effect, a federation of first-aiders, (a) who meet
regularly for combined ambulance-practice, and (b) who are
prepared to render first aid on public occasions.   The organization comprises "divisions" and "corps."   A division consists
of not less than 8, and a corps of not less than 72 members.
A corps is frequently formed by the union of several divisions.

The features which are calculated to make it specially applicable to mines are:—(a) The great attention given to methods of
carrying the injured, and (b) the confidence which comes from
combined practice, enabling a division or corps to work in
thorough unison on occasions of exceptional disaster.

It has been decided to form a branch of the Brigade in connection with the Cumberland Mines and Quarries Centre, and the writer hopes at an early date to see it established.

Ambulance being one of the subjects taught under the Technical Instruction Act, assistance is given, more or less, by most of the County Councils to the formation and conduct of ambulance-classes. The County Councils of Durham and Northumberland make a liberal capitation-grant; Cumberland adopted this course originally, but about five years ago, it materially reduced its contribution to this important work, and now merely pays the expenses of the examination. All the classes held, at mines, under the auspices of the Cumberland Mines and Quarries Centre, receive assistance from the County Council.

The writer recently came across a letter to the *Newcastle Chronicle*, written about half a century ago, and preserved in the Library of the Institute, describing, with illustrations, an invention for conveying injured persons from the working-place to the hospital or their home. The appliance consisted of a shallow wooden box without a lid, the injured person being placed in the box. On reaching the shaft, the box was put in an outer case, which was then slung to the pit-rope or chain and conveyed to the surface. In the light of modern experience, the appliance has a crude appearance, but it stands as the expression of a desire to lessen human suffering. In these days, the Low Moor jacket and Furley stretcher make an effective combination, which enables injured persons to be conveyed out of the mine with the minimum of pain.

The Coal-mines Regulation Act provides that "ambulances or stretchers, with splints and bandages, shall be kept at the mine ready for immediate use in case of accident." It would certainly be advantageous, if in addition to those appliances a Low Moor jacket, and a copy of the *Emergency Book* of the St. John Ambulance Association, were also provided at every mine.

The writer hopes that he has succeeded in proving that the work of ambulance at mines is worthy of the attention of all persons engaged in mining, comprising the workman, the manager, the mine-owner and royalty-owner. He has always found that mine-owners and managers are both willing and anxious to encourage in every way the formation of ambulance-classes amongst their workmen, and he has no doubt that the owners of mining royalties throughout the country, once they fully recognize its intrinsic importance, will be equally ready to subscribe liberally for the furtherance of this humanitarian work.

Mr. R. D. Bain stated in 1900 that " several collieries have been provided with wheeled ambulance of the most modern type, thereby causing the minimum of suffering to injured persons on the way to their homes or hospitals ; "[*] and the writer suggests that royalty-owners might equip the mines in which they are interested with the latest appliances for the conveyance of injured persons.

Some managers now decline to appoint any person as an official, unless he is an ambulance-man ; this principle might be carried further, and managers might make it known that no young man would be considered qualified for promotion to the rank of a miner, unless he is in possession of the certificate or medallion, the qualifying badge of the St. John Ambulance Association. This has been done with happy results in at least one of the hæmatite-mines of Cumberland.

The names of qualified ambulance-men are conspicuously posted at some mines, so that everyone may know where to find a qualified man in case of emergency. The writer cordially approves of this commendable practice, pending the period when there will be an ambulance-man in every working-place.

In conclusion, some striking remarks on this subject by the Secretary of State for the Colonies may be quoted. Mr. J. Chamberlain, at Birmingham, in presenting South African medals to ambulance-men who had served in the war, said : " I do not think that I need dwell upon the importance of such an Association as that to which you belong, but its value depends upon the ubiquity of its members. They must be everywhere, at all times. It is necessary, therefore, that you should be a very numerous body, so that if any accident befalls any member of the community, he may not be far out of the reach of your assistance ; and, if that end should be achieved, then I am quite certain that by your instrumentality many valuable lives will be saved, and the dangers which result from the accidents, which, after every precaution is taken, are still inevitable in every community, will be reduced to a minimum."[+]

[*] *Annual Report*, 1900, page 36.
[+] *The Times*, September 1st, 1902, page 6.

24

Mr. JOHN GERRARD (H.M. Inspec
wrote that he had read Mr. Leck's
He had so many times seen where li
aid, by the stopping of bleeding, t
handling of persons with serious frac
after being recovered from beneath a
person being carried out for dead,
and laid in the road, and after several
that, it therefore, enabled him to test

Dr. C. LE NEVE FOSTER (London)
ambulance work were summed up in
mental Committee upon Merionethshi
Parliamentary paper [C.-7692] in 18
had reported as follows:—

II.—*Care of Injured Persons.* 72.—Our
evidence of the medical men, fully warrant us
of rendering "first aid" to injured persons sh
boy employed in mines. We consider that the
give an impulse to such teaching by affording
ance work, by taking a personal interest in the
establishing ambulance-corps, such as alread
collieries. We need hardly add that we should
of the Coal-mines Regulation Act incorporated
the working of slate-mines. This rule compels
splints and bandages, ready for immediate use.
this, and not allow any person to be a manager
of the St. John Ambulance Society, or some
equal standing. The certificate is so easily ob
inflicting a hardship upon the working populati
of his employées to pass the necessary examina

The old method of conveying injured
unwieldy litters should be abandoned, and cor
provided in their place.*

In the eight years which had el
issued his views had undergone no
much convinced of the value of sy
aid to the injured " as he was when
classes about 20 years ago.

On being appointed Professor at
in 1889, he at once induced the late
then at the head of the Science and
ambulance-classes for the students.

* Page xxvii.

a diploma in Mining or Metallurgy until he has received the certificate of the St. John Ambulance Association that he is qualified to render "first aid."

Mr. M. H. HABERSHON (Sheffield) wrote suggesting that the training of suitable men in the use of apparatus for breathing in noxious atmospheres was a branch of ambulance-work which might very well be undertaken by such an organization as the one in Cumberland referred to in this paper. With the most recent form of the Giersberg apparatus, selected men could breathe for 2 hours and do light work without discomfort, after being properly trained. This apparatus contained several important improvements which he thought made it superior to other apparatus previously tried in this country. He thought also that the success obtained with it by Mr. Meyer at the Shamrock Colliery, Westphalia, would now justify any organized centre for mining ambulance-work in taking steps towards the training of men in its use. Owing to the absolute necessity of having the apparatus always kept in perfect order, and of having men thoroughly trained and instructed (which could not be done without some expense), it seemed unlikely that progress in this matter would be attained by individual collieries; and therefore it was, he thought, particularly suitable for the joint centre or organization to which Mr. Leck had referred.

Mr. W. H. HEPPLEWHITE (Nottingham) wrote that he had read Mr. Leck's paper on ambulance-instruction at mines with great interest, and, as he had always associated himself whenever possible with the furtherance of first aid to the injured in all its branches, he congratulated Mr. Leck in bringing the subject before the members.

The Midland mines-inspection district was probably the pioneer in introducing hand ambulance-carriages and stretchers at some of the large collieries. He well remembered their introduction, preceding the instruction department by many years, so that injuries happening underground had to await the attention of a qualified medical man either at home or at the hospital. A workman with a broken leg was put into a pit-tub without in any way receiving attention towards preventing the foot from wobbling about. He had known a case where a pit-pony's saddle was placed under the broken part to assist in keeping it rigid.

The use of splints and bandages was unknown 25 years ago. Since that time, the development of first-aid instruction had been rapid and general throughout the Midland district.

It was worthy of record that, at nearly every colliery, classes had been formed by the managers, and continued year by year until many of the stallmen and all officials held either certificates or medallions. At one large colliery, it was one of the conditions of the employment of a stallman that he must hold either a certificate or a medallion, and he quite agreed with this condition. There were in the Midland district many ambulance-corps established in connection with the classes, and as they all wore the regulation uniform they presented, when on parade, or under review, a smart military appearance.

As a typical example, he would describe the Babbington colliery-corps. The Tibshelf centre, which includes Birchwood, had for president Sir Charles Seeley, and Deputy Commissioner Mr. S. C. Wardell of No. 5 division as chairman. The Babbington centre had for president Mr. C. Hilton Seeley, M.P., and Mr. George Fowler as chief superintendent.

The terms usually commenced at the beginning of each year, and the examinations were generally held in the following March or April. The number of members attending the different classes were generally about 100, which comprised persons going in for their first, second or final examinations. The number of persons that had obtained certificates since the formation of the classes in 1878, was 1,912 and the number receiving medallions 822. There was, in addition to the men's class, a women's class, many of whom studied nursing, as well as first aid.

Ambulance-stations were placed at convenient spots. Those inbye were generally near the deputy's cabin, and were provided with stretchers, splints, blankets, and the requisite material for dressing wounds. In case of an accident at the coal-face, one or two men were despatched for the ambulance-material, while others got the patient in a position to be attended to; and if the case was serious a message was sent by telephone from the site of the accident to have the roadways cleared of tubs, and to have the ambulance-van at the pit-top ready for removing the patient to his home or to the hospital. In mines, where the accident might happen a considerable distance inbye from the shaft, and the roadways might be blocked with tubs, the use of the telephone could not be too highly estimated. Almost without exception, the patient could be taken from the workings to the hospital without a stoppage, and in several instances the medical attendant had been summoned by telephone and was on the pit-bank awaiting the arrival of the patient.

The Babbington ambulance-corps was formed in 1889, and was now composed of 8 officers, 11 sergeants and 134 privates, making an effective force of 153 men. All wore a quiet-coloured, but neat uniform. They met regularly in the summer evenings for drill, and the members attended the ambulance-classes and assisted the new beginners. About September in each year there was a review of the corps, and an ambulance-display.

He (Mr. Hepplewhite) differed entirely from Mr. Leck in suggesting that H.M. inspectors of mines should assume the rôle of managing any of the centres, or should take any position as an officer; but he should in every way give his moral support and advice. He (Mr. Hepplewhite) was often called upon to distribute prizes at ambulance competitions and to give a short address on the advantage of the knowledge of being able to alleviate the suffering of a fellow-workman by rendering first aid.

There could be no doubt that ambulance-parades and the study of ambulance-work had a marked tendency to humanize the men and brought out a strong feeling of good fellowship with each other, in short, ambulance-work was a capital education for the collier.

Dr. T. OLIVER (Newcastle-upon-Tyne) wrote that Mr. Leck had rendered yeoman service to all engaged in the mining industry of this country, by the clear and pleasing manner in which he had urged the claims of ambulance-instruction at mines. With most of all that he had said he (Dr. Oliver) quite agreed. There were advantages in having ambulance-centres under the wing of the St. John Ambulance Association. Not only had this society been in existence for several years, it was well organized, and had numerous centres all through the country; it undertook the formation of detached classes and its main object was the instruction of persons in rendering first aid in case of accident. It was besides, not only an examining body giving certificates and medallions to successful students, but was under Royal patronage, and this circumstance to some extent threw a glamour around the association which it otherwise would not possess. For many reasons, therefore, it was desirable, if ambulance-instruction was to be given at mines, no matter the directing body (whether the employers alone, the miners themselves or the county council), that any centre that was formed should work in union with the St. John Ambulance Association.

Mr. Leck said that a course of not less than five lectures should be given. This was fewer than in most instances would be found desirable, since the lectures proper must be prefaced by other instructions dealing in a general way with the human skeleton, the heart and lungs, circulation, the position of the main blood-vessels of the body, and a few details concerning the nervous system. At the most only a small number of lectures could be given for (1) it was not every medical man who cared to undertake the kind of teaching necessary, (2) his professional calling made too many and often very irregular demands upon his time, and (3) it was undesirable to multiply the lectures and demonstrations too much, in case of wearying the audiences.

As to the value and importance of ambulance-instruction there could be no doubt. Knowledge was power. The acquisition of that special kind of knowledge which enabled one to render first aid to the injured was of the greatest possible advantage to every man, no matter what his rank in life might be. Especially valuable was it to the miner whose vocation was a dangerous one and who perforce was often separated from his fellow-men by a considerable distance.

By a miner engaged hewing coals for a great part of the day, attendance on a winter's evening at a class for instruction in first aid should be looked upon in the light of recreation. It was an entire change to him from his ordinary work, and, besides, as already stated, he was gaining information that could be turned to good account for other people. The Home Office recognized the value of the instruction and urged the formation of classes. Ambulance-instruction was like higher education: it stimulated the miner to read and think: it gave him a larger interest in life, encouraged in him, were it needed, feelings of mutual respect for his fellow-workmen and supplied a particular kind of knowledge which would never come amiss.

Apart from arresting severe hæmorrhage, one of the most important things to note, when an accident that is not immediately fatal has occurred, is the manner in which an injured person is lifted and carried by his fellow-workmen. While a strong arm is necessary, this is quite in keeping with a gentle touch and delicate handling. Knowing how to place a wounded limb in the easiest position and how to lift it when required robs an accident of much of its terror, and often is the means of preventing a simple fracture from becoming compound. From experience, he (Dr. Oliver) could bear testimony to the excellent work done by men at the mines who had rendered first aid. The injured often came to the infirmary properly splinted, and with their wounds well dressed.

It is an interesting fact, of which probably many members of the Institute are aware, that miners' wounds usually heal remarkably well. After a lacerated injury to the soft parts of a limb, the wound may look black and be much discoloured, yet such a wound will in most instances heal very readily. Owing to this fact it is often contended that coal-dust is an antiseptic. Without going quite that length he (Dr. Oliver) was in a position to state that in making bacteriological cultures of coal, microorganisms are, practically speaking, non-existent. In other words, coal at the working face is perfectly sterile, and does not contain microbes. It is to this circumstance of the coal-dust carried into a wound not being germ-laden, and not to any assumed antiseptic action which it possesses, that must be attributed the rapid healing of miners' wounds. What he had said of the healthy character of the wounds applied to those injuries received at the mine-face. In the main ways close to the foot of the shaft and near the stables, the air is not free from germs, so that wounds received at this particular part of a pit may become septic. Owing to the proximity of the stables also, wounds may become, as he had known them, infected with a tetanus bacillus. These are details which, although small of themselves, become of importance to the wounded so far as immediate treatment and future results are concerned, and where such instruction as that gained in ambulance-lectures might be most useful.

During a course of ambulance-instruction, in addition to the application of first aid to the injured, the attention of those attending the class should be directed to the effect produced by inhalation of the various gases that may be present in coalmines, and how to treat persons that have been overcome by them.

Seeing too that electricity is becoming more and more employed in mines, the men should also be instructed as to how electricity kills, and how workmen apparently killed by electricity may be restored to life. Severe electrical shock causes death by immediately paralysing the heart. The muscles of respiration are thrown into a state of temporary spasm, but the heart stops beating, even though the breathing continues for some time after death. The point to remember, therefore, is that death in electric shock is from the heart, and that in order to resuscitate persons apparently killed the same methods of artificial respiration must be adopted as those employed in restoring the apparently drowned.

The illustrations that he (Dr. Oliver) had given shewed that there must be many ways in which such knowledge as that gained in ambulance-lectures might come in very helpfully. Apart from the utilitarian aspect of the question, he knew of nothing which could bring greater satisfaction or pleasure to a man, than to know that by any service that he had rendered to a fellow-workman, he had relieved pain or been the means of saving life. For these reasons, he warmly advocated " Ambulance Instruction at Mines."

Mr. W. WATKYN-THOMAS (Workington) wrote that the members were very much indebted to Mr. Leck for his paper on this very important subject; and it was specially fitting for an inspector of mines to advocate this question, as it was his first duty to encourage everything which tended to the greater security of life; and " first aid " did so materially, directly and indirectly, as no one doubted who had experience of it. He wanted to emphasize what Mr. Leck had said of the results of the combination in the West Cumberland district of H.M. inspectors of mines and the mine-managers in ambulance-work; and mineowners, officials and workmen are unitedly in earnest in trying not only to maintain the already widespread interest therein, but to extend its operations.

The fact, reported by the late Mr. J. L. Hedley, that " conveying a complete course of instruction in ' first aid ' to upwards of 1,400 workmen, representing every mine and important quarry in West Cumberland," during a period of six years, was proof of the excellence of the organization. He was convinced that this success had been attained chiefly through the able, active and sympathetic work of Mr. W. Leck and his colleagues; and, in his (Mr. Watkyn-Thomas') opinion, the good done through the direct influence of H.M. inspectors of mines was not limited to the cause of humanity, in lessening suffering in accidents, which were inseparable from mining work, since the whole scheme elevated and educated officials and workmen, and cultivated more kindly feeling between them, and a greater sense of discipline and method, generally. These results, and the hearty co-operation of all classes, were largely, if not chiefly, due to the fact that the organization is officered by H.M. inspectors of mines; and the memory of the late inspector of mines, Mr. Hedley, and of his ever-ready sympathy in ambulance-work, which endeared him to them, will long remain with the whole of the mining community.

It would not, however, be expedient to take it as a matter of course, from the experience in this district, that H.M. inspectors of mines should be expected to act in an official capacity in all such organizations, as he (Mr. Watkyn-Thomas) thought that each district must decide for itself, and the widest discretion should be allowed to H.M. inspectors of mines in their relations thereto.

Mr. W. LECK (H.M. Inspector of Mines, Cleator Moor), replying to the discussion, wrote that he felt exceedingly gratified at the profitable discussion which his paper had called forth. The succinct and practical remarks of Mr. John Gerrard and Dr. C. Le Neve Foster, and the specially interesting notes of Dr. Thomas Oliver, were an encouragement to all who were spending time and energy in promoting ambulance-instruction at mines.

With reference to the number of lectures, the St. John Ambulance Association prescribes a minimum course of five; but, as Dr. Oliver remarks, more are usually needed. In connection with the Cumberland-mines classes an extra lecture is nearly always given, and frequently the course is extended to seven or eight separate lectures.

He cordially agreed with the admirable suggestion of Mr. M. H. Habershon that an organization like the Cumberland Mines and Quarries Centre might very well undertake to promote practical instruction in the use of apparatus for enabling persons to work in noxious atmospheres. A general suggestion of a similar kind was mentioned to him (Mr. Leck) some little time ago, but although the idea was not altogether new, he was obliged to Mr. Habershon for mentioning it in this connection, and he hoped that his proposal would ere long crystallize into actual realization.

Mr. W. H. Hepplewhite's opportune remarks constitute an important historical contribution to ambulance-literature, and it is clear that the Midland mines-inspection district is well to the fore in ambulance-work.*

The PRESIDENT (Sir Lindsay Wood, Bart.), in moving a vote of thanks to Mr. Leck for his paper, stated that no doubt many lives had been saved, both in mines and various works, since ambulance-instruction had been given. The members could not be too high in their praise of ambulance-classes.

Mr. T. E. FORSTER seconded the resolution, which was warmly approved.

——

Mr. E. A. NEWELL ARBER's paper on "The Use of Carboniferous Plants as Zonal Indices" was read as follows:—

* An interesting incident has just occurred in connection with the annual ambulance-competition among Cumberland miners, held on June 27th, 1903. The challenge-shield was won by a team representing Whitehaven colliery, and to mark the merit of the performance, the Miners' Association have awarded a grant of £5, to which the Whitehaven Colliery Company have added another £5, for the purpose of supplying each of the four members of the winning team with a gold medal. This commendable example of co-operation between employers and workmen forms a striking illustration of one of the important side-issues of the ambulance-movement in connexion with West Cumberland mines.

THE USE OF CARBONIFEROUS PLANTS AS ZONAL INDICES.

By E. A. NEWELL ARBER, M.A., F.L.S., F.G.S.,

TRINITY COLLEGE, CAMBRIDGE ; UNIVERSITY DEMONSTRATOR IN PALÆOBOTANY.

INTRODUCTION.

When the writer had the honour to receive a communication inviting him to contribute a paper on fossil plants to the Keswick meeting of the North of England Institute of Mining and Mechanical Engineers, some doubt was felt as to the possibility of doing so with advantage to the members. Palæobotany, even when illustrated by actual specimens or by figures of fossil plants, is, in many of its aspects, too technical to appeal largely to those who have not previously taken some active interest in the subject. Yet, in regard to Carboniferous rocks, mining engineers have much in common with those who are making a special study of the distribution of the fossil plants. There exists a common desire to be able to recognize their " whereabouts," if one may so state it, in this great formation. The recognition of land-marks, in the shape of definite horizons, is a matter which the writer imagines is of vital importance to the engineer, and one which has long been the aim of the geologist and palæontologist. It has seemed, therefore, that this opportunity might perhaps be employed with advantage in giving a brief account of the present position of the study of the distribution of Carboniferous plants, in discussing the principles and methods by which advances have been recently effected, and in describing some of the difficulties which bar the road to a more rapid progress.

The Carboniferous system is the thickest of all the great divisions of the British stratigraphical series. It is also one which, in its upper portion, presents comparatively little lithological variety as traced vertically and laterally. As practically all the workable coals of England belong to the Upper Carboniferous, lithological character is usually of little help for the purpose which we are here considering. Thus in the Somerset and Bristol coal-field, we find two sets of carbonaceous deposits resting on rocks which are termed Millstone Grit. The general character of these deposits is not, as a whole, dissimilar from the single series of measures in Yorkshire, also resting on Millstone Grit (Rough Rock).* Yet it has been shown by palæobotanical evidence that the Coal-measures of the southern coal-field are a newer set of deposits than those in Yorkshire, belonging to what are known as the Upper and Upper Transition Coal-measures,† whereas those of the northern coal-field belong to the Middle and Lower Coal-measures.

There are three types of evidence to which we must look for help in zoning the Carboniferous. These are the fossil plants, the fossil invertebrates, and the fossil fishes.

With regard to the fossil fishes, their occurrence is hardly numerous enough to be regarded as a main line of evidence. Fish-remains do, however, occur in many different beds of the Carboniferous, and it is now possible to distinguish clearly between an Upper and a Lower Carboniferous fauna of this nature.‡ We find, therefore, in the fossil fishes, often valuable support for results obtained by other means.

At the present time, efforts are being made to zone the Carboniferous along two distinct lines of evidence. The study of Carboniferous mollusca has had for several years the advantage of a special share of the attention of a Committee appointed by the British Association for the Advancement of Science to assist in the progress of zoning the Carboniferous.§ Although perhaps the results achieved in this direction have not entirely fulfilled the hopes which have been entertained, the time has not yet come to gather how far we may depend upon this type of

---

* "The Yorkshire Carboniferous Flora," first report, by Mr. Robert Kidston, *Transactions of the Yorkshire Naturalists' Union*, 1888, part 14, page 6.

† "On the Fossil Flora of the Radstock Series of the Somerset and Bristol Coal-field (Upper Coal-measures)," by Mr. Robert Kidston, *Transactions of the Royal Society of Edinburgh*, 1887, vol. xxxiii., page 335.

‡ Dr. R. H. Traquair, *Quarterly Journal of the Geological Society of London*, 1903, vol. lix. (discussion), page 24.

§ "Life-zones in the British Carboniferous Rocks : Report of the Committee," *Report of the Seventieth Meeting of the British Association for the Advancement of Science, Bradford*, 1900, pages 340-342.

evidence. It must, at present, be conceded that the study of plant-remains has attained a more assured position in this respect, despite the valuable assistance which the close examination of the Carboniferous mollusca has afforded to palæontology.

We may turn now to a short explanation of the progress which has been made in this country with regard to the use of Carboniferous plants as zonal indices. It is not proposed here to trace out the history of this progress, in regard to either the British Carboniferous rocks or to those of the Continent.* We may rather devote such time as may be at our disposal to a consideration of the principles and methods which are adopted, and the results which have so far been attained.

PRINCIPLES AND METHODS.

The student of Carboniferous plants has long ago realized that the kind of evidence which is drawn successfully from the distribution of marine invertebrata is inapplicable and inaccurate in the case of fossil plants. For instance, as is well known, the Jurassic rocks are divided into a number of definite zones on the occurrence of a particular species of Ammonite, confined or almost wholly confined to that zone. Apparently in regard to the Carboniferous mollusca, the same principle is being applied. Efforts are being made† to obtain a characteristic and common mollusc, which occurs in one subdivision of the Coal-measures, but which is absent or almost entirely absent from others; and to use such species as zonal indices. How far this will prove possible in the case of a fauna which is not truly marine, but largely littoral, estuarine, or even freshwater, remains to be seen. The discovery of restricted species of plants is not, however, the primary object of the palæobotanist. Some geologists, realizing that fossil plants do not commonly afford this type of evidence, have rather hastily concluded that such remains are, therefore, useless as zonal indices. It has been shown, however, that this is not the case. It is true that, in British rocks, a number of plants, so far as our

* For the use of plants as zonal indices in the Carboniferous rocks of the Continent, the works of Geinitz, Weiss and Stur (Germany), and of Messrs. Grand'Eury and Zeiller (France) should be consulted.

† "Life-zones in the British Carboniferous Rocks: Report of the Committee," *Report of the Seventieth Meeting of the British Association for the Advancement of Science, Bradford*, 1900, pages 340-342.

knowledge extends, are confined to one of the minor divisions of the Carboniferous, such as the Middle Coal-measures. This is the case with *Zeilleria delicatula*, Sternb., and *Sigillaria ovata*, Sauveur, plants which have recently been found in the Cumberland coal-field.* The evidence of such restricted species is not, however, the foundation of any method of zoning by means of fossil plants, although it is often important as affording confirmatory support to conclusions gained on entirely different grounds.

In order to establish the position of any bed in the Carboniferous, it is necessary to collect and to study a number of different species from it, or from the associated rocks; in other words, we must know not one or two species, but a flora. The number of species need not, however, be very large. Usually twenty species or even less will suffice, if they belong to diversified types of plants; but the larger the number, the better. It is the relative abundance of certain types of plants at any one horizon, and the absence of other types, rather than the occurrence of particular species, which gives the solution to the problem of the horizon of the bed in question. By taking into account the aggregate or assemblage of plant-types, the common occurrence of certain classes, genera, subgenera or species, and the absence or rare occurrence of others, species which have a wide range in time in Carboniferous rocks can be made to yield evidence. Such species, despite their range, are found to be much more abundant in some subdivisions of the Carboniferous than in others.

Thus the common occurrence of *Lepidodendron aculeatum*, Sternb., in the Cumberland coal-field points to these beds being of Middle or Lower Coal-measure age, rather than Upper; since this species has been found to occur most abundantly on these horizons in other coal-fields, and less abundantly in the Upper Coal-measures. From a number of separate small conclusions of this nature, a general conclusion can be arrived at, for which support can often be found from other evidence, such as the occurrence of restricted species. Again, an abundance of such types of plants as *Calamites* and

---

* "The Fossil Flora of the Cumberland Coal-field, and the Palæobotanical Evidence with Regard to the Age of the Beds," by Mr. E. A. Newell Arber, *Quarterly Journal of the Geological Society of London*, 1903, vol. lix., page 1.

Sigillaria in association with *Sphenopteris*, and an absence of particular types of *Pecopteris*, *Alethopteris* and *Cordaites*, will help to distinguish a Middle from an Upper Coal-measure flora.

Ocassionally, small points of disagreement with a general result are found. Thus *Alethopteris Serli*, Brongt., a characteristic Upper Coal-measure fern-like plant, is occasionally found in the Middle Coal-measures, as for instance in Cumberland. The disagreement of a single character does not, however, invalidate a conclusion drawn from an aggregate of characters. Such disagreements occur among recent plants, which are classified on exactly the same principles as those applied here. In the recent family Scrophulariaceæ, the presence of five stamens in the flower is a single character contributing towards an aggregate of characters, which distinguishes this family or natural order from others. But many, perhaps the majority of genera belonging to this family, possess only four or two stamens, whereas their other characters, as a whole, clearly point to close affinity with those members of the Scrophulariaceæ which possess five stamens. It need hardly be pointed out that if all the plants, which possess five stamens, were thrown into a group on this character alone, that group would not be a natural one, since it would include a large number of genera in no way related one to the other.

Let us now endeavour to see with what success the principles, which we have discussed, have and are being applied towards zoning the Carboniferous rocks.

It should, however, be pointed out that we owe our present knowledge of the distribution of fossil plants in British rocks almost entirely to one palæobotanist, Mr. Robert Kidston. For several years, Mr. Kidston has devoted his attention to the examination of the Carboniferous plants of many British coalfields, and in a remarkable paper published in 1894,* he gathered together the results of years of patient and critical research, and gave us the first clear enunciation of the methods by which Carboniferous plants may be used as zonal indices. The principles explained here are those which Mr. Kidston then initiated. As a direct result of Mr. Kidston's work, it is now

---

* "On the Various Divisions of British Carboniferous Rocks as determined by their Fossil Flora," by Mr. Robert Kidston, *Proceedings of the Royal Physical Society of Edinburgh*, 1893, vol. xii., page 183.

possible to ascertain, within certain limits, the age of Carboniferous beds, by an examination of their plant-remains in cases where the position of such beds has been regarded as uncertain, or is in dispute.

PALÆOBOTANICAL SUBDIVISIONS OF THE CARBONIFEROUS.

The following are the subdivisions of the English Carboniferous rocks on lithological grounds:—

Upper Carboniferous:—
> Coal-measures.
> Millstone Grit.

Lower Carboniferous:—
> Yoredales and Upper Limestone Shales.
> Mountain Limestone.
> Lower Limestone Shales, etc.

In Scotland, the Lower Carboniferous rocks are, as a whole, of a somewhat different type, and are subdivided as follows:—

Lower Carboniferous:—
> Carboniferous Limestone Series.
> Calciferous Sandstone Series.

(1) *Lower Carboniferous.*

It is an easy task for anyone who is at all familiar with Carboniferous plants, to distinguish between an Upper and a Lower Carboniferous flora. The general facies of a flora is in itself sufficient to determine the division to which the beds should be assigned. With regard to subdivisions in the Lower Carboniferous, comparatively little information has so far been published, and we are at present chiefly familiar with plant-remains from the Scottish beds. It is believed, however, that Mr. Kidston, who has devoted considerable attention to this subject, is about to greatly extend our knowledge of both the Lower Carboniferous plants of Scotland, and those of the North of England.

The Lower Carboniferous flora of Britain, as a whole, agrees closely with that of the Culm and other Lower Carboniferous floras of Germany, and elsewhere on the Continent. The whole aspect or facies of this flora differs markedly from any yet found in undoubted Upper Carboniferous rocks. The fern-like plants (*Adiantites, Rhacopteris* and *Sphenopteris*)* differ remarkably from the Pecopterids, Alethopterids and Neuro-pterids of the Upper Carboniferous. The well known *Calamites* of the English Coal-measures are extremely rare in the Lower Carboniferous, and their place is taken by another member of the Equisetales, *Asterocalamites*.† *Lepidodendron*, an abund-ant type of Lycopod in the Upper division, occurs also in the Lower, but the species are distinct in each division. Lastly, *Sigillaria* is much rarer in the Lower Carboniferous than in the Upper. So far as the writer is aware, with the exception of one or two doubtful cases, and of such composite species as *Stigmaria ficoides*, no species of Carboniferous plant is common to both the Upper and Lower divisions.

Two subdivisions of the Scottish Lower Carboniferous rocks, the Carboniferous Limestone and the Calciferous Sandstone, can be recognized by differences in their fossil flora, as well as by their general lithological character.

(2) *Upper Carboniferous.*

With regard to the Upper Carboniferous, progress has natur-ally been more rapid, on account of the relatively greater abundance of plant-remains, and of workable coal-seams in this series. At the present time, we can distinguish in it four palæobotanical subdivisions.

Upper Carboniferous :—

> Upper Coal-measures.
> Upper Transition Series.
> Middle Coal-measures.
> Lower Coal-measures and Millstone Grit.

The Millstone Grit, which is usually regarded as the lowest division of the Upper Carboniferous, is not very prolific in plant-remains. But our comparative ignorance of its flora is prob-ably due to the small attention which has so far been devoted to this horizon, rather than to a general barrenness of the rocks, as is often supposed. A more thorough collection of the

---

* *Sphenopteris*, a very artificial form-genus, occurs both in the Upper and Lower Carboniferous, but no species, so far as the writer is aware, is common to both.

† Also known as *Archæocalamites.*

25

plant-remains from these beds is urgently needed. Such species
as are known from the Millstone Grit have been found to be in
every case identical with plants from the Lower Coal-measures.
Thus, while we are at present unable to distinguish the flora of
the Millstone Grit from that of the beds immediately succeeding it, we do, however, know that this horizon is rightly regarded
as a member of the Upper Carboniferous.

In the Lower Coal-measures, fossil-plants are fairly common,
but not so abundant as in the higher series. Coal-bearing rocks
of this age occur in many of the coal-fields of England and
Scotland. Their fossil flora is distinguished chiefly on negative
evidence. All the commoner Lower Coal-measure plants are
found in the Middle Coal-measures, but the latter flora is
characterized by the common occurrence of plants which have so
far not been found in the lower series. Certain plants are,
it is true, only known from the Lower Coal-measures,* but these
are unfortunately of rare occurrence, and are, therefore, not to
be relied upon as zonal indices. On the whole, our knowledge
of the flora of the Lower Coal-measures is fairly satisfactory, but
there is still room for more research as to the distribution of
plants at this horizon. It is not improbable that, in the upper
beds of the Lower Coal-measures, a transition-flora will eventually be detected. Such a transition-flora is already known to
exist in the beds immediately above the Middle Coal-measures,
which yield a mixture of typical Upper and Middle Coal-measure
plants.

The Middle Coal-measures are the most abundant coal-
bearing rocks in England ; and, in beds associated with the coal,
fossil plants are, as a rule, extremely common. This flora is distinguished, as we have seen, from the subordinate series by the
common occurrence of species restricted to this division. It
marks a period during which several plant-types, such as
Sigillaria among Lycopods, and *Neuropteris* and *Sphenopteris*
among fern-like plants, attained their maximum development.

Intermediate between the Middle and Upper Coal-measures,
there is an horizon known as the Upper Transition series. The

---

* For a list of these, and for an enumeration of the characteristic plants of the
different subdivisions of the Carboniferous system, see " On the Various Divisions
of British Carboniferous Rocks as determined by their Fossil Flora," by Mr. Robert
Kidston, *Proceedings of the Royal Physical Society of Edinburgh*, 1893, vol. xii.,
pages 223-232.

general character of the flora of these beds has been already indicated. These rocks are present, and have been recognized by their fossil flora, in many coal-fields, including the South Wales[*] (Lower Pennant rocks), Somerset[†] (New Rock and Vobster series), Potteries,[‡] South Lancashire (Ardwick series), and possibly also in the Cumberland coal-field; though in the last mentioned, few fossil plants have yet been found in the Upper division of the Sandstone Series of Whitehaven.[§]

The flora of the Upper Coal-measures, an horizon which only occurs (so far as our present knowledge extends) in the three southern coal-fields of South Wales, Somerset and the Forest of Dean, is, as a whole, very distinct from that of the Middle Coal-measures. It is especially characterised by an abundance of certain types of *Pecopteris* and *Alethopteris* (*A. Serli*, Brongniart), and by a comparative rarity of *Lepidodendra* and some other genera, which are abundant in the lower series. The Upper Coal-measures are the highest series of Carboniferous deposits which occur in this country, though still higher sub-divisions are found in the Carboniferous beds of the Continent.

———

Such is an account, necessarily brief, of the use to which British fossil plants of Carboniferous age have been put as tests of succession, and of the progress which has attended the study of the distribution of such remains. There is not time here to mention the invaluable services that such plants have rendered to botany in the elucidation of the phylogeny of plants, and in extending our knowledge of the habit and structure of members of the vegetable kingdom in past times.

The writer would wish to point out in conclusion the great need for, and importance of, still further work among Carboni-ferous plants to both the botanist and geologist; and, he might add also, to those engineers who are directly connected with the working of our carbonaceous deposits. Several of our coal-fields are as yet unexplored botanically, and in most of them

---

[*] Mr. R. Kidston, *Ibid.*, 1894, pages 228 and 229.   [†] *Ibid.*

[‡] "Additional Records and Notes on the Fossil Flora of the Staffordshire Potteries Coal-field," by Mr. Robert Kidston, *Transactions of the North Stafford-hire Naturalists' Field Club*, etc., 1897, vol. xxxi., page 128.

[§] "The Fossil Flora of the Cumberland Coal-field, and the Palæobotanical Evidence with Regard to the Age of the Beds," by Mr. E. A. Newell Arber, *Quarterly Journal of the Geological Society of London*, 1903, vol. lix., page 17.

more detailed work than has yet been possible is necessary on many horizons, if we are to increase the number of landmarks in the Carboniferous system. The workers in this particular field of research are at present few, and thus the task of detailed collection of plant-remains over such wide areas is beyond their powers. Such work can, in most cases, be only carried out successfully by local geologists. The writer would, therefore, make a special appeal to members for help in the collection of specimens, and for records of the occurrence of plant-remains in the rocks with which they may be familiar. If only a few members, who are engaged in mining operations in rocks of Carboniferous age, would take an active interest in the subject, by "keeping a look-out" for plant-remains, and by collecting and forwarding them as they may be found to those* who are specially working on the distribution of Carboniferous plants, it would soon be possible to extend considerably the use of such plants as zonal indices beyond the limits which are here described.

————

Mr. J. J. H. TEALL (Director of the Geological Survey, London) wrote that the work of determining the distribution of fossil Carboniferous plants in space and time is undoubtedly one of great importance, both from a scientific and from an economic point of view. It is, moreover, a work which can only be effectively performed by the co-operation of those who are actively engaged in raising coal and in sinking shafts, with palæobotanists like Mr. Newell Arber and Mr. R. Kidston. The results already obtained by these gentlemen are most important, and constitute a guarantee that any assistance which members of the Institute can render will not be thrown away.

Whenever borings are made for coal in undeveloped fields, as for example in the concealed part of the great Yorkshire and Nottinghamshire coal-field, every trace of a plant or other fossil should be most carefully preserved and submitted to competent specialists for determination. But the full value of the evidence furnished by fossils found under such circumstances can only

* Specimens may be forwarded either to Mr. R. Kidston, F.R.S., 12, Clarendon Place, Stirling, N.B., or to the writer at the Sedgwick Museum, Cambridge. It is of the utmost importance that each specimen should be labelled with a record of the exact bed, pit, and locality from which it was obtained. Such information is often much more valuable than the specimen to which it applies.

be obtained, when the distribution of fossils has been accurately determined in areas where the relations of the strata are well known. Every effort should therefore be made to increase our knowledge in this respect.

As Mr. Arber pointed out, there is no probability that the Carboniferous rocks will ever be zoned by fossil plants as minutely as the Jurassic rocks have been zoned by ammonites and the Lower Palæozoic rocks by graptolites, but it has already been proved that, by taking into account assemblages of plants and noting the relative abundance of particular forms, it is possible to establish a chronological classification of great value; to correlate, within certain limits, the strata of different coal-basins in this country; and to bring our measures into relation with their equivalents in other parts of the world.

Mr. R. Kidston had kindly undertaken to prepare a memoir on the horizons of the British coal-measures, as determined by fossil plants for the Geological Survey, and as Director of that survey, he (Mr. Teall) desired most strongly to support Mr. Newell Arber's appeal for assistance from members of this Institute for the work which he and Mr. R. Kidston were doing.

Mr. E. Leonard Gill (Newcastle-upon-Tyne) wrote that Mr. Newell Arber had given a clear and interesting account of the broad results attained through a study of the fossil-remains of the Coal-measure floras. His paper showed that the labours of the naturalist may have a practical value, even for the mining engineer; and on the other hand it made it evident that only such thorough and patient work as that in which Mr. Robert Kidston and the author were engaged could really have any such value. But anyone having access to collieries and shale-heaps was in a position to help with this work, in the way pointed out by Mr. Arber; and now that the practical and systematic nature of the investigation had been demonstrated before this Institute, it was much to be hoped that the members would contribute to the attainment of a better knowledge of the fossil flora of the Northumberland and Durham coal-field. He (Mr. Gill) might add that the Natural History Museum at Barras Bridge contains a collection which will be found useful for reference by any who may become interested in the subject, and could be used as a temporary deposit for specimens. It also contains the Hutton collection of coal-plants (presented by this Institute), which is a classic example of what may be done towards forwarding this particular branch of knowledge by anyone who makes that use of his opportunities which the author and Mr. Kidston so strongly urge.

Mr. AUBREY STRAHAN (Geological Survey, London) fully concurred with the author in attaching great importance to the search for definite recognizable horizons in the Carboniferous rocks. In the exploration of unproved coal-fields, and especially of those concealed under newer rocks, the need for such landmarks was now being felt daily, and would be felt more in the near future. Lithological character was an untrustworthy basis for correlation in such rocks as the Coal-measures. Sandstones and shales replace one another rapidly even in the same basin, and can only be matched in a most general way in adjoining basins. In correlating the measures of coal-fields so far apart as those of the south-west and those of the Midlands or the North of England, reliance would have to be placed on the fossils.

It was unfortunate that neither of the two lines of investigation referred to by the author had yet yielded results so precise as were desirable for practical purposes. Among a large number of molluscan species, scarcely any could be trusted as indices of horizon, and in the case of the plants the enormous proportions attained by "transition-series" shewed the difficulty of classifying strata by fossils with so indefinite a vertical range. There were other fossils, however, not only in the Lower Carboniferous rocks, but at several horizons in the Coal-measures, which probably would be a surer guide to the zoning, if they were not unfortunately extremely rare. The fossils referred to were of purely marine types, and belonged to genera which ranged through a large part of the Carboniferous system. Among them Cephalopoda were generally represented, and would probably furnish a better palæontological time-table than either the Lamellibranchs or plants. But though attention was being directed to them, with the result of shewing that they were more frequent than was supposed, the fossiliferous bands were still too rarely recognized to be of much practical use.

Notwithstanding the difficulties referred to, the study of the Coal-measure plants had yielded, and would yield valuable re-

·sults.   The importance of being able to determine whether a
boring carried down through Secondary rocks into Carboniferous
strata had struck those strata below or above the productive
measures could hardly be overestimated, and the researches referred to in this paper had gone far towards rendering such determinations possible.   Their progress in the future must depend
largely on the amount of material available: for the relative
abundance of certain types of plants at any one horizon, on which
the author rightly placed much reliance, could be ascertained
only by the collection of many specimens.   In this work the co-
operation of mining-engineers was urgently needed.

Dr. WHEELTON HIND (Stoke-upon-Trent) wrote that there
was of course nothing like leather, but while leather was a very
useful article in its place, its usefulness was distinctly limited.
It had always seemed to him that palæobotanists had not done
enough field-work, and, therefore, were too much inclined to
magnify the value of plant-remains as a factor in determining
horizons in the Carboniferous series.   At the outset, it must be
remembered that a very large portion of the great thickness of
Carboniferous rocks was deposited under marine conditions and,
therefore, if plants were found in them, they were there only
casually.   Again, in the Lower Carboniferous rocks, plants were
absent through many hundred feet of strata in certain districts,
and, therefore, afforded no help whatever to the identification of
horizon.   Carboniferous plants were undoubtedly land-organisms,
and could rarely, if ever, be in the exact spot where they grew, but
had drifted to a greater or less extent.   The mainland of the
Carboniferous period was not far away, probably occupying the
North of Scotland, while part of the Lake country was also land.
Plants would be laid down only in strata near to land, and hence
were not distributed over wide areas.

He (Dr. Wheelton Hind) quite agreed that in attempting to
zone the Carboniferous rocks, it was the fauna or the flora as
a whole which was important, and not so much the presence of one
or two species.   This was certainly true of the mollusca, and he
gladly learnt that the flora also bore out this fact.   As a matter
of fact, in the Upper Carboniferous, where plants were fairly
common, the mollusca gave a much more exact and definite
index to the actual bed than the plants, which he understood

only as a whole, illustrated artificial subdivisions rather than beds. The table of subdivisions of the Carboniferous sequence was to his mind somewhat unsatisfactory, because no definition of the Yoredale series was given, and many writers confused two distinct series under this name. He was not aware that plants had been found to any extent in this series; but, as the true Yoredale series was only the equivalent of the upper part of the thick limestone of the Midland counties and contained the same fauna, and the series was probably marine from top to the base, the question of plant-remains was quite subordinate to the mollusca as an index of horizons.

If the Yoredale series was intended to include the shales and black limestones of the Midland counties, then he emphatically joined issue, for these beds were characterized by a rich fauna, which unmistakably indicated a different condition of deposit from that which obtained in the true Yoredale country. In this series, the Pendleside series, plants occurred at a few horizons, but they were very rare, and too unfrequent for much to be based on them. At present, therefore, plants were only plentiful enough in the Calciferous Sandstone series (a series of detrital material, laid down not far from the shore), and in the productive Coal-measures, to give any help, and it behoved all those who came across specimens to have them identified and properly labelled for reference. He might say, in conclusion, that his own researches did not lead him to accept the subdivisions of the Coal-measures proposed by the palæobotanists, but he was open to conviction. At present he was of opinion that the plants and the fauna of the Carboniferous series told the same great tale, as nearly similar, as organisms with such different habits could do; and the only scientific attitude was not to neglect the evidence of either, or to belittle the results obtained from the study of a single group. It was most improbable that fossil plants and mollusca would give absolutely the same results.

Prof. G. A. Lebour (Durham College of Science) congratulated Mr. Arber on his valuable paper, on the possible zoning of the Carboniferous rocks by means of plants. Such attempts were the more to be encouraged, because of their extreme difficulty and of the almost hopeless nature of the enquiry. He (Prof. Lebour) had no desire to throw cold water on work of this kind, but he could not shut his eyes to the many obstacles which must be met with in carrying it on. Zoning in the Carboniferous series by means of animal-remains was only now beginning to enter the sphere of practical geology, and that chiefly in the Midland counties, owing to the excellent detailed investigations of Dr. Wheelton Hind, Mr. Walcot Gibson and Mr. J. T. Stobbs. Animal-zones were. however, far more easily determined than plant-zones. They were also more useful, because the life of a genus or of a species was, in general, shorter in the animal kingdom, and, therefore, restricted to a smaller range of strata than in the vegetable kingdom. Thus a single plant-zone might be and often was, represented by several animal-zones. Again, those portions of the organism which enabled one to name a genus or a species with certainty were much more often preserved in animal-remains than in those of plants. The literature of fossil botany showed, by the terrible bulk of its synonymy, the want of agreement as to nomenclature that existing among experts, and this was especially true as regards Carboniferous plants. This evil, though it existed also with respect to animals, was, in this case, far less rampant and more easily remedied. As against these objections to plant-zoning, it must be admitted that whereas in the non-calcareous divisions of the Carboniferous series animal-remains were comparatively scarce and represented by a small number of forms, plant-remains were abundant both individually and specifically. It was probable that when some standard of nomenclature—say Mr. Kidston's— came to be adopted more generally than at present, and when (if such a degree of knowledge ever be possible) the associated plants of each district at each moment of time during the Carboniferous period came to be known, so that the flora of the Lower Coal-measures (for example) of South Wales could be intelligently compared with that of the synchronous period in the Newcastle coal-field—it was probable that then, but not till then, plant-zoning in these rocks might become of practical value. It was as a valiant attempt to take part in this forlorn hope that he, as a member of the slow-moving British Association Committee to which the writer of the paper referred welcomed Mr. Arber's very interesting communication.

Mr. J. T. STOBBS (Stoke-upon-Trent) wrote, cordially welcoming Mr. Arber's contribution to the *Transactions*. He perused the paper with mixed feelings: on the one hand, it was another indication of the growing interest of the geologist and the specialist in the Coal-measures and it would probably result in an increase of the number of workers in the subject; but, on the other hand, it made an unfair estimate of the work that had been done along other lines. Whilst he would like to consider himself just as anxious for the greater use of fossil plants as of fossil mollusca or fishes, yet many years' devotion to Coal-measure geology had compelled the acceptance of views which did not harmonize with those expressed in the paper.

Dealing more particularly with Coal-measures, in addition to the three types of zoning enumerated by the author, namely: —Plants, fishes and invertebrates, there was a fourth method which he (Mr. Stobbs) believed capable of more exact and far-reaching results, namely:—The discovery and the tracing of marine beds known to exist in most of our coal-fields. The latter method enlisted the knowledge of specialists of the above-named three types of organic life, and it might be seen at a glance that a combination of these forces must be more powerful than their isolation and, should it be said, their mutual negation and neutralization?

As to the nomenclature of the Carboniferous system, adopted by Mr. Arber, it was unwieldy and was a compromise with a system that had now fallen out of use. In seeking to initiate a scheme of classification based on palæontological data, why cling to one based on lithological characters? The artificial divisions advocated by Mr. Arber were objectionable for the following reasons:—(1) There was no actual break in the succession or sequence of the Carboniferous rocks; and (2) no palæobotanical line of demarcation capable of being mapped had been pointed out in any of our coal-fields.

With regard to the actual zoning by fossil plants indicated by Mr. Arber, the limits were somewhat nebulous, and it seemed to him (Mr. Stobbs) to be almost as easy by other criteria to recognize single beds, as the scientific compartments devised by palæobotanists. Further, it was necessary that the zones should be numerous enough to ensure the identification of coal-seams in more than one district; for, it must be remembered that palæontology dealing with other organisms than palæobotany had already achieved this. It was more than doubtful if fossil plants could ever be relied upon for work of such exactness, although it might well be that certain broad divisions might be taken, which, as a whole, might materially promote more exhaustive zoning by other life-forms.

The inferiority of plant-remains for the object in consideration seemed to arise from : —(1) The difficulty in determining the species with accuracy. Sir R. I. Murchison related how Mr. Robert Brown, the botanist, would never name the " genus," much less the " species," of a fossil plant. (2) The fact of their presence in a stratum being no evidence of the condition under which the bed was deposited. He (Mr. Stobbs) had carefully worked ten different marine beds in the North Staffordshire coal-field and had found plant-remains in every one of them, and obviously the plants would not indicate marine conditions. In his opinion, the time was not ripe for zoning by fossil plants (as evidenced by the breadth of those zones specified by Mr. Arber) nor would it be, till one typical coal-field had been exhaustively worked. In the interests of the science of the Coal-measures, it was to be hoped that specialists would desist from framing classifications which had no regard to the evidence afforded by branches of knowledge other than their own.

The Rev. J. F. Blake (London) wrote that he recognized the great interest and truth of Mr. Arber's paper, and believed that on the principles laid down certain zones could be indicated. In comparison with the indications by animal-remains it had been claimed that plants are superior as zonal indices. Thus, Mr. R. Kidston* had stated that at the Hamstead colliery, Great Barr, near Birmingham, some strata which had been classed as Permian, were found to contain animal fossils which would refer them to Lower Coal-measures, but the plant-remains shewed them to be Upper Coal-measures, which was much nearer the mark. Plants, however, are much more dependent on climate, and this might be the main cause of any difference between them. We learnt that Lower Coal-measures, as determined by the plants characterized Scotland, transition-zones the North of England, Middle Coal-measures the Midlands, Upper Coal-measures the South of England, and still higher measures the coal-fields of France. This zoning by latitude seemed significant. All these

* *Proceedings of the Royal Physical Society of Edinburgh*, 1893, vol. xii., page 186.

floras might be co-existent, and they were not entirely distinct; thus only the hardiest reached the northern coal-fields, while the tenderer for climate (but more prolific), flourished in the south, and as the climate became warmer the latter gradually reached a little farther northward, so as to characterize the upper part of the Coal-measures there, and so we got the sequence.   But it was not like the zones of the Chalk or the Lias—where the occurrence of any one zone led one to expect all the lower zones beneath it. If the flora of the Upper Coal-measures was found in any sinking we could not, speaking generally, assume that the Middle and Lower Coal-measure floras and their corresponding coals would follow at a lower level, but only if such coals were known to occur in the district.   The recognition of the zonal character of a flora might assist in the extension of a known coal-field, but it would not determine the value of a new one.

Prof. R. ZEILLER (Paris, France) wrote that he heartily congratulated Mr. Newell Arber on having so clearly summarized the characters of the flora of the different stages of the Coal-measures. He hoped that those, to whose lot it fell to work mines, would increasingly appreciate the interest of palæo-botanical investigations and the practical advantages to be derived from these in determining the relative ages of the workable coal-groups. Already well nigh half a century had passed since the late Prof. H. B. Geinitz had shown the variations in the flora of the Saxon Coal-measures on passing from one horizon to the next; but the distinctions pointed out by the learned professor were hardly susceptible of generalization, as the Saxon measures comprized only a limited portion o1 the Carboniferous series. It was not until 1877 that Mr. Grand'Eury indicated with precision the successive modifications undergone by the flora during the Carboniferous period, and the characteristics exhibited by it at each successive stage. The conclusions at which he arrived were, moreover, confirmed by the researches carried on at about the same time by other geologists, such as Messrs. Weiss, Stur and Lesquereux, and the present writer. If the Carboniferous formation be regarded as a whole, three great divisions may be distinguished therein, each so sharply characterized by its flora, that it is sufficient to identify a few species from one or the other in order to assign to it the particular deposit which one may be investigating.

The first of these divisions, starting from the base, is the Lower Carboniferous or Lower Coal-measures, or Culm stage, with which may be correlated the Lower Carboniferous of Great Britain, especially characterized (as Mr. Arber observes) by a group of species of the genera *Adiantites, Rhacopteris, Cardiopteris, Rhodea, Diplotmema, Archæocalamites*, and by certain particular forms of *Lepidodendron*.

Above this come the Middle Coal-measures, or Westphalian, to which belong the greater portion of the deposits included in the great coal-bearing belt that stretches across Europe from England to Russia, comprizing the Upper Carboniferous of Great Britain, the Franco-Belgian coal-fields, and the Westphalian and Silesian coal-fields. The flora of these is characterized by species of the genera *Sphenopteris, Mariopteris, Alethopteris, Lonchopteris, Neuropteris, Sphenophyllum*, and by the abundance of ribbed *Sigillaria*.

The topmost division, Upper Coal-measures, or Stephanian, (unrepresented, as it would seem, in England) more especially includes the coal-fields of Central and Southern France, the highest zone of the Saar coal-field, the uppermost horizons of Saxony, etc. The predominant forms are ferns or Cycadofilicinæ of the genera *Pecopteris, Callipteridium*, and *Odontopteris*, ribless *Sigillariæ*, and *Cordaites*. Thence we pass, without observing any notable variations of flora, up to the Permian, which is characterized, however, by certain special generic or specific types, such as *Callipteris, Tæniopteris* and *Walchia*.

If we enter into details, and endeavour in each of these great divisions to distinguish subsidiary divisions by means of their respective floras, it becomes necessary to compare a greater number of specimens, and to determine with precision the specific forms represented among them. Mr. Newell Arber has indicated, in a few words, the characteristics which differentiate the several stages, from the Lower Coal-measures up to the Upper Coal-measures, in which latter the species of the Stephanian flora begin to appear in fairly large number, though still associated with a considerable proportion of Westphalian forms. Variations of much the same order allow us (as Mr. Grand'Eury had shown) to distinguish a series of groups in the

Stephanian formation, the lowermost beds of which still contain a few species of the Westphalian flora, the Cordaitæae dominating towards the middle, while the uppermost beds are characterized by the abundance and variety of ferns, and by the incoming of some forms which are destined to prevail in the Permian.

It may not be altogether uninteresting to remind the members that on more than one occasion have palæobotanical investigations led to practical results of real importance, from the point of view of the working of mines. Thus, for instance, at La Grand'Combe, in the Gard, Mr. Grand'Eury and the present writer were enabled to recognize as being of manifestly distinct ages two groups of measures separated by a fault; these were originally thought to belong to the same horizon, but, after palæobotanical studies had revealed the unconformity of their age, a boring was put down below the younger group and struck at a depth of about 2,460 feet the seams of the older group.* Conversely, the present writer was enabled to determine as contemporary, and as belonging to one and the same group, the seams worked in different districts of the Graissessac collieries, and originally thought to be of different ages. In the Allier coal-field, moreover, the investigation of the flora had enabled Mr. Grand'Eury to assert that it was needful to look, at a greater depth than had been hitherto attempted so far, for the main seam of St. Eloi, thrown out by a fault; and the exploration-work carried out in accordance with his predictions revealed this seam at a depth of 951 feet, with a thickness of 45 feet of coal. There is little doubt that in many cases the study of vegetable remains, if these be collected with the precautions recommended by Mr. Arber, will continue to render similarly useful service to the mining industry.

What lends particular importance to the characteristics which we have been considering is their remarkable constancy from one locality to another on the globe. The flora, which appears to have been then uniform all over the earth, at least up to about the end of the Westphalian age, underwent everywhere the same modifications in the same order of succession. Thus, in the United States and in Canada on the one hand,

* *Bulletin de la Société Géologique de France*, 1884, series 3, vol. xiii., pages 131 to 149 ; and 1885, vol. xiv., pages 32 to 37.

and in Asia Minor on the other, the same groups are found as in Europe, containing the same specific forms associated in a similar manner, with the sole exception of an extremely small number of types which appear to have been restricted to certain limited areas.

Without entering into minute detail, it may be permissible to remind the members that the palæobotanical characteristics described by the present writer in 1888 as distinguishing the several stages of the North-of-France coal-field were subsequently observed by Mr. R. Kidston in Great Britain and by Mr. Cremer in Westphalia, exactly as the present writer had noted them. It has thus been possible to correlate horizon by horizon the successive strata of these different coal-fields.* The lower and the middle zones of the Valenciennes coal-field correspond almost exactly to the Lower and Middle Coal-measures of England, while the upper zone may be assimilated to the Upper Transition series as defined by Mr. R. Kidston. On the other hand, the Upper Coal-measures of England are not represented in the North of France, but their equivalents occur in various other parts of Europe, as, for example, in the Saar coal-field, in the United States, and in the Heraclea coal-field in Asia Minor, forming there, as in England, the summit of the Westphalian formation.

So complete an agreement between the results obtained by different observers must induce confidence in the characteristics noted by them, as admitting of the precise determination of the particular horizon of the beds which form the subject of their investigations.

Prof. C. GRAND'EURY (St. Etienne, France) wrote that he had read with extreme interest Mr. Newell Arber's paper on the results to which he had been led by the application of fossil-plants to the classification of the British Coal-measures: the determinations appeared to be absolutely accurate. The Edinburgh coal-field does indeed belong to the Culm; and the great English coal-fields correspond exactly to the Westphalian; while the

---

* "Sur les Subdivisions du Westphalien du Nord de la France, d'après les Caractères de la Flore," *Bulletin de la Société Géologique de France*, 1894, series 3, vol. xxii., pages 483 to 501; and "Correlation of French Northern Coal-measures with those of England and Westphalia," *Colliery Guardian*, 1895, vol. lxx., pages 59 to 61.

Millstone Grit, which he (Prof. Grand'Eury) had seen near Manchester, is undoubtedly associated with the Middle Coal-measures. But in England the Stephanian is incomplete, and between the Upper Coal-measures and the Permian there is a great gap, represented by the coal-fields of Central France.

A more detailed study of the species will most certainly lead to the subdivision of British groups into zones. He (Prof. Grand'Eury) doubted, for instance, whether *Alethopteris Serlii* ranges up into the Upper Coal-measures. In France, it is replaced by *Alethopteris Grandini*, which is near to *Serlii*, and if Mr. Newell Arber desired, he would most willingly send him an example of the French species, in order that he might compare it with the British species.

Mr. E. A. NEWELL ARBER, replying to the discussion, wrote that some of the points raised were of great interest, but that he could only attempt to reply briefly to a few of the objections which were urged in regard to the value of fossil-plants as zonal indices. In the first place, he wished to gratefully acknowledge the sympathy with the aims and objects of the paper expressed by several writers, and more especially by the Director of the Geological Survey of the United Kingdom.

In reply to Dr. Wheelton Hind's interesting remarks, he (Mr. Newell Arber) was glad to learn that, in zoning by means of mollusca, the aggregate of the fauna was regarded as important. Judging by the most recently published list of molluscan zones, to which a reference was given in the paper, this did not appear to be always the case. In more than one instance, the occurrence of a single species, such as *Anthracomya Phillipsii*, seemed to be solely relied upon as marking a particular zone. With regard to field-work, there was no doubt that a great deal remained to be done, more perhaps than was in the power of those engaged in the subject at the present time: hence the appeal for local help which he had ventured to make in this paper. At the same time, a considerable number of British coal-fields had been most carefully examined and collected from by palæobotanists, who were constantly engaged in field-work of this nature. The results, which had been attained, had not all been placed on record as yet.

He (Mr. Newell Arber) was glad to see that Prof. Lebour had

called attention to some of the difficulties of the subject, but he was far from agreeing with the conclusion that the task was of an "almost hopeless nature." On the contrary, the application of the knowledge acquired in areas where the age of the beds was clear on stratigraphical grounds to other areas where their age was in dispute had given results which bore a distinct promise of the successful application of the principles explained here. It was true that, in the past, palæobotany had suffered severely at the hands of those who had insisted on the identification of remains, often too fragmentary or too ill preserved for such a purpose. It was to this that the confusion in the synonomy to which Prof. Lebour referred was largely due. The far higher standard of preservation and completeness of detail, which was nowadays demanded before identification is attempted, had tended to eliminate much of this confusion. The dictum of Robert Brown, quoted by Mr. J. T. Stobbs, was a wise one at the time, for the study of fossil-plants was then in a chaotic state. Within the last forty years, palæobotany had been raised to the dignity of a science, and the detailed knowledge which we possess of several Carboniferous types would no doubt astonish some who may not be familiar with the present position to which this branch of palæontology had now attained. These facts alone had tended to remove many sources of error in identification.

He (Mr. Newell Arber) was inclined to agree with Mr. J. T. Stobbs that it might some day be necessary to put forward a scheme of zonal classification in terms of palæobotanical evidence. As, however, the work was still in progress, and it was hoped that many further additions to our knowledge would be forthcoming, there seemed to be no immediate necessity for such a scheme at present. He (Mr. Newell Arber) was, however, unable to agree with Mr. Stobbs that the zones proposed here were invalid because an exact line could not in some cases be drawn between them. Just as there was usually no actual break in the sequence of Carboniferous rocks, so, in regard to the Coal-measures, there was no break in the flora, but merely a gradual change. This was one of the most interesting facts which had been revealed by the study of Carboniferous plants. In the neighbourhood of the hypothetical line between two Coal-measure zones, a mixture of types typical of the rocks above and below occurred, which at least afforded some evidence as to the position of the bed, while there was a marked contrast between the two floras on either side of the line at some little distance from it. Breaks in the flora did, however, occur, for (as had been shown here) there was a marked break between the flora of the Upper and Lower Carboniferous rocks. It might be also pointed out that the number of zones proposed here was quite as numerous as those put forward on the evidence of the mollusca in the paper already referred to.

The question of the influence of climate which Mr. Blake raised was an interesting one, and one on which the evidence of fossil plants was not a little puzzling. Such evidence as was available rather pointed to the fact that local variations did not appreciably affect the flora. This was shown by the close identity of the flora in Upper Carboniferous times in Europe, North America, and in certain portions of Asia. Among Mesozoic plants, this uniformity was even more marked in several cases, many, if not the majority, of the species possessing an almost world-wide distribution.

He (Mr. Newell Arber) gladly welcomed the remarks of Prof. Grand'Eury, whose magnificent work on the palæobotanical subdivisions of the French Carboniferous rocks was well known in this country. With regard to the occurrence of *Alethopteris Serli* in the English Upper Coal-measures, the author believed that all those who had worked at the flora of these beds were agreed that this species was very abundant in them, and far more so than at any other horizon in the British coal-fields. Possibly what was known as Upper Coal-measures in this country might not exactly correspond to the zone in which *Alethopteris Grandini* was common in France.

He (Mr. Newell Arber) also desired to express his thanks to Prof. Zeiller for his valuable and interesting remarks, which, coming from so high an authority on this subject, would have great influence in establishing confidence in the value of fossil-plants as zonal indices in this country.

The PRESIDENT (Sir Lindsay Wood, Bart.) moved a vote of thanks to Mr. Arber for his valuable paper.

Mr. M. WALTON BROWN seconded the resolution, which was cordially approved.

DISCUSSION OF DR. F. SCHNIEWIND'S PAPER ON "THE PRODUCTION OF ILLUMINATING-GAS FROM COKE-OVENS."*

Dr. P. P. BEDSON (Newcastle-upon-Tyne) wrote that Dr. Schniewind's account of his experiments with Otto-Hoffmann bye-product coke-ovens should prove of great interest at the present juncture, when there is evidently a revived interest in the manufacturing of coke in retort-ovens. This paper contained, so far as he was aware, the most complete and exhaustive account of the different phases of the process of coking yet published. It dealt not alone with the quantity and composition of the coke produced, but gave also data as to the hourly variation in the composition and volume of the gas formed during the process of distillation: data not alone of interest from the insight afforded into the nature of the chemical changes involved, but of considerable practical importance as demonstrating at what period of coking the gas produced could with advantage be separated into a richer portion and a poorer portion, the latter to be used to supply heat for the distillation of the coal and the former to be employed as an illuminating-gas. This "surplus gas" might also be utilized in other ways: for example, as a power-gas.

The perusal of the paper and the study of the numerous experimental data were most assuredly convincing of the complete utilization of coal, which the carbonization in the bye-product oven made possible, and substantiated many of the advantages claimed for this mode of treatment. Regarded simply from the point of view of coke-production, there was a distinct increase in the yield of coke, and that, as shown by the analysis, in the form of dry coke.

Most instructive was the examination into the heat-distribution in the products of distillation of the coal and the comparison which Dr. Schniewind instituted between the working of the bye-product coke-oven and the results of the practice in the distillation of coal in gas-retorts, which placed the efficiency of the bye-product oven in a most favourable light. Again it was shown that the proportion of the nitrogen of the coal obtained in the form of ammonium sulphate was higher than usual.

* *Trans. Inst. M.E.*, 1901, vol. xxii., page 619.

As one who had always taken keen interest in the utilization of coal, he had read Dr. Schniewind's paper with very great
profit, and he felt that the facts cited in it put the question
of the bye-product oven outside the domain of mere theoretical advocacy, leaving a burden of considerable responsibility
upon the practice of coking with an expenditure of some 30 to
35 per cent. of the heat-value of the coal.   But then the use of
raw coal in our domestic fires under boilers, etc., was responsible
for a great waste of the heat-value of coal, and further for the
production of the dust and smoke-laden pall which hovered over
our large towns; a condition of affairs which would be greatly
improved were smokeless fuels, such as coke and gas, more
extensively employed.

DISCUSSION OF MR. T. ADAMSON'S PAPER ON "WORKING A THICK COAL-SEAM IN BENGAL, INDIA."[*]

Mr. THOMAS ADAMSON (Giridih, India) wrote that the value
of a system of working should be judged by the death-rate which
it entailed; and for the last 12 years, there had not been a
single death arising from the system described.   The collieries
land 600,000 tons of coal per annum, and the death-rate per
1,000 persons employed underground and aboveground was
0·38 and 0·23 for the years 1901 and 1902 respectively.   The
British record under the Coal-mines Regulation Act did not
compare favourably with these figures, for in 1900 the death-
rate was 1·30 per 1,000 persons employed.   The overmen had
all had experience of mining in Great Britain, and all labour
was performed by natives.   At least 90 per cent. of the coal
in the seam was wrought, and this compared somewhat favourably with the instance given by Mr. J. B. Atkinson, of the
Fifeshire seam, of which only 60 per cent. was worked.

The seam occurs in the Lower Coal-series of the Gondwana
system of Permian age.

In the Joktiabad mine of the East Indian Railway Company, visited by Mr. R. R. Simpson, the seam is 17 feet thick,
including a band of stone 2 feet.   The lower portion of the
seam (8 feet thick) up to the stone-band, is taken out in the

    [*] Trans. Inst. M.E., 1903, vol. xxv., pages 10 and 192.

first working over an area of 50 feet by 50 feet, about one-fourth of the area of a pillar, 100 feet square. The band of stone and the upper and remaining 7 feet of coal are then dropped in the same way as the 13 feet of top-coal at the Komaljore mine.

Four *chowkidars*, or tell-tales, are left in an area of 100 feet square at the Joktiabad mine, while four are left in an area of 80 feet square at the Komaljore mine. Consequently, as stated by Mr. Simpson, less coal was lost at the Joktiabad mine than as stated in his (Mr. Adamson's) paper at the Komaljore mine. The thickness of strata over the seam at the Joktiabad mine is 450 feet. There is a goaf standing now, measuring 400 feet by 250 feet, with nothing but chowkidars (tell-tales) left standing; these tell-tales are 10 feet square, and are spaced from 40 to 50 feet apart. The system of working, as stated by Mr. Kirkup, was introduced by Dr. Saise about 16 years ago, and, under Dr. Saise's superintendence, the working under varying conditions had perfected the system.

Mr. F. L. G. SIMPSON (Mohpani Mines, Central Provinces, India) wrote that the system of preparing the top-coal for dropping seems as perfect as it well could be, but the possibility of entering the goaf at all to cut up and fill the fallen pillar of coal evidently depends upon having a very exceptionally good roof such as that described by Mr. Adamson; and a description with sketches shewing the mode of attacking this block of fallen coal, and the method of timbering would be highly interesting, as this part of the operation does not seem to be so easily carried out as the preliminary drivings. With most roofs, and even with a good post-roof after a time, a heavy fall of top-stone seems pretty sure to follow the dropping of such an area of top-coal as that described, and this fallen stuff would probably fill up the spaces on the two sides farthest from the goaf, and from which the coal has to be again approached. The time required for getting at the dropped coal, in addition to the 24 hours during which the place is fenced off, would seem to give time for the main roof in many cases to fall, especially when, as mentioned by Mr. Kirkup, a goaf of any considerable size had been formed and the condition of the roof impaired. With the roof unsettled or partly fallen or damaged, as it appears likely in many cases to be in practice, it would seem highly impracticable to re-timber the goaf, especially where logs 21 feet high are to be built and trees 21 feet long set up, for as the filling and timbering advance into the goaf, the weighting of the roof is becoming more imminent. The very strong character of the roof, which is so great an advantage in allowing the dropped coal to be reached, has its disadvantages in not allowing it to fall when desired, and thus throwing extra weight on the pillars next to the goaf, and this would seem likely to destroy the chowkidars or tell-tales, if only 8 or 10 feet square and 21 feet high, almost as soon as they were formed, and would also prevent the nine small knobs or stooks from being reduced as much as required. One can hardly suppose that such conditions as the above could always be avoided, even with the splendid roof described at Komaljore colliery, and it would be useful to learn how they are met when encountered. It would be interesting to know the cost of timber, the cost of labour for setting and withdrawing same, and how many cutters and fillers can usually be engaged in clearing away the 40 feet square of dropped coal at one time; whether they go at it on the two entire sides of the block farthest from the goaf, or in what manner, and how long it takes to remove the 850 tons of coal which it contains. It would also be most useful to know whether trouble has been experienced with surface-water during the monsoons, as the great settlement experienced in wholly removing thick seams will lead to the formation of lakes in the rainy season, which might be suddenly emptied into the workings by such a fall as that described by Mr. Adamson. An area of depression, about 600 feet square or over 8 acres, if 1 foot deep, would contain about 3,500,000 gallons of water, and with a rainfall amounting to say 6 inches in as many hours, such a lake would be collecting feeders at the rate of 5,000 gallons per minute. As to fires in the coal, it would be useful to know what happens when from any cause it has been found impossible to remove all the fallen coal.

At the Mohpani mines, the writer had worked out some 10 acres of coal in a seam 25 feet thick lying under a river-bed, with a cover of 60 to 100 feet of strong posts and shales, but not always a post-roof, next to the seam. This seam was worked on the bord-and-pillar system to the boundary, with the bords and headways first driven next the roof, 6 feet wide, and then cut right down to the thill at that width, leaving pillars 40 feet square with no idea of entirely removing them. Afterwards, however, finding it possible to remove them, the "brokens" were taken clean out by a system of modified longwall, taking three to four lifts one above the other, the first being next the thill, and packing and following up. This packing was sent down from the surface, and hauled to the coal-face by a self-acting incline. All the lifts were packed, except the top one, and that only partly. This system formed three to four sets of faces in steps, with hewers about 10 feet apart along each face, and packing following within 10 or 15 feet; and the only limitation to the output was the rate at which the packing could be kept up. This system was profitable under the special circumstances of the case, as it gave perfect control of the settlement of the roof, and the coal could not have been removed below the river by any other method, but the cost of packing would in ordinary circumstances prohibit its adoption. The seam was inclined about 1 in 3, rising both eastward and westward from the bottom of the basin. The cost of packing, including working at the surface-quarry, filling into trams, sending down, tipping and filling into goaf underground, with supervision, etc., was 6 annas per ton; the cost of pit-timber, 3 annas per ton; the cost of timbering and drawing timber, 1 anna per ton; and the total cost amounted to 10 annas (10d.) per ton. The system of driving next the roof and then cutting down is not desirable, and in new seams at Mohpani mines of about the same thickness, the first workings are all driven in the bottom-coal next the thill.

The Komaljore system appears to be adapted only to special circumstances; and, to render the principle applicable to ordinary cases, the top coal would probably have to be dropped in smaller lifts, or really in vertical slices across the end of the pillar. The *chatnies* might be cut in such a position that the slice would drop close to the fast side, thus rendering it easier to get at, and enabling the coal to be filled without having to enter very far into the goaf for that purpose.

Mr. JAMES GRUNDY (H.M. Inspector of Mines in India) wrote that members freely criticized the system of working before they had been able to fully understand it, and even went so far as to state that it seemed very dangerous " and one which would not be allowed by H.M. inspectors of mines in this district,"[*] and one of H.M. inspectors of mines endorsed this opinion. It was quite natural, when such opinions were expressed, that he, who had been the only inspector of mines in India for several years, during which time he was cognizant of the working of this system, should ask himself whether he, and the colliery-managers who were most closely concerned in this case, were less careful of workmen's lives and general safety than they were in some parts of Great Britain.

Let them now at once turn from surmise to facts, and these should settle the question. He (Mr. Grundy) was told by the officials that for a period of 12 years this system had been responsible for 70 per cent. of the total coal raised, and about 90 per cent. of the present output at this large colliery was worked by this system, yet there had not been a single death in this time that could in any way be attributed to this particular system of working. He (Mr. Grundy) did not mean to infer that they had such a remarkable colliery, etc., that no accidents had happened at it, because there had been 15 deaths during the past five years; still the brilliant record remained that there had been none attributable to the system of working under discussion.

To avoid wrong impressions, he would detail the causes of the 15 deaths above mentioned:—On surface tram-line, 3; by run-away tubs, underground, 3; by falling from cage in shaft, 2; in a sinking shaft, 1; by a piece of coal falling down shaft, 1; by falls of roof: (a) Went into a goaf, immediately after the timber had been drawn, and a piece of roof-stone fell on him: this mine was worked on the old system; (b) a piece of coal fell from the side of a pillar worked on the old system; (c, d and e) a piece of roof-stone fell in a place worked on the old system of pillar working, 5; a total of 15 deaths.

Reference had also been made to the system of "leaving the little knobs to come down at random."[†] He (Mr. Grundy) might state that the above degree of safety had not been brought about, and maintained for so long a time, by allowing anything to happen at random. It would not be easy to find anything that was allowed so to happen; and, on the other hand, the excellent results obtained might rather be attributed to the management having very earnestly endeavoured to understand

* *Trans. Inst. M.E.*, 1903, vol. xxv., page 13.   † *Ibid.*, page 14.

the nature and behaviour of the excellent sandstone-roof, and to deal with it as they believed to be best; and the same might be said of the coal.

Much was due to very careful supervision and inspection, both of which he (Mr. Grundy) believed to be fully equal to the best at any colliery in any district in Great Britain; also to careful selection and training of the European overmen, the work and supervision that these men must carry out themselves, and the kinds of work that a native was not allowed to do; and finally the innumerable precautions to avoid accidents that were based on experience and forethought.

The description of the heavy fall of roof was very graphic: it did not, however, represent what usually took place, but rather an extraordinary occurrence. As to this matter, he (Mr. Grundy) was only speaking of the past, and would not surmise as to the future, any more than to hint that as time went on the size of a standing area before a fall might be such as to give quite a new experience when it did happen to fall. In this respect he might also state that great reliance was placed on the *chowkidars*, "policemen," or "tell-tales." No one expected these comparatively small pillars to keep up the roof, but it was really wonderful how much of the tale they did tell to an observant official when the roof began to move and he wanted all the information that anything and everything could give him. The *chowkidars* were of value because:—They gave valuable and timely warning; the workpeople were then removed; when the *chowkidar* had done most of its duty the roof fell; but the value of the *chowkidar* did not cease until the fall was complete, and it did much to modify the suddenness and force of the fall.

Again, one or two systems of longwall were recommended. The section showed that the seam was 21 feet thick, with a strong sandstone-roof and floor. To work the seam safely, by any ordinary system of longwall, would require very thorough packing. Where was the packing-material to come from, and what company would go to the trouble and expense of packing when the coal could be worked so well and safely without it?

It would indeed be very interesting to have some particulars of any colliery where so large a proportion of the total coal was extracted, the coal sent out of the mines in such good condition, with so few accidents, and at so small a cost.

DISCUSSION OF MR. S. J. POLLITZER'S PAPER ON "THE UNDERLAY-TABLE."*

Mr. H. D. HOSKOLD (Buenos Aires) wrote that Mr. Pollitzer had produced a very ingenious and curious original invention, and, in his (Mr. Hoskold's) opinion, when speed was not a question and it was immaterial how much time was expended upon an underground survey, then no objection should be raised against the use of the underlay-table as an additional means for connecting an underground survey to the surface.

However, further investigations might enable Mr. Pollitzer to devise a second instrument, or apparatus, which could be used on, or parallel, to the underlay-table, and so command a much greater distance at each setting up of the instrument. If this could be done, the length measured (or hypothenusal line) would be a constant quantity, and, with a suitable clinometer, the vertical angle for each length could be measured and the perpendicular and base determined. This, however, was only a suggestion for the consideration of Mr. Pollitzer, and not intended to convey an adverse opinion against the underlay-table for the use for which its inventor intended it. It was to be hoped that he would be equally successful in his further efforts. This was an age of electric speed, as also of rigid competition in doing things; and, for these reasons, generally speaking, mine-surveyors would, naturally, select such an instrument or instruments for their use as would not only ensure the greatest possible accuracy and speed, but offer the greatest facility in saving time and labour.

Mr. A. LUPTON had correctly estimated the capabilities of the Hedley improved mine-surveying dial. We had, also, the Thornton dial-circumferentor, and more recently, the newly introduced Grubb-Davis miners' dial-circumferentor, each being constructed with a vertical semicircle capable of reading vertical angles to 90 degrees. These instruments were very useful and handy for surveying narrow inclined tortuous roads in mines. Undoubtedly the best and most speedy practice would be to use three low tripod-stands with these instruments, as indeed with all other classes.

The common, or plain Y theodolite, in its improved form, with

* *Trans. Inst. M.E.*, 1903, vol. xxv., pages 24 and 196.

divided circles of from 4 to 4½ inches in diameter, with a vertical semicircle reading to 80 or 85 degrees and mounted upon a low stand, might be employed for such surveys as they were discussing with great advantage. The triangular levelling-frame, screwed to the tripod-head, and into which the three levelling-screws of the theodolite are locked, carries a small and convenient traversing or centreing apparatus, enabling the surveyor to move the theodolite in all directions for a certain distance, or until the vertical axis and plumb-line are in the same vertical plane as the station-point. There was nothing new in this traversing apparatus, as it was supplied with all the best instruments. If, however, the greatest possible accuracy were to be sought, then, for general mine-surveying, a small specially constructed and light transit-theodolite, with from 4 to 4½ inches circles, reading to either 30 or 20 seconds of arc would give satisfactory results. To render such an instrument capable of being employed for observing vertical angles from the nadir or perpendicular line to the zenith, he (Mr. Hoskold) had devised an improved form of adjustable prism-reflector attachment.

The experience of the writer (Mr. Hoskold) was that the connection of a surface line with an underground survey and made through an underlay-shaft, was one of the commonest and most facile operations to be found in the whole art of mine-surveying; and, further, it was beyond all doubt that the greatest accuracy and speed could be obtained by the proper use of a handy theodolite of moderate weight, although the roads to be surveyed might be tortuous and have a great dip such as those described by Mr. Pollitzer. More than 50 years ago, when he (Mr. Hoskold) occupied himself largely with mine-surveying operations under all conditions, and without possessing such facile appliances as now exist, he encountered no insuperable difficulty in completing such surveys satisfactorily. Some 25 years ago, when he made valuations of various calamine- and blende-mines situated in Picos de Europa, Spain, it became necessary to make an instrumental survey. The underlay-roads had all manner of zig-zag directions, and, in various places, dipped at high angles up to 80 degrees. The underlay-rock consisted of limestone and was very smooth in places; still, with great caution and a proper selection of places as stations of observation, the survey was closed in a satisfactory manner. At that time he possessed his angleometer mine-surveying instrument, represented in Figs. 15 and 16, illustrating his "Notes upon Ancient and Modern Surveying and Surveying-Instruments,"[*] and as the vertical circle attached to that instrument read vertical angles to 90 degrees, it was a matter of indifference what amount of dip underlay-roads in mines had.

The silver- and copper-mines of this country exist under conditions similar to those above mentioned, and it would be difficult to find mines with less facility for surveying operations: still when he (Mr. Hoskold) had occasion to commission the engineers of his official staff to make surface and underground surveys, and to connect the same through underlay-shafts, all obstacles were overcome. It was true that more labour, care, and time were expended than in more ordinary cases.

If any survey, such as that described by Mr. Pollitzer, were to be made for the object of setting out and conducting the execution of some long and expensive drift or tunnel through solid rock from one or two points at the same time, no moderate amount of trouble and cost should be spared in order to ensure accurate results, and in order to guarantee this, it might sometimes be necessary to repeat the survey. If, therefore, wooden platforms were really required for the surveyor's use and to be fixed at various points in the underlay-road, the small comparative cost connected therewith certainly ought not to be considered. It was out of all reason to urge that such platforms could not be easily and properly fixed by the use of short blocks of wood and wedges, by means of which the platform could be secured to the sides of the roads or to the upright timbers, if any, as the case might be. Apart from the question of platforms, he (Mr. Hoskold) believed that no more difficulty would be experienced in setting up a theodolite tripod-stand upon a given inclined road, than would be attached to the tripod-stand used by Mr. Pollitzer with his underlay-table. No doubt it would be preferable to use a theodolite in such surveys, and to adopt wooden platforms to facilitate the work, and then it would be an easier matter to fix the tripod-stand in such a convenient position that the line of sight to a forward station would not be impeded by either of the legs of the stand.

---

[*] *Trans. Inst. M.E.*, 1900, vol. xix., pages 232 and 233.

Mr. Hoskold felt great pleasure in being able to endorse the sentiments expressed in the statement of Mr. Henry Jepson in reference to the connection of an underground survey with a surface survey,* and upon this question, Mr. Pollitzer had shown no good and sufficient reasons for preferring the uncertain plumb-line plan for such a connection as indicated, to the more precise, scientific and highly practical system associated with a properly-constructed optical instrument.

Upon this same point, Mr. Bennett H. Brough, had also fallen into a great mistake in his statement, " as the optical methods present serious difficulties." The theoretical opinion conveyed in that expression was certainly contrary to the evidence, as the successful practice of Messrs. Bourns, Beanlands, Liveing, Richards, Jepson, the writer and various other British and North American engineers proved.

However, it was true that, at the expenditure of much time, labour and inconvenience, and under certain particular conditions in comparatively shallow pits, the plumb-line mode had given fairly good results. But there was always much uncertainty and great risk attending it, when employed to give the direction of long tunnels to be driven from two points; and, for these reasons, the plan had been abandoned by men of great experience in favour of the surer system by such optical instruments as offered every advantage for this process.

The information offered to us in reference to the exploits of "Hero, of Alexandria, 2,000 years ago," would have been more acceptable and interesting if a footnote had been added giving a reliable authority in its support.

Referring to mistakes sometimes made in the use of surveying instruments, he (Mr. Hoskold) had frequently indicated that the accuracy of a horizontal angle might always be known before leaving a station by observing the supplementary angle, which, together with the angle itself first observed should, theoretically speaking, make up an entire circle. But, as small imperceptible errors arose from incorrect setting up of the instrument, sighting, coarse divisions of the instrument, personal defects and booking, the sum of the two angles would, sometimes, be in excess or defect of 360 degrees. However, with a good instrument and proper manipulation, the differential error due to such defects should

* *Trans. Inst. M.E.*, 1903, vol. xxv., page 22.

never exceed the amount indicated by the smallest subdivision obtainable from the verniers. This method of observing was much to be recommended, because it proved the amount of accuracy obtained at each station, and also gave the surveyor the opportunity of making a correction for any error which might have occurred. There would, however, always exist a very small angle amount, which could not be read, no matter how finely the instrument and verniers might be divided, and this common error became an accumulated amount at the end of the survey; but it could not be known unless the survey finished at the point of commencement. In such a case it became a question of balancing the survey, or distributing the total error between the whole of the angles. It had, however, been stated that the errors likely to arise, in the manner indicated, were counterbalanced: but this could not happen, unless the length of the lines between the stations were equal, and the errors could, by some unknown influence or law, equalize themselves in plus and minus quantities.

In order to provide for observing short lengths of lines not commanded by the telescope, to which Mr. Pollitzer seemed to refer, he (Mr. Hoskold) placed two plain sights on the top of the telescope more than 40 years ago, and he always found it a great convenience for the object intended. The error mentioned by Mr. Pollitzer as having occurred in his survey, and derived from the use of his theodolite, must have been due to some inherent defects in the instrument itself; otherwise, with good manipulation, such errors ought not to have existed.

He (Mr. Hoskold) would like to know how Mr. Pollitzer obtained the horizontal and vertical angles by the use of his underlay-table in the tortuous underground-roads that he had described. If such necessary elements could not be so obtained, then it was to be inferred that it would be necessary to employ some other class of angle-measuring instrument.

Referring to a remark made in a former discussion upon Mr. Pollitzer's paper, it might be stated that various North American instruments, with a second or interchangeable auxiliary telescope, had been introduced from time to time. When the second telescope was required for use it was placed on the top of the permanent telescope, or attached to the side of the horizontal axis of a transit-theodolite, and in either position it was believed that sights could be taken down a perpendicular or inclined shaft. All these instruments might be considered to be of the same type, and in a recent American scientific publication discussing the Shattuck double-reflecting solar attachment, those transits with auxiliary telescopes were referred to in the following terms:—" Every engineer knows of the extreme difficulty in carrying an accurate traverse-line up or down a shaft whose dip is greater than 50 to 60 degrees. Beyond this point, small instrumental errors such as collimation, inequality in height of standards, errors in plate-levels, etc., become very exaggerated. By the use of the top or side telescope attached to an ordinary transit, some one or all of these errors—according to the method used—affect the work and cannot be eliminated or easily allowed for, throwing uncertainties about the results, not mentioning the great care required to avoid disturbing the adjustment of the attached telescope between stations, especially in narrow or crooked shafts."

Such was the dictum of a well-informed North American writer, from which we were led to infer that something better and more certain than a transit-theodolite with auxiliary telescope was really needed, or was already in existence, for sighting down perpendicular and inclined shafts and making a connection between underground and surface-surveys.

DISCUSSION OF PROF. A. RATEAU'S PAPER ON " THE UTILIZATION OF EXHAUST-STEAM BY THE COMBINED APPLICATION OF STEAM-ACCUMULATORS AND CONDENSING TURBINES."[*]

Mr. B. WOODWORTH (Baxterley) wrote that the system of utilization of exhaust-steam devised by Prof. A. Rateau was an extremely interesting development of the regeneration-principle applied to exhaust-steam; and if the variations necessary for free absorption and re-evaporation of intermittent exhaust-steam could be kept within the limits of 15 to 19 pounds of absolute pressure there was every prospect of a successful future for the system, if the actual costs of the apparatus and its working did not absorb all the profits of the working. It became specially applicable in the case of central condensing-plants for

* Trans. Inst. M.E., 1902, vol. xxiv., page 322 ; and vol. xxv., page 40.

groups of engines (where the original motive power could deal with its own work without the aid of a vacuum) as it avoided the trouble from excessive cooling of the engine-cylinders, etc., when not at work, and leakages of air at the joints to injure the vacuum, except for the few parts connecting the accumulator and turbines. All the other appliances were subject to pressures slightly above the atmosphere, and consequently shewed up any defects by leakage of steam; but if the vacuum power was required to help the original motors to do their work the system would not be applicable in such cases.

He (Mr. Woodworth) was not enough of a scientist to determine exactly the amount of material necessary to form an accumulator for absorbing and giving out the amount of heat that should be dealt with during intermittent working in the very short spaces of time allowed for these actions to take place : but if it was necessary to allow much surplus steam to escape, so as to avoid abnormal back pressure on the prime movers, or to use much live steam to assist the accumulator in working the turbine, he would doubt very much that the results would be profitable, and for continuously working engines he would hardly consider the system a profitable substitute for the ordinary low-pressure cylinder of the prime mover.

Prof. A. RATEAU (Paris, France) wrote that the diagrams which were attached to his paper showed conclusively that the variations of pressure necessary for the absorption and regeneration of the steam could be restricted to 15 or 19 pounds per square inch, and to less even than that, if the weight of the substances accumulating the heat were sufficiently increased.

The cost of installation of the accumulator-apparatus could not be regarded as high. Moreover he (Prof. Rateau) had recently found a means of dispensing with the cast-iron basins by using water alone for accumulating the steam. On an average, 3 or 4 tons of water would suffice to keep going an engine of 100 electric horsepower. The apparatus became then very inexpensive and very practical.

As Mr. B. Woodworth had very justly remarked, the most obvious application of the system was to the case of central condensation-plants, and it was precisely in a case of that kind that the first plant was erected, at the Bruay collieries. But they should not run away with the idea that it was needful to allow the steam of the primary motors to escape at atmospheric pressure, in order to ensure the success of the system. The desired advantage was gained, even if the low-pressure turbine worked at a pressure inferior to that of the atmosphere. Thus, for instance, in a case where the steam would enter at an absolute pressure of half an atmosphere, and the condenser would yield a vacuum of 27 inches (an easily attainable vacuum), the turbine would still produce one electric horsepower at the poles of the dynamo for a steam-consumption of only 40 pounds per hour.

For cases where the turbine would receive the exhaust-steam during a part only of its working time, and where consequently it must needs be supplied with live steam during a comparatively long interval, he (Prof. Rateau) had indicated in his paper a solution that would satisfy the most exacting conditions. This consisted in providing a turbine to work at high pressure and at low pressure simultaneously, in such wise that it could be supplied at one and the same time both with steam at high pressure (unrelieved by preliminary expansion) and with steam at low pressure. In this fashion, a plant was got together that could be erected no more expensively than an ordinary group of high-pressure piston-engines, and yet capable of utilizing under the most favourable conditions imaginable either high-pressure or low-pressure steam. Engines of this kind, of 300 electric horse-power, were now being built for a mining company, and several others were being designed for steel-works.

Mr. DAVID BURNS' paper on " The Gypsum of the Eden Valley " was read as follows : —

THE GYPSUM OF THE EDEN VALLEY.

By DAVID BURNS.

On the line of the Midland railway, for some miles south of Carlisle, there is a considerable industry in the manufacture of plaster-of-paris and allied cements, which has been steadily on the increase for a couple of generations. Turning to *Mines and Quarries: General Report and Statistics for 1901*, we find that the total output of gypsum in the United Kingdom in 1873 was 66,124 tons, while in 1901 it was 200,766 tons.* There is probably no better index of the wealth, comfort and refinement of a nation than the amount of gypsum that it consumes, seeing that it enters so largely into ornament and improved house-property. Of the above output in 1901, the Eden valley produced 26,655 tons, or over 13 per cent. of the whole. This report adds the following note: " The principal gypsum-deposits worked in the United Kingdom are those of Cumberland, Nottinghamshire and Staffordshire, where the mineral occurs in irregular seams and in spheroidal and lenticular masses in the Keuper division of the Trias. The gypsum of Derbyshire and Westmorland occurs in rocks of the same age, whilst the seam worked in Sussex is considered to belong to the Purbeck Beds."†

Gypsum is at present worked at five different points in the Eden valley, three of which are in Cumberland and two in Westmorland‡ (Fig. 1, Plate XV.).

The northernmost is situated at Cocklakes, near Cumwhinton station, and is owned by Messrs. John Howe & Company. There are here extensive works for the manufacture of plaster-

---

* Page 203.   † Page 202.

‡ The writer desires to express his thanks to the proprietors and managers of these works, for granting him every information and facility in their power. Especially is he indebted to the late Mr. Ch. J. Howe, who took the greatest interest in the investigation, and but for whose sudden death while this paper was in preparation it might have been more conclusive on some points still unsettled.

of-paris, parian and other cements. There are several abandoned quarries near the works, but now all the rock is raised from shafts.

The next mine, at Knothill, is nearly 2 miles south of Cocklakes, and is about due west of Cotehill station. It is owned by Messrs. Joseph Robinson & Company, Limited. There are here also extensive works, and, besides manufacturing plaster-of-paris, they make a fire-proof cement known as " Robinson cement." Here again the outcrop-rock had all gone, and mining had begun, but a steam-navvy has been introduced and removes the 50 feet of soft sandstone and marls, which overlie the bed of gypsum. In this way it is possible to obtain the whole of the rock, and this would not have been practicable in underground workings, owing to the very bad roof which covers the gypsum.

Each of these works is connected to the Midland railway by branch-lines. They are the largest works and the pioneer gypsum-mines of the Eden valley. Before the Midland railway was prolonged from Settle to Carlisle, the gypsum was carted into Carlisle, a distance of 5 to 7 miles, and manufactured there into plaster-of-paris. The first person to use gypsum from this district was an old woman who dug out pieces, baked it and sold it to her sister-cottagers to make geometrical patterns on their hearths and doorsteps.

The third mine is that of the Long Meg Plaster Company (named after a celebrated prehistoric circle close by), situated between the Lazonby and Little Salkeld stations. It is smaller and of more recent origin than the two works already mentioned. It is connected by a branch-line, worked by water-power, with the Midland railway, and is well equipped for the manufacture of plaster-of-paris. Originally it was worked as a quarry, but as the rock dips slightly into the steep south-eastern bank of the Eden, it has had to be mined for many years.

The fourth mine is situated between Culgaith station on the Midland railway, and Temple Sowerby station on the Eden valley branch of the North-Eastern railway. It is owned by Mr. H. Boazman of Acorn Bank. There are no works here, the rock being sold as such, and carted to the station. This, like the

others, was first worked as a quarry,
worked as a mine.

The fifth mine* is near Kirkby
valley railway, and belongs to Mess
pany, Limited. There are no works
to the station. This deposit, like
from the outcrop, and is still work
overburden, of soft marls and clay, i
It is removed with pick and shov
hauled in bogies by a steam-engine,

There is only one other gypsum-
England, and though it does not c
paper, its position may be pointed o
It is situated close by the sea-shor
near Whitehaven.

When we remember that gypsu
ordinary rain-water, and that probab
that comes down our rivers to t
burden of gypsum and there leaves
tribution must be wide and its fo
perplexing. We accordingly find g
rocks and in the newest formations,
and in crystals. In proportion to
its treatment by recognized authori
hit on an explanation that accord
locality, they have generalized it,
repeated unthinkingly by book-ma

The Eden-valley gypsum, ho
original treatment. Mr. J. G. Go
Survey, in the course of his survey
valley, and took a special interest
Among other writings on the subje
able paper to the Geologists' Associ
with truth that current theories a

* A quarry.
† " Some Observations upon the Natura
of the Geologists' Association, 1888, vol. x., pa

along with much valuable information, proceeds to elaborate a theory of his own. Mr. Goodchild had noticed that the surface of the gypsum-bed had a very irregular hummocky form, and he explained the presence of the bed and its outline in the following manner:—

When the marls, in which the bed of gypsum lies, were deposited in shallow, inland seas, they carried a considerable proportion of gypsum which got deposited as crystals of selenite throughout the body of these marls, sometimes in greater quantity and sometimes lesser. At the horizon of the bottom of the present bed was a particularly strong layer of selenite. When surface-waters began to circulate they dissolved these crystals of selenite, and carried them down to the bottom-bed where the gypsum was deposited. In this way, the bed grew upwards, and vertically under the part where the marls were most loaded with crystals, the solid oed grew fastest and thickest. In short, the present thickness of the bed of gypsum at any point is the measure of the extent to which the marls over it were filled with selenite before the accumulation of that bed. Just as the hump of the growing dromedary rises against the pressure of the superincumbent atmosphere by the animal inhaling and appropriating part of it, so Mr. Goodchild supposes that the molecule of gypsum thrust up in some semi-vital way, the superincumbent rock to find room for itself upon its fellows. But it may be better to state Mr. Goodchild's views on this point in his own words:—

From this phase of development it is often easy to trace further gradations, up to the point where nodular masses of various dimensions consist entirely of mixtures in variable proportions of selenite, satin-spar and granular gypsum – the clay at that point having been, it would seem, entirely extruded by the growth of the nodule. I have seen many such cases, especially in Staffordshire, and in both Westmorland and Cumberland, where large nodular masses of gypsum have steadily grown *in situ*, and have clearly extruded the clay whose position they now occupy. In such cases the original bedding of the marls can be distinctly traced through the nodules themselves, just as in the case of the nodular masses of ironstone, or of calcareous matter, concreted in many of our Carboniferous sandstones, or as the bedding of the Chalk can be traced through flint, or of the Carboniferous Limestone through chert, or, again, of the same limestone through the pockets of hæmatite occurring in it in Cumberland and elsewhere. In each and all of these cases, gypsum, flint, hæmatite, it is clearly a case of replacement, only that in the case of the gypsum molecular change has played a less important part, and mechanical forces have been at work equally with the chemical.[*]

[*] "Some Observations upon the Natural History of Gypsum," *Proceedings of the Geologists' Association*, 1888, vol. x., page 435.

The writer has no hesitation in saying that no feature of the marls is traceable into the gypsum of the main beds of the Eden valley, but such may possibly have been observed in certain gypsum-beds occurring over the principal bed, and presently to be noticed. Again:—

In association with the nodular gypsum just described, one occasionally meets with deposits of a much more persistent character. These differ from the nodular deposits only in regard to their horizontal extension. . . . . These have long been worked in open quarries in the lower parts of the basin of the Eden, between Appleby and Carlisle. A careful examination of these in the field teaches us that much the same mode of occurrence as that just described characterizes these deposits also. But, on the other hand, their stratiform features are often strongly marked, and it is difficult to believe that in these cases we are not dealing with a veritable stratum contemporaneous with the beds amongst which it now occurs. Still, there is present the same evidence of passage into nodular gypsum, which can be traced into fibrous gypsum, or into selenitic marls; and it is quite certain in many cases that the lamination seen in the gypsum can be traced into the marls, as in the other cases noted. And there is, again, abundant evidence of irregular upgrowth of the surface, as if the same kind of action that has been at work concreting the nodules had in this case simply united, or added on, gypsum that had been drained out of the clays above, to a pre-existing sheet or bed occurring below. . . . . This view will enable us to account satisfactorily for such of the marked irregularities of the upper surface of the gypsum as are not referable to the dissolving action of surface-waters. . . . . It will also serve to explain the strange contortions so commonly seen in clays overlying gypsum in deposits of all ages all the world over.[*]

It is scarcely necessary to add that on dynamical grounds this theory is quite untenable, and protracted study of these formations has quite satisfied the writer that it is not necessary. There has been no addition to the Eden-valley gypsum since it ceased to be a land-surface and became covered by the shales and marls, except the bed presently to be referred to, and it occurs in the marls, and there has been little loss, except along certain joints, and between the mass and the enclosing clay at the sides, and except, of course, where these beds have been swept away by denudation in common with the other beds of the district.

Nor is there any call for a special theory so far as the origin of this bed is concerned. It has been formed as the limestones of the Carboniferous period were formed, by the accumulation of calcareous organisms. Whether these lived in salt or fresh water there is probably no means of telling, as, so far as the writer has been able to learn, every trace of organic structure is

* "Some Observations upon the Natural History of Gypsum," *Proceedings of the Geologists' Association*, 1888, vol. x., page 436.

gone. The white gypsum is in striking contrast to the red beds both above and below, and the gypsum must have been formed in some sea or lake from which the red water then prevailing was all but completely shut out. This would serve the double purpose of allowing the shells to thrive, and of keeping the gypsum pure. Near Whitehaven, there is a bed of pink gypsum, showing that molluscan life was possible with a certain small amount of iron in the water. A bed of calcareous matter having accumulated, as had so often happened before in the previous period, sulphuric acid was washed into it and converted it into a sulphate. Whether shells, lying at the bottom of a deep sea acidulated with sulphuric acid, would gradually become converted into a solid crystalline mass of gypsum without any other covering, the writer could not pretend to say, and possibly experiment may be necessary to settle that point, and certainly the evidence tends to this effect: but, under the one circumstance or the other, the bed became a great crystalline mass of sulphate of lime.

There is nothing new in this theory. Prof. James D. Dana says:—"There is a noted 'acid-spring' in Byron, Genesee County, New York . . . besides others in the town of Alabama. This sulphuric acid acting on limestone (carbonate of lime) drives off its carbonic acid and makes sulphate of lime, or gypsum; and this is the true theory of its formation in New York."[*]

Sir Archibald Geikie states that gypsum may be formed "(2) through the decomposition of sulphides and the action of the resultant sulphuric acid upon limestone; (3) through the mutual decomposition of carbonate of lime and sulphates of iron, copper, magnesia, etc.; . . . (5) through the action of the sulphureous vapours and solutions of volcanic orifices upon limestone and calcareous rocks."[†]

Sir Charles Lyell says:—

The crater of Taschem, at the eastern extremity of Java, contains a lake strongly impregnated with sulphuric acid, ¼ mile long, from which a river of acid-water issues, which supports no living creature, nor can fish live in the sea near its confluence . . . . Near the volcano of Talaga Bodas, ... sulphureous exhalations have killed tigers, birds, and innumerable insects.[‡]

[*] *Manual of Geology*, 1863, page 248 ; and third edition, 1880, page 234.

[†] *Text-book of Geology*, 1882, page 115 ; second edition, 1885, page 121 ; and third edition, 1893, page 152.

[‡] *Principles of Geology*, eleventh edition, 1872, vol. i., page 590.

The Pusanibio, or Vinegar river of Colombia, which rises at the foot of Puracé . . . . is strongly impregnated with sulphuric and muriatic acids, and with oxide of iron.[*]

It is probably more interesting than profitable to speculate on the source of the sulphuric acid that formed the gypsum of the Eden valley. It may have accompanied the Whin Sill when it spread among the limestones of the North-east of England if indeed that intrusion was not of later date; or it may have come from the overflow of the mineral veins of Alston Moor, which were still later, abounding as they do in sulphides; or it may have come from volcanic vents in the South of Scotland or elsewhere.

After the bed had become a compact crystalline mass, it became exposed to subaërial denudation, at any rate at its northern end, its present surface having been unequivocally a land, or at the lowest a littoral surface. At one time Messrs. John Howe & Company had ¼ acre of it exposed, and unwrought in their quarry at Cocklakes, where this feature was seen beyond all dispute. Lines of little runnels were clearly visible across it, and its appearance was rather suggestive of the weathered surface of a fine-grained crystalline limestone in the bed of a river. Along one or two lines there had been some decay of gypsum since the overlying marls had been deposited, and the bottom-beds were there slightly broken; but, over most of the area, the surface of gypsum, exposed by the workmen, was the identical surface that had sunk below water-level, and had been covered up by dark mud followed by red shales and sandstones.

If this be the second time that the gypsum has been covered by strata since its accumulation as calcareous matter, it seems rather strange that the whole of the first covering has been swept away and a perishable rock like gypsum left to stand the brunt of the denuding forces. But, a solid bed of gypsum would be able to withstand force such as that of running water, or the waves of the sea or lake for a limited time, much better than loose marls such as overlie it now. Gypsum perishes more readily by chemical action than by mechanical abrasion. As it is, the areas known and those remaining to be discovered are probably only isolated patches of what was a continuous bed that may have extended over a large area. Still, though the

[*] *Principles of Geology*, third edition, 1834, vol. iv., page 125.

writer has made persistent inquiry, he has never seen nor has he had reliable verbal evidence from managers or foremen-quarrymen of gypsum having been worked up against a face of stratified marl. In every case that has come under his observation, he has found the rock cut off by Boulder-clay or valley-gravel. Were he to dogmatize on his limited field of observation, he would say that the bed had been continuous till Glacial times, but this is extremely unlikely. Undoubtedly, however, there has been a great sacrifice of gypsum by Glacial action. Mr. Howe thought, from the many borings that he had put down at Cocklakes, that he could trace a line of denudation, as if it had been the course of a river.

The gypsum-rock shows posts as limestone does, but the joint between the posts is sometimes so close that it takes the practised eye of the miner to detect it. The writer once saw on a beautiful face of rock in the Knothill quarry a faint contortion highly suggestive of lateral pressure, but such is not a common feature, although the contorted junction of posts in the Long Meg mine would sometimes bear such a construction.

Gypsum is a hydrated sulphate of lime, and consists of, in round numbers:—Calcium, 33 per cent.; sulphuric acid, 46 per cent.; and water of crystallization, 21 per cent.

The rock is ground to a fine powder between millstones, and then placed in circular pans under which is a fire. In this way the water is "boiled" off the powdered rock, which is being continually agitated by stirrers. The water comes off in little puffs of steam, and has much of the appearance of a fluid boiling. The dehydrated powder is commercial plaster-of-paris.

A varying proportion of the rock is anhydrite,[*] and has the same chemical composition as commercial plaster-of-paris, but none of its other qualities. Anhydrite is much harder than gypsum, is harsher to the feel, and is at present of no commercial value. It usually occupies in some proportion the centre of the deposit. The prevailing idea is that the whole bed was at one time anhydrous and has gradually, through geological ages, been converted into gypsum, a process which is still going

* Called "cobble" by the miners, while gypsum is styled "rock."

on. This conclusion is founded on the following observations:
—(1) Near the outcrop of the bed, where the circulation of water
has been freer, the anhydrite entirely disappears, as it also does
at the edges of the bed where it comes up against the clay; (2)
following a deposit into the hill, in the centre of its breadth, as
the cover increases so usually does the proportion of anhydrite:
and (3) wherever a feeder of water has been found to traverse
the bed, the rock has been found hydrated in that neighbourhood.

Were anhydrite a more perishable rock than gypsum, the
above observations would be very convincing, but the opposite
is to some extent the case, and one might with some consistency
reason as follows :—

Before the bed was broken up by denudation, the parts that
were pure gypsum got filled with swallow-holes, and this weak-
ened those areas and all the beds that were over them, and hence
they fell more readily than the anhydrite-areas to the ravages
of denudation. The pure gypsum round the areas left is on the
edges of the pure areas that have been removed. The reason why
a feeder of water is in a particular position is because the absence
of anhydrite allowed the water to cut its way through at this
spot. We might even go further, and say that the whole mass
was originally gypsum, but where it has had most cover and
been kept driest and warmest the water of crystallization has
been slowly passing off, as it does in the pan of the manufacturer.

Seeing that the bulk of the rock when gypsum, is about
33 per cent. more than when anhydrite, it is difficult to see how
a ball of anhydrite, in the midst of gypsum, could be converted
into gypsum without bursting the surrounding rock. The very
different texture of the two rocks is another difficulty, but possibly both could be overcome by supposing that the balance of
the anhydrite for which there was no room was carried away
by the circulating water.

It is a pity that the anhydrite, of which there is a considerable
quantity in the Eden-valley mines, cannot be turned to commercial account, containing, as it does, nearly 59 per cent. of
sulphuric acid. Any sulphur or acid derived from this source
would probably be chemically pure, and seeing that sulphur
free from arsenic is a desideratum, there is here a problem worth
the attention of inventive members.

Mr. Goodchild states that " sulphate of lime precipitated from solution under a pressure of ten atmospheres falls as anhydrite."[*] He does not state his authority, but if this be the case, it may have an important bearing on the presence of anhydrite in these beds. Almost universally in the Cocklakes district and to some extent in the others, in the top-post, the rock consists largely of crystals of gypsum, that seem to have crystallized out of a magma of the same substance, for these crystals lie among crystalline gypsum of nearly equal purity. Mr. Howe informed the writer that these crystals were about equally prevalent in the bottom-post, which also bore pieces of selenite as clear as glass. These imbedded crystals are occasionally met with in both the top and bottom posts of the quarry at Kirkby Thore. Now, clearly these external posts were never anhydrite, and it might be that the top and bottom posts were first attacked by the sulphuric acid and converted into gypsum. Then the acid, eating its way in, either from pressure or heat, left anhydrite in the central posts instead of gypsum.

It is entirely a subject for the experimental chemist, and further speculation without experiment may not be of much value. The writer would say, in leaving the subject of anhydrite, that in his opinion the balance of evidence is to the following effect :—(1) The top-post has always been gypsum, and the bottom-post has nearly always been the same ; (2) the central posts, where now anhydrite, have always been anhydrite : and where now gypsum they have generally been gypsum ; and (3) in exceptional cases, and to a moderate degree, by the free access of water and air, anhydrite has been converted into gypsum.

These deposits of gypsum are often spoken of as "lenticular." They certainly sometimes become a little thinner at the edges, owing to the greater decay there : but they are no more lenticular than any bed of pure and perishable limestone is that has decayed somewhat near its outcrop.

Fig. 2 (Plate XV.) shows how a bed of anhydrite was cut off sharply as it approached a feeder of water running from the top to the bottom of the gypsum-bed at Cocklakes. Fig. 3

[*] *Proceedings of the Geologists' Association*, 1888, vol. x., page 429.

(Plate XV.) shows that about 1 foot of the gypsum has been washed away over an area of several square feet and red sandstone deposited in its place. Exactly over the point, where this wash dies out the overlying anhydrite thickens, about 2 feet, on its lower margin.

The bed is usually over 20 feet thick, and it has been quarried at Cocklakes as much as 28½ feet thick, and probably the undenuded thickness in that district may be about 29 feet. It is dangerous drawing conclusions from isolated instances, and the total areas mined form but a very small proportion of the original deposit. But, judging from what the writer has been able to ascertain, the bed thins steadily from north to south, and scarcely attains 20 feet, when complete, at Kirkby Thore.

Some 3 feet over the top of the gypsum at Cocklakes comes a thin greenish bed some 2 inches thick. It is so constant throughout the Eden valley, where gypsum is known to exist below, as to challenge attention. The writer examined it under a microscope, and found that it consisted, seemingly, of the ordinary shale, mixed with a large number of minute crystals. He is indebted to his colleague at the time, Mr. McCreath, of Aspatria Agricultural College, for the following analysis :—

	Per cent.
Water and other volatile ingredients	8·30
Oxide of iron	6·00
Alumina	32·50
Carbonate of lime	3·18
Sulphate of lime	7·88
Magnesia	3·06
Potash and soda	0·10
Insoluble silicate and sand	38·98

This analysis and the appearance under the microscope show that when the shale had been deposited up to this point, sufficient mineral waters invaded the area to cause a slight deposition of crystals of sulphate and carbonate of lime. The thinness and general uniformity of this bed throughout the area mined show a corresponding uniformity in the conditions when the shales were being deposited.

At one point in Knothill quarry, the section between the top of the ordinary gypsum-bed and the green bed is largely made up of impure gypsum, alternating with thin bands of shales, clearly the result of deposition. As far as can be seen, this extra gypsum thins rapidly and only covers a small area of a few hundred square feet, and has not been considered of any commercial value.

In the Acorn Bank mine, however, this top bed has been recently found in a remarkable state of development. Along the face of the old quarry, the green bed can be seen at one or two places with its usual development. From the quarry face, the bottom rock was followed into the hill by an adit-level and worked for many years, but to the north it was found to pass largely into anhydrite. A new adit-level was, therefore, driven in such a position as to explore this higher bed, which showed in places balls of gypsum. It was found to be about 10 feet above the lower bed and to be about 6 feet thick. About 3 feet of the top of this bed is mixed with impurities, and has been left and forms an excellent roof. This leaves from 3 to 5 feet of good gypsum, which has been proved north and south, for a length of 600 feet. Part of this rock is not distinguishable from the best of the rock in the main bed below, but much of it has a banded appearance, shewing different shades of colour, which betrays its sedimentary origin. The bottom post, about 1 foot thick, is in places an anhydrite, so that possibly the determining conditions that have made gypsum here and anhydrite there have yet to be discovered.

The writer was confidently informed by Mr. Boazman and his mine-foreman that plant impressions, described as " ferns," had been seen in the top layers of this bed: this is the first intimation of fossil evidence of which the writer has heard.

There can be little doubt that the source of the gypsum of the upper bed was the decay of the lower bed. The following table shows a comparison of the gypseous deposits at Cocklakes and in parts of the workings at Knothill and Acorn Bank:—

COCKLAKES.		KNOTHILL.		ACORN BANK.	
Description of Strata.	Thickness. Feet.	Description of Strata.	Thickness. Feet.	Description of Strata.	Thickness. Feet.
Green bed.		Green bed.		Green bed.	
Dark brown shale ...	3	Impure gypsum ...	2	Impure gypsum ...	2
Extreme limit of good		Marls predominating	2	Pure gypsum	4
gypsum ...	29	Gypsum predominat-		Marls	10
		ing in thin layers ...	6	Probable thickness of	
		Probable amount of		main bed under the	
		good gypsum ...	22	good upper bed ...	16
Total	32	Total	32	Total	32

Not only does the main bed thicken towards the north, but it has been more exposed as a land-surface and more wasted, and quite sufficient material has been taken from it to form all the supplementary beds yet discovered. The distance from the bottom of the main bed to the green bed is about constant. To explain the higher beds it would be necessary to suppose, after the close of the formation of the main bed, that the gypseous area was divided into three groups:—(1) Areas that were raised above the water-level, and in which denudation was going on; (2) areas into which the gypseous waters from the first areas drained and were evaporated, sometimes with an admixture of mud, as at Knothill, and sometimes without it as at Acorn Bank; and (3) areas into which muddy water from a distance flowed forming only shales and marls.

The boundaries of these areas have frequently changed so that an area that was marl-forming at one time became gypsum-forming later, and the reverse. But, for a brief period, when the green bed was being formed practically the same conditions obtained throughout the whole area, and water carrying sulphate of lime, carbonate of lime, magnesia and mud spread uniformly over most of the area, leaving a marl filled with crystals; and then the gypseous period came to an end for the time being. Further developments may reveal further local variations, but this in the main was the succession of events in the Eden valley.

Probably the impure gypsum that forms the roof in the present Acorn Bank mine is the result of the deposition of mud on the yet soft gypsum below, and the top of the main bed at Kirkby Thore seems to show the same conditions. This is quite different to the gypsum of extra purity found in the form of imbedded crystals at Cocklakes. The evidence seems to prove that in the Cocklakes district the waters of the acid sea were cut off and it gradually dried up: the uppermost 6 inches or so of the deposit being the result of the gypsum crystallising out of the mother-liquor. When this liquor completely evaporated and the surface became dry land the surface of the gypsum assumed the weathered form now found.

This still leaves the crystals in the bottom bed at Cocklakes to be explained. These are similar to those in the top bed, but are generally smaller and lie more in horizontal bands. The very bottom is a mixture of gypsum and marl, very similar to the mixture in the roof of the mine at Acorn Bank. The ribs of satin-spar running into the red beds below, both at Cocklakes and Long Meg, have suggested to the writer a dried and sun-cracked surface invaded by gypseous waters. It may be as well, however, to suspend judgment on this point, but the following is conceivable : —

(1) The introduction of calcareous covered life into the area was at first very fragmentary.

(2) When it had made moderate accumulations of calcareous matter, it was attacked and all but exterminated by an inrush of acid water, accompanied no doubt by some readjustment of levels, and this washed the dissolved sulphate of lime into areas hitherto dry and which now show masses of crystals.

(3) The supply of acid water was then stopped and the invaded areas dried up, the cracks were filled with sulphate of lime and mixed the top layers of marl with it. Then after a time, when the final drying up was approaching, gypsum crystallized out of the mother-liquor as it did later when the top bed was forming.

(4) Water again invaded the whole area, this time without acid, and shell life flourished uninterruptedly until the calcium of the whole bed had accumulated.

But what makes this thin line of discoloration more interesting is the fact that it rises and falls with the rise and fall of the solid gypsum below. This gives quite the appearance of the beds just above the gypsum having been thrust up by the prominences of the latter, and no doubt suggested to Mr. Goodchild his theory of their origin. But a careful study of the excellent freshly-cut section, some 40 feet high, which bounded the bared surface at Cocklakes, already referred to, made quite clear the meaning of this peculiar phenomenon (Fig. 4, Plate XV.).

The first bed to be deposited on the billowy surface of the weathered gypsum was a rather dark shale. This filled up the hollows, first to the level of the prominences, and then spread evenly all across, and this went on till the green bed was formed. Then, either owing to the lapse of time, or by pressure, or by the area having become dry land or nearly so for a short time, which is quite consistent with the previous supposition as to the cause of the green bed, the black mud shrank. Where it was deepest it shrank most, namely, over the hollows ; and where it was shallowest, over the prominences, it shrank least. In this

way, the rises and falls in the surface of the gypsum were reproduced in a somewhat modified degree in the overlying green
bed. By the time a thickness in the covers of less than 6 feet
is reached all trace of the undulations is lost, and the strata run
on undisturbed, except at one or two points where there has been
further decay of the gypsum along certain joints by circulating
water since it was covered up, and the marls have fallen down
a little.

The gypsum-bed is situated a short distance above the Penrith Sandstone, a red rock largely developed in the Eden valley.
Mr. Goodchild claims to have discovered signs of unconformity
between the gypsum and the Penrith Sandstone, but the writer
has never been able to detect any evidence of this. Indeed, into
some of the cavities in the gypsum near the bottom of the bed
at Cocklakes, sand had washed and had been formed into a
sandstone identical in texture and colour with the Penrith
Sandstone, showing as far as such evidence can be trusted, that
after the gypsum had been accumulated and consolidated into
crystalline rock, the Penrith Sandstone had not yet become strong
enough to resist the attack of underground circulating waters,
which carried it, particle by particle, into the vacant space in
the gypsum. There is other evidence that sandstone is slower
in consolidating than limestones. No sandstone of the texture
of Penrith Sandstone is known to occur above the gypsum-bed.

The quotation at the beginning of this paper places this
gypsum-deposit in the Keuper division of the Trias and Mr.
Goodchild appears to take the same view.* Without entering
on that prolific source of discussion, the classification of strata,
the writer would point out that in the Eden valley the continuity
of the beds from the Penrith Sandstone (so-called Permian) to
the top of the marls with the gypsum (so-called Trias) is most
distinct and unbroken, and at the deposition of the gypsum-
bed, which no doubt marked the greatest change in the physical
condition of the district in the period embraced, there is not
the slightest indication of an unconformity. It seems rather
remarkable to the mere stratigraphical geologist to be told that
in the middle of red and similar strata, without any break in

* " Some Observations upon the Natural History of Gypsum," *Proceedings of
the Geologists' Association*, 1888, vol. x., page 423.

the continuity of deposition, or evidence of violent disturbance of any kind, the greatest transformation in life, from its first creation to the creation of man then took place, namely, the change from the Palæozoic to the Mesozoic! Indeed, in the infiltration of Penrith Sand into the cavity of the gypsum-bed alluded to above, there is the intrusion of Palæozoic sedimentary strata into Mesozoic rocks.

The writer would point out that Magnesian Limestone may have been formed by the influx of solutions of magnesian salts among accumulations of calcareous matter, just as gypsum was formed by the addition of sulphuric acid, and that the same bed may be dolomite in one area and gypsum in another according to the chemical nature of the volcanic products that have reached it. The 3·06 per cent. of magnesia, in the analysis* of the green bed, equal to the carbonate of lime present, and to half of the sulphate of lime may be significant and would represent the magnesia of about 14 per cent. of dolomite.

The following are analyses of Eden-valley gypsum, in each case from the main bed:—

	Acorn Bank.		Cocklakes.
	I.	II.	
Sulphate of lime	79·09	78·57	78·88
Water of crystallization	20·85	21·32	20·77
Oxide of iron (Fe₂O₃)...	—	0·11	—
Silica	0·06	—	0·35

An analysis of the water from Cocklakes mine is as follows:—

	Grains per gallon.
Chloride of potash (KCl)	3·13
Chloride of soda (NaCl)	1·79
Sulphate of lime (CaSO₄, 2 H₂O)	149·76
Carbonate of lime (CaCO₃)	7·96
Carbonate of magnesia (MgCO₃)	7·76
Silica (SiO₂)	0·18

It is desirable that the features of the Eden-valley deposit be briefly considered as compared with those of Nottinghamshire and Derbyshire as set forth in the *Transactions* in an instructive paper by Mr. A. T. Metcalfe.† Wash-holes are mentioned as being frequent in that district. The writer has seen one in the

* *Trans. Inst. M.E.*, 1903, vol. xxv., page 420.

† *Ibid.*, 1896, vol. xii., page 107.

Cocklakes mine : it consisted of a round symmetrical chamber about 5 feet in diameter and about 10 feet high to the top of a gothic roof. The roof and upper parts of the sides above the level of the stagnant water were smooth and ribbed; the lower parts of the wall were also smooth and covered with very fine mud that had been washed in : and the middle part of the walls towards the surface of the water was covered with crystals of selenite. The surface water getting in had dissolved the gypsum of the roof, and some of it had been deposited as crystals against the walls possibly during periods of drought. There is much satin-spar about the bottom part of the bed, and in the red beds just below, at the Cocklakes and Long Meg mines, but that described by Mr. Metcalfe must be harder and finer grained. It does not occur above the main mass that the writer has seen.

Mr. Metcalfe says " The innumerable inosculations of minute veins of gypsum frequently bind beds of marl into extremely hard rock : "* this is similar to what occurs at Acorn Bank on the upper bed of gypsum. He further refers to the " saddle back " formation over the balls in the marls and like Mr. Good-child ascribes it to the " increase in bulk of the concretionary mass below ; " also that it has been suggested as due to the swelling of the mass by the passage of anhydrite into gypsum. He further notices that around these balls the marls are destitute of gypsum, while it mixes with the marls in the hollows. The late Mr. Howe pointed out to the writer that at Cocklakes on the top of the bed in the hollows there was a scale of gypseous marls, whereas the prominences were smooth and polished and made a clean contact with the shale. The con-ditions in Nottinghamshire seem, therefore, to be very similar to what they are at Cocklakes. The writer's view is that in each case probably, and certainly at Cocklakes, the prominences have still been exposed and weathering, while muddy and gypseous water invaded at intervals the depressions and then formed by evaporation the hard scale. The ordinary alternation of rain and dry weather would be sufficient to accomplish this.

The writer is not aware that any of the gypsum of the north has ever been sold as " alabaster " for statuary purposes, but sometimes the workmen in their leisure time make very pretty little objects such as letter weights and the like. Mr. Metcalfe

* *Trans. Inst. M.E.*, 1896, vol. xii., page 108.

adds:—" Apparently there are no grounds for believing that the
gypsum deposit of Nottinghamshire and Derbyshire are in any
way whatever connected with volcanic agencies."[*] He adopts
the evaporation and precipitation theory.

The writer is indebted to Prof. Lebour for the following
abstract of Prof. Ochsenius's views brought forward in 1885, and
since, owing to the high reputation of that chemist, held in
great esteem by geologists. According to Prof. Ochsenius, the
conditions essential to the formation of a natural salt pan, like
those of the Caspian Sea, are :—(1) The presence of a shallowing
or "barrage" which impedes without completely interrupting
the communication between the sea and the concentration basin ;
then (2) a climate dry enough for active evaporation, whilst the
basin can receive but a feeble proportion of fresh water. In
these circumstances the water which flows into the basin is heated
and evaporates in part. What is left becomes more dense, falls
to the bottom, and determines little by little the death of
organisms or the departure of such of them as can pass the bar.
Deposition of gypsum begins when the percentage of salt is
quintupled and salt itself is precipitated when its proportion,
multiplied eleven times, has raised the density of the water
to 1·22. Above the deposit of salt floats a mother-liquor in which
are present, in ascending order, sulphate of magnesia, chloride of
potassium, chloride of magnesium, borates, bromides, salts of
lithium and compounds of iodine. When the density of this
liquor permits it to surmount the force of the current flowing
into the basin, the dissolved salts overflow into the sea over the
bar, just touching its crest, whilst above them the sea-water continues to arrive in the contrary direction. But the latter on mixing
with the saline liquor, abandons its sulphate of lime, which, in
traversing the salts of the liquor in the act of precipitation,
becomes anhydrous and tends to form a cap of anhydrite above
the sea-salt already precipitated.

If at this moment communication with the sea becomes
interrupted by a rise of the bar, the mother-liquor evaporates
in place, attacks the anhydrite while forming palyhalite and may
afterwards give rise to an entire series of deliquescent salts, such
as are observed in Germany in the Stassfurt deposit. Moreover,

[*] *Trans. Inst. M.E.*, 1896, vol. xii., page 113.

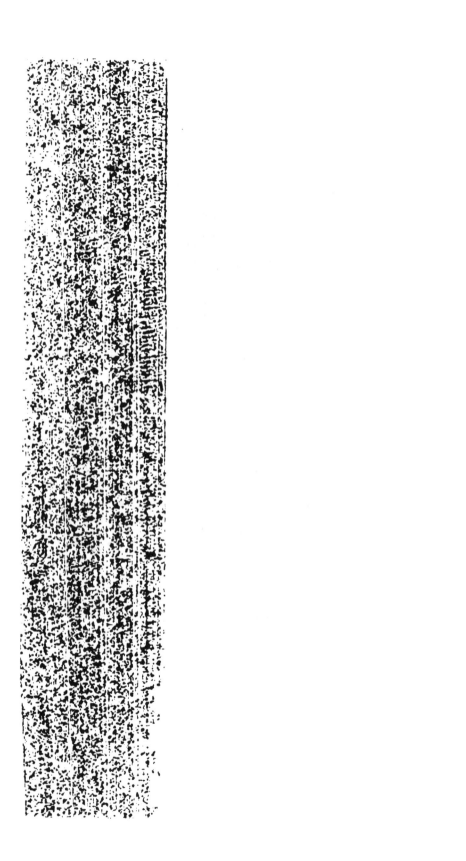

to him) might be cleared up by the writer of the paper, or by some other speaker. With regard to crystallization, in some measure it seemed to him possible to look upon that as confirmatory of Mr. Goodchild's position, in which the writer of the paper did not concur. Mr. Goodchild treated the deposition of gypsum as due to the wasting away of the upper marls. Mr. Burns had not, he thought, said anything which clearly proved that the marls might not have lost the crystals which went to form the gypsum below. With a great deposit of gypsum, a great feeding stratum above would be required, and he did not know whether Mr. Goodchild's theory dealt with that point.

As to the difference of origin between gypsum and anhydrite, and the supposed flow of water through the anhydrite to bring about the hydration of the gypsum, might not that be due to the anhydrite lying more or less along the middle horizon of the gypsum-deposit? Might not the occurrence of anhydrite (the absence of water of crystallization making it so) be due to the fact that the greatest crystallizing-activity took place along the central line of the deposit, and so eliminated the water? It seemed to him difficult to conceive that a flow of water through the central bed of gypsum or anhydrite should cause the rest of it to become hydrated. It appeared to him that the anhydrite had been more probably gypsum to start with, and that the water of crystallization had been driven towards the top and bottom of the gypsum-deposit.

Mr. D. A. Louis (London) said that it was not necessary to attribute a volcanic origin to the sulphur in deposits of gypsum. The immense deposits in the neighbourhood of Paris were obviously deposited from sea-water and a similar occurrence was taking place at the present day in Karabugaz bay on the eastern side of the Caspian Sea. This bay was connected to the Caspian Sea by a very narrow channel, so that in the dry season the loss of water by evaporation greatly exceeded the supply of water; consequently, concentration ensued, and a considerable deposition of the salts from solution took place.

Prof. HENRY LOUIS (Newcastle-upon-Tyne) said that it was only fair to the writer of the paper to point out that it was well ascertained that there were two very different ways in which gypsum could be formed. A typical deposit of the one kind

was found at Stassfurt, which had certainly been formed by
evaporation.   He had, on the other hand, seen several gypsum-
deposits, which had as certainly been formed by the action
of sulphuric acid.   He pointed this out in 1878, when he
examined some extensive deposits in Nova Scotia, and he based
his argument on the fact that the gypsum, in certain places,
enclosed small crystals of sulphur.   Another important point
was that a number of rare borates occurred; and he thought
that the occurrence of boric acid was a crucial test as to this
method of formation.   In Almeria, in the south of Spain, also,
there was a similar occurrence covering many square miles, and
these two were typical instances which showed that Mr. Burns's
theory was, at any rate, possible.   The possible hydration or
dehydration was not a matter of difficulty.

Mr. D. A. Louis (London) said that the chemical method of
formation, volcanic or otherwise, was well-known, but he had not
seen any very large deposits of chemically-formed gypsum.
In volcanic districts, it was plain that the gypsum formed there
was due to chemical action, but then it occurred only in local
bunches.   The Paris beds were very extensive, and undoubtedly
they were formed by evaporation.

Mr. D. Burns said that in his paper he spoke only of the
formation of gypsum in the Eden valley.   There are two principal ways in which gypsum could have been formed in any
position in which it was now found, and most theories were
refinements or modifications of one of these.   The first was the
deposition from sea-water of gypsum that had been leached
out of the land by rain-water, or had been dissolved off some
pre-existing beds of gypsum.   The other was the formation of
gypsum out of other material, and the only substances out of
which it could be formed were carbonate of lime and sulphuric
acid.   The former was the method by which it was renewed
and redistributed, but the latter was the means by which its
amount had been added to.   No one surely would doubt that
there was a great addition to the stock of gypsum all the world
over in New Red Sandstone times, and to suppose that the shell
life that had been so abundant, and had built up such thick
limestones in Carboniferous times still lived on, and formed the
calcareous basis of the beds of Magnesian Limestone and

gypsum of the earlier Red Beds, and that magnesian salts and sulphuric acid were in a peculiar degree products of that particular phase of volcanic activity which characterised Permian times, is the most natural and direct way of accounting for what we now find. The onus of proof lies with those who would base the theory of the origin of a world-wide deposit on subtle laboratory experiments or the discovery of crystals of selenite or isomorphs of salt in the marls.

It is easy to reason from anhydration to hydration or the reverse, and in his view the special conditions that gave rise to gypsum and anhydrite respectively formed the principal difficulty that remained and on which further discussion would be valuable.

Prof. G. A. LEBOUR (Durham College of Science) wrote that Mr. Burns had brought forward, as he always did, a theory of great ingenuity. This theory depended, however, upon an assumption which few geologists would be willing to accept without much stronger evidence of its probable truth than Mr. Burns had offered in his paper. It was obvious that gypsum might result from the occurrence of sulphuric acid in the presence of limestone. It was quite another thing to say that great deposits of gypsum such as those of the Eden valley had been formed in this manner, especially since there were several other modes of gypsum-production known, some of which would seem, in the absence of proof to the contrary, to be more applicable to the cases in question than the one selected by Mr. Burns. There was, as he quoted from Prof. Dana, "a noted 'acid spring' in Byron, Genesee County, New York," but there was no such spring in the region under consideration, nor was there anything to show that anything of the kind ever existed there. In fact Mr. Burns had still to show that "the sulphuric acid that," he says, "formed the gypsum of the Eden valley" was at the proper time available in that part of England. He suggested that this acid "may have accompanied the Whin Sill," but beyond the fact that it must be later than a portion of the Lower Carboniferous series, the age of the Whin Sill was, and must in all likelihood remain, unknown. It might be of much earlier date than the gypsum-bearing beds of Cumberland and Westmorland. The sulphides of the Alston Moor veins and volcanic vents in the South of Scotland "or elsewhere" were mentioned by him as possible sources of the acid necessary for his theory. The first of these possible sources could only be seriously considered after the age of the infilling of the Alston veins had been proved. There were certainly Permian volcanoes in Ayrshire, Nithsdale, and Annandale, and their aid might possibly be invoked in accounting for gypsum-deposits in or near their immediate neighbourhood, if there were any. To look to them for explanations of the Vale of Eden beds of gypsum could scarcely be seriously proposed. If the Whin Sill injection was accompanied by the occurrence of sulphuric acid, and if the action of this acid on limestone were such as Mr. Burns supposed, where were the huge gypsum-masses which there must have been such ample opportunities of producing from the Carboniferous Limestone strata traversed by the Whin Sill? Also, where were the similarly easily-formed masses in the vicinity of the Alston metalliferous veins? It came to this, that where there was some slight evidence (though no proof) of possible acid-emanations of adequate magnitude, gypsum in large masses was unaccountably absent. It was, however, as Mr. Burns well showed in that valuable portion of his paper which dealt with actual observed facts, present in considerable quantities among those red sandstones and marls (the natural home of gypsum, anhydrite and salt all over the world) which, in the North-west of England, yield no tittle of evidence of such conditions as might favour the production of sulphuric-acid springs and the like.

Mr. PHILIP ALLAN (Carlisle) wrote that there were many features connected with the deposits of the Eden valley requiring further study and investigation. Among these might perhaps be mentioned the beds of anhydrite (or "cobble" as it was locally called) found associated with the gypsum. It was not an uncommon occurrence for two separate and distinct beds of anhydrite to extend throughout the greater area of the deposit. On the underside of these beds of anhydrite, there was generally a distinct parting between the gypsum, while on the top side the gypsum and anhydrite was somewhat grown together. Between the two bottom beds of gypsum, there was also a very distinct division. Fig. 5 (Plate XV.) might perhaps more fully explain the occurrence. It would appear, therefore, that the gypsum had been deposited in layers, and that alternate changes had taken place in the calcareous matter as the gypsum was being formed.   On the other hand, the anhydrite was sometimes found in rolls or balls, enveloped entirely in gypsum, with no distinguishing division whatever.

Mr. DAVID BURNS (Carlisle), replying to the discussion, wrote that it was possibly an indiscretion, indulged in with evident hesitation, to refer to the Whin Sill and Alston veins as the source of the sulphuric acid, which formed the gypsum of the Eden valley. If they were such, it was merely because they were local phases of a world-wide outburst in late Permian times. But, taking the metalliferous veins, no one could deny that they had been filled with mineral matter that was largely a compound of sulphur and various metals. What was now left were the truncated remnants of the original veins as formed and filled; and the measure of the action of these sulphides on the limestone was the width of the veins in limestone as compared to what they were in other rocks, which was considerable. The limestones were probably not turned into gypsum, because a hot and dry sulphide had not the ability to perform this change. But when the upwelling stream overflowed at the surface, and came into contact with water and the air, free sulphuric acid would be formed, and this would change carbonate of lime into sulphate of lime, wherever it was washed, if it were a hundred miles away. Prof. Lebour was rather unreasonable in asking the writer to point out the exact " acid spring " from which any particular bed of gypsum had been formed. A geologist could seldom point out with any certainty the position of the land-area from which a particular bed of shale had been washed, or even the position of a volcano from which a bed of ash had, for a certainty, been scattered. Then how could it be expected that any one could point out, among perishable beds, the particular solfatara from which came some special supply of sulphuric acid? We are dealing, as had been pointed out, with the " natural home of gypsum . . . all over the world," and it is sufficient that we can point to an equally wide-spread contemporaneous volcanic activity, which was competent to supply the necessary sulphur. Prof. Lebour knew of other modes of gypsum-production, but the writer did not know of any.   As pointed out, elsewhere, the evidence of re-distribution brought forward by several observers was no explanation whatever of how the " natural home of gypsum   . . all over the world " was filled with that remarkable mineral.

He (Mr. Burns) agreed with Mr. P. Allan that there is much that is unsatisfactory in our knowledge of what has determined the relative distribution of gypsum and anhydrite; and his paper will have served a good purpose if it induced the managers of gypsum-mines to record accurately observed facts in this connection.   Mr. Allan's sketch (Fig. 5, Plate XV.) shewed that there are distinct posts (usually about four) in the gypsum, just as there is in a bed of limestone, and that the division between the anhydrite and the gypsum often follows the joints. But it does not always do so; and in one case, at least, in Cock-lakes mine, observed by the late Mr. Howe and the writer, the junction-line between the gypsum and the anhydrite was as shewn in Fig. 6 (Plate XV.).   As indicated by Mr. Allan, the rounded appearance of the masses of anhydrite is often suggestive of something of the nature of concretionary action.

The PRESIDENT (Sir Lindsay Wood, Bart.) moved a vote of thanks to Mr. Burns for his interesting paper.

Mr. T. E. FORSTER seconded the resolution, which was cordially approved.

———

Mr. J. MALCOLM MACLAREN's paper on "The Occurrence of Gold in Great Britain and Ireland " was read as follows :   *

---

* This paper was read at a General Meeting held on October 11th, 1902.

THE OCCURRENCE OF GOLD IN GREAT BRITAIN AND IRELAND.

By J. MALCOLM MACLAREN, B.Sc., F.G.S.

INTRODUCTION.

The following outline of the conditions under which gold occurs in Great Britain and Ireland is compiled from the many references, historical and otherwise, to the subject to be found scattered throughout the various papers brought together in the appended bibliography, together with extracts from notes made at the auriferous localities themselves, all of which have been visited by the present writer.

In dealing with the historical references, none but those which are connected with veritable gold-occurrences are included. Many reported gold-discoveries in the fifteenth, sixteenth and later centuries no doubt, arose, as indeed was the case as late as 1852, in Fifeshire, and in 1901, at Leith, from the ignorance of those who mistook pyrites or mica, or other minerals with yellow lustre, for the precious metal.

The subjoined references in the works of Latin authors to the occurrence of the precious metals in Britain afford but little definite information; and it may indeed be considered a moot point whether gold and silver in quantity were obtained from these islands in ancient times.

Strabo, writing about 19 A.D., and dealing with Britain's trade and relations with Rome, mentions gold and silver as occurring among its products.* Tacitus, in his biography of his father-in-law, Agricola, relates that the latter in his oration to his soldiers before the battle at the foot of the Grampians, in 84 A.D., where Galgacus was defeated with great slaughter, heartened them in this wise: "*Fert Britannia aurum et argentum, et alia metalla, pretium victoriæ.*"† On the other hand, Galgacus is

* Book iv., cap. 279. † *Vita Agricola.*

reported to have indicated the inevitable issue of defeat, as servitude in the mines "*Ibi tributa, et metalla, et cetera servientium poena.*"* And as is shown by Diodorus Siculus (8 B.C.), the Romans were indeed hard taskmasters. Again Glaucus in his speech (with indirect reference to Britain) says, "*Neque sunt nobis arva, aut metalla, aut portus, quibus exercendis reservemur,*"† thereby inferring that all these were to be found in Britain. The elder Pliny (77 A.D.), states that gold (*auraris metallis*) was found in the stream-works (*elutia*), the stream of water washing out (*eluens*) black pebbles, a little varied with white, and of the same weight as the gold.‡ Cæsar also, in his *Commentaries*, accuses the Britons of having assisted the Gauls with their treasures.

Of the early Roman gold-workings, there are no authentic remains; but it has generally been supposed that the old workings of Ogofau, near the village of Pumpsant, some 12 miles west of Llandovery, are evidences of Roman occupation and of their search for gold. The name Ogofau or Gogofau is probably Ogofawr, which means a large cave or large disused workings, *Ogo* being a generic term for such old excavations.§ At this spot, the remains of Roman pottery, ornaments and baths have been found. Some of the ornaments are of gold, and are of considerable artistic merit. The workings are extensive, and have evidently been opened first along the cap of the lodes. When these open cuts became too deep, levels 170 feet long, 6 feet high, and 5 feet to 6 feet wide were driven through the country-rock to cut the lode. The upper level communicated with the opencast workings by a rise, and the lower and upper levels are similarly connected as shewn in plan and section (Figs. 2 and 3, Plate XVI.).‖ The workings are in Lower Silurian rocks, which here dip slightly to the northward. The lodes are of quartz, and vary both in dip and strike. The quartz is massive and somewhat opaque, showing in places a tendency to form interpenetrating growths of crystals. The accompanying minerals are iron-pyrites, in cubes and pyritohedra, and a little galena. A white sericitic mica and inclusions of slate are not uncommon. The slates, when fine-grained,

* *Vita Agricolæ.*   † *Ibid.*
‡ Pliny, *Historia Naturalis* (Natural History , book xxxiv., cap. 47 to 49.
§ *British Mining*, by Mr. Robert Hunt, page 40.
‖ *Memoirs of the Geological Survey of Great Britain*, 1846, vol. i., plate VIII.

are very dark and very fissile, and through them run occasional thin veins of greenish-blue serpentinous mineral.

Some idea of the primitive method of crushing employed is furnished by an interesting relic. Between the old workings and the entrance to Dolau Cothi is a rudely prismatic block of very coarse sandstone, the grains of which are not by any means well rounded. The block is now set up vertically on a small mound, and in the course of time local traditions have gathered round it furnishing miraculous explanations of its present shape. On each of the four longer sides is a depression, from 2 to 3 feet long and about 8 to 10 inches wide, caused by the successive formation of elliptical grooves, the minor axis of the ellipses being successively moved to a parallel position along the line of the major axis. The latter axis was probably about 15 inches, and the former about 9 inches, in length, and the grooves vary from 4 to 6 inches in depth. It is quite clear that the stone was used as an anvil or mortar on which the quartz from the adjacent workings was crushed. From the gradual shallowing of the depression to its edge, it would appear that the groovings were made by a suspended pestle or block of stone to which a compounded horizontal and vertical reciprocating motion had been given—something indeed akin to the old Australian dolly, where the heavy pestle is raised for the next blow by the spring of a bent pole. In this case, instead of the simple vertical motion given to the Australian machine, the motion was in all probability horizontal, the point of suspension being over the minor axis of the ellipse. The result certainly could not have been produced without the suspension of the pestle, for a single workman operating a heavy pestle by mere attrition would form a depression crescent-shaped in plan as he moved the stone to and fro. The crushing might indeed have been performed after the manner practised, even at the present time, in remote parts of the world, where two natives facing each other and seated astride a large stone alternately draw a large stone over a groove in the lower stone in which the material to be crushed is placed.

Pliny, though he gives the sequence of operations in gold-mining in Roman times, unfortunately supplies no details of the actual processes (*quod effossum est, tunditur, lavatur, uritur, molitur in farinam, ac pilis cuditur*, which is mined, crushed, washed, burned, ground to powder and pounded with pestles).[*] It

* Pliny, *Historia Naturalis,*

may be here noted that Pliny omits the final washing, which is necessary to separate the gold from the products of the oxidation of the pyrites and other concentrates produced by the initial washing process.

Gold was first noted at Ogofau in modern times by Sir W. Warington Smyth[*] and Dr. Percy, though Sir Roderick Murchison had some years previously, submitted the quartz to assay without result.[†]

The Gogofau veins were worked for a short time (during 1889 to 1891) by the South Wales Gold-mining Company, but the results were extremely discouraging, the total yield being 4 ozs. 19 dwts. The whole of the milling machinery has now been removed, leaving only a large heap of tailings to mark the scene of this company's operations.

ENGLAND.

Cornwall.—Of the production of gold in Cornwall and Devon in early times there are no records, but that from time to time the gold-mines of these counties were considered sufficiently remunerative to be worked is evidenced by the numerous writs and grants from Henry III., and from his successors down to Elizabeth, to various grantees. With all these, however, not a single ounce of gold is recorded as having been obtained.

In 1564, a patent or monopoly was granted to William Humphreys, Cornelius Devos, Daniel Hochstetter and Thomas Thurland, to seek for gold, silver and quicksilver in certain counties in England, Wales and Ireland within the Pale. This patent was confirmed and amplified by James I., and became the charter of the Mines Royal Company, which existed and claimed the right to all royal metals until after the middle of the nineteenth century. It does not appear, however, that their operations at any time met with any degree of success.

Notwithstanding the fact that gold in quantity has not been recorded from Cornwall, there is no doubt that the tin-streamers from time to time in the course of their work obtained small quantities. Indeed, Carew[‡] says "Tynners doe, also, find little hoppes of gold among their owre, which they keep in quils, and

* *Memoirs of the Geological Survey of Great Britain*, 1846, vol. i., page 480.
† *Siluria*, page 450.
‡ Carew's *Survey of Cornwall*, 1602, book i., page 7.

sell to the goldsmithes oftentimes with little better gaine then Glaucus exchange."

In the early part of the last century, gold was obtained in small quantities at Ladock by Sir Christopher Hawkins. A specimen presented by him to the Royal Geological Society of Cornwall was enclosed in its quartz-matrix,[*] indicating, therefore, a local origin.

Native gold has been found in most of the Cornish tin-streams flowing to the south. Of these the Carnon stream, at the head of Restronget creek in the Falmouth estuary, has perhaps yielded most abundantly. Small nuggets are not uncommon here. One specimen, found at Carnon, is reported to have weighed more than 10 guineas, and was probably about 2 ounces in weight.

In 1753, certain tin-streamers in the parish of Creed, near Grampound, met with some grains of gold, and "in one stone, a vein of gold, as thick as a goose-quill was found." Shortly after, gold was discovered at Luny, in the parish of St. Ewe, in blue sandy slate. "Some of it," to use the miners' phrase, "was kerned about spar." Gold-ore is reported to have been worked in 1846, at Wheal Samson, in St. Teath, but little was obtained.[†] In 1852, gold was discovered in quartz-veins at Davidstowe, North Cornwall.[‡]

Borlase mentions that he had seen a nugget from the parish of Creed, near Grampound, weighing 15 dwts. 3 grains, which was then in the possession of Mr. Lemon, of Carclew. Gold was also found in the Crow Hill stream-works, at Trewarda, at Kenwyn, and at Llanlivery, near Lostwithiel. In the Natural History Museum at South Kensington there is exhibited a small water-worn nugget from Wendron, near Helston. Gold is also reported from Cornwall in the matrix from a cross-course in Huel Sparnon, and in the gossan of the Nargiles mine.[§] Mr. D. Forbes[||] records the presence of gold in the argentiferous tetrahedrite, chalcopyrite and galena of a lode at Bound's Cliff, near St. Teath.

The alluvial gold is generally found associated with stream-tin, and such gold as has been recovered has been obtained during the

---

[*] *Transactions of the Royal Geological Society of Cornwall*, vol. i., page 235.

[†] *Mining Journal*, 1847, page 222.

[‡] *Report of the Geological Society of Cornwall*, 1852, quoted in *Quarterly Journal of the Geological Society of London*, 1854, vol. x., page 247.

[§] *Transactions of the Royal Geological Society of Cornwall*, vol. vii., page 90.

[||] *Philosophical Magazine*, 1869, vol. xxxvii., page 322.

course of streaming operations. Analysis of several grains of gold from St. Austell Moor, the largest of which was only 2·1 grains in weight, gave Mr. D. Forbes[*] the following result:—Gold, 90·12; silver, 9·05; and silica and iron, 0·83 per cent. The specific gravity of the gold was 15·62. Gold from Ladock was analysed by Mr. A. H. Church, showing it to be slightly finer in quality than the above:—Gold, 92·34; silver, 6·06; and silica, 1·60 per cent. The former specimen was worth about £3 16s. 6d. and the latter about £3 18s. 6d. per ounce.

Devon.— In this county the existence of gold has been known, or assumed, for many centuries. In the beginning of the nineteenth century, a miner named Wellington is reported to have found gold at Sheepstor, on South Dartmoor. At different times he brought to a silversmith at Plymouth quantities which in the aggregate were valued at about £40.[†]

The principal auriferous locality in Devon, is, however, at North Molton (Fig. 4, Plate XVI.). Here in 1852, the gossan-ores of the Britannia and Poltimore mines were discovered to be payably auriferous. This discovery, coming immediately after the world-wide excitement and unrest caused by the discovery of the Californian diggings, attracted an extraordinary amount of interest. The first trial of the gossan yielded 26½ ounces from 20 tons of ore, and the average yield of further trials of 50 and 75 tons was 16 dwts. per ton. The gold was of very good quality, and was worth nearly £4 4s. per ounce. The total value of the gold produced from the Poltimore mines up to November 2nd, 1853, was £581 5s. 1d.[‡]

The North Molton copper-mines are situated in an area of Devonian rocks, some distance away from their contact at the surface with the overlying Carboniferous sandstones (Fig. 4, Plate XVI.). Both the Devonian and Carboniferous strata are very highly inclined, in many cases approximating to verticality, and the lodes appear to dip with the country-rock. The auriferous gossan-lode is from 4 to 10 feet wide, and dips to the north. There is considerable evidence of this mine having been worked, prob-ably for copper, in very remote times. The auriferous gossan

[*] *Philosophical Magazine*, 1869, vol. xxxvii., page 322.
[†] *Transactions of the Royal Geological Society of Cornwall*, vol. v., page 141.
[‡] *Mining Journal*, 1853, page 690.

is a friable ironstone highly mineralized, and containing copper. It is brown on the western side of the Mole and reddish on the eastern bank. The latter is reputed to be twice as valuable in auriferous content as the former (assaying 17 dwts. and 8 dwts. respectively).

The Britannia mine is ¾ mile north of the Poltimore. Gold was found there, in grains and small plates, by Mr. Flexman of South Molton, prior to 1822. It is also a gossan-ore, but more siliceous than the Poltimore.

These gossans arise from the decomposition of slightly auriferous metallic sulphides, mainly iron-pyrites. In a specimen from North Molton, in the Natural History Museum at South Kensington, small particles of gold are clearly visible in the brown and somewhat siliceous ironstone.

Cumberland.—In the tenth year of the reign of Elizabeth, after a famous lawsuit with the Earl of Northumberland, resumption was made by the Crown of the rich copper-mines at Goldscope, Newlands, Keswick, Cumberland (where gold was reported to have been first found in the reign of Henry III.). The mines were claimed in respect of the gold and silver which the copper-ore contained. For as set forth in that trial:—" Where the Oar, which is digged from any Mine doth not yield, according to the Rules of Art, so much Gold or Silver, as that the value thereof doth exceed the charge of Refining, and the loss of the baser Metal in which it is contained and from whence it is extracted, it is called a Poor Mine, but when the Oar doth yield such Gold or Silver as exceedeth the charge of Refining and loss of the Baser Metal, it is called a rich Oar, or a *Mine Royal*, this appertaining to the King by his Prerogative."*

Though these mines were resumed by the Crown, it would appear that it was rather on account of the argentiferous content than for any gold that they may have contained, for a writer of 1709† remarks " The ore got by the gin under level was so rich in *silver* that Queen Elizabeth sued for it, and recovered it from the Earl of Percy for a royal vein." The name Goldscope was formerly Gowd-scalp and would suggest the occurrence of gold at

* *Plowden's Reports*, page 301.

† *An Essay towards a Natural History of Westmorland and Cumberland*, by — Robinson, London, 1709.

that place. These mines were worked extensively by Germans
introduced into this country, for their mining knowledge, by
Elizabeth, who, according to Sir John Pettus[*] was moved to do
so " from Her observation of the inartificialness of former Ages
in this concern, which may be collected from Her sending for
and imploying so many *Germans* and other Foreiners (where *Mines*
were plentiful and the *Arts* belonging to them) who might put
us into the tract of managing ours, in *finding* and *digging*, and
in *smelting* and *refining* Metals."

Somerset.—Gold has been recorded from the Carboniferous
Limestone near Bristol.[†] Messrs. W. W. Stoddart and Pass
found appreciable quantities of both metals in the limestone at
Whalton, near Clevedon. Analyses of the dried limestone gave:—

Alumina (Al_2O_3)	...	...	...	...	0·8777
Ferric oxide (Fe_2O_3)	...	...	...	...	4·8000
Calcium carbonate ($CaCO_3$)	...	...	...	94·3000
Silica (SiO_2)	...	...	...	...	0·0200
Silver (Ag)	...	...	...	...	0·0023
Gold (Au)	...	...	...	...	traces

Mr. J. P. Merry, of Swansea, found in one sample 94 grains of
silver and in another nearly 1 ounce of silver; while both contained 3 to 5 grains of gold per ton. In the absence of proof of the
absolute purity of the fluxes used (and especially of the litharge),
these results must be received with some degree of caution.

WALES.

The auriferous veins of Merionethshire appear to have been
discovered in 1843. There are three claimants for the honour of
the discovery—Mr. Arthur Dean, who read a paper, reporting the
existence of gold at Dol-y-frwynog, before the 1844 meeting of the
British Association for the Advancement of Science;[‡] Mr. James
Harvey, owner of the Cwm Eisen mine : and Mr. Robert Roberts
of Dolgelly, who claimed to have been cognizant of the existence
of gold in 1836, and to have several years later pointed out the
locality to Mr. Dean.[§] The locality of the discovery, according to
Mr. Roberts was Cwm Eisen. Mr. Roberts says :—" In the year

---

[*] *Fodinæ Regales.*

[†] Ure's *Dictionary of Arts*, etc., seventh edition, vol. iv., page 419.

[‡] *Report of the British Association*, etc., 1844, page 56.

[§] *Mining Journal*, 1844, page 383 ; 1845, pages 6, 37 and 38.

1836, I had samples of ore assayed which proved to be very rich in gold and silver, but not being aware that the working of such ores would be allowable in North Wales, I left off further search until 1843 . . . . I certainly informed him (Mr. Dean) 12 months ago (October, 1843) of the existence of precious metals in this country."[*] Mr. Dean appears to have been called in by Mr. Harvey, the proprietor of the Cwm Eisen mine, to furnish an expert report on the mine, in order to form a company to work the gold-ores. To which claimant belongs the honour it is now impossible to say, but it is at least certain that Mr. Dean's paper, read before the British Association in 1844, contained the first published notice of the discovery.

In 1846, an attempt was made to raise capital to work the gold-mines, but, owing to the ridicule cast on the project, the attempt resulted in failure. Early in 1847, however, the North Wales Silver, Lead, Copper and Gold-mining Company was floated with a capital of £125,000 in 12,500 shares of £10 each to work the veins at Vigra, Clogau, Tyddyn-gwladys and Dol-y-frwynog. Vigra and Clogau were then being worked for copper alone. The Cwm Eisen lode during this year was worked vigorously, but only for copper-ore and for argentiferous galena. Dol-y-frwynog lode yielded a little gold during 1847; but, being in places 12 feet wide with good copper-ore, it was worked almost entirely for the latter metal.[†]

Before January, 1849, the first extensive trials of Welsh auriferous veins had been made at Cwm Eisen, and 7 pounds of gold of the approximate value of £350 had been obtained from $10\frac{3}{4}$ tons of concentrates, the produce of 300 tons of ore.[‡] Gold-mining operations were at this time much hindered by the claims of the Mines Royal Corporation, to which, as we have already seen, the Crown had granted, by patents of Elizabeth, its royal prerogative in Wales. The matter was finally settled by the Crown requiring a royalty of 5 per cent. on private property and of 10 per cent. on Crown land.

In 1853, an impetus was given to gold-mining in Wales by the introduction of the Berdan machine for gold-recovery, and coming as it did at the height of the excitement caused by the gold-discoveries in California it created a mild boom, of which the

[*] *Mining Journal*, 1845, pages 37 and 38.

[†] *Ibid.*, 1847, page 23.   [‡] *Ibid.*, 1849, page 94.

usual advantage was taken by unscrupulous persons. At that time, gold was reported from all parts of England and Wales, nearly all the alleged discoveries being, of course, fictitious. The gold-mines worked during this boom were all about the upper waters of the Afon Mawddach, in the vicinity of the Rhaiddr Mawddach.

On August 16th, 1853, gold was discovered at the Prince of Wales mine (now the Voel mine, Merioneth) about ½ mile west of the junction of the Mawddach with the Afon Wen, and in the same week a similar discovery in an old dump was made at Vigra (Clogau?) by Messrs. Goodman and Parry of Dolgelly.* In 1854, a single piece of stone worth £25 was crushed from Clogau, and 2 years later, 100 pounds of quartz from the same mine yielded 14½ ounces of gold.†

It was not, however, until 1860, that the St. David's lode of the Clogau mine gave any indication of its exceedingly rich bonanzas. On May 21st, 1860, a mass of 15 cwts. of gold-quartz of the estimated value of £500 to £600 was broken down. During the first 6 months of 1861, 983 ounces of the value of £3,664 were obtained. This rich discovery naturally stimulated enterprise in the vicinity; and, in 1863, the Clogau, Cefn Coch, Dol-y-frwynog and Cwm Eisen mines were being vigorously worked, and visible gold was discovered at Garth-gell, Cambrian, Cae Mawr, Prince of Wales, Moel Offryn, Glasdir, Tyddyn-gwladys and Ganllwyd mines. In April, 1862, gold was met with *in situ* in the Berthllwyd mine, near Tyn-y-groes, and a crushing of 333½ tons from the adjacent Welsh Gold-mining Company's mines gave a yield of 282½ ounces of gold.‡ The gold of the Gwyn-fynydd lode, which yielded so handsomely a quarter of a century later, was discovered early in 1864 by Capt. Griffith Williams, but the discovery was kept secret until February 23rd, 1864.

During 1865, many thousands of pounds were expended on useless metallurgical machinery: Berdan pans, Mitchel pans and Mosheimer machines being erected and condemned in rapid succession. Considering, however, the world-wide ignorance of the principles of gold-recovery at that period, this result is not surprising. In 1865, the Clogau mine paid £22,575 in dividends

* *Mining Journal*, 1865, page 134.

† *Ibid.*, 1860, page 670; and *Transactions of the Manchester Geological Society*, vol. ii., page 97.

‡ *Mining Journal*, 1864, page 674.

and had, in little more than three years, produced gold to the value of £43,783.* Table I. contains an estimate of the total produce of gold, of an average value of £3 4s. per ounce, obtained from the Mawddach district from 1844 to 1866.†

TABLE I.—GOLD-PRODUCTION OF THE MAWDDACH DISTRICT FROM 1844 TO 1866.

	Ounces.
Berthllwyd, Welsh Gold-mining Company *	282½
Castell Carn Dochan, Llanuwohllyn†	1,104
Cefn Coch	478
Cwm Eisen	176
Gwynfrwynog (Gwyn-fynydd ?)	6
Prince of Wales, Voel Mines	63
Vigra and Clogau	11,788
	13,897½

* *Mining Journal*, 1866, page 168.
† *Ibid.*, 1864, page 674.

Table I. is certainly incomplete; and Table II. contains another and probably more accurate return to the end of 1865,‡ compiled from official records (1865).

TABLE II.—GOLD-PRODUCTION OF THE MAWDDACH DISTRICT TO 1865.

	Ore.			Gold.		
	Tons.	Cwts.	Qrs.	Ozs.	Dwts.	Grs.
Alluvium	100	0	0	5	0	0
Cambrian	50	0	0	30	0	0
Castell Carn Dochan, to June 30th ...	3,500	0	0	1,606	0	0
Cefn Coch, Welsh Gold-mining Company	1,982	8	1	666	6	2
Cefn Dewddwr	5	0	0	8	0	0
Clogau, prior to 186.	50	0	0	200	0	0
Do., from 1860	5,140	0	0	11,850	0	0
Cwm Eisen	487	0	0	334	0	0
Dol-y-frwynog...	312	1	2	167	5	5
Gwyn-fynydd	5	1	3	15	15	12
Mawddach River	1	4	0	2	0	0
Prince of Wales	20	0	0	63	0	0
Totals	11,652	15	2	14,947	6	19

After 1866, gold-mining languished for nearly 20 years, and there is little of importance to note in that period. The Vigra and Clogau mines had worked out their bonanzas, and in 1868, produced only 490 ounces of gold.§ In 1870, the total yield

* *Mining Journal*, 1865, page 518.

† *Treatise on Ore-deposits*, by·Mr. J. Arthur Phillips, page 204; second edition, by Prof. H. Louis, 1896, page 294.

‡ Ure's *Dictionary of Arts*, etc., seventh edition, vol. ii., page 699.

§ *Mining Journal*, 1869, page 137.

from Wales was only 191 ounces, of which Gwyn-fynydd contributed 165 ounces. During the following year not a single ounce of gold was produced.

TABLE III.—GOLD-PRODUCTION OF MERIONETHSHIRE.

Year.	Gold-ore.		Gold.			Estimated Value at the Mine.			Actual Value of Gold.		
	Tons.	Cwts.	Ozs.	Dwts.	Grs.	£	s.	d.	£	s.	d.
1861	—		2,886	3	0		...		10,816	17	0
1862	803	14	5,299	1	12		—		20,390	15	5
1863	385	15	552	12	19		—		1,747	0	0
1864	2,336	10	2,887	0	0		—		9,991	0	0
1865	4,280	15	1,664	11	0		—		6,408	10	0*
1866	2,928	0	742	16	10		—		2,859	7	10*
1867	3,241	4	1,520	6	21		—		5,853	3	5*
1868	1,191	10	435	14	23		—		1,677	12	9*
1869	—		—				—		—		
1870	— ·		191	0	0		—		735	7	0*
1871	—		—				—		—		
1872	—		—						—		
1873	—		—				—		—		
1874	—		385	0	12		—		1,477	6	11*
1875	—		548	1	21		—		2,105	17	6
1876	—		288	18	6		—		1,119	10	9
1877	—		139	4	13	638	10	6 (?)	536	0	4*
1878	—		697	12	16		—		2,825	8	6
1879	2	0	447	7	21		—		1,790	0	0
1880	—		5	0	0		—		19	5	0
1881	—		—				—		—		
1882	—		226	0	0		—		863	0	0
1883	869	0	66	0	0	100	0	0	254	2	0*
1884	—		—				—		—		
1885	35	0	3	10	0	7	0	0	13	9	6*
1886	—		—				—		—		
1887	0	17	58	0	0		—		209	0	0
1888	3,844	0	8,745	0	0	27,300	0	0	29,982	0	0
1889	6,226	0	3,890	0	0	10,746	0	0	13,277	0	0
1890	575	0	206	0	0	434	0	0	675	0	0
1891	14,067	0	4,002	7	0	12,200	0	0	13,700	0	0
1892	9,990	0	2,835	0	0	9,168	0	0	10,511	0	0
1893	4,489	0	2,309	0	0	7,657	0	0	8,619	0	0
1894	6,603	0	4,235	0	0	13,573	0	0	14,811	0	0
1895	13,266	0	6,600	0	0	16,584	0	0	18,528	0	0
1896	2,765	0	1,352	10	0	4,257	0	0	5,035	0	0
1897	4,517	0	2,032	0	0	6,282	0	0	7,185	0	0
1898	703	10	395	0	0	1,158	0	0	1,299	0	0
1899	3,047	0	3,327	0	0	10,170	0	0	12,086	0	0
1900	20,802	0	14,004	0	0	42,925	0	0	52,147	0	0
1901	16,374	0	5,900	0	0†	13,920	0	0	21,000	0	0†
Totals since 1867	107,268	10‡	59,832	17	0‡	176,374	0	0‡	280,547	13	11‡

* Estimated at £3 17s. per ounce.
† Estimated from private sources, as the official returns are not yet available.
‡ The value of the gold from 1838 to 1901 inclusive is £208,855.

In 1881, a low-level tunnel was driven to intersect the St. David's lode (Clogau mine); and, shortly after it was intersected, a small patch of 225 ounces was obtained. Nothing of importance was recorded from the district until 1888, when a rich shoot was discovered in the Gwyn-fynydd lode. The Morgan Company was floated to work this mine, which in 2 years produced over £35,000 worth of gold. After the exhaustion of the shoot, the company suspended operations; but a new company was formed, and carried on operations for many years, with varying success. The two most important mines, the Clogau and the Gwyn-fynydd, are now united as the St. David's Gold and Copper-mines, Limited.

Table III., compiled from official records, shows with some degree of accuracy the total produce of the Merionethshire gold-field since its discovery. The column "estimated value" of Table III., requires some explanation. It is the " estimated value of the ore at the mine " as furnished to the Mines Department of the Home Office, presumably by the mine officials, and means merely the estimated profit that should accrue from the treatment of the ore, the estimated expenses for the year having been deducted. These figures are, therefore, almost valueless, and yet it is these that are added in the official returns to obtain the total value of the gold-ore raised since 1873. This total value is now, however, accurately shown for the first time in the last column of Table III.

It will be seen that the returns, at times, have been so large as to admit of a handsome profit. The net profits of the St. David's Gold and Copper-mines, Limited, for the year 1900, were £39,729, which admitted of the payment of dividends at the rate of 60 per cent. on the capital. While the gross receipts for that year were £51,344 4s. 10d., the total expenses were only £8,423 9s. 7d., or 8s. 7¾d. per ton. The royalties paid to the Crown were £2,038 7s. 7d., or, at the rate of 2s. 1d. per ton of ore crushed. During 1901, 15,500 tons of ore were treated for a yield of 5,537 ounces of bullion, worth together with the concentrates £19,710. The total expenses for the year were £6,730, or about 8s. 8d. per ton, thus leaving a net profit sufficient to pay a very handsome dividend.

Geology.—The rocks (Fig. 5, Plate XVII.) of the auriferous area of North Wales may be grouped as shewn in Table IV. The

Harlech grits, the oldest members of the sequence, occupy the most westerly position in the area under discussion. They extend from Barmouth to Harlech, and for a considerable distance eastward. They are composed in the main of greenish-grey grits, with, towards their upper horizon, interstratified bands of green and purple slates. Their dip is to the east at fairly low angles, and their estimated thickness is 6,000 feet. Their junction with the overlying Menevian beds is best seen about ½ mile east of Barmouth.

TABLE IV.

Upper Cambrian or Ordovician	Bala Series	...	... Castell Carn Dochan Slates.
	Bala Age	...	... Felstone-porphyries and felspathic ashes.
	Arenig Age	...	... Igneous intrusive rocks. Diabases (greenstone).
Middle Cambrian	... *Lingula*-flag Series		... Dolgelly Beds. Ffestiniog Beds. Maentwrog Beds.
Lower Cambrian	... Menevian Series	...	
	Harlech Series ...		... Harlech Grits.

The Menevian series overlies conformably the Harlech grits, and is composed of dark sandstones and flags. These are as a rule felspathic, and at times are somewhat coarse-grained. In weathering, these Lower Cambrian rocks fail to furnish any soil, and the outlook northward from Y-Garn across their exposure is barren in the extreme, offering a wide expanse of succeeding ridge and scree. Neither in the Harlech grits nor in the Menevian beds are metalliferous veins developed.

The horizon of the auriferous veins of Wales is that of the *Lingula*-flags. These are divided into three groups. The lowest, the Maentwrog, rests in this area directly and without unconformity on the Menevian beds, and like them, dips south-east and east at angles varying from 45 degrees near Barmouth to 10 degrees near Gwyn-fynydd. They are fossiliferous at Tyddyn-gwladys and Cwm Eisen. At Tyddyn-gwladys, the following fossils have been found on this horizon[*]:—*Anopolenus Henriçi, A. Salteri, Paradoxides Hicksii, Microdiscus punctatus, Agnostus Davidis, A. princeps, A. nodosus, Paradoxides Davidis, Theca corrugata,* and others. From Cwm Eisen, *Olenus cataractes, O. gibbosus, Agnostus princeps* and *A. trisectus* are recorded. From

[*] *Quarterly Journal of the Geological Society of London,* vol. xxii., page 505.

Penrhos, *Olenus micrurus*, one of the most characteristic fossils of this horizon, has been obtained. The most richly auriferous lodes are in these beds, and include Gwyn-fynydd, Cwm Eisen, Cefn-dewddwr, Ganllwyd, Berthllwyd, Cefn Coch, Voel mines and Clogau. The junction between the Maentwrog and the under-lying Menevian beds is, especially in the Llechau and Mynach valleys, clearly traceable on the surface, the slates of the Maent-wrog beds yielding a fair soil which supports an abundant vegeta-tion. This junction-line runs northward from the estuary of the Mawddach to near Llyn-bodlyn, and passing round the head of the valley, turns south-eastward to near Clogau, whence it crosses

FIG. 10.—PONTDDU AND CWM LLECHAN VALLEY.

the Cwm-llechau and Cwm-mynach valleys in an easterly direc-tion, gradually turning, however, to the north-east. At Cefn Coch it coincides with the large quartz-lode, and, continuing in a north-easterly direction, crosses the Eden opposite the high bluff at Cefn-dewddwr, and thence pursues a northerly course passing within a short distance of Gwyn-fynydd. The rocks of the Maentwrog beds are, on the whole, grey and dark-coloured slates, sometimes highly ferruginous, together with occasional bands of sandstone.

The Ffestiniog beds, which overlie conformably the Maent-wrog beds are developed from Moel-Hafod-Owen through Glas-dir to Penmaenpool. The auriferous veins on this horizon are

those at Dol-y-frwynog and at Glasdir. The slates are highly
fossiliferous near Moel-Hafod-Owen, yielding *Lingulella Davisii*,
Hymenocaris vermicauda, *Bellerophon cambrensis* and others.
The Ffestiniog beds, in this neighbourhood, have been very considerably altered by dynamic stress, occasioned possibly by the ex-
trusion of the great neighbouring igneous mass of Rhobell Fawr.
The ordinary slaty rocks of the *Lingula*-flags give place to a
hard massive rock, indistinguishable in many cases from the intrusive felspathic igneous rocks of the area. In places, it contains
a large quantity of talc, becoming a talcose schist, weathering
along fissure-planes to a somewhat kaolinic clay.

The Dolgelly beds are exposed near the town of that name,
and also north of Rhobell Fawr, and are, as a rule, soft black
slates. No auriferous veins have been discovered on this horizon.

Igneous rocks are well developed in the area under discussion,
the intrusions, especially north of the Mawddach river, between
the Barmouth estuary and Llanelltyd, running parallel with the
strike of the lower beds of the *Lingula*-flag series. They occasionally occupy fault-lines, a remarkable instance of which is seen
between Tyn-y-groes and the Clogau. Here, the large Cefn Coch
quartz-lode occupies the plane of contact between the Menevian
and Maentwrog beds. Farther south-west, the fissure, which
runs into the head of the Mynach valley, is occupied by an
intrusive diabase, at times again giving place to quartz. No less
than 150 of these intrusions, varying from a few feet to nearly a
mile in length, have been mapped by the officers of the Geological
Survey. Many of the dyke-rocks are light in colour, exhibiting
imperfect crystallization due to rapid cooling. Some are calcareous, showing effervescence on treatment with acid. They
are, as far as may be seen from hand-specimens, dolerites and
diabases.

The Welsh gold-mines are, with the exception of Castell Carn
Dochan, disposed along the northern and western slopes of the
watershed of the Afon Mawddach, a stream flowing into St.
George's Channel, at Barmouth. The auriferous belt extends
from near Pontddu, midway between Barmouth and Dolgelly, in
an easterly direction to 1 mile beyond the falls at Rhaiddr Mawddach, beyond which no discoveries of importance have been made.

The two most productive lodes are located one at each end of the already proved auriferous belt—the Clogau on the south-west and the Gwyn-fynydd on the north-east. They are now both under the control of the St. David's Gold and Copper-mines, Limited. The Clogau mine is situated some distance from Pontddu up the Cwm-llechau valley. Midway between Pontddu and the mine is the crushing-mill, the ore being conveyed by an aerial tramway from the mouth of the main level. The mountainous nature of the country permits of the lode being worked level-free, and at the same time furnishes abundant fall for the use of the water of the Llechau as a source of motive-power. The St. David's lode lies in the Middle Cambrian slaty rocks (*Lingula*-flags) a short distance south of their line of surface-contact with the coarse greenish-grey underlying Lower Cambrian or Menevian grits and sandstones. The vein, which has a nearly east-and-west strike, or parallel with the line of contact mentioned, is almost perpendicular, any dip being towards the north. It varies in width from 2 to 9 feet, but it is much split in places, forming occasionally large horses. The matrix of the vein is quartz, somewhat white and chalcedonic in appearance, especially near and at the surface, to which the lode has been sloped. Calcite is not uncommon, and occasionally contains gold. In general character, the auriferous matrix is a fairly clean opaque white quartz, occasionally stained with iron-oxides, but in depth it contains undecomposed sulphides. Of the sulphide-ores, blende is by far the most abundant, but iron-pyrites and chalcopyrite also occur in quantity. An uncommon associate of gold is met with at Clogau, namely, the silvery white telluride of bismuth, tetradymite. The gold itself is occasionally in the clean white quartz, where it is shotty, but is more often associated with blende or with a darker veinstone, probably rendered so by the contemporaneous deposition of sulphides in a state of extremely fine division.

The Gwyn-fynydd (White Mountain) mine lies a short distance above Rhaiddr Mawddach, and like the Clogau has the advantage of an ample supply of water under a good head, and also is worked level-free. This mine was originally worked as a lead-mine, but in 1870 a small rich pocket of gold-ore was discovered a few feet below the surface, portions of which yielded at the rate of 7 to 16 ounces to the ton. The auriferous character of the lode was first discovered in 1864. The Gwyn-fynydd lode, like the St. David's, from which it is distant about 8 miles, is close to the contact between the Maentwrog slates and the Menevian sandstones. The former, in this area, dip to the east at angles varying from 10 to 60 degrees. The latter also dip in the same direction, but at much lower angles.

The Gwyn-fynydd lode strikes east-and-west, dipping to the north at about 80 degrees. It branches in several places forming numerous small horses of slate. As a natural consequence, its width varies considerably, 2 feet to 20 feet being the extreme limits. The matrix of the gold is a white and opaque quartz. In places, it is much mineralized, the most abundant sulphide

FIG. 11.—OLD STOPES, GLASDIR.

being blende: but pyrite, mispickel, galena and chalcopyrite are also present. The gold here is, as a rule, much finer than that from Clogau: indeed, in some cases, it is so finely divided that it imparts a yellow stain to the stone, with which it is obviously of contemporaneous origin. In other cases, the gold is of subsequent deposition, occurring in vughs in blende, and infiltrating the somewhat cavernous quartz. In the latter case, the gold is often leaflike and wiry.

Since the discovery, in 1888, of the rich shoot, which has been traced for more than 300 feet, this lode has yielded consistently, and with vigorous prospecting for new shoots, should yield equally well in the future.   For many years it has furnished the greater proportion of the Welsh gold-yield.

In the vicinity of Gwyn-fynydd, the mines which have yielded good specimens but have never been sufficiently rich in gold to pay for working expenses, are the Cwm Eisen (Cwm-heisian), Dol-y-frwynog, Cefn-dewddwr, and Tyddyn-gwladys.

Of these, as we have already seen, Cwm Eisen and Dol-y-frwynog were among the earliest worked, and though never remunerative, the gold produced from them has been of considerable quantity.   Cwm Eisen, in the early days of gold-mining, yielded two large returns, of 170 ounces from 300 tons and 148 ounces from $157\frac{1}{2}$ tons respectively.   There are several specimens (wrongly labelled Cwm-y-swm) from this mine in the Jermyn Street museum.   The quartz is on the whole rather clear, and the invariable associate of the gold is zinc-blende, the latter being sometimes contemporaneous and sometimes prior in point of deposition.   Galena and pyrites also occur, and indeed the veins were originally worked as a silver-lead mine.

The Dolfrwynog (or Dol-y-frwynog) mine, about 1 mile east of Cwm Eisen, has produced some very rich specimens.   The gold here is somewhat fine, at times staining the quartz.   It is also associated with blende and with pyrites.   The main lode averages about 5 feet in width, strikes west-north-west and east-south-east, and dips towards the north at about 40 degrees.   At a depth of 200 feet, very rich ore was met with in this mine.

The Tyddyn-gwladys silver-lead-mine has yielded a small quantity of gold, as also has the Cefn-dewddwr.   Both are situated almost at the junction of the Menevian and the Maentwrog beds.

On the west of the river Mawddach, below its junction with the river Eden, gold has been obtained in small quantities from Ganllwyd, Coed-cy-fair, Berthllwyd, Goitref, Cae-gwernog, Cefn Coch and Cae-mawr.   These are all in the Maentwrog beds or, as in the case of Cefn Coch, are at the junction with the underlying Menevian beds.   East of the Mawddach, and opposite the above are Penrhos, Tyn-y-Penrhos, and Glasdir.

The last is situated opposite the Tyn-y-Groes Hotel, a short distance up the Afon Pabi.   The country-rock here is of bedded slate (Ffestinoig beds) striking about north-east and south-west,

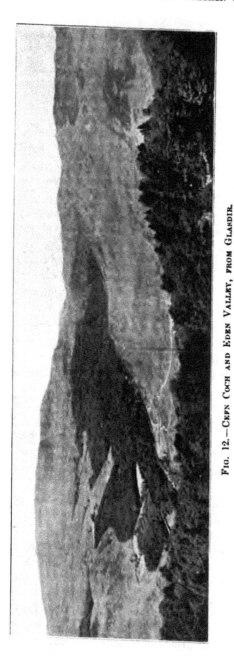

FIG. 12.—CEFN COCH AND EDEN VALLEY, FROM GLASDIR.

and dipping south-east-ward. The ore-body is not a defined vein, but appears to be an impregnation of the country-rock along a line of faulting, and is contained between two fairly well-defined walls, which are usually slickensided. The auriferous pyrites (pyrite and chalcopyrite) is distributed in irregular patches throughout the ore-body. The general tenour of the ore-body is about 1·1 per cent. of copper, with a very small proportion of gold—less than 1 ounce per ton of concentrates.

The only other lodes to be noted in this area are those included in the Voel mines near Llanelltyd. So far as can be seen, there are here two main lodes, both striking north-northeast and south-southwest, and dipping east-southeastward with the country - rock. Throughout this section are numerous intrusive sheets of diabase, generally conforming to the strike and dip of the slates. The lodes in places occupy the plane of contact between the diabases and the slates, the igneous rock, in one case, forming the hanging-wall and the slates the foot-wall of the vein. The auriferous quartz is generally stained with green chloritic matter, and is associated with the usual *indicator* for gold, namely, zinc-blende. The gold is sometimes contained in the quartz, but is more often deposited on the accompanying blende.

The only important auriferous occurrence outside the water-shed of the river Mawddach is that of Castell-carn-dochan, about 5 miles from Bala and 2 miles from the small village of Llanuwch-llyn. The main auriferous vein strikes north-east and south-west, dips southward, and is composed of extremely clean quartz, completely free as a rule, from sulphide-ores. The gold occurs in specks disseminated throughout the quartz. The lodes are in soft black shaly rocks, dipping eastward at about 45 degrees, very near their junction with the felspathic ash-beds and lavas which form the summit of Castell-carn-dochan.

Complete reduction-works were erected in 1864, and up to the end of 1865, about 3,500 tons had been treated for a yield of 1,606 ounces. The lode has since been worked spasmodically, yielding 12½ ounces from 50 tons, in 1889; and, during the years 1895 to 1898 inclusive, 393 ounces of gold from 2,638 tons crushed.

The gold or electrum of the Welsh auriferous region, when met with *in situ*, is scattered throughout a quartz-matrix, or occurs deposited on blende or pyrites in vughs and cavities. It rarely shows any approach to crystallization. The following are aver-age percentage analyses of vein-gold from Clogau :—

No. of Sample.		Gold.		Silver.		Quartz.		Loss.
1	...	90·16	..	9·26	...	0·32	...	0·26
2	...	29·83	...	9·24	...	0·74	...	0·19

These samples represent a value of £3 16s. to £3 16s. 6d. per ounce.

The alluvial gold of the river Mawddach is found mainly in the bed of the stream, but a fair prospect may be washed in many places from the soil on the slopes of the valley. It occurs in small flattened grains, often coated with a hæmatitic film, and is as-sociated with galena, blende, titanic iron-ore, marcasite and pyrite. Its specific gravity is low, namely, 15·79, due, however, not so much to impurities as to the presence of numerous small air-cavities. As a general rule, the Mawddach alluvial gold is

worth about 5s. per ounce more than vein-gold.   It is also lighter
in colour than the Clogau gold, owing to the admixture in the
latter, of copper with the ordinarily prevailing silver.

The earliest recorded attempt to obtain gold from the sands of
the river Mawddach was that of Mr. Frederick Walpole and Sir
Augustus Webster, who obtained an appreciable quantity in the
summer of 1852.   In 1870, owing to the unprecedented lowness
of the river Mawddach, several Australians and Californians
worked its bed with good results.   One sample of about 1 ounce
weight was taken to Liverpool and there assayed at the rate of
23¾ carats (nearly 990 fine).*   Above Gwyn-fynydd, no nuggets
have been found, but they occur along the whole course of the
river Mawddach from Rhaiddr Mawddach to Cymmer Abbey, the
gold gradually becoming finer as the latter place is approached.

An analysis of the alluvial gold of the river Mawddach made by
Mr. David Forbes† gave the following results:—

Gold.		Silver.		Iron.		Quartz.		Loss.		Specific Gravity.
84·89	...	13·99	...	0·34	...	0·43	...	0·35	...	15·79

It will be noted that this analysis shows a much lower value
than those of the vein-gold from Clogau cited above.   This is
due to the fact that none of the alluvial gold of the river Mawddach
is derived from the Clogau lode, but in all probability arises from
the degradation of the Gwyn-fynydd or neighbouring lodes, the
gold of which is worth much less than that from Clogau, 8 miles to
the south-east.

SCOTLAND.

The earliest recorded notice of the occurrence of gold in Scotland is found in a grant of 1153 A.D. to the Abbey of Dunferm-
line of a tithe of all the gold which should accrue to David I. from
Fife and Fothrif.‡   Gilbert de Moravia is said to have discovered
gold at Duriness (Durness), in the north-west of Sutherland,
in 1245.   By an Act of May 26th, 1424, it was enacted in the Parliament of James I. that if " ony myne of golde or siluer be fundyn
in ony lordis landis of the realme and it may be prowyt that thre
halfpennys of siluer may be fynit owt of the punde of leide The

* *Mining Journal*, 1870, page 699.

† *Philosophical Magazine*, 1867, vol. xxxiv., page 344.

‡ *Registrum de Dunfermelyn*, page 16 ; and *Hailes Annals*, 1819, vol. ii.,
page 461.

lordis of parliament consentis that sik myne be the kyngis as is vsuale in vthir realmys,"[*] an enactment which would indicate that the produce of precious metals in Scotland had hitherto been negligible.

With the discovery of the gold-mines of Crawford Moor in the reign of James IV. (1488 to 1513), we pass from the region of speculation to that of fact, for in the Treasurer's accounts for 1511, 1512 and 1513, are found many payments to Sir James Pettigrew for working the gold-mines of that region.[†] As early as 1513, an expert report was furnished to the King by John Damiane, Abbot of Tungland, whose expenses were met by the royal exchequer. After Flodden (1513), in which James IV. perished, the mines were controlled by the Queen Regent. Peculation and evasion of royalties was, even in those remote days, not unknown, for in 1542 a justice-ayre was held in Edinburgh to take cognizance of those who had broken the mining ordinances and had carried gold out of the kingdom.[‡] In 1524, it was also enacted that the gold from Crawford Moor should be minted at the Cunyie House (the Scottish Mint). The Albany medal, struck in the same year, was made from gold found on Crawford Moor, as no doubt was much of the coinage of that period.

In July, 1526, a lease of all the mines of gold, silver and other metals was granted for 43 years to certain Germans and Dutchmen, Joachim Hochstetter, Gerard Sterk, Antony de Nikets, and others. To the same grantees, a license to coin was issued in the following year.[§] But the results could not have been encouraging, for in 1531, a payment is recorded to "the Dutchmen quhill cam here for the myndis, at their departing hamewart."[||] In 1535, a commission was appointed to enquire into the workings of the gold-mines, with the result that miners were imported from Lorraine in 1539.[¶]

It was to hunt over the bleak uninviting country of the Lowthers in the Crawford Moor district that James V., in 1537,

---

[*] *Early Records relating to Mining in Scotland*, by R. W. Cochran-Patrick, page 2.

[†] *Loc. cit.*, page xiii.

[‡] *Loc. cit.*, page xiv.

[§] *Records of the Coinage of Scotland*, 1878, by R. W. Cochran-Patrick, vol. i., page 64.

[||] *Early Records relating to Mining in Scotland*, by R. W. Cochran-Patrick, page xv.

[¶] *Loc. cit.*, page xv.

brought his court, largely composed of the train of his first wife,
Madeleine, the youngest daughter of the King of France. From
this visit arose an interesting story, illustrative of the then production of gold. Newly arrived as the courtiers were, from the
sunny valleys of the Seine and the Loire, their disparagement of
the bare uplands of the Glengonnar and Wanlock was loud and
unsparing. Nettled by these complaints, the Scottish King
wagered that at a forthcoming banquet he would produce from
that unpromising countryside fairer fruit than ever grew on the
smiling slopes of France—a wager, needless to say, immediately
accepted. At the banquet, a covered dish was brought in, which,
on the removal of the cover by the Queen, was seen to be filled with
newly-minted gold " bonnet-pieces," a fruit held to be sufficiently
goodly to justify the award of the wager to the King.

During the following 4 years, 41¼ ounces of native gold were
used in making a crown for the King and 35 ounces for a crown
for the Queen; 17 ounces were added to the King's great chain,
and a belt made for the Queen (Mary of Guise) weighed 19¼
ounces. Other articles of jewellery made at the same time from
native gold were a "bairtuithe" (a boar's tusk mounted in gold),
a shrine for "ane bane of St. Audrian of May," a gold whistle for
the King, and "ane dragoun, an amulet, and ane target of the
King's awin gold for his Majesty."[*]

Though some of the early coins of Mary's reign are of native
gold, there seems to have been nothing worthy of record about
the mines until 1567 when one Cornelius de Vois, otherwise known
as Cornelius De Vos, or Devosse, who pursued, among other avoca-
tions, that of artist and painter to Queen Elizabeth, arrived in
Scotland, with a recommendation from that Queen to the Scottish
Court. "And then Cornelius went to view the mountains in
Clydesdale and Nydesdale, upon which mountains he got a small
taste of small gold. This was a whetstone to sharpen his knife
on," says the old historian (Stephen Atkinson), "and this natural
gold tasted sweet as honeycomb to the mouth." Cornelius consulted with his friends in Edinburgh, and "by his persuasions
provoked them to adventure with him." He showed them some
samples of the gold, "it was in stems and some like unto bird's
eyes and eggs." Finally, the adventurers raised a capital of

---

[*] *Early Records relating to Mining in Scotland*, by R. W. Cochran-Patrick,
page xv.

about 5,000 pounds Scots or £416 sterling. The principal share-holders were the Earl of Morton, Secretary Ballantine and Abraham Peterson or Greybeard. Soon Cornelius had " six score men at work in the valleys and dales " (at fourpence sterling per day, or at the rate of a merk to twenty shillings the ounce for gold, if paid by results): " he employed both lads and lasses, idle men and women, which afore went a-begging, and he profited by their work, and they lived well and contented." On one occasion, he sent to Edinburgh 8 pounds of gold worth £450—the produce of 30 days' work. This was there converted into the current gold coin of the realm—twenty pounds Scots gold pieces.

Within the next two or three years, the same Abraham Peter-son or Abraham Greybeard, a Dutchman, obtained sufficient gold to make " ' a verie faire deepe bason ' which contained by estima-tion within the brim thereof, an English gallon of liquor. The same bason was of clean, neat, natural gold ; itself was then filled up to the brim with coined pieces of gold called unicorns ; which bason and pieces both were presented unto the French King by the said regent, the Earl of Morton, who signified on his honour unto the King, saying ' My Lord, behold this bason and all that therein is, it is natural gold, gotten within the kingdom of Scot-land, by a Dutchman named Abraham Greybeard ' ; and Abraham Greybeard, who was standing by, affirmed it on solemn oath, but he said unto the King that ' he thought it did engender and in-crease within the earth, and that he observed it soe to do by the in-fluence of the heavens.' "

Sir Walter Scott, however, tells the story in somewhat differ-ent fashion. He says that King James presented the vessel filled with gold bonnet-pieces to the French and Spanish ambassadors —a story which has the greater air of truth, for the Regent Morton is not likely to have met the French King.

During the next thirteen or fourteen years, licenses to work the gold-mines of Scotland were successively assigned to Arnold von Bronchhorst, to Abraham Peterson (or Greybeard) and to Eusta-chius Roche (1583). The royalty demanded varied from 6 to 7 ounces per 100 ounces obtained, the remainder to be brought to the Cunyie House (the Scottish Mint) where £22 Scots was paid for the ounce of fine gold and 40 shillings Scots for the ounce of fine silver.

About 1578, Sir Bevis Bulmer appeared on the scene, who

was destined to play a great part in the development and working of the Crawford Moor deposits. This "ingenious gentleman" was engaged by Thomas Foullis, a goldsmith in Edinburgh, to work his lead-mines in Lanarkshire, but appears to have devoted his attention entirely to the gold-deposits. The scenes of his operations lay principally on Mannock Moor, and Wanlock Water in Nithsdale, and on Friar's Moor and Crawford Moor, and the district in the Leadhills. He worked the deposits very systematically, constructing head-races and tail-races, and appears to have been fairly successful.

Atkinson says of Bulmer in this respect:—"On Wanlock Water, in Robert's Moor, he caused search diligently for natural gold, of which he got a pretty good quantity, and made water-courses to wash it . . . . Some say, he found out the vein of gold Mr. Bowes had discovered. In Frier Moor in Glengonnar Water in Clydesdale Mr. Bulmer got store of gold. . . . By help of a watercourse he got much straggling gold, on the skirts of the hills; and in the valleys, but none in solid places. which kept him in great pomp, keeping open house for all comers, as is reported."

"On Short Cleuch Water, in Crawford Moor, he built another goodly watercourse, and intended to make several dams there, to contain water for the buddles, and sources, and for washing gold. of which he found store. . . . From Short Cleuch, he removed up the great hill to Lang Cleuch Head, to seek gold in solid places: whereof he discovered a spring, but there he wanted a water-course to help him. This vein had the sapper stone plentifully in it, which sometimes held natural gold; but the salmoneer stones in that vein at Lang Cleuch Head held much silver and may prove a rich mine, if followed by such as know the nature of minerals. It is said that vein was powdered with gold, called small powdered gold. It was a vein and not a bed."

Bulmer's dumps or waste-heaps are still to be traced about the Gold Scaur, a rivulet falling into the Elvan Water. He built for himself a large house in Glengonnar, over the lintel of which he carved.

> In Wanlock, Elwand and Glengonnar
> I won my riches and my honour.

The largest nuggets recorded by him are of 6 ounces and 5 ounces respectively, of pure gold, found within 2 feet of the moss

at Lang Cleuch Head. At the same place he found a piece of
" sapper stone " (quartz) weighing 2 pounds from which 1 ounce
of gold was obtained. (There is now in the Edinburgh Museum
of Science and Art, a piece of white quartz from Lang Cleuch
Burn showing free gold). At the place, Bulmer erected a stamp-
ing mill, " called abroad *anacanago*," from which he obtained
"much small mealy gold." From his own records, he appears
to have been most successful on Henderland moor, in the Ettrick
Forest, where he obtained much gold—"the like to it in no other
place of Scotland." It was "like Indian wheat, or pearls, and
black-eyed like beans."

On his return to England about 1595, he presented to Queen
Elizabeth a porringer made of native Scottish gold on which was
engraved : —

> I dare not give, nor yet present,
> But render part of that's thy owne :
> My mind and heart shall still invent
> To seek out treasures yet unknowne !

Sir Bevis Bulmer appears to have been a man of indomitable
energy. Not only did he thoroughly exploit the gold of Lanarkshire, but also the silver mines of Hilderstone, Scotland, the silver-
lead-mines at Chewton, in the Mendips, and also of Combe Martin
in Devon. His remarks anent mining are shrewd. "How
long it " (the silver mine named God's Blessing at Hilderstone)
" will continue is known unto God, for mines are as uncertain as
the life is to man, which is like a bubble on the waters to-day,
to-morrow none." Again, " a mineral man should be a hazard
adventurer, not much esteeming whether he hit or miss."
Bulmer, from lavish hospitality, fell on evil days, but was
appointed by James I., in 1605, to be " Chief Governor of the
King's Mines " which post he held until his death in 1613.

In 1593, James VI. granted to Thomas Foullis, goldsmith,
burgess of Edinburgh, in consideration of " the grite soumes of
money restand awand be his Majestie and his darrest spous to
him and for sindrie advancementis maid by thame to him in the
quhilk nocht onlie hes he imployit ane grite parte of his awne
substance bot alswa remains addebtit to sindrie personis alsweill
in England as in Scotland off the quhilkis his Majesties officiaris
ar nocht able to mak him presentlie payment be his Hienes
fynanceis being swa exhaustit and swa spoillit be the moyane of

inordinat suitaris . . . . in speciall . . . . the mynis on Glengounnar."   The sum advanced by Foulis was £14,594. The mines so granted to him are the property of his descendants at the present day, Sir Robert Hope, an ancestor of the present Marquis of Linlithgow, having married a descendant of Thomas Foulis.

In 1603, a sum of £200 was granted to Sir Bevis Bulmer, and in 1604, £300 to George Bowes to search for gold and other metals on Crawford Moor.   Bowes' letters to the authorities are of considerable interest.   He describes the progress of his search with some detail, although it was fruitless.   His chief endeavour was to find a vein of auriferous quartz in Glangrese gill, the existence of which had been reported to him by one who had seen specimens obtained from it by "the Dutchmen" (Cornelius de Vois and party), many years before.   His mode of operation appears to have been hushing, or washing off the soil and exposing the underlying rock by sudden outbursts of water.   He reported the discovery of an auriferous vein, but later was doubtful of it being so.   Bowes gave up his work in 1604.   A curious light is thrown on former methods of working by one passage in his letters:—"though I have not got the value of 30s. in gold, yet I doe assure yoᵣ loᵖ I have wrought more work than Mr. Bulmer hath done hethertoo, resolving what gold I shall gett, faithfullie to deliver it to his Maᵗⁱᵉ use, boulding it inconvenient to this service to bye gold and make shew thereof as gotten in these workes: my travill onelye tending for discouerie of a vaine of gold." . . .•

After Bowes' retirement, little appears to have been done till 1616, when a grant of the Scottish mines was made to Stephen Atkinson, an Englishman and a refiner in the Mint of the Tower of London.   This man, it was, who endeavoured to persuade James VI. to adventure once again in the Scottish gold-mines. But James had spent about £3,000 (a vast sum in those days) in searching for gold on Carnworth Moor; and had obtained no more than 3 ounces, worth £12, an unfortunate speculation for a King whose regard for money was so notorious.   To Atkinson's representations, although attacking the King at his weakest point, that of cupidity, he therefore turned a deaf ear.   Atkinson assured the King that if only the search were carried out properly the

* *Early Records relating to Mining in Scotland*, by R. W. Cochran-Patrick, pages 103, 107, 108, 109 and 111.

result would make " his majesty the richest monarch in Europe,
yea, in all the world."   A subtle attack was also made in another
quarter—"In respect," says Atkinson, "of the wonderful
resemblance which many of his majesty's gracious deeds have with
the prophet David, and Solomon the wisest."   Astute as was
Atkinson, the previous lesson had been too severe and James
sought gold on Carnworth Moor no more.   At the same time, he
gave some countenance to Atkinson's projects, for in 1616 we
find an Order in Council instructing Atkinson " to move twenty-
four gentlemen of England, of sufficient land, to disburse £300
each " for which service they were to be created knights of a new
order, entitled " The Knights of the Golden Mines."   Atkinson's
plan appears to have failed, for Sir Bevis Bulmer and Sir John
Claypool were the only two knights so created.   By Atkinson's
licence he was to render to the Crown a tenth part of all the metal
obtained.

The operations were apparently unsuccessful, for in 1621, a
lease was granted to John Hendlie, physician, for a period of 21
years, and another in 1631 for 7 years to James, Marquis of
Hamilton.

In 1633, a rare gold medal was struck to commemorate the
coronation of Charles I. of Scotland, which bears the inscription
" EX AVRO VT IN SCOTIA REPERITVR."

In 1649, grants are recorded in favour of Sir James Hope in
respect of the Crawford Moor mines.

Certain manuscripts in the British Museum, written probably
by George Bowes, throw considerable light on the subject* and
show the author to have been a man of keen perception.   He
argues that the original sources of the gold " lyenge dispersed in
Chevore rockes neeare the topes and heighte of the mountaynes."
He says that it was " in King James the Fourth his tyme about
80 years sythence the Scotishe mene did begyne to wash golde."
(A date probably about 1520, if Bowes be the author.)   Also that
in the time of James V., some 300 people maintained themselves
for several summers by washing and that the "golde gotten
thearin " is " of greater vallue than one hondred thousand poundes,
yet in so many years and so many people workynge for goulde

* *Early Records Relating to Mining in Scotland*, by R. W. Cochran-Patrick,
pages 103 *et seq.*

no vaynes of gold have byne knowne to be founde." He further states that "gold hath byne gottene by washinge by the Lord of Markiston (Merchistoun?) in Pentley Hyles distant from Leadhile house in Craforde More (28) myles, and great plente of gold hath byne gottene in Langham Water (14) myles, and Megget Watere (12 myles) and over Phinland (16) myles distante from Leadshill house . . . . golde hathe byne found in more then (40) severalle gylles falynge in to the fowere greatere waters of Alwyn [Elvan], Glangonar, Wanlock and Mannocke."

The author also states that one piece of 30 ounces and some of greater weight had been found, which were "flatte and myxed with spar and some with keele [earthy oxide of iron] and some wth brimstone . . . . which shewe thear are vaines of gold from whence thos peices weare torne by the force of wateres eithere at Noes flood or by the vyolence of wateres synce that tyme." He also records that though he made many trials he found no gold on the higher slopes or tops of the hills, but only in or near the streams and that therefore "the same muste eithere growe theraboutes or by vyolente wateres be drivene out of higher places wheare they did growe wthin the circumference of those places wheare the gold is founde."

The following Scottish gold-localities, though they are undoubtedly unreliable in every respect, are nevertheless interesting. They are derived from a memorandum written during the reign of James V. and have been copied by Atkinson* and by Cochran-Patrick.† They were here inserted. on account of the comparative rarity of the works of these authors.

Ane Memorandum left by Robert Seton, commonly designed of Mexico, anent the Metals in Scotland, especially Gold.

Gold is found in the following places in Scotland :—

[1] In the Boggs of New-Lesly, upon the burn side ; and at Drumgavan, where George Lesly did dwell, two miles from Drumdeer [Dunideer].‡

[2] And in Northfideil hill, in one John Keith's land, beside Reivenshevin.

[3] In the Over hill, beneath the kyln, in the In-town, in the parish of Belhelvies.

* Stephen Atkinson's *Discoverie and Historie of the Gold Mynes in Scotland,* Bannatyne Club, 1825.

† *Early Records relating to Mining in Scotland,* by R. W. Cochran-Patrick, page xxii.

‡ *Ibid.*, page xxvii.

[4] And at Menzies, in the Golden-bank there, in the parish of Foveran [10 miles north-east of Aberdeen], and at the hill at South Fardin.*

[5] And in ane fould, called the Peltones.

[6] And in Dinrey hill in Carrick, not far from Mayboll.

[7] And in Caylies moor within the burn, that is betwixt the Sorne and Machlin place.†

[8] And in Henderland, Glengaber burn there.

[9] And in Monbengar braes and burn there.

[10] And in Dowglass braes, and at Dowglass craig.

[11] And in Windie-Neil in Tweddale. It marches with the Black-house in Yarrow.‡

[12] And in Borthwick hill, betwixt Hawick and Branxome.§

[13] And in Longlie burn, in the north side of Selkirk.

[14] And also at the New Town in Augen [Annan ?] Cait burn. in Annandale.

[15] And in Over-Lochan burn. *Alto y baxo, lacus aureus.*

[16] And in Bonarte hill in Fife, at Sarus Arrius.

[17] Solway-sands, near to the new town of Annand, not far from Dumfries, *Micie chiltir.*

[18] Durreness in Stranaver ; it belongs to the Lord Ray. *Ally ay una pedra muy relucente de noche tambien an ay Metall muy bueno.*

[19] In Glen-Yla, at Cassels, and there at Calderhall, and Over-Glen above St. Bride's Kirk. *Plata de azur.*

[20] And at Normandhill, on the side of Camps water.

[21] And in Annan water, and Cherries-braes. *Lauola,* Hoksay, and Long-cleuch.

[22] And at Bellyes baith and Jervies Mayr.

[23] And in Glen-naip betwixt Carrick and Galloway.

[24] And in Ruberlaw Hill, a mile from Bothwell.

[25] And in Galloway, in the Barony of Tareagles, and in an hill called Colochen hill.||

[26] And in Largo Law in Fife. *Plata.*¶

[27] And in Hara in Caithness, in the Laird of Rater's land.

[28] Dumpender law. It belongs to Bothwell.

[29] And in the Moir, or in Airlaw, beside Crichton dean. *Oro.*

[30] And in the Laird of Down's land in Caithness.

[31] And in my Lord Brotherstown's land : lead two miles from the sea.**

[32] And in Courtoit burn.

[33] In Galloway, in the hill of Skrill, in a strype of water. *Mucho oro y grandes pedaços.*

[34] And in the water above Threpland miln, and in sundry other parts thereabout.

---

* This is noted by Col. Borthwick as a locality for silver and not for gold, *Early Records relating to Mining in Scotland,* page xxvi.

† Silver only, according to Col. Borthwick's Memorandum, *ibid.,* page xxviii.

‡ Silver only, according to Col. Borthwick's Memorandum, *ibid.,* page xxvii.

§ Copper only, according to Col. Borthwick's Memorandum, *ibid.,* page xxviii.

|| Silver only in Col. Borthwick's Memorandum, *ibid.,* page xxvii.

¶ Copper enough, Col. Borthwick, *ibid.,* page xxvii.

** Lead only, Colonel Borthwick, *ibid.,* page xxviii.

[35] And in a burn that runs from the head of Moffat water, or Annand water in Annandale. *Oro.*

[36] And in Glencloucht, where the miners did find much gold, long since at the Kirkhill [. . . . toward the east side thereof].[*]

[37] And in Long-Forglan moor, three miles from Dundee.

[38] And at Stains hill.

[39] And in James Crawford's at Muckert. *Millen unelto.* Not far from Culross, three or four miles above Torie burn.

Further memoranda transmitted by "Colonell Borthwick who had the direction of the mines," to Sir Robert Sibbald,[†] in 1683, declare that:—

Gold is found in severall places in Scotland; the most famous place is Craw-ford-moor, where it was found by King James the Fourth, and King James the Fifth, and is yet found by passing the earth through searches, and the same brought down with speats of raine. I have seen pieces of it as big as a cherry. It is exceeding fine Gold. The Ore, as it was tryed at the King's Mint in London, afforded eleven parts of Gold, and the refuse was Silver.

A place called Dunideer is famous for Gold [several miles beyond Aberdeen]

Some report, that at Clovo, at the head of South Esk, some eight miles from Killiemuir, there is found gold and silver.

There is a gold Mine very rich, in a husband toun, called Overhill, in the parish of Belhelvie, that belongs to my Lord Glames, three fathoms, beneath the kyln that is at the head of the In-town.

A gold Mine was found in King James the Fifth's time, in Lamington burn. Gold is found at Kersop, upon Yarrow water, in Philiphaugh's ground.

At the present time, it is, of course, impossible to estimate with any approach to accuracy the quantity of gold yielded by the Crawford Moor alluvial deposits during the sixteenth and seventeenth centuries. Pennant, on what authority it does not appear, says "In the reigns of James IV. and James V., vast wealth was procured in the Leadhills, from the gold found in the sands washed from the mountains; in the reign of the latter not less than £300,000 sterling." Dr. Lauder Lindsay places the yield as still higher, namely, £500,000, but his authorities for this high sum are equally obscure. In a matter which, from the conditions of working, and from the lapse of time, must necessarily be of considerable uncertainty, it is advisable to follow those authorities who speak with the most immediate knowledge of the subject. In this case, the authority is the letter of George Bowes, quoted above. Bowes, speaking of the total produce of the Crawford Moor district, which took place during his own and part of

* Col. Borthwick's Memorandum, *Early Records relating to Mining in Scotland*, page xxvii.

† *Early Records relating to Mining in Scotland*, pages xxv., xxvii and xxviii.

the preceding generation, places the yield at £100,000 sterling, and even this is probably overstating rather than understating the amount.

Leadhills.—The district of Leadhills, southern Lanarkshire, (Fig. 6, Plate XVIII.), lies about 44 miles south-east by south from Glasgow. The village itself has an elevation of 1,300 feet, and is reputed to be the highest in Scotland. The surrounding country is barren and uninviting, running up on the south to the Lowther Hills (2,403 feet) and stretching away in the north to Crawford Moor. The high land in the immediate vicinity, which at its greatest elevation has a north-easterly trend from Cairn Hill (1,471 feet), in the Nith valley to Watchman Hill (1,487 feet) above Elvanfoot on the Clyde, serves as the main watershed of the district, from which flow branches both to the rivers Clyde and Nith. The two highest peaks on this range are Green Lowther (2,403 feet) and Lowther Hill (2,377 feet). Flowing to the north and north east, are the Potrail Water (which, with Daer Water, really forms the river Clyde), Elvan Water, Glengonnar (flowing through the village of Leadhills) and Snar Water (a tributary of Duneaton Water). Joining the Nith, and running as a general rule to the west and south-west are Wanlock Water, Menock Water, Enterkin Burn, and Carron Water. Access to the metalliferous district is now rendered easy by the recent construction of a light railway-line from Elvanfoot, running up the Elvan valley to the villages of Leadhills and Wanlockhead.

The auriferous area of the Leadhills lies almost entirely in rocks of Lower Silurian age—of Llandovery, Caradoc-Llandeilo, and Arenig time. The surface-contact line of the Llandovery and the older underlying Caradoc-Llandeilo beds runs in this area approximately north-east and south-west, parallel with the course, and some little distance to the north, of the Potrail Water.

The oldest rocks in the district are pillowy diabase-lavas[*] which, with the overlying radiolarian cherts, are exposed in rapidly recurring folds wherever denudation has proceeded sufficiently far to remove the younger rocks. The folds are generally isoclinal, and relief is often obtained by the development of thrust-planes. Overlying the radiolarian chert (Lower Llandeilo) is a well defined but thin band of black shales—the Glenkiln shales (Upper

[*] *Memoirs of the Geological Survey of Scotland*, 1899, vol. i., page 656.

FIG. 13.—WANLOCKHEAD VILLAGE, LANARKSHIRE.

Llandeilo). The fossils of these beds are graptolites (*Cœnograptus gracilis, Didymograptus* spp., *Climacograptus* spp., etc.).

Overlying the Glenkiln shales, at a short interval, and without any stratigraphical break, are the Hartfell shales (Caradoc), also characterized by their graptolites (*Pleurograptus linearis, Dicranograptus Clinguni, Climacograptus Wilsoni*, etc.). The Hartfell shales at the Leadhills occasionally give place to coarse greywackes, grits and conglomerates. It is in these arneaceous sediments alone that the metalliferous (galena) veins of the Leadhills are developed. As these veins approach the black shales, either laterally or in depth, they gradually become poorer, and finally, with contact, the galena disappears from the vein. In all probability, this peculiar distribution is due to an upward movement of precipitant solutions derived from the carbonaceous shales.

The gold of the Leadhills area is found in the streams, into which it has been washed from a gravelly clay, locally known as "till," which lies disposed on the slopes of the hills. The gold generally occurs as fine dust, but small nuggets of varying size have from time to time been observed. The largest on record weighed 27 ounces, and is, or was, in the collection of the Marquis of Linlithgow.[*] This nugget is said to have been discovered about 1502. It is bigger than the Wicklow nugget of 22 ounces, and is therefore the biggest recorded British nugget. Another also in the same cabinet weighs 1½ ounces. Gold-washing as an industry has been abandoned at Leadhills for many years, such gold as has been obtained during the last century having been collected for the purpose of making jewellery for wedding-presents, etc., to the proprietors. Dr. Lindsay[†] records that 975 grains of gold were collected, in 1862, for the Countess of Hopetoun, and that on another occasion 600 grains were collected in 6 weeks by 30 men working in their leisure-time. Between May and October 1863, three miners working at intervals collected 33 grains of gold from the " till " at a point about 120 feet above the bed of the stream and halfway down the Langcleuch burn, between Leadhills and Elvanfoot. They sold it to Dr. Lindsay at the rate of 7½d. per grain. In the summer of 1862, the miners, by way of holiday work, frequently collected quantities of from 14 to 54 grains, for which the average price charged was 6d. per grain.[‡]

The gold from Wanlockhead is of the average quality of British gold, the following being an assay[§]:—Gold, 86·60; silver, 12·39; copper and iron, 0·35; loss, 0·66. The specific gravity was 16·50.

An analysis made by the late Prof. Heddle[||] of alluvial gold in small hackly scales and nuggets from Langcleugh burn, in the Leadhills district, differed but little from the above:—Gold, 87·32; silver, 11·80; ferric oxide, 0·41; silica, 0·22; and loss, 0·25. The specific gravity was 17·52. The value of the above gold ranged, therefore, from £3 13s. 6d. to £3 14s. per ounce.

In the Edinburgh Museum of Science and Art lies the " Gemmell nugget," over the nativity of which there has been a

---

[*] *Report of the British Association*, 1867, Dr. W. Lauder Lindsay, page 64.

[†] *Transactions of the Geological Society of Edinburgh*, 1868, vol. i., page 109.

[‡] *Ibid.*, page 110.

[§] *Chemical News*, by Prof. Church, vol. xxix., page 209.

[||] Edinburgh Museum of Science and Art.

long, and at times acrimonious, newspaper discussion. It is not
a nugget in the true sense, but a large fragment weighing
probably about 3 pounds, of highly auriferous quartz. It was
picked up by a miner named Gemmell on the roadside at Wanlockhead in 1872. The finder broke the specimen into several
pieces and sold the fragments to various collectors. Owing to
the exertions of Mr. Dudgeon of Cargen, the several pieces were
collected, replaced in position and presented to the above
museum.

The quartz sufficiently resembles that of the Leadhills to
lend some colour to the theory of local origin, but the evidence

FIG. 14.--LEADHILLS, LANARKSHIRE.

brought out by the discussion referred to would on the whole
favour an Australian origin. It appears that on the day prior
to the discovery a handbarrow containing—among other effects
of a returned Australian miner who was "flitting" from one
house to another in Wanlockhead—a number of specimens of
Australian gold-quartz, was wheeled along the road. The pre-
sumption therefore is that the Gemmell nugget is one of these
unnumbered and unmarked Australian specimens. The quartz
certainly resembles that from many Victorian gold-fields.

Gold has also been found *in situ* in the Leadhills district.
In 1803, Prof. Traill records gold from a vein of quartz at Wan-

lockhead, and in the Edinburgh Museum of Science and Art there is a specimen of clean white quartz slightly waterworn, containing gold which shows a tendency to wiriness.

A specimen in the Natural History Museum, South Kensington, labelled " Leadhills," is composed of a clean white quartz with a few grains of shotty gold. On the authenticity of this specimen, Dr. Lauder Lindsay throws some doubt, for, as he points out, in the case of the Leadhills gold, which commanded nearly more than double its intrinsic value, there was every incentive for the substitution of similar Australian and other auriferous quartz for that of local origin. It would further appear

FIG. 15.—LEAD-MINES, WANLOCK BURN.

that the South Kensington specimen was purchased from a London mineral dealer who had secured the whole cabinet of a gentleman " who had collected extensively in the Leadhills district." The quartz differs little in appearance from that of the Leadhills district, and the present writer is disposed to accept it as a genuine Leadhills specimen. It is a large piece of quartz containing visible gold in one corner only, and it seems very improbable indeed that an Australian miner would have carried for any length of time or for any distance so large a piece of valueless gangue when the corner containing the very small quantity of gold might easily have been broken from the specimen, and could then have been placed in the waistcoat-pocket.

A specimen of auriferous quartz in the Edinburgh Museum of Science and Art, from Wingate burn, Leadhills, shows somewhat wiry gold but little waterworn, and is associated with a clean, milky-white quartz. It is evidently derived from an adjacent vein. Another specimen from Stake burn, Wanlockhead, in the same museum, shows native gold disseminated throughout limonite and quartz.

Sutherland.—Though as far back as 1853,* a nugget weighing 1½ ounces† is recorded as having been picked up in 1840 in the Kildonan stream, and though, as we have already seen, gold was reputed to have been obtained in 1245 by Gilbert de Moravia at Durness (a few miles south-east of Cape Wrath) gold was not known to occur in any quantity in Sutherland until November, 1868. As first suggested by the Rev. J. M. Joass, it is not at all improbable that the remnants of the numerous Pictish towers or brochs scattered over the hills of Strath Ullie and forming, when complete, a chain of strongholds from Dunnaine on the coast to Ben Ghriam-beg at the sources of the Ullie, indicate a means of defence for a digging population against the attacks of marauding bands of maritime rovers, attracted to the spot by the report of gold. It is noteworthy that the remains of these small forts are particularly numerous in the immediate vicinity of the Suisgill and Kildonan streams, which are, so far as is at present known, the most highly auriferous in the district.

In November, 1868, Mr. R. N. Gilchrist, a native of the county, was induced to try the various streams for gold. He had been for 17 years in Australia, where he had met with success as a gold-digger. He was led to the search, it is said, by the likely appearance of the gravels of the Ullie, but it is much more probable that traditions of the gold already found there, were in existence. His search soon resulted in the discovery of gold in Kildonan burn, a small tributary of the Ullie. Following up the discovery, gold was found in the neighbouring burns, and a rush to the neighbourhood took place. At one time, in 1869, no less than 400

---

* *Gold Rocks of Great Britain, etc.*, 1853, by John Calvert, page 163.

† More than 1 ounce, according to David Forbes, *Philosophical Magazine*, 1869, vol. xxxvii., page 327.

men were employed in the diggings. That the work was remunerative, for the time being, is evidenced by the continued payment
during a year of the license-fee for each digger of £1 per month,
besides the royalty of 10 per cent. demanded by the Government.

During the short period that these gravels were worked after
the discovery of their auriferous character, royalty was paid on
£3,000 worth of gold; but, as the temptation to conceal the greater
portion of the gold discovered must have been almost irresistible,
it is probable, as estimated by Dr. Joass, that the total amount
recovered was not less than £12,000.[*]

Much of the gold from the Sutherland gold-field was purchased
by Mr. P. G. Wilson of Inverness. In three weeks in February,
1869, he purchased respectively £28, £18 14s., and £14 worth
of gold. By April 8th, he had purchased £431 worth, and by
August 24th, between £5,000 and £10,000 worth.[†] The price
was at first £4 per ounce, but it afterwards dropped to £3 10s.
According to Dr. Lauder Lindsay, £15,000 worth of gold was obtained from Sutherland in 1869.

About the same time, gold was also discovered, but in smaller
quantities in the Allt-Smeoral, or Gordon-bush burn, and in the
Uisge Duibh or Blackwater, two streams falling into the head of
Loch Brora. These were, however, worked for a very short time.
since the license-fees obtained did not by any means compensate
for the damage occasioned by the diggers to pastoral interests, by
driving sheep away from the sheltered valley to the bleak moorland. Digging was therefore prohibited in the Brora district from
January 1st, 1870, and has never been resumed.

The gold-fields of Sutherland are restricted to two main localities—to the tributary streams flowing from the north into the
Ullie or Helmsdale, and to the two streams, already mentioned,
flowing into Loch Brora.

All the former have their sources in the high lands running
along the boundary between the counties of Sutherland and Caithness. The auriferous streams are, in order from the mouth of the
Helmsdale upward, the Allt Torrish, Allt Breacich, Allt Duibh,
Ildonan, Allt Ant' Fionnaraidh, Suisgill and Kinbrace (Cn
reas). The Craggie, flowing from the west into the Ullie, has
also yielded alluvial gold.

[*] *Geology and Scenery of Sutherland*, by H. M. Cadell, page 95.
[†] *Transactions of the Geological Society of Edinburgh*, vol. ii., page 27.

The whole country through which these streams run is typical moorland, with heatherclad lower hills and with extensive marshy ground at the sources of the streams in the high lands.   The valleys of the streams have been cut down rapidly, and are narrow and fairly straight.   Alluvial flats of any size are wanting along their course, and it is only in the main stream, the Ullie or Helmsdale, that such are developed.

The rocks of this district have been mapped by officers of the Geological Survey, and are, in the main, granites and schists (Fig. 7, Plate XIX.).   The granites occur in mass only on the north-western and on the eastern boundaries of the area, the east-

FIG. 16.—BLACKWATER STREAM, AT THE HEAD OF LOCH BRORA, LOOKING
NORTH-WESTWARD.

ern development being prolonged to and forming the Ord of Caithness.   The auriferous district is therefore almost wholly in the schistose rocks.   These have been divided by the officers of the Geological Survey into quartz-schists, flaser-mica schists and granulitic biotite-schists, clearly representing original sandstones and shales, probably of Lower Silurian age.   The quartz-schists occasionally pass in the mass into quartzite, of which there is an extensive development at Cnoc-an-Eiranneach, a hill 1,700 feet in height, at the head of the Suisgill and Kildonan burns. The schists are well exposed in the bed of the Ullie, just below Kildonan Lodge, and also in the bed of Kildonan burn, about 300 feet above the bridge. Their general dip is east north-eastward at a high angle. They are here thinly-bedded flaggy schists, with occasional bands more or less quartzose. Some of the latter pass into true quartzites. With the quartzose beds are thinner bands (4 to 6 inches thick) of highly micaceous (muscovite) schists. The flaser-mica-schists, which occur as very thin bands through the quartz-schists, in the lower courses of the Kildonan and Suisgill, have a much greater development toward the heads of these streams. Their mica is muscovite. The granulitic biotite-schists are grey rocks, made up of quartz and felspar-grains rudely oriented, together with an occasional grain of biotite-mica. The Kildonan, in the upper part, and the Suisgill for the main part of its course, flow over these rocks.

Overlying the metamorphic schists, and rendering it at all times difficult and in some cases impossible, to map out the boundaries of the rocks, is a heavy deposit of Glacial Drift, overlain in its turn by thick beds of peat. The Glacial Drift shows but little sign of stratification, and fine-grained sandy or clay-beds, are, as a rule, absent. The general constituent of the drift is a coarse gravel, through which are interspersed numerous small stones and boulders ranging from 6 inches to 2 feet in diameter. These boulders, so far as can be seen, are fragments of rocks which are all represented *in situ* in the immediate neighbourhood, and a comparatively local origin for the drift may therefore be assumed.

The gold is found as small grains in the beds of the streams, and in the gravel-banks along their courses. It is naturally most abundant in the rocky potholes and in the crevices afforded by the upturned edges of the flaggy schists, across the strike of which the streams run; but it appears to be also disseminated throughout the drift. Although most abundant in the lower courses of the streams it is not found there alone, but occurs right up to the heads of the burns, clearly demonstrating, either long-continued denudation, or more probably a concentration of the gold in the drift which caps all but the highest hills. The grains of gold are generally flattened, and, except in the case of the larger nuggets, present very little evidence of rolling or attrition by the action of water. The heaviest nugget yet discovered here weighed 2 ounces 17 dwts. and was picked up by a man named Rutherford in April, 1869, shortly after the discovery of the gold-field. A model of this nugget is shown in

the Jermyn Street museum.   Generally speaking, the gold
becomes finer from north-west to south-east, indicating, therefore,
a north-western origin for the gold of this area.   The alluvial
gold of Kinbrace burn is coarse and shotty, as also is that, but
in a less degree, of the Suisgill, while Kildonan gold farther to
the south-east is very much finer than either of the above.   The
richest deposits yet found have been in the Gold burn, a stream
flowing from the east into the Suisgill.   Here, indeed, several
colours or specks of gold may be obtained from nearly every dish.

FIG. 17.—LOCH BRORA FROM THE SAME POINT AS FIG. 16, LOOKING
SOUTH-EASTWARD.

The matrix of the gold is to be sought for in the quartz-veins
in the local schists, and possibly in similar veins in the granites
to the north-west; but, in the latter case only where they are
adjacent to, or intersect the schists.   A careful examination of
the beds of the streams disclosed several quartz-veins, apparently
striking and dipping with the country-rock.   The quartz of
these veins is white and opaque, resembling very much in character the quartz-veins in the phyllites and similar metamorphic
schists of Otago, New Zealand, from the denudation of which and
the manifold concentrations of the residual débris have resulted
the extremely rich fluviatile auriferous deposits of that region.

The following percentage analyses of the gold of Sutherland
have been made:—

No. of Sample.	Authorities.	Gold.	Silver.	Silica.	Iron.	Specific Gravity.
1	Forbes	81·11	18·45	0·44	—	15·79
2	Do.	81·27	18·47	0·26	—	15·79
3	Makins	79·22	20·78	—	—	16·62
4*	Heddle	80·34	19·86	—	0·12	15·612

* Specimen in Edinburgh Museum of Science and Art.

In the early days of the gold-field, small nuggets weighing 1, 2, 3, and even up to 5 dwts., were not uncommon. Mr. Gilchrist, the original discoverer, had in his possession five nuggets varying from ¼ ounce up to 1 ounce troy weight.

The minerals which accompany the gold, and are left behind in the dish on panning off, are almandine-garnets, muscovite-mica, ilmenite, magnetite and iron-pyrites. According to David Forbes* the magnetic particles have the following composition:— Oxide of iron, 91·26; titanic acid, 8·03; and silica, 0·70; while their specific gravity is 5·08.

The Kildonan veins, according to the late Prof. Heddle, have indeed yielded gold, and there is in the Edinburgh Museum of Science and Art a small bead, said to have been obtained from quartz-veins in the Kildonan area. This bead represented a yield of 39·2 grains of gold per ton, and was contained in an electrum of 28·57 per cent. of gold and 71·43 per cent. of silver. It would appear, however, that the method of separation used was that of amalgamation, and it is difficult to account for the extremely high proportion of silver, except on the supposition that it came from the mercury used; more particularly as it does not appear that special precautions were taken to ensure the absolute purity of the mercury nor the freedom from gold of the materials used for crushing. Nor does it appear that the vein itself contained galena (the only mineral likely to furnish silver). Taking everything into consideration, this particular result must be regarded with some suspicion. As already seen, the alluvial gold of the district is about 800 fine.

Judging from analogy with alluvial gold derived from similar rocks in other parts of the world, the richness of the Sutherland alluvia will depend almost entirely on the conditions of concentration, and but little on the comparative richness of the parent quartz-veins. As a case in point, the extremely rich alluvial deposits of the Clutha river in New Zealand have been derived from veins, the remnants of which are far too poor to work at the present time. Numerous examples of a similar kind

* *Philosophical Magazine*, 1869, vol. xxxvii., page 327.

Suisgill Burn.   Cnoc-an-Eiranneach.   Gold Burn.

FIG. 18.—SUISGILL AND GOLD BURNS, KILDONAN.

might be adduced to show that, without an absolute knowledge of the conditions of alluvial deposition, no inference can be drawn as to the richness or poverty of the parent-reefs of the Kildonan area.

As already stated, the petrological and lithological characters of the schists leave but little doubt of their sedimentary origin ; and, such being the case, it would seem that it is in the metamorphic schists that the parent auriferous veins must be sought for. Only on such an assumption can the local distribution of the gold be satisfactorily explained, for the gold would naturally be concentrated in leads (in this case littoral rather than fluviatile, and corresponding on the whole to modern beach-sands) in the old sediments as they are in those of the present day. Notwithstanding the fact that gold appears to have been obtained attached to granite, the source of the vein-gold of Kildonan is to be ascribed not to the granites of the area, but to the schists. At the same time it must be stated that an occurrence of gold in the quartz of granite as a primary constituent is reported from Sonora, Mexico.*

* George P. Merrill, *American Journal of Science*, 1896, vol. i., page 309.

As will be seen from Appendix II., numerous assays have been made of the rocks of Sutherland in the hope of determining the source of the gold. These were conducted with large samples, no less than 200 grammes (3,086·4 grains) of each rock being taken. Extreme care was taken to ensure the absolute purity of the fluxes used, and especially of the litharge. The beads obtained being far too minute to be weighed, or in some cases to be seen with the naked eye, were measured with a

Fig. 19.—KILDONAN BURN.

standardized eye-piece micrometer in a powerful microscope, and their weight was calculated from their diameter. By this means, beads representing 0·01 grain per ton, or 1 part in 1,568,000,000 may easily be determined, certain necessary refinements in the methods of treating the minute beads having been adopted. Though the results set forth herein are negative, they have yet a certain value, and are by no means useless.

With reference to the profitable working of the alluvial deposits of Kildonan, it must be said that the prospects of success

do not appear alluring.   Neither could a head of water sufficiently
great to command the gravels be obtained, nor is the water in the
immediate neighbourhood sufficiently plentiful to work the gravels
on such a scale as the economics of the scheme would demand.
The over-burden (large stones and boulders) of the Glacial Drift
is very heavy, and sluicing operations of any kind would probably
be impracticable during 2 or 3 months in the year, owing to the
inclemency of the winter season.   But probably, the greatest obstacle to working such a deposit on a large scale lies in the fact
that the salmon-fishings in the lower portion of the Helmsdale
would be injured—if not altogether destroyed—and the prospec-
tive revenue from royalties, etc., on gold would certainly not compensate for that loss.

TABLE V.—GOLD-PRODUCTION OF SUTHERLAND IN 1896.

No. of Party.	Time at work.		No. of Days worked.	Gold.		
	From	To		Ozs.	Dwts.	Grs.
1	May 6	August 6	92	4	3	12
2	May 6	August 6	92	3	11	0
3	May 7	August 6	91	0	15	5
4	June 3	August 6	64	0	14	9
5	June 3	August 6	64	0	14	9
6	June 15	August 6	52	2	12	11
7	July 16	August 6	21	0	5	2
		Totals	476	12	16	0

The Sutherland County Council in 1896, endeavoured to re-
open the diggings, and with the consent of the owner set several
men to work in Suisgill burn.   Table V. shows the quantities
of gold obtained.*   This result would indicate a yield of 13
grains of gold per *diem*, or nearly 2s. in value, a result clearly
unprofitable.   The experiment naturally led to no further trials.
In this connection, however, it must be stated that local opinion
as to the experiment (on apparently no better ground than that of
innate belief in the richness of the field) explains this poverty
of return by the assumption that not all the gold obtained was
reported to the authorities.

The two auriferous localities at Loch Brora are the Allt
Smeoral, or the Gordon-bush burn, and the Uisge Duibh, or
Blackwater.   The former flows from the north into the loch
about ¾ mile from its head.   In the ravine, where crossed by the

* *A Treatise on Ore-Deposits*, by Mr. J. Arthur Phillips ; second edition,
1896, by Prof. H. Louis, page 320.

bridge, a good section of the rocks is displayed. They are Lower Silurian flaggy quartzites and micaceous schists dipping south-eastward from 40 to 60 degrees. Though the whole course of the stream lies in the above rocks, the eastern boundary of its water-shed lies along the Old Red Sandstone, here represented in the mass by conglomerates through which run gritty bands of sand-stone. Ben Smeoral (1,592 feet) is composed of this rock, which here dips south-eastward at low angles from 10 to 20 degrees. Granite-dykes and quartz-veins are common in the upper waters of the Allt Smeoral. The gold is found both in terraces and in the flats of the stream, in the bottom-stratum of coarse grit lying

FIG. 20. — BAILE N'OIR, KILDONAN BURN.

on the rock, and is overlain by a deposit of reddish clay and sand, much of which has been obviously derived from the Old Red Sandstone area.

The Uisge Duibh, or Blackwater, flows into the head of Loch Brora, in the upper part of its course, over Lower Silurian rocks precisely similar in composition to those noted above, and through an alluvial flat for more than 2 miles of its lower course. A short distance above its junction with the Brora river, 2 miles from the Loch, gold has been found. Here the burn runs across the strike of the rocks, which dip south-eastward at angles of about 20 degrees. The micaceous schists and quartzites are seamed by numerous narrow dykes of granite. The gold occurs in a bluish sandy clay, together with rolled fragments of red granite and quartz, and is somewhat coarse in character. There is, however, little alluvium in the stream after it leaves the flat formed by the filling of Loch Brora.

Other Localities.—The other undoubted Scottish localities in which gold has been discovered may be grouped into three divisions :—

(*a*) Occurrences which may be associated with the Leadhills alluvial deposits. These are, besides the streams already mentioned as flowing from the high land in the vicinity of Leadhills (Shortcleuch, Leadburn, Elvan, Langcleuch, Glengonnar, Wanlock), those flowing into the Tweed (Manor Water, Meggat, Yarrow and Glengaber), and those flowing into the Annan (Moffat Water and Dobbs Linn). All the above occurrences are alluvial, but auriferous pyrites is recorded from Torbockhill, near Annan.* This on analysis yielded 4 dwts. of gold and 10 ounces of silver per ton. The auriferous pyrites was taken from an old working called "the cave," which was worked in the eighteenth century by Germans.

(*b*) Perthshire occurrences (Breadalbane area), about Loch Tay and the headwaters of the Tay. According to Dr. Lauder Lindsay, a nugget found here in former times weighed 2 ounces. He also records gold in its matrix from Tyndrum, at the head of Strathfillan, western Perthshire, where argentiferous galena occurs in mica-slate near its junction with quartzite. In 1861, James Tennant found gold in quartz, associated with iron-pyrites at Taymouth (Fig. 8, Plate XIX.).

Gold has also been recorded by various observers from other parts of Perthshire; from galena-veins at Lochearnhead near the railway station† where arsenical pyrites yields at the rate of 6 ounces to the ton, and where particles of native gold have been found in the gossan;‡ Glen Lednoch;§ Ardvorlich, south side of Loch Earn, in mining for argentiferous galena; Cornebruchill, on the southern side of Loch Tay, opposite Ben Lawers;‖ Glenturret;¶ Glenalmond;** and Glenquaich, near Loch Freuchie.††

---

* *St. James' Gazette*, June 15th, 1901.   † Murchison.

‡ *Quarterly Journal of the Geological Society of London*, vol. xvi., page 424.
§ Ritchie.

‖ C. H. G. Thost, *Quarterly Journal of the Geological Society of London*, 1860, vol. xvi., page 424.   ¶ Anstead.   ** Mercer.   †† Nicoll.

There is in the Natural History Museum at South Kensington, a nugget weighing 1,010 grains from Turrerich, Glenquaich, Breadalbane. It is of a brassy-yellow colour, and is apparently of very poor quality. It contains about one-third of its weight of quartz. The gold is extremely cavernous, and shows a tendency to crystallization, though no distinct crystal-faces are to be seen.

Small quantities of alluvial gold have been recorded from the tributaries of the Dee, at Braemar and Invercauld, and in the sea-sand of the coast, about Aberdeen.[*]

In 1869, gold-dust in considerable quantity was found in the alluvium of the headwaters of the Errick (Ericht?) and Nairn rivers in Inverness[†] and were also washed from the granites there by Dr. Bryce in 1870.

Dr. W. Lauder Lindsay, from ancient manuscripts, infers that gold has been worked at times in the Clova district and the Braes of Angus in Forfarshire.

Gold was reported to have been found in Unst (Shetland) at Ureh, near Braewick, but a close examination of the locality failed to discover the slightest traces of that metal.[‡]

In May, 1852, considerable excitement was caused in Fifeshire by the reported discovery of gold in that county, and no less than 300 men were engaged on these diggings for the better part of a month. The rush had its origin in the letter of a convict in Australia to his friends in Kinnesswood, stating that certain rocks near his native village were very similar to those being worked for gold in Australia. The chief scene of their labours appears to have been in a quarry of Carboniferous Limestone on the south base of the West Lomond hill, overlooking Loch Leven. Over-lying the fossiliferous limestone is a bed of hæmatitic clay containing huge globular masses of iron-pyrites, which were carried away by the diggers in the belief that they were nuggets of gold.

IRELAND.

There is no actual knowledge of the discovery or working of gold-deposits in Ireland before 1765. Inferentially, however, from the great numbers of golden ornaments known to have been in existence in the early Christian ages, it may, with

* Leask.
† Dr. James Bryce, *Report of the British Association*, 1870, page 70.
‡ *Mining Magazine*, 1878, vol. ii., page 18.

some reason, be assumed that the deposits discovered and worked during the past century, together with others at present unknown, had been discovered in remote ages, and with impoverishment of auriferous content had been allowed to drop into oblivion. At the Bog of Cullen, near Limerick, and at the fords of the Shannon, not only have golden ornaments been found, but also the implements for smelting the metal. Here and there, scattered throughout the ancient annals, are more or less reliable records of the presence of gold in Ireland. In the *Annals of the Four Masters*, and also in the *Book of Leacan*, it is recorded that Tighernmas (A.M. 3656) first smelted gold in Ireland at Foithre-aithir-liffe. This is the main ridge of the Wicklow mountains, and its proximity to the Ovoca auriferous district would suggest a local source for the gold. Keating, in his *History of Ireland*, states also that this king was "the first who discovered gold-ore in Ireland." The inhabitants of Leinster were called "Laighnigh-an-Oir," *i.e.*, the Lagenians of the Gold, "because it was in their country that gold was first discovered." Giraldus Cambrensis (Gerald Barry), writing in the thirteenth century, states that Ireland abounded in gold. Gerard Boate, in his *Natural History of Ireland*, written in 1652, mentions the occurrence of alluvial gold in the Mayola (Miola) river, which flows into Lough Neagh through a portion of Londonderry county. Since the rocks through which this river runs are mica-slates of Silurian age—already seen to be exceptionally favourable for the occurrence of gold in the British area—some credence may be placed in the statement. The more so, since he explicitly states the amount (1 dram) gathered from the sands. But if auriferous deposits were known in the early and middle ages, it is quite certain that their exact positions were forgotten in the eighteenth century—a fact not surprising, when the troublous times through which Ireland had passed are taken into consideration.

It would appear that the first well-authenticated discovery of gold in Ireland was made about 1765. This was a small nugget picked up from the Ballinvalley brook, flowing into the Aughrim river, near its junction with the Ovoca. Five years later another small nugget was found in the same stream by a boy while fishing, but it was not until September, 1795, that it became generally known that the gravels of this stream (then called the

Aughatinavought, but afterwards called by Mr. Thomas Weaver the Ballinvalley or Gold-mine river) were more or less auriferous throughout its whole course. A rush to the spot naturally followed, and in a very short time a great concourse of peasants (men, women and children) were engaged in the arduous and unaccustomed work of gold-washing, using the crudest of appliances. In October, 1795, when the news of the discovery came to the ears of the authorities, a strong force of Kildare Militia was sent to turn away the peasants, who, driven from Ballinvalley and Ballinasilloge, the richest spots on the Aughatinavought, flocked to the neighbouring streams, but these appar-

FIG. 21.—EASTERN STREAM, GOLD-MINE RIVER, LOOKING SOUTHWARD.

ently did not prove as rich as that first exploited, for work in them ceased after a time.

The Treasury, having received an Enabling Act from the Irish Parliament, proceeded to work the auriferous deposits at Ballinvalley, rather, it would seem, for the purpose of exhausting the easily available gold in order that no temptation might be left for the peasants to assemble and congregate in great numbers, than in the hope of remunerative working; for the directors, Messrs. King, Weaver & Mills were instructed to continue to work the deposits only until the great depth of overlying gravels would preclude the possibility of success from individual and spasmodic efforts.

Government operations were suspended in May, 1798, when the works were destroyed by the rebels and the workings were deserted for more than 2 years. During the period of working, the directors had obtained 555 ounces 17 dwts. 22¼ grains of gold valued at £2,146 15s. The cost was £1,815 16s. 5d., and thus, as Mr. Weaver says,[*] " Government had fully reimbursed its advances, the produce of the undertaking having defrayed its own expenses, and left a surplus in hand." On the resumption of work in 1801, the directors recommended a search for gold in its matrix, and extensive exploratory works were carried out by trenching and costeaning, and by driving levels, but these failed in every case to reveal the existence of an auriferous vein. Streaming operations, carried on at the same time in the branches of the Gold-mine river and in adjacent streams, yielded 388 ounces 6 dwts. 16¾ grains, valued at £1,528 12s. 11¼d. so that the total quantity of gold recovered by the Government operations was 944 ounces 4 dwts. 15 grains, of the value of £3,675 7s. 11½d.

In 1803, the difficulty of working increasing and the auriferous gravels becoming poorer, the directors advised an abandonment of operations, to which recommendation effect was immediately given. For some time afterwards, a guard of militia was stationed at the spot, and on their withdrawal, the peasants again invaded the stream and for a time vigorously carried on sluicing and washing operations. It has been estimated that not less than £2,000 were annually obtained by them for some years, but this is probably far too high an estimate, since the carefully considered exploitation by Messrs. King, Weaver & Mills in earlier years had failed to yield anything like this amount.

In 1840, Messrs. Crockford & Company obtained the rights to the auriferous deposits and worked them energetically for a period of nearly 4 months, during which time they obtained no less than £1,800 worth of gold, including one nugget of 11 ounces and another of 4 ounces 12 dwts. 12 grains. It seems curious that, notwithstanding this apparently profitable return, the enterprise should have been abandoned so quickly, and that, if we except more or less surreptitious working by the peasants, no

* _Transactions of the Geological Society_, 1819, vol. v., page 207.

attempt was made to recover the gold of the Gold-mine river until 1862, when the Carysfort Mining Company leased the gold-royalties. This company appears to have devoted its attention rather to the discovery of auriferous veins in the neighbourhood, than to the working of the stream-gravels; but its operations in the former respect were no more successful than were those of Messrs. King, Weaver & Mills in the early years of the century. The company ceased active work in 1865, having obtained only £203 5s. worth of gold. Since that time, with the exception of some desultory streaming by Mr. F. Acheson from 1876 to 1879, no work has been done on these gravels.

Fig. 22.—Eastern Stream, Gold-mine River, looking Northward.

Table VI., prepared by Mr. G. H. Kinahan,[*] shows the gold-production of the Ovoca gravels. The total produce since 1795 is, therefore, estimated at from 7,440 to 9,390 ounces; and a value between £28,855 and £36,185; but, as will be seen, the estimated amount bulks very largely in these totals and the total produce is probably much less than that stated.

* "The Mode of Occurrence and Winning of Gold in Ireland," *Journal of the Royal Geological Society of Ireland*, vol. vi., page 147.

TABLE VI.—GOLD-PRODUCTION OF THE OVOCA GRAVELS.

Worked by.	Duration of Works.	Gold.			Authorities.
		Ounces.	Value.	Largest Nuggets.	
Peasants .	To October 15th, 1795	From 800 to 2,666	£ s. d. 3,000 0 0 to 10,000 0 0	ozs. dwts. grs. 22 0 0; 20 2 21; 5 0 0; 2 17 0	Mills. R. Molesworth. Mills. Do.
Government	August 12th, 1796, to May 26th, 1798	ozs. dwts. grs. 555 17 22¼	2,146 15 0	18 0 0; 9 0 0	Reports to the Government by Messrs. King, Weaver & Mills.
	Sept. 8th, 1800, to June 24th, 1801	43 9 10	112 14 11½	7 0 0; 2 10 0	
	June, 1801, to 1803	344 17 6¾	1,415 18 0	3 2 19	Royal Dublin Society, nugget.
Peasants ...	To 1839 ...	5,000 0 0	20,000 0 0	4 10 0	Estimated in 1840.
Crockford & Company	4 months in 1840	600 0 0	1,800 0 0	11 0 0	Returns to H.M. Commissioners of Woods and Forests.
Peasants	To 1857 ...	?	?	24 0 0; 6 0 0	Hugh McDermott, Arklow.
Carysfort Mining Company	1862 to 1866 ...	52 11 5	203 6 0	0 13 8; 0 4 9; 0 3 11	Geological Survey, Dublin, Director's Report.
Mr. F. Acheson	1876 to 1879	14 3 0	60 0 0	?	Mineral Statistics.

Though, so far as is yet known, no other auriferous deposits of economic value occur in Ireland, the presence of gold has been detected in various places, both in veins and in alluvial sands. Under the latter head are the sands of the Glendun river, county Antrim, which enters the sea at Cushendun, and flows from the flanks of Slieve-an-Orra;* the sands of the Dodder river above Rathfarnham, which yielded the two small nuggets picked up many years ago on Stephen's Green, Dublin;† Balliscorney Gap, county Dublin;‡ and the " black sand " deposit near Greystones, county Wicklow. This last deposit appears to have resulted from the concentration, by wind and by wave-action, of the heavier constituents of the drift-sands that are here exposed on the beach. It extended along the beach for several hundreds of feet and was,

* "The Mode of Occurrence and Winning of Gold in Ireland," *Journal of the Royal Geological Society of Ireland*, vol. vi., page 140.

† *Ibid.*   ‡ *Ibid.*

when examined, several inches in thickness.   It contained 21·5 per cent. of magnetic material (magnetite, chromite and ilmenite), together with red and brown hæmatite, iron-pyrites, rutile, cassiterite and garnets.   On washing and panning 7½ pounds of black sand, 37 colours of very finely divided gold were left in the dish. Indeed, "gold was found in small quantities in all the specimens of black sand taken from the beach."[*]

Gold, *in situ*, has also been reported from several localities, namely, from Bray Head, county Wicklow,[†] from the gossan of the Dhurode copper-lode, Carrigacat, county Cork;[‡] and from

FIG. 23.—LOOKING UP THE EASTERN STREAM, FROM THE JUNCTION.

the pyrites and gossan of the mineral lodes in the Ballymurtagh, Cronebane and Connary mines in the Vale of Ovoca, several miles to the north of the Gold-mine river.   The pyrites-lodes of the last mentioned district have long been known to carry a small quantity of gold.   Mr. Thomas Weaver[§] states that the silver from Cronebane contained 6¼ per cent. of gold, and recent assays of that complex sulphide (kilmacooite) from Connary, and of copper-pyrites

* " Black Sand in the Drift north of Greystone, County Wicklow," by Mr. Gerrard A. Kinahan, *Journal of the Royal Geological Society of Ireland*, 1881, vol. vi., page 113.

† " The Mode of Occurrence and Winning of Gold in Ireland," by Mr. Gerrard A. Kinahan, *Journal of the Royal Geological Society of Ireland*, 1882, vol. vi., page 138.

‡ *Ibid.*

§ *Transactions of the Geological Society of London*, 1819, vol. v., page 207.

from the Ballygahan copper-mine show in each case 12 grains of gold per ton.*

Assays for gold in the rocks in the neighbourhood of Dublin have been made by, or rather for, Prof. J. P. O'Reilly from time to time, with somewhat astonishing results. A greenstone, or diorite, from the quarries about Montpelier and Bohernabreena, county Dublin, and used for the paving of Dublin streets yielded at the rate of 2 dwts. of gold per ton.† From this single assay, it is ingenuously estimated that the quantity of gold laid down since 1851, in the streets of Dublin is worth £196,224. At the same time, it must not be forgotten that these are the rocks over which lies the course of the Dodder river, from the sands of which, as has already been seen, two small nuggets were obtained. It is also of importance to note that these rocks "nearly always contain pyrites," and therefore have been subjected to a secondary impregnation. Iron-pyrites from lenticular diorite masses in the Lower Silurian clay-slate of Crookling Hill, Brittas valley, county Dublin, yielded to the same assayer 6 grains of gold per ton. The diorite-rock itself, on analysis, yielded a maximum of 2 grains of gold per ton. Traces of gold were also found in the rocks from various other places. In the absence of details as to the method of conducting the assay and of ensuring the purity of the fluxes used, and regard being had to the nature of the rock analysed, it would appear advisable to receive the above results with some degree of caution.

From the foregoing it will be at once apparent that the only occurrence worthy of detailed description is that of the Gold-mine and adjacent valleys in Wicklow (Fig. 9, Plate XIX.).

The Gold-mine river flows into the Aughrim river, at Woodenbridge, just above the junction of the latter with the Ovoca. Its sources are on the southern and eastern slopes of Croghan Kinshelagh mountain, 1987 feet high, and the highest eminence in the vicinity. For the greater part of its course it flows through a miniature ravine, with steep, well-wooded sides, excavated for itself in the slaty rocks. These narrow trench-like valleys are characteristic, not only of the tributary streams, but also of the

---

* "Prospecting for Gold, etc.," by Mr. E. St. J. Lyburn, *Scientific Proceedings. of the Royal Dublin Society*, 1901, new series, vol. ix.. page 422.

† *Scientific Proceedings of the Royal Dublin Society*, 1890, new series, vol. vi., pages 450 to 459.

main rivers—the Aughrim and the Ovoca. About ¾ mile above the confluence of the Gold-mine river with the Aughrim, the former is augmented in volume by the Eastern stream, also auriferous. All the other auriferous streams lie to the west of the Gold-mine river and are tributaries of the Aughrim. They are the Ballin-temple, 1 mile above Woodenbridge; the Clone; and the Cool-bawn, which flows also from the slopes of Croghan Kinshelagh but to the north-west. Gold in small quantities has also been reported from the Ballythomas stream, still further to the west.

All these streams run mainly through an area of Lower Silurian (or Cambro-Silurian) grey, green and dark slates; sandy shales;

FIG. 24.— LYRA, AT THE FORKS OF THE EASTERN AND WESTERN STREAMS.

and grits, belonging probably both to the Caradoc and to the Llandeilo beds. They have, in the district under discussion, a general north-east to south-west strike and a dip southeastward of 70 to 80 degrees. One of the best exposures of the sedimentary rocks is seen in an abandoned quarry on the Eastern auriferous stream. Here the rocks are very fissile, dark blue and brown slates, showing a thickness in section varying from 40 to 50 feet. In places, the slates show evidences of extensive exposure to meta-morphic agencies.

To the west and north-west of the Gold-mine river, and form-ing the high lands of the Croghan Kinshelagh (1,987 feet),

Monayteigue (1,892 feet) and Ballycoog (1,169 feet) hills, occurs a great development of plutonic and volcanic rocks. Both appear at the surface with outcrops elongated in a general north-easterly and south-westerly direction. The plutonic rocks occur as narrow dykes or masses, and are essentially microgranites. They are best developed on the western and south-western flanks of Croghan Kinshelagh. According to Dr. Hatch* "they are essentially granitoid rocks of rather fine grain. In colour the specimens from the large masses vary from a creamy-white to a yellowish-brown; but those obtained from the dykes are of a more bluish-white or grey. They are usually spotted over with small aggregations of a dark mineral (biotite, sometimes altered to chlorite). Examination under the microscope shows that these rocks consist essentially of quartz, orthoclase and plagio-clase, together with a small quantity of mica, in holocrystalline and granitic aggregation. Sphene occurs as an accessory con-stituent in small grains." In the microgranite of Croghan Kinshelagh there appears a micropegmatitic or granophyric arrangement of the quartz and felspar.

The volcanic rocks are developed farther to the north-east than the microgranites, but preserve in the outline of their exposures the same general north-east to south-west elongation noted in the case of the microgranites. They are mainly epi-diorites, quartz- and augite-diorites, and dolerites. The main mass of Croghan Kinshelagh is an epidiorite†—dolerite altered by contact or dynamic metamorphism or perhaps by a combina-tion of both. The special feature of the epidiorite is the com-plete paramorphism of the augites of the dolerite to hornblende.

The gold of the Croghan Kinshelagh area is in all cases found in the gravels in the beds of the streams. Since, in Wicklow, the river-valleys are extremely narrow and correspondingly deep, it follows that concentration of the gravels has been restricted, at any rate, since the initiation of the present valley-system, to the well-defined lines represented by the present courses of the streams; and also that the fluviatile plains, or even the small flats, so characteristic of the alluvial deposits of other countries, are here absent. Under these conditions, the maximum of mechanical concentration may be expected. The gravels con-

* *Memoirs of the Geological Survey of Ireland, Explanatory Memoir to accompany Sheets 138 and 139*, page 53.   † *Ibid.*

tain numerous rounded boulders of the plutonic and volcanic rocks—the microgranites and the epidiorites—of Croghan Kinshelagh. The richest deposit is here, as in most alluvial goldfields, at the base of the gravels, and rests immediately on the underlying Lower Silurian slates.

The concentration of heavier metals and minerals at the base of such deposits is not due to the gravels of the stream having been deposited over already concentrated sands, but to the natural tendency of the heavier minerals to settle in obedience to the force of gravity.* According to Mr. G. A. Kinahan, who saw those workings when exposed by Mr. Acheson, the auriferous

FIG. 25.—LOOKING UP BALLINVALLEY STREAM, FROM THE RED HOLE.

black sands were disposed in parallel lines along the master-joints of the slate, which apparently were coincident with the course of the stream.† Where cross-joints, transverse to the course of the stream, occurred, there was generally a concentration of black sand ; and as both master-joints and cross-joints are regular and

* A striking illustration of this property recently came under the writer's observation. A large phial, containing osmiridium, platinum and gold, mixed with sand, from the Hunter river, New South Wales, had lain for many years in a mineral cabinet in London, and subject to repeated vibrations from passing traffic. On inspection, it was found that the heavier minerals had concentrated on the lower side of the phial to such an extent as to form a clearly defined zone much darker in colour than the overlying sand. In water, this action is, of course, facilitated by the buoyant action of the water on the sand-grains.

† *Manual of the Geology of Ireland*, by Mr. G. H. Kinahan, page 344.

frequent in occurrence, the gold is found at points disposed after a regular pattern, at the intersections of the lines of a chessboard.   These joints also acted as natural riffles, as is apparent from the fact that the gravels were richest where the Ballin-valley stream crossed the strike of the rocks.   The gravels were naturally also more productive where, from any cause, such as a bend of the stream, a reduction of its slope, or the confluence of two streams, the velocity of the current was checked.   In the very early days of the workings, it was recognized that the slack water behind a large boulder was favourable to the deposition of heavy minerals; and when such were found in the course of

Fig. 26. – The Red Hole and Ballinvalley Stream, looking Northward.

working the gravels, rich deposits were generally expected on the lower or downstream side.

The black sand is composed mainly of magnetite, ilmenite, hæmatite and iron-pyrites, but cassiterite, galena, wolfram, molybdenite, gold, copper-pyrites and oxides of manganese also occur in the sand.   The late Mr. W. Mallet* records having obtained from a washing of 150 pounds no less than $3\frac{1}{2}$ pounds of stream-tin, in all sizes from small grains up to pebbles $\frac{1}{2}$ inch in diameter, and of the variety known as wood-tin.

The gold of the gravels is generally in fine grains, presenting evidences of considerable attrition, especially in the lower por-

* *Philosophical Magazine*, 1850, vol. xxxvii., page 392.

tions of the streams. Mr. Thomas Weaver,[*] however, noted gold "crystallized in octahedrons, and also in elongated garnet dodecahedrons," and "frosted" or crystallized gold has been remarked by various observers from the upper portions of the valleys.

The biggest nugget found at Wicklow was picked up by a party of peasants, in or about September, 1795, and weighed 22 ounces. Its subsequent history is somewhat obscure, but it would appear that it was bought by Mr. Turner Carnac for £80 12s. for presentation to George III., who had the nugget, or part of it, made into a snuff-box. A smaller nugget of 4 ounces 8 dwts., was for long in the Royal Dublin Society's museum, but was stolen in 1865. Fourteen other nuggets from Wicklow, ranging in weight from 3 ounces 2 dwts. 14½ grains, to 1 dwt. 4 grains are recorded from the Dublin Museum of Science and Art, from the Edinburgh Museum of Science and Art, and from the Jermyn Street museum.

The following are various assays of Wicklow gold:—

No of Sample.	Gold.	Silver.	Iron.	Copper.	Silica.	Totals.	Specific Gravity.	Authority.
1	94·06	5·94	—	—	—	100·00	12	Weaver.
2	90·62	7·82	1·56	—	—	100·00	—	Alchorne, Mint-master.
3	92·32	6·17	0·78	—	—	99·27	16·342	Small grains by Wm. Mallet.
4	91·01	8·85	—	—	0·14	100·00	14·34 } 15·07 }	D. Forbes.
5	89·00	8·10	2·10	trace	—	99·20	—	Scott.

Disregarding No. 1 sample, the low specific gravity of which was due to the cellular nature of the nugget, the average fineness is 90·74 in gold and 7·73 in silver, representing at the present time a value of £3 17s. 3d. per ounce.

The richest deposit appears to have occurred in the upper course of the Western auriferous stream (also called, by Mr. Thomas Weaver, the Ballinvalley stream) about ½ mile below Ballinagore bridge. Here, at the Red Hole, and for some 1,200 feet below, the most remunerative results were obtained by the peasants, by Messrs. King, Weaver & Mills, and by all later workers. From the Red Hole, the old workings extend down stream nearly to the junction with the Eastern stream. The

[*] *Transactions of the Geological Society of London*, 1819, vol. v., page 207.

Eastern stream, especially below the confluence of the Killa-
hurler brook and the stream running down from Slieve Foore
(1,356 feet), carries a fair quantity of gold, as also do the streams
coming in on the right bank below the deserted quarry. At Lyra,
the junction of the Eastern and Western streams, a rich deposit
was found containing much coarse gold. This was perhaps only
to be expected, for at Lyra is to be found the first alluvial flat
that occurs in the downward course of either stream. Below
Lyra, and as far as Rostigah, the gravels of the main stream were
productive; but, below Rostigah, they became too poor, and the
overburden proved too heavy to work.

Fig. 27.—Ballinvalley Stream, showing the Ravine.

The Ballintemple brook, flowing into the Aughrim from the
north-western flank of Croghan Kinshelagh, was worked by
Messrs. Crockford & Company, and subsequently by the Carysfort
Mining Company, in both cases yielding gold, both fine and
coarse. In the Coolbawn stream, flowing northward to the
Aughrim from Croghan Kinshelagh, Mr. Thomas Weaver found
a 2½ ounces nugget—the largest discovered anywhere except in
the Ballinvalley stream.

On the source of the alluvial gold of Wicklow, there is much
room for speculation. The results of all Mr. Thomas Weaver's,
Messrs. Crockford & Company's and the Carysfort Mining Company's examinations of the quartz-veins of the vicinity would

point to a source far removed from Croghan Kinshelagh as a matrix for the gold. Mr. Thomas Weaver, as has already been indicated, vigorously prosecuted a search in the vicinity for auriferous veins. Struck by the fact that on the three main branches of the Gold-mine river, the gold ceased at points which were in a straight line transversely crossing the streams, and pointing to the existence of the vein-matrix along that line, he trenched, costeaned and drove levels in various directions, but without being rewarded by the glimpse of the smallest speck of gold; and that although some fifty or sixty quartz-veins were intersected in a single level, driven 1,068 feet to the north-west at Ballinagore. Costeaning by Messrs. Crockford & Company and by the Carysfort Mining Company was no more successful, the latter company, according to Capt. Philip Argall, having crushed 300 or 400 tons of quartz from Monayteigue and Ballintemple, without obtaining a single particle of gold.

Notwithstanding the above evidence, it appears to the writer that the matrix of the gold is, or more probably, was, in the immediate vicinity of Croghan Kinshelagh, and that the present auriferous deposit in fact represents the concentrates of a pyritous lode which has suffered degradation. It need not necessarily have been, and indeed was probably not, rich. The enormous amount of concentration that has taken place in these valleys—their narrowness and the condition of the bottom admirably adapting them to play the part of natural sluices—would certainly point either to poverty, or to scarcity of auriferous quartz. A further piece of evidence, though not an important one, as to an immediately local origin is the occurrence of fragments of quartz containing gold. By far the most important corroboration, however, is that afforded by Mr. E. St. John Lyburn, who in April, 1899, assayed quartz from a vein, 8 inches wide, and in the immediate vicinity of the old Government workings, with a yield at the rate of 4 dwts. of gold to the ton.* The assays appear to have been performed with every care, the fluxes —a fruitful source of error—having been previously assayed with negative results. The assays were also performed in this case in triplicate. Further assays from the quartz-veins in the immediate vicinity of the Gold-mine valley yielded, in every case, traces of gold.

* *Scientific Proceedings of the Royal Dublin Society*, 1901, new series, vol. ix., page 426.

It has been hinted, rather than suggested by various writers, that the matrix of the alluvial gold may have been the outcrops of the pyrites-lodes of Ballymurtagh, Cronebane and Connary, which are known to be auriferous;[*] and that, either by fluviatile or by glacial action the precious metal has been transported to its present position.   But a consideration of the physiographical features of the district would appear to preclude this source and these modes of transport.   From the fact that the observed glacial striae are, generally speaking, parallel with the courses of the present streams, it would appear that the present valley-system was initiated anterior to the occupation of the valleys by

FIG. 28.—THE GOLD-MINE RIVER, AT LYRA.

glaciers.   The natural course, therefore, at Woodenbridge, of fluviatile or glacial débris from the Upper Ovoca was, not to the south-west to the Gold-mine river, but to the east with the valley of the Ovoca; and it is indeed difficult to see how, in any case, detritus from the Ovoca could be deposited in the Gold-mine valley.   Moreover, assuming that the sources of the gold were the pyrites-lodes of Cronebane, auriferous deposits intermediate in position between Cronebane and Woodenbridge would naturally be expected.

[*] At a Swansea copper-works, 3 tons of residues from Wicklow pyrites, after treatment with sulphuric acid, assayed 8 ounces of gold and 154 ounces of silver to the ton, or a value of £74 per ton. The Wicklow pyrites furnished by Messrs. Williams, Foster & Co., contains ¼ to ½ ounce of gold per ton.—*Mining Journal*, 1849, page   .

Consideration of the conditions under which gold is found in Wicklow does not favour the assumption of yet undiscovered rich deposits, to be profitably worked in the future.   As has already been pointed out, the parent-veins, if portions of them still exist, are in all probability too poor to be worked.   The narrow, trench-like nature of the auriferous valleys would also indicate the improbability and indeed the impossibility of extensive high-level auriferous gravels, for high-level terraces demand a broad valley with more or less sloping sides.   The hope of payable gravels on the broad hills between the valleys is rendered slender by the presence of the glacial striae indicating, of course, a complete sweeping of the country by ice. Finally, it would appear that the best hope of future profitable working lies in the deep gravels that cover the floor of the lower portion of the Gold-mine valley and of the Ovoca below Woodenbridge. So far as the writer is aware, no attempts have been made in this direction and the potentialities of these gravels are absolutely unknown.

CONCLUSION.

In reviewing the geological distribution of the known auriferous veins of Great Britain (Merioneth, Leadhills, North Molton and Cornwall) the most striking feature is their more or less intimate connection with the older Palæozoic rocks. Further, in the case of the alluvial auriferous deposits of Sutherland and Wicklow, where the parent-veins have not been located; the available evidence leads to the inference that these veins also are, or have been, located in Lower Silurian areas. And an inspection of the various auriferous localities described has forced the writer, somewhat against his preconceived ideas, to the belief that it is in the older Palæozoic sedimentary rocks, and not in the often associated igneous intrusions, that the former locus of the gold, which is now found in veins or in alluvial deposits, must be placed.   On no other supposition can the restriction of the auriferous veins in geological time be so satisfactorily accounted for, as by the assumption that many of the sandstones and shales of the Cambrian and Silurian seas were derived from the denudation of an auriferous area, the metallic contents of the derived sedimentary rocks being subsequently leached into, and concentrated in convenient fissures.

To connect the various auriferous occurrences with the presence of local igneous rocks would necessitate the granting of so many postulates, themselves of dubious value in the present state of knowledge of the conditions of ore-deposition, as entirely to deprive the suggestion of any scientific value. Moreover, the igneous rocks present in the various areas differ widely in petrological character and there are many areas in which the same conjunction of igneous and sedimentary rock may be found, but which are yet not auriferous.

Although it cannot be granted that the igneous rocks have in this case yielded the gold of the veins, yet it is very probable that the local heat furnished by their intrusion caused the formation and circulation of the solvent agencies that leached the gold from the sedimentary rocks and deposited it in the fissures.

The total yield of gold in Great Britain and Ireland may, with the exception of that from the Leadhills, be computed with sufficient approach to accuracy, to give a considerable degree of value to the estimation. The yield for each country has already been dealt with under its respective heading and is as follows : —

			£
England : North Molton	...	...	581
Wales : since 1844	... ...	...	280,547
Scotland : Leadhills	... ...	...	100,000
Sutherland (1868-1869)		...	3,000
Ireland	... ...	...	28,855
	Total	...	£412,983

Of this sum, almost all, except that from the Leadhills, has been obtained during the nineteenth century, and nearly half of it (£208,855) during the last 14 years. The outlook for the present century is by no means so promising. Little or nothing may be, for reasons already set forth, expected from Wicklow, Sutherland, Leadhills or Cornwall. From the auriferous district of North Wales, the outlook is much more hopeful. The veins here yield somewhat high-grade bonanzas, and while they must be regarded always as speculations rather than as commercial investments, yet there is no reason why (provided always that a large reserve-fund is set aside when a bonanza is encountered, in order to carry on that vigorous prospecting which should be the distinctive feature of " patchy " mines) their working should not result in commercial success. With water-power, cheap labour, the absence of serious difficulties in metallurgical treatment, and exceptional facilities for mining, there seems no reason why the present low cost of treatment, already under 9s. per ton, should not be yet further reduced. But the prime factor in the successful working of the Welsh gold-mines must be the recognition of the axiom that the more patchy the vein, the greater the necessity for building up a large reserve-fund to be devoted to prospecting for further bonanzas; and also that the exploratory work must be carried on at the same time as the working of the bonanza.

APPENDIX I.—BIBLIOGRAPHY.

England and Wales.

1602. Carew, Richard, of Antonie, *Survey of Cornwall*, book i., London.

1707. Wyat, John, *Some Account of the Mines, and the Advantages of them to this Kingdom*, printed by W. B., for John Wyat, at the Rose, in St. Paul's Churchyard, London.

1818. Hawkins, Sir Christopher, Bart., "Observations on Gold found in the Tin-stream Works in Cornwall," communicated in a letter to John Ayrton, Paris, *Transactions of the Royal Geological Society of Cornwall*, vol. i., page 235.

1839. De la Beche, Sir Henry Thomas, *Report on the Geology of Devon, Cornwall and West Somerset*, page 613.

1839. Murchison, Sir Roderick Impey, *The Silurian System*, part i., page 368.

1844. Dean, Arthur, "Note respecting the Discovery of Gold-ores in Merionethshire, North Wales," *Report of the British Association for the Advancement of Science*, Transactions of the Sections, York, page 56.

1846. Smyth, Sir Warington W., "Note on the Gogofau or Ogofau Mine," *Memoirs of the Geological Survey of Great Britain and of the Museum of Economic Geology of London*, vol. i., page 480.

1853. Pattison, S. R., "A Day in the North Devon Mineral District," *Transactions of the Royal Geological Society of Cornwall*, vol. vii., page 223.

1853. Calvert, John, *The Gold-rocks of Great Britain and Ireland, and a General Outline of the Gold-regions of the World, with a Treatise on the Geology of Gold*, octavo, London.

1854. Ramsay, Sir Andrew Crombie, "On the Geology of the Gold-bearing District of Merionethshire," *Quarterly Journal of the Geological Society of London*, vol. x., page 242.

1854. Pattison, S. R., "On Auriferous Quartz-rock in North Cornwall," *Quarterly Journal of the Geological Society of London*, vol. x., page 247.

1854. Murchison, Sir Roderick Impey, "Gold-deposits in Great Britain," *Mining Journal*, London, vol. xxiv., page 717.

1858. Ansted, D. T., "Gold in Wales," *Mining Journal*, London, vol. xxviii., page 241.

1858. Ramsay, Sir Andrew C., "The Physical Structure of Merionethshire and Caernarvonshire," *The Geologist*, vol. i., page 169.

1860. Readwin, T. A., "The Gold Discoveries in Merionethshire," *Transactions of the Manchester Geological Society*, vol. ii., part 9, page 97.

1861. Smyth, Sir Warington W., "Gold-mining at Clogau, North Wales," *Mining and Smelting Magazine*, vol. i., page 359.

1861. Readwin, T. A., "On the Gold of North Wales," *Report of the British Association for the Advancement of Science*, Transactions of the Sections, page 129.

1862. —, "On the Gold-bearing Strata of Merionethshire," *Report of the British Association for the Advancement of Science*, Transactions of the Sections, page 87.

1863. —, "On the Recent Discovery of Gold near Bala Lake, Merionethshire," *Report of the British Association for the Advancement of Science*, Transactions of the Sections, page 86.

1865. Hunt, Robert, "British Gold," *Quarterly Journal of Science*, vol. ii., page 635.

1865. Salter, J. W., "Explanation of a Map of the Faults in the Gold-district of Dolgelly," *Report of the British Association for the Advancement of Science*, Transactions of the Sections, page 73.

1866. Ramsay, Sir Andrew C., "The Geology of North Wales." *Memoirs of the Geological Survey of the United Kingdom and of the Museum of Practical Geology*, vol. iii.

1866. Plant, J., and E. Williamson, "Fossils of the *Lingula*-Flags," *Transactions of the Manchester Geological Society*, vol. v., pages 76 and 220.

1867. Forbes, D., "Researches in British Mineralogy," *London, Edinburgh and Dublin Philosophical Magazine, and Journal of Science*, fourth series, vol. xxxiv., page 338.

1867. Murchison, Sir Roderick Impey, *Siluria*, fourth edition, pages 449 and 546.

1868. Forbes, D., "Gold from the Clogau Quartz-lode, No. 2," *Geological Magazine*, vol. v., page 224.

1868. —, "Stream-gold from the River Mawddach," *Ibid.*, page 225.

1869. —, "Researches in British Mineralogy," *London, Edinburgh and Dublin Philosophical Magazine, and Journal of Science*, fourth series, vol. xxxvii., page 321.

1870. Readwin, T. A., "Notes on a Merionethshire Gold-quartz Crystal, and some Stream-gold recently found in the River Mawddach," *Report of the British Association for the Advancement of Science*, Transactions of the Sections, page 84.

1870. Smyth, Sir Warrington W., "Occurrence of Gold in North Wales," *Transactions of the Royal Geological Society of Cornwall*, vol.   , page   ; and *Mining Journal*, vol. xl., page 940.

1875. Readwin, T. A., *Dr. Ure's Dictionary of Arts, Manufactures and Mines*, seventh edition, vol. ii., page 690.

1875. —, "Gold in Wales," *Mining Journal*, vol. xlv., pages 844, 929, 1042, 1096, 1208, 1236, 1292, 1319, 1347, 1404 and 1431.

1876. —, "Gold in Wales," *Mining Journal*, vol. xlvi., pages 20, 48, 73, 126, 152, 180, 233, 261, 289, 318, 345, 374 and 401.

1877. Hall, T. M., *Geology and Mineralogy of North Devon*.

1879. Readwin, T. A., "Notes on some Minerals of the Mawddach Valley," *Mineralogical Magazine and Journal of the Mineralogical Society of Great Britain and Ireland*, vol. iii., page 122.

1880. —, *Gold in Wales: the Mawddach Valley, Merionethshire*, octavo.

1880. —, "Gold in Wales," *Mining Journal*, vol. l., pages 848, 877, 904, 931 and 988.

1884. —, "Note on Welsh Gold," *Mineralogical Magazine and Journal of the Mineralogical Society*, vol. vi., page 108.

1888. Vanderbilt, A. T., *Gold not only in Wales, but in Great Britain and Ireland*, London, 1888.

1895. Smith, E. A., "Gold in the British Isles," *Knowledge*, London, vol. xviii., page 33.

1896. Louis, Henry, *A Treatise on Ore-deposits*, by J. A. Phillips, second edition, rewritten and greatly enlarged, pages 190 and 292.

Frequent references to the gold-mines of Great Britain and Ireland are to be found in the *Mining Journal*, from 1840 to the present time.

SCOTLAND.

1769-1772. Pennant, Thomas, *A Tour in Scotland*, 1769, *and Voyage to the Hebrides*, 1772, Warrington, 1774-1776.

1805. Jameson, R., *A System of Mineralogy*, vol. ii., page 108.

1825. Atkinsoune, Stevine, *Discoverie and Historie of the Gold Mynes in Scotland*, Edinburgh (Reprint by the Bannatyne Club, 1825).

1852. Harkness, R., "On the Silurian Rocks of the South of Scotland, and on the Gold-districts of Wanlockhead and the Leadhills," *Quarterly Journal of the Geological Society of London*, vol. viii., page 393.

1860. Thost, C. H. G., "On the Rocks, Ores, and Other Minerals on the Property of the Marquess of Breadalbane in the Highlands of Scotland," *Quarterly Journal of the Geological Society of London*, vol. xvi., page 424.

1865. Irving, G. V., *The Upper Ward of Lanarkshire*.

1867. Lindsay, Dr. W. Lauder, "On the Gold-fields of Scotland," *Report of the British Association for the Advancement of Science*, Transactions of the Sections, page 64.

1867. —, "The Gold and Gold-fields of Scotland," *Transactions of the Edinburgh Geological Society*, vol. i., page 105.

1869. —, "The Gold and Gold-fields of Scotland," *Journal of the Royal Geological Society of Ireland*, new series, vol. ii., page 176.

1869. —, "The Fifeshire Gold-diggings of 1852," *Transactions of the Edinburgh Geological Society*, vol. i., page 272.

1869. —, "Preliminary Report on the Sutherlandshire Gold," *Transactions of the Edinburgh Geological Society*, vol i., page 273.

1869. Campbell, John, of Islay, "Something from the Gold-diggings in Sutherlandshire," by the Author of "Frost and Fire," *Odds and Ends Series*, No. 22, duodecimo, Edinburgh.

1869. Forbes, David, "Researches in British Mineralogy," *London, Edinburgh and Dublin Philosophical Magazine*, fourth series, vol. xxxvii., page 321.

1869. Joass, Rev. John M., "Notes on the Sutherlandshire Gold-field, with an introduction by Sir R. I. Murchison," *Quarterly Journal of the Geological Society of London*, vol. xxv., page 314.

1870. Lindsay, Dr. W. Lauder, "On the Gold-fields of Forfarshire," *Transactions of the Edinburgh Geological Society*, vol. ii., page 27.

1870. —, "The Gold and Gold-fields of Perthshire," *Proceedings of the Perthshire Society of Natural Science*, 1869-1870, page 35.

1870. Bryce, Dr. James, "On the Matrix of the Gold in the Scottish Gold-fields," *Report of the British Association for the Advancement of Science*, Transactions of the Sections, page 70.

1870. Cameron, W., "On the Sutherlandshire Gold-fields," *Transactions of the Geological Society of Glasgow*, vol. iv., page 1 ; abstract, *Geological Magazine*, vol. vii., page 139.

1874. Church, A. H., "Analyses of Scotch Gold [from Wanlockhead and Sutherlandshire]," *Chemical News*, vol. xxix., page 209.

1874. Lindsay, Dr. W. Lauder, "Recent Gold Discoveries in Scotland," *Scottish Naturalist*, vol. , page .

1875. —, "Auriferous Quartzites of Scotland," *Scottish Naturalist*, vol. iii., pages 46 and 177.

1876. Porteous, Dr. Moir, *God's Treasure House in Scotland.*

1876. Dudgeon, Patrick, of Cargen, "Historical Notes on the Occurrence of Gold in the South of Scotland," *Mineralogical Magazine and Journal of the Mineralogical Society of Great Britain and Ireland*, vol. i., page 21.

1877. Lindsay, Dr. W. Lauder, "Museum Specimens of Native Gold," *Transactions of the Edinburgh Geological Society*, vol. iii, page 153.

1878. —, "Australian Gold-quartz in Scotland," *Transactions of the Geological Society of Glasgow*, vol. vi., pages 68 and 131.

1878. Cochran-Patrick, R. W., of Woodside, *Early Records relating to Mining in Scotland*, Edinburgh.

1878. Lindsay, Dr. W. Lauder, "Gold-field and Gold-diggings of Crawford Lindsay," *Scottish Naturalist*, vol iv., pages 208, 211, 256 and 355.

1880. French, A., "On a peculiar pasty Form of Silica, from a Cavity in Gold-bearing Quartz," *Mineralogical Magazine and Journal of the Mineralogical Society of Great Britain and Ireland*, vol. iv., page 42.

1884. Heddle, M. Forster, "The Geognosy and Mineralogy of Scotland," part vi., *Ibid.*, vol. v., page 271.

1884. Hunter, J. R. S., "The Silurian Districts of Leadhills and Wanlockhead, and their early and recent Mining History," *Transactions of the Geological Society of Glasgow*, vol. vii., pages 373 and 433.

1895. Greenly, E., "Notes on the Sutherland Gold-field," *Transactions of the Edinburgh Geological Society*, vol. vii., page 100.

1896. Cadell, H. M., *The Geology of Sutherland*, Edinburgh, second edition, page 95.

1899. Peach, B. N., and J. Horne, *Memoirs of the Geological Survey of the United Kingdom*, "Silurian Rocks of Great Britain," vol. i., Scotland, page 656.

1901. Maclaren, J. M., "The Source of the Alluvial Gold of the Kildonan Field, Sutherland," *Report of the British Association for the Advancement of Science*, Transactions of the Sections, Bradford, page 651.

1901. Heddle, M. Forster, *Mineralogy of Scotland*, vol. ii.

IRELAND.

B.C. 348. *Annals of the Four Masters.*

1652. Boate, Gerard, M.D., *Ireland's Natural History*, S. Hartlib, London.

1749. Simon, James, *An Essay towards an Historical Account of Irish Coins*, Dublin, page 2.

1786-1790. Vallancey, Colonel Charles, *Collectanea de Rebus Hibernicis*, vol. vi.

1795-1796. Coquebert, Charles, "Notice sur les Mines de Cuivre de Cronbane et Ballymurtagh dans le Comté de Wicklow," *Journal des Mines*, vol. iii., part xvi., page 82.

1795. Lloyd, J., "An Account of the Late Discovery of Native Gold in Ireland," *Philosophical Transactions of the Royal Society of London*, volume for 1796, page 34.

1795. Mills. A., "A Mineralogical Account of the Native Gold lately discovered in Ireland," *Ibid.*, page 38.

1796. Anon., "Gold-nugget from Wicklow," *Gentleman's Magazine*, part i., page 8.

1797. An Act To enable the Lords Commissioners of His Majesty's Treasury to conduct the Workings of a Gold-mine in the County of Wicklow, received the Royal Assent on April 24th, 1797.

1800. Mills, A., "Mineralogical Account of the Native Gold lately discovered in Ireland," with a Map, *Transactions of the Royal Dublin Society*, vol. ii., part i., page 454.

1801. Kirwan, R., "Report on the Gold-mines in the County of Wicklow, with observations thereon," *Ibid.*, vol. ii., part ii., page 129.

1801. Fraser, Robert, "General View of the Agriculture, Mineralogy, etc., of Wicklow," drawn up for the Dublin Society, Dublin.

1802. Mills, A., "Second Report on the Wicklow Gold-mines," *Transactions of the Royal Dublin Society*, vol. iii., page 81.

? Lloyd, Dr. Bartholomew, "Presidential Address," *Transactions of the Royal Dublin Society*, vol. ., page .

1809. Lloyd, — and A. Mills, *Discovery of Gold in Ireland.*

1819. Weaver, Thomas, "Memoir on the Geological Relations of the East of Ireland, with Map," *Transactions of the Geological Society of London*, vol. v., pages 117 and 207.

1827. Griffith, Sir Richard, "Report on the Metallic Mines in the Province of Leinster," *Transactions of the Royal Dublin Society*, vol. ., page .

1832. Geisecke, —, *Catalogue*, 1832.

1835. Weaver, Thomas, "On the Gold-workings formerly conducted in the County of Wicklow, Ireland," *London and Edinburgh Philosophical Magazine, and Journal of Science*, vol. vii., page 1.

1840. Anon., *Mining Journal*, vol. x., pages 30 and 326.

1841. Anon., *Ibid.*, vol. xi., pages 47 and 213.

1844. Kane, Sir Robert, *Industrial Resources of Ireland*, page 208.

1849. Anon., "The Gold-mines of Wicklow," *Mining Journal*, vol. xix., page 15.

1850. Mallet, William, "On the Minerals of the Auriferous Districts of Wicklow," *London, Edinburgh and Dublin Philosophical Magazine, and Journal of Science*, vol. xxxvii., page 392; and *Journal of the Geological Society of Dublin*, vol. iv., page 269.

1853. Smyth, Sir Warington W., "On the Mines of Wicklow and Wexford," *Records of the School of Mines and of Science applied to the Arts*, vol. i., page 389.

1855. Anon., *Mining Journal*, vol. xxv., page 782.

1856. Anon., *Mining Journal*, vol. xxvi., pages 585 and 653.

1860. Griffith, Sir Richard, "Catalogue of Irish Mines and Metalliferous Indications," *Journal of the Geological Society of Dublin*, vol. ix., page 145.

1865. Sanders, G. and J. K. Boswell, "The Gold-valleys of the County of Wicklow" (Discussion), *Transactions of the Royal Geological Society of Ireland*, new series, vol. i., pages 98 and 99.

1874. *Watt's Dictionary of Chemistry*, second supplement, page 572.

1878. Kinahan, G. H., *Manual of the Geology of Ireland*, pages 339 and 363; and *Quarterly Journal of Science*, vol. xv., page 189.

1879. Argall, Philip H., "Notes on the Ancient and Recent Mining Operations in the East Ovoca District," *Scientific Proceedings of the Royal Dublin Society*, new series, vol. ii., page 211.

1881. Kinahan, Gerrard A., "'Black and' in the Drift North of Greystones, County Wicklow," *Journal of the Royal Geological Society of Ireland*, new series, vol. vi., page 111.

1882. Kinahan, Gerrard A., "On the Mode of Occurrence and Winning of Gold in Ireland," *Ibid.*, new series, vol. vi., page 135.

1883. Kinahan, G. H., "On the Possibility of Gold being found in Quantity in County Wicklow," *Scientific Proceedings of the Royal Dublin Society*, new series, vol. iv., page 39.

1883. Joyce, P. W., *The Origin and History of Irish Names of Places*, vol. ii., page 360.

1886. Kinahan, G. H., "Economic Geology of Ireland: I.—Metal-mining," *Journal of the Royal Geological Society of Ireland*, new series, vol. viii., page 3.

1888. Hull, E., *Memoirs of the Geological Survey: Explanatory Memoir to accompany Sheets 138 and 139 of the Map of the Geological Survey of Ireland*, page 24.

1890. O'Reilly, J. P., "Notes on Some Assays for Gold of Rocks occurring in the Neighbourhood of Dublin," *Scientific Proceedings of the Royal Dublin Society*, new series, vol. vi., page 450.

1895. Ball, V., "On the Gold-nuggets hitherto found in the County Wicklow," *Ibid.*, new series, vol. viii., page 311.

1901. Lyburn, E. St. John, "Prospecting for Gold in County Wicklow and an Examination of Irish Rocks for Gold and Silver," *Ibid.*, new series, vol. ix., page 422.

APPENDIX II. Rock-assays.

No. of Label.	Locality.	Nature of Rock.	Gold per Ton.	Remarks.
1	Annie Farm, foot of Loch Lubnaig. Callander. Perthshire	Pebbly grit	Grains. Nil.	On panning-off, it yielded a few grains of pyromorphite, but no pyrites.
2	Rudha Ban, ½ mile south of Ardlui, Head of Loch Lomond, sheet 38	Albite-gneiss	Nil.	0.8 per cent. of magnetite left on panning. No visible sulphides.
3	Meall Garbh, Pass of Leny, Callander, Perthshire	—	Nil.	
4	Green beds, Sron Aonach, Ben Lomond	—	0·0.60	
5	Green beds, Ardlui, Loch Lomond	—	Nil.	
6	Gold-mine, near Lochearnhead Railway Station	Metamorphosed slate	0·0060	Originally a fine grained slate, now highly siliceous by replacement.
7	Do. do.	Quartz	Nil.	Lode-matter, the quartz including brecciated fragments of the country-rock.
8	Near Farmeston, Kilmahog, Perthshire.	Grit	Nil.	
9	About 4½ miles south-west of Aberfoyle Bridge, sheet 38	Hornblende-schist or epidiorite	Nil.	
10	West of Strathyre, Perthshire	Epidiorite	Nil.	
11	Locality-ticket missing ...	Quartz	Nil.	

APPENDIX II. — ROCK-ASSAYS. — *Continued.*

No. of Label	Locality.	Nature of Rock.	Gold per Ton.	Remarks.
			Grains.	
12	One mile north-east of Corriecoich, Berriedale Water, loc. 11	Quartz	Traces	Quartz apparently derived from a vein.
13	North-eastern slope of Morven, near the 1,250 feet contour line, sheet 109, loc. 10.	Ferruginous sandstone	Nil.	Old Red Sandstone.
14	Suisgill Burn, loc. 5 ...	Decomposed granulitic biotite-schist	Nil.	
15	Do. loc. 3 ...	Pegmatitic dyke-rock	Nil.	
16	Do. loc. 5a ...	Flaser mica-schist	Nil.	
17	Do. loc. 4 ...	Granulitic biotite-schist	0·0007	Hard undecomposed rock.
18	Do. loc. 4 ..	Do.	0·0007	
19	Do. loc. 5 ...	Do.	Nil	
20	Kildonan Burn, loc. 1 ...	Quartz - pegmatite	Nil.	
21	Opposite Signal Box, Kildonan, loc. 6 ...	Quartz-schist	Nil.	
22	Do. do.	Granitic rock	Nil.	Little decomposed.
23	Kildonan Burn, loc. 2 ...	Granulitic biotite-schist	Nil.	
24	Albridge, Kildonan, loc. 7	Quartz-schist	Nil.	
25	Upper conglomerate, Beinn Ratha, east of summit, 5 feet from granite, 2 miles south of Reay, sheet 115	Conglomerate	Nil.	
26	Streamlet, west of Loch Dubh, 1 mile south-east of Ben Alisky, sheet 109	Sandstone	Nil.	Old Red Sandstone.
27	At a point in stream, 1 mile north-east of Morven, sheet 109, loc. 8	Earthy fissile slate	Nil.	Originally micaceous, very fissile, taken from surface, and somewhat ochreous.
28	Eastern part of Small Mount, 1 mile west of Morven, sheet 109, loc. 9	Quartzite	Nil.	Passing at one end into true vein-quartz.
29	Top of Ben Alisky, a few feet west of Cairn, north-east of Strathmore Water, sheet 109.	Conglomerate	Nil.	Ochreous and much decomposed.
30	Stream, west of Boreland, 1 mile north of road, 1¼ miles west of Loch Tay	—	0·0400	Gangue, obtained by panning off pyrites.
31	Do. do.		0·0060	Pyrites, sample yielded 21·2 per cent. of clean quartz, obtained by panning off the pyrites.
32	Do. do.	--	0·0200	Pyrites obtained by panning off, equalled 11·4 per cent. of the whole.

APPENDIX II. —Rock-assays.—*Continued.*

No. of Label.	Locality.	Nature of Rock.	Gold per Ton.	Remarks.
			Grains.	
33	Stream, west of Boreland, 1 mile north of road, 1¾ miles west of Loch Tay	—	0·0007	
34	No. 1, Fridd level, Voel Mines, Dolgelly, Wales	Intrusive	Nil.	From mouth of level.
35	St. David's Lode, Clogau, Dolgelly	Dyke-rock	Nil.	From beside vein.
36	Do. do.	Carbonaceous shaly flucan	Nil.	Do.
37	Holy Cross-cut, Voel Mines, Dolgelly	Dyke-rock	0·0060	
38	Trawsfynydd Road, above Tyddyngwladys, near Dolgelly	Sandstone	0·0007	Lower Cambrian Sandstone.

———

The PRESIDENT (Sir Lindsay Wood, Bart.) moved a vote of thanks to Mr. Maclaren for his interesting paper.

Mr. W. O. WOOD seconded the resolution, which was cordially approved.

———

The CHAIRMAN (Sir Lindsay Wood, Bart.) moved that the thanks of the members of the North of England Institute of Mining and Mechanical Engineers be accorded to the Cumberland Coal-owners Association; to the owners of the collieries, mines, quarries and works which had been thrown open for their inspection; to the Cockermouth, Keswick and Penrith, and London and North-Western Railway companies; and to the members of the Committee, especially those resident in Cumberland, for making such excellent arrangements for that meeting.

Mr. T. E. FORSTER seconded the resolution, which was most cordially approved.

'd in Great Britain and Ireland"

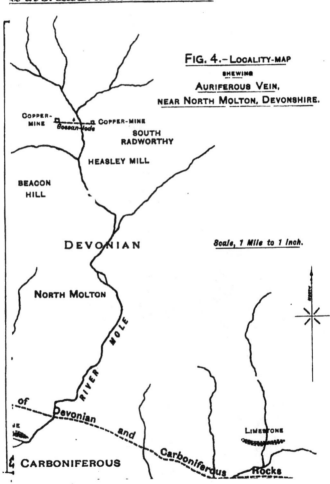

FIG. 4.– LOCALITY-MAP
SHEWING
AURIFEROUS VEIN,
NEAR NORTH MOLTON, DEVONSHIRE.

COPPER-MINE   COPPER-MINE
Gossan-lode

SOUTH RADWORTHY

HEASLEY MILL

BEACON HILL

DEVONIAN

NORTH MOLTON

Scale, 1 Mile to 1 Inch.

RIVER MOLE

of
Devonian   and
Carboniferous

LIMESTONE

CARBONIFEROUS   Rocks

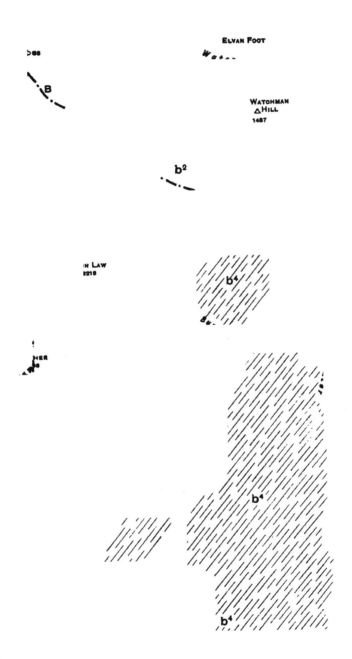

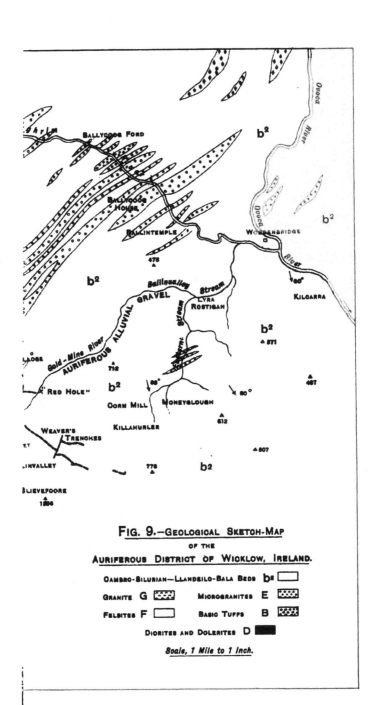

FIG. 9.—GEOLOGICAL SKETCH-MAP
OF THE
AURIFEROUS DISTRICT OF WICKLOW, IRELAND.

CAMBRO-SILURIAN—LLANDEILO-BALA BEDS b²

GRANITE G   MICROGRANITES E

FELSITES F   BASIC TUFFS B

DIORITES AND DOLERITES D

Scale, 1 Mile to 1 Inch.

THE INSTITUTION OF MINING ENGINEERS.

GENERAL MEETING,

HELD IN THE ROOMS OF THE GEOLOGICAL SOCIETY, BURLINGTON HOUSE, LONDON, JULY 2ND, 1903.

SIR LINDSAY WOOD, BART., PRESIDENT, IN THE CHAIR.

REPORT OF THE DELEGATE TO THE CONFERENCE OF DELEGATES OF CORRESPONDING SOCIETIES OF THE BRITISH ASSOCIATION FOR THE ADVANCEMENT OF SCIENCE, BELFAST, 1902.

The Report of the Delegate (Mr. M. Walton Brown) representing the Institution was read as follows:—

NEWCASTLE-UPON-TYNE, *October 24th,* 1902.

TO THE PRESIDENT AND COUNCIL OF THE INSTITUTION OF MINING ENGINEERS.

GENTLEMEN,

The Conferences were held at Belfast on September 11th and 16th, 1902.

The papers and reports of interest to the members include:—

"The Definition of the Unit of Heat," comprized in the Report of the Committee for making "Experiments for Improving the Construction of Practical Standards for Use in Electrical Measurements." *

Seventh Report of the Committee on "Seismological Investigations."†

Final Report of the Committee upon "The Nature of Alloys [Constitution of Copper-tin Alloys]."‡

Report of the Committee on "Life-zones in the British Carboniferous Rocks."§

Third Report of the Committee on "The Movements of Underground Waters of North-west Yorkshire."‖

"The so-called 'Fossil' Water of Sedimentary Strata." By Dr. William Mackie.¶

I am, Gentlemen,

Yours faithfully,

M. WALTON BROWN.

Mr. H. KILBURN SCOTT'S paper on "The Mineral Resources of the State of Rio Grande do Sul, Brazil," was read as follows:—

* *Report of the British Association for the Advancement of Science, Belfast,* 1902, page 55.

† *Ibid.,* page 59.

‡ *Ibid.,* page 175, and *Proceedings of the Royal Society of London,* 1902, vol. lxix., page 320.

§ *Ibid.,* page 210.   ‖ *Ibid.,* page 224.   ¶ *Ibid.,* page 608.

THE MINERAL RESOURCES OF THE STATE OF RIO GRANDE DO SUL, BRAZIL.

By H. KILBURN SCOTT.

Introduction.—During the last few years an increasing amount of interest has been taken in mining work in Brazil, and this is partly accounted for by :—(1) The depreciation in the value of the national currency, making the opening of any mining property relatively easy, as compared with the conditions obtaining when exchange is near par, as in 1890. This depreciation was most pronounced during a part of 1899, when the milreis, which has a par value of 2s. 3d., fell to $5\frac{1}{2}$d. And (2) the profitable development of the manganese-ore industry, which was initiated in 1895, and the successful re-opening of the famous Morro Velho gold-mine, which gave a distinct fillip to mining work.

In no part of Brazil was the interest in mining felt more keenly than in the state of Rio Grande do Sul, and during the last few years, several companies have been formed to work metalliferous and other deposits in this state.

General Remarks.—The state of Rio Grande do Sul is the southernmost of the United States of Brazil, and it is bounded on three sides by the state of Santa Catharina and the republics of Uruguay and Argentina, and on the fourth side by the Atlantic Ocean. Its geographical position is between 27 degrees 5 minutes and 35 degrees 45 minutes of south latitude, and between 6 degrees 22 minutes and 14 degrees 16 minutes of west longitude from Rio de Janeiro (Fig. 1, Plate XX.).

The language spoken is Portuguese, although, owing to the contiguity of the state to the Argentine and Uruguayan republics, many Spanish words are in general use.

The climate is most healthy, especially in the highlands of the interior, where the temperature seldom passes below zero, or above 35° Cent. It is very suitable for Europeans, as is proved by the successful colonization of the district, north and west of Porto Alegre (the capital), by Germans and Italians: there being about 100,000 of each nationality.

As often obtains, the natives are intelligent but very indolent, and the presence in the country of the German and Italian colonists has had a most beneficial influence on them. The larger proportion of the natives earn their living directly or indirectly by the breeding of stock and the preparation of dried beef—principally for export.

The recent financial crisis brought about by the burning of paper-money, which was one of the obligations that the country took upon itself in 1899 in order to keep faith with its foreign creditors, has made itself felt in the state of Rio Grande do Sul, and most severely in those parts of it in which the fresh and dried meat-trades were hitherto the principal industries.

Efforts are being made throughout Brazil to establish new industries, and mining, because it develops the natural resources of the country, is having its share of attention; although it must be explained that latterly the abnormal rise in the gold value of the national currency from 6d. to 1s. per milreis has caused a temporary set-back to mining work, by reason of the consequent increased cost, in gold, of labour and materials bought in the country. Already the currency value of labour, etc., has fallen, and it may be taken that once the transition-stage is passed, not only will the country be able to satisfy its foreign bondholders, but the native industries, being supported by a solvent country, will become more profitable than heretofore.

The great recuperative power of Brazil is proved by its present financial stability as compared with 1899, when universal bankruptcy appeared to be inevitable; and the country is so rich in natural resources that three years has sufficed to accomplish what would have taken a decade in other countries.

The three principal towns are Rio Grande, Pelotas, and Porto Alegre, the state capital. The first-named is situated immediately inside the "bar," where the estuary of the Rio Grande empties into the Atlantic Ocean. The Rio Grande bar is somewhat shallow, but, if carefully navigated, steamers up to 5,000 tons burthen can reach the town. Pelotas and Porto Alegre are at the southern and northern extremities respectively of the Lagôa dos Patos (Ducks' Lagoon), a large and shallow lake which drains

the greater part of the state.   Pelotas is a flourishing commercial town, the centre of the cattle-trade, whereas Porto Alegre
is more official in character.

The important towns in the interior are Bagé, San Gabriel,
Santa Maria and Cachoeira, the two former being in the centre
of the cattle-breeding district, and all of them are situate
within the limits of the metalliferous region.

The railway-system passes through the more important towns
and consists of a line of 1 metre-gauge, part of it from Rio
Grande to Bagé being worked by a British company, known as
the Brazilian Great Southern Railway of Rio Grande do Sul;
while the rest of the system from Bagé by San Gabriel to
Margem on the Jacuhy river (where it is joined to Porto Alegre
by river-steamers) is leased from the government by a Belgian
company.   Another line, built by a Belgian company, runs
from Santa Maria in a northerly direction, and is intended
eventually to join the railway-system of the northern states.
As most of the lines have a guaranteed interest, the Brazilian
government is reducing its foreign obligations by gradually buy-
ing out the foreign companies, and leasing the lines again at a
remunerative rate ; and it is likely that it will eventually acquire
the British railway above mentioned, and lease it to the Belgian
company, which works the rest of the line.   None of the railways have, as yet, been very remunerative, but the British line
has been the most successful, and it is to be presumed that the
Belgian company will make an effort to obtain a lease of it with
a view to more economical working being possible by having the
through system under one management.   The gradients on the
railways are very heavy, in some places as much as 3 per cent.,
and the curves are severe, as small as 328 feet (100 metres)
radius, but improvements could be made by proper contouring.
In the vicinity of the coast-towns, and for some distance into the
interior, the country is very flat : this being especially the case
near the Lagõa dos Patos.

Generally speaking, the country is not mountainous, the
higher lands consisting of low, rounded hills, except in the con-
glomerate-region where some of the hills are very steep.   In
consequence of the large area of eruptive rocks, and the absence
of the decomposed metamorphic schists, which are so prominent
a feature in other parts of Brazil, the roads are easy of transit.

Geology.—The geology of the state of Rio Grande differs considerably from that of the state of Minas Geraes.[*]

In the southern part of the state, there is a chain of hills rising to 2,000 feet (600 metres) above sea-level, known as the Serra do Herval, and consisting in great part of granite and eruptive rocks. In the western part of the range, the granite gives place to gneiss, fine and coarse porphyries, metamorphosed chloritic schists, etc., and these extend into the Republic of Uruguay. The metamorphic schists are intersected by veins of auriferous quartz, in which the gold is sometimes visible, but it is generally associated with iron-pyrites.

The Carboniferous formation extends along the northern side of the Serra do Herval, and consists of soft sandstones, shales and coal-seams, producing a coal of a semi-bituminous and schistose character at the principal mine, known as São Jeronymo.

In the south, huge deposits of conglomerate are found over an extensive area, which have been disturbed by a large eruption of melaphyre. They are made up, for the most part, of granitic material with a small proportion of the débris of metamorphic rocks, quartzites, micaceous and chloritic schists, and the constituents vary in size from large boulders to the finest particles forming a sandstone. Thin layers of materials of even size, alternate with each other, and the fine conglomerates and sandstones seem to be by preference in the lower, and the coarser conglomerates in the upper portion of the series. The beds generally lie horizontally, but they have been broken in every direction; and parts have fallen down, while masses stand 175 to 200 feet (50 to 60 metres) higher with sharply-defined escarpments.

In many places, the eruptive rock not only appears on the surface, but has covered the hills with a lava-like flow. Besides this main movement, the conglomerates have been subjected to local upheavals: the principal one, of which the Camaquam copper-mine is the centre, is 12½ miles (20 kilometres) long and 4 to 5 miles (6 to 8 kilometres) wide. The beds are inclined on both sides at an angle of 40 degrees, and in many parts the melaphyre appears on the surface. At the Cerro Martino, the

[*] "The Manganese-ores of Brazil," by Mr. H. Kilburn Scott, *Journal of the Iron and Steel Institute*, 1900, vol. lvii., page 179; and "The Iron-ores of Brazil," by Mr. H. Kilburn Scott, *Ibid.*, 1902, vol. lxi., page 237.

disturbance of the conglomerate and sandstone-beds is much greater; and pieces of them, surrounded by the melaphyre, form islands of sedimentary rock.

Copper occurs frequently in the whole region, which has been affected by the melaphyre-eruption, and there seems to be a close connection between the metalliferous deposits and the eruptive rock.

On the south-west of the conglomerate-formation, syenite covers an extensive area. It is very often auriferous, the gold occurring either in quartz-veins, or as impregnations in the decomposed syenite-rock. The village of Lavras, at the eastern extremity of this formation, is at present the centre of the gold-mining industry, although work is being carried on intermittently at other places, such as São Sepé to the north and Bom Retiro, near São Gabriel, to the north-west.

To the south of the conglomerate-formation, the Carboniferous rocks extend as on the north. The principal deposits of this basin are known as Candiota and Santa Rosa, and are very extensive.

History.—Mining in this state has been carried on more or less intermittently for the past 50 years. In 1835, gold was discovered in veins at São Antonio das Lavras, the metal having previously been located in the neighbouring streams.

The gold-deposits at São Antonio das Lavras and the copper-deposits near the Camaquam river were originally worked on a small scale by Brazilian owners, but the concessions were subsequently acquired by a British company, known as the Rio Grande do Sul Gold-mining Company, Limited. A certain amount of exploratory work was carried on in the vicinity of Lavras and on the copper-deposits of Camaquam, but, for some reason which it is difficult to discover, owing to the lapse of time, work was gradually suspended and the mines abandoned. In 1898, attention was again directed to these metalliferous deposits, and three companies were formed to work the auriferous deposits in the vicinity of the village of Lavras, known as the :—(1) Vista Alegre Gold-mining Estate Company, Limited; (2) Société des Mines d'Or du Cerrito; and (3) Omnium Minier Lopes-Talhouarne: the first being a British company, and the other two, Belgian. Messrs. Belfort & Company, a local company, are opening up a mine on the Bom Retiro estate, near São Gabriel, while the copper-mine of Camaquam, which had been worked for some years by Germans, has now passed into the hands of the Société des Mines de Cuivre de Camaquam. This latter is a Belgian company organized by the Banque d'Outre Mer, which holds the majority of the shares.

Of the Carboniferous deposits, the two best-known are São Jeronymo and Candiota, the former being the most extensively worked. So long ago as 1841, Dr. Perigot was engaged by the Brazilian government to report upon these deposits, and the São Jeronymo mine has been irregularly worked since about 1870. At the present time it is owned by a native company, known as the Companhia das Minas de São Jeronymo.

Auriferous Deposits.—As already remarked, the principal known gold-deposits occur in the vicinity of the village of Lavras, which is situate 37 miles (60 kilometres) north of Bagé and about 800 feet (250 metres) above sea-level.

The gold occurs, principally, in the syenite-rock in small stringers of quartz and impregnations (stockworks) in the enclosing rock on each side of them. The syenite, at its contact with the quartz-stringers, seems to be decomposed to a varying thickness, which amounts in some places to 6 feet (2 metres) or more, the quartz having presumably disappeared as a result of the action of the solutions carrying the gold, etc. The mineralization is irregular, and generally limited in area, the rock being found to be very rich in places but often pockety.

Galena, iron pyrites and zinc-blende occur associated with the gold, the first-named being considered a good guide for auriferous ground, although as a matter of fact no rule can be established, as non-auriferous galena is common.

Ordinary iron-pyrites is often very rich in gold, and parcels of pyrites, weighing a few tons, were shipped to England for reduction in the early stages of the exploration, containing from 500 to 700 grammes of gold per ton.

At São Sepé, 22 miles (35 kilometres) north-west of Caçapava, auriferous quartz-veins are plentiful over an area 12½ miles (20 kilometres) long by 4 to 6 miles (6 to 10 kilometres) broad. At one place, an auriferous quartz-vein cuts the granite and metamorphic rocks, close to their contact with the Carboniferous formation. In fact, coal, mined about ½ mile (1 kilometre) distant from the auriferous deposit, is used to drive a 5 heads stamp-mill at the mine.

The quartz-veins vary in thickness from 1½ to 6½ feet (0·50 to 2·00 metres) and are fairly easy to follow, as compared with the stringers found in the syenite at Lavras. The gold is sometimes visible, and, as at Lavras, often found in rich pockets, so that it is difficult to estimate the average gold-contents of the lodes. Near the village of Dom Pedrito, situate 37 miles (60 kilo-metres) to the south-west of Lavras, the Barcellos Gold-mining Company worked some auriferous deposits about 15 years ago. This company suspended operations shortly after its formation, presumably owing to payable bodies of ore not having been found.

In the Republic of Uruguay, near Taquarembó, several deposits are being worked, which have much in common with those in Rio Grande, and are said to belong to the same forma-tion. The gold occurs in thin quartz-veins and impregnations in the syenite and metamorphic rocks as at Lavras and São Sepé; and, as far as can be gathered, the deposits, which are known as São Gregorio and Cuñapirú, are being profitably worked.

Speaking generally, it must be confessed that, up to the present time, the working of the different auriferous deposits in the State of Rio Grande, has not been particularly success-ful, the difficulty being the small size of the quartz-veins and the irregularity of their gold-contents, owing to their pockety tendency. Mining operations are being continued by the several companies, and it is anticipated that results as good as those shown in Uruguay will be obtained before long.

Cupriferous Deposits.—The principal copper-deposit as yet known, namely, that of Camaquam, is about 50 miles (80 kilo-metres) from the Rio Negro station of the Southern Brazilian railway and 2 miles (3 kilometres) from the Camaquam river. The mine is situated on a large upheaval of conglomerate, on the south-eastern point of the formation, caused by the large melaphyre eruption which affected the whole country: the mela-phyre appears in many parts on the surface, but is not found in the mine (Figs. 2 and 3, Plate XX.).

The zone in which mining is expected to be most profitable

is 1,600 by 2,600 feet (500 metres by 800 metres), and the average
height above the lower adit is 250 to 300 feet (80 to 90 metres).
The upper adit was driven 520 feet (160 metres) in a north-
easterly direction, and cuts a dyke of gabbro, which does not
come to the surface.   Most probably, this eruption caused the
opening of the veins which diverge from the dyke and supplied
the materials (quartz and copper and iron sulphides) which fill
the lodes.   In some cases, small particles of copper-glance have
been found in the gabbro, and dendrites of native copper have
been found in the fissures of the conglomerate, close to the

FIG. 4.—VIEW OF THE SURFACE-WORKS AT THE CAMAQUAM MINE.

former rock.   It is interesting to note that these dendrites present
a similar appearance to copper that has been deposited by
electrolysis.

The lodes have not proved to be so good in the sandstone-beds
as in the hard conglomerate, as through the softer rock they lack
regularity or even continuity.   Of the four lodes at present
being worked, three are cut by two adits, respectively 200 and
380 feet (60 and 115 metres) below the top of the hill, and the
other by a level opening directly on the outcrop.   This latter
only contains copper-pyrites, the gangue being quartz with heavy

spar.   The three former are similar, with copper-glance near the surface, and bornite, copper-pyrites and some iron-pyrites lower down.

The first lode, cut by the lower adit, shows only copper-glance, the vein being thin but rich, giving excellent ore very easy to concentrate by hand-sorting.   The two other veins yield a mixed ore of copper-glance and copper-pyrites.   The average thickness of the lodes worked during 2½ years has been 4 feet (1·25 metres), and the ore assayed 6½ per cent. of metallic copper, with about 19 pennyweights (30 grammes) of gold per ton.

FIG. 5.—ORE-DUMP, ETC., AT THE CAMAQUAM MINE.

The cars of mineral coming out of the mine are tumbled over a grizzly and perforated plate with 1 inch (25 millimetres) mesh. The mineral is sprayed with water and sorted by hand, the picked rich ore assaying 30 per cent. of copper.   The barren rock is thrown away, and the remainder, which averages only 7 per cent. of copper, is dressed in a small concentrating plant with breaker, roll-mill, trommels, hand-jigs and round-buddles. The resulting concentrate averages 28 per cent. of copper.

The actual monthly output of the mine is 90 to 100 tons of ore containing 28 to 30 per cent. of copper, which is packed in canvas-bags and sent to England. The ore is free from such impurities as arsenic, lead and zinc.

The Camaquam Copper-mines Company is, at present, erecting a plant for dressing 80 tons per day. The concentrates will be smelted so as to obtain a matte containing 50 to 60 per cent. of copper; and the monthly output is expected to be 220 to 250 tons of matte. The plant will be run by hydraulic power. The dam, built for the purpose, is 400 feet (130 metres) long and 52 feet (16 metres) high at the deepest point. It will hold a reserve of water sufficient to run the mill during the whole year, so that the plant will not be stopped by the long droughts which sometimes occur.

FIG. 6.—VIEW OF DAM IN CONSTRUCTION AT THE CAMAQUAM MINE.

With regard to the continuation of these lodes in depth, as bearing not only on the future of these deposits but also on others in the same district and found under somewhat similar conditions, the author saw no evidence of any probability of their giving out. Although, on being opened, the principal lode along its strike proved to be somewhat irregular, its variations in value and width are only local. Mr. Jules Jadot, the late manager of the mine, informed the author that he had noted no evidence of deterioration of the lode, although the present adit-level had only come upon the apex of the eruptive-mass of gabbro, and therefore its influence on and relation to the ore-bodies could not well be noted. When the lower adit-level has been holed into the principal lode, it will be possible to determine with certainty its size and its value in depth. The low adit is now driven for a length of 750 feet (230 metres) and has cut two lodes showing only copper-glance; it will reach the third (the most important) lode before the end of 1903. The development workings in this lode (São Luiz) are 650 feet (200 metres) long. It is intended to continue the adit-level to a length of about 3,300 feet (1,000 metres) and it is calculated that it will cut several other lodes, which crop out very promisingly on the surface.

The Cerro Martino mine is situated about 86 miles (142 kilometres) from the Rio Negro station, already referred to, and 81 miles (130 kilometres) from Cachoeira on the Uruguayana railway, which runs from Margem on the Jacuhy river. As already remarked, the geological conditions are somewhat different from those which obtain at the Camaquam deposit, for the conglomerates seem to have suffered very much more from melaphyre, which has spread itself over the two hills, in which the copper-ores are found. The conglomerates, between the two hills, are very fine and may be considered sandstones. They have a dip of about 40 degrees, and the lodes cross them approximately at right angles, as at Camaquam. Most of the copper occurs as glance, and the other constituent minerals of the lodes are iron-pyrites and heavy spar. Analyses made by the author of samples taken by himself gave copper-contents varying between 7 and 25 per cent.

The Primavera mine is situated near Caçapava, in the central chain of hills formed by the Cambrian rocks and granite. The vein of quartz, containing copper-glance and pyrites, and showing sometimes signs of native silver, is situated at the contact of granite and micaceous schists. The copper-glance occurs disseminated through the ore-mass in small nodules, and the total copper-contents may be taken on an average sample as about 7 per cent.

The Cerro do Geraldo deposit is situated about 2½ miles (4 kilometres) in a north-easterly direction from Caçapava. The ore crops out on the summit of a steep hill, and a shaft which has been sunk to a depth of 66 feet (20 metres) shows copper-ore impregnations in the micaceous schists. At the foot of the hill, some copper-glance impregnations have also been noticed, but the mineral is nowhere of any commercial value.

Copper-ores have been found at Lavras and Quarahim, but the author has no reliable information available concerning them.

Carboniferous Deposits.—The two principal coal-deposits are known as São Jeronymo and Candiota, the first situated to the north-east and the other to the south-west of the Serra do Herval.

The São Jeronymo deposit lies about 11 miles (18 kilometres) from the Jacuhy river, with which it is connected by a metre-gauge railway. The coal-seam is 6½ feet (2 metres) thick, and is mined by the pillar-and-stall method, the conditions for successful and economical mining being most favourable. The roof is good, and the quantity of water in the mine very small: no pumps being required as the water can be taken from the sump by a bucket hauled by the winding-engine, working 2 or 3 hours a week. The coal is of a semi-bituminous and schistose character, and must be carefully picked in order to obtain good results.

In 1899, a careful examination of the São Jeronymo coal was made by Mr. Jules Koeber, of Rio de Janeiro, on behalf of the Rio Grande Gas Company, in order to determine its value, as compared with good English coal, for gas-making (Tables I., II. and III.).

TABLE I.—ANALYSES OF SÃO JERONYMO AND ENGLISH COALS.

	São Jeronymo Coal. Gas-work's Sample. Per cent.	São Jeronymo Coal. Coal Co.'s Sample. Per cent.	English Coal. Gas-work's Sample. Per cent.
Hygroscopic water	10·68	15·14	2·70
Volatile matter	23·72	38·78	31·33
Fixed carbon	38·86	37·34	62·16
Ash	26·82	8·78	4·23
Sulphur	1·87	1·43	1·05
Calorific power in calories ...	5,323	6,043	7,769
Specific gravity	1,580	1,580	1,638

These analyses prove that the mineral is not satisfactory as a gas-coal. The great variation in the results of the analyses of the different samples may be explained by the fact that the Coal Company's sample was doubtless made up of picked mineral, whilst that of the Gas Company may be taken as representing the class of coal generally furnished to consumers.

TABLE II. - TRIALS OF SÃO JERONYMO AND ENGLISH COALS IN AN
EXPERIMENTAL PLANT AT THE GAS-WORKS.

	São Jeronymo Coal. Gas-work's Sample.	São Jeronymo Coal. Coal Co.'s Sample.	English Coal. Gas-work's Sample.
Furnace-temperature, degrees Cent.	1,000	1,000	1,000
Retort-charge, kilogrammes	1	1	1
Duration of experiment, minutes	20	20	20
Gas distilled from 100 kilogrammes, cubic metres	22·60	27·21	31·53
Coke, kilogrammes	64·95	51·45	62·85
Candle-power of gas produced	3·59	10·26	15·17

TABLE III. -ANALYSES OF THE COKES PRODUCED BY DISTILLATION OF
SÃO JERONYMO AND ENGLISH COALS.

	São Jeronymo Coal. Gas-work's Sample. Per cent.	São Jeronymo Coal. Coal Co.'s Sample. Per cent.	English Coal. Gas-work's Sample. Per cent.
Hygroscopic water	3.72	2·66	1·29
Fixed carbon	51·30	78·47	92·80
Ash	44·66	19·48	5·86
Sulphur	1·22	1·16	0·90
Calorific power, calories ...	2,333	5,480	7,560

NOTE.—The ash of the São Jeronymo coal was white and calcareous, and that of the English coal was yellow and siliceous.

The quality of the coal can undoubtedly be much improved by picking, and when that is done its utility must be very great in a country where imported coal is so very expensive. It can be sold for 20s. a ton in competition with English coal, which costs from 45s. to 50s. per ton at Porto Alegre.

The output of the mines has averaged between 10,000 and 20,000 tons during the last few years, and the greater part has been used by the Uruguayan railway and native coasting-steamers. Owing to the high percentage of ash in the coal, specially wide grate-bars must be fitted to the boiler-furnaces in which it is used.

An analysis of the São Jeronymo coal made by Mr. Vandecapelle gave the following results:—

	Per cent.
Hygroscopic water	10·4
Volatile matter	23·3
Fixed carbon	40·8
Ash	25·5

The São Jeronymo Carboniferous basin has a very large area, and outcrops of coal occur along the valley of the Jacuhy river.

Exploration-work has been done at Irapuá, near Cachoeira; and the coal has been analysed by Mr. Vandecapelle with the following results:—

	Per cent.
Hygroscopic water	13·85
Volatile matter	19·91
Fixed carbon	34·19
Ash	32·03

The Carboniferous formation in the valley of the rivers Candiota and Jaguarão is very extensive, being about 54 miles (90 kilometres) long by 31 miles (50 kilometres) wide. Many attempts have been made during the last century to work these deposits; and, as far back as 1863, a concession was given by the Imperial Government for the construction of a railway to connect the deposits with a port.

The coal occurs in a bed about 7 feet thick, resting on mica-schists and overlain by a ferruginous sandstone, and it has an inclination of about 10 degrees towards the south-west. The outcrop being very extensive along the sides of the valleys, the conditions for economical working are even better than at São Jeronymo. The bed of coal consists of thin alternating layers, of almost infinitesimal thickness, of carbonaceous and argillaceous matter. Owing to this peculiar structure and the swelling of the argillaceous matter on exposure, the mineral will not bear transport or storage.

The following are analyses of samples of Candiota coal by Mr. Vandecapelle:—

	1 Per cent.	2 Per cent.
Hygroscopic water	15·2	21·8
Volatile matter	28·5	29·2
Fixed carbon	38·1	41·5
Ash	18·2	7·3

NOTE.—No. 1 was a general sample, while No. 2 was a picked sample.

The analyses show that the quality of the mineral is better than that of São Jeronymo, but the grave defect of its friability on exposure causes it to have little commercial value. Some of the mineral has been used from time to time for the electric-light plant at Bagé, but the results have not been satisfactory.

To use either the São Jeronymo or Candiota coals, a *sine-quâ-non* is the adaptation of the furnaces of the boilers to this poorer coal, by the fitting of a special type of grate-bar and the provision of forced draught.

Miscellaneous Minerals.—Wolfram has been found near Encruzilhada in a vertical quartz-vein, in the granite-rock which could be traced for some distance. On both sides of the quartz-vein, and between it and the granite, small bands of crystalline white mica (muscovite) were found ¾ inch to 1¼ inches (2 to 3 centimetres) thick.

A few feet (3 or 4 metres) below the surface, the wolfram was gradually replaced by copper and iron-pyrites, and black tin has been found in the streams near by, but not in the veins. A parcel of this wolfram-ore, sent to England, contained 69·20 per cent. of tungstic acid and was sold for 9s. 6d. per unit.

Agate is found, in quantity, at Uruguayana, near the boundary between the republic of Uruguay and the state of Rio Grande. Originally, the industry was very prosperous, but latterly it has been in a decadent condition owing to the unremunerative prices ruling for the mineral.

Conclusion.—The author has given as complete and reliable an account as possible of the known mineral-resources of the state of Rio Grande. So little exploration has, however, been carried out, and so exceptionally favourable are both the physical and economical conditions for successful mining, that there is every possibility of a great increase in mining work in the future.

The author gratefully acknowledges the help of several gentlemen in the preparation of this paper, his special thanks being due to Messrs. Jules Jadot and Albert Vandecapelle.

———

Mr. ROBERT NORMANTON (Beverley, Yorkshire) wrote that he had read with great interest Mr. H. Kilburn Scott's paper on "The Mineral Resources of the State of Rio Grande do Sul, Brazil," and having travelled over the country, he could fully endorse and appreciate Mr. Scott's communication with slight exceptions. It is doubtful whether the "bar" of Rio Grande will allow steamers to pass with a tonnage of 5,000 tons; only under very exceptional tides could a vessel of this tonnage pass the bar, but specially constructed German steamers of 2,000 tons go up monthly to the city of Rio Grande. He could fully endorse Mr. Scott's observations, wherein he states that the metalliferous deposits seem to have a close connection with the eruptive rock.

3. 2.—PLAN OF THE COPPER-MINES OF CAMAQUAM.

SITE OF
PROPOSED
REDUCTION-
WORKS

In the Serra Santa Barbara, possibly not visited by Mr. Scott, black basalts and obsidian of a peculiar structure occur. The Serra Santa Barbara is at an elevation of about 1,600 feet (500 metres) above sea-level, trends in a general westerly direction, and is about mid-way from the Serra do Herval and the copper-mines of Camaquam, but at a much higher level than those mines. The latest advice from this zone is that rich copper-ores have been discovered, and he was led to believe that the quantity was large. The Camaquam copper-mines, now worked by Belgians, were visited by him during the exploration-stage, the depth of the excavation being about 16 feet. This property was formerly held on lease by a British company, who executed a certain amount of work, and then abandoned the lease, and something of this nature also took place with the Barcellos Gold-mining Company. About 3 miles from the Barcellos mine are other fine gold-properties, and should the Barcellos mine commence operations again, their mill could be supplied with ore from outside mines and used as a customs mill. By the opening of the Great Southern Railway, the position of various properties has been greatly improved. Mr. Scott was perfectly correct when he stated that but little exploration had been carried out, barely a scratching of the ground, with the exception of the Camaquam copper-mine, although for mining, there were most favourable conditions, and as regards climatic conditions the Rio Grande do Sul is a white man's country.

Mr. LOUIS COUSIN (Brussels) wrote that the description of the Camaquam copper-mines is accurate, with the exception of the tenour of gold in the ore: some time ago, the ore contained 15 grains (1 gramme) of gold per ton per unit of copper, and con-centrates of 30 per cent. of copper contained 19 pennyweights (30 grammes), but that percentage had not been maintained since Mr. Scott's visit to the mine. The continuity of the lode in depth had now been proved by the lower adit (cross-measures drift) which had met the São Luiz lode at a depth of 394 feet (120 metres) below the surface.

Mr. JAMES BARROWMAN (Hamilton) pointed out that the specific gravity of the coal appeared to be exceptionally high.

The PRESIDENT (Sir Lindsay Wood, Bart.) remarked that it was probably a slaty coal.

Mr. H. D. HOSKOLD (Buenos Aires) wrote that Mr. H. Kilburn Scott's paper was opportune, interesting and welcome, because it dealt with a country, the mineral resources of which, in all probability, may be large. He had, therefore, rendered a great service in bringing the subject of his paper before the members.

Doubtless Brazil is, more or less, in the same position as various of the other South American republics, that is, the country is not thoroughly explored, the mines are not much developed, and there is a want of accurate, detailed, official mining statistics from the time when mining was first commenced in a formal manner. No doubt Mr. Scott must have felt this, for it is to be supposed that if it had been otherwise he would have been able to include such data in his paper, not only from Rio Grande do Sul, but also from the other mining states of Brazil. He (Mr. Hoskold) had taken considerable pains to procure a complete copy of the official mining statistics of Brazil, but had failed to do so. It must be evident that the value of a mine or mines depends altogether upon the annual production and the net benefit derived over a series of years in the past, as well as on future prospects. For the object of determining the comparative advantages to be derived from the expenditure of capital which may be devoted to the exploitation of mines and the reduction of the ores in various countries; and, consequently, to aid in selecting mines in a country with the best conditions, mining statistics are not only important but absolutely necessary.

The nature of the mining laws of a foreign country may, also, affect the profitable working of mines to a considerable extent. Before, therefore, any serious mining operations are entered upon, such legislative measures as may exist should be thoroughly examined. Mr. Oswald Walmesley, the author of an excellent *Guide to the Mining Laws of the World*, states that Brazil possesses no code of mining laws; but that various decrees or isolated legal dispositions have been made at different periods, regulating mining in that country.

It is curious to note that in various of the South American republics anything and everything of a black or carbonaceous nature found deposited in the rocks is considered, by those who are not practical geologists and mining engineers, to belong to the Carboniferous formation. But this is only true in a limited sense, that is, that certain more recent formations may contain coal, as is the case in the Argentine and Chilian republics, without belonging to the great Carboniferous formation, or of the age of that found in Great Britain and other countries. It is, therefore, highly probable that the coal-deposits mentioned by Mr. Scott as existing and extending " along the northern side of the Serra do Herval," where the principal coal-mine, São Jeronymo, is being worked, do not belong to the Carboniferous system. However, the fossil fauna and flora, if any exist in the district referred to, would settle the question definitely. Mr. Scott has not referred to this interesting and important palæontological evidence; but he does say, " At one place, an auriferous quartz-vein cuts the granite and metamorphic rocks, close to their contact with the Carboniferous formation.   In fact, coal, mined about $\frac{1}{2}$ a mile (1 kilometre) distant from the auriferous deposit, is used to drive a 5 heads stamp-mill at the mine." Doubtless this statement would appear curious to many and, without more explanation, would not be well understood, for it is not a common thing to find " granite and metamorphic rocks, close to their contact with the Carboniferous formation."

When officially representing the mining and metallurgical industries of the Argentine Republic at the Chicago Exhibition in 1893, and in the character of a member of the International Jury of Awards, he (Mr. Hoskold) had frequent occasion to inspect and report upon the exhibits in the sections of various nations for awards.   On visiting the Brazilian mining section, his attention was particularly directed to a series of beautifully drawn maps and sections of the São Jeronymo coal-mines, and to his great surprise discovered that the seam of coal had been deposited upon a thin band of micaceous schist, and that this latter rested immediately upon a granite-base. Not being satisfied with the first casual examination of the sections, other inspections of them were made, and finally a copy of the whole procured.   The coal-seam was represented as existing in basin-form, of which there were three in continuous order on the section and under the same conditions; but they were disconnected one from the other by denuding influences which had, in former geological epochs, carried away the more elevated parts exposed to its effects. Unfortunately, the copies of these plans and sections have been either lost or mislaid, so that, at present, it is not possible to refer to them for verification. The data now offered confirm, in a general manner, the indications made by Mr. Scott.

At the time when he (Mr. Hoskold) examined the geological

section of these coal-measures, it appeared to him that the micaceous-schist, under the coal-seam, was much too thin to allow of the growth of an immense tropical forest of trees and other vegetable products such as would have been sufficient to form a deposit of coal like that now found in the mine of São Jeronymo, which, according to Mr. Scott, has a thickness of 6½ feet. One of the greatest authorities upon the time necessary to form a bed of coal was Sir J. W. Dawson, and in one of his scientific discussions he said "We may safely assert that every foot of thickness of pure bituminous coal implies the quiet growth, and fall of at least 50 generations of *Sigillaria*, and therefore an undisturbed condition of forest-growth enduring through many centuries." Taking this estimate as a basis, the bed of coal in the São Jeronymo mine must have required 325 generations of forest-growth. It is, therefore, evident that the vegetable-matter, originally forming the coal-seam under consideration, must have been derived from forest-growths situated at a distance from the locality of the coal-seam, and that it must have been transported by floods and deposited in the locality and under the conditions in which it is now found. Doubtless, such a process is, in a great part, the reason why the coal-seam is found in a schistose condition and, consequently, much inferior to British coal dating from the great Carboniferous period. Considering, therefore, all the conditions under which the coal in the São Jeronymo district exists, he (Mr. Hoskold) is of the opinion that it does not belong to the Carboniferous system.

After having studied this question for many years, he is also convinced that the same remarks apply to the coal-deposits existing in the Argentine, Chilian and Uruguayan republics.

Mr. M. H. Mills (Mansfield), in moving a vote of thanks to Mr. Scott for his interesting paper, remarked that there was plenty of work for their children, and for their children's children, if not in Great Britain, then in Brazil and other countries. They might also feel secure that their own coal-fields would continue to hold their own as to quality, if not in quantity.

Mr. C. C. Leach seconded the resolution, which was cordially approved.

———

Mr. A. S. E. Ackermann's paper on "Pneumatic and Electric Locomotives in and about Coal-mines," was read as follows:—

PNEUMATIC AND ELECTRIC LOCOMOTIVES IN AND ABOUT COAL-MINES.

By A. S. E. ACKERMANN, A.C.G.I., A.M.Inst.C.E.

The subject of traction, especially by electric means, is one which is very much before the public and engineers in particular, just at present. Consequently, many papers are being written, either to be read before such a meeting as this, or for publication in one or other of the technical journals. The author feels that he may be wearisome to the members, by repeating matter that they already know, but having made a tour in the United States extending over 5,000 miles and lasting four months, chiefly for the purpose of reporting on American mechanical methods of coal-mining, he ventures to hope that some of the information about to be given may be of use and interest to the members, and he will endeavour to reply to any questions that may arise in the discussion.

In describing the systems of pneumatic and electric haulage as they exist in America, it is extremely important to realize the different local conditions that exist there, and which consequently have largely determined the systems in use. The differences between British and American conditions were dealt with by the author in a short article in the *Engineering Magazine*,* and those specially interested are referred to that journal. Briefly, it may be stated that the seams which are being worked in America are much thicker and nearer the surface than they are in Britain, the dip is very much less, 5 per cent. being about the maximum, while faults are almost unknown. The roofs are usually good, fire-damp is rare, and a large percentage of the mines are drift ones. This latter fact is perhaps the most important of all in connection with pneumatic and electric haulage as practised in the United States, for the electric haulage there

* 1902, vol. xxiii., page 357.

is not in the least like that which is in use in this country. Here, the electrical energy is transmitted to the pit-bottom, where it is used for driving one of the rope-haulage systems; in America all the cases that the author saw (he inspected 21 coal-mines) were on the overhead-trolley system, very similar in fact to most of the electric trams in this country, but the locomotive is quite distinct from the tubs, that is, the system is not that known as the multiple unit. Similarly, the pneumatic system uses a single locomotive. Pneumatic locomotives are much more cumbersome than electric locomotives, owing to the two large steel compressed-air vessels which they carry, and hence they can only be used on main haulage-lines and where there are no sharp curves. These storage-cylinders are (in the case of a 13 tons locomotive) 2 feet 9 inches in diameter and 16 feet long, with a capacity of about 160 cubic feet each. Occasionally, there is a tender to the locomotive carrying three more such storage-cylinders, so as to increase the distance that the locomotive can travel without having to re-charge.

For a 13 tons locomotive, makers recommend that the radius of the sharpest curve should not be less than 35 feet, and state that the least radius practicable for it is 20 feet. In one mine visited all curves had a 54 feet radius, although electric, and not pneumatic, locomotives were in use.

The air-storage vessels are charged with air at a pressure of 700 to 1,000 pounds per square inch, but this is reduced to about 140 pounds before it enters the cylinders. In one case, the locomotives had compound engines, and then the initial cylinder-pressure was 225 pounds per square inch. The overall dimensions of a 13 tons pneumatic locomotive are about 5 feet 3 inches high, 6 feet wide and 16 feet long, the wheel-base being 5 feet 3 inches long, the gauge 3 feet 6 inches, and the drawbar-pull 4,500 pounds. The smallest size of pneumatic locomotive given in a maker's list is 4 feet 4 inches high, 4 feet 5 inches wide, 10 feet long, weighing 5 tons, and having a draw-bar-pull of 1,500 pounds.

The charging is done both in and out of the mine, depending on the length of the haul, the gradient, and the load brought out. When charging is done, whether in or out of the mine, the air-vessels on the locomotive are not coupled up direct to the pneumatic main, but to a larger air-storage vessel, which in turn is connected to the air-main from the power-house. In one mine, where pneumatic coal-cutters were used, as well as pneumatic locomotives, the air for the cutters was taken from the same stationary air-storage vessel in the mine as for the locomotives, but a reducing valve was used to reduce the pressure to about 70 pounds per square inch, the pressure at which most of the cutters work. This strikes one as a very inefficient, though no doubt convenient method, as the loss due to compressing the air to 700 pounds per square inch and then reducing down to 70 pounds per square inch must be very great. With steam, the loss is not so great, as the wire-drawing dries the steam. Where the conditions of hauling are such that one charge will take the locomotive in and out again, there is of course no need to carry a high-pressure main, as well as a low one, into the mine. In this case, and also usually where pneumatic cutters and haulage are used, a separate compressor is used to supply the coal-cutters. The more usual arrangement is, however, to have electric haulage, even where the cutting is done by pneumatic machines.

There is one great advantage, and one that perhaps is not sufficiently kept in mind, in connection with mechanical haulage, and that is fewer or even no horses are required in the mine. Thus one source of fouling of the atmosphere is removed, for 1 horse must be equivalent to at least 5 men in the generation of carbon dioxide, in addition to fæcal matter; while if the system is pneumatic haulage, a considerable addition is made to the fresh air in the mine, and the expansion of the exhaust-air from the locomotives and coal-cutters also has a cooling effect. In the case of one mine, using both pneumatic haulage and cutters, and with a daily output of 1,000 tons of screened coal, the combined capacity of the air-compressors was 2,500 cubic feet of free air per minute.

The air-mains are usually carried along the ground, but in one case they were slung overhead from the cross-timbers by which the trolley-wires were also supported, the reason for this being that the water from the mine was very corrosive. The air-mains are of wrought-iron, with screwed joints caulked with lead. For high-pressure air for the pneumatic locomotives, the main is usually 4 or 5 inches in diameter, and for low-pressure, for the coal-cutters and pumps, the mains are 6 to 8 inches in diameter.

With regard to the comparative cost of the plant for pneumatic and electric haulage, Mr. A. de Gennes stated that the cost of an electric installation is only one quarter that of a pneumatic plant, though the author does not think the difference can be quite as much as this. The same authority states that a rope-haulage costs, in the first instance, three times as much as an electric-trolley one.* In the power-house, in the one case, boilers, steam-engines and dynamos are required; in the other, boilers, steam-engines and air-compressors, so that no very great differ-

FIG. 1. - A DRIFT-MINE AND AN ELECTRIC LOCOMOTIVE.

ence in cost can occur here. In conveying the energy into the mines, there is considerably more room for difference, for the air-mains are expensive and take much longer to lay, on account of their weight and the numerous joints which have to be made air-tight. A straight bed has also to be prepared for them. On the other hand, the electric trolley-wire is very simply fixed from the roof where it is good, or from the cross-timbers where it is bad, and the rail is used for the return; though in the case of the distribution of electrical energy to coal-

* *Annales des Mines*, 1900, vol. xviii., page 244.

cutters, there are usually two bare wires, both supported from one set of brackets fixed to the roof and with the earthed wire on the outside, so that it slightly protects the positive or live wire. It has been said that a man can lay a greater length of electric cable in a day than he can pipes in a week. Lastly, there is the cost of the electric locomotive to be compared with that of the pneumatic, and here the electric one has some advantage. In upkeep the electric system also has the advantage, for there are fewer moving parts to be oiled and for dirt to work into, and no reciprocating parts, while the efficiency of electrical transmission of energy is considerably higher than that by air. It has been said, however, that if as much attention were given to improving and perfecting pneumatic transmission as there is to electrical, the former would run the latter very close. The track for either system is the same, the rails being of the flat-bottom type, spiked to rough sleepers about 6 inches square. The weight of the rails is usually 40 pounds per yard, though in one case 60 pounds was in use, this being old main-line metal. In nearly every mine they wished that they had heavier metals, whatever their actual ones happened to be. In most cases, secondary haulage of some sort is used for collecting the loaded tubs to make a trip, mules being the most common means; but in one case, a group of ten drift-mines belonging to the Berwind White Coal-mining Company at Windber, Pennsylvania, not a horse or mule is employed. The mines are extremely systematically worked on the pillar-and-room system, electric-trolley haulage is used in the headings, and the men who load the tubs in the rooms push them to the entries, where they are picked up by the electric locomotives. This same group of mines has no doors, all the ventilation being carried out by means of overcasts. Sand is used freely, and is quite an item in the running expenses, for increasing the adhesion of the driving wheels to the metals.

In the case of the pneumatic locomotives, oil head-lights are used, but the electric ones use incandescent electric lamps, and in one instance an arc head-light was used.

The electric locomotives are much smaller than the pneumatic ones of the same weight: the overall dimensions of a 13 tons electric locomotive being about 3 feet high, 4 feet 8 inches wide and 12 feet 6 inches long, with a wheel-base 4 feet 8 inches long, gauge 3 feet 6 inches (the gauges vary in different mines, but 3 feet 6 inches may be taken as the average), and capable of exerting a drawbar-pull of 4,300 pounds.   Much smaller electric loco-motives can however be obtained, as for example one which is only 2 feet 5 inches high, 3 feet 6 inches wide, and 9 feet 3 inches long, with a wheel-base 3 feet long, weighing 4 tons, and having a drawbar-pull of 1,000 pounds.   This is, of course, a great advantage, and they run at the same speed, namely, about 8 miles per hour.   At first, there used to be considerable trouble

FIG. 2.—A DRIFT-MINE AND AN ELECTRIC LOCOMOTIVE.

on account of the armatures being burnt, owing to overloading and the heavy starting current.   Now, even an 11 tons electric locomotive is fitted with two 40 horsepower motors, one on each axle.

Drift-mines, of course, lend themselves particularly well to pneumatic and electric locomotive-haulage, because the tubs are run without a stop, or intermediate handling, right from the collecting-place in the mine to the tippler and screens over the railway-trucks on the siding, or barges on the river.   Sometimes, the distance from the mine-entrance to the tippler is several hundred feet. This external part of the system is almost exactly like a small overhead trolley-tram system, the conductor being attached to cross-wires supported by two rough wooden poles, one on either side of the line. The trolley-wire in such cases is usually, but not always, kept at such a height as to just clear one's head.

Both electric and pneumatic haulage are, however, used in cases where the conditions are not so favourable. For example, in two or three cases, the entrance to the mine was down an incline too steep to be worked by a locomotive, and the tubs had to be brought out by wire-rope haulage, but they were brought to the foot of the incline by locomotives. Similarly, in a couple of cases of shaft-mines, locomotives were used to bring the tubs to the pit-bottom. The length of any single haul was also considerable, being frequently a mile, and in one case (pneumatic) $2\frac{1}{2}$ miles. An interesting case was met with in which the mine-owners wanted an electric locomotive of about 20 tons in weight, but this was at a time when 13 tons locomotives were the heaviest made. They therefore obtained two 13 tons ones, and coupled these together, both mechanically and electrically, so that only one collector or trolley-pole was used. This device acted admirably, and with it they have brought out 31 tubs, each carrying 2 tons, against a 6 per cent. grade, and have continued to use these coupled, even now that they have an 18 tons single electric locomotive.

Electric storage-battery locomotives have also been made, but the author did not meet with any.

As to the pressure of the current, 250 and 500 volts were the only pressures met with, and all were direct current, though in one mine three-phase current was used for working the pumps. It looks probable that three-phase current will be much more used in mines in the future, for there are no commutators used with the system, and consequently one of the chief sources of sparking is absent. On the other hand, it is more dangerous.

Of the mines visited which used electric haulage, 7 used a pressure of 250 volts and 5 used 500 volts. The latter pressure is used by the previously mentioned large and important group of 10 drift-mines at Windber, Pennslyvania. Sparking between the trolley-wire and the collector is, of course, common with either pressure. As to the danger of 500 volts, Mr. A. P. Trotter,

direct current, the writer saw one of the pit-hands strike his
cheek against the trolley-wire which had just been brought down
by a fall of roof in the main haulage-heading.   The shock
knocked the man down, and he seemed dazed for a few seconds,
but was up and off again in ¼ minute.   Mules and horses are said
to be much more susceptible to electric shock than man, but they
are also said to learn how to avoid the wires.

The tendency in America is undoubtedly to use electric locomotives and coal-cutters more and more.   Of the mines visited,
12 used electric locomotives, and 4 pneumatic locomotives, while
a fifth was changing from rope to pneumatic haulage, and the
remaining four had rope or mule-haulage.   As to the cost of
electric haulage, this has already been compared with that of
pneumatic.   In comparing it with the cost of haulage by mules,
the saving over the latter is most marked.

Mr. H. W. Hughes, in his book on *Coal-mining*, states that
instances have been quoted where the cost per ton hauled by
electric locomotives is only one-tenth of that which prevailed
when mules were employed.   To again refer to the Windber
mines, the seven power-houses, which were running when the
author was there, have a total capacity of 2,500 kilowatts.   There
are 75 miles of trolley-line, worked by forty-three 13 tons electric locomotives, of which the combined power is about 3,400
horsepower.   There are 14 dynamos, and, in addition to this, the
plant consists of 14,000 tubs, each having a capacity of 2,500
pounds, with a tare of about ½ ton, and 220 pneumatic percussive
coal-cutters supplied with air at a pressure of 70 pounds per
square inch from 18 air-compressors.   The output from one of
these mines is 2,000 tons per day, and the others have about the
same output each.

As to the reduction in cost that resulted from replacing
rope-haulage and all the mules in one of these mines by
electric traction, Mr. A. de Gennes stated that, in addition to
rope-haulage, they had 8 mules and 8 drivers, costing £4 per day
for an output of 200 tons, that is 6d. per ton.   When two 13
tons electric locomotives were installed, the output increased to
1,000 tons per day at a cost for haulage of £1 12s.—or less than
½d. per ton.*

* *Annales des Mines*, 1900, vol. xviii., page 244.

The maximum gradient that can be worked satisfactorily by electric or pneumatic locomotives is apparently 8 or 9 per cent.

As to applying either of the systems here described to our mines, the author feels that there will be great diversity of opinion among mining engineers, and it will be instructive and interesting to hear these opinions. Of course, there is no doubt that the electric system, at least in its present form, is debarred from many mines on account of the sparking, but all our mines are not fiery, and again the main haulage-headings of many others are free from gas, and consequently electric locomotives might be used. On the other hand, the pneumatic locomotives do not possess this danger, so that argument falls to the ground : and as to the cost of the latter, while it is pretty obvious that it is better than several other systems or it would not be used, it may be that it might be cheaper, under the conditions that hold in this country, to transmit the energy into the mine electrically, and then at the pit-bottom or other safe place to convert it into compressed air for pneumatic haulage and coal-cutters.

———

Mr. C. C. Leach (Seghill) asked whether, leaving out the question of horse-haulage altogether, there was any advantage in adopting locomotive haulage in preference to rope-haulage underground.

Mr. Henry Hall (H.M. Inspector of Mines, Rainhill) said he was glad to hear that Mr. Ackermann did not appear to agree with his own authorities that electric was very much more economical than compressed-air haulage. No doubt in the United States, as was pointed out in the paper, conditions were very different from what they were in this country, and the difficulty of introducing electric haulage and machine coal-cutters was much less there. After thinking somewhat about this matter, he was coming to the conclusion that compressed air, if properly applied, would be almost as cheap as electricity, and, if that were so, they need not trouble with enquiries as to whether it was safe or unsafe to introduce electric coal-cutters into British mines. Mining engineers had removed a large number of dangers from mines at considerable expenditure; why should they now think of using electricity which might prove highly dangerous, if there was no substantial advantage to be gained?

Mr. T. E. FORSTER (Newcastle-upon-Tyne) remarked that pneumatic haulage by means of locomotives could not be regarded, in the North of England at any rate, as a novelty, for they were used some 14 or 15 years ago in one of the Lambton collieries.   He did not know whether the system was still in use, but it proved only partially successful.   Its use did not become more general because, owing to the size of the roads and the general conditions, they could not get locomotives to travel a sufficient distance.   Compressed air was not then used at the same pressure as it is now, and the section, the height of the seam, and the width of the roads in America were much more favourable to its introduction.   He was much impressed with the idea of electric-locomotive haulage some 3 or 4 years ago, when he wished to make an engine-plane to run a considerable distance under the sea, but the danger of overhead wires had to be considered, and he arrived at the conclusion that they did not know how far they might go without legislative interference: so, principally for that reason, the use of ropes was adopted.   The calculations as to difference of cost were really very unimportant.   It was of course possible to introduce electric haulage into some mines—the iron-mines of Cleveland, for instance—which had seams of a large section, and they would possibly get a larger output, if the roads were sufficiently wide, by means of electric locomotives than with ordinary rope-haulage.

Mr. PHILIP KIRKUP (Birtley) asked, with respect to the conveyance of air underground for the compressed-air locomotives, whether the air-receivers were placed near the face of the workings.   As regards the electrical installation, it seemed to him that the wagonway must be fairly high, or otherwise it would be extremely dangerous to those travelling on the engine-plane to have a live wire overhead and to use the rails as the return.   Were the mine wet, this would be an important consideration where electric haulage was adopted.   He suggested that the relative cost of the system should be given " per ton per mile."

Mr. S. MAVOR (Glasgow) remarked that Mr. Ackermann spoke of three-phase *versus* direct current for machines for underground work.   Did he know of any mining locomotive driven by three-phase current, or had he any suggestions from his own experience as to the application of the three-phase motor for the purpose? With regard to the adoption of three-phase current generally for mining purposes, he was strongly of opinion that, unless the distance to be transmitted was so great that the additional cost of cables would be overwhelming, direct current was without doubt the best to employ. This would be suitable for rope-haulage and for coal-cutters, and there was no doubt whatever that direct current was the most suitable for driving locomotives. The low head-room compelled the use of a motor of small diameter, and this imposed limitations on the speed of rotation. There must also be greater clearance, on account of the vibration of a loco-motor as compared with a fixed machine. These and other requirements were difficult to combine with a three-phase motor. Reference was made in the paper to the advantage of compressed-air plant, owing to the fact that the fresh air would cool the atmosphere. It would be interesting to learn any data upon this point, as the volume ejected by the coal-cutters would be only a very small proportion of the air passing through the mine, and the result on the temperature would, he thought, not be noticeable.

Mr. M. H. MILLS (Mansfield) regretted to hear Mr. Hall express the opinion that the difference in cost between electricity and compressed air was not of sufficient importance to be taken into consideration. He felt sure that, as time went on, it would be found that electricity would be used throughout the mines, and very little compressed air or any other power would be employed. It was necessary to compete with foreign methods of working in the most economical manner, and everything must be done in our schools and colleges to further the use of electricity in mines.

Mr. M. W. WATERHOUSE (Bedworth) said that, in looking for a reason why locomotives were adopted in the United States while in Great Britain they scarcely thought of using them, it occurred to him that probably they had an unusually large number of curves in American mines. It was stated that rope-haulage cost in the first instance about three times as much as an electric plant: and it would be interesting to

have some figures shewing how this was calculated, as it seemed
a very high proportion.   As the rails were used for the return
in the same manner that tram-rails were used in this country,
he asked whether any steps were taken to bond the rails?
Referring to the remarks of Mr. H. Hall, he was strongly of
opinion that, if a certain given power was to be transmitted by
air and electricity, it would be found that greater power was
consumed in the compressed-air than in the electrical plant.

Mr. C. C. LEACH (Seghill) remarked that compressed-air
plant was by no means obsolete yet, and it had been improved
very much, but when compressed-air machinery was out of
order, the waste was enormous.   He had taken out a compressed-air pump, which required 70 horsepower, and replaced
it by an electric pump, which used 12 horsepower, and it was
doing three or four times the work.

Mr. SYDNEY F. WALKER (London) said that there was no
difficulty whatever in using two-phase or three-phase motors,
if they liked, on electric locomotives.   It had been done on
the Valtellina railway in Italy with 3,000 volts taken directly
to the cars, but he did not think that it had ever been used in
the case of small electric locomotives for use in mines.   There
was, however, no reason why it should not be: it was a
question of expediency, and of ultimate cost.   They would
have to consider the question of repairs and attendance, etc.,
and it must not be forgotten that three-phase current required
three cables whereas the continuous only required two.   There
were advantages, of course, to compensate that, but undoubtedly
it was a distinct disadvantage.   Mr. Ackermann had given
the members a good insight into American mines, and it
must be remembered that they had now to compete with
American mines, which are in the position that we were in
50 years ago in regard to the working of upper measures and
thick seams, and working levels and drifts very largely.   They
had not much gas, and there was not that regard for human life
which was insisted upon in this country. · Mr. A. B. Markham, in
his evidence before the Departmental Committee of the Home
Office, stated that in a certain district there were 14 deaths in
one year from electric trolley-wires, and there may have been
others.   If compressed-air plant was developed in the same way

as electric plant had been developed there was no doubt a great future for it, and electricity would find it difficult to hold its own, especially in fiery mines. A compromise suggested, and used in Germany, was where electricity was taken from the pit-bottom some distance inbye and there used to compress air. He did not see any reason why air should not be compressed to 1,000 pounds per square inch and carried underground, using a reducing valve, if necessary. Perhaps the writer would say where the heavy loss occurred, in converting from high to low pressure. The cost of compressing to 1,000 pounds per square inch would be greater relatively than to 50 pounds, but there would be a smaller quantity to transmit. There would be some loss in reducing the pressure, but the greatest loss arose because they could not use the full expansive effort in the motor as they could with steam. He would also like to know how they overcame the breaking of the compressed-air pipes in the American mines, for undoubtedly a large proportion of the loss arose from leakage. The statement that three-phase current was more dangerous than continuous current was correct in the abstract, but it should be qualified, for in practical application he doubted whether, all things considered, 500 volts alternating current was more dangerous than 500 volts continuous current. It was stated that, at Windber, a quantity of 2,500 kilowatts was supplied by seven power-houses; this was very small for so large a number, and should have been provided from one house. Power-stations could not supply power at the same price as that at which a colliery could supply it, especially if two or three collieries were supplied from a central plant.

Mr. PHILIP KIRKUP (Birtley) asked whether there was not great difficulty in starting machines at full load with alternating current. How were they going to overcome the difficulty of making the rotor sufficiently small to traverse a seam 18 inches thick? The question of additional cost was important, for there would be 50 per cent. extra for the cables at the same voltage.

Mr. JOHN GERRARD (H.M. Inspector of Mines, Manchester) thought that compressed-air locomotives were tried more than 20 years ago at Messrs Briggs collieries in Yorkshire.

Mr. ISAAC HODGES (Normanton) remarked that, about 20 years ago, 13 compressed-air locomotives were in use at the Whitwood

collieries, near Normanton, Yorkshire. The locomotives were
of two sizes, with cylinders 3 inches and 4 inches respectively in
diameter, worked by compressed-air at a pressure of 350 pounds
per square inch, but the installation had to be abandoned owing
to their ineffectiveness on even moderate gradients and their
general unreliability. Generally speaking, he thought that the
reason why locomotives had not found greater favour in Great
Britain was that they had much heavier gradients than in
America. With steep gradients it would be necessary to use
a rack-railway. Another reason why rope-haulage was so
largely adopted was that they knew it would do its work. It was
instructive to know that mules and horses subject to shock in
American mines learned, in course of time, to avoid the wires.
In speaking of locomotives, there could be no question that the
electric was superior to the compressed-air locomotive on account
of its smaller size. Mr. Ackermann stated that, for an output
of 1,000 tons per day, 2,500 cubic feet of free air was compressed
per minute, but that was almost a negligible quantity so far as
regards any advantage in ventilation. They did not calculate
their ventilation to such exactitude that 2,500 cubic feet for such
a mine was of any particular moment. It might be so in a
smaller mine, but even there a smaller quantity of air would
be compressed. Mr Ackermann did not give comparative figures
of the capital-cost of the plant for air and electricity, but there
was something wrong in stating that this was as four to one,
and rope-haulage as three to one.

The Rev. G. M. CAPELL (Passenham) said it might be
interesting for the members to know that within an 8 hours'
journey from London, at the Marles collieries in the Pas de
Calais, they had had a large number of small electric locomotives in use for about 10 years. They were from 6 to 7 feet long
and traversed the whole mine. He heard that their use had
been increasing and that locomotives were being supplied to
other collieries in the northern district of France.

Mr. W. H. BORLASE (Greenside Mines) wrote that there
could be no doubt of the adaptability and suitability of the
electric locomotive for underground haulage, where they could be
used in non-gaseous mines, nor of the superior advantages they
afford in metalliferous mines over rope-haulage and pneumatic

locomotives. The initial cost was very much less and the cost of maintenance very small, in comparison with the two latter systems; and it had long been a surprise to him that electricity had not been adopted by some of the leading ironstone-mines in this country. The sparking of the trolleys no doubt is sufficient to prevent its adoption in fiery mines, but he thought that this fault could be reduced to a minimum, and these useful machines made suitable for the average coal-mine by employing, when the return is through the rails, two positive trolleys, one of which would be always in sufficient contact with the wire to carry the whole of the curernt, while the other is passing under the insulator, the jar at the insulators being the chief source of sparking. At the Greenside lead-mines, Westmorland, there had been working for the last 10 years a 12 horsepower electric locomotive, running a load of 18 tons, at a cost of 1 farthing per ton per mile, for miles travelled. The power is generated by water-turbines, and is conveyed $1\frac{1}{2}$ miles at a pressure of 600 volts into the mine to the distributing station, where it is converted by a motor-transformer to a pressure of 250 volts for the locomotive-wires. The wires are placed $5\frac{1}{2}$ feet above the rail and about 20 inches apart. The adit-level, being the main roadway into the mine, is traversed daily by horses employed in the lower levels and by men going to and from their work, without fear or inconvenience. The plant has given very little trouble, and breakdowns are not numerous.

Mr. R. HOOD HAGGIE (Derby) wrote that Mr. Ackermann's paper opened up a field which is eminently suitable for discussion, as there are so few collieries in this country where any save steam-driven locomotives are used. The question of compressed air *versus* electricity in a coal-mine resolves itself into the consideration of the risks attendant upon either rather than of the cost; and the prevention of accidents to those at work underground must be the first consideration. With pneumatic locomotives there is the possibility of a joint bursting, as in the accident near Normanton (Yorkshire): this is the only real danger to life, and would no doubt be confined to a small area. With electricity, if bare overhead trolleys are used, there is danger in all parts of the pit should a fall from the roof occur. The main roads are seldom more than 6 feet high, and in most pits considerably less, so that a man may touch the wires with his head when walking, and as the most generally-used pressure is 500 volts he may be seriously injured. In thin seams and low roads, it is often the case that it is almost impossible for the man to withdraw from a contact, and the majority of the accidents that have occurred have been through want of proper care in the thorough insulation of both poles throughout the installation. There is another danger, and that is, the sparks from the trolley, which cannot be avoided. Of course storage-batteries might be used on the locomotives, but they are seldom attended to properly at a mine, and so might cause a greater expense than any other kind of haulage. Taking Mr. Ackermann's figures, it would be interesting to know what distance the pneumatic locomotives would run when fully charged to their capacity of 160 cubic feet and doing their full load on the level, and against a 5 per cent. gradient. This would help to compare the efficiency of pneumatic and electric locomotives. The reference to three-phase motors for use underground should not in the meantime be made in connection with locomotives, as so far, they have not been successfully applied, owing to the low-starting torque of these motors without slip-rings; and the use of slip-rings on a motor on so shaky a foundation as a locomotive would be anything but satisfactory.

Mr. H. F. BULMAN (Barcus Close) wrote that at Nanaimo colliery, Vancouver Island, when he visited it in December, 1897, some 500 tons of coal in an 8 hours' shift were being hauled along an underground plane 2 miles in length by an electric locomotive on the trolley-system. The road was fairly level, the gradient varying from 1 in 125 to 1 in 200. There was plenty of height, and the roof was strongly timbered to preclude any danger from falls of stone. Curves were numerous, some of them very sharp, only 30 feet in radius. The gauge of way was 2½ feet, and the steel-rails were of the ordinary flat-bottomed type, fished at the joints, and fastened to the wooden sleepers by spikes. They formed the return for the current. The tubs were made of wood, with steel wheels, 10 inches in diameter, weighing about 4½ cwts., and holding 14 cwts. of coal. From 70 to 80 tubs were run in a set, and they were connected together with the usual chain-couplings. They travelled at a speed of 6 to 8 miles an hour. The bare conducting wire overhead was hung from the timbers so as to be in a line with one of the rails, but

it was hung so as to allow about 5 inches of play on it from side
to side. On straight lengths of the road, the insulated supports
were fixed at intervals of about 25 feet, but round the curves
they were spaced closer together and there were also horizontal
radial wires fastened to the timber on the inside of the curve,
and at their other end to the conducting wire. The set of tubs
ran round these curves easily, and seldom got off the way,
although there was no check-rail nor side-supports. The pressure of the continuous current was 250 volts. The generator
was on the surface; and the depth of the shaft 650 feet. The
cost of working was said to be 3d. to 3½d. (6 to 7 cents) per ton
on the coal hauled along the plane, amounting to about 1,000
tons daily in two shifts of 8 hours each. It had been in use for
5 years without any accident. The cost of repairs was small,
being confined mainly to the rewinding of armatures and the
refacing of the commutators. This electric haulage had superseded the work of mules, which are used at this colliery. No
bits or bridles are employed, the only control being by the voice.

The PRESIDENT (Sir Lindsay Wood, Bart.) said that, with
regard to the statement made by Mr. S. F. Walker, it should not
be allowed to go forth that the members agreed with the statement
that Americans did not consider seriously the question of the
loss of a few lives, if they effected a saving of 1d. or 2d. per ton.

Mr. A. S. E. ACKERMANN, replying to the discussion, said
that as regards the danger from bare wires, and machinery in
mines generally, he had gone into the question, and would probably read a paper on this subject to the Institute. In reports
which he had published on coal-mining by machinery in
America[*] he had shewn that the accidental deaths per 1,000,000
tons raised were for Great Britain about 4·5, and for the
United States about 5·9 (this included one particularly bad
accident); but, after making certain more or less justifiable
allowances, the difference was small. It was very remarkable
that in those States in which large quantities of coal were cut
by machinery, and doing practically all the haulage by
machinery, the accidental death-rate per 1,000,000 tons was
much less than in the States not using machinery. It had been
mentioned that pneumatic or electric haulage might be used

* *Coal-cutting by Machinery in America*, London, 1902.

where there was a single track, but in the cases described in his paper they all had double tracks. A question had been asked with reference to air-receivers being placed near the working-face. No doubt they were initially placed near, but they were taken inbye as the workings extended. In one case, a pressure of 700 pounds was reduced to 70 pounds, and carried from the air-receiver to the headings. The height of the bare wires was so low in some instances that it was necessary to lower the head to avoid contact. In the case mentioned in his paper, the rope-haulage was taken out and locomotives were used in its place, so that the distance hauled was exactly the same. With regard to the use of three-phase electric locomotives, he did not see any three-phase current in use except at the Davies mine, where it was being used for driving pumping-machinery. The wires were carried in the air-way. The cooling and renewing of the air by the exhaust from pneumatic machinery was a small matter, but it occurred at the very place where the men were working. There was not a large number of curves in the mines, and there were probably fewer in American mines than in our own. The rails were not bonded in order to secure a good return for the electricity. A question had been asked with regard to the breakage of compressed-air pipes, but he did not think that this had given much trouble; the pipes were usually in remarkably good condition; they were very substantial and looked thoroughly tight and sound, and considerably better than many steam-pipes which one frequently saw. In the particular case, where he saw a man struck down by a current of 250 volts pressure, the wire was hanging in a deep loop about 2 feet from the ground; the ground was damp and the man's face was moist with perspiration and formed a good contact. He agreed with the President that Americans were not so bad as they were painted, and he could prove by figures that their accidental death-rate in coal-mines was not high.

The PRESIDENT moved a vote of thanks to Mr. Ackermann for his interesting paper.

Mr. G. S. CORLETT seconded the resolution, which was cordially approved.

————

Mr. A. C. CORMACK's paper on "Electrical Plant-failures, their Origin and Prevention," was read as follows:—

ELECTRICAL PLANT-FAILURES, THEIR ORIGIN AND PREVENTION.

By A. CAMPBELL CORMACK.

INTRODUCTION.

The title of this paper is somewhat misleading, as a very wide subject is indicated; but within the limits less can be included than was anticipated. Attention can be directed only to continuous-current dynamos and motors; and even then merely the general aspects of the subject can be considered.

The writer does not wish to give the idea that all dynamos and electric motors are liable to an excessive number of breakdowns in comparison with the breakdowns met with in non-electric machines. Unfortunately, a large number of bad and unsuitable plants have been installed in this country, and the results met with in such plants have in many cases been looked upon as inseparable from and characteristic of all electric plants, but such judgment is inaccurate and unfair. With reasonable care in design, manufacture and working, electric machinery can be, and has been, made more reliable than steam, gas or oil power-plants. This remark should be kept in view when dealing for a time with a somewhat blacker aspect of the case.

At the end of the paper, an analysis is given of the breakdowns which have occurred to several thousand machines, ranging in size from an 800 kilowatts dynamo to a $\frac{1}{2}$ horsepower motor, the observations extending back for a period of about four years. The group of machines from which these statistics are prepared are under somewhat better than average conditions, so far as superintendence is concerned. Some causes of breakdown have been entirely eliminated, and others have been very much diminished. Further, owing to the elimination of a number of very bad machines, plants from which these statistics are prepared are considerably better than the average. No coal-cutting machines have been included, as the statistics from these are far from complete. Three tables have been prepared:—Table A is descriptive of the nature of the accidents. When an accident was of a compound nature, several parts being injured, each is included in the table, excepting those cases where the failure of one part is certain to follow on the failure of another. In Table B, the point of origin of each breakdown enumerated in Table A is traced: and in Table C an analysis of the causes of the breakdowns is given.

The preparation of these tables has been somewhat complicated by the complex nature of some of the derangements. In some cases, the derangements were not caused by any single fault, and for this reason in preparing Table C, it was necessary to adopt a system of marks allocating the same number to each breakdown, but distributing them if more than one fault was the cause of the breakdown.

DAMAGE TO DYNAMOS AND MOTORS.

Table A is descriptive of the kind of accident, but as the others will be of greater interest, this table is not dealt with in detail.

Damage to Mechanical Portions.—In 25·05 per cent. of the total number of breakdowns, purely mechanical parts of the machines were damaged, and the first striking feature is the small number of broken shafts met with. This is due to the fact that, in order to provide the requisite stiffness, shafts are as a general rule made much stronger than is necessary to bear safely the bending and twisting stresses which they must undergo. Shaft-breakages seem to be almost confined to a few makes of machines.

Damage to Binders and Fastenings.—In 9·4 per cent. of the breakdowns, binders and fastenings have been damaged, such damage usually consisting in a bursting of the binder from purely mechanical stress.

Damage to Bearings.—In 8·6 per cent. of the breakdowns, the bearings were damaged. Such damage consisted in seizing of the brasses, or in melting out of the white metal. This latter accident accounts for a number of damaged binders.

Various Mechanical Parts Damaged.—Of the other mechanical parts damaged, 4·7 per cent. consisted mainly of fractured beds, brackets, pulleys, etc., and were almost all of the nature of consequent damage.

Damage to Brush-gear, Connections and Terminals.—In 7·01 per cent. of the accidents, brush-gear was damaged, 1·56 per cent. being purely mechanical. In 2·35 per cent. of the failures of brush-gear, the insulation had given way. These failures were mainly of the kind where arcing takes place across an insulating ridge separating a live metallic portion from the frame of the machine, and have been caused mainly by dust and oil collecting on these ridges. The remaining 3·1 per cent. of accidents to the brush-gear were really due to connections and terminals of the machine, but it has been convenient to place these under this head.

Damage to Field-magnet Coils.—Of the accidents in which the field-magnet coils were damaged, 10·16 per cent. were mainly cases in which the insulation between the coil and coil-former or cores had given way. The 1·56 per cent. representing coils gone altogether (that is, arcing from series to shunt on the same coil or arcing between two separate coils), show that this is not altogether an uncommon accident. The latter accident is most frequently met with in bi-polar machines, where the coils often touch. In this case, unless the outer layers are at the same potential, there is considerable risk of accident.

Damage to Armatures.—The failure of armatures constitutes by far the most serious item on the table. The form which such failure most frequently assumes is the burning-out of coils. This occurred in the case of 29 per cent. of the total number of accidents observed to machines. In many cases the entire armature was burned out; and in others only a few coils.

The percentage of coils gone to earth, 19·5 per cent., is also a large item, and most of these failures were due to destruction of insulation. The accumulation of dust, or other conducting matter, on the insulating ridges, such as is met with at the end of the slot-insulation or at the end of the coil-supports of a barrel-wound armature, accounts for the majority of the remainder.

Breakages of wire, occurring in 15·6 per cent. of the derangements, were confined mainly to the smaller sizes of wire-wound armatures, and usually took place at or about the points where the wires leave the coils or join the commutator.

The joint-failures, 8·6 per cent., are more common than they ought to be, although some were undoubtedly caused by short-circuiting in another portion of the armature.

The 3·1 per cent. in which binders were burnt are those cases where fusing took place by arcing between the binder and armature-wires, and did not arise from purely mechanical stresses.

Damage to Commutators.—The commutators are frequently damaged, this having occurred in 33·6 per cent. of the breakdowns. The failures of commutator-insulation have for the most part occurred at the ends, causing leakages to the earth-connected portions of the commutator or between bars or radial connections. In the majority of cases, these failures are caused by the presence of dirt, or other foreign matter, on the insulating ridges or between the radial connections. The mechanical failure of commutators, amounting to 10·95 per cent., consists of burst commutators, fracture of commutator-lugs, and failure of keying of commutators. The 12·5 per cent. of the cases where surface-fusion of the commutator has occurred are cases where the surface of the commutator has been destroyed by violent sparking, sometimes between segments, the latter fault being the usual accompaniment of a broken wire.

ORIGIN OF BREAKDOWNS.

Table B is the most interesting, as here the breakdowns mentioned in Table A have been analysed with a view to showing the portions of the machines at which the breakdown originated or the portions which by being faulty caused the breakdown.

Faulty Shafts.—The breakdowns originating in fracture of the armature-shaft amounted to 0·78 per cent. This, as already stated, is a somewhat rare accident; and is, as might be expected, usually accompanied by destruction of the armature-windings. But in one case under consideration, the fracture occurred outside the pulley-end bearing. This was caused by weighting the driving pulley with a heavy fly-wheel, which overhung a good deal. The other breakages in shafts took place, as might be expected, at those points where oil-grooves were turned in the shaft, or where collars were formed for the purpose of supporting the core-plates, etc. Breakdowns to shafts have been caused by the armature coming out of the centre and being considerably nearer to the bottom of the pole-bore than to the top, with the result that the magnetic pull on the armature which was formerly balanced and equal was then in a downward direction, and produced a severe bending moment on the shaft. In these cases, of course, the trouble originated at the bearings, not at the shaft.

Faults caused by Bearings.—This same magnetic pull is no doubt responsible for aggravating some of the 8·6 per cent. of the accidents which originated in the bearings. If a bearing heats and the armature gets below the centre, worse heating is sure to follow.

Most of the defects in bearings are due to improper lubrication. In cases, where use is made of automatic lubrication of the type having a loose ring running on the shaft dipping into an oil-well below, defects in bearings have been caused by the breakage or sticking of these rings.

An interesting defect in dynamo- and motor-bearings, which sometimes occurs and requires watching to prevent damage, is the wearing of the bearing upwards. This sometimes occurs in cases where the armature has been placed nearer the top of the race than the bottom, in order that the upward pull of the magnets may relieve the pressure on the bearings. When the upward pull of the magnets is greater than the gravitational attraction, the bearings wear upward, and allow the armature-core to approach the pole-tips when working. This, owing to its liability to escape detection, is a somewhat dangerous defect; for when the machine is examined at rest the armature-clearance will appear satisfactory.

The failure of bearings by fusion of white metal is usually accompanied by stripping of the armature-windings, and sometimes also by damage to the armature-conductors, this latter being most frequent in the case of smooth-cored armatures. In slot-wound armatures, damage is often done by the core-plates cutting through the insulation of the slots. Such damage may not be apparent at first, but frequently shows up some time later.

The percentage of defects originating in bearings would, the writer believes, have been much higher if the machines had not been inspected periodically.

Failures due to Mechanical Portions.—Breakdowns traceable to failures of the mechanical portions of the machines other than those specially mentioned amounted to 1·65 per cent. These were almost all of an accidental nature, not being preventable by reasonable care in manufacture or working.

Failures due to Brush-gear.—Brush-gear failures from purely mechanical causes originated 3·9 per cent. of the breakdowns. They do not form a serious item. In the case of metal brushes, accidents were not uncommonly caused by loosening of the pinching-pin securing the brush in position. This usually resulted only in damage to the commutator-surface, but one case is included in which a shunt field-magnet coil went to earth from this cause.

In the case of carbon-brushes, a similar accident usually results in more serious damage. When a brush drops out, the metal portion of the holder comes down on the commutator-surface, and (being broader than the brush) touches more segments than the brush did: also, having a lower resistance, it allows very heavy currents to flow across it, and in the coils connecting the commutator-segments immediately below it. Such an accident frequently results in fusion of the joints of the armature-wire with the commutator. A number of the mechanical defects in brushes were due entirely to faulty or flimsy design. A fault, which happily is not so common as it used to be, is that of allowing springs to carry part of the current to the brushes. This results in overheating and loss of temper of the spring and of the attendant. The tempers can usually be kept by providing metallic connections direct to the brushes, and by insulating one end of the spring.

Breakages of brush-holder spindles are usually due to carelessness in screwing up.

In the analysis here made only 2·35 per cent. of the breakdowns are due to brush-insulation. This is one of the frequent causes of accidents, which efficient periodical inspection practically eliminates. Breakdowns, here, frequently lead to complete burning-out of armatures. They were usually caused by the accumulation of oil on the insulating ridges separating the live portions of the brush-gear from the rockers or from the frame of the machines; such accumulations temporarily bridge over the ridge and allow a flash to pass, after which the arc is easily maintained across the gap. Where the insulating ridges are of ebonite or other brittle material, they are often cracked by mechanical stresses caused in screwing up. Such cracks usually fill up with oil, and are sure to cause trouble. In many cases, the insulating ridges are too small; though large enough to prevent cross-sparking in a laboratory, or in a new machine on the maker's test plate, they do not allow any margin for the small accumulations of dust which in many cases are all but unavoidable in practical working.

Failures due to Leads and Terminals.—2·35 per cent. of the breakdowns were due to the failure of leads and terminals. Terminal failures are practically of the same nature as those met with in brush-insulation. The failures due to defective leads are almost all owing to their being improperly protected. and allowed to come into contact with the commutator or with an earth-connected portion of the machine. Brush-leads in particular are frequently too long, and the danger from the leads touching the machine-bed is aggravated if they are at all oily. The connecting leads between the field-magnet coils are occasionally of a flimsy nature, and cause trouble by breaking when being cleaned, or by touching the frame of the machine and burning-out their insulation. The latter defect has resulted in the burning-out of field-coils.

Failures due to Field-magnet Coils.—7·8 per cent. of the total number of accidents were due to failure of the insulation of the field-magnet coils. No distinction has been made between cases in which the insulation separating coils from their formers and those cases in which insulation between wires failed. Most belong to the first class. About half of the breakdowns were caused by perishing of the insulation due to overheating, this overheating being caused by working the coils at higher temperatures than is proper.

A somewhat common cause of field-magnet coil breakdowns in compound and shunt machines is the improper arrangement for switching off the current from the shunt-coils, or lifting the dynamo-brushes before the magnets are demagnetized. When the current from a field-magnet coil is switched off in such a manner that the core demagnetizes rapidly, the coils surrounding the core are caused to cut magnetic lines of force very rapidly. This may produce a voltage at the coil-terminals many times greater than the dynamo-voltage, and frequently sufficient to puncture the insulation. When the machine is next run, it will either refuse to work, or burn out a field-magnet coil, perhaps also an armature. The defects traceable to this cause have, of course, been credited to starting switches.

Another frequent cause of the failure of field-magnet coils is the presence of lubricating-oil. The effect of oil on insulation will be more fully dealt with in a later part of this paper.

Breakages of field-magnet wire are only responsible for 0·78 per cent. of the total number of the accidents. Only in one case has further damage been done than breakage of the wire itself. This was with a motor where the fuse failed to act, and the armature was burnt out.

Failures due to Armature-core Insulation.—47·80 per cent. of the breakdowns originated in the armatures, and this is clearly the most vulnerable part of a dynamo or motor. The failure of the insulation between the windings and core (which includes slot-insulation, insulation of core-supports, etc.) is responsible for 18 per cent. of the total number of machine-failures. Failures of this nature are often accompanied by partial burning-out of the armature-coils, but in more than half of the cases, the damage has been confined to the conductors immediately over the point which failed, these being fused and burnt by an arc passing between the windings and core. Failures of insulation to core have been divided into two classes: —(a) Those in which flashing has taken place through the insulation; and (b) those in which flashing has taken place past or over the insulation.

In class (a), flashing through the insulation causes 12·5 per cent. of the breakdowns. The damage is usually more extensive in class (a) than in class (b), and more difficult to repair. The great majority of failures have been due to the insulating materials used being unsuitable or insufficient to meet the working conditions. It is here that the greatest temptation seems to be given to designers, as a small reduction in the thickness of the insulation and the size of the conductors makes a considerable reduction in the overall dimensions and the cost of the machine. Almost all the failures in this class were due to the insulating-material having lost some of its original mechanical strength and electrical disruptive strength; sometimes also its insulating properties.

It may be desirable to explain that by disruptive strength is meant resistance to the passage of an electric spark; by insulating properties, the resistance to the passage of a small leakage current (insulators being only bad conductors). A false security is often felt when a test indicating a high-insulation resistance between the windings and the frame is obtained. Unfortunately, high electric disruptive strength is not always accompanied by high-insulation resistance; and sometimes an increase in resistance means a decrease in disruptive strength.

An impression seems to prevail that insulations consisting of organic substances are comparatively safe, short of the point at which they become carbonized, but this is not so, and many failures have been caused by the fracture or pulverization of insulation, owing to its disruptive strength being so lowered that it can be punctured by accidental rises of pressure which appear to occur in normal working. It is worthy of note that few break-downs have occurred where machines have the armature-insulation or a portion of the thickness of the armature-insulation made of mica.

Some of the failures in class (a) have been due to the insulation being destroyed by rough handling in the construction of the machine. Some of the trouble occurs in the length of the slot, but it is of more frequent occurrence where the conductors leave the slot, particularly if a bend occurs there. Another frequent cause of this class of breakdown is the presence of moisture and decomposed oil in the insulating material.

In class (b), which is responsible for 5·5 per cent. of the total number of breakdowns, flashing takes place from the armature-conductors to the armature-coil supports, or other earthed part of the armature where the conductors approach the earthed metal work, and are separated from it by the prolongation of the insulation. This flashing over, which results at least in the destruction of the ridge, is for the most part caused by the presence of damp and dust on the ridges. As in the case of commutator-ridges these were frequently too short to meet

practical working conditions. Some of the armature insulating
ridges are difficult to get at for cleaning, notably those on the
commutator-end coil-support, where they should be made much
larger than is the general practice. The failures in class (b)
are probably much below the percentage of failures due to this
class of defect in ordinary practice; for this point received
special attention in the machines under consideration.

Failures from Armature-coil Insulation.—Failure of coil-
insulation, that is, failure of the insulation between the coils,
was responsible for 11·7 per cent. of the total number of
accidents. A number of accidents were in slot-wound armatures.
A few of the cases occurred between conductors in the slots.
In slot-wound armatures, with cylindrical end-connections, the
fault occurs with greater frequency in the slot itself than in that
portion of the wire outside the slots. A frequent point of
failure is the insulating-belt between the cylindrical end-connections in a barrel-wound armature. In an armature with
evolute end-connections, the failures between the coils take
place frequently at a point in the windings nearest to the shaft.
The arrangement for supporting the cross-over here is sometimes very inefficient mechanically. In drum-armatures, the
insulation burnt is usually that which separates neighbouring
conductors at maximum potential difference, although the
insulation is usually supposed to be proportioned to meet this
increase of pressure. Comparatively few breakdowns occur
through the failure of the insulation of the wires themselves.

The reasons for the failures of the insulation are practically
the same as those given for failure of the insulation to the core,
except that in the case of smooth-cored armatures abrasions of
the conductor-insulation due to relative motion play a more
important part. As a general rule, owing to the end-connections being better ventilated than the conductors on the core,
the insulation, here, seems to suffer less from the effects of high
temperature.

The failures of insulation above enumerated frequently
result in the entire burning-out of the armature.

Failures due to Joints.—Joint-failures are responsible for
7 per cent. of the total number of accidents, and most of these

joint failures are caused by bad workmanship. They have been, for the most part, confined to one or two makes of machines. Unfortunately, if a faulty joint is nicely finished outside, it is very difficult to detect until it causes trouble. This usually begins on the commutator-surface. A number of joint-failures are due to improper fastening of the commutator, but such have not been included in the percentage.

Failures from Binders and Fasteners.—In 1·65 per cent. of the failures, the damage originated in the mechanical bursting of the binders and fastenings. This is a smaller percentage than would be attributed to this cause in ordinary practice. As a general rule, binders are of sufficient strength to resist the ordinary stresses due to centrifugal force and the magnetic pull on the conductors. Those cases where binders have merely partly loosened out by unwinding have not been considered accidents, if no further damage resulted. One cause of the bursting of binders in large machines, with high temperature-rise, is the stress in the binder set up by the expansion of the heated conductors. This can be overcome by placing a cushion of soft tape below the binder, in addition to the usual mica-insulation. In a number of slot-wound machines, binders in the pole-tunnel are dispensed with, wedges of wood or fibre being depended on to hold the conductors in place. Owing to shrinkage and the liability of the material so used to become brittle, the writer does not look upon such fasteners with favour, unless they are supported by binders. The number of break-downs from this cause is small, but it has only been kept so by careful supervision.

Failures from Binder-insulation.—7·8 per cent. of the total number of breakdowns were due to the failure of the insulation of the binders. The writer feels sure that many breakdowns commonly attributed to bursting of binders from mechanical weakness have really been caused by a portion of the binder being burnt by arcing between the binder and the armature. Most of these accidents have occurred on smooth-cored armatures, although slot-wound armatures appear to be just as liable to this defect in the region outside the pole-tunnel. The largest number of defects seem to have been caused by the binders becoming loose, and shifting to the edges of the insulation. A number are apparently due to dampness of the insulation, and to the neglect of small tears in the edges of the insulation. Some defects are directly due to careless handling of the armature when being put in position. Dust does not appear to cause much trouble with binder-insulation. There is little lodgment for it, and it appears to be swept away by the rotation of the armature.

Failures from the Driving Horns.—Defect, or absence, of driving horns is responsible for 1·65 per cent. of the breakdowns. This applies only to smooth-core machines, and comparatively few of these are now being built. The failure, or want, of driving horns produces trouble by allowing the conductors to be dragged round by the pull of the magnets, causing cutting and tearing of the insulation.

Ring-wound armatures are now little used, which seems, in the case of small motors for high pressures, rather a pity. They may come into fashion again for such machines, and perhaps also for very large machines. Where such armatures have smooth cores, they certainly do require driving horns on the outside of the ring, the interior spiders not being sufficient to prevent slipping of the windings without causing undue stresses on the insulation of the core-ends.

Failures from Commutator-insulation.—Commutators are responsible for 14·9 per cent. of the breakdowns, and these may be divided into two classes. The first, in which the insulation has broken down, is responsible for 9·4 per cent. The failure most frequently occurs between the bars and supporting rings, the parts subject to the greatest difference of pressure. Such breakdowns often result in the burning out of the armature-windings. To a much slighter extent, breakdown of insulation between the segments occasionally occurs. This seldom results in more than a few coils being damaged. Only in a very few cases does a breakdown occur in the body of the insulation, permitting flashing through it. These few cases have all occurred where a material which absorbs moisture has been substituted for mica. Almost all the accidents to commutator-insulation are due to the collection of oil, dust, etc., on the insulating ridges separating the segments from the clamping-nuts. In many cases, these ridges are made much too short. Sometimes there are no ridges at all, the insulation being flush or even sunk below the level of both bars and rings. The insulation at the outside end is usually most exposed to such accumulations, but being more easily cleaned, fewer breakdowns appear to occur here than at the back end where cleaning is very difficult, sometimes impossible.

Very few cases have occurred where the insulation has broken down, owing to the commutator-wedging arrangements slacking back and permitting sufficient chattering to pulverize the insulation. This usually causes sufficient sparking at the brushes to draw attention to the defect, before more serious mischief occurs.

Faults from Commutator-fastenings.—Defective fastening and keying of the commutator has caused 5·5 per cent. of the total number of accidents. Defective fastenings, sometimes, although rarely, cause breakdowns of the insulation as described in the previous paragraphs. Bursting of the commutator by centrifugal force is not now so common as it used to be, and few cases are met with. As a general rule, the proportions of the parts are sufficient to resist the simple centrifugal stresses, and most of the bursts usually met with have been caused by improper locking arrangements. A cause of bursting of commutators which does not figure at all in this table, is the bursting of the commutator owing to the bars being allowed to wear down too far.

The defect or absence of keying of the commutators has caused in this class the largest number of breakdowns. When the keying is defective, the commutator is driven by the armature-wires, and sometimes the commutator and the armature have slight relative motion on the shaft. This has caused a large number of breakages of the wires joining the armature to the commutator. In commutators of the type having separate radials, fracture often takes place at the joints, these breaks causing flashing at the commutator. Cases were not infrequent where, owing to bad keying, the whole of the commutator-radials or of the wires joining the coils to the commutator-segments required renewal, as they had been rendered very brittle by bearing variable mechanical stresses. The remedy is the secure keying of the commutator-sleeve to the shaft; or, in the case of large machines, the bolting of the commutator to the core-sleeve.

Defects from Motor-starters.—Starting switches of motors have been the cause of 4·7 per cent. of the breakdowns in the table, and the writer may say that this figure has been kept low by timely alterations to a number of dangerous switches. In some cases in colliery-work, no starting switches were used at all, but the use of these was insisted on. The most frequent trouble has occurred with the starting switches of shunt-wound motors. These are sometimes so arranged that the shunt-circuit is broken in such a manner as to allow of high inductive rise in pressure. The simple expedient of arranging a conducting circuit across the terminals of the shunt before it is opened, effectually prevents trouble from this cause. If the armature be left across the shunt, it is sufficient.

In a number of cases, no arrangement has been made for disconnecting the shunt-circuit of the machine, this being left on always. While in damp situations this may contribute to keeping the machine dry, this advantage does not warrant the danger that is introduced. In the event of the circuit being switched off from the engine-house, when the machine is at rest, the shunt-circuit is often broken in a dangerous manner. This has been a cause of breakdown of field-magnet coils, and not infrequently of armature-insulation also. The breakdown of armature-insulation can easily be understood when it is remembered that one end of the shunt is always left connected to the armature (and with an earth on the main connected to the other terminal of the shunt), the point at which the pressure to earth would be highest would be in the armature, and the flash to earth would take place there. The writer may here say that he attributes a number of faults in mains to the same cause.

While he believes that improper proportioning of the resistance of starters has shortened the life of many motors, there are few cases where breakdowns can be attributed solely to this cause. A not uncommon fault in starters which has caused burning out of armatures is a breakage of the wire near the last stop. In such cases, a careless attendant occasionally continues the use of the starter subjecting the armature to very heavy starting currents by applying practically the full voltage to a standing armature.

In this connection may be mentioned another very common fault which is found in starters for shunt motors, that is, taking the connection from the last stop of the resistance. This results in the voltage applied to the shunt at the moment of switching-on being lowered to the same extent as the voltage applied to the armature, with the result that to get the necessary starting torque, the full-load current has to be much exceeded. The starting torque can be kept up by attaching the shunt-wire to the first stop in the resistance. When the switch is in the " on " position, the current for the shunt requires to flow back along the resistance, but as a general rule this is a matter of little importance, and where it is desirable to arrange otherwise, an additional stop can be put in the switch to cut out this resistance after the lever has reached the last stop.

Untraced Damage.—Origins.—In 3·13 per cent. of the breakdowns, the point of origin could not be traced. In some cases this was owing to there being no opportunity of making an investigation. In other cases, the damage was too extensive to permit of any definite conclusion being arrived at as to which part gave way first.

Derangement from Various Causes.—The 1·56 per cent., various, is made up of those cases where the damage originated in some portion of the machine other than those above specified.

CAUSES OF DERANGEMENTS.

Table C sets forth as far as possible the conclusions reached as to the cause of breakdowns. Some of the headings in this table mean practically the same thing, but it seemed desirable to make out separate categories. For example, the heading "Dust and damp" might have been merged in the " Defective attention " heading; but a number of breakdowns have been caused by dirt and damp, that could hardly have been attributed to defective attention on the part of the attendant.

Constructional Defects.—The first of the main headings, " Constructional defects," is responsible for 39·36 per cent. of the total number of breakdowns. With greater laxity in supervision, resulting in a greater number of breakdowns, this heading would probably have been a smaller percentage. A number of the defects in design, construction, etc., were known to exist when the machines came under observation, but they were usually of such a nature that little could be done to remedy them without entirely reconstructing the machines.

Bad Design and Perishing Insulation.—Almost half of the 39·36 per cent., that is 18·36 per cent. of the total failures, were due to bad designs, and 7·4 per cent. to premature deterioration of the insulation under normal conditions of working. The writer looks upon this latter as bad design, because it is the use of unsuitable materials, or of materials under unsuitable conditions.

Within the limits of this paper it is impossible to enter fully into the defects in design. The criticism of design might occupy many papers. Care has been taken not to attribute a failure to this cause, unless the fault has been perfectly clear. Comparatively few of the faults were found in the mechanical portions of the machines. Those usually took the form of insufficient keying, bad arrangement of bearings, and other defects familiar to all mechanical engineers. The great majority of the faults in design appear to be in the electrical portions of the machine, and for the most part take the form of supplying too little insulation, insufficient insulating ridges, and improper arrangements for supporting the conductors and their insulation. The writer is afraid that, owing to the stress of competition, a number of designers sacrifice to a great extent reliability to cheapness. Particularly in the case of small and medium-sized motors the tendency is to design machines having higher rises of temperature than experience has shown to be compatible with reasonable durability of the machine. No margin appears to be left for the slightly abnormal conditions that almost every machine sooner or later meets with in its working life.

Bad Workmanship.—Bad workmanship is responsible for 13·6 per cent. of the breakdowns. A number of these defects occurred in the mechanical portions of the machine. A considerable number were due to carelessness in winding, resulting in tearing or abrasion of the insulation. A large proportion was caused by bad and defective jointing of the conductors, and insufficient and careless soldering.

Conditional Defects.—Overloading, over-rating and unsuitability for the purpose are responsible to the extent of 3·71 per cent. of all the breakdowns. These conditional defects occur with less frequency than would have been met with in usual circumstances. Many conditional defects were removed at the time that the machines came under observation.

Overloading.—1·37 per cent. are due to wilful overloading; 1·56 per cent. to over-rating—which practically amounts to the same thing, except that this is due to the seller and not to the user of the machine.

Unsuitable Machines.—The 0·78 per cent. refers to those cases in which trouble arose through using a machine which, however good for other employment, was not suitable for the purposes and circumstances in which it was running.

Maintenance-defects.—Maintenance-defects are responsible for breakdowns, to the extent of 31·46 per cent.

Damp and Dust.—Damp and dust, 7·4 per cent., refers to those cases where a reasonable amount of attention was given to the machines by the attendant, but where owing to unfavourable conditions it was not possible to keep the machine sufficiently clean and dry to prevent trouble. Dust is frequently somewhat hygroscopic, absorbing quantities of moisture. Defects from dust occur most frequently in semi-enclosed motors. These are splendid automatic dust-bins; the provision for cleaning usually is insufficient; many of them cannot be cleaned properly without being taken apart, and it would in most cases be impracticable to do this every time the machine required cleaning. The defects in semi-enclosed motors might quite well be attributed to bad design or unsuitability, but the writer leaves the reader to place it as he thinks fit, with the advice to adopt, where necessary, wholly enclosed motors, large enough to run cool; and easily accessible open-type machines in all other situations.

Rough Usage.—The 1·56 per cent. explains itself. It has usually been merely a form of carelessness.

Defective Attention.—Defective attention, 22·5 per cent., refers principally to those cases where the machines have not had the attendance necessary to ensure satisfactory working.

The inattention takes the form of neglect of cleaning, allowing oil to gain access to the windings, setting the brushes badly, want of lubrication, carelessness in starting and stopping, and the neglect to perform the adjustments necessary for the proper working of the machine.

Damage by Oil.—The main source of trouble is the presence of dust and oil on the machines; the latter is most destructive; windings that have been allowed to absorb any appreciable quantity of lubricating oil come to grief sooner or later. The remark is often made that oil is a good insulator. This is so in its usual condition, but oil in the windings of electrical machines at their ordinary working temperatures, not merely rots the insulators, but usually ceases to be a good insulator. This appears commonly to be attributed to the high temperatures; but while it may be admitted that these exercise influence, examination of a large number of faulty windings where oil was present seems to point to the conclusion that electrolysis plays a most important part in causing the trouble.

Owing to the weakness of the currents passed by the insulating oil in the early stages, the change appears to be slow, but becomes more rapid as the oil begins to conduct better, possibly through portions of it combining with the metal of the wires to form conducting metallic salts. If electrolytic action plays an important part in the breakdown of oily insulation, derangements due to this cause will be less frequent in alternating-current machinery than in continuous-current machinery. This is exactly what is found, even when the temperature-rise in alternating-current coils is greater than with continuous current. The presence of oil on windings is usually due to improper oil-catching arrangements. Some of the oil-catching arrangements, while sufficient to meet normal conditions, are insufficient to deal with the abnormal rush of oil caused by the flooding of bearings. This defect, due to carelessness, is most frequently met with in automatic lubricators where no arrangements have been made to prevent the oil-wells from being filled too full.

Deficient Lubrication.—Deficient lubrication is a frequent cause of accident, and with white-metal bearings the result is often disastrous.

Unknown Causes.—The breakdowns, for reasons already explained, attributed to unknown causes amount to 13·1 per cent.

Accessories.—The faults in accessories, 5·48 per cent., may seem low. This is accounted for by the fact that, in some cases, an accident may have been due to the combined defects of a fault in the machine and a fault in the accessories, if the machine-defect was bad enough to have ultimately caused the accident, the accessories were not condemned.

SUMMARY.

The significance of the preceding part of the paper relating to breakdowns is only as that of wrecks and rocks to a pilot. They teach how disaster may be avoided. In the case of breakdowns, there is one good, deep, safe channel. There are often many very tempting side-branches from this channel, usually not easily discriminated from the channel itself. They may be all that is desirable at first, but they all sooner or later lead to wreck. The only safe way is found in installing suitable first-class electrical machinery of ample rating for the purpose, and in looking after it well. This, although safe, is not always a very easy channel to find and keep to, so if you are not quite sure of it yourself you ought to get a pilot.

To enter fully into the detail-question which should decide the adoption of any electrical scheme is beyond the scope of this paper, but the writer feels compelled to touch on the general aspect so as to justify the plea for good plant. The root of every electrical undertaking is financial, and financial considerations rightly rank first in the arrangement of any such scheme. The problem to be solved in every case is how to attain satisfactorily the desired end in the cheapest possible manner. This involves a careful balancing of capital-expenditure, standing charges (interest, depreciation and sinking fund) and running charges, including employers' liability, the latter being the financial aspect of safety.

Class and Rating of Plant.—A doctrine now much preached in relation to machinery is that of the merry and short life, or high rating. Probed to the bottom this only means low capital-expenditure and high upkeep-charges. This is one of the dangerous side-channels. Such a scheme employing cheap, highly-rated plant can sometimes, on paper, be made to look more favourable than one entailing more capital-expenditure and providing rationally-rated long-life plant; but, on sifting out, it often appears that the standing charges are insufficient to meet the more rapid wearing out of machinery. Even if satisfactory allowance is made for these charges, there is another point which affects mining work more than any other industry, that is, the necessity for reliability. High rating of machinery only means working nearer the limits of safety, and, therefore, it entails a greater liability to derangement (a phase of the merry life). Reliability is only another point in the financial aspect, and if one could foresee and estimate the losses due to the stoppage of plant, and put those charges in the balance-sheet, the high-rating sheet would in the majority of cases be at once discarded.

If it is found, on considering the desirability of adopting any electrical scheme, that to make it compare favourably with a satisfactory non-electrical alternative method, the permissible capital-expenditure and standing charges do not allow for the best obtainable rationally-rated plant, leave electricity alone and adopt the alternative method. In the great majority of such cases (at least 99 per cent.), owners would be acting in their interests and in the interests of electrical engineering in not using electrical plant. In the remaining 1 per cent. of the cases, the course is very doubtful. Having decided to obtain high-class plant of the type suited to the requirements, the next step is to see that this is obtained.

Plant-tests and Guarantees.—The percentage of breakdowns attributed in the preceding pages to defective design and workmanship show how important it is to make sure that the article supplied is actually first-class; very often the only condition attached to the supply of the dynamos and motors is that they shall drive a given load. Occasionally high efficiency is specified and obtained, but the user of electrical plant requires more than this. He ought to ensure as far as possible that the plant would be likely to continue the satisfactory performance of its duties. Unfortunately there is no simple test or set of tests that can ensure this. A satisfactory temperature-test helps. An insulation resistance-test is of little service. A guarantee of twelve months, while being useful, does not ensure continued satisfaction, as it is often only at the expiry of this period that many defects in design, particularly in relation to insulation, appear.

Inspection of Plant.—The only way to obtain reasonable expectations of the continued satisfactory working of the machine is to combine a careful examination with the above-mentioned tests. This examination should show whether, in the light of past experience, the proportions and arrangement of parts and the selection of materials are such as could be expected to give satisfactory results. Unfortunately, this cannot always be ascertained by examination of finished machinery. There are also many defects in workmanship which may not be discovered after the completion of the machinery.

Details of Design requiring Attention.—The safest course appears to be as follows:—First obtaining suitable, satisfactory guarantees of efficiency, output, freedom from sparking, and of general satisfactory working; then the detailed design of the machine should be gone over on the plans to see that all points are satisfactory; and afterwards the machine should be examined at several stages in the course of manufacture, to ensure as far as possible that the workmanship is satisfactory.

It is not proposed to enter into the question of detailed design. A number of the points which require watching are given on the pages dealing with breakdowns. The most important are:—Use of suitable insulating-materials in the armature; the provision of sufficient insulating-ridges where bared, or thinly insulated, portions at considerable difference of potential abut on one another; the general mechanical design of the machine, and particularly the keying and locking of the various parts; the provision of suitable lubrication of bearings; and the prevention of lubrication where it is not wanted.

Inefficient Insulation.—The writer would like to say, in connection with the subjects of insulation and temperature-rise, that consideration of a large number of breakdowns leads to the conclusion that, owing to the stress of competition, insulation is in many cases being cut down to smaller dimensions than experience warrants, and is frequently subjected to higher temperatures than is desirable.

Temperature-rises.—It is certain that under ordinary circumstances, where organic insulating-materials, paper, cotton-tape, fibre, etc., are solely used, ultimate economy is sacrificed in allowing temperature-rises of more than 60° Fahr. Where a portion of the insulation of the machine is of suitable mineral matter, such as mica, micanite, 70° Fahr. appears the highest desirable maximum limit. While slight difficulties of construction and increase in overall dimensions are sometimes entailed by having a portion of the insulation of mica, it would appear, after due allowance for its friability, that its use is much to be desired in the slot-windings of armatures and all warm portions of dynamos and motors where thinness of insulation has a considerable bearing on the economy of construction of the machine.

Erection and Attention.—Having obtained suitable machinery, it is necessary for the attainment of the desired end that it should be well treated. In installing, every possible precaution should be taken to guard against the two great enemies, damp and dust. Afterwards, in working, it should be kept free from damp, dust, and, except in the bearings, oil. Where enclosed motors are used, these ought to be taken apart for thorough cleaning, at least once a year, and oftener if possible.

It must not be assumed that this implies that electrical machinery requires careful nursing and highly skilled attention. With machinery of the character here advocated, this is far from being the case. In almost all but very complicated plants, all that is required is the periodical cleaning, lubrication, and attention to the simple adjustments of brushes, etc., with due observations of a few simple precautions in starting and stopping. This unskilled attendance is of such essential importance that it is advisable to place a check on it; and while the continuous attendance of a highly-skilled electrician is not necessary, it is desirable that such a person should be available to give occasional advice to the attendant and to test and examine periodically the machine for incipient defects that can be detected by an expert. If this be done, the proverbial stitch may be made in time. The writer feels sure that those who have taken, or will take in the future, the safe course herein indicated, will agree in maintaining the high opinion expressed in the beginning of this paper regarding the reliability of electrical plant.

Conclusion.—In conclusion, the writer desires to say that the main facts on which his paper is based are derived from the records of the National Boiler and General Insurance Company, Limited, for which company he acts as electrical engineer, and the records have been kindly placed at his disposal for the preparation of this paper by Mr. Edward G. Hiller, the chief engineer.

TABLE A.—SHOWING THE RELATIVE FREQUENCY WITH WHICH VARIOUS PORTIONS OF DYNAMOS AND ELECTRIC MOTORS WERE DAMAGED WHEN BREAKDOWNS OCCURRED.

	Per cent.	Per cent.
Mechanical portions		25·05
Broken shafts	2·35	
Binders and fasteners	9·40	
Bearings	8·60	
Other mechanical parts	4·70	
Brush-gear, connections, terminals, etc.		7·01
Failure of brush-insulation	2·35	
Connections and terminals	3·10	
Mechanical portion of brush-gear	1·56	
Field-magnet coils		10·16
Coils burnt out	5·50	
Coils earthed	3·10	
Coils gone altogether	1·56	
Armatures		78·15
Coils short-circuited and burnt out	29·00	
Coils earthed (that is, failure of insulation of frame)	19·50	
Breakage of wire	15·60	
Joint-failures	8·60	
Binders burnt	3·10	
Driving horns	2·35	
Commutators		33·60
Failure of insulation	10·15	
Mechanical failures	10·95	
Surface fusion	12·50	

TABLE B.—SHOWING POINTS OF ORIGIN OF BREAKDOWNS. THE PERCENTAGES GIVEN SHOW THE FREQUENCY WITH WHICH THE VARIOUS PARTS CAUSED BREAKDOWNS.

	Per cent.	Per cent.
Mechanical portions		11·03
Shafts	0·78	
Bearings	8·60	
General	1·65	
Brush-gear, leads and terminals		8·60
Mechanical	3·90	
Insulation	2·35	
Leads and terminals	2·35	
Field-magnets		8·58
Failure of coil-insulation	7·80	
Breakage of wire	0·78	
Armatures		47·80
Core and slot-insulation ... ... 12·50		
End-insulation (that is separating ridges) ... 5·50		
	—— 18·00	
Coil-insulation ... ... 11·70		
Joints ... ... 7·00		
Fasteners and binders ... ... 1·65		
Binder-insulation ... ... 7·80		
Driving horns . .. . 1·65		
Commutators		14·90
Insulation ... ... 9·40		
Fastening and keying ... ... 5·50		
Starters		4·70
Unknown		3·13
Various		1·56

TABLE C.—SHOWING THE CAUSES TO WHICH THE DERANGEMENTS WERE DUE.

	Per cent.	Per cent.
Constructional		39·36
Bad design ... ... 18·36		
Perishing of insulation ... ... 7·40		
Bad workmanship ... ... 13·60		
Conditional		3·71
Overloading ... ... 1·37		
Over-rating ... ... 1·56		
Unsuitable ... ... 0·78		
Maintenance		31·46
Dust and damp ... ... 7·40		
Rough usage ... ... 1·56		
Defective attention other than above ... ... 22·50		
Accidental and unavoidable, with reasonable care in construction and working		7·05
Unknown		13·10
Caused by faults in accessories		5·48

Mr. A. C. CORMACK, replying to a question, stated that breakdowns of coal-cutting plant formed a distinct class by themselves, and he had not been able to gather sufficient data to give reliable averages as a separate class.

Mr. C. C. LEACH (Seghill) said it was misleading to state that breakdowns of dynamos were less numerous than that of engines. When they had an engine driving it, it did not matter to the colliery-owner whether the engine or the dynamo was at fault; the result was the same.

Mr. G. S. CORLETT (Wigan) said it would be very valuable if they had the percentage of machines that had actually failed: there was nothing to indicate in the paper whether it was 1 per cent. or 50 per cent. of the total machines examined. Further, there was nothing to indicate the life of the machines: whether they had been running 1 or 10 years; nor what class of work they were doing: whether in an extremely dry and clean engineering-shop, or in a coal-mine or chemical works, etc. He had tried to get some figures as to the percentages of failures of machines actually at work, and he found that there was a great difficulty in a business like his own in keeping in touch with the history of a machine that had been at work for a number of years; but he had taken a group of 65 machines running for periods of not less than 5 and not more than 15 years, and the average was $8\frac{1}{2}$ years. They were all continuous-current and working in collieries in different parts of the world, and he had been able to trace closely their history from the start to the present time. He found that there had been 17 failures, or an average of 2 per annum. He had included one case where the motors were laying under water for 2 months, but he had not included legitimate renewals to brushes and commutators. He did not know whether he was right in interpreting one paragraph, in which the course that a customer should adopt to ensure getting a reliable plant was described, but so far as he could follow it Mr. Cormack postulated that if a colliery-manager wanted to purchase a 10 horsepower machine, he must (1) obtain all sorts of guarantees of efficiency, (2) discuss carefully the designs of the motors, (3) visit the works where the machinery was being manufactured at various stages of the manufacture, and (4) finally make all sorts of tests; but if a manager wanted to do that, when installing a considerable number of motors, he would be inspecting electric machinery instead of managing his colliery. He thought that so elaborate a procedure was quite unnecessary, as there were many manufacturers from whom reliable machines could be obtained.

Mr. W. C. MOUNTAIN (Newcastle-upon-Tyne) said that the paper seemed to deal firstly with the construction of dynamos and motors, and secondly with their upkeep, etc. Speaking of the manufacture of dynamos and motors, the purchaser was justified, and was quite right, in seeing that the machinery was up to its work and thoroughly good, but too much interference was not good either for the manufacturer or for the customer. If the purchaser employed an inspector to look after his interests and if the inspector, or the firm that he represented, insisted upon some particular design of his own being adopted and not the well-tried designs adopted by the manufacturer, then it was only fair that the purchaser should take a certain amount of responsibility as regards the satisfactory working of the machinery. At the same time, manufacturers would be only too pleased to carry out any suggestions which it was to the purchaser's interest to adopt. The question of maintenance was one of the probable causes of breakdowns. Colliery-machinery had heavier duties, and worse treatment, than machinery put into a clean engine-house or engineering-shop. In the case of one large installation, costing £14,000, a man was appointed to look after it, who originally drove a hauling-engine of 8 horsepower; it was said that the new machinery justified a better class of man, but the reply was to the effect that the electric plant ought to run without any attention, or it was not worth while adopting. If electric machinery was properly looked after by competent workmen, who would keep it clean, etc., a large number of the accidents which appeared to arise from structural defects would not arise; and the great percentage should be charged against insufficient maintenance, or want of attention. Everybody knew that coal-cutting was hard work; to some extent motors had been blamed when the coal-cutters should have been, as the shock and jar to the machinery was bound to damage the motor. During the last 18 years, machines had altered largely in design, and it was not fair to compare modern machines with machines erected many years ago, especially in the case of an advancing science like electricity.

Mr. S. F. WALKER (London) said that the paper might be summarized under four heads:—(1) Faults due to defective insulation; (2) faults which have grown up since town-lighting and power-stations had so much developed the idea of using machines to their utmost limit and sometimes beyond; (3) want of attention; and (4) want of an expert to give advice. Insulation he thought had been the bugbear of electrical engineers as long as electricity had been a practical subject, and the larger the machines became, the higher the efficiency, the higher the pressure, and the more important was the insulation. He thought that the author of the paper was correct in saying that in some cases, in the cheaper classes of machines, insulation had been "cut financially:" it had thus enabled the maker to produce a cheaper machine, and the repairing bill had been considerably enhanced. He had always maintained that efficiency was not the 90 per cent. which they got out of the machine on test, but it was what they got out of it at work; and they must take into account the repairing bill. In colliery-plant, it was always desirable to have a large margin; if they wanted 50 horsepower he advised them to instal 100 horsepower, and as much more as they could, and the repairing bill would go down in proportion. Insulation was a difficult question; the material was damaged by heat, and there was also a certain current going through it which affected especially organic insulators. There were sparks passing across, and the insulation was broken down. In the case of slotted armatures, perhaps there was a small pin-point left; a spark passed across, and the motor broke down. As regards attention, 25 or 30 years ago, he was told that unless he could erect plant that would work by itself it was of no use. His rule in those days was to reckon everything of which he could think, and then multiply the power required by 10. As regards an expert, who was independent and not connected with any works or colliery, the colliery-owner did not want him, and the manufacturer did not want him, for he knew either too much or too little; but he quite agreed with Mr. Cormack that, in hundreds of cases, an expert, backed up by a thorough knowledge of the work, would save many hundreds of pounds at very small cost.

Mr. S. MAVOR (Glasgow) thought that colliery-engineers, who had most experience of modern electric plant, would be the first to concede that electric machines were less liable to derangement than any class of machine with which they had to deal. From that generalization, however, coal-cutters might be excluded. Some types of coal-cutters were more difficult to drive than others, as they imposed much more severe conditions on the motors, which were therefore more liable to breakdown. An important point with mining work was the rating of the electric motor and nothing was said as to the time-limit. In ordering a 10 or 20 horsepower motor, they should always state the nature of the work required of it. A motor which was quite large enough to do such work as a single haulage, for which a maximum of 50 horsepower was required, would be too small a machine to do the work of a pump requiring 50 horsepower. In one case it had to run continuously, and in the other with varying loads and periods of rest. A time-clause was therefore as necessary as a temperature-clause.

Mr. W. MAURICE (Hucknall) wrote that an examination of Mr. Cormack's figures correctly prove that half the breakdowns which occurred to electrical machinery were due to neglect of one kind or another. The advertising phrase, " electrical plant will run without any attention," had, he feared, been in the past too literally accepted. If well-made electrical machinery were given reasonable and intelligent attention, it would, without doubt, prove as good as any possible substitute. The damage caused to armatures, etc., by oil was serious and should, he thought, be pointed out more frequently to workmen in charge of machinery liable to be so injured. Mr. Cormack's remarks on motor-starters also deserved special attention.

Mr. R. HOOD HAGGIE (Derby) wrote that one statement worthy of remark in Mr. Cormack's paper was that, out of several thousand accidents that had occurred in electrical plants, there were only 13·1 per cent. to which no cause could be assigned. As a rule, if the accident be in the armature, the cause is burnt out and it is put down to the insulation being poor or carelessly done. All the men attending to the machine at the time know nothing unusual that occurred before the accident, and so the truth remains hidden. It would be interesting to hear whether all the reasons given in the tables were correct, or were they largely arrived at by guesswork; if the former, it should certainly be a source of satisfaction to the insurance-company.

Mr. A. H. STOKES (H.M. Inspector of Mines, Derby) wrote that Mr. Cormack had correctly described a large number of electrical plants when he wrote that they were " bad and unsuitable," as a large number of failures of electrical installations were due to bad arrangements and under-power machines. Many plants had been put down of just sufficient size to do the work required when all the machinery was moving and working smoothly, and no surplus of power was provided to overcome the inertia at starting. Such motors when called upon to start, after standing for some time, refused to answer the call made upon them and a breakdown occurred. If electricians, when told the horsepower required, would double the figures and supply accordingly, there would be less numerous breakdowns and better accounts of electric machinery. Mr. Cormack had given the members valuable records of failures, and it would greatly enhance the value of his paper if he would tabulate a code of rules, and standard figures or tests, whereby the high efficiency at which he aimed might be secured.

Mr. J. CLIFTON ROBINSON (London) wrote that where a high-speed motor was driving direct another shaft which had not been properly lined up, a breakage had been known to occur. The danger of carbon-brush accidents can be entirely eliminated by having fixed holders with a sliding brush. Occasionally, it had been found that where only a small clearance had been provided between the brush-holder and the radial connections to the commutator in the case of a motor driving a fan or any machine having an end-thrust, when at work, and also in crane-motors, a side-play on the armature had developed and caused the brush-holder to short-circuit and burn-out the armature-coils. Where a number of motors and dynamos were placed within a comparatively small space, it was desirable to instal a small air-compressor, by means of which dust and dirt could be removed from parts otherwise inaccessible, and it was surprising how much dirt could be removed from an apparently clean machine. Where motors were used for driving toothed gearing, especially in pump-gearing, the incessant vibration rendered breakdowns of the armature-coil insulation more prevalent, and it was advisable to have an armature with the coils made specially rigid in the slots. Where the motors were small and of a high voltage (say 500), the wires being of small diameter were liable to be broken by the vibration, and it had been found advisable to use series-wound armatures in preference to the ordinary parallel type, so as to be able to use a larger gauge of wire than otherwise. Also it had been found advantageous in particular cases to have a few series-wound turns on the field-magnets of motors, when the load was of a pulsating nature, and the armature-life had been increased considerably by this means. Sometimes it was preferable to have a shallow channel turned in the core-discs so that the binders might be flush with the surface of the armature, and thus less accessible to damage due to the wear of the bearings and other causes. Some of the substitutes for mica as commutator-insulation had given much trouble, and the material used very much resembled mica in appearance. It did not apparently absorb moisture, and seemed to lose its properties at a comparatively low temperature. In high-voltage motors, it was not always sufficient to provide for the armature being left across the shunt, as the high induction of the armature itself greatly retarded the immediate flow of current on breaking circuit. Where current was taken by a motor from a supply in which the negative was earthed, the insulation of the starting-switch needed careful watching. It was frequently advisable not to connect the shunt-lead to the first stop of the starting resistance, but to the negative side of the motor-switch, the positive being connected to the positive armature-terminal on the motor. By this means the field of the motor was made so soon as the motor-switch was closed, and had attained its full value before the armature-current began to flow. Also, if the starting resistance released automatically on no load, a nasty spark was avoided by so connecting, as the arm swung back off the first contact and opened the armature-circuit. With the field connected to the first contact, a reduced field-voltage had to be broken as well as the armature-circuit, which sometimes caused a rapid burning-away of the first contact. He agreed with Mr. Cormack that the semi-enclosed motor was often a snare. The amount of enclosure was not sufficient to keep out the harmful damp and dust, and yet was quite capable of making it a very awkward matter to clean, overhaul and inspect the machine in question. In choosing an enclosed motor, perhaps one of the most important features to be considered was its accessibility, particularly on the commutator side; for one had known of motors where it was a practicable impossibility to get at the underneath brushes, without taking the motor adrift. Where copper-brushes were used on the commutator of a motor, careful attention should be paid to the angle of contact made by the brushes on the commutator in order to avoid undue sparking. Where a motor was to be used in a situation where the atmosphere was at a high temperature, sufficient allowance should be made for this in the size of the machine and the temperature-guarantee when ordering.

Mr. A. CAMPBELL CORMACK, replying to the discussion, understood Mr. Leach to say, that it was unnecessary to have a dynamo more reliable than the engine. When both were equally reliable, they would expect to have the same number of breakdowns in ten years. If the engine broke down at the same time as the dynamo, and *vice versa*, it would be convenient; but they would not, and stoppages would be multiplied. In addition to the dynamo, there was the motor, and if motors were not more reliable than the engine, it would entail a great number of stoppages. It therefore seemed clear that the standard of reliability should not be curtailed by the limitations of the engine. The percentage of machines that had actually failed, so far as he could remember, taken over four years, had been about $7\frac{1}{2}$ per cent. per annum. As to the age of the machines, they varied, but making a rough estimate he would say that the average age of the machines under consideration was between four and five years, certainly not more. The class of work being performed was various, and roughly it approximated to mining work; a large number were in mines, a number at iron-works, and a few in central stations. As to the interpretation of the paragraph which dealt with the examination of machines during construction, it was not meant to imply that every 10 horsepower motor should have a separate specification drawn up for it; the remark was intended to apply to the case where a large plant was being put down, and there it would pay to carry out the steps indicated. Even with single small machines, a certain amount of checking was desirable; and this should be done by an expert, and not a colliery manager. Interference with manufacturers was not suggested, provided that their standards represented good practice, and that the work was satisfactory. In electrical machines, standardization had gone so far that individual specifications could not improve much on good standard patterns; but they were of use in that they ensured adherence to proper standards to meet the case. In the cases, which had come within his own knowledge, where these plans had been adopted, they had been able to work on a satisfactory footing with the manufacturer, who had only been too pleased to adopt the suggestions made. He agreed with the remarks as to the very inefficient attention given to electrical machines.

He was very glad that Mr. S. F. Walker agreed so thoroughly with him in advocating a high class plant. The temperature-limit of which he spoke was the maximum temperature to which the machine was likely to be subjected; and this would probably be arrived at in the case of a small continuous-current motor at the end of 5 hours. Where they only wanted intermittent work, it would be cheaper to specify a shorter time for reaching the temperature-limit, but it was necessary to exercise great care in doing this, as in the second and succeeding runs the machine usually started from a higher temperature.

The PRESIDENT (Sir Lindsay Wood, Bart.) moved a vote of thanks to Mr. Cormack for his valuable paper.

Mr. C. C. LEACH seconded the resolution, which was cordially approved.

———

Mr. J. WALTER PEARSE's paper on " Luxemburg and its Iron. ore Deposits " was read as follows : —

LUXEMBURG AND ITS IRON-ORE DEPOSITS.[*]

By J. WALTER PEARSE.

Inasmuch as Luxemburg is in intimate relation with the countries by which it is surrounded—forming part of the German Zollverein, having formerly extended into the Belgian province of the same name, and its main railway-system having been worked by the French Eastern Railway Administration—it partakes more or less of all these three countries, and consequently a great deal of misconception prevails as to the status of the little country, so that the following particulars may perhaps not be out of place.

By the treaty of London in 1867, the independence of the Grand Duchy of Luxemburg was guaranteed by the Powers; and since that period it has enjoyed constitutional government, Duke Adolf of Nassau (who has now abdicated in favour of his eldest son), having succeeded the late King of Holland as Grand Duke of Luxemburg, in default of heirs male, owing to the Salic law prevailing in Luxemburg though not in Holland.

The country, which takes the form of an irregular isosceles triangle with the apex towards the north, is wedged in, as it were, between the Rhine Province of Prussia on the east and the Belgian province of Luxemburg on the west, while the eastern and larger portion of the base is bounded by Lothringen, or Lorraine, and the remainder by the French Department of Meurthe-et-Moselle.   Situate between 49° 35' and 50° 16' north latitude, and between 5° 45' and 60° 30' longitude east of Greenwich, the country has an area of nearly 1,000 square miles, being therefore equal in size to a small English county. If small in area, however, and containing a population ever increasing, but never exceeding 225,000 owing to constant emigration, Luxemburg is remarkable in many respects.

---

[*] A map of the iron-ore district, showing the mines, ironworks, steelworks and the railway-communications, that had been prepared to accompany this paper, has unfortunately been lost in the post; and it would be difficult to replace as the original map, prepared by the engineers of the Prince Henri Railway for the Luxemburg Government, is now out of print.

Its railway-system, fostered by iron-mining both directly and indirectly, is more extensive in proportion to area and population than that of any other country. It possesses a thermal spring more highly mineralized than any in Europe. Its iron-ore, richer and more easily mined than that in neighbouring countries, is also more readily smelted owing to the intimate admixture of lime, so that with a suitable mixture of the charge no extraneous flux is required, while its mining laws differ from those of any other country. Moreover, the capital, long known as the stronghold of Europe, is now endowed with a bridge of special and economical construction, having the largest single span of any stone bridge existing. A peculiarity is that, while the native dialect, claimed by some to be a separate language, more nearly resembles English than any other, the inhabitants are accustomed to speak French and German with equal facility, and localities have generally names in both languages.

The surface of the country is very diversified, being flat or undulating with small abrupt hills or *kops*, in the south, but wild and mountainous in the north. The Wilhelm-Luxemburg or Guillaume-Luxembourg railway, the main line of which traverses the country longitudinally, affords communication with Liége and Aix-la-Chapelle on the north, and with Thionville and Metz on the South. This line is intersected almost perpendicularly at the capital by its branches, affording communication with Trèves on the east, and Brussels on the west. Other branches also give direct access to the most important mining centres of the country, terminating westward however at Esch-sur-l'Alzette, in the centre of the iron-ore district. This railway is generally credited with having been the means of discovering the iron-ore deposits; but such is not strictly correct, because iron-ore was worked to a slight extent before the railway was made, although the line was certainly the means of proving the beds and facilitating their working.

From Esch, the Prince-Henri railway was made in a north-westerly direction, purposely with many sinuosities for easy access to the ore in the central and south-western districts, and was continued northward and then eastward, intersecting the Wilhelm-Luxemburg railway at Ettelbruck, whence it proceeds eastward, and along the frontier southward, to its present terminus at Grevenmacher. A branch put out from Rodange, near the French frontier, affords communication, near Longwy, with the French Eastern system; and, on account of the tortuous nature of the line between Esch and Pétange, a new and direct railway has been made between that station and Luxemburg City. An independent line of the Prince-Henri railway also starts from Kautenbach on the Wilhelm-Luxemburg railway, and proceeds westward, until it communicates at the frontier with the Belgian State railway.

In addition to these two standard-gauge railways, there are four light lines of 3 feet 3⅜ inches (1 metre) gauge, one from the capital to Remich on the eastern frontier; another from Cruchten on the Wilhelm-Luxemburg line to Fels, or Larochette; a third from Bettembourg, also on the Wilhelm-Luxemburg railway near the southern frontier, to the Aspelt station of the Luxemburg-Remich railway; and lastly one from Noerdange on the Prince-Henri railway to Martelange, on the Belgian frontier. Independent of the mineral traffic, all these railways, except the first-named, are intimately connected with the iron-industry, because they were subsidized by ore-concessions, granted to promote enterprise and to compensate for possible working at a loss.

The Luxemburg-Remich railway passes through Mondorf where, in 1841, the engineer Kind started a bore-hole for a Belgian company with the object of proving a deposit of rock-salt, which was thought to exist, because saline springs occur in the neighbourhood. This expectation was not realized, and a depth of 2,415 feet (736 metres) was reached without rock-salt being struck, while the non-existence of coal and Permian deposits was proved. At a depth of about 1,640 feet (500 metres), however, a thermal spring was tapped, the water of which issued at a temperature of 79° Fahr. (26° Cent.) and at the rate of 154 gallons (700 litres) per minute. Analysis shows that 1 litre of this water contains 8·6774 grammes of sodium chloride, 0·1909 gramme of potassium chloride, 3·2323 grammes of calcium chloride, 0·4096 gramme of magnesium chloride, 0·0077 gramme of lithium chloride, 0·00005 gramme of magnesium iodide, 0·1409 gramme of magnesium bromide, 1·4669 grammes of calcium sulphate, 0·0986 gramme of strontium sulphate, 0·0974 gramme of calcium bicarbonate, 0·0026 gramme of magnesium bicarbonate, 0·018 gramme of bicarbonated iron protoxide, 0·0087 gramme of silicic acid and 0·0007 gramme of arsenic acid, while also containing in solution 30·8 cubic centimetres (1·88 cubic inches) of carbon dioxide and an approximatively equal volume of nitrogen. Owing to its remarkable richness in bromine, the water of the Mondorf spring, now owned and worked by the Luxemburg Government, has proved to be a valuable therapeutic agent, becoming more and more appreciated every year. The following section of the bore-hole will afford some idea of the geological formation of the country : —

No.	Description of Strata.	Thickness of Strata. Metres.	Ft.	In.	Depth from Surface. Ft.	In.
1.	Liassic limestone and marl	41·50	136	2	136	2
2.	Sandstone overlying conglomerate ...	12·60	41	4	177	6
3.	Keuper	206·20	676	7	854	1
4.	Shelly limestone	79·91	262	1	1,116	2
5.	Gypsum and salt-clay	32·39	106	4	1,222	6
6.	Green and red marl, with gypsum and limestone	21·5℔	70	6	1,293	0
7.	Bunter marl, with gypsum and sandstone...	56·30	184	9	1,477	9
8.	Bunter sandstone	260·70	855	4	2,333	1
9.	Grauwacke	15·00	49	2	2,382	3

The geological formation in the northern part of the country is Palæozoic, comprising the lower series of the Devonian rocks, while those in the south form part of the Mesozoic group, consisting of the Triassic from the variegated sandstone upwards, and of the Jurassic up to and including the Oolites. The Jurassic comprises the two lower divisions in their entirety, while the upper division is only represented by the Bajocian group; but these strata are rarely found more complete than in Luxemburg. The bituminous shales of this formation, which are so rich in some places that they burn with a slight flame, are in most cases directly overlain by the iron-beds, which generally begin as a hard and tenacious marl, insensibly merging into a sandy limestone, and containing a few isolated beds of pure sandstone, while in other cases the lower beds consist of a friable sandstone. The composition of the iron-ore presents many variations; and layers of true sandstone are often met with, sometimes at the bottom and sometimes in the middle of the measures.

Minerals, other than iron-ore, occur at various points in the north, and have been worked at different times. The lead-mines at Oberwampach extend over the frontier to Longwilly in Belgium, where they are worked extensively. There is a copper-mine at Stolzenburg on the Prussian frontier; and antimony is found at Goesdorf, where both lead and copper also occur. There is a valuable deposit of dolomite at Grevenmacher, the eastern terminus of the Prince-Henri railway; and slate is quarried extensively at Martelange on the Belgian frontier.

The iron-ore deposits occur in the lower districts of the country, extending along the western half of the southern frontier, being overlain by a slight alluvial cover. The iron-ore, stratified as in Cleveland, is a ferruginous oolite, or limonite, known as *minette*, and forms the small but very important northern portion of the great iron-field which extends southward a little beyond Nancy, with a width of about 62 miles (100 kilometres), the eastern and western boundaries being generally parallel with the Moselle and Meuse respectively, while the seams dip slightly to the south-west.

The Luxemburg portion of the deposit is divided by a fault, bearing nearly north and south, into two districts—the Esch-Rumelange-Dudelange on the east, and the Belvaux-Differdange-Lamadelaine on the west, the difference in geological level between the two being about 135 feet. In the eastern district, there are from four to six rich beds, separated by unproductive rocks, the lowest or grey seam consisting of ferruginous limestone, while the upper series, comprising the red and yellow seams, contain more or less sand agglomerated by a calcareous cement; but in the western district the seams are closer together, forming practically a single bed of ore, which is worked with great facility. In the western district, the yellow seam disappears, or becomes unworkable; on the other hand, the brown and black seams make their appearance below the grey seam.

Whereas the iron-content of the ore does not vary greatly over the Luxemburg portion of the deposit, averaging about 39·5 per cent., the lime-content increases gradually from west to east, while that of silica diminishes in a similar ratio. The Luxemburg portion is the most important of the whole field, on account of the iron-content being higher, the mixture of lime more intimate, and the deposit more easily worked. At Dude-lange, in the eastern extremity (where all the surface, and also the concessible ore is in the hands of the Dudlelange Smelting Company, which was the first establishment to make steel in Luxemburg), the red seam is 4¼ feet (1˙25 metres) thick; the cal-careous seam, 5 feet (1˙5 metres); the yellow seam, 6½ feet (2 metres), and the grey seam, 7¼ feet (2˙25 metres), making a total thickness of 23 feet. At Rumelange, a little towards the west, all the above-named seams occur, but with about double the thickness; and they are all worked together—the red seam to great advantage, notwithstanding its high silica-content, because the others are calcareous.

The Galgenberg Hill, still further to the west, contains three workable seams, the red calcareous, the grey calcareous and the brown siliceous, making a total thickness of 28¼ feet (8˙65 metres); but there is rather a thick cover, consisting of limestone, marl, calcareous rock, and argillaceous and siliceous iron-ores that are not workable.

At Esch-sur-l'Alzette, the centre of the mining district, no less than eight workable seams are encountered: namely (with their iron-content), the siliceous (23˙6 per cent.), the third minor (38˙9 per cent.), the second minor (25˙7 per cent.), the first minor (33˙5 per cent.), the main red (42˙8 per cent.), the yellow (33˙5 per cent.), the main grey (32˙5 per cent.), and the brown seam (41˙6 per cent.), making a total thickness of more than 50 feet. At Belvaux, towards the western extremity, the cover is not so thick; and the three calcareous seams are worked, having a total thickness of about 31 feet. At Differdange, farther westward, the ore is generally very friable, and therefore, instead of being sent away, it is more suitable for smelting on the spot, as it now is by the Differdange-Dannenbaum Company, which has erected steel-works as well as blast-furnaces; and the two siliceous seams, the grey and the black, afford a total thickness of nearly 19 feet. At Rodange, on the extreme west, but north of Differdange, a total thickness of 34½ feet comprizes the two red calcareous seams, the upper and lower band of the grey seam, divided by 3 feet of sterile rock, and a poor seam of marly iron-ore.

Table I. has been compiled from the results of analyses made from iron-ores at the various points.

TABLE I.—ANALYSES OF LUXEMBURG IRON-ORES.

District.	Sesquioxide of Iron. Fe_2O_3	Silica. SiO_2	Alumina. Al_2O_3	Phosphoric Acid. H_3PO_4	Lime. CaO	Carbon Dioxide. CO_2	Sulphuric Acid. H_2SO_4
EAST—							
Dudelange, Upper Seam	56·32	—	5·05	1·49	11·23	—	—
do. Lower Seam	51·67	—	4·69	2·03	13·13	—	—
Rumelange, Red and Yellow Seams ...	51·40	—	6·00	1·80	14·50	—	—
Rumelange, Yellow Seam (screened) ...	55·70	—	7·00	1·90	10·00	—	—
Galgenberg, Red Seam...	56·20	—	5·75	1·50	9·96	—	—
do. Grey Seam	50·07	—	3·30	1·32	17·70	—	—
do. Brown Seam	58·63	—	3·90	2·36	7·00	—	0·09
Esch-sur-l'Alzette, Red Seam	57·39	8·57	5·27	1·72	10·74	8·43	—
Esch-sur-l'Alzette, Grey Seam	42·60	4·84	2·91	1·90	22·10	17·37	—
Esch-sur-l'Alzette, screened	56·40	7·70	7·35		9·96	—	—
Esch-sur-l'Alzette, unscreened	51·54	7·75	3·94		14·80	—	—
Belvaux, Grey Seam ...	58·95	—	5·00	1·27	4·21	—	—
do. Red Seam ...	56·35	—	9·75	1·20	6·25	—	—
Differdange, Yellow Seam	49·77	—	7·65		10·40	—	—
do. Grey Seam..	49·73	—	7·57		9·50	—	—
do. Black Seam	46·27	—	5·88		11·00	—	—
Rodange, Lamadeleine (average)	54·60	10·05	7·60		8·40	—	—
WEST—							

At these seven points, the mean iron-content of the various seams are:—39·42, 36·2, 36, 39, 39·47, 35·45, 45, 40·17, 29·82, 39·475, 36, 41·38, 39·57, 34·84, 34·81, 32·37 and 38·2, per cent. giving the above-named average of 39½ per cent. The mean water-content of the Galgenberg ores has been found to be:— Red seam 10·5 per cent., grey seam 8·3 per cent., and brown seam 12·2 per cent., with respective specific gravities of 2·77, 2·65 and 2·56, determined at a temperature of 59° Fahr. (15° Cent.). The chemist at the Rumelange iron-works considers that phosphorus exists in the ore in the state of iron phosphate, because all analyses of Luxemburg ore show the phosphorus-content to be in direct ratio to that of the iron. This content varies from 1·7 to 1·8 per cent., so that the ore is suitable for producing the fine-grained greyish special pig, between white and grey, required for basic steel-making.

The French mining law of 1810, which forms the basis of mining legislation in Luxemburg as in most other Continental countries, enacts that, so soon as the surface-owner can no longer work his ore opencast to advantage, it belongs to the State and can be conceded to third parties; but the original obligation of mine-owners to supply neighbouring blast-furnaces with ore has long been obsolete in Luxemburg, and is now definitely abrogated. The Grand-ducal law of March 15th, 1870, extended the surface-owner's right by empowering him to continue working, even with the aid of shafts or adits, when the cover does not exceed a certain height above the uppermost seam; but all the ore below this depth may be conceded by the State. The actual depth differs in the eastern and western districts owing to the dislocation of the measures above referred to; but it is fixed so as to place the surface-owners in each on an equal footing.

This " Little National Fortune," as it is called by the Director-general of Finance extends over 4,695 acres (1,900 hectares)—not a bad proportion, considering that the area of the whole country is only 639,393 acres (258,744 hectares); and some indications as to the thickness of the deposit have already been given.

The erection of a blast-furnace carries with it the right to an ore-concession, each concession forming the subject of a separate law; but ironmasters are obliged to smelt their ore on the spot. Ore-concessions may also be granted, by way of subvention, to those who carry out works of public utility, although hitherto such grants have only been made to railway-companies, with the right, however, of selling their ore. The Wilhelm-Luxemburg Railway Company received a money subvention; but the Prince-Henri Railway Company, part of whose system was unremunerative at first, received an ore-concession in the western district and also the valuable Galgenberg Hill, which, extending over 141 acres (57 hectares), is let to a mining company.

The method of working, almost invariably the same, varies only with the nature of the ground. Trenches are cut in the hill-side so as to lay bare the rock below, descending parallel with one another from ridge to base in the case of a crest, but radiating from apex to periphery in an isolated hill. The surface-soil, between the trenches, is then turned over downward on to the hill-side, thus forming a series of terraces suitable for cultivation. The ore cropping to the surface is then taken out, the sterile rock being stacked to serve for packing future mine-workings. The same method is continued, until a depth of about 65½ feet (20 metres) from the original surface is attained, below which opencast-working would become too expensive. Mining by adits and drifts is then resorted to, the roof being allowed to fall when there is not sufficient material for packing.

In Luxemburg, working by shafts is the exception; but there is a small pit, 82 feet (25 metres) deep, affording communication between the red, grey and brown seams, worked together, at the Heintzenberg mine, which with some furnaces at Esch has been taken over by the Aachener-Hütten-Actienverein (Rothe-Erde Company). By means of 90 horsepower electric loco-motives, hauling 60 tons at a speed of 10 feet per second on a rising gradient of 1 in 33, the ore of the grey and brown seams is taken to this shaft, where it is raised by a 45 horse-power electric engine, which can wind 1,500 tons in 24 hours; and from the mouth, in the red seam, the ore is conveyed directly to the furnaces, also by electric locomotives. In a damp adit, dipping towards the mine, a fan, driven directly by a dynamo, delivers 5,500 cubic feet (155 cubic metres) of air per minute; and rotary electric rock-drills, with a 250 volts current, make a hole nearly 2 inches in diameter and more than 3 feet deep, in from 6 to 8 minutes, permitting an advance of 229 feet (70 metres) per month by three shifts of seven men.

The general practice is to break up the ore, first by men with sledges and afterwards by boys with smaller hammers, but in some cases by stone-breakers. In this manner, the calcareous concretions can be readily separated, and a more uniform fur-nace-charge obtained. At some of the larger blast-furnaces, the experiment was tried of charging in the ore as it is mined, when it was found that the coke-consumption did not materially increase; but, at some older and smaller furnaces owned by the same company, this experiment was followed by a decided in-crease in the coke-consumption.

The first furnace was built of ashlar stone, 44 feet (13·5 metres) high, at La Sauvage on the French frontier, about 1790; and another, adjoining it, of rubble-masonry, about 49 feet (15 metres) high, was worked so late as 1880. The early furnaces were erected to smelt the alluvial ore in their neighbourhood by means of charcoal; and it so happens that there are still two sets of furnaces, far removed from the limonite, that are still working. These furnaces, and also those erected on the great body of ore, turn out an increasing proportion of basic pig every year; and, as will have been gathered, steel-works have been added to three of the iron-works, for converting the pig-iron into steel by the Thomas-Gilchrist process.

In 1901, owing to the depression of trade, the production of iron-ore diminished by nearly one-third; and the 75 mines only yielded 4,455,179 tons, against 6,171,229 tons in 1900. With an ore-consumption of 2,878,150 tons, 2,771 workpeople, and 25 furnaces in blast, pig-iron was produced to the amount of 916,404 tons (against 970,885 tons in 1900), comprising 111,593 tons of forge, 132,438 tons of foundry, 672,075 tons of basic, and 297 tons of transition-pig; and with 426,376 tons of pig-iron the three steel-works turned out 257,055 tons of ingots, blooms, slabs, billets and finished steel, against 184,714 tons in 1900.

—————

Mr. OSWALD WALMESLEY (Lincoln's Inn) said that the Luxemburg iron-ore deposit described in the paper was interesting, both as being very valuable and as being worked to a large extent under exceptional mining legislation. The principal feature of the latter was that it recognized the principle of *domainiality;* that is, although the State reserved from the surface-owner all mines, except such as could be worked by open-cast working, it did not grant concessions to the first finder or applicant, or at discretion subject to certain conditions, as in many foreign countries, but when making a grant of a mine did so subject to a special contract made by law with the grantee, reserving what the mine could fairly pay to the State. This was a special feature of mining legislation, although not confined entirely to the Luxemburg law; it seemed at any rate to have answered extremely well in Luxemburg, both for the State and for the grantees. For this, and other perhaps more practical matters, referred to in the paper, it seemed deserving of special commendation.

Mr. BENNETT H. BROUGH (London) said that the thanks of the members were due to Mr. Pearse for an account of an iron-ore field regarding which there was but little information in the *Transactions;* and the paper was the more valuable as coming from the pen of one who had lived so long in the district. Although the Luxemburg field covers but one twenty-fifth of the area of the great district of Jurassic iron-ore, yet the author made it clear that the thickness of the beds and their satisfactory nature rendered this portion of it of conspicuous importance, and it was gratifying from a patriotic point of view to remember that it was entirely owing to a British invention (that of the basic-steel process by Messrs. Thomas & Gilchrist) that the development of these enormous deposits had been rendered possible. There appeared to be some confusion in the analyses given. A recent paper by Mr. W. Kohlmann gave the following average composition of the ores:—Brown ore: iron, 38 to 42 per cent.; silica, 14 to 15 per cent.; alumina, 8 per cent.; and lime, 3 to 5 per cent. The grey seam contains: iron, 30 to 32 per cent; silica, 8 to 9 per cent.; alumina, 4 to 5 per cent.; and lime, 16 to 18 per cent.

Mr. J. S. JEANS, having visited the field described in the paper, and examined several of the leading mines and furnaces, said that not only was the Luxemburg field very important as such, but so were those also of Nancy and Alsace-Lorraine, which were practically a continuation of the same deposit in that part of Europe. The Luxemburg field was one of the first to be developed, and to the best of his recollection it had not been further developed to any extent within the last quarter of a century. The ironworks in Belgium and many in Germany were dependent on this field. The Alsace-Lorraine deposits, however, were, he believed, now worked to a much more considerable extent, and were to be deemed among the most important fields of any kind in the whole of Europe, both with regard to their extent and duration.

Mr. J. WALTER PEARSE, replying to the discussion, wrote that, at least 20 years ago, it had been predicted that Luxemburg would eventually become the centre of pig-iron production in Europe; and this prediction now appeared in a fair way of fulfilment. Not only had the number and size of the blast-furnaces greatly increased during the period named, with a corresponding increase in the output, but also the proportion of basic steel-pig had increased, at the expense of forge and foundry pig. The figures given of the iron and other contents were derived from actual analyses made at different times of samples of ore taken from various seams and localities. Of course, the general result depended upon the selection and proportion of the samples taken. He desired to refer members interested in the subject to a report* on the mine-concessions of Luxemburg, by Mr. H. Neuman, Secretary of the Grand-ducal Mining Council.

The PRESIDENT (Sir Lindsay Wood, Bart.) moved a vote of thanks to Mr. Pearse for his interesting paper.

Mr. BENNETT H. BROUGH seconded the resolution, which was cordially approved.

———

Mr. F. HIRD's paper on "The Electrical Driving of Winding-gears" was read as follows:—

* *Rapport Les Concessions Minières dans le Grand-duché de Luxembourg sous le Rapport du Prix et du Mode d'Aliénation*, by Mr. H. Neuman, 1894.

THE ELECTRICAL DRIVING OF WINDING-GEARS.

By F. HIRD.

The application of electricity to the driving of winding-gears forms one of the most recent and important developments in electrical mining work; and as the subject is now attracting a good deal of attention, the writer hopes that a sketch of the progress made and the advantages gained may prove of interest and value.

The improvements, which have made electric winding a serious rival to steam, are of quite recent date, nevertheless many electrical winding-gears have already been installed and have been in satisfactory operation for several years.

The earliest applications of electricity have, however, been confined to cases where fuel was costly and water-power was available, within a reasonable distance. The advantages of electricity as a means of efficiently transmitting power were sufficient in such cases to determine its adoption for driving the winding-gears as well as the pumps and other machinery. In the mountainous mining districts of America and the Continent of Europe, such conditions are frequent, and a considerable number of electrical winding-gears have been manufactured, in America, by the General Electric Company, the Westinghouse Company, and others; and, in Europe, by Messrs. Siemens & Halske of Berlin, Messrs Siemens & Halske of Vienna, Messrs. Schuckert & Company of Nuremberg, the Allgemeine Elektricitäts Gesellschaft, and others, for work under these conditions (Figs. 17, 18 and 19). Time will not allow of our dwelling on these early applications of electricity, for we are concerned rather with the improvements which have widened the field of electric winding and made it applicable to the conditions usual in this country.

It will readily be understood that the chief problem for the electrician lies in finding the most suitable method of control. The motor itself is a well-tried machine, and presents no difficulty; but the controlling apparatus must be designed to meet the arduous conditions of frequent starting and stopping, must provide the greatest accuracy of control, and must permit of steady running at slow speeds for the purpose of repairs, inspection, etc., in the shaft.

The usual method of starting an electric motor is by inserting a variable resistance in the circuit of its armature or rotor, and gradually cutting it out as the speed rises. This method is known as the rheostatic method of control, and is illustrated diagramatically in Figs. 1 and 2 (Plate XXI). It is the

FIG. 17.—ELECTRIC WINDING-ENGINE AT THE HOHENEGGER SHAFT, KARWIN.

simplest method for general purposes, but it is objectionable in cases where the starts are numerous, on account of the heavy loss of energy in the resistance when starting. In Fig. 9 (Plate XXII.) the thick, black line represents the horsepower required in the case of a particular lift. The area within the curve represents the total energy consumed, while the shaded portion represents the energy wasted in resistance during a rheostatic start. The loss amounts in this case to about 30 per cent. of the total energy supplied for the trip.

Another objection to this method of control lies in the instability of the speed when much resistance is inserted. The speed of a motor becomes very sensitive to changes of load, when there is considerable resistance in circuit, and it is extremely difficult to maintain a steady slow speed under these conditions. This difficulty can be partly overcome by providing a very fine subdivision of the resistance and by careful attention to the speed on the part of the operator.

In spite of its disadvantages, the simplicity of this method of control determined its adoption in the early applications

FIG. 18.—UNDERGROUND ELECTRIC WINDING-ENGINE OF THE HERCYNIA MINING COMPANY, VIENENBURG.

of electricity to winding-gears. It was, however, evident to electrical engineers that a more suitable method must be devised if the full advantage of electrical driving was to be secured, and accordingly we find that a great deal of attention has been paid to this problem, especially by German engineers who have done much careful work on the subject.

One of the first attempts to improve on the rheostatic control was made by the Allgemeine Elektricitäts Gesellschaft, first at a mine at Hollertzug in Siegerland, and later at a coal-mine at Zwickau, in Saxony. In these cases, the speed of the motor was controlled by varying the voltage of the supply-current. This was accomplished by varying the exciting current of a special generator, whose only function was to supply the winding-motor. The general arrangement is shewn in Fig. 3 (Plate XXL). The speed of the motor, which is practically proportional to the electric pressure of its supply, is under complete control, and there is no instability at low speeds: further the heavy rheostatic losses during acceleration are avoided, and the controlling apparatus has to deal only with small exciting

FIG. 19.—UNDERGROUND WINDING-ENGINE AT THE THIEDERHALL
SALT-MINE.

currents instead of with the heavy main current. This method, however, necessitates a special generator for each winding-engine, and is not very suitable if the generating station is situated at any considerable distance.

In another system which has been used, the pressure of the supply-main is, as it were, subdivided by means of a series of balancing motors, coupled together. Fig. 4 (Plate XXI.) shows a case of this kind in which five motors, of 100 volts each, are connected in series between the 500 volts mains and coupled together so that they must always run at the same speed.

By tapping the points between the motors, the following series of pressures are available:—100, 200, 300, 400 and 500 volts. The control thus obtained is not so good as that in the method shewn in Fig. 3 (Plate XXL), since the pressure can only be varied by steps of 100 volts instead of by an imperceptible gradation: it is, nevertheless, much better than a rheostatic control, and does not necessitate a special generator.

Another method, which may be regarded as a modification of that shewn in Fig. 3 (Plate XXI.), would be to use a motor-

Fig. 20.—Electric Winding-Engine for Zollern II. Colliery.

generator instead of a special generator; this would enable the system to be employed, even if the generating-station were situated at some distance from the winding-engine. The writer is not aware that this system has ever been applied in practice, but with some very important additions, it forms an essential part of the Siemens-Ilgner system presently to be described.

All these systems, in various degrees, meet the objections to a rheostatic control, avoiding the rheostatic losses and providing a perfect control, which enables the attendant to drive the motor steadily at any desired speed.

The writer now comes to an advance of a different kind, in which an attempt has been made to secure these advantages, and, in addition, to equalize the load on the generating-station, and thus secure economy in steam-consumption as well as simplified working conditions of steam-generation.

Messrs. Siemens & Halske of Berlin, were, the writer believes, the first to devise an electric winding-system which efficiently equalizes the power-demand by means of power-storage. A winding-engine built by the Bergwerksverein Friedrichs-Wilhelmshütte at Mülheim-on-the-Ruhr, and electrically equipped by Messrs. Siemens & Halske of Berlin, on this system, was shown at the Düsseldorf Exhibition of 1902. This plant has been fully described,[*] and it has no doubt been seen by many of the members at the Düsseldorf Exhibition (Fig. 20). Every one who has seen it must have been struck by the perfection of the control and the ease of its manipulation.

In this plant, two motors are used which are directly coupled to the winding-drum without the use of intermediate gearing. A battery of electric accumulators furnishes the means of storing power during intervals of no load or of light load, and of assisting the generating-station during the times of heaviest demand. During retardation at the end of a trip, the kinetic energy of the moving parts is paid back into the accumulators, instead of being wasted in the brakes. The battery fulfils yet another function, namely, that of furnishing a series of voltages for effecting the speed-control. It is divided into four groups, and the two motors are capable of being connected in series or parallel. The various electrical groupings thus rendered possible, give a very satisfactory and efficient speed-control. A rheostat is only used to smooth out the steps between successive groupings.

The various electrical connections, which have to be made during each journey, necessitate the moving of heavy switching-gear, and it is necessary that the correct connections for starting, stopping, paying back energy to the accumulators, and so on, should be effected safely and reliably without any thought or discretion on the part of the attendant. All this has been effected by means of a pneumatic system of control, and the

* *Engineering*, 1902, vol. lxxiv., page 12, and in the *Colliery Guardian*, 1903, vol. lxxxv., page 845.

compressed-air is also utilized for operating the emergency-brakes.

The safety-appliances are most complete and carefully thought out, and whatever complexity there may be in the switching arrangements is entirely dealt with by the mechanism, and leaves the actual handling unaffected. All the necessary operations are effected in the correct order by the motion of a single handle.

The credit of having clearly demonstrated in actual practice the possibility of securing an equalized load on the primary generator, and the great advantages resulting therefrom, belong to Messrs. Siemens & Halske. It must, however, be admitted that these advantages have been secured at a somewhat heavy price. Apart from the complexity of the heavy switching-gear, a battery of accumulators must be provided. Now, the first cost of a battery is high and its maintenance is heavy and somewhat uncertain; and, altogether, capital-expenditure on electric accumulators is a form of investment which one would gladly avoid, if it were possible to obtain the advantages of power-storage in any simpler manner.

The plant exhibited at the Düsseldorf Exhibition is now erected at the Zollern II. mine, the property of the Gelsenkirchen Mining Company and it is interesting to note that the advantages of the Siemens-Ilgner system have so impressed themselves on the responsible authorities that an alternative method of control on the Siemens-Ilgner plan has already been put in hand for this plant, and is expected to be in operation by the end of July.

Mr. Ilgner, chief engineer of the Donnersmarkhütte, should be credited with the suggestion of providing for the necessary storage of energy, by a suitable high-speed flywheel. The combination of this flywheel with a motor-generator, which furnishes the means of control, constitutes the essential feature of the Siemens-Ilgner system. Messrs. Siemens & Halske, co-operating with Mr. Ilgner, have had the benefit of this gentleman's valuable suggestions and also of his great experience as a practical mining-engineer; they have, therefore, been able to produce a plant which secures all the advantages attained by that shown at the Düsseldorf Exhibition in a simpler and more effective manner. Fig. 5 (Plate XXI.). represents the arrangement diagrammatically. The control is provided by a motor-generator which consists of the following parts:—(1) A motor capable of furnishing the average power required by the winding-gear. This motor may be a three-phase induction-motor, or a continuous-current motor, according to circumstances. (2) A continuous-current generator, the function of which is to provide the current for the winding-gear motor. This machine is arranged so that the excitation of its field is variable through wide limits, in fact, from its full maximum through zero to its maximum-value in the opposite direction. By this means, the voltage or pressure of supply to the motor can be varied through the same range. (3) A small continuous-current machine, the function of which is to provide the exciting current for the generator, just mentioned, and also for the winding-gear motor. This special exciter is necessary, in order to ensure stability of the field and, therefore, of the pressure, which could not be attained through such a wide range, if the generator were allowed to furnish its own exciting current. And (4) a heavy steel flywheel, specially designed for high circumferential speeds, so as to store a large amount of energy in proportion to its weight.

The current from the generator is taken straight to the armature of the motor, which drives the winding-drum, without passing through any switches, fuses or controlling gear, for it is one of the special advantages of this system that no manipulation of the heavy main current is required, the whole of the control being effected by light switching-gear dealing only with the small currents used for exciting the field-magnets.

The motor for actually driving the winding-drum is of the continuous-current type. The drums may be driven through gearing, but for powers over 100 horsepower, it is considered better practice to couple the motor directly to the drum without any intermediate gearing. The slow speed thus necessitated makes the motor somewhat larger and more expensive in the first place, but the disadvantages of gear for heavy powers, together with the noise and the wear-and-tear, which are inseparable from its use, more than set off the extra cost.

The necessary control is effected by the motion of a single lever, which is arranged to work in either of two slots provided in the supporting-frame, one for each direction of motion.

When the lever is drawn back to its extreme position, and there is therefore, no current on, it can be passed from one slot to the other.   The effect of this motion is to throw over a reversing-switch, which controls the direction of the current in the field-circuit of the generator.   This determines the direction of the main current furnished to the winding-motor, and, consequently, the direction of its rotation.

The motion of the lever in either of the slots has the same effect, it gradually cuts out the resistance which is connected in the field-circuit of the generator, which thus acquires a stronger field and, therefore, provides a proportionally greater electric pressure to the winding-motor.   The excitation of the field of the motor, however, is kept constant and, since its speed is proportional to the pressure of its supply, the forward motion of the lever determines a corresponding increase in the speed of the winding-motor.   For each position of the controlling-lever, there is a definite speed of the winding-motor; and this is quite stable and independent of the load, and can have any desired value from nothing up to the maximum rated speed.

Too rapid motion of the controlling-lever is prevented by means of an adjustable oil dash-pot, which may be seen in Figs. 6, 7 and 8 (Plate XXL).   These figures also shew the various safety-appliances and brakes, which prevent over-running and, generally, guard against the effects of mishandling the plant.

A depth-indicator is provided which consists of two vertical screws suitably geared to the drum, each having a travelling-nut, representing one of the winding-cages.   These nuts are provided with curved steel fingers, which come into operation near the end of their travel and push back the control-lever to a position which allows only of a greatly reduced speed.   The attendant can then no longer increase the speed, but, as the lever is free to go farther back, he has complete control as regards further slowing down or stopping.   Should he, however, through any accident or negligence miss stopping at the right moment, a second finger on the nut pushes the control-lever into the off-position and applies the emergency-brakes.

There are usually two sets of brakes provided, the working brakes and the emergency-brakes.   The former are applied by the action of a piston working in a cylinder to which compressed air can be admitted. The valve for admitting the compressed air is operated by a handle, placed conveniently near the controlling-lever, and connected with it in such a manner that when it occupies the off-position the brakes are applied, and they are released when it is moved to the starting position.

It should be noted that normally the brakes are only used to hold the load, and that there is no absorption of power, any excess of energy in the moving system at the end of the journey being taken up electrically. Thus, when the winding-cages are travelling at full speed, if the control-handle be moved back nearly to the off-position, the pressure of the generator is reduced below that which the winding-motor can produce. The latter, therefore, works as a generator, pays back energy into the motor-generator, accelerates its speed, and thus stores energy in its fly-wheel; and meanwhile, it draws its generating power from the energy of the moving system and thus produces the necessary braking effect. After a little practice, the right moment for drawing back the control-lever is ascertained, and the winding-cage is brought to rest at the right place without using the compressed-air brake until the last moment.

The emergency-brakes are entirely distinct from the working brakes, and are applied by means of a counterbalance-weight, which is normally supported by the pressure of the compressed air on a piston, and by this means the brakes are held off the drums during working.

The brakes are, therefore, applied, if through any accident the air-pressure fails; they are also applied by a special trigger on the depth-indicator should the cage overrun the bank as mentioned already; and they can also, in the case of emergency, be applied by hand. In any case, the fall of the weight is arranged to open a special emergency-switch which entirely disconnects the winding-motor. It will be seen that very ample provision has been made for security, and to guard against misuse or negligence.

The writer will now consider, in a little more detail, the exact action of the flywheel on the motor-generator, as this forms a most important feature of the system. In Figs. 9 and 10 (Plate XXII.) the thick, black lines represent the horsepower momentarily required on the drum or pulley. Fig. 9 represents the case of an unbalanced rope, and the fluctuations of horsepower are naturally much greater than in Fig. 10, where there is an endless rope passing over pulleys at the top and bottom of the shaft; the average horsepower, however, is the same in both cases. If we succeed, therefore, in completely equalizing the demand for power by means of flywheel storage, the demand for power will be the same in both cases; we shall, however, require a larger flywheel in the first than in the second case. Taking the first case, we find from consideration of the curve that the amount of stored energy available must be 9,000 horsepower-seconds, or 4,950,000 foot-pounds. To store so large an amount, the flywheel must be made entirely of steel, and run at a peripheral speed of about 16,000 feet per minute. Under these conditions, a 14 tons flywheel would give out 9,000 horsepower-seconds with a reduction of speed of about 10 per cent. In practice, the flywheel would be made of a weight of about 20 tons, so as to allow for the inevitable loss of energy in the double conversion.

The total amount of energy stored is, of course, much greater than the amount mentioned above, but it is only available by admitting of a greater reduction of speed than is advisable in normal working. Should, however, any accident occur, which involves the entire cutting-off of the electric supply, the energy stored in the flywheel would all be available, allowing only for electrical losses. The net stored energy in such a flywheel is about 380,000,000 foot-pounds, of which at least 75 per cent. is available for effective work on the net useful load. The useful work to be done in one lift is 5,500 pounds by 1,650 feet, or 90,750,000 foot-pounds. Consequently, it will be recognized that there is enough energy stored for three lifts after the total cessation of the electric supply.

At the beginning of a journey, the motor-generator may be assumed to be running at nearly its maximum speed. When the lift commences, the power is at first supplied entirely by the generating-station, but as soon as the demand has become greater than the average, the motor-generator begins to slow down and its flywheel furnishes the excess of power required; as soon, however, as the demand for power has fallen again below the average, the motor-generator begins to speed up again, storing power once more in its flywheel. It continues to do so after the completion of the trip and before the next lift begins, drawing this power from the generating-station and providing it with a load during the period of rest. If the plant is well proportioned for the conditions of the work, the motor-generator will have nearly reached its maximum speed when the next call for power comes upon it. Should an unusually short rest occur, it will not have reached so high a speed and will run somewhat more slowly for one or two trips; while, if an unusually long stop occur, it will reach its full speed before the next lift commences, and its demand on the generating-station will begin to fall, so that the equalization of load will not be complete. In practice, however, with a fair margin in the flywheel, the equalization will usually be pretty complete, and if the generating-station be used for other work, or if there are several electric winding-engines at work, the actual fluctuations in the demand on the generating-station will be practically negligible.

The writer has now described a system which unites in the simplest and most practical way, yet suggested, the advantages of power-storage and the consequent equalization of the load on the generating-station, the absence of rheostatic losses, and the most perfect speed-control. It remains to shew the advantages which such a plant offers over steam winding-engines.

The advantage of this system of winding is clear from the fact that, although the scheme has nowhere yet been put into practical operation, there are actually in course of construction for various mines, winding-engines on the Siemens-Ilgner system of a total capacity of over 17,000 horsepower; the installers need not be considered very venturesome in adopting an untried scheme, for winding by electricity has already established itself as a proved success in numerous instances; and the novelties of this system consist only in the combination of known and tried devices, the effect of which is easily and certainly calculable beforehand.

One of the principal considerations which has led to such a rapid adoption of the Siemens-Ilgner system is the great economy in steam-consumption which results from it. In certain quarters, the opinion prevails that such economy is of little or no importance at the pit-mouth, where the fuel burned is often of so inferior a grade as to find no market. The economic fallacy of this opinion is, however, obvious to engineers from a practical experience with their own accounts, but it may be worth while to say a few words on the subject:—(1) Improvements in boilers, furnaces, and stokers are constantly facilitating the use of lower grades of fuel, which are therefore acquiring a certain market-value. (2) It often happens that the fuel actually used on a mine is of a fairly high grade, because the use of a lower grade would necessitate capital expenditure in increased boiler-plant to obtain the necessary power; and (3) the cost of fuel is only one item out of many in the cost of producing steam. Labour, repairs, and maintenance and interest on cost of boilers, stokers, steam-pipes, etc., are heavy items, most of which are wholly or in part proportional to the quantity of steam produced. It is evident, therefore, that steam costs a definite and by no means negligible sum per pound, even if the fuel be supposed to cost absolutely nothing; if, therefore, we can do with two-thirds or half of the steam, we shall effect a real saving in money, provided of course that we do not introduce compensating expenses in other directions.

The steam-consumption of winding-engines varies through an enormous range, according to the nature of the work, of the engine itself, and of the attention that it receives; but, in any case, it is extremely high as compared with ordinary steam-practice on fairly steady work. It is clear, therefore, that the conditions of steam-winding are unfavourable for economy. The engine has to be capable of very great though short overloads, the load itself is extremely intermittent, and there are intervals of standing idle during which steam is being condensed in the pipes and, very possibly, it is blowing off from the boilers in great quantities as well. Again, the engine must be capable of running in either direction and of starting from any position. These conditions are in themselves unfavourable for economical valve-setting; but they have the still greater disadvantage that the steam-distribution has to be largely under the control of the operator, and this in itself is a most prolific source of waste.

The writer is aware that there are many members who can furnish actual figures of steam-consumption from their own experience, and he hopes that they will not fail to do so in the discussion, as information on this point is very valuable. The writer has been able to obtain a few figures from actual tests, which he thinks are very instructive.

A trial has recently been made on the large twin-tandem winding-engine of the Rheinelbe III. colliery, the property of the Gelsenkirchen Mining Company. The trial extended over one shift of 9 hours, and the consumption of steam was found to be 114 pounds per effective horsepower-hour. (The writer may remark, in order to save repetition, that all the steam-consumptions given in this paper are per horsepower-hour on the nett useful load lifted in the shaft). During this trial, the winding-engine was handled in the usual manner by its ordinary attendant, and the conditions were those ordinarily obtaining. A second trial of 9 hours was next taken, in which the only difference was that the operator was strictly supervised during the whole extent of the trial by an expert engineer. The result was a consumption of 70 pounds per effective horsepower-hour. These figures speak for themselves, but they have a special significance from our point of view, as will presently be seen. In the preceding trials, the steam exhausted into the atmosphere; and a further trial, under condensing conditions, with the same careful supervision, gave a steam-consumption of 57 pounds per effective horsepower-hour.

Other searching trials, extending over long periods, have been made by Mr. Buschmann on the winding-engine at the Heilbronn salt-mines, and this engineer concludes that with such care and attention as can be realized in practice, 77 pounds of steam per effective horsepower-hour can be attained, though with special supervision of the operator, he reached as low a figure as 68 pounds.

We shall however, err greatly, if we suppose that such figures represent a fair average of what is taking place in the winding-engines actually in operation. These results were obtained with modern engines of the best and most economical type, and care was taken that the engines should have full duty during the trials; but if we take into account the losses incident to times of slack lifting or of total cessation of work, a figure of about 95 pounds will be nearer the mark; while if we consider cases such as are unfortunately too common, where very low pressures, late cut-off, and generally obsolete and inefficient machinery and methods are in use, it is hard to say what extravagant figure may not be reached.

We are now in a position to understand the full importance of an equalized load on the prime-mover. For it may be observed that the extravagant steam-consumptions, just given, are almost entirely attributable to the intermittent nature of the work. On the other hand, given a fairly uniform load, at a constant speed, there is clearly nothing to prevent us from availing ourselves of the best modern practice in the economical generation of power. In making a comparison with the tests just quoted, however, we will not assume the very highest attainable economy but a moderate working figure. A fair average steam-consumption for a modern generating plant under the conditions of daily work with moderately fluctuating loads, such as we may expect in this case, is 20 pounds of steam per electric horsepower-hour, and this value will be taken as the basis of our comparison.

Experience with existing electric winding-gears and a careful consideration of all the working conditions with the Siemens-Ilgner system leads one to expect a total efficiency of 60 per cent. under daily working conditions. That is, to say, of the total electric energy supplied to the terminals of the motor-generator, 60 per cent. appears as nett work in raising the useful load in the shaft. To be on the safe side, however, an efficiency of only 50 per cent. will be taken, from which we deduce a steam-consumption of 40 pounds of steam per effective horsepower-hour on the useful load, a great deal less than the best result with steam and less than half what may be regarded as an average working result with steam-winding.

No negligence or mishandling on the part of the operator can appreciably affect this figure. The steam-distribution in the engine-cylinders is entirely out of his control, and the chief cause of waste is thus removed. Practically, the only harm that he can do is to make a bad stop with an undue amount of braking; yet even this has little effect on the economy, for with this system the energy of the moving parts is not wasted in friction but stored in the flywheel and at least 75 per cent. of it may be recovered. It is in the writer's opinion, one of the chief advantages of the system that it removes from the attendant the dangerous power of controlling the steam-consumption.

It is of course advisable in this, as in any system, that attention should be paid to the correct operation of the winding-engine, for only in this way can rapid and uniform handling of the output be attained, but any lack of care is not attended with the direct and disastrous losses which occur in steam-plant.

In this connection the writer would like to draw the atten- tion of the members to Figs. 11 to 16 (Plate XXII.), which are generally similar to the curves plotted in Figs. 9 and 10. These curves, however, are not calculated, but are careful copies of actual diagrams taken automatically from an electric winding- gear at Thiederhall. Instead of horsepower, electric current in ampères, is plotted, being, of course, much easier to measure : for all practical purposes, however, the current is proportional to the horsepower, the electric pressure being (as it was) kept constant, and this may be considered to be a horsepower-curve. Now, it is particularly interesting to notice the variations which these curves show from each other and from the theo- retical diagrams. So far as starting and running are concerned, they are practically the same, the differences occur in the ends of the diagrams, representing the retardation ; and they are due to the differences in judgment of the attendant as to the right moment for shutting off current and beginning braking.

It is true that men acquire great dexterity in this matter by long practice, but the writer thinks that these diagrams reveal the fact that this dexterity is not quite so great as may commonly be imagined. The writer doubts whether any one observing the actual lifts represented, would have suspected such great differ- ences in the stops as are shewn in the diagrams. It must, at any rate, be a great comfort to a mine-owner or manager to feel that he has a plant in which little unavoidable errors of judgment are not attended with a direct and heavy loss.

The writer must not close this paper without briefly referring to the subject of the first cost involved in electrical plant. Naturally it is not possible to say anything very definite on this point, since it greatly depends upon the particular circumstances of each case; but, broadly speaking, one may say that the cost of an electrical winding-plant with motor-generator and all safety and controlling-devices complete, would be about the same as that of a good modern steam winding-plant of the same capacity.

We have, of course, to consider in addition the means of generating the electricity. The cost of this will depend greatly upon circumstances, but it will not usually be a heavy item. In most cases, a generating-station would be erected for other work besides winding, and the average power only has to be supplied. In such cases, the fair share of the cost attributable to the winding-gear would probably not exceed the sum saved in boilers, stokers and steam-pipes, which would be dispensed with owing to the smaller steam-consumption.

From what has now been said, the writer hopes it will be clear that the application of electricity to winding-gears has distinct advantages and merits, and has now been sufficiently developed to be considered as within the range of practical possibilities for mines and collieries in this country. The writer will be satisfied with the result of his paper, if it induces some of those actually concerned in the manufacture, purchase, or use of winding-engines to give serious consideration to the economic possibilities of electrically-driven winding-gears.

———

Mr. D. SELBY BIGGE said that this paper opened up an enormous field to the electrical engineers of this country. As yet there had been no applications on any large scale of electric winding. This was not so in other countries, and they must look to what the other parts of the world were doing. In Germany, Belgium and Westphalia, they were applying this new departure in electricity, that is, to main winding-engines. The question to decide, of course, was one absolutely of cost, as to whether it would pay to adopt electricity in preference to steam. The writer considered that there would be a saving as against a ordinary steam winding-plant of nearly 50 per cent. in the c of working. He (Mr. Bigge) had recently been considering t matter very closely, and he could not put the saving nearly high as that, but he thought that 16 to 26 per cent. would be actual saving by using an electric winding-gear as compared wi the best modern steam-engine. He might cite a case in whi electric winding would commend itself to mining engineers: th was the case of a group of collieries where they had coke-ove the waste-gases from which could be utilized. If they we sinking a new colliery, they would be able to generate power

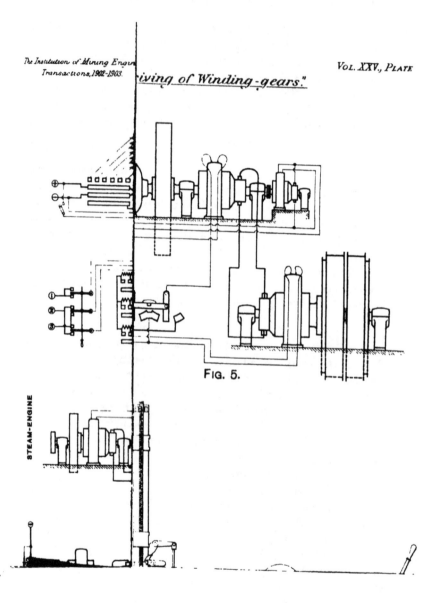

FIG. 5.

STEAM-ENGINE

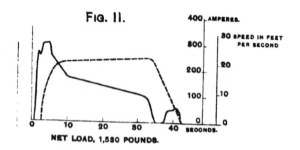

FIG. 11.

400 AMPERES.

300

30 SPEED IN FEET
PER SECOND

200

20

100

10

0   0

0   10   20   30   40 SECONDS.

NET LOAD, 1,530 POUNDS.

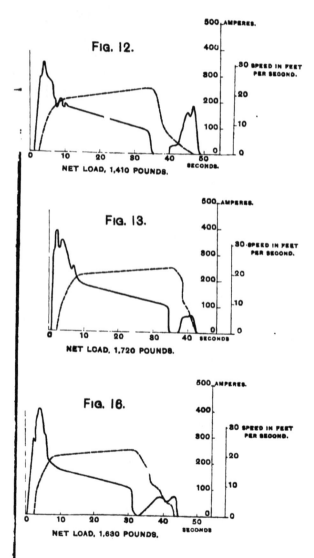

500 AMPERES.

FIG. 12.

400

30 SPEED IN FEET
PER SECOND.

300

200

20

100

10

0   0

0   10   20   30   40   50 SECONDS.

NET LOAD, 1,410 POUNDS.

500 AMPERES.

400

FIG. 13.

30 SPEED IN FEET
PER SECOND.

300

200   20

100   10

0   0

0   10   20   30   40 SECONDS

NET LOAD, 1,720 POUNDS.

500 AMPERES.

400

FIG. 16.

300

30 SPEED IN FEET
PER SECOND.

200   20

100   10

0   0

0   10   20   30   40   50 SECONDS

NET LOAD, 1,630 POUNDS.

the coke-ovens at a minimum of cost, and transmit it electrically to the new winning. This immediately obviated all initial expense in steam-engines and boilers; and in such a case, main electric winding-engines should certainly be considered. The system described was very ingenious, and if it worked out in practice as well as it appeared in theory in the paper, it should be a great success. He had not seen the method except at the Düsseldorf Exhibition, but the results should be very carefully watched.

The Rev. G. M. CAPELL (Passenham) said that at Grand Hornu colliery, near Mons, a double tandem engine, with a wheel-generator of 4,000 horsepower was driven by superheated steam, and the consumption of water was guaranteed not to exceed 10 pounds per indicated horsepower. The plant, with the condenser, worked three mines at a considerable distance apart; one was 2 miles off, another $1\frac{1}{2}$ miles, while the generating plant was at the central station. The winding would all be effected by electrical tri-phase motors; the whole of the surface-transport of coal from the three winding-shafts to a central screening-station, and the carriage of the coal to the canal and to the railway, would also be worked from the central station. He thought that it would be the first complete electric installation at a colliery in Belgium, and it had entirely displaced the old system.

Mr. A. H. STOKES (H.M. Inspector of Mines, Derby) wrote that Mr. F. Hird attached considerable importance to the saving of fuel, and although the fuel burned at a colliery was often of very inferior quality, yet it would be an economic fallacy to ignore the amount of coal consumed. The paper was very bare of actual details and cost, and the author appeared to forget that electricity must be generated, or produced mechanically, and that the primary source of the electrical power at a colliery was the steam-engine. As regards the economy of plant, they required for comparison the cost of the steam-engine and generating plant, plus the winding arrangement, and then the comparison of that cost with the cost of a winding-engine under present conditions of winding. This would supply data upon which an opinion could be formed, as to the cost of a steam *versus* an electric winding-plant. The actual fuel consumed by both plants was required under similar conditions of winding, and this would give ground upon which a comparison might be made of the cost of winding.   In the absence of such information, it would be unwise either to condemn the present system, or to rush into unknown and untried electrical installations for winding coal.

Mr. S. MAVOR (Glasgow) congratulated Mr. F. Hird on being the first to bring this important subject before the members. The difficulty in this country was to shew a sufficient saving to warrant the enormous capital expenditure.   In the case of isolated collieries, it would be impracticable, unless or until the power-supply from some of the big electric-supply companies was available at a sufficiently low rate.   In the case of new collieries it was different, especially where more than one shaft could be included.   He believed that there were cases where groups of existing collieries might be economically dealt with in this manner.   He thought that Mr. Hird had not overstated the case when he said that the cost in steam-consumption would be about one-half of that of the ordinary method, but he had gone into one or two cases where he found that a saving of one-half the coal would not be nearly enough to balance the depreciation on the electrical plant; in other words, coal was too cheap. Of course the district in which coal was much dearer was the one in which the system was the most likely to find favour.

Mr. M. W. WATERHOUSE (Bedworth) said that the paper opened out a wide field for investigation.   The method described was very ingenious, but it struck him that they would require at least five dynamos or motors to work one set of winding apparatus, and this seemed rather complicated, notwithstanding the obvious advantages from an economic point of view.

Mr. H. PERKINS (Leeds) said that only within the last few years had it been possible to talk of electric winding.   The generating station generally would not bear the demand, which would vary from nothing up to 1,500 horsepower; but, by the ingenious system of averaging the load, electric winding was rendered possible.

Mr. JOHN ROBINSON (Haydock) asked what was the contemplated life of the new winding apparatus.   Would it last as long as the steam-engine, or, in other words, would the cost in repairs not be greater than in the case of the steam-engine? He ventured to think that there was a bright future for electric winding-plant, especially where the colliery was of great depth. As an instance of the possibility of doing without boilers, he mentioned the case of the Mond gas-plant. It was one of the schemes of the future, but he was sorry that Mr. Hird had not given more practical information as to the saving that could be effected, as 50 per cent., he feared, was altogether an erroneous figure. If they could effect 25 per cent., he felt sure that mining engineers would not have any hesitation in adopting electric winding, provided that the cost of repairs did not exceed those of the steam-engine.

Mr. S. F. WALKER (London) said that Germans with their system of protection were prepared to go in for experimental plant, and we should reap the benefit of their work. There were two methods of adapting electricity to winding-engines: (1) by means of an electric accumulator, and (2) by a mechanical accumulator such as the heavy flywheel; and, so far as he understood, it was a question of reliability. One heard of electric accumulators in central stations that worked beautifully, but they have a staff of trained men constantly watching them, and they have no coal-dust to contend with. It would be possible to have one in an engineer's shop, properly screened and looked after, which would give a very good account of itself. Electromobile engineers were working at the problem, and mining engineers would reap the benefit. On the other hand, could they depend on an enormous steel-casting, 20 tons in weight, revolving at a high speed, and be sure that it would not burst into fragments? If they could be sure of the flywheel, it was perhaps the better accumulator of the two, because it was simpler. Although there might be five dynamos, there was not anything like the same amount of complications as there were about an ordinary winding-engine.

Mr. T. P. OSBORNE YALE (Blaenau Festiniog) said that he had under his observation nine electric winding-engines, some of which were working on the three-phase system and some on the continuous and varying from 15 to 200 horsepower. These engines were working absolutely without a hitch, and he knew of many metalliferous mines where steam-plant was being taken out and electrical plant substituted. In the use of low pressures, the three-phase could not contend from a financial point for a moment with continuous current, but when it came to a question of high pressures it was useless discussing the relative merits of the two systems, for the present state of the law forbids the use of three-phase current under conditions which would pay to use it.

Mr. F. HIRD, replying to the discussion, said that he did not agree with the remark that electric winding would not prove advantageous in the case of an isolated colliery, nor that the interest on capital should be paid out of the saving effected. Of course it would not pay in every case to take out steam, and put in electricity, but he did claim that in the case of a new plant it would pay to install electric winding, even at an isolated colliery. The five dynamos appeared more complicated than they were in reality. There was no difficulty in designing motor-generators to run at 2,000 to 3,000 revolutions per minute, and the capital invested need not be large. With regard to the danger of a burst flywheel, there were many ways in which the flywheel could be built so as to be absolutely safe; say, with boiler-plates for instance. With regard to modern improvements in accumulators, he had been told by experts in accumulators that the improvements were in the price-lists and not in the accumulators themselves. It was futile to discuss the general question of three-phase *versus* continuous current. Any consulting electrical engineer had an open mind on the subject, and would consider that there were cases for which each was best adapted. It was doing harm to give the impression that electricians were not agreed on the point, as each case must be considered on its merits.

Several members had remarked on the absence of actual figures of the cost of electrical and steam winding, he was very conscious how much the practical value of his paper would have been enhanced by such figures, but it must be remembered that the only electrical system which by equalizing the load could compete with steam in coal winding, was still quite new and absolutely no figures were as yet forthcoming. Several plants, however, would soon be at work, and no doubt actual comparative costs would be available at an early date.

The PRESIDENT (Sir Lindsay Wood, Bart.) moved a vote of thanks to Mr. Hird for his very interesting paper.

Mr. M. WALTON BROWN seconded the resolution, which was cordially approved.

THE INSTITUTION OF MINING ENGINEERS.

GENERAL MEETING.

HELD IN THE ROOMS OF THE GEOLOGICAL SOCIETY, BURLINGTON HOUSE, LONDON,
JULY 3RD, 1903.

SIR LINDSAY WOOD, BART., PRESIDENT, IN THE CHAIR.

DISCUSSION OF MR. M. R. KIRBY'S PAPER ON "STEAM-GENERATION BY THE GASES FROM BEEHIVE COKE-OVENS."*

Mr. J. C. B. HENDY (Etherley) wrote that it was not clear whether the amount of air stated by Mr. Kirby as being used per ton of coal in coking was the amount arrived at by experiment at the ovens forming the subject of his paper. He thought that for purposes of comparison this was important, as the amount of air necessary would vary considerably, according to the condition of the coal when put into the ovens, the composition of the coal, and the temperature of the air-supply. Mr. Kirby did not state whether the coal had been mechanically treated in any way before being put into the ovens, but as by Table I. the amount of water in the coal is 7·3 per cent., he presumed that it had been washed. This large amount of water was of course very much against the coal. For washed coals it was perhaps not excessive, as in many instances he believed that as much as 10 per cent. and more of water was put into the ovens after washing, through imperfect draining. The loss of heat and useful effect, however, with such an amount of water was enormous, for not only was the calorific power of the coal reduced in the same proportion as the amount of water that it contained, but a larger quantity of heat was practically wasted in evaporating the water in the ovens. In justice to the coal, he thought that Mr. Kirby should have made special allowance in his paper for the loss due to this excess of water. The paper did

* Trans. Inst. M.E., 1902, vol. xxiv., page 441.

not give any information as to the shape or area of the main flue for the coke-ovens, which was a very important item in the matter of radiation. He thought that in many instances too little attention had been given to the proportions of the main flue, as it was not uncommon to see a flue almost of the same size for the whole of its length, the area being practically the same for 10 ovens as for 50 ovens or more, which manifestly must be wrong. The fact that a circular flue would present less radiating surface than any other shape of equal area was also in many cases entirely ignored. He believed that great economy could be effected by making the flues of proper dimensions, and with areas graduated according to the varying number of ovens along the whole length of the range. He noticed in the analysis of the coke in Table I. that there were no traces of sulphur or phosphorus, and would like to ask whether this was exceptional and correct. Some years ago, he made some experiments as to the steam-generating power of two Lancashire boilers, the results of which were recorded in the *Transactions*, and in those experiments he always found that the best results were obtained when the temperature at the bottom of the chimney varied from 450° to 500° Fahr., and this conclusion seemed to agree with the results given by Mr. Kirby.

————

DISCUSSION OF MR. F. J. NORMAN'S PAPER ON "BORING IN JAPAN."*

Mr. F. J. NORMAN (Calcutta) wrote that he had been engaged on behalf of the Government of India in putting down an experimental bore-hole by the Kazusa system. The site chosen afforded about as hard a test as could well be imagined, being situated on a rocky ridge of quartzite, with veins of particularly hard quartz running through it in all directions, but mostly east and west. Yet, despite this, the bore-hole was put down to a depth of 40 feet in less than twenty days, with a miserable gang of coolies, who had never done any other work than digging earth with a hoe, and carrying the same on their heads in a basket. Not one of the men weighed 9 stones, and their average weight was 7 stones 5 pounds. An improved drill

* *Trans. Inst. M.E.*, 1902, vol. xxiii., page 685.

was used, weighing 120 pounds, and therefore quite a manageable affair in the hands of six coolies. At least, it ought to have been, but when the writer's back was turned upon them one day they managed to let the chisel drop down the bore-hole, with the consequence that it got jammed. And then, instead of informing the writer, they set to work tugging away at it, and so got it jammed crosswise in the bore-hole, for all endeavours to withdraw it have hitherto been ineffectual.

Mr. Tom David, after only a three days' attendance at the boring operations in Calcutta, had just succeeded in putting down a bore-hole to a depth of 300 feet. He only used one of the tools, namely, the sludge-pump and cutting-chisel combined. This cost him £1 2s. 8d. (17 rupees); the bamboos he cut down from among those growing wild on the site of the bore-hole. The rings were made by an old *mistri*, or blacksmith, from scraps of iron, and cost next to nothing. The cost of labour also came to little for at this slack season of the year, there was no work going on in jute-works.

He (Mr. Norman) enclosed a small quantity of the débris removed from out of the bore-hole in Purulia, and a specimen (model size) of a bamboo-joint as described in his paper.

———

Mr. T. P. Osborne Yale's paper on "Electric-power Distribution by Continuous Current, for Mining and General Purposes in North Wales" was read as follows:—

ELECTRIC-POWER DISTRIBUTION BY CONTINUOUS CURRENT, FOR MINING AND GENERAL PURPOSES IN NORTH WALES.

By T. P. OSBORNE YALE.

Introduction.—Any claim that this paper may have to interest is not based on the magnitude of the plant concerned, but on the fact that it illustrates the way in which the supply of electricity, either public or private, in this country, is hampered by existing laws; and how, under exceptional circumstances, a supply may be given to the public without statutory powers.

The writer ventures to think that these points may be of some interest to the members, as there are many places in Great Britain where several mines adjoin, or where the same firm has works and villages within a comparatively short distance of each other, where it would be well worth while to erect a private central supply-station. The paper also deals with some of the points to be considered in deciding on the use of tri-phase or continuous current in this country.

Lastly, small as it is, the writer believes that the plant to be described is the largest hydro-electric plant for public supply in Great Britain, entirely dependent on water all the year round.

Description of District.—The district with which this paper is concerned lies in the north-eastern corner of Merionethshire, and has long been one of the great centres of the slate-industry in the world. The output of finished slates from this district is about 100,000 tons per annum, and the value is about £500,000.

Blaenau Festiniog has a mining population of about 13,000 and lies on the average about 700 feet above sea-level. It consists chiefly of a very long narrow street, following the semi-circular contour of the mountains, with a radius of about 6,000 feet.

All the slate-mines and quarries lie above the town, and, roughly speaking, their surface-workings follow the same general contour-line at a height varying from about 1,000 to 1,900 feet above sea-level, that is about 300 to 1,200 feet higher than the town.

The slate-veins dip at an angle of about 1 in 3, and have been followed downward, so that, in most instances, what was developed originally as an open quarry, has long since become a deep and extensive mine, worked partly by means of adits and partly by inclined shafts following the dip of the veins. The whole process of getting and manufacturing the slate for the market is most interesting, and is continuous. That is to say, the miners open up new ground and are followed by the rockmen, who get the blocks, which are then sent up to the surface, where they immediately go into the mills and are there cut into various sizes and shapes required for the market. Almost all this work is performed by contract, and with the exception of the mining, which goes on more or less day and night, is only carried on in the day-time.

Power is chiefly required for haulage on the inclines, for sinking and for pumping; and also for driving the mills and repairing-shops. Hauling and mill-driving are only carried on during the day, unless under very exceptional circumstances.

Haulage.—Haulage on the surface is largely performed by means of gravity, for it is apparent from consideration of the above facts that the finished slate has to descend from the mills, and this difference of weight is used to haul up the empty wagons and necessary mine-supplies. Another method of utilizing gravity for haulage, which is largely employed aboveground, is by means of funicular railways: that is to say, two water-tanks, fitted with suitable platforms and attached to opposite ends of a rope passing round a drum at the top of the incline, are employed, and water is admitted to the tank which happens to be at the top, to haul up the weight carried on the other platform.

A large vertical lift of this kind is in use in one of the mines. This is a very wasteful application of water. With these exceptions, all haulage is performed by means of steam or electricity. The engines operate from one to four or more roads by means of single ropes on independent drums, which are thrown in and out of gear. The full wagons going down an incline do not assist the ascent of the empty trucks, but are let down on the brake.

The loads must be started gently, as frequently they consist of stone-blocks of great weight, up to 3 tons, and being carried on a trolley with a short wheel-base, there is considerable risk of them being tipped, at the bottom of the incline, where they come on to it from the level. In many instances, the landing must be very smartly effected, if overwinding is to be avoided, as the loaded waggons on reaching the top are not always run straight under the drums, as is usual, but are immediately turned off at a sharp angle, and, in these cases, the control over the engine must be absolutely perfect.

Mill-driving.—Steam or electricity is employed for mill-driving, but where water-power is available, water-wheels and turbines are employed to a considerable extent. Mill-machinery consists chiefly of slate-shears and saw-tables, but there are other machines also to be driven, such as planes and saw-sharpeners, as well as the tools in the repairing-shops, etc.

Pumping.—Water-wheels and turbines are largely used for pumping water, in which the turbine or wheel on the surface is connected by means of rods to the pumps. In one case, an auxiliary steam-engine is installed for use, in case of failure of the water-power. Where water-power is not available, with very few and trifling exceptions the pumping is now effected by electricity.

Rock-drilling.—Rock-drilling is almost entirely performed by hand, but a certain amount of this work is done by compressed-air drills. Experiments have been made with electric drills, but while fairly successful, they have not yet shown themselves superior to compressed-air drills.

The writer is about to try an electric drill, which has just been introduced, and is stated to yield excellent results.

Introduction of Electricity.—About 12 years ago, a small electric pumping-and-lighting plant, working at 100 volts, was erected at one of the mines: it is operated by turbines, and has always given satisfaction. Beyond this, however, nothing more in the district was done for many years, although the writer commenced his efforts to extend its use about 8 or 9 years ago.

Finally about 5 years ago, he succeeded in forming an independent company for the purpose of supplying electricity for general purposes to the public. Arrangements were made by which certain water-power was secured about 10,000 feet from the mines, where the available head of water is about 194 feet, and there is a constant supply of water all the year round, without the necessity of undertaking any heavy expenditure on water-works. Among other points in favour of this water-power were:—(1) It is close to an existing market for its power, and it could be acquired at a price which made it compare favourably with coal at 18s. per ton, the average cost in the district for coal delivered at the boilers (bought in large quantities), while for small quantities the price is somewhat higher. (2) The cables from the power-house to the mines would pass through the centre of the town, where a most suitable site for a substation could be obtained, with the Festiniog railway (which is in direct communication with every mine and works in the district) on one side, and a road on the other, thereby favouring the establishment of general stores and repairing-shops as well as having many other advantages. This site is about 7,500 feet distant from the power-house and 3,000 feet from the mines. The supply of light to the town was also included in the undertaking, so as to secure a good load-factor, as the power-load is almost entirely a day-load. Of course, this is the reverse of the usual experience of electric-lighting companies, as they generally have an insignificant day-load. The gas-works belong to the local authorities, who supply gas at 4s. 2d. per 1,000 cubic feet. (3) The supply of power for pumping is almost automatic, and the more power required for pumping, the more water-power there is available to do it.

System of Supply.—The next point for consideration was whether continuous current or three-phase should be adopted. The writer examined carefully into this question, and advised the adoption of the continuous-current system, for the following reasons:—

The state of the law with regard to electric supply is such that it is almost impossible to move about outside the strictest boundaries of private property without statutory powers, and with statutory powers, overhead wires are expressly forbidden without the consent of the local authorities.   In fact, statutory powers cannot be obtained without the consent of the local authorities.   In any case, parties obtaining statutory powers are placed in a very disadvantageous position, both from a commercial and from an engineering point of view.   The local authorities were most hostile to the scheme, at that time, and endeavoured in every possible way to thwart its initiation.

Consequently, great care was necessitated in complying with the regulations of the Board of Trade and of the Postmaster General, and the regulations of these authorities are much more exacting in every way with regard to the use of alternating current than with regard to continuous current.   The regulations of the Board of Trade are such that, if the pressure exceeds 500 volts, it is impossible to convey any reasonably large amount of power, even for private purposes, beyond the boundaries of private property, as defined by the public roads, without special permission from that authority.   Another regulation limits the pressure of the supply to other parties to 250 volts, except by special permission.   Even if this special permission be granted, the plant is liable to inspection and such extra regulations as the authorities may see fit to apply.   The mine-owners, at that time, were not particularly anxious to introduce electricity, and they absolutely declined even to consider the question, if further Government inspection and regulations were forced upon them. As it was, they only consented to use electric power on condition that the plant was provided and erected at their mine, free of expense, with the option of purchase, which has been exercised. The hauling and pumping was taken by contract, including the attendance and maintenance of the machinery: this pleased the mine-owners, and suited the electric-supply company, as it placed them in the position of parties supplying themselves, thereby enabling them to use a pressure of 500 volts, without any special permission.   Otherwise, the limit would have been 250 volts, for as pointed out above, the mines refused to have any extra inspection forced upon them.   Had the pressure been restricted to 250 volts, the scheme, so far as utilizing the water-power was concerned, would probably have been abandoned; and it was only by making special arrangements that the comparatively low pressure of 500 volts was used.   Bearing this limitation in mind, therefore, it remained to consider the respective merits of the three-phase and continuous-current systems.

Before going any farther, the writer may state that he is entirely in favour of the use of three-phase current when the limits of low pressure can profitably be exceeded; but it must be understood that, although the Board of Trade declare the limits of low pressure for continuous current to be 500 volts, they place the limit for three-phase current at 250 volts. Otherwise, the writer considers that each system must be considered on its merits in relation to any particular case, for instance, in a fiery mine, three-phase current has merits of its own, which give it an undoubted advantage.

With regard to the case under discussion, the reasons for adopting continuous current were as follows:—(1) As pointed out above, the Board of Trade limits for low pressure are only 250 volts for three-phase, while they allow 500 volts for continuous current. This fact alone was sufficient to decide in favour of the latter. (2) Even if it had been permissible to employ three-phase current at the same voltage as continuous current, namely, 500 volts, then the weight of copper in the line would have been about 50 per cent. greater, for the same efficiency. Mr. Gisbert Kapp has pointed out* that if the copper in a line weighs 100 tons for a continuous-current plant, and if another line be worked of the same length and giving the same efficiency of transmission and the same stress on the insulation with a three-phase plant, 150 tons of copper would be required. This would have ultimately meant probably 30 or 40 tons extra in our case, while this would also have entailed extra cost for carriage and erection, for increased strength of supports, insulators, etc., for an overhead line, or for insulation, trenching, etc., for an underground line; and under the circumstances this was a convincing argument in favour of the use of continuous current. (3) The regulations, generally, are more favourable to continuous current. (4) There was less likelihood of disturbance to telegraph and telephone-lines from continuous than from three-phase current; and at all events if complaints were made, we should be in a much stronger position. The writer may state that the Postmaster General has the power, and rightly so, to stop any line from working (even, the writer, believes, on private property) if it interferes with the working of his lines. As the power-lines run for considerable distances more or less in proximity to, and often parallel

* *Journal of the Institution of Electrical Engineers*, vol. xxii., page 555.

with, the Post Office lines, besides crossing the same at various points, this was a very important consideration. The writer may mention that one of the most important telegraph-lines in the world runs through the district, namely, the line from London to Dublin and New York. The same considerations governed the crossing and running parallel of the power-lines with the wires of the railway-companies, and the writer may state that having examined our proposals, both of the companies concerned—the Great Western and the Festiniog railway companies—afforded every assistance, and both companies adopted electric lighting of their passenger and goods-stations and yards. The preceding considerations also applied to private telephone-lines, with some of which the power-lines run parallel for long distances. (5) Continuous current lent itself more to our purposes. For instance, at the substation we have motor-generators, which reduce the pressure for lighting purposes to 230 volts. The main generators can, however, be run, when required, at reduced voltage, and by means of suitable switchgear, the lighting system can be coupled direct to the power-mains. This practically duplicates the plant, and enables us to dispense with many expensive spare parts which would otherwise have to be stocked at the substation for the lighting-plant.

Description of Plant.—The mine-owner, to whom power was first supplied, stipulated among other things, that:—(1) No existing machinery was to be stopped during working hours, while the change over from steam to electric driving was being effected; (2) existing starting-levers, etc., were to be retained or imitated as far as possible; and (3) the steam-engines and plant were not to be moved, and the electric driving-arrangements must be so designed that a change back to steam could be readily effected.

The writer had no hesitation in accepting these conditions, but it must be admitted that they were a heavy handicap. At the same time, there appeared to be compensating advantages in this arrangement both for the mine and for ourselves. namely:—In view of the fact that the change back to steam could be so readily effected, it would not be necessary to stock spare electrical parts for the winding-motors such as armatures, etc., as it would practically take no longer to change over to steam than to change the armature, and the damage could be quietly repaired on the spot, thereby saving a great deal of handling and carriage, besides which it had the advantage of not absolutely scrapping the steam-plant, which would form a very valuable standby in case of accident.

Haulage.—The haulage-installation, at present, consists of three motors, one of 120 brake-horsepower and two of 80 brake-horsepower each. In each case, the starting and regulating switches are actuated by the same levers that previously controlled the stop-valves, when the engines were steam driven, and are worked by the same men, who had no previous electrical training whatsoever. One of these motors has been working for about three years, and the last of them for nearly two years. Another motor of 80 brake-horsepower is being erected. These are all series-motors, and since being installed have worked without a hitch electrically. In each case, a combined magnetic brake and flexible coupling of the Sandycroft type is provided between the motor and the gearing, and is found to work very satisfactorily.

On one or two occasions, owing to purely mechanical causes, which might have occurred to any engine, such as a hot bearing or a broken pinion, there has been occasion to revert to the use of steam, temporarily, and the whole arrangement has worked admirably.

The drivers, who are as conservative in their habits as most men of their class, much prefer the electric driving, as there is less cleaning, etc., to be done after they shut down, consequently they get shorter hours of work whilst getting the same pay.

Pumping.—There are four sets of three-throw pumps, namely:—One set capable of lifting as normal duty 350 gallons per minute to a height of 200 feet; one with a normal duty of 160 gallons per minute, to a height of 350 feet; one to lift 50 gallons per minute, to a height of 400 feet; and a sinking pump, lifting 25 gallons per minute against a head of 120 feet. The mine has been entirely dependent on electricity for pumping purposes during the last three years.

One possible advantage in connection with leaving the steam winding-engines, *in situ*, is seen in connection with the pumping-plant. Should the mine, through some cause, be cut off from the power-station for any length of time, which is, of course, most improbable, one of the winding motors might temporarily be run as a dynamo by the steam-engine for the purpose of pumping.

The largest and smallest of the above sets are worked by shunt-wound motors, and the other two by series-wound motors. In the writer's opinion, shunt-wound motors are, generally, the best type to use for pumping, but as he has already pointed out, in the case of a breakdown of the lighting plant, it might be desirable to change over on to the main generators, running them at a slower speed, so as to reduce their pressure to about 230 volts. At this pressure, it would not be possible, in practice, to work a shunt-wound motor for 500 volts, whereas a series-wound motor would simply run at half speed. Their great drawback is that they run away, if they lose their load, whereas shunt-wound motors run at practically constant speed whether at full or no load. For this reason, in the writer's opinion, the latter are decidedly the best type, as a rule, to use for pumping purposes. It would seem, from the above, that series-wound motors must have an attendant while running. As a matter of fact, a runaway can be guarded against by means of a minimum cut-out, in which case the same care is not so necessary on the part of the attendant; but this method should only be adopted in special cases.

In the case under consideration, there is a man whose sole duty it is to attend to the pumping, so that he has time if necessary to start the shunt-wound motor pumps, to leave them, and then to go and do what pumping is required from the series-wound motor-pumps.

The great advantage to the electric company of the pumping load, and of having the contract for it, is that they are able to level their load with it at the times that suit them best.

Mill-driving.—At present, mill-driving is not much developed, owing to the fact that in the first mines to take electric supply, the mills were almost all driven by water-wheels or turbines, as already described. The whole of the mill-driving done by steam in the two mines referred to only amounted to 15 horse-power and this is being done by two shunt-wound motors: one of

10 and the other of 5 horsepower, driving respectively a small mill and a fitting-shop. The use of electricity for this purpose is, however, about to be extended to adjoining mines, less favourably situated with regard to water-power on their premises.

Lighting.—Underground current is taken from the power-circuits, supplying five 100 volts lamps in series, at the pumps and engines, and a few fixed points. It is of course liable to great fluctuations, but under the circumstances these are of no importance. The cost of the supply of light is included in the contract.

For the public supply, two motor-generators are at present installed at the substation: One of 70 kilowatts normal output and one of 37½ kilowatts. There is also a small Tudor battery to take the day-lighting load. A special lighting dynamo of 43 kilowatts output is being erected at the power-house, for the purpose of supplying an outlying part of the district, which can be more conveniently supplied direct from the power-house than from the substation.

Generating-plant.—The generating-plant, at present, consists of two 6-pole compound-wound continuous-current generators, each capable of giving an output of 90 kilowatts at 560 volts. They are direct-coupled, one on each side, by means of flexible couplings, to a Pelton wheel. There is a 43 kilowatts lighting dynamo, mentioned above, driven by a turbine, and an auxiliary high-pressure dynamo of 15 kilowatts. There is also a differentially-wound booster working in conjunction with a Tudor battery, which is a very satisfactory arrangement, and tends to a great extent to level the load. These two last machines can be driven by the turbine, which drives the 43 kilowatts lighting dynamo; or, if desired, the auxiliary high-pressure dynamo, mentioned above, can be run as a motor to drive the booster. This arrangement permits the turbine to drive the booster during the day, and to take the lighting-load at night. Also, when it is desired to shut down the main generators owing to the load going off, the auxiliary dynamo can be thrown in to finish the day's work. In order to effect this, the auxiliary dynamo and the booster are coupled together upon the same shaft, which can be connected to one side of the turbine by a clutch; and on the other side, the same arrangement can be applied to the 43 kilowatts lighting dynamo. This arrangement ensures a fairly good load on the lighting turbine, which otherwise would have a very poor load-factor.

A small exciter, driven by a Pelton wheel, is used for separately exciting the shunt-windings of the high-pressure dynamos. This affords greater security against lightning troubles, and it also tends to steady the voltage under the great fluctuations of the winding load.

Water-power.—The fall is 194 feet in 2,400 feet, and the catchment-basin has an area of about 10 square miles or 6,400 acres, with an average rainfall of about 90 inches per annum. The catchment-area contains some fine natural lakes, as well as many reservoirs made by the mines to store water for their own purposes. This is a great advantage, from our point of view, as mine-owners look after the water in their own interests, which in this respect are the same as our own. Should it become desirable at any time, however, the storage-capacity can be greatly increased. So far, it has not been necessary to take any steps in this direction, and only one dam has been constructed at the head of the pipe-line, across a deep and narrow gully. This dam was finished in 1900, and is nearly 30 feet high. It is built in a natural hole formed by the action of the water in the solid rock of the stream-bed, and is constructed of solid masonry. Considerable difficulty was experienced in getting in the foundations, owing to the gully being so narrow at the bottom, that it did not permit of the water being diverted from one side to the other, except during very dry weather, and three great floods were experienced in August one after the other, during the building, causing great damage. However, the dam was completed the same autumn, and it has on several occasions since had 4 to 5 feet of water over its crest. Owing to the steepness of the sides of the gully, it has been possible by means of this dam, which was built in the first place as a head-works for the pipe-line, to store 1 acre of water with an average depth of about 15 feet at the head of the pipe. The overflow is taken over the top of the dam. Two spare pipes of 30 inches diameter are provided for flushing out the pool, or emptying it if desired; and one or both of these pipes can be used, if desired, for further extensions.

Switchboards, Overhead-mains, etc.—The switchboards are of the usual type for this class of work.

The overhead-mains consist of stranded cables supported almost entirely by double poles of the A frame type, so far as the main line is concerned. This form of pole will permit of extensions when required, and ensures a very wide margin of safety even under the worst conditions to be met with in this district, which is subject to very violent storms and gales. A liberal supply of lightning-arresters and conductors is provided.

Price of Electricity, etc.—Power is supplied to large consumers at 1d. per unit, and in one special case at ½d. per unit. Electricity is supplied to the Urban District Council for public and private lighting at a maximum rate of 4d. per unit.

At the present time, motors to the amount of about 500 horse-power and about 5,000 lamps of 8 candlepower, or their equivalent, are actually connected to the mains. Electric power is in one instance supplied for heating a chapel.

Telephones.—Telephones are extensively employed, and are arranged so that the substation is the telephone-exchange, which is a very convenient arrangement, as it enables the staff to keep in close touch with every part of the system. The telephone-wires are carried on the same poles as the power-and-lighting mains, and no disturbance arises from this cause. The company undertakes the maintenance of telephones in the district.

Conclusion.—The writer hopes that the members will recollect, that he expressly stated at the commencement of his paper that this plant is not dependent on its size (which is quite a secondary consideration) for its interest. At the same time, on a small scale, it is an example of what can profitably be done by electricity, whether by continuous or three-phase current, and he trusts that his paper may be of assistance to mining engineers.

The writer's main object in venturing to submit his paper has been to enlist their sympathy and support against the present laws and regulations dealing with the production and use of electricity in this country. It is generally agreed that high-pressure currents are desirable in many cases, and that the best way of conveying them is by overhead transmission. In Italy, recently, the writer saw many overhead lines working at pressures up to 15,000 volts. These lines are perfectly safe when properly constructed, yet the cry of "danger to the public" is continually raised in this country.

The writer understands that on the Milan transmission-line, conveying about 13,000 horsepower for 30 miles from the Adda falls at Paderno, at a pressure of 13,500 volts (which is sometimes raised to 15,000 volts), with 550 line-standards and 10,000 porcelain-insulators in operation, there has not been a single case of an insulator breaking down, or of a line falling, since the system was set to work some years ago. Of course, every care was taken in erection, and all insulators were tested under severe conditions to 30,000 volts at least. The writer does not say that it is desirable to carry these overhead trunk-lines through our cities. The currents can be transformed for use in the cities, but outside, the transmission-lines should be taken at high-pressures as straight as possible across country. The Board of Trade are well able to see that suitable earthing devices are provided, arranged to act automatically in the exceptional case of a broken wire.

In the writer's opinion, the chief reforms needed are:— (1) The No. 3 regulation of the Board of Trade, limiting the pressure of supply to a consumer to 250 volts, should be modified very much in favour of higher pressure for power, as distinct from house-lighting purposes, without having to obtain special permission, which as shown above, complicates matters, and may even stop progress altogether in many cases. Of course, such higher pressure would be subject to regulations for the public safety, the same as all existing lines. (2) No. 14 regulation requiring that "no high-pressure aerial line shall be used to transmit more than 50 kilowatts (that is about 67 horsepower)" without the consent of the Board of Trade, should be abolished. And (3) other regulations should also be amended in accordance with the preceding suggestions.

The recommendation of Lord Cross' Committee of 1898, namely:—"That the provisions of the Act of 1888, which require the consent of the Local Authority as a condition precedent to the granting of a Provisional Order, should be amended," does not appear to have materially influenced the

Board of Trade towards a wider exercise of their dispensing powers; and, in the writer's opinion, practical effect should be given to this recommendation as soon as possible.

The fact of obtaining statutory powers should not place an absolute veto with regard to overhead wires in the hands of the local authority, who are often most unreasonable in this matter. For instance, if they could have exercised it at Blaenau Festiniog, where the wires run principally over a rocky and thinly populated district, the writer believed that they certainly would have done so, for no good reason; but possibly from an idea that it sounded better to insist on putting the wires underground. In that case, the current could not possibly have been supplied at the price now charged per unit, and even possibly not at any reasonable price.

––––––

Mr. A. H. STOKES (H.M. Inspector of Mines) wrote that Mr. Yale appeared to take strong objection to the Board of Trade regulations which were promulgated in the interest of safety, but if these regulations be compared with the conditions imposed by the mine-owner,[*] there can be little doubt that the former were not to be compared with the latter in respect to onerous terms. With regard to high-pressure currents, the question was one of safety and economy. With respect to safety, it must be clear that 5,000 volts is a far more dangerous current than 500 volts, and if 500 volts will satisfy our requirements, why go to 5,000 volts? With reference to economy, it would probably be found that for any distance under 1 mile the cost of cable, carrying a current of 500 volts, would be less than that for a smaller cable and transformer using a current of 5,000 volts, or that it would be only where a large horsepower is required at 1 mile or more in-bye that high pressure would be economically used.

Mr. G. HOOGHWINKEL (London) wrote that electric rock-drills are extensively used on the Continent and also, though more recently, in the United States of America. He had installed more than 125 electric rock-drills of the percussion and also of the rotating type. They were driven by electric

[*] *Trans. Inst. M.E.*, 1903, vol. xxv., page 620.

motors, either by means of a flexible shaft, or directly mounted on to the drills. In hard holing seams and quarries, percussion-drills were used, while in the soft iron-ores of Luxemburg and Sliegen rotating drills were usually employed. They had been at work for over 4 years and were superior to air-driven drills. They require $\frac{1}{5}$ to $\frac{1}{4}$ of the power, had more power to withdraw the drill when jammed, were less cumbersome and noisy (exhaust of air under high-pressure), and of course had all the advantages of the electric distributing system as against air-plant. More than 600 machines were now at work in different mines on the Continent, and were giving every satisfaction.

Mr. D. SELBY BIGGE (Newcastle-upon-Tyne) thought that it was fortunate that there was divergence of opinion, for it was only through divergence of opinion that they arrived at the best results. He had gone very minutely into the question, with a perfectly open mind, he had inspected some 80 to 100 mining installations on different systems, and he had come to the conclusion that the only practical system, and the best for mining work, transmission, and driving, was the three-phase system. In the first place, the machinery itself was of simpler construction, and there were no commutators or brushes. The generators were also of the simplest construction that it was possible to devise; they could be mounted on the shaft direct, and run at speeds as low as 50 to 60 revolutions per minute. They could manipulate the three-phase current to-day, in as perfect a manner as steam used to be: he did not think that this was generally known. It was chiefly on the score of simplicity that he recommended the use of three-phase machines for mining work, and his opinion was endorsed by that of the leading firms in Belgium, Germany, Switzerland and Italy.

Mr. J. GERRARD (H.M. Inspector of Mines, Manchester) said that, as a mining engineer having only a little acquaintance with electricity, he regretted that electrical engineers could not agree more nearly as to what was safe in the use of electricity. One electrical engineer recommended one form of application, and another recommended exactly the opposite, and it became difficult for the mining engineer to decide between them. There was a large field for the application of electricity in mining, as it was so very convenient; but in introducing it they wanted to be assured that they were not interfering with all that had been done hitherto in regard to lighting and ventilating the mines and the use of explosives.

Mr. S. F. WALKER (London) remarked that the divergence of opinion between electricians was more apparent than real, and alternating current, properly applied, would do all that direct current would do, and *vice versa*. The advantage of alternating current became apparent where the distances were great, and there was then considerable saving in cables.

Mr. W. C. MOUNTAIN (Newcastle-upon-Tyne) said that Mr. Selby Bigge had the same desire that they all had to further the use of electricity in mines, but he entirely differed from the view that the three-phase system was the only one to be used in mines. Haulage-gears required a certain amount of adjustment of speed, and the pumps also, and with three-phase current it was impossible to get them adjusted economically. If there was a sudden inrush of water, it was necessary to increase the speed. At Tredegar collieries, there was over 2,000 horsepower of electric machinery, including 11 sets of haulage-gear and 13 sets of pumps, and the plant was working very satisfactorily; but at the same time it did not work more satisfactorily than continuous-current plant at other collieries. As to the cost of producing three-phase current, speaking of a colliery employing only electric plant, continuous current could be produced more cheaply because it had the better power-factor. At Tredegar, the cost was 0·4d. to 0·5d. per kilowatt per hour, and at the Broughton and Plas-power collieries, with 1,200 horse-power, the cost was 0·29d. As regards large generating plants for a group of collieries, where the current had to be conveyed over long distances, or where the machinery was placed at a distance beyond, say, 1 mile from the generator, he agreed with Mr. Selby Bigge, but where they were erecting a small plant, or even plants sufficient for one particular colliery, the three-phase system had no advantage over the other. In fact, he was inclined to think that, except in the case of fiery mines, the continuous current was the best. It is very difficult, however, to lay down any general rule, as each colliery-installation must be considered on its own merits, as there are circumstances such as the size and arrangement of the haulage-gears, the pumps and other machinery which it is required to drive, together with the distances at which this machinery is placed from the generator, which necessitates each installation being considered on its own merits. An engineer, who had recently erected a large three-phase plant, stated that if he had to erect another plant he would adopt continuous current. In using armoured cables, these must be three-core, and, when working ¾ mile inbye, the high potential became a saving.

Mr. T. P. OSBORNE YALE remarked that there were many objections to the present state of the law. The chief object of his paper was to enlist the sympathy and support of the members, for if public opinion was brought to bear on the matter the Government would be more likely to attend to it.

The PRESIDENT (Sir Lindsay Wood, Bart.) remarked that Mr. Yale's paper was a valuable one for the members, as it pointed out many of the restrictions imposed on the use of electricity. He thought that it would be well if the attention of the Government or the Board of Trade was drawn to these. No doubt these restrictions were limiting the use of electricity in this country, and enabling other countries to develop their resources to a greater extent, so that our manufacturers were severely handicapped.

Mr. J. C. FORREST (Essington) said that he was connected with the slate-quarries in a district neighbouring that described by Mr. Yale in his paper. He could confirm what had been said by the writer, but they had other restrictions to contend with besides those of the Board of Trade. When they contemplated erecting the plant, they were told that they could have the water-power for "a mere nothing," but when they proceeded to secure it they found the "nominal" cost had become so great that the benefits which would otherwise have resulted from electricity had disappeared. Coal now costs from 15s. to 16s. per ton, and although as a coal-getter he was pleased to supply it, as a slate-getter he would like to see some cheaper and better method of obtaining power, and to be in a position to use the water which was so plentiful. One could not but be struck with the amount of water-power coming through the valleys of Wales.

Dr. C. Le Neve Foster (London) said that he had listened to the paper with extreme interest, as it was practically the fulfilment of a dream of his own of 23 years ago. In his first annual report upon that district, he pointed out the large amount of water-power going to waste, and the desirability of utilizing it and transmitting it electrically. With regard to the restrictions of the Board of Trade, would it not be advisable for this Institution to appoint a Committee to consider the question? The Board of Trade was composed of reasonable men, and if the Institution, after careful deliberation, propounded a scheme by which the difficulties would be met, the Board of Trade would be likely to listen to them.

Mr. S. F. Walker (London), referring to the large amount of water-power going to waste, said that we were apt to say that other countries were more favoured in this matter of water-power, but this was not exactly correct, though nature had in some instances placed an accidental dam. During the recent rains, it was stated that there was a rainfall of 9,000,000,000 gallons within a few days over the London area. If that had been stored at a height of 500 feet above sea-level, and allowed to run down to sea-level, it would have supplied 25,000,000 horse-power hours. He thought that Parliament would eventually have to consider the question of water-rents; and if we had to compete with other countries, land-owners must not be allowed to demand prohibitive rents. The United States Government promoted anything that would facilitate their trade, while if our own Government did anything at all it was in the direction of hampering the development of manufactures. Nobody would blame the Government for the regulations which had been made to stop the dribbling away of lives in mines, but mining was an industrial enterprise which also had to live. The Government, however, did nothing to help the development of this, or any other industry, but, on the other hand, by adding to the cost of working, very greatly reduced the mineral industry of the Kingdom.

Mr. Henry Hall (H.M. Inspector of Mines, Rainhill) said that he had noticed a tendency on the part of electrical engineers to belittle the dangers of electricity, and probably if they were a little more reasonable, the Board of Trade would be more accommodating. He had recently been at a quarry where the water came over a large wheel at the top and drove a part of the machinery, the over-flow running down hill again drove two dynamos, and the electricity was conveyed back to the top of the hill to finish the work. They could not expect landlords to give them the water for nothing, even electrical engineers would not give their services for nothing, and of course if power was wanted it must be bought.

Mr. M. H. HABERSHON (Thorncliffe) said the writer remarked that for fiery mines the three-phase system was the only one which should be adopted. He thought that the writer had made the statement without considering its full import. There was a difficulty as to the definition of a fiery mine and electrical engineers should be very careful before making statements of so sweeping a nature, seeing that in many collieries they were using continuous current for driving coal-cutting plant, where three-phase current could not be adopted.

Mr. T. P. OSBORNE YALE, replying to the discussion, said that he readily accepted the correction of the last speaker. He had spoken hastily about the three-phase current and fiery mines, and only mentioned it as an instance to shew that he was not biassed either way, as the merits of each particular case should be taken into consideration. He did not agree with the remark that electrical engineers belittled the dangers of electricity. It was possible for a person to be killed as the result of a shock from 100 volts; it might cause him to fall off a wall, for instance, and break his neck. But he had himself often touched 500 and even 600 volts: it was painful and under exceptional circumstances might have been fatal. His point about the Board of Trade regulations was that it was generally admitted throughout the civilized world, by people competent to judge, that experience proved that there were certain pressures which might be applied under certain conditions and with certain restrictions which were perfectly safe. Mining itself was a dangerous occupation, unless it was carried on under certain regulations: if the regulations were observed it ceased to be dangerous in the ordinary sense of the word. The pressure of steam was dangerous, and if a boiler burst at a pressure of 50 pounds it was practically just as dangerous as if it burst at 200 pounds, and in each case certain precautions against bursting should be observed. The same thing applied exactly in the case of electricity; it was perfectly harmless at 500 or at 5,000 volts, if it were applied properly. Throughout the civilized world, except in Great Britain, electricity at high pressure was permitted under certain conditions imposed by the respective Governments, and therefore, owing to the absence of vexatious restrictions, it had become of more general everyday use than in this country. The suggestion that the Institution should appoint a Committee to approach the Board of Trade was a valuable one, and he would be pleased if the discussion of the paper should lead to anything of that kind being adopted. Since the paper was written, he had heard that the Board of Trade were considering seriously the question of amending their regulations, but he did not know whether this was a fact or not. An interesting article had recently appeared in *Engineering** on this question.

Mr. J. GERRARD (H.M. Inspector of Mines, Manchester) said it seemed to be of primary importance that the members should have a clear view of the question as to the danger of the transmission of electric currents in mines. Did the presence of water on roadways in the mine, or the fact of the man being in a state of perspiration, render dangerous a shock of 500 volts?

Mr. T. P. OSBORNE YALE, referring to the use of 500 volts, said that he was speaking of its use in an ordinary way. He did not advocate the use of wires conveying a current of 500 volts being placed next the colliery-roof so that the men would be liable at any moment to touch them. . It might not be fatal electrically, but it might knock a man down, and allow him to be run over.

Mr. JOHN GERRARD said that fatalities had occurred above ground in the case of persons standing in water: would the interposition of water on a roadway make the shock more dangerous?

Mr. T. P. OSBORNE YALE said that the presence of water did make the shock more dangerous. Electric current at any pressure might, under exceptional circumstances, be dangerous. It would be within the memory of many of the members that a man was killed by 200 volts in his bath; and under the same conditions 100 volts would be dangerous. It was all a question of how much current might be forced through a man: it was the quantity that killed and not the pressure of the current.

* 1903, vol. lxxv., pages 861 and 862.

Under certain conditions, it might take 10,000 volts to force the fatal amount of current through a man, or it might only take 1,000 volts, or 100 volts. Of course precautions should be taken with 500 volts, so that persons could not touch the wires without deliberately trying to do so.

Dr. C. LE NEVE FOSTER (London) thought that they could not get any further to-day, and the only course appeared to be to appoint the proposed Committee to enquire into the matter. He therefore moved "that the Council be requested to appoint a Committee to consider the present restrictions imposed by law on the transmission of electrical power, and to take such steps as they deem desirable."

Mr. G. A. MITCHESON (Longton) seconded the proposal, and suggested that the proposer might add to it a recommendation that the Committee should make some enquiry as to what constituted a fiery mine so far as electricity was concerned.

Dr. C. LE NEVE FOSTER thought that would be interfering with what was already in the hands of the Departmental Committee of the Home Office. He wished to abstain from trespassing in any way on the functions of that committee.

Mr. G. A. MITCHESON thereupon withdrew his suggestion.

Mr. W. N. ATKINSON (H.M. Inspector of Mines, Stoke-upon-Trent) remarked that the enquiry of the Departmental Committee of the Home Office was confined to the use of electricity in mines. It had always been felt by H.M. inspectors of mines and mining engineers in this country that it was practically impossible to define a fiery mine, with reference to the use of electricity or explosives or any other subject.

The resolution was unanimously adopted.

The PRESIDENT (Sir Lindsay Wood, Bart.) moved a vote of thanks to Mr. Yale for his interesting paper.

Mr. G. A. MITCHESON seconded the resolution, which was cordially approved.

Mr. A. R. SAWYER'S "Further Remarks on the Portuguese Manica Gold-field," were read as follows:—

FURTHER REMARKS ON THE PORTUGUESE MANICA GOLD-FIELD.

By A. R. SAWYER.

A further examination of the Manica gold-field towards the end of 1902, has enabled the writer to add to the geological map (Fig. 1, Plate XXIII.) of this area begun in 1899 and referred to in his paper on this gold-field.*

The greater part of the map is as complete as surface-indications will allow. The north-western corner and the western extension of the Isitaca range, were not visited by the writer. Although he believes that granite-masses extend farther in these directions, he prefers not to indicate them on the map. The mountainous and partly wooded nature of the country made it difficult to follow up the various rock-masses shown; this can only be done by a prolonged and extended geological survey.

The writer has endeavoured to show the main geological features of this gold-field. In many respects it is a type of all the Mashonaland gold-fields, some of which the writer has described in his book on *The Gold-fields of Mashonaland*. He is of opinion that the geological features of these gold-fields have so important a bearing on the values of the reefs occurring in them, that they should be considered in their entirety previous to launching out into large mining operations.

With regard to this particular gold-field, it is of interest and importance as being situated closer to Beira, the natural seaport of Rhodesia, than any of the Rhodesian gold-fields. In 1893, when reporting on all the Mashonaland gold-fields to the late Mr. Cecil Rhodes, the writer drew his special attention to the importance of the Manica gold-field.

It will be noticed on the map (Fig. 1, Plate XXIII.) that the gold-bearing formation, previously described as having a trend from west to east, is bordered on the north and south by large granite-masses. The northern mass of granite culminates in

* *Trans. Inst. M.E.*, 1900, vol. xix., page 265.

Mount Gorongoe, about 6,000 feet high, which stands alone, like a huge pointed monument, about 18 miles north of Macequece, and in a few other heights ranging up to 6,970 feet farther north. The southern mass culminates in Mount Vumba, over 5,000 feet high, about 3 miles from Macequece. Figs. 2 and 3 show these two granite-masses, viewed from the eastern and southern slopes of Chiromiro mountain, respectively. The intrusion of these masses of granite has compressed the gold-bearing formation, with the result that cleavage-planes have been produced, running mostly north and south at many points, as already described. In places, it would be impossible to state which is the true plane of stratification were it not for other facts, such as the general trend of the formation from west to east.

FIG. 2.—VIEW, LOOKING NORTHWARD FROM THE EASTERN SLOPE OF CHIRO-MIRO MOUNTAIN, WITH MOUNT GORONGOE IN THE DISTANCE.

In the Murchison Range, in the Transvaal, as shown in the writer's book on *Mining in the Murchison Range*, similar rocks also have a trend from west to east. This is moreover pretty general in the Mashonaland gold-fields and in most other South and East African gold-fields of the same age. In these gold-fields, there has been a similar compression of the gold-bearing strata (crystalline schists, etc.) by huge masses of granite-intrusions on the north and south.

Although granite, doubtless, underlies all these formations, forming the foundation upon which they all rest, the granite-masses, thus associated with the crystalline schists, are of more recent date than these schists and sedimentary rocks. They are intrusive masses, which have disturbed the continuity and regular deposition of certain sedimentary deposits, isolated attentuated remains of which may still be found.

Fig. 4 gives an idea of the nature of the prominent Chiromiro range. The conglomerate-beds appear to be lenticular, though a definite statement cannot be made, owing to the formation being hidden by the large amount of vegetation which covers the southern slope of the range.

The crystalline dolomitic limestone already alluded to as resembling the extensive Dolomite series in the Transvaal called by the Boers "Olifant Klip" (Elephant-rock) owing to its

FIG. 3.—VIEW, LOOKING SOUTHWARD FROM THE SOUTHERN SLOPE OF CHIROMIRO MOUNTAIN, WITH THE VUMBA RANGE IN THE DISTANCE. THE FIGURE IS STANDING ON THE OUTCROP OF THE DOLOMITIC LIMESTONE.

FIG. 4.—VIEW, LOOKING WESTWARD, FROM THE SOUTH-EASTERN SLOPE OF CHIROMIRO MOUNTAIN, WITH MOUNT VENGA IN THE DISTANCE. MOUNT CHIROMIRO IS SEEN ON THE RIGHT-HAND, AND, WITH THE TWO PEAKS, FORMS "NEKS," ON WHICH THE DOLOMITIC LIMESTONE IS SEEN DISTINCTLY.

likeness on the surface to an elephant's skin, is shown. It is 200 feet thick in the line of section, and there forms a " nek."* A similar nek, in which the limestone is also exposed occurs farther west. The preservation of the limestone, at these two points, is due to denudation having been there retarded by the presence of the two peaks, which with the main range form these neks.

The cleavage-planes, referred to above, are so developed in this locality, both in the dolomitic limestone and in the neighbouring rocks, that they have influenced the process of denudation; and at first sight it appears as if they had a north-and-south strike with an easterly dip of 45 to 50 degrees. The presence in this dolomitic limestone of numerous thin quartz-veins, along cleavage and bedding-planes, also, at first sight, adds to the mysti-

---

* In the Boer, and not in the geological, meaning of the word.

fication. The writer has traced this bed of dolomitic limestone, as shown on the plan, for a distance of over 1 mile. Considering its thickness, it is a very important deposit, which amidst other surroundings might have considerable commercial value. It is the more interesting, because the writer has not seen so important a limestone-deposit in any of the similar formations in the Transvaal or in Rhodesia.*

The copper-ore deposits, south of Dillon's camp, previously mentioned, have been explored, and several reefs of good ore have since been discovered. It would appear that the copper-ores occur both in the cleavage and in the bedding-planes.

The reefs found in this gold-field are similar to all those seen by the

FIG. 5.—VIEW, LOOKING EASTWARD, FROM CHIMESI CAMP, WITH MOUNT DABSI IN THE DISTANCE, ON THE LEFT-HAND.

* The only limestone-deposits seen by the writer in Mashonaland are described in his *Gold-fields of Mashonaland*, page 16. One is a dolomitic "limestone," north of the Penhalonga Range in Manica ; and the other is a pure limestone-deposit, near Salisbury.

writer, in South and East Africa, in this formation, and are mostly bedded veins, as stated by him in *The Gold-fields of Mashonaland.*

An opinion having been expressed that these veins are "true fissure-veins," a term which implies a course different from that of the formation in which they occur, it is interesting to note that Dr. G. A. F. Molengraaff, state geologist under the late Transvaal Government, in his paper on the "Géologie de la République Sud-Africaine du Transvaal,"[*] states with regard to the occurrence of minerals other than gold in reefs in the Primary System, in which he includes the Hospital Hill and Witwatersrand series:—"On les trouve généralement dans des veines qui ont la même direction et le même pendage que les strates encaissantes."[†]

Exceptions, no doubt may occur, and the writer pointed out one of these, namely, possibly the Beatrice Reef[‡] in Mashonaland. Dr. Molengraaff also states that a reef, near Steynsdorp, which has not yet been properly studied, appears to be a true fissure-vein ("un vrai filon de fracture").[§]

It is possible that reefs, which assume a course parallel to the formation, may deviate into a cleavage-plane, as may be the case with the Edmundian copper-deposits.

The writer has had an analysis made of the dark mineral, already described, of which a large quantity is found in the Chimesi valley. Half its weight is talc (a hydrated silicate of magnesia), the remainder consists of oxides of iron and manganese (wad), and of sandy and clayey matter. It contains traces of gold.

———

The PRESIDENT (Sir Lindsay Wood, Bart.) moved a vote of thanks to Mr. A. R. Sawyer for his paper.

Mr. W. N. ATKINSON seconded the resolution, which was cordially approved.

———

Dr. J. S. HALDANE read the following paper on "Miners' Anæmia, or Ankylostomiasis":—

[*] *Bulletin de la Société Géologique de France*, 1901, series 4, vol. i., page 13.
[†] *Ibid*, page 28.
[‡] *Gold-fields of Mashonaland*, pages 13 and 80.
[§] *Bulletin de la Société Géologique de France*, 1901, series 4, vol. i., page 29.

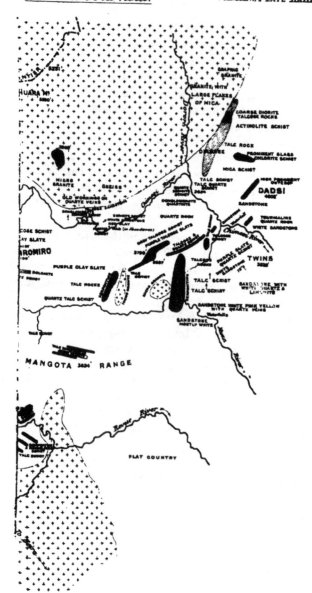

REFERENCES.

Granite, Felsite, Pegmatite and other Acid Igneous Rocks : Gneissoid Granite

Rocks corresponding to the Lower Member of the Transvaal Hospital Hill Series :
Crystalline Schists, Sericite Schists, Conglomerates, Sandstones, Quartzites, etc.

Banded and Ferruginous Quartzites, the most prominent and distinctive members
of the above Series, which form indices. Three distinct Beds of Banded
Quartzite occur, where undisturbed. Frequently they are closely associated
with Conglomerate and Breccia, with pebbles of Banded Quartzite.

Prominent Dolomitic Limestone-bed in the above Series.

Diorite : Intermediate Igneous Rocks.

MINERS' ANÆMIA, OR ANKYLOSTOMIASIS.

By J. S. HALDANE, M.D., F.R.S., FELLOW OF NEW COLLEGE, OXFORD.

A few months ago, the writer was asked by the Home Secretary to make an enquiry, in conjunction with Mr. J. S. Martin, H.M. inspector of mines, into the health of Cornish miners. One of the first results of this enquiry (in which Dr. A. E. Boycott, and later Dr. S. G. Scott, co-operated) was to show that an illness which affected the miners at Dolcoath and other mines in the Camborne district of Cornwall was the tropical disease known to medical men as ankylostomiasis.[*] As ankylostomiasis may easily spread in either metalliferous mines or coal-mines, the disease and its manner of communication are well worth the attention of mining engineers.

The main symptoms are:—(1) Pallor, best seen on the inside of the lips and eyelids; and (2) palpitations and great shortness of breath on any exertion. In well-marked cases, the man affected is unable to work; and, in severe cases, the weakness is so extreme that death may result in the absence of suitable treatment. Besides the pallor, palpitations and shortness of breath, there is often, though not always, a good deal of pain or discomfort in the stomach, and a capricious appetite. The Cornish epidemic has been accompanied by the prevalence among the miners of well-marked skin-affections, locally known as "bunches." These are small irritable inflamed swellings, which appear usually in groups, on almost any part of the skin, and frequently suppurate. Vesicles containing clear fluid often cap these "bunches." In other cases, there is only a localised eruption of urticara or nettle-rash, accompanied by much itching. The "bunches" only occurred after work underground.

[*] *Report to the Home Secretary on An Outbreak of Ankylostomiasis in a Cornish Mine*, by the writer, 1903. Fuller details as to symptoms, etc., are given in a paper by Dr. Boycott and the writer in *The Journal of Hygiene*, 1903, page 95.

The disease had come to be well known locally as "Dolcoath anæmia," and many of the more seriously-affected men resorted for treatment to the Miners' Convalescent Home at Redruth. Table I., compiled from the hospital-books, gives some idea of the course of the epidemic.

TABLE I.—MINERS' CONVALESCENT HOME.

Year.	Cases admitted for Anæmia.	Year.	Cases admitted for Anæmia.
1893	1	1898	23
1894	3	1899	12
1895	9	1900	11
1896	13	1901	7
1897	29	1902	8

It will be seen that the first cases appear to have occurred about 1894 or 1895, and that there was a great increase during the next two or three years, followed by a marked decline, which will be referred to later. A considerable number of miners are still, however, more or less affected ; 61 per cent. of the hospital-cases were of miners who came directly from Dolcoath, which is much the largest mine in the district. The remainder were from other neighbouring mines, and to judge from the records certain of these seem to have been affected quite as early as Dolcoath.

TABLE II.—AIR-ANALYSES AT DOLCOATH MINE.

Samples taken at	Oxygen.	Carbonic Acid.	Oxygen diminished.	Carbonic Acid increased.
	Per cent.	Per cent.	Per cent.	Per cent.
A. Surface	20·94	0·030	0·000	0·000
B. 302 fathoms level in eastern (down-cast) shaft	20·94	0·030	0·000	0·000
C. Same level at end of cross-cut through granite : air being blown in ...	20·82	0·080	0·120	0·050
D. 375 fathoms level on north lode, at stoping	20·80	0·110 / 0·115	0·140	0·080
E. 412 fathoms level, large open gunnis	20·85	0·095	0·090	0·065
F. 440 fathoms level, in gunnis	20·83	0·090	0·110	0·060
G. 470 fathoms level, rise in cross-cut : drill idle	20·66 / 20·67	0·220 / 0·230	0·275	0·195
H. 455 fathoms level. Bottom of engine (upcast) shaft	20·85	0·095	0·090	0·065
I. 254 fathoms level, taken from gig in engine-shaft	20·81	0·115 / 0·115	0·130	0·085

The anæmia at Dolcoath mine had been very generally attributed to impurities in the air of the mine, the presence in noxious amount of carbonic oxide or some other gas from explosives being specially suspected. The analyses (Table II.) made by the writer on his first visit to the mine showed, however,

that the air was on the whole unusually pure; and, no trace of
carbonic oxide could be found. The points at which the samples
were taken, and the temperatures observed, are marked on the
accompanying section (Fig. 7, Plate XXIV.). The total upcast
ventilation is about 80,000 cubic feet per minute.

In spite of the great depth and extent of the workings the air at
Dolcoath mine was, on the whole, much purer, chemically speaking, than is usually the case in coal-mines. The average increase
of carbonic acid in seven samples of upcast air in coal-mines, as
found by the writer* was 0·255 per cent., the average deficiency
of oxygen (partly due to presence of fire-damp) being 0·67 per
cent.; whereas, at Dolcoath mine, the average increase of
carbonic acid was only 0·072 per cent., and the deficiency of
oxygen 0·125 per cent. in the upcast air. This result was, at
the time, somewhat surprising, as the air was warmer and moister
than that usually met with in coal-mines, and the general ventilation was dependent solely on natural air-currents. The
problem of ventilation in a metalliferous mine differs very
greatly, however, from what it is in a coal-mine.

The intestinal worm known as *Ankylostoma duodenale* (sometimes also called *Anchylostomum duodenale, Dochmius duodenalis,*
or *Uncinaria duodenalis*) was discovered in post-mortem examinations by Dubini in Italy in 1838; but its causal relation to
anæmia was only demonstrated in 1854 by Griesinger, who showed
that the disease known as Egyptian chlorosis, or anæmia, is due to
its presence. A few years later, Wucherer proved that the anæmia
common in Brazil is ankylostomiasis; and ankylostomiasis has
since been identified in many other tropical and sub-tropical
countries as a common, widespread and very troublesome disease.
It seems somewhat doubtful whether any tropical or sub-tropical
country is free from it. About twenty years ago, there occurred
among the men engaged in constructing the St. Gothard tunnel,
a very serious outbreak of anæmia, associated with great loss
of life. This disease was shown by Perroncito and others to
be ankylostomiasis. Perroncito also discovered the presence of
ankylostomiasis among the metalliferous miners at Schemnitz,
in Hungary. The disease was, in consequence, much more
fully studied, and the present efficient mode of treatment intro-

* *Trans. Inst. M.E.,* 1895, vol. viii., page 549; and 1896, vol. xi., page 272.

duced. In recent years, ankylostomiasis has spread to an alarming extent in some of the Continental colliery-districts—particularly in collieries in Austria, near Liége in Belgium, and in Westphalia, where it is at present causing much trouble and anxiety. It has also been discovered in the St. Étienne collieries, among brick-makers near Cologne and Bonn, and among sulphur-miners in Sicily. The disease is, to a slight extent, endemic among the agricultural population in parts of Italy, and possibly elsewhere in Europe.

Ankylostomiasis has been known under a large variety of synonyms in different countries and languages, since it is only recently that the identity of the disease has been established. Among British synonyms may be mentioned "tropical anæmia," "Egyptian chlorosis," " miners' anæmia," " dirt-eating," " earth-eating," or " geophagy" (the three latter from the fact that one symptom, particularly among children and coolies, is an abnormal craving for earth or clay), " ground-itch " or " water-itch " (from the skin-symptoms), and, in Cornwall, the name " Dolcoath anæmia " was invented.

Fig. 1.—Posterior-end of Adult Male Ankylostoma, magnified 40 Diameters.

It is not possible to discover, now, the original source of the Cornish epidemic, but it is evident that Cornish mines are particularly exposed to infection, as Cornish miners are constantly going to and coming from metalliferous mines in tropical countries all over the world. The infection may easily have been brought from West Africa, or Brazil, or Mysore, or Kimberley, or various other places where ankylostomiasis exists. Infection from any European source seems much less probable.

The adult worm [specimens shown] is about ½ inch long and

of a yellowish-white colour, and blood is frequently seen showing through from the
intestinal canal (Figs.
1 and 2). Under the
microscope, the copu-
latory bursa of the
male has a very char-
acteristic appearance
(Fig. 1) The worm
is not found in the
fæces, unless a vermi-
cide remedy, such as
thymol or extract of
male fern, has been
administered. The
eggs are, however,
always found in
abundance in the
fæces. To find them,
a trace of the

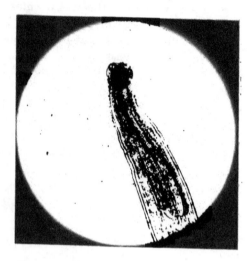

FIG. 2.—HEAD OF ADULT ANKYLOSTOMA, MAG-
NIFIED 30 DIAMETERS.

fæces is spread on a glass slide with a drop of water, and the
eggs searched for with
a low power Their
appearance and size are
shown in Figs. 3 and 4.

Pallor of the lips,
etc., in a miner, may,
of course, be symp-
tomatic of other dis-
eases besides ankylos-
tomiasis; but when
unusual pallor is seen
in several men, the
presence of the dis-
ease should be sus-
pected, and arrange-
ments made for micro-
scopic examination of
the fæces. In exam-
ining a number of

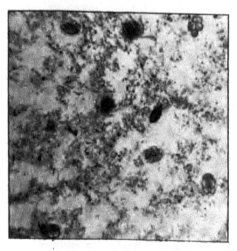

FIG. 3.—ANKYLOSTOMA-OVA AT DIFFERENT STAGES :
NEAR THE CENTRE ARE OVUM OF TRICHO-
CEPHALUS DISPAR, FROM NEARLY FRESH FÆCES,
MAGNIFIED 100 DIAMETERS.

hæmoglobin is present in the red corpuscles of the blood, and as might be expected, they also are present in diminished proportion. One characteristic of the disease is, however, that the diminution in the proportion of corpuscles is not so great as in the proportion of the hæmoglobin. Each corpuscle has thus a diminished amount (about 67 per cent. of the normal, on an average) of hæmoglobin. The proportion of corpuscles can be counted under the microscope, by means of the instrument known as the hæmacytometer.

As hæmoglobin is the substance which carries oxygen from the lungs to the tissues of the body, it is only what might be expected that the disease should cause serious symptoms, and particularly shortness of breath. In carbonic-oxide poisoning, which is, unfortunately, only too familiar to miners in connection with underground fires and colliery-explosions, the oxygen-carrying function of the blood is also interfered with. If, however, the symptoms and the interference with the oxygen-carrying function be compared in the two conditions, a marked difference is found. In the case of ankylostomiasis it is not uncommon to find miners at work with only 35 per cent. of the normal proportion of hæmoglobin in their blood, whereas a man whose blood is saturated with carbonic oxide to such an extent that only 35 per cent. of the hæmoglobin remains free for respiratory purposes will certainly be unable to stand, and probably unconscious and in imminent danger. Moreover, in carbonic-oxide poisoning, shortness of breath is not nearly so prominent a symptom as a tendency to faint on the slightest exertion. To what is this difference due? The explanation is probably as follows:—The hæmoglobin has, in reality, two functions to perform, one being to carry oxygen to the tissues, and the other to assist in carrying off carbonic acid. In the case of anæmia, both functions are impaired, since the hæmoglobin is greatly diminished in quantity, whereas in the case of carbonic-oxide poisoning and in asphyxia of any kind from simple want of oxygen, only the oxygen-carrying function is interfered with. Now, experiments made recently, by Mr. J. G. Priestley and the writer, have shown that excess of carbonic acid in the blood is the normal stimulus to respiration and circulation; and as this excess is necessarily present in anæmia, the respiration and particularly the circulation are increased to such an extent as to compensate to a large extent both for the excess of carbonic acid and the deficiency of oxygen. In carbonic-oxide poisoning, there is no such compensation, as want of oxygen by itself is not an efficient stimulus to respiration and circulation. Hence the relative absence of shortness of breath and the far greater tendency to fainting, etc., in the latter condition.

The anæmia of ankylostomiasis naturally tends to cause a variety of other symptoms besides pallor and shortness of breath, since the aëration of the tissues in every part of the body is more or less defective, particularly in severe cases. One of these symptoms is general want of energy, disinclination for work, and depression. In severe cases there may be dropsy, and a tendency to rupture of small blood-vessels. The heart is also apt to become flabby and dilated. Naturally also, the digestion suffers, though it is difficult to say how far the disturbances of digestion are due directly to the irritant action of the worms in the alimentary canal. The curious craving for earth or clay seen so markedly in children, and even among adults in the lower races, is perhaps caused in some way by the anæmia. Somewhat similar cravings are sometimes met with in the ordinary anæmia or chlorosis so common among young women. The pain or discomfort in the region of the stomach in ankylostomiasis is probably due directly to irritation produced by the worms, although the latter are present, not in the stomach, but in the upper part of the small intestine.

How do the worms produce the anæmia? To this question it is impossible at present to give a definite answer. The current explanation is that the worms suck so much blood as to cause the anæmia. The experiments made in Cornwall showed, however, that the "anæmia" is not due to any deficiency in the total amount of hæmoglobin or red corpuscles in the body, but to a great dilution of the blood. In anæmia from bleeding, on the other hand, similar experiments have shown that there is a very marked deficiency in the total quantity of hæmoglobin and corpuscles. It may be added that anæmia is commonly produced by tape-worms, which are known not to suck any blood. The cause of the anæmia is thus, for the present, a mystery.

The skin-eruptions observed in the Cornish epidemic are of

great interest in view of recent investigations. Curiously enough, these symptoms do not seem to have been described in connection with previous outbreaks underground, though possibly they were present and well-known to the miners themselves. Some three or four years ago, Prof. Looss, now of Cairo, spilt on his hand a few drops of water containing fully-grown *Ankylostoma*-larvæ. Believing the larvæ to be harmless unless introduced into the mouth he did not trouble to wipe off the drop: but he was surprised to find that after a short time the place became inflamed. Some time later he found that he was seriously affected by ankylostomiasis. He then made the experiment of placing a drop of a similar culture on the leg of a patient shortly before the leg had to be amputated. On examining, microscopically, a piece of the infected skin, he found that the larvæ had penetrated it, apparently entering round the roots of the hairs. He has just published a further and far more conclusive experiment. A man, who was carefully ascertained to be free from ankylostomiasis, was infected by a drop of culture on the arm; and 71 days later ova began to appear in the fæces. At the same time, two puppies (animals which are susceptible to ankylostomiasis) were smeared along the back with a culture. They both died at the end of 9 days, and large numbers of young immature worms were found in the upper part of the small intestine. These experiments seem to leave no doubt that the larval worms are capable of entering the body through the skin, causing considerable local irritation in doing so, and that they penetrate to the intestine within a short time, where they reach full maturity in about two months. Previous experiments by Leichtenstern and others had shown that they also reach the intestine when they enter through the mouth.

Further important evidence on the subject of skin-infection was furnished by Dr. Bentley about a year ago. He found that a disease, which is known as "ground-itch" or "water-itch," and affects coolies in Assam during the rainy season, is associated with ankylostomiasis. The skin-affection appears on the feet of coolies working barefoot in wet soil polluted by human fæces. The symptoms are itching, followed by an eruption of vesicles and sometimes of pustules—apparently the same skin-symptoms as were so often seen associated with "Dolcoath anæmia." Dr. Bentley made the experiment of incubating two portions of sterile soil for a week with human fæces, the first portion being mixed with fæces containing *Ankylostoma*-ova, and the second with fæces free from ova. A small portion of both samples was smeared on the arm of a man, with the result that the skin-eruption was produced by the first portion, but not by the second. He also found that when the infected portion of the culture had been previously allowed to dry in the air for 8 hours, it produced no effect even when re-moistened. The *Ankylostoma*-larvæ were dead.

The life-history of the worm is as follows:—The male and female worm live in the upper part of the small intestine, attached to the lining membrane by means of four hooks round the mouth-capsule. Occasionally they seem to penetrate beneath the epithelium. They can detach themselves temporarily, but are unable to reproduce themselves within the body. At any rate embryos have never been found in the intestine, and the eggs found in the fæces are only at the very earliest stages of development.

Fig. 5.—Recently Hatched Larval-worm.
[Illustration reproduced by permission from *The Journal of Hygiene*, 1903, vol. iii., plate V., Fig. 8.]

The worms seem to be able to live for several years. In Cornwall, there were several cases of men who had been away from all possible sources of re-infection for two or three years, and yet still continued to discharge ova in their fæces. Nevertheless, the worms evidently die off in time, if their number is not added to by re-infection; and men removed from the source of infection gradually recover.

The females produce ova in enormous numbers. Experiments made outside the body show that these ova will hatch as larval worms (Fig. 5)* within a few hours if kept at or near the body-temperature: yet the ova discharged are quite immature. The reason why they do not develop in the intestine is apparently that there is no free oxygen, and they cannot develop without oxygen. In fæces, deposited on the ground, the ova gradually develop under suitable conditions. The writer has found that they will hatch at any temperature between about 60° and 90° Fahr., provided that the fæces are not dried up too soon, and yet are not so wet as to deprive the ova of oxygen. The lower the temperature the slower is the process of hatching. The stages of development are shown in Fig. 6. At one stage, the embryo-worm may be seen moving actively within the shell. After hatching, the larvæ cast their cuticles once or twice in the course of development, and finally reach a stage at which they are capable of infecting a man if they are swallowed, or brought into contact with the skin. As already stated, it appears that they do not become mature till about two months after the infection. The larvæ are very quickly killed by drying, and various disinfectants will also kill them, and it is probable that they do not develop

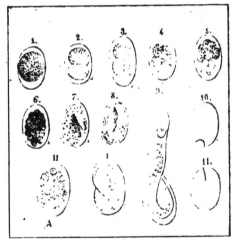

FIG. 6.—STAGES IN THE DEVELOPMENT OF THE LARVAL-WORM (AFTER PERRONCITO).

to the infective stage, except at temperatures varying from 70° to 85° Fahr.

Closely-allied species of worms have been found in the intestine of the horse and of the dog, fox, and cat. At one time, horses were believed to be a source of infection in mines, but this has been disproved, as the worm found in the horse is of a distinct species.

* Figs. 1 to 5 are reproduced from micro-photographs made by Dr. C. A. Coventon, Oxford, of specimens from Cornwall.

As might be expected, considering the mode of infection, other human intestinal worms are often associated with the *Ankylostoma.* Among the Cornish miners the *Trichocephalus dispar* (whip-worm) and the *Anguillula intestinalis* were frequently present in the intestine. Both of these worms are, however, relatively harmless.

Ankylostomiasis is, fortunately, very susceptible to treatment, and no one need die of the disease if only its nature is recognized in time. The worm can be expelled by means of large doses of thymol or extract of male fern. These should, however, only be administered under medical supervision, as the doses required are somewhat formidable. As a rule the improvement is very rapid.

The important question of the prevention of the spread of ankylostomiasis in mines must now be discussed. Unfortunately, the disease has obtained a firm hold in Cornish mines, which will thus certainly continue for some time to be a possible source of infection to British collieries. Men returning from tropical countries, or coming from mines on the Continent, might also at any time introduce infection. As ankylostomiasis has gradually spread in Austrian, Belgian and Westphalian collieries, there is no doubt that it is capable of spreading in British collieries, given the necessary conditions; and for all we know the infection may already be present. At any rate it seems safest to assume that the disease may appear at any time. It should also be remembered that a manager, who knows the symptoms, will be almost more likely to detect the first cases than a doctor.

As regards prevention, the chief point to bear in mind is that infection is spread solely by pollution of the ground with infected fæces, and by men coming in contact with the polluted ground. The disease is not in the slightest degree directly contagious from person to person, and there is no danger of its spread aboveground, provided that the most elementary sanitary precautions are taken. If we can stop the pollution of the ground, or prevent men from coming in contact with the polluted material, we can absolutely prevent the spread of the disease. At present, there is a great lack of attention to the prevention of contamination of the ground in mines by human fæces, and there appears to be no reason why this state of matters should not be reformed, and all mines in the country be made absolutely proof against the spread of ankylostomiasis and other diseases propagated in the same way. The men should, in the first place, be given to understand that the habit of emptying the bowels underground is a source of trouble and danger; and those who do so, except in case of necessity, and at certain specified places, might well be subjected to some penalty. Provision ought, also, to be made in some definite form for cases where the necessity arises. At Dolcoath mine, a system of metal receptacles for use underground has been introduced. A tin of disinfectant for use with each pail is provided, and the pails are sent up to the surface at intervals. In coal-mines, however, it would probably be quite sufficient to dispose of the soil by burying it in dry coal-dust with a disinfectant in some dry place so separated off that there would be no risk of infective material coming in contact with the men, or being carried by water.

In parts of a mine, where the ground and the air are dry, there is not the same risk as at wet places, since the larvæ cannot withstand ordinary drying. In Cornish mines, the air is nearly saturated, and the ground moist in most parts; but this is not the case in coal-mines, where an enormously greater ventilation is needed and the air usually reaches the bottom of the downcast heated by compression, and, therefore, relatively dry; while the relative dryness may further increase as the air is still further warmed in its passage through the mine. The effect of moist ground in favouring the development of the larval worm is strikingly illustrated by the facts mentioned above with regard to coolies' water-itch, and by the following statistics as to the prevalence of ankylostomiasis in the Westphalian collieries before and after the Government regulations with regard to watering of coal-dust were introduced.[*]

TABLE III.—CASES OF ANKYLOSTOMIASIS IN WESTPHALIA.

Years	1896	1897	1898	1899	1900[*]	1901	1902
Number of cases	107	113	99	94	275	1,030	1,355[†]

[*] Watering was introduced in 1900.   [†] Till October.

It will be seen from Table I. that since 1898 the epidemic of ankylostomiasis at Dolcoath has considerably diminished.

[*] *Glückauf*, 1902, vol. xxxviii., page 1250.

This is probably attributable to two causes. In the first place the manager, Mr. R. Arthur Thomas, was led to suspect that the skin-affections were somehow connected with the pollution of the mine by fæcal deposits. He, therefore, requested the men to be more cleanly, and ordered chloride of lime and permanganate of potash to be spread on the more specially-polluted places. At the same time, he took measures for distributing the air from the downcast shaft more equally, the result being that the workings were kept drier, and the temperature probably reduced somewhat. It is, perhaps, too early yet to say what the effects have been of introducing the pail-system; but he informs me that the skin-affections which were so troublesome before seem to have almost disappeared, and that he has not heard of any new cases of anæmia.

In Westphalia there is now an elaborate system of medical examination of all men working, and before they are engaged at a mine, or re-engaged. Men who harbour the worm are rigidly excluded from work underground until they are cured, special hospitals being also provided at the expense of the colliery-owners. Considering that, wherever the disease exists, a large number of healthy men harbour the parasite, this system would seem to entail a great amount of hardship, inconvenience and expense, and would certainly be very unpopular in Britain. It appears to the writer that there ought to be no necessity for such a system, and that the proper aim is to make the sanitary condition of the mines such that the disease, even if introduced, can never give any real trouble. A few isolated cases are of little account, provided they are duly recognized and properly treated.

It is evident that the proper carrying out of the necessary sanitary precautions will require not only careful consideration on the part of colliery-managers, but also intelligent co-operation by the men. The writer hopes, however, that as a result of the discussion of the subject by the members of The Institution of Mining Engineers, efficient action may be taken towards keeping British collieries free from the disease.

———

Mr. W. H. PICKERING (H.M. Inspector of Mines, Doncaster) expressed the indebtedness of the members to Dr. Haldane for his important paper. The writer stated that the disease was

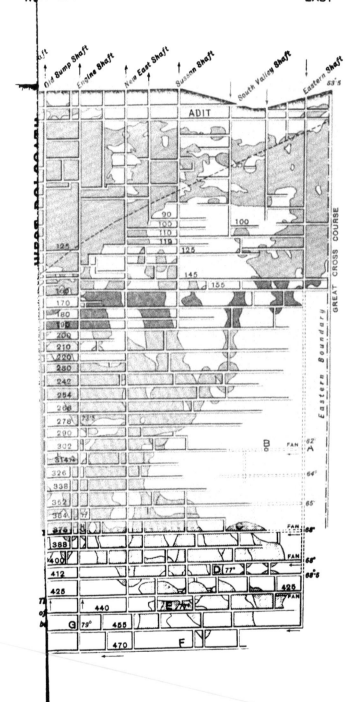

susceptible of cure, but according to the report of Dr. König, of Düsseldorf, the cure was severe, and blindness sometimes resulted. It was also stated that in Germany, where wash-houses were provided, these were a source of danger and infection. It was impossible to emphasize too much the importance of superabundant ventilation. The air in Dolcoath mine was chemically pure, but it was hot and saturated with moisture. About 800 men were employed underground, so that on the average each man was supplied with about 100 cubic feet of air per minute, but in metalliferous mines the air was not coursed throughout the mine as in the case of coal-mines; and in many places the air was practically stagnant. There would be no difficulty in providing the ordure-receptacles mentioned by the author in certain parts of the mine, but the goaf, a convenient and proper place, was used by men working at the face.

Dr. C. LE NEVE FOSTER (London) said that he was glad to hear Dr. Haldane state that the ventilation of Dolcoath mine was so good. In some analyses of air that he had himself taken from the working-places in other mines, there was only 19 per cent. of oxygen, and as much as $1\frac{1}{2}$ per cent. of carbonic acid gas. He was specially interested to hear Dr. Haldane's remarks upon ankylostomiasis, as the first tests for the worm in the case of British mines had been made at his instigation. Suspecting the existence of the disease at the Cwmystwyth mine, Cardigan-shire, in 1898, he obtained permission from the Home Secretary to have the fæces of some of the miners sent to Liége to be investigated. Dr. Haldane had not mentioned the number of worms; but, if his (Dr. Foster's) memory served him aright, he had read that several thousand worms had been found in the intestines of a man. It was stated that the number of female worms was always larger than that of males; and, for this reason, the male worm in order to perform the act of copulation had to unhook himself, and was often washed away by the flood of matter descending into the intestine. The point that there was infection through the skin was new to most people. In Germany, it was stated that the worm was invariably taken in through the mouth, and not through the skin. The figures published in German journals as to the increase of the disease since watering the mine was introduced were very important. A dusty colliery must be watered by government regulation, the danger of explosion was lessened, but the danger of this disease was increased. It occurred to him, not only with regard to this worm, but also with regard to the dust, that it might be well to remove the dust pneumatically, by exhaustion in the same way that houses were now cleaned by the vacuum-process. If once the dust was removed, then the danger would also be removed.

The point as to wash-houses required explanation: if a man got into a bath containing dirty water, he might catch the disease; if, on the other hand, the man had a shower-bath, he thought that the dangerous worm would be washed away, consequently, it should be distinctly stated what kind of wash-house had been in use at collieries where the disease had been propagated. While agreeing thoroughly with Dr. Haldane as to the desirability of restrictions, he was not sure that a mine-manager or any of H.M. inspectors of mines would propose so drastic a remedy as that suggested by Dr. Haldane, namely, that the men should be liable to a statutory penalty if they deposited their fæces except in certain places or at certain hours. A good deal had been said to-day about the restriction imposed by the Board of Trade with regard to electricity, and if the Home Office were to get an Act passed containing a restriction of the kind suggested with regard to the depositing of fæces, there might be a considerable amount of trouble.

Mr. E. W. HAWKER (London) said that in the large silver-lead mines of Australia there were many cases of supposed lead-poisoning, which had not been cured, and it was quite possible that the men had been suffering from this disease. Doctors, however, like miners, jumped to conclusions too rapidly, and in one instance, where the manager of a lead-smelting works was suffering from intense headache, several doctors stated that he was suffering from lead-poisoning. As a matter of fact he was suffering from astigmatism.

Mr. J. C. FORREST (Essington) said that when trade was good in the coal-mines a number of men were employed at the collieries from the metalliferous mines. In Spain, where he had come across the disease, it was probably due to this, and no doubt the introduction into Austria and Germany had resulted from the introduction of men from metalliferous mines. In Cornwall, it had certainly been derived from workmen who had been employed in foreign mines. In collieries, the waste could be used for the purpose mentioned in the paper, but there was no danger of infection in coal-mines so long as the men were prevented from using the main roads for such deposits.

Mr. ISAAC H. SAXTON (Grassmoor Colliery) said that at the Grassmoor colliery receptacles had been provided for many years at the pit-bottom, but since the publication of Dr. Haldane's blue-book, metal-receptacles had been placed at all points where 5 or 6 men or boys congregated together, such as the junctions on haulage-roads, etc. A plentiful supply of disinfectant-powder was also provided at each place.

Mr. PHILIP KIRKUP (Birtley) said that the method proposed of removing the dust instead of damping it, would of course mean an increase of cost. Would not a solution of permanganate of potash in the water solve the difficulty?

Mr. H. R. HEWITT (H.M. Inspector of Mines, Derby) asked in what state the worms entered the body; was it as a worm or as an egg on the point of breaking?

Dr. HALDANE: As free larvæ.

Mr. HEWITT said that in metalliferous mines the men were using the same roads, year after year, travelling up ladders, using alternately hands and feet, and the dirt was thus transferred from the roads to the rungs of the ladders, and then to the hands of the men.

Mr. W. N. ATKINSON (H.M. Inspector of Mines, Stoke-upon-Trent) said that this subject demanded the most serious attention of mine-owners in this country, in view of what had taken place in the Westphalian coal-field. The disease had been introduced into this country, and so far as was known at present was confined to certain mines in Cornwall. The suggestions made by Dr. Haldane as to keeping the coal-mines in such a sanitary condition that they would be kept free from this disease was no doubt a very valuable suggestion, and one worthy of the attention of colliery-managers. It seemed to him, however, considering the large number of collieries working under all sorts of different conditions, that it was hardly likely that such measures would be adopted until they became essential. That was, until the disease actually appeared in a district, it was hardly likely that the mines would be so carefully guarded as to render the propagation of the disease impossible. It seemed to him that, considering the disease was at present confined to a few mines, it would be worthy of consideration whether something in the nature of isolation or quarantine for those mines could not be adopted, so as to confine the disease to the mines where it already existed, and so stamp it out.

Mr. HENRY HALL (H.M. Inspector of Mines, Rainhill) thought that Dr. Haldane would agree with him that the enquiry into this subject had not gone far enough to warrant the suggestion that definite action should be taken. He had read both Dr. Haldane's and foreign reports, and there was a great difference of opinion among the scientific gentlemen who had taken the matter in hand. It could not be questioned that coal-mines were better ventilated than metalliferous mines. In the latter, all the ventilation was produced by diffusion, whereas in coal-mines the air was constantly moving. He was surprised to hear that 80,000 cubic feet of air per minute were passing into Dolcoath mine without any mechanical means of ventilation, and Dr. Haldane must have hit upon a most favourable day for his visit. There would no doubt be times when there would be very little air passing, especially when the wind was blowing in certain directions. If the worms did not hatch out in the human body, how did it happen that as many as 30,000 worms were found in the intestines? According to foreign reports, one of the scientists in Westphalia had tried to find these worms on the thill of the mines. He had failed to do so, but he had found them in large numbers on the bars higher up. If they were developed on the ground, how was it that none were found there?

Mr. J. GERRARD (Inspector of Mines, Manchester) said it was obvious that in Westphalia this disease was a serious matter indeed. If they could keep it away from their extensive coal-mines, Dr. Haldane would have accomplished very good work, and if he would give some simple plain directions, the members would probably be willing to put them into operation. If temperature did affect the development of this horrible worm, then as the mines were getting deeper and deeper, and the temperature of coal-mines was increasing, it was well that they should know it. It was his good fortune to go to Westphalia last autumn, and at several of the mines he saw that the baths, in which a number of the workers would immerse themselves, were being removed, and, sprays or shower-baths, at intervals, in the building, were taking their place. In April last, he began to make enquiries in the Manchester mines-inspection district as to the arrangements existing, and he found only three closets in operation underground. Generally, the men at the shaft-bottoms, incline-stations, etc., relieved themselves by the side of the roads, in a few cases they went into a disused roadway, and in a number of cases the men used water-channels. The currents, in some of these, being very sluggish, the excrement would lie at the bottom of the channel. The investigation had led many managers to make provision, and now a number of closets were established near shaft-bottoms, boxes and pails at important junctions on engine-planes and inclines, and notices posted threatening dismissal if the workmen fail to use the places provided. From a sanitary point of view, many managers have stated that these measures were necessary. Dr. Haldane appeared to consider that it required an expert to discover the presence of the disease, but in Lancashire they were relying upon their medical men for information as to the disease.

Dr. J. S. HALDANE said that it required an expert to make the special test mentioned on the blood, but it did not require an expert to recognize the ova in the fæces.

Mr. JOHN GERRARD said that Dr. Haldane recommended a simple form of bucket or pail. If there was one form of pail which recommended itself, perhaps Dr. Haldane would say which it was.

Mr. HENRY HALL asked whether the men in Cornish mines took water or tea to drink underground.

Dr. ARTHUR E. BOYCOTT (S. Thomas's Hospital, London) wrote that Dr. Haldane had already dealt with everything in connection with ankylostomiasis which was of practical interest to those who had mines in Great Britain under their control, and it was a reasonable expectation that any fresh outbreak which might occur in this country would receive that immediate recognition, the want of which was the sole difficulty in the way of the adequate treatment of the disease. Anæmia is comparatively so uncommon an affection among the adult males who form a mining population that a suspicion that there is infection by this worm should be at once suggested by the occurrence of any such symptoms as pallor and shortness of breath.

The natural history of an epidemic of ankylostomiasis presents several features which may be of material assistance. As has long been known from the experience of medical men in the tropics, only a rather small proportion of those who are infected ever suffer from severe symptoms; in other words, comparatively few individuals are susceptible to the disease in any great degree. It follows that, at the outbreak of any epidemic, there will be a larger number of such susceptible individuals than at any later period and a corresponding number of acutely severe cases. At a later period, some of these will have improved, either by removal to some kind of work where re-infection is impossible, or by a process of natural recovery, which may take place without treatment or change of occupation, and while the individual is still subject to the chances of daily infection. The more severe cases have probably been compelled to give up mining altogether. Under such circumstances, there may be few men who are actually very ill, and it would be an extremely suggestive fact were such cases found to occur in those who had but recently come into the mine. An epidemic may in this way seem to wear itself out, while as a matter of fact there are still a very large number of infected individuals working in the mine. These men are an obvious source of danger both to fresh workers and to other mines, and, though not suffering from marked symptoms, are probably not in perfect health, and will, in consequence, not be quite such efficient workers. At the time of the investigations at Dolcoath in 1902, there were certainly not more than twenty individuals who at the time shewed any serious deterioration of their working capacity, while in 1897-1898 as many as 52 men had been sufficiently ill to be admitted to the Redruth Hospital for anæmia. Yet of some 60 underground workers who were examined, every one was found to be infected, and of those who had no definite symptoms the majority had a distinct diminution of hæmoglobin (about 10 per cent. less than normal). Some of these gave a history of severe anæmic symptoms, from which they had almost recovered. It is necessary, therefore, to view everyone in and from an infected mine with suspicion, as a possible source of infection, without regard to any symptoms of anæmia which may be present at the moment or to any history of apparent recovery from such symptoms.

It is fortunate that any suspicion of ankylostomiasis which may be aroused is readily verified or refuted. In most cases, the eggs occur in such numbers in the stools that a very brief search will reveal them. As they are distributed uniformly throughout the whole mass, it is unnecessary to examine more than one portion of each sample. In cases where they are not found at once, a more prolonged search must be made on more than one occasion, before pronouncing definitely that they are absent. In such cases, an examination of the blood may be of assistance; this, however, cannot give the absolutely positive information which is afforded by the discovery of the eggs, and requires a certain amount of training and experience before it can be utilized to advantage. The eggs are easy to find, and there cannot be any serious difficulty in their identification by anyone who has seen them before.

The suppression of any outbreak of ankylostomiasis embodies two stages of procedure: (1) The cure of those who are urgently ill, by their removal from any chance of fresh infection, and by the expulsion of the worms already present; and (2) the cleansing of contaminated areas, and the establishment of some efficient and compulsory system of underground privies, whereby further contamination may be prevented. In this country, the conditions of temperature, etc., which prevail, combined with the systematic disposal of sewage, will effectually prevent any spread of the disease among the general population. In mines, the tropical circumstances are practically irremediable so far as the temperature is concerned, though something may occasionally be done in the way of diminishing excess of moisture. There is, however, no reason why any fouling of the mine by excrement should occur. It may be stated with a considerable degree of certainty that contact with infected fæcal matter is an absolutely necessary condition of infection, so that it should not be a matter of great difficulty to render this impossible.

It is, of course, better to prevent the introduction of the

disease into any mine than to have to cure it when it has once
obtained a footing. But the difficulties and expense involved
in any adequate examination of all fresh hands (especially those
coming from abroad) are practically prohibitive, especially in
view of the fact that under ordinary conditions of civilized
cleanliness any infected man who may come into a mine is no
material danger to his fellow workers. If, on the contrary,
the filthy habits which prevail in certain mines are permitted to
exist, a single worker who harbours *Ankylostoma* may place
the whole of an underground community in a most undesirable
position.

Dr. THOMAS OLIVER (Newcastle-upon-Tyne) wrote that
the miners of the North of England had fortunately, so far,
remained free from ankylostomiasis, and therefore had not
suffered from that form of bloodlessness which Dr. Haldane
had described and which in Cornwall was known as " Dolcoath
anæmia." The absence of ankylostomiasis in Northumberland and Durham was remarkable, for many of the miners
of the North had travelled abroad and worked in other
countries. The disease was not unknown, however, in Scotland.
Prof. Stockman, of Glasgow, had reported two cases of
ankylostomiasis in miners. Dr. Haldane had so fully discussed
the subject in all its aspects that little remained for him to
add. It is satisfactory to learn that the cause of " Dolcoath
anæmia " has been ascertained, for once the cause of a malady is
known, treatment of the same is usually easy, comparatively
speaking.

In Dr. Haldane's paper, attention is drawn to one very
important point, that is, the absence of any connection of the
Dolcoath miners' anæmia with imperfect ventilation. There is,
for example, a form of anæmia which is very prevalent at the
lead-mines at Linares, in Spain, a condition of things which is
attributed to the absence of ventilation in the mines; but
knowing how widely ankylostomiasis is distributed, also the
dirty habits of the Spanish miner, and that the anæmia of
the Dolcoath miner was likewise believed to be due to imperfect ventilation (until Dr. Haldane showed by analysis how pure
the air in the mine was), it is more than probable that the anæmia
of Spanish miners, like that of the sulphur-miners of Sicily, is

due to the ravages of the intestinal parasite *Ankylostomum duodenale.*

Ankylostomiasis played sad havoc with the miners during the construction of the St. Gothard tunnel, but at the Simplon works, owing to the excellent hygienic measures adopted by the two medical men at Brigue and Iselle, no case of ankylostomiasis so far had occurred during the tunnelling of the Simplon, a circumstance which reflected the greatest credit upon Drs. Pommata and Volante, especially when they knew how prevalent the disease was among the Italian peasantry and miners. So severe have the epidemics of the disease been in Belgium that the malady properly formed the subject of discussion at the recent Conference of Miners in Brussels. It was in 1884 that the parasite, so far as Belgium is concerned, was first discovered in a miner in the neighbourhood of Liége. Since then so widely distributed had the disease become that during last year a medical commission of the province was appointed to examine nearly 10,000 of the 24,000 colliers living in the district. This commission reported that 55 per cent. of the workmen were suffering from ankylostomiasis. Of 134 miners examined in 1901 at the bacteriological laboratory at Mons, 55 had ankylostomiasis. The disease is also widespread through the mining districts of Westphalia. According to the *Deutsche Bergarbeiter-Zeitung,* the numbers of miners in the Ruhr coal-field suffering from ankylostomiasis from 1896 to 1902 were as follows:—

1896.	1897.	1898.	1899.	1900.	1901.	1902.
107	113	99	94	275	1,130	1,355

As moisture favoured the development of the larvæ, the rapid increase in the above numbers had followed the compulsory watering of the mines, introduced in 1900.

Dr. Haldane was the first, so far as he (Dr. Oliver) was aware, to describe the skin-affections observed in miners who are the subjects of ankylostomiasis. Hitherto it has generally been believed that the parasite only gains admittance into the human body by the alimentary canal, but there is now sufficient evidence to show that the ova are capable of penetrating the skin and of ultimately making their way into the alimentary canal, since in persons infected through the skin the fully-developed parasite is subsequently found in the duodenum. All this evidence tends to show how great is the need for the exercise of the greatest care on the part of the affected miners, for it is by the fæces that the ova leave the body, and careless miners by defæcating anywhere in the mine may spread the disease far and wide. The ova when discharged are immature. They cannot develop into the fully formed parasite in the intestine of their host, owing to want of oxygen. The disease fortunately tends to disappear spontaneously, unless it is reinforced by fresh infection. The larvæ are killed by drying, and the parasite itself by such a germicide as thymol internally administered. It is of the utmost importance that the disease, when imported into a mining district, should be at once recognized, for its spread can be prevented by attention to hygienic details, personal cleanliness on the part of the miners, and by medical treatment.

Dr. J. S. HALDANE, replying to the discussion, referred first of all to Mr. W. H. Pickering's question as to the cure of this disease being severe and sometimes dangerous. According to the consular reports, that apparently was so. Of course, they were not medical reports, but no doubt excessive doses of male fern would be attended by very unpleasant results. Male fern was an exceedingly unpleasant medicine to take, but he had never seen it administered in such doses as they appeared to give to the men in Germany. The only time that he had himself administered it, the man said it was as bad as if he had swallowed at least a pint of castor-oil. When a man had a dose of thymol, which had been used in Cornwall, he was told to keep his bed during the morning on which it was taken, but he did not know of any case where the man was seriously inconvenienced after a dose. Evidently the idea of danger was based on unreliable information, or it was due to excessive doses of extract of male fern having been administered. With regard to washhouses, these were the places where the men, after a shower-bath, hung their dirty clothes, and left them until the next morning. These were supposed by some German authorities to be a source of danger, but he himself did not believe that they were. His reason was that where the men changed and had their clothes washed at home the disease did not spread amongst their families. The drying of the clothes also killed the small young larvæ. With regard to the question of the air at Dolcoath mine not being good, he felt bound

to stand up in defence of the air in Cornish mines, although
Dr. Le Neve Foster knew more about that matter than he did.
He had found that there were enormous spaces in these mines;
and if they could imagine in coal-mines the whole of the goaf
being left open, they would not feel much air-current; although
the air might be moving, it would not be so noticeable as where
it was passing through narrow passages. If they observed the
temperature at the top and bottom of the mine, there was considerable ventilating force, but he did not deny that some
particular parts were badly ventilated. The temperature underground was lower in these mines, proportionately to the depth,
than in the best ventilated coal-mines. If the heat-production
from oxidation in coal-mines was calculated, it would be found
that the ventilation in most Cornish mines had a cooling effect,
whereas the effect in coal-mines was to raise the temperature.
In one case in Cornwall, where the ventilation was small, the
mine had been heated up gradually, as the heat produced by
oxidation in the mine was much greater than the heat carried
away by the air. His impression was, comparing the reports
made on this subject some time ago, that there had been much
improvement in recent years. In the report of the Royal
Commission of 1864, the published analysis was very bad.
They could only find such air, nowadays, in very exceptional
places, such as the one referred to in the table of analyses given
in his paper. In one case, where he tried to get a sample, the air
was quite stagnant, and a candle would not burn, the temperature was 105° Fahr., and the place was not being worked.
Eventually air-pipes were introduced, and the temperature was
very rapidly reduced. With regard to the number of worms
found in the intestines, he had never heard of so many as 30,000,
the number which had been mentioned, but it was common to
find 500 or 1,000 in a bad case. With regard to the multiplication of the germs, the evidence was that they did not multiply in
the intestines. They had just followed, one by one, either through
the skin or in some other way, and the reason for this assumption
was that if any man went away from the source of infection he
gradually got better. If he had been infected, say, with typhoid
fever, the germs would multiply enormously wherever he was,
but in this case it was altogether different. The worms would
live a long time: 2 or 3 years. He cordially agreed with Dr.

Le Neve Foster that it would be absolutely impossible to re-gulate the hours at which a certain natural operation should occur, but he did think that it was possible to regulate the places; it was only a matter of introducing order underground. They required a system which should provide the means of disposing of the fæces in such a way that the men would never come into contact with them. He hoped that some-thing would be done to keep the mines clean, as anyone could keep the ground clean on the surface. He was interested in the remark about silver- and lead-miners, and the anæmia among them. It would be difficult to distinguish at first sight, but any-one who had seen a good deal of both could readily distinguish them, from the appearance of the man. The first question that he asked at Dolcoath was whether lead-ore was being worked.

Mr. E. W. HAWKER asked whether the presence of carbonate of lead or sulphide of lead affected the matter.

Dr. J. S. HALDANE said he believed that carbonate of lead was poisonous, but there was no danger with galena. He was inte-rested in Mr. Saxton's remarks as to the precautions taken in Derbyshire, and they were just the kind of precautions that should be adopted in other mines. The men should understand that they ought to behave with common decency in a mine, as they would anywhere else. He did not think that the use of anti-septics, such as permanganate of potash, was likely to be advan-tageous in mines. Mr. H. Hall asked why the larvæ found on the timber were not found on the roads. He had grave doubts as to whether the larvæ found on the timbers had anything to do with ankylostomiasis. Many worms were very similar, it was difficult to distinguish them, but he did not believe that they were the same; and they were not found on the roads, because they did not look for them in the fæces. The one thing needful was to get rid of fæcal deposit, and to prevent it getting upon the men's boots and being carried about the mine. He did not think that the worms ate the blood, as suggested by Mr. H. R. Hewitt, for, according to experiments, the affected man had more blood, though of inferior quality. The question of quarantine raised by Mr. W. N. Atkinson was an important element in the dis-cussion; and he thought that this suggestion ought to be very carefully considered. A good deal could be done in the way of watching any new men coming to the mine; and any one looking anæmic, and coming from a suspicious part of the country, should not be employed without a medical certificate of his freedom from ankylostomiais; but he did not think that the quarantine system would do away with the disease. A few days ago, a man came to the Glasgow Infirmary; his history was that he had been a soldier in India; he had been ill for some months, and had been treated for some other disease. Prof. Stockman discovered that this man had ankylostomiasis. There must be a number of such men coming back from abroad, and unless they could make the conditions underground such that the disease would not spread, he did not know how they were to guard against it. They could not institute a minute investigation of the inside of every man who applied for work. As regards the question of men taking water underground, in Brenn-berg in Hungary, one of the most infected collieries, they took down small kegs of water, just as they did in Cornwall. The keg probably got very dirty inside, at any rate, nobody ever saw the inside of it. In Westphalia, he did not know what they drank: in Cornwall, so far as he could ascertain, the men did not drink underground water, as it was impregnated with copper.

Mr. HENRY HALL (H.M. Inspector of Mines, Rainhill) said that in Lancashire mines the men drank tea or coffee, anything except water.

Dr. HALDANE replied that it was unusual for the miners, anywhere, to drink underground water.

Mr. J. GERRARD (H.M. Inspector of Mines, Manchester) said that at Pendleton colliery the bulk of the men carried cans of water into the mine.

The PRESIDENT (Sir Lindsay Wood, Bart.), in moving a hearty vote of thanks to Dr. Haldane, said that this was only one of the very valuable series of papers that he had contributed, and he hoped that the members would be favoured with the results of his investigations from time to time.

Mr. JOHN GERRARD seconded the resolution, which was warmly approved.

Mr. G. H. KINAHAN's paper on "The Re-development of the Slate-trade in Ireland," was read as follows:—

THE RE-DEVELOPMENT OF THE SLATE-TRADE IN IRELAND.

By G. H. KINAHAN, M.R.I.A., District Surveyor (Retired), Geological Survey of Ireland.

Introduction.—Of late, there has been a movement to develop the resources of Ireland and among others the mining and quarrying industries. In connection with mining and quarrying, some of the proposals are not on such sound business and commercial principles as to give prospects of success, and any failure is necessarily detrimental to the country at large.

In connection with the mineral veins, some of the proposals are to work old mining setts, solely because, at one time, they were in " riches; " quite ignoring the fact that the original adventurers sold their interest when they had no hopes of further gain, and that their successors had to abandon the undertaking after having lost more or less money. Of course, there are some old mines, which, on account of the new methods for treating poor ores, may have a future before them : to these, however, the writer does not refer.

Similarly, with the coal-fields, it appears inexpedient to expend vast sums of money in looking for coal, when it is highly improbable that even when found the seam will not be of sufficient dimensions to admit of being profitably worked.

There are some beautiful and good granites in Ireland, that have been worked in a small way; but as a general rule the veins have only been holed, no good deep quarries having been opened; consequently they do not pay as they ought to do; these might be developed. But, instead of doing so, it has been proposed to start companies in wild districts, where there are now no quarries, no quarry-men, no trade, or other necessary facilities.

Slates.—Speculations that are quite ignored, although they seem quite legitimate, are the regeneration of the now abandoned slate-quarries that formerly were a considerable source of revenue to the country. To these it is proposed to call attention in this paper.

In Ireland, the best slates almost invariably occur in the upper division of the Ordovician (Lower Silurian), which from its fossils seems to be equivalent to the May Hill Sandstone of England. In the Devonian (Old Red Sandstone of the Geological Survey Maps) of south-western Ireland, there are, however, some good slate-veins, and also inferior slates in the Carboniferous Slate, useful for a local trade. At the present time, Ireland is greatly dependent for its slate upon Wales, the few supplied from the home-quarries at present at work, scarcely supplying the local trade; yet, as will be seen in the following records, at one time the slates commanded in places a considerable export trade.

Irish slates, in general, are heavier and not of as good colour as the Welsh. This in a great measure is due to the quarries not being deep enough, because similarly as in Wales, the deeper the working is on a vein the better are the metal and colour, besides being more easily wrought and split. In the majority of cases, before the Irish quarries were abandoned (on account of the depression of trade in 1850 and subsequent years), the "eyes" only were taken out; while all detritus was allowed to accumulate, and consequently the works are now more or less smothered. This filling-stuff should be removed before the quarry could be in proper working order, more especially so, because in many cases the débris have been dumped on the floor of the working, thereby preventing a lower and better bench from being worked. It must be said that from appearances it would seem as if many of the quarries were worked on a "lazy bed" system, the waste not at once being carried away, but shifted about from one part of the workings to another, as a lower bench was cut into and quarried. In some cases, as mentioned hereafter, when plenty of power was available, it was proposed to grind up the waste and make bricks, pottery and tiles at the works, or to export the clay.

Mr. S. Lewis, in his *Topographical Dictionary* (1835), from information supposed to have been supplied by that well-known miner and geologist, Mr. Thomas Weaver, gives in each county

history very exhaustive lists and information of the numerous
slate-quarries scattered about the island; some being on veins
of more or less considerable size, while others were of small
dimensions and suitable only for a local trade. All of these,
when the trade-depression came in 1850 and subsequent years,
were abandoned, and the export-trade died out; while since
then only a few of them have been worked. The information
collected by Mr. S. Lewis has been republished in the *Economic
Geology of Ireland.** On this account it seems superfluous to
mention here all the different localities; the writer therefore proposes to confine himself to the principal quarries—those that seem
to be so circumstanced that a slate-trade might be developed
with a prospect of being remunerative to the adventurers.

As already stated, Ireland now is largely dependent on Wales,
as the few Irish quarries at work have more orders than they
can supply, and a very inferior French slate is on the market.
After the late storm in Dublin, on Feburary 27th, 1903, slates ran
up to famine-prices, so that in some cases French slates had to be
used.

The quarries will be mentioned as far as possible under the
different counties, arranged in alphabetical order.

Cork.—At Audley Cove and Tilemuck, Ballydehob, in 1834,
quarries were opened by the West Cork Mining Company, who
had 500 hands employed. The slate was of good quality, hard,
compact, durable, and had a ready sale in London and other
English markets.

In Sherkin Island, Baltimore Harbour, there is an excellent
hard and durable slate. In 1835, 100 hands were employed, and
many cargoes were shipped to England, where they had a ready
sale. This slate since then has been highly approved of, and
some time since a company was formed to work the vein, but
on account of one of the " unrests " to which Ireland is unfortunately subject, the enterprise was abandoned. This quarry ought
to be of special interest to the Department of Agriculture
and Technical Education, on account of the fishery at Baltimore.
All fishermen ought to have " two strings to their bow," therefore, besides being fishermen and women, they ought also to be
capable of working at the slates in slack times; more especially

* Page 327 *et seq.*

as fishes are so erratic in their movements, that at any time they may leave their usual feeding grounds, as they have done on former occasions. If this took place, the employees at the fishery would be without work, if they had not the slate-quarry to fall back on till the fishes returned.

At Roberts Cove, parish of Ballyfoil, 10 miles north-east of Kinsale, there is a valuable slate-quarry, from which slates were exported to England in the ships that brought coal to the bay.

All these quarries, on or near the southern coast of county Cork, have great facilities for shipping to England, as so many of the colliers and other trading vessels to this and the west coast of Ireland have to return in ballast for want of cargoes.

Donegal.—Two miles west of St. Johnstone is the Glentown quarry. Hereabouts, there are two or more veins of slate that have been holed in places. Tradition has it that slates have been produced here for more than 100 years. The oldest workings are about ¼ mile from the present quarry; they are mentioned in Dr. MacFarlane's *History of Donegal,* published about 1805. The main vein, from 20 to 30 feet wide, ranges about north 60 degrees east, and hades northward, about 50 to 70 degrees; while the cleavage strikes north 20 degrees west, and dips eastward at about 55 degrees. For over ¼ mile there are old workings along its outcrop. About 1820 a deep quarry was opened to the westward at the margin of the drift-slope, the stuff, waste and water being raised by an engine. About 1845, after it had fallen into the hands of the landlord, the Duke of Abercorn, a tunnel 1,140 feet long was run from the valley on the east; this cut the vein 84 feet below the surface. The tunnel drained the quarry, besides being used for a tramway by which the stuff and rubbish were removed. The vein was worked to the level of the tunnel, and at the eastern end to a further depth of 40 feet below it. The work was continued until 1879, when, on account of depression in trade, it ceased to pay, and had to be abandoned. At that time, there were about 30 hands employed, all natives, who were used to work slate from their childhood. The slate is good "metal," of a durable grey colour, and light, so that roofs require timber of but small scantling. As in other veins, the stuff improves in quality in depth, but at the same time it is remarkable what good metal can be procured close to the surface. The vein, however, is crossed by joints which prevent the slates from being raised of very large dimensions. There was a ready sale in Derry and neighbourhood, the largest being sent to Glasgow.

The work might have been more judiciously planned, as all the waste, instead of being run out of the quarry, was dumped on the old floor of the portion worked. An apparently advantageous method of resuming the work would be to remove all the débris of the ancient workings and any bad stuff on the back of the main vein, into the valley to the westward, and from the western brow to carry a breast eastward along the vein. The removal of the back of the vein would not be altogether unprofitable, on account of the good quality of some of the surface-stone, which could be wrought into slate.

Kerry.—In Valentia Island, in the eastern slope of Geokaun is the wellknown quarry—as this may be considered a working quarry we may refer for its history to the writer's *Economic Geology of Ireland.*[*] The writer should, however, mention that it has principally been worked for slabs, which are not to be excelled for beauty, strength or size. These were largely exported to London, and also to America. At present the trade is not pushed so vigorously as it ought to be.

Kilkenny.—In the valley of the Lingaun at the mearing of county Kilkenny—are the Ormond quarries. The slate is of excellent quality, but, unfortunately, near the river where the quarries are situated, the strata are very much cut up by faults, which add considerably to the working expenses, as the slate-veins are constantly cut out when a quarry is in good working order. Away from the river, the veins seem to be more continuous, and therefore ought to be more suitable for quarrying. These quarries were worked to supply a large quantity of slates for export, and a local trade in Carrick and elsewhere.

Tipperary.—On the opposite side of the river in county Tipperary, are the Victoria quarries, 6 miles from Carrick-on-Suir and 14 miles from Waterford. As these quarries are at

[*] Pages 342 and 343.

present being worked we may refer for further information as to their history to the writer's *Economic Geology of Ireland.*[*]

One-and-a-half miles south of Ballynacourty or New Forest House, on the southern side of the glen of Aherlow, county Tipperary, there was an extensive quarry. The slate is of a greenish colour, sound and good, but rather heavy; it was, however, extensively used in the vicinity and neighbouring towns.

On the north side of this glen of Aherlow in the Ahaphuca valley, county Limerick, there is a thick vein of very similar slate that was also extensively quarried. It is situated 4 miles south-south-west of Ballylanders.

In county Tipperary east of the northern arm of Lough Derg, and north-north-east of Killaloe, county Clare, there are various quarries that have been called after that town. Now, all the "Killaloe slate-quarries" are idle, except that at Corbally, near Portroe, now being worked by the Killaloe Slate Company. For its history members may be referred to the writer's *Economic Geology of Ireland.*[†] Of the other workings, the largest was at Derry Castle. Here the slate was not considered as good at at Corbally and the Imperial Mining Company, who were working it, abandoned the undertaking when they bought the Corbally quarry. These Killaloe quarries are most advantageously situated, as from Killaloe there is water carriage to Limerick, Dublin, and most parts of Ireland.

Waterford.—In county Waterford, 4½ miles east of Kilmacthomas, are the Ross quarries, which were worked up to 1863, The slate is good, and of a dark-grey colour.

In Glenpatrick, 6 miles from both Carrick-on-Suir and Clonmel, there were, before 1850, rather extensive quarries working on veins of good slate. The largest are:—Toor to the north, and a quarry 900 feet to the south, of Glenpatrick bridge. In the first, the slate is fine-grained, of a bright bluish-grey colour, and in the second of a lighter grey: both the veins are of very good widths. A third is situated in Clondonnel, at the base of the Reeks of Glenpatrick, on a vein of finely cleaved, earthy grey slate.

The quarries in this glen seem well worthy of being reworked, but the last operators left them rather smothered with

---

[*] Pages 346 and 347.　　　　　　　　[†] Pages 348 and 349.

waste. Some time ago, a syndicate was formed, which proposed to work a double trade—as they intended to grind up the waste and make pottery, tiles and bricks, or to export the clay to England, besides working the slate-veins. As the glen is situated on the navigable river Suir, there are considerable facilities for an export trade. The syndicate did not succeed in forming a company, on account of one of the Irish " unrests."

Wexford.—In county Wexford, 1½ miles south-south-west of Bunclody or Newtonbarry, in the townlands of Ballyprecas and Glaslacken, north and south of a glen, there are peculiar slate-veins; because, although the quarries are opposite to one another and the veins seem to have the same strike, yet those to the north in Ballyprecas are blue and ribboned, while those to the south, in Glaslacken, are grey and greenish-grey—a grey vein in the latter is of excellent quality. At the present time, these quarries are more or less smothered, yet there are great facilities for opening a good quarry, that is, if both the quarries were worked as one, and a deep cut made up the glen from the westward. By this means the veins could be drained to a considerable depth, and a way made for running the waste into the valley to the west.

Wicklow.—Near Ashford and the Devil's Glen, county Wicklow, a slate-vein was being worked in 1845, for both roofing-slates and slabs. Sir R. Kane and Mr. Wilkinson wrote approving of it, and stated that it strongly resembled Welsh slates, being about intermediate in character between those of Bangor and Llanberis, while like those the veins are of Cambrian age. This quarry was worked for a few years with every prospect of success, but the proprietor, like many others, suffered from the great depression of trade in 1850 and subsequent years, during which the works were abandoned.

Conclusion.—The foregoing notes are merely intended to draw attention to Irish slate-quarries now idle that seem to be capable of being hereafter worked at a profit: and it seems to be outside the writer's object to describe working quarries, these are therefore only mentioned and for their brief histories the reader may refer to the writer's *Economic Geology of Ireland*. Some of those mentioned, as the quarries have not recently been visited, may possibly be at work, but as there are no slates from them in the Dublin market, any such works must needs be of small dimensions.

In different places, there are good slate-veins which have only been holed to supply slates for local purposes. Some of these afford facilities which suggest that they might be profitably opened up, but as such new adventures are foreign to his present object, the writer has not mentioned these localities. It should be pointed out that almost invariably in Ireland where there are slates, even if no big quarry has been worked, will be found persons capable of working slate; so that if works were commenced all the hands would not have to be imported.

———

The PRESIDENT (Sir Lindsay Wood, Bart.) moved a vote of thanks to Mr. Kinahan for his valuable paper.

Mr. M. WALTON BROWN seconded the resolution, which was cordially approved.

———

Mr. W. DENHAM VERSCHOYLE's paper on "The Smelters of British Columbia," was read as follows:—

THE SMELTERS OF BRITISH COLUMBIA.

By W. DENHAM VERSCHOYLE.

INTRODUCTION.

Mining in British Columbia is at present in a rather unsatisfactory state, owing perhaps mainly to the vast extent of the country, to the want of transport-facilities, and to the generally complex nature of the ores met with, which renders it almost impossible to procure successful treatment except under the most careful and experienced management. Unfortunately, in a great number of cases this has been wanting, and the result has been a long list of failures. But a change is now apparent, the days of "wild-catting" are to a great extent passed, under experienced management many of the mines are coming to the front, and the value of the mineral-production is expanding every year at a rapid and satisfactory rate.

TABLE I.—TOTAL VALUES OF THE METAL AND MINERAL PRODUCTION OF BRITISH COLUMBIA.

Years.					Dollars.
1890	...	...	...	...	2,608,803
1895	...	..	..	...	5,643,042
1896	...	...	...	...	7,507,956
1897	...	...	...	...	10,455,268
1898	...	...	...	...	10,9'6,861
1899	...	...	..	...	12,393,131
1900	...	...	...	...	16,344,751
1901	...	...	...	...	20,086,780
1902	...	...	...	...	17,486,550

Table I. shows this expansion for a few recent years. Table II. shows the value of the metal and mineral production for the past 4 years, as distinct from other materials, such as lime, bricks, etc. Table III. shows the total production from lode-mines. It was found necessary to include small gold-and-silver values that have not passed through the smelter, on account of the difficulty in accurately segregating these.

TABLE II.—Quantities and Values of the Metal and Mineral Production of British Columbia.

	1899.		1900.		1901.		1902.	
	Quantity.	Value.	Quantity.	Value.	Quantity.	Value.	Quantity.	Value.
		Dollars.		Dollars.		Dollars.		Dollars.
Gold, placer ...ounces	67,245	1,344,900	63,936	1,278,724	48,505	970,100	53,657	1,073,140
„ lode ... „	138,315	2,857,573	167,153	3,453,381	210,384	4,348,603	236,491	4,888,269
Silver ... „	2,939,413	1,663,708	3,958,175	2,309,200	5,151,333	2,884,745	3,917,917	1,941,328
Copper ... pounds	7,722,591	1,351,453	9,997,080	1,615,289	27,603,746	4,446,963	29,636,057	3,446,673
Lead ... „	21,862,436	878,870	63,358,621	2,891,887	51,582,906	2,002,733	22,536,381	924,832
Coal tons of 2,240 lbs.	1,306,324	3,918,972	1,439,595	4,318,785	1,460,331	4,380,993	1,397,394	4,192,182
Coke „ „	34,251	171,255	85,149	425,745	127,081	635,405	128,015	640,075
Other materials	...	206,400	...	251,740	...	417,238	...	480,051
Totals		12,393,131		16,344,751		20,086,780		17,486,550

TABLE III.—Quantities and Values of the Metal-production of British Columbia.

Years.	Gold.		Silver.		Lead.		Copper.		Total Values.
	Ounces.	Value.	Ounces.	Value.	Pounds.	Value.	Pounds.	Value.	
		Dollars.		Dollars.		Dollars.		Dollars.	Dollars.
1894	6,252	125,014	746,379	470,219	5,662,523	169,875	324,680	16,234	781,342
1895	39,284	785,271	1,496,522	977,229	16,475,464	532,255	952,840	47,642	2,342,397
1896	62,259	1,244,180	3,135,343	2,100,689	24,199,977	721,384	3,818,556	190,926	4,257,179
1897	106,141	2,122,820	5,472,971	3,272,836	38,841,135	1,390,517	5,325,180	266,258	7,052,431
1898	110,061	2,201,217	4,292,401	2,375,841	31,693,559	1,077,581	7,271,678	874,781	6,529,420
1899	138,315	2,857,573	2,939,413	1,663,708	21,862,436	878,870	7,722,591	1,351,453	6,751,604
1900	167,153	3,453,381	3,958,175	2,309,200	63,358,621	2,691,887	9,997,080	1,615,289	10,069,757
1901	210,384	4,348,603	5,151,333	2,884,745	51,582,906	2,002,733	27,603,746	4,446,963	13,683,044
1902	236,491	4,888,269	3,917,917	1,941,328	22,536,381	824,832	29,636,057	3,446,673	11,101,102

From these tables it will be seen that the smelting business of the Province represents a large proportion of the total mining output, and present indications seem to show that before many years, the returns from all other branches of mining, will be insignificant in comparison with those obtained from smelting.

SMELTING.

The following are the names and some particulars of the operations of the smelters of British Columbia. On the accompanying map (Plate XXV.), the position of each is marked, and the district from which their ore-supply is drawn, is where possible roughly outlined.

FIG. 1.—TRAIL SMELTER: CANADIAN SMELTING AND REFINING COMPANY.

(1) The works of the Hall Mining and Smelting Company, at Nelson, do a custom-smelting business in both lead and copper-ores, and during 1901, treated 22,220 tons purchased from various mines in East and West Kootenay. During the latter part of the year they treated 20,679 tons from the Silver King (their own) mine: 4,840 tons of bullion were produced, containing 9,370,207 pounds of lead, 1,080,079 ounces of silver and 7,933 ounces of gold. The bullion and 1,708 tons of copper-matte were shipped for refining to Newark, New Jersey.

(2) The works of the Canadian Smelting and Refining Company, known as the Trail smelter, are at Trail. This is a custom-smelter, entirely, and buys and treats Rossland lead and copper-ore principally. During 1901, about $65,000 were expended in extending and improving the works (Fig. 1).

The present plant consists of 3 copper blast-furnaces, 3 lead blast-furnaces, 1 softening furnace, 2 O'Hara roasters, 6 Bruckner roasters, 10 hand-roasters for lead-ores, 24 roasting stalls, 2 lime-kilns and 2 briquetting plants. There are two separate sampling-mills, one for copper-ores and the other for lead-ores. There are transformers, motors and other electrical

FIG. 2.—GRAND FORKS SMELTER: GRANBY CONSOLIDATED MINING, SMELTING AND POWER COMPANY, LIMITED.

appliances for 1,000 horsepower. There are also complete assay-offices and laboratories, machine and blacksmith shops, etc. The present capacity of the works is about 1,200 tons of ore daily. The value of the property with its water-works, electric-lighting plant (the smelter generates its own electricity, and lights the town of Trail besides lighting its own works), limestone-quarries, etc., is close upon $1,000,000.

During 1901, they smelted about 100,000 tons of copper-ore and 50,000 tons of lead-ore. The products of copper-matte, 4,500 tons, and bullion 6,300 tons, contained approximately:—

The power by which the blowers, sampling-works, converter, etc., are operated, is derived from three 16 inches turbine-wheels, which develop 1,200 horsepower, the motor-water being taken from the northern fork of the Kettle river (Fig. 3). Each wheel is directly connected to a Westinghouse rotating arm, alternating-current generator (Fig. 4). Another wheel drives two pumps, each having a daily capacity of 750,000 gallons (Fig. 5).

The smelter consists of two double-decked, steel-jacketted furnaces, 160 inches by 44 inches at the tuyeres, made by the Gates Ironworks, of Chicago. The gases pass off from the

FIG. 4.—INTERIOR OF POWER-HOUSE AT THE GRAND FORKS SMELTER.

top in a downtake pipe 4 feet in diameter, which is connected with the big-flue dust-chamber leading to the stack; the flue-chamber measures 10 feet by 10 feet on the outside and 300 feet in length. The stack is 11 feet by 11 feet inside measurement and 152½ feet high. There are three blowers, one for each furnace and one in reserve. These are connected with the furnaces by a pipe, 54 inches in diameter, all blowers being connected with the one main pipe. Each of these blowers is driven by a 75 horsepower Westinghouse induction-motor, which is belted direct to the blower.

The ore is delivered in bins from the mine-cars, and thence it goes to a No. 5 Gates gyratory crusher, with a capacity of 1,000 tons per day, from which it is delivered into a continuous steel bucket-elevator, lifting it to the top of the building. It is then sampled in a 60 inches Snyder automatic sampler, and is thereupon delivered to the smelter-bins. The work of enlargement was begun in May, 1902. The power has been increased by a 250 horsepower horizontal turbine, directly connected to an electric generator. The water-supply was also increased by 750,000 gallons daily, for the water-jackets and for granulating the slag.

FIG. 5.--POWER-HOUSE AT THE GRAND FORKS SMELTER.

An additional No. 8 Connersville blower, 2 new furnaces, and crushing and sampling machinery have increased the capacity of this department to 2,000 tons daily, while the storage-capacity of the bins has been increased to 10,000 tons. From the furnaces, the copper-matte after settling goes to the converters, where it is blown up to about 98 per cent. of pure copper. The converter also treats matte from the smelters of the Greenwood and Hall mines.

The following table shows the monthly tonnage of ore treated by the Granby smelter during 1901. The amount of ore treated in 1900 was 62,387 tons. A noticeable feature of the operations is the increased efficiency of the plant. In November and December, 1900, two furnaces treated 36,517 tons of ore; and in the same two months of 1901, the same two furnaces treated 42,677 tons, an increase of 101 tons per day. The coke and fluxes used are not included in the foregoing statistics.

FIG. 6.—GREENWOOD SMELTER: BRITISH COLUMBIA COPPER COMPANY, LIMITED.

Months.						1901. Tons.	Daily average. Tons.
January	...	...	...	...	...	... 17,640	569
February	...	...	...	...	...	... 17,708	632
March	...	...	...	...	...	... 19,713	636
April	...	...	...	...	...	... 18,995	633
May ...	...	...	...	...	...	... 19,075	615
June ...	...	...	...	...	...	... 18,510	617
July ...	...	...	...	...	...	... 18,176	586
August	...	...	...	...	...	... 18,028	581
September	...	...	...	...	...	... 20,059	668
October	...	...	...	...	...	... 20,347	656
November	...	...	...	...	...	... 20,706	690
December	...	...	...	...	...	... 21,971	708
Totals	...	...	...	...	...	... 230,928	632

(4) The Greenwood smelter of The British Columbia Copper Company treats chiefly ore from the company's Mother-lode mine and does a custom copper-smelting business. Two furnaces are running, one since February, 1901. and one only lately completed. They are stack-furnaces, 42 inches wide by 150 inches long at the tuyeres, of which there are ten on each side, each being 3½ inches in diameter.   The engine-and-blower house, 60 feet by 45 feet, contains two No. 7½ Connersville blowers, and the high-pressure cylinder of a compound condensing Reynolds-Corliss engine, 16 inches in diameter by 35 inches stroke, rated at 150 horsepower; the other cylinder will be added later. In the boiler-room, there are 3 horizontal return-tubular boilers, 66 inches in diameter by 16 feet long; each of 100 horsepower (Fig. 6).

The following table shows the tonnage treated by this smelter :—

| | | | | | | 1901. | |
Months.						Tons.	Daily average Tons.
February	...	...	...	...	...	3,026	274
March	...	..	...	...	...	10,519	330
April	...	...	...	...	..	11,322	377
May ...	...	...	...	...	...	11,830	381
June ...	...	...	...	...	...	11,206	373
July ...	...	...	...	...	...	11,943	385
August	...	...	...	...	...	7,386	350
September	...	...	...	...	...	11,823	394
October	...	...	...	...	...	12,660	408
November	...	..	...	...	...	12,264	408
December	...	...	...	...	...	13,098	422
Totals	...	...	...	...	...	117,077	320

The increased efficiency of this smelter is as noticeable as in the Granby smelter. The figures of the table represent the work of one furnace, and are exclusive of coke and flux. During 24 hours of one day last January, 460 tons of ore were put through this furnace, which record affords evidence of the favourable nature of Boundary ores, and the skill that has been shown in the construction and working of the furnaces.

(5) The plant of the Standard Pyritic Smelting Company, situated at Boundary Creek, 3 miles south of Greenwood, was erected to treat ores from certain mines, as well as custom-smelting.

The large main building—the smelter-house—is 182 feet
in length by 120 feet in width: measuring from the feed-floor
in the centre of the building, the height is 64 feet; and from
the furnace-floor to the roof nearly 80 feet. In the sampling
department, there are two 36 inches and two 48 inches automatic samplers, a 7 inches by 10 inches Blake rock-crusher,
2 sets of rolls, 12 inches by 20 inches, and 2 belt-elevators; and
east of the sampling department the bins for the sample discard are placed. Next are placed two parallel rows of ore-
storage bins, eight in a row, and each bin is 34 feet by 16 feet.

Fig. 7.—Van Anda Smelter.

Farther east, are lime and coke-storage bins, the whole group
of bins occupying the central portion of the building from
north to south, and over them run double railway-tracks. At a
lower level, the furnace-floor extends eastward from the stone retaining wall, 60 feet, and has a length of 140 feet. The dust-flue,
of stone-walls with an arched brick roof, runs about 200 feet to
the steel smokestack, which is 9½ feet in diameter and 112 feet
high, above a 14 feet brick-base. On the furnace floor-level
are two 75 horsepower engines, one to run a No. 7 Connersville
blower and the other to run the sampling machinery.

(7) The Crofton smelter, situated at Osborne bay, Vancouver Island, has been erected to treat Lenora (Mount Sicker) ores. Two blast-furnaces have been erected, the largest with an estimated capacity of 400 tons per day; and the furnace-room is built large enough to accommodate three of these stacks. A modern bessemerizing plant has also been installed. The power-plant consists of a compound condensing engine of 500 horsepower, but only the high pressure cylinder has been erected. Two Connersville blowers supply the air for the furnaces and converter. Three 200 horsepower boilers supply steam for all purposes (Fig. 8.).

FIG. 9.—TYEE SMELTER: TYEE COPPER COMPANY, LIMITED.

(8) The Tyee smelter has been erected at Ladysmith, Vancouver Island, to treat Tyee (Mount Sicker) ores. The plant consists of a 150 tons water-jacketted furnace, with a complete sampling-plant and a storage-capacity for 1,600 tons (Fig. 9).

Several other small works have been built in the Province, but have been closed for various reasons. There are several others in contemplation, although no definite information is obtainable about them. A large tonnage of British Columbian ores is sent across the boundary to the Tacoma and Northport smelters in the State of Washington, U.S.A. The former receives ores from Vancouver Island, Texada Island, and the coast, and the latter treats ores from Rossland and adjoining districts.

MINES.

It hardly comes within the province of this paper to describe the mines which are the source of the smelting-ore supply, but a short description of the principal mines may prove interesting, as showing upon how solid a basis the smelting-industry rests. As a rule, each smelter is built in proximity to some large mine or group of mines; and, although a large custom-business may be operated, generally speaking, the chief ore-supply comes from that mine or group of mines. Because only a few mines are here described, it must not be inferred that they are the only good ones, for there are many large mines, particularly in the Boundary country, which ship merely a small tonnage, but are quite capable of sending out a very large one. Such mines are either now doing development-work alone, and any ore shipped is simply in the nature of a trial-shipment; or else they are limited or entirely prohibited by lack of transport-facilities.

(1) The Nelson smelter, during 1901, treated 20,679 tons from (its own) the Silver King mine, and this represents almost half its total tonnage. There are a number of good mines shipping to this smelter, but at present their output is limited by the high freight-charges. The Silver King mine was supposed to be worked out when the 500 feet level was reached; but during 1900 and 1901, nearly 5,000 feet of development-work, done below that level and down to the 1,000 feet level, has opened up new ore-bodies of great value. The average assay of the ore treated yields 16·1 ounces of silver and 3·8 per cent. of copper. The typical ore of the Nelson and adjoining districts is a high-grade argentiferous galena, intermixed as a rule with more or less auriferous pyrites and chalcopyrite.

(2) The Trail smelter derives its ore-supply principally from the neighbourhood of Rossland, but it also draws a considerable tonnage from ports on the lakes and rivers in the country extending from the U.S.A. boundary, north to the main line of the Canadian Pacific railway. The tonnage shipped in 1901 from various mines in the Rossland district to the Trail and Northport smelters, aggregated 283,307 tons valued at $4,621,299 and the principal shippers in that district were as follows :—Le Roi, 158,598 tons: Le Roi No. 2, 35,956 tons ; Rossland Great Western, 10,046 tons ;

Centre Star, 53,590 tons; and War Eagle, 19,863 tons. Labour troubles interfered very much with the work in this district during 1901, and curtailed to a considerable extent the tonnage shipped. The Le Roi is the principal mine in this district, and its geological characteristics and methods of mining and treatment are to a certain extent characteristic of the district.

At the Combination shaft there is a fine massive head-gear, 85 feet high, with first-class crushing, conveying, sorting and sampling machinery, all driven by electricity. There is also an automatically-working aerial tramline leading to the ore-

FIG. 10.—ELECTRIC-DRIVEN AIR-COMPRESSOR: WAR EAGLE MINE.

bins on the Great Northern railway. There are two double-cylinder modern-type winding-engines, one of 1,000 nominal horsepower and the other of 500 nominal horsepower. The larger engine is used exclusively for hoisting ore, skips of 4 tons capacity being used. The smaller engine is used for the purpose of raising and lowering men, tools, timber, etc. The boiler and air-compressor plant, together with fitting-shop, timber framing-shed, blacksmith-shop, store, etc., are situated on the Black Bear claim, about 800 feet distant from the shaft. There are 11 boilers with a capacity of 2,000 nominal horsepower, and 2 large cross-compound air-compressors, are each capable of running forty 3¼ inches drills at sea-level (Fig. 11).

(3) The Grand Forks smelter derives its principal ore-supply from the Old Ironsides and Knob Hill mines, situated in what is known as the Boundary district, which includes some very valuable mines, and is one of the most important mining districts in British Columbia. The main lode in this camp runs through the Knob Hill, Old Ironsides and other claims. A large amount of development-work has been done in these mines, which up to April, 1902, consisted of about 2,689 feet of sinking and rising and 12,082 feet of cross-cutting and drifting, making a total of 14,771 feet; but this does not include work done in opening

FIG. 11.—STEAM-DRIVEN AIR-COMPRESSOR: BLACK BEAR CLAIM
OF THE LE ROI MINE.

out the great stopes, characteristic of these mines. The workings in the Old Ironsides mine are connected with those in the Knob Hill mine by drifts, and at the 300 feet level are 550 feet below the outcrop. The main lode has been opened on the surface for a distance of over 3,000 feet, by open-cuts and quarries, and the width has been demonstrated to be about 400 feet. The majority of the ore is now mined by open-quarry methods, and arrangements are being made for loading the railroad cars, that transport the ore to the main bins, by means of steam-shovels (Fig. 12).

The total shipments from these mines to the Grand Forks smelter was 231,762 tons in 1901. The capacity of the mine and smelter is being increased as rapidly as possible.

(4) The principal source of ore-supply to the Greenwood smelter is the Mother-lode mine, situated in the Deadwood camp. The main lode has been traced for over 1,000 feet on the surface and has been found up to 160 feet wide. The development-work done up to April, 1902, consists of about 1,450 feet of sinking and rising, and 5,350 feet of drifting and cross-cutting, totalling to 6,800 lineal feet. The main shaft is down 325 feet and at the 300 feet level a cross-cut has shown the width of the lode to be over 100 feet, while a drift has proved it for 350 feet. A large ore-quarry is being opened on the surface. The ore

FIG. 12.—SHAFT-HOUSE AND ORE-BINS AT THE OLD IRONSIDES MINE.

shipped in 1901 was 99,548 tons, and 45,000 tons during the year 1902.

(5) The Sunset is the principal mine proposed to be worked by the Standard Pyritic Smelting Company. The lode has been proved to be over 100 feet wide and has been followed to a depth of 400 feet. About 5,000 feet of development-work has been done.

(6) In the Cornell mine, owned by the Van Anda Company, there have been very satisfactory developments during 1902. The main shaft is being sunk to the 560 feet level and considerable exploration work is projected before stoping commences.

(7) The Crofton smelter proposes to treat principally ores from the Lenora mine in the Mount Sicker camp, Vancouver Island. The development, up to date, has proved a large body of high-grade ore, on this and many other properties in the camp (Fig. 13).

(8) The Tyee smelter will treat ore from the Tyee mine in the Mount Sicker camp. The lode has been proved to a depth of 340 feet, and is shown to be about 40 feet wide, while its con-

FIG. 13.—TRAMWAY FROM THE LENORA MINE TO THE ESQUIMALT AND NANAIMO RAILWAY.

tinuity has been proved for about 500 feet. A smelt of 200 tons returned 8 per cent. of copper, 5 dollars of gold and 5 ounces of silver per ton. The Mount Sicker camp is likely to prove, in the near future, a most important mining centre.

THE BOUNDARY DISTRICT.

The Boundary district, from its importance as a mining centre and its many interesting geological features, might well afford material for a separate article, but without increasing very largely the scope of the present paper, the writer cannot do more than notice superficially the most salient features. Mining and treatment-charges of the ores from this district have been reduced probably as low as anywhere in the world, and therefore the subject-matter of the following notes may reasonably be included in this paper.

The district may be roughly defined as that lying along the international boundary-line in the neighbourhood of the Kettle river and Boundary creek, British Columbia. The mountains, while not especially rugged, and often clear of timber, are covered with drift, and prospecting is therefore rendered difficult.

Eruptive rocks, such as granites, lavas, tuffs and greenstones are widely distributed, and sedimentary and highly altered meta-morphic rocks, including serpentine, are often met with, being generally much broken up and found as inclusions in the intruded rocks. Crystalline (mica and hornblende) schists and limestones are among the oldest rocks of the district. Overlying these are volcanic rocks, probably of Tertiary age, consisting of tuffs, ashes, basalts, andesites, etc.

The ore-bodies occur in all the older rocks, the Tertiary volcanic and younger granite and porphyritic dykes being generally barren. They may be classed as follows:—(1) Small quartz-veins carrying gold and silver; (2) large oxidized veins; and (3) large sulphide-veins.

The third class is the most important, and is the only one that will be described in this paper. In this class, there is practically no surface-oxidized zone, the unaltered ore being met with on or within a few feet of the surface. The most striking feature of these deposits is their size, which is generally very great. They have been formed by the circulation of mineralized thermal waters along fissures, contact- or shear-zones, and by the replacement of the country-rock met with and the contemporaneous deposition of the contained minerals. Pyrrho-tite, chalcopyrite and pyrite are the principal minerals.

The average value of the ore in the Granby company's mines may be taken as follows:—Copper 1·7 per cent., gold 1·60 dollars and silver 33 cents per ton. With copper at 11⅜ cents per pound, the total values would be 5·93 dollars per ton. A safe estimate for the available ore in the whole district would be 25 to 35 pounds of copper, 25 to 40 cents of silver and 1·50 to 2·50 dollars in gold per ton of 2,000 pounds.

pyrrhotite (Fe_7S_8 or $6FeS,FeS_2$) and generally sufficient silica to slag off the iron protoxide formed in the reaction :
$$FeS + O_3 = FeO + SO_2.$$

Economic Factors.

Working on ores where there is such a narrow margin of profit as those of the Boundary district, it is obvious that the prices of material used and of the main product are of very serious import. The principal factors that determine the profits of production, and may be said even to place limitations on production (since production will never be carried on for long without a reasonable profit), are, in the district referred to, the price of coke and copper. In the north-eastern district, the determining factors are the price of lead and silver.

Fig. 15.—Coke-ovens at Fernie.

The coke used in the Nelson, Rossland and Boundary districts, comes from the Crow's Nest coal-mines at Fernie, British Columbia, and passes over the Canadian Pacific railway and steamboats to the smelters. Under great pressure, the price of coke has been recently reduced from 4·50 to 4 dollars and coal from 2·50 to 2 dollars per ton. Considering that the prices which rule in Pennsylvania are :—Coal 80 cents and coke 1·75 dollars, under somewhat similar labour-conditions, it seems not unreasonable to assume that with competition, coal should be sold at 1·25 dollars and coke at 2·50 dollars per ton.

The total coal produced from the Crow's Nest collieries during 1901 was 379,355 tons of 2,240 pounds, an increase over the previous year of 172,552 tons. Of this coal, 180,768 tons were converted into coke, producing 111,683 tons; and of this coke, 32,121 tons were sold into the United States and 77,241 tons in Canada. The 300 ovens were kept busy during the year; 102 new ones were built at Fernie and 224 at Michael; and foundations have been laid for 200 ovens more.

The recent withdrawal of American purchasers of lead-ore for flux and the fall in the price of lead rendered it unprofitable

FIG. 16.—LOADING COKE-OVENS WITH AN ELECTRIC-TROLLEY AT FERNIE.

to operate a large number of British Columbian mines. The mines affected are low in silver-value. The rise in the silver-production is largely due to the opening up of new mines, the extension of operations in old ones having high-silver values, and to increased production from the copper-mines. In 1900, 90 per cent. of the silver-production was derived from silver-lead ore, but in 1901, owing to the situation of the lead-market, the production from these ores has declined, while 30 per cent. has been derived from the copper-silver ores. A noticeable feature in the year's operations is the rapid increase in the production of copper, as compared with that of silver and lead.

The Institution of Mining Engineers.   To illustrate Mr W. Denham Verschoyle's Paper on "The Smelters of British Columbia."   Vol. XXV., Plate XXV.

Transactions 1902-1903.

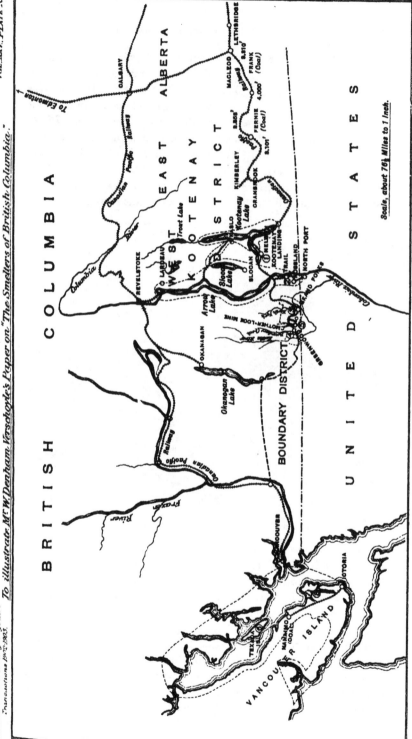

Scale, about 76½ Miles to 1 Inch.

To illustrate the effect of the prices of coke and copper on the low-grade ores of the Boundary district, the following figures may be given. In one case, the ore contained:—Copper 1·66 per cent. or 33·2 pounds per ton, worth 2·99 dollars with copper at 9 cents per pound; silver, worth 0·28 dollars; gold, worth 1·53 dollars; and a total value of 4·80 dollars per ton. Using 12 per cent. of coke at 4·50 dollars per ton, the cost of mining and treatment was:—Mining, 1·72 dollars; freights, etc., 1·10 dollars; smelting, 2·20 dollars, or a total cost of 5·02 dollars per ton. If coke had been procurable at 2·50 dollars per ton, the total cost would have been 4·78 dollars, and it would have been just possible to continue operations without actual loss, although no provision could have been made for depreciation and dividends. With coke at 2·50 dollars per ton and copper at 12 cents per pound (latest New York quotation), the operation of this property would result as follows:—Value of ore, 5·69 dollars per ton; less total cost of treatment, 4·73 dollars per ton, leaving a profit of 96 cents per ton.

SUMMARY.

During 1901, 920,416 tons were treated from the lode-mines of British Columbia. The values saved in gold, silver, lead and copper totalling to 13,683,044 dollars. Of this quantity, the smelters of British Columbia treated 540,904 tons. And part of the balance, to which the Le Roi mine contributed 158,598 tons, was smelted outside the province at the Northport and Tacoma smelters, and part was treated at various large and small quartz-mills within the province.

———

The PRESIDENT (Sir Lindsay Wood, Bart.) moved a vote of thanks to Mr. Verschoyle for his valuable paper.

Mr. M. WALTON BROWN seconded the resolution, which was cordially approved.

———

Mr. H. W. HALBAUM's paper on " The Common-sense Doctrine of Furnace-draught " was read as follows:—

THE COMMON-SENSE DOCTRINE OF FURNACE-DRAUGHT.

By H. W. HALBAUM.

Introduction.—The original purpose of the writer was merely to point out that certain factors of chimney-draught and their simultaneous fluctuations could be more easily correlated and measured by geometrical than by arithmetical methods. A look into various engineering and mining text-books, however, showed that some of the supposed certain factors were, in reality, very uncertain. The authorities were found to be at variance with each other in some important respects, and also far more given to stating their propositions as *ipse dixit* than to giving the why and wherefore of their pronouncements. The result of this observation was to convince the writer that definitions were necessary, as well as means of measurement. The present paper, therefore, covers rather more ground, and ground of a rather different nature, than was at first contemplated. The aim is to set forth the theory of the case to some extent, and show the why and wherefore of the matter. No doubt, practice will always diverge more or less from theory; it will never approach perfection so nearly as theory does; but, in this case, as in all others, it is obvious that the ideal standard of reference and comparison is the theoretic standard. The ground to be traversed in this paper is very debatable, and some of the positions occupied therein may not be absolutely impregnable. But the writer believes with all his heart that they are worth reconnoitring.

Motive Column and Volume.—Taking any given duct, the volume of the flow per unit of time is a constant multiple of the velocity. If, therefore, the volume be denoted by V, and the motive column, or the head producing velocity by h, we may write

$$\frac{V^2}{h} = c \qquad \qquad \qquad (A)$$

where c is a numerical constant. Thus, as the writer pointed out in a previous paper, the volumes V_1, V_2, V_3, etc., are ordinates at the abscissae, h_1, h_2, h_3, etc., in the parabola whose parameter or *latus rectum* is the duct-constant, c. Observe that this is not the relation of the mass and the pressure, nor yet that of the mass and the head, but it is the relation of the volume and the head; and the distinction is by no means of a purely academic kind.

In Peclet's well-known formula for the determination of the dimensions of a boiler-chimney, the given duct comprises the flues and chimney taken together. The formula assumes that the head, h, of the same density as the flowing air in the chimney, varies directly as the square of the velocity in the chimney. This assumption is scarcely warranted—the writer hopes to show the reason in a future paper—but it is an instance of a wellknown and much-quoted formula, and a formula much relied on in practice; and the writer will accept it provisionally for the nonce, as it will enable him to meet some possible objectors on their own ground.

Pressure and Mass.—It is evident, however, that we are more concerned with the weight, or mass, of air supplied to the furnaces than with the volume that it happens to occupy in the chimney. [For the present purpose, we may consider weight and mass as equivalent terms.] Now, let p be the pressure in pounds per square foot: let m be the mass of the whole volume, V, in pounds and let D be the density (or weight in pounds of 1 cubic foot of the volume), then by formula (A):—

$$h \; : \; V^2,$$
$$hD \; : \; V^2D,$$
but
$$hD \; = \; p,$$
and
$$VD \; = \; m,$$
hence
$$p \; : \; mV \quad \ldots \ldots \ldots \ldots \ldots \quad (B)$$

Thus, since the volume, V, is a constant multiple of the velocity, v, we may write:—

$$p \; : \; mv, \quad \ldots \ldots \ldots \ldots \ldots \quad (C)$$

Hence, the pressure varies as the mass multiplied by the velocity, which is a fundamental law of mechanics applicable to the motion of fluids and solids alike.

It will be noted that the pressure varies as the square of the velocity, only when the mass and the velocity vary together at equal rates. In any furnace-shaft, however, the mass and the velocity never vary together at equal rates. Until a certain point of temperature is attained, the velocity in the chimney always varies at a faster rate than the weight of air supplied to the furnaces, while beyond this point of temperature, the mass and the velocity actually vary in opposite directions. It is evident, therefore, that the popular axiom which states that the pressure varies as the square of the velocity is not true in the case of chimney-draught. Of course, the law (C) is well enough known to most engineers: but why should text-books not plainly enunciate the law which is always true, instead of constantly quoting an antiquated axiom which, at its best, is only approximately true within a very limited radius? The writers of text-books usually qualify the axiom in either one of two ways. With respect to mine-ventilation by furnace, they say that " neglecting shaft-friction," p varies as v^2. But, if we neglect shaft-friction in such cases, we shall often be neglecting that which absorbs the greater part of the total pressure. Then we find another kind of qualification employed by the authorities. They say that the pressure varies as the square of the velocity, provided " the densities remain constant." Surely no language could be more absurd. For the entire draught is nothing but a constant multiple of the difference of the densities, and every fluctuation in the amount of draught is due to a corresponding fluctuation in the difference of the densities. Hence if the " densities remain constant," the pressure and the velocity will also be constant, and any law of their mutual variation becomes, for the time being, an absolutely dead letter. It is therefore suggested by the writer that, instead of quoting the popular axiom which purports to connect pressure and velocity, text-books should generally adopt and invariably enunciate the infinitely more satisfactory rule which connects pressure, mass and velocity—or pressure, mass and volume, as the case may be—as set forth in equations (C) and (B) respectively. The pressure, then, in any duct, is in ratio to the momentum, mv; and the work or power is in ratio to the pressure multiplied by the distance through which it acts in the unit of time. In other words, the power varies as the kinetic energy, mv^2.

Relation of Densities and Temperatures.—In an enquiry of this kind, it is convenient to assume a standard barometer, and this procedure will presently be adopted. But a word may be said concerning the general case. In order to find the density (or weight in pounds of 1 cubic foot) of air, the mining text-books usually employ a formula in which 1·3253 is the coefficient of the barometer, and 459 the complement of the Fahrenheit thermometric scale in the absolute temperature. But it is now generally believed that the absolute zero is 460·66 say 461, degrees below the zero of Fahrenheit. If, however, we substitute this number in the formula in place of 459, it has the effect of putting the barometric co-efficient out of truth. The present writer looked into this case of the density some time ago, and recon-structed the formula.* Using 461 as the complement of the Fahrenheit reading, the coefficient of the barometer comes out at a little over 1·33, which is so nearly equal to 4/3 that this improper fraction may be used for all practical purposes. Now let D be the density of dry air (pounds per cubic foot); F, the temperature in degrees Fahr.; t, the absolute temperature; and B, the barometrical reading in inches of mercury; then:--

$$t = F + 461, \quad \ldots \ldots \ldots \ldots (D)$$
and
$$tD = 1·33B. \quad \ldots \ldots \ldots \ldots (E)$$

All the formulæ lettered from (A) to (E) are perfectly general.

For the purpose of this paper, we may assume a standard barometer of 30 inches, although the generally accepted standard is a tenth of an inch or so less. The density of the external air will then be affected only by the absolute temperature, t. This density will be called the normal density, D, and t is the normal temperature. The absolute temperature of the air within the chimney is greater than t; let it be T. But the density of this heated air is less than D, let it be d. Then under the standard barometer of 30 inches we have from the general formula (E):--

$$tD = Td = 39·9 = \text{say } 40 \quad \ldots \ldots \ldots (1)$$

[The numeral, 40, is much more convenient than the numeral that it displaces, and the difference is hardly appreciable, especially as in this paper we are more concerned with the differences of two densities, and the square roots of densities, than with any single density the whole value of which is taken by itself.] From equation (1), it will be evident that t and D are co-ordinates

* *Colliery Guardian*, 1897, vol. lxxiv., page 285.

in a rectangular hyperbola, and that T and d are co-ordinates in the same hyperbola, the condition of the said hyperbola being that the product of the co-ordinates to any point on the curve is the constant number, 40.

Relation of the Pressure, the Mass and the Densities.—Consider the case of a boiler-chimney, f feet high, and assume for the present one of two hypotheses. We assume, either that the flue-resistances are small enough to be neglected, or else we assume with Peclet that the head, h, consisting of air of the temperature T, varies directly as the square of the velocity of the chimney-gases whose temperature is also T. The normal volume of air drawn from the external atmosphere is V, and its mass is m. Within the chimney, the mass is still practically m, but its volume is now V_0, its density d, and its absolute temperature T. We have, then a certain ratio, call it R, and it has three values, namely:—

$$R = \frac{V_0}{V} = \frac{D}{d} = \frac{T}{t} \qquad . \qquad . \qquad . \qquad . \qquad (2.$$

But whatever volume or velocity we consider, it is evident that the total pressure is equal to the difference of the densities multiplied by f. But f is constant, hence:

$$p \; : \; (D - d) \qquad . \qquad . \qquad . \qquad . \qquad . \qquad (3,$$

But the general formula (B) affords another value for p, namely mV_0: for V_0 is a constant multiple of the velocity of the heated gases in the chimney. Hence we may write from (3) and (B):

$$mV_0 \; : \; (D - d).$$

Now multiply each side by the inferior density, d. Then, since $V_0 d$ equals m, we obtain for the value of the mass of air supporting the combustion:

$$m \; : \; \sqrt{(D - d)d} \qquad . \qquad . \qquad . \qquad . \qquad (4)$$

This ratio of m (a much more practical entity than any ratio of V_0) is very easily plotted by one of Euclid's methods, and the simultaneous fluctuations of the temperatures, densities, pressures and masses may all be shown in a single diagram.

The Diagram of Chimney-draught. Under the standard barometer of 30 inches, the absolute temperatures and the densities are, as already stated, co-ordinates to all points on the curve of a rectangular hyperbola whose constant is 40. But,

since the actual values of the co-ordinates differ so greatly in
magnitude, the ordinates of the temperatures must be plotted to
a far smaller scale than the ordinates of the densities. In the
diagram under consideration, therefore, the scale of the densities
is 10,000 times greater than the scale of the temperatures. For
since : —

$$tD = 40,$$

we may write also that : - -

$$\frac{t}{100} \times 100D = 40,$$

which gives the scales of the diagram. The actual absolute
temperature is obtained by moving the decimal point of the

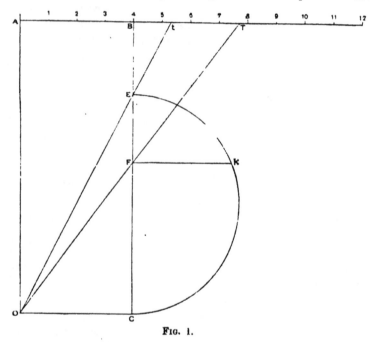

Fig. 1.

diagrammatic temperature two places to the right. Thus, an
absolute temperature which appears on the diagram as 5·3 is in
reality 530. Similarly, the actual density is obtained by moving
the decimal point of the diagrammatic density two places to the
left. Thus, a density which appears on the diagram to be 7·7 is
in reality 0·077, and so on.

Then, in any diagram of the type illustrated by Fig. 1,

$AOCB$ may be any rectangle, the area of which is 40. In Fig. 1, AB equals 4, and BC equals 10; and their product is 40. If any line such as OT be drawn, the horizontal length AT and the vertical height CF will fix the position of one point on the hyperbolic curve whose asymptotes are OA and OC produced. The construction is no doubt familiar to all engineers having had to do with indicator-diagrams, and it is therefore unnecessary to draw the curve itself.

In the application of the rectangular hyperbola to the present study, the line AB produced is the line of absolute temperatures, and a line drawn from O to any temperature, t (the geometrical value of which is At), possessed by the external air will intersect the line, CB, at a point E. Now CB is the line of densities, and just as At is the geometrical measure of the algebraic value t, even so is CE the geometrical measure of the algebraic value D.

And similarly, if AT be the geometrical measure of the chimney-temperature, the algebraic value of which is T, then CF is the geometrical measure of the inferior density, the algebraic symbol of which is d. For OT intersects the line of densities at F.

Again, since CE equals the normal density, D, and since CF equals the reduced density, d, it is clear that $CE - CF$ is the geometrical equivalent of the difference of the densities. But $CE - CF$ equals FE. Hence FE is the geometrical measure of the algebraic value p, the ratio of which by equation (3) is $(D - d)$.

Finally, CE, the superior density, is the diameter of the semicircle, CKE; and by Euclid (Book II., Proposition 14) we have FK equals $\sqrt{FE \times CF}$. But the algebraic value of FE is $(D - d)$, and that of CF is d. Hence FK is the geometrical measure of the algebraic value denoted by $\sqrt{(D - d)d}$. But by equation (4), this is the ratio of the mass of air, the symbol of which is m. So that FK is the geometrical ratio of the mass or weight of air supplied to the furnaces.

Thus, all the factors of the mass are measured as straight lines in the hyperbolic diagram, whose constant is equal to 40; and the mass itself is measured by a straight line erected on the diameter of a semicircle at a point fixed by the hyperbolic factors of temperature and density. To sum up the features of the diagram in a compact form, we write its properties thus:—

	Algebraic Symbol.		Geometrical Ratio.
When	t	:	At,
and	T	:	AT,
then	D	:	CE,
and	d	:	CF,
Also	p	:	FE,
and	m	:	FK,

$$ \dots \dots \dots \quad (5) $$

Thus, by merely drawing two straight lines from O to the absolute temperatures, t and T, the whole of the remaining values, D, d, p and m, straightway plot their ratios automatically, so to speak. It is evident that the diagram also furnishes a ready method of working back from a given mass to the temperature, T, that will produce it.

The Maximum of the Mass.—In order to do a given amount of work, it is necessary to liberate a given amount of energy. To obtain this energy requires the combustion of a given amount of coal, and the combustion of this coal requires a given weight of air. Hence any increase in the volume, V_0, can only be valuable in so far as increment of volume is associated with increment of mass. Sometimes, indeed, increment of volume and mass are alike undesirable, but that aspect of the case need not be considered just now. It may at present suffice to point out that while the volume, V_0, may be expanded indefinitely, and that while the pressure, p, may be theoretically augmented until $(D-d)$ equals D, the mass, m, is limited by a very definite maximum.

Referring again to Fig. 1, the maximum of the mass is obviously attained when FK becomes equal to the radius of the semicircle, CKE. Now, whatever be the condition of the external atmosphere—standard barometer or otherwise—the diameter of CKE is the superior density, D. The ordinate, FK, is removed from one extremity of that diameter by the inferior density, d, and from the other extremity by the difference, $(D-d)$, of the densities. But when FK is equal to the radius, the inferior density is also equal to the radius, and the difference of the densities is again equal to the radius. At this juncture, therefore, d equals $(D-d)$. So that the mass, m, the geometrical ratio of which is FK, is a maximum when D equals $2d$. Now, by equation (2), when D equals $2d$, we have T equals $2t$.

AOCB may be any rectangle, the area of which
AB equals 4, and *BC* equals 10; and thei...
any line such as *OT* be drawn, the horiz
the vertical height *CF* will fix the posi'
hyperbolic curve whose asymptotes a·
The construction is no doubt fami'
had to do with indicator-diagrams,
to draw the curve itself.

In the application of the re·
study, the line *AB* produced
and a line drawn from *O* t·
value of which is *At*), po·
the line, *CB*, at a point
just as *At* is the geo·
even so is *CE* the g·

And similarly
chimney-temper·
is the geometr·
symbol of w'
at *F*.

Agai·
CF e·
the '
Bu
t'

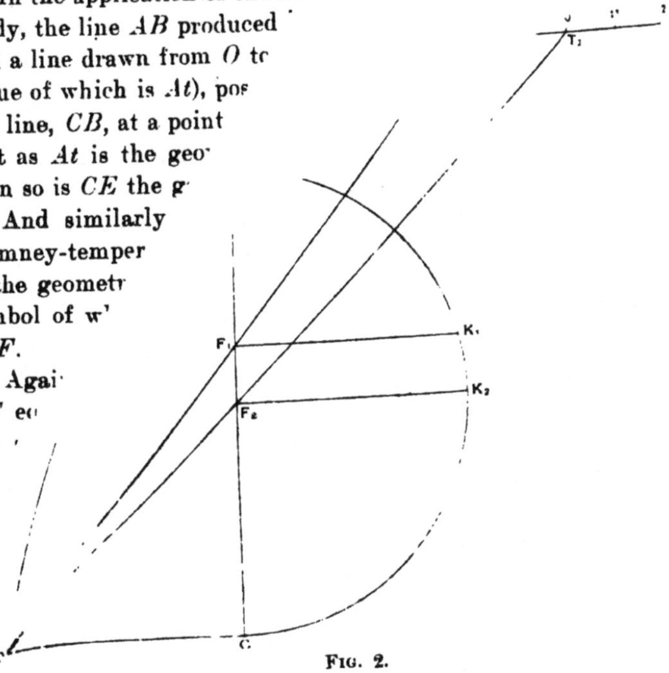

FIG. 2.

development of energy. Maxima of this class are always impos-
ing, but they are frequently apt to cost more than they are
worth. Is it not possible that the maximum mass is one of this
class? It is certain that the maximum is much more costly
than masses which are only slightly less.

Fig. 2 has been drawn to illustrate this point. The numerical
values have been chosen only because they are whole numbers, the
significance of which is manifest at sight. Two cases are illustrated in the diagram. In the first case the external temperature

'), the normal density CE equals 8 (hundredths).

ature is T_1, equals 800, absolute of course,

is F_1E, equals 3, the ratio of the mass

best chimney-temperature is $2t$.

since t equals 500, we make

rises to F_2E equals 4, which

nt.; and the ratio of the

equal to $\sqrt{16}$, or about

question naturally arises

ent. of mass is really worth the

33 per cent. of pressure.

cost of this extra 33 per cent. of pressure?

, a mass of air, the ratio of which is $\sqrt{15}$ is raised

range of temperature amounting to 300 degrees. In

second case, which is the case of the maximum mass, a weight of air, the ratio of which is $\sqrt{16}$ is raised through a range of 500 degrees of temperature. [The specific heat, being the same for both cases, may be left out of account.] The comparative cost in the first case, therefore, is represented by $\sqrt{15} \times 300$ equals 1,162, and the comparative cost in the second case is as $\sqrt{16} \times 500$ equals 2,000. Now 2,000 is 72 per cent. greater than 1,162, so that the cost of the maximum mass is 72 per cent. greater than the cost of a mass which is only 3 per cent. less than the maximum itself. Is the maximum mass, then, worth so much? The text-books affirm that it is, for they lay it down as an axiom that the best ratio of the temperatures is that which secures the maximum mass of air and the maximum consumption of coal. It may be of interest to examine this case a little more fully.

The Maxima of Energies.—The writer would respectfully suggest that the maximum mass may not, after all, be the best mass in any commercial sense. It has already been pointed out that increment of volume can only be valuable in so far as it coincides with increment of mass. But we may now go a step further, and say that increment of mass can only be valuable in so far as increment of mass is associated with increment of useful energy. The object of the chimney is not the passing of volume, nor the circulation of mass, nor even the consumption of coal, except in so far as these are means to a certain end. That end is the raising of steam, say; and if so, it is clear that the best ratio of temperatures is not necessarily that which will circulate the greatest mass of air, or that which will ensure the greatest consumption of coal—the best ratio is that which will develop the largest amount of energy available for raising steam.

First, then, let us agree on an expression to represent the total energy developed by the combustion. In order to effect the combustion, we must supply a sufficient amount of air. Let that amount of air be a pounds for each pound of coal. Then, the total weight of the air-supply will be a times as great as the total weight of coal consumed. Hence, since the weight of air is m, the weight of the coal consumed in the unit of time will be m/a pounds. And if the combustion of each pound of coal develop, on the average, u units of heat, the total energy developed will be E units of heat: and

$$E = u \frac{m}{a} \quad . \quad . \quad . \quad . \quad . \quad . \quad . \quad . \quad . \quad (6)$$

And since u and a are presumably constant values in a given case, we have the rule that E varies as m. Hence, the maxima of E and of m occur simultaneously, both occurring when R equals 2.

The total energy, E, is expended in various ways, but in this paper, it will be divided into two portions only. One of these portions is expended between the fire-doors and the chimney-base, and the remainder is expended beyond the base of the chimney. This latter portion will be considered as being expended on draught alone, and it will be symbolized by e_0. Then :--

$$e_0 = ms\,(T - t)\ . \quad . \quad . \quad . \quad . \quad . \quad . \quad . \quad (7)$$

where s is the specific heat of air at constant pressure. The quantity, e_0, may be called the lost energy, and it is pretty certain that the energy expended between the base and the top of the chimney is always greater than its theoretical value as defined by equation (7). For the absolute temperature at the chimney-base is certainly somewhat greater than T, which is evidently the mean absolute chimney-temperature.

The other portion of the total energy—that expended between the fire-doors and the chimney-base, may be called the effectual energy; for, although the whole of it is not effective in the best sense of the word, it represents a ratio of the real efficiency. Its value may be denoted by the letter ϵ: then the effectual energy is:—

$$e = E - e_o.$$

Putting in the values of E and e_0, as per equations (6) and (7), we obtain:—

$$e = m \left[\frac{u}{a} - s\,(T-t) \right] \quad . \quad . \quad . \quad . \quad . \quad . \quad (8)$$

The form of this equation suggests that e is limited by a maximum, which must of necessity occur at a lower value of R than that associated with the maxima of m and E. And, if so, it is of importance to enquire at what value of T that maximum is attained.

In any enquiry of this nature, one is obliged to depend almost solely upon theoretical values, but it is highly probable that any argument based on such values will lose just as much force in some directions as it gains in other directions. Hence, it is probable that such an argument will, on the whole, be practically sound. Equation (8) may be reduced to a more convenient form for use in this connection. First divide each side by st, and substitute for T/t its value R. Then substitute for m its ratio, which is $\sqrt{(D-d)\,d}$. Then substitute for d its value D/R, as denoted by equation (2). Next divide both sides by D. We then obtain:—

$$\frac{e}{stD} \quad : \quad \frac{\sqrt{R}-1}{R} \left[\frac{u}{ast} - (R-1) \right];$$

but s is a numerical constant, and tD equals 40 (under the standard barometer). We may therefore cast out these constants from the lefthand member, and write:—

$$e \quad : \quad \frac{\sqrt{R}-1}{R} \left[\frac{u}{ast} - (R-1) \right] \quad . \quad . \quad . \quad . \quad (9)$$

An average case may now be examined. The following are probably fairly average values for the annexed symbols:—

$u = 14{,}000$ (units of heat per pound of coal),

$a = 24$ (pounds of air per pound of coal),

$s = 0{\cdot}2379$, and

$t = 510$ (degrees absolute).

This value of t corresponds to 49° Fahr., which is approximately equal to the mean annual temperature of Great Britain. Substituting these numerical values in equation (9), we obtain:— -

$$e \; : \; \frac{\sqrt{R}-1}{R} \left[4\cdot808 - (R-1) \right] \quad . \quad . \quad . \quad 10)$$

The quantity outside the brackets is in ratio to the total mass, m; the numeral is in ratio to the total energy per unit of mass, and the quantity between the inner brackets is in ratio to the energy lost in the draught per unit of mass, m. Table I. has been prepared by applying equation (10) to the various values of R indicated in the first column.

TABLE I.

Absolute Temperatures.			Chimney-temperature in ordinary reading Fahr.	Ratio of Total Energy developed or Ratio of Total Coal-consumption: or Ratio of R.	Ratio of the Effectual Energy: or Ratio of e.	Loss of Energy in the Draught.	
R	t	T				Ratio of e.	Percentage of the Total
1·009	510	510	49	nil.	nil.	nil.	nil.
1·100	510	561	100	1·3822	1·3535	0·0287	2·1
1·200	510	612	151	1·7918	1·7173	0·0745	4·2
1·300	510	663	202	2·0257	1·8993	0·1264	6·2
1·400	510	714	253	2·1720	1·9913	0·1807	8·3
1·500	510	765	304	2·2665	2·0308	0·2357	10·4
1·569	510	800	339	2·3116	2·0379	0·2737	11·9
1·600	510	816	355	2·3276	2·0371	0·2905	12·5
1·700	510	867	406	2·3662	2·0217	0·3145	14·6
1·800	510	918	457	2·3891	1·9916	0·3975	16·6
1·900	510	969	508	2·4006	1·9512	0·4494	18·7
2·000	510	1,020	559	2·4039	1·9039	0·5000	20·8
2·080	510	1.061	600	2·4021	1·8623	0·5398	22·5

Now whether e be an exact ratio of the energy employed in raising steam or not, it is the ratio of the energy expended in the vicinity of the boilers, provided that the absolute temperature of the gases at the chimney-base be equal to, without being greater than T. And it is evident that if that temperature be greater than T, the energy expended within the flues will be less than its theoretical value, e, while the energy actually lost in the chimney will be greater than its theoretical value, e_0. In such a case, it will be granted that the actual maximum of e must occur at a lower value of R than that associated with the theoretical maximum. Hence, it is reasonable to suppose that the argument against the text-books as furnished by the table is understated if anything. But, even when we accept these theoretical values, it is seen by Table I. that the maximum of e occurs at a value of R, which is approximately equal to 1·569. At this point, T equals 800, and F_0 equals 339. (F_0 is the ordinary Fahr. reading

of the chimney-temperature.)   Comparing the cases of the
maximum of e when R equals 1·569, and the maximum of E
when R equals 2, it is seen that in the latter case we burn more
coal, and obtain less effective power in the boiler-flues.   A still
more striking contrast is furnished by the cases of R equals 2·08
and R equals 1·3.   In the case where R equals 2·08, the
temperature F_0 equals 600° Fahr., and many of the text-books
assert that 600° Fahr. is the best chimney-temperature.   But
if we adopt this temperature, it is seen from the table that
we actually obtain a less value of e (although we burn 18 per
cent. more coal) than if we had adopted a temperature some
400 degrees lower.   The reason is easily found from the table.
By making F_0 equal 600 instead of 202 degrees, the total energy
is increased 18 per cent., but the energy expended on the draught
is increased by 427 per cent.   Or taking the figures from the
fifth column, the total energy is increased by a positive amount,
the ratio of which is (2·4021 − 2·0257 equals) 0·3764; while, taking
the figures from the seventh column, the lost energy is increased by
a positive amount, the ratio of which is (0·5398 − 0·1264 equals)
0·4134.   The net loss is therefore (0·4134 − 0·3764 equals) 0·0370;
and the effectual energy in the second case is less by that amount.

In the foregoing table, the value of t is based on the mean
annual temperature.   But if we take values of t corresponding
to the summer-temperature or the winter-temperature, the result,
so far as the maximum of e is concerned, is pretty much the
same.   In any of these three cases, summer, winter, or mean,
the maximum of e is attained at some value of R which lies in
the neighbourhood of 1·57.

For other values of u, a and t than those here assumed, the
maxima of e may be ascertained with sufficient accuracy by
means of an ordinary slide-rule, for which the following formula
is intended. Let any desired numerical values be taken, and put : —

$$x = 1 + \frac{u}{u\,s\,t}.$$

Then x is a whole number, and E, e and e_0 being the ratios
respectively of the total energy, the energy expended within the
flues, and the energy expended on the draught, we may write
for slide-rule calculations : —

$$\frac{R}{\sqrt{R}-1} \; : \; \frac{R-1}{e_0} \; : \; \frac{x-R}{e} \; : \; \frac{x-1}{E} \quad . \quad (11)$$

In Table I. the value of x is taken at 5·808. If x be greater than this value, the maximum of e will be greater; and it will be associated with a somewhat higher value of R. But in no practical set of circumstances can the maximum of e occur at a value of R so high as 2. This value of R is associated with the maxima of the mass, m, and the total energy, E. And these maxima occur when R equals 2, irrespective of the value of x. But by the very nature of the case, the loss, e_0, of energy caused by the draught must always increase at a much more rapid rate than E increases, so that the maximum of the effectual energy, e, is bound, in all cases, to occur sooner than that of E.

Let us suppose, however, that by " some happy combination of fortuitous circumstances," the numerical value of x is raised to 10, although it is difficult to imagine what those circumstances should be. Then a little trifling with the slide-rule will show that even in that impossible case, the maximum of e would occur when R equals 1·7, or thereabouts. The setting up of any chimney-temperature greater than T equals 1·7 t would therefore, even in that case, only result in reducing the effective power.

We may conclude, then, that in all circumstances of an ordinary kind, the maximum of e occurs between R equals 1·5 and R equals 1·6; and the temperatures associated with these values of R are therefore the best to adopt—if by the best we mean that temperature which will give the maximum effective energy within the flues. [It will, of course, be evident from the table that a still lower temperature is more economical, provided that a sufficient combustion is also forthcoming.] When R is about 1·5 or 1·6 the effective power is a maximum.

A possible objection may arise here, although it hardly seems necessary to discuss it. It may be said that when R equals 1·57 or so, the temperature of the gases escaping from the flues will only be about 340° Fahr., which is the temperature of steam at 100 pounds pressure per square inch. Hence, wherever steam of a moderately high pressure is required, the temperature of the gases escaping into the chimney cannot be reduced to 340° Fahr. or anything like it. No doubt this is true enough, but the answer is obvious. Let the temperature of the escaping gases be anything that may be preferred, but let the excess of temperature be absorbed by a feed-heater or economizer placed at or near the base of the chimney. In fact, a very pretty argument, based on the science of radiation, could be furnished to show that, apart from the subject of draught altogether, the feed-heater or economizer is a very essential detail of any boiler-plant.

The writer has endeavoured to state this case as moderately as possible, and will be glad if those who perceive any flaw in the argument—and there are probably many of them—will kindly draw attention to the same. There are one or two other matters upon which he would have desired to touch, but he hopes to return to the subject at an early date. It will perhaps be better to close with a brief recapitulation.

Summary.—It is argued in this paper that:—

1. The pressure varies, not as the square of the velocity, but as the mass multiplied by the velocity (in a given duct, of course).

2. The pressure varies as the difference of the densities. If the densities be constant, the pressure, the volume, the mass and the velocity are constant also.

3. The mass of air circulated by the draught varies as the square root of the product of the inferior density, d, and the difference $(D-d)$ of the densities.

4. All the factors of the mass are hyperbolic quantities.

5. The maximum of the mass occurs when the ratio of the densities, which is also the ratio of the absolute temperatures, is 2.

6. The maximum total energy, or the maximum consumption of coal, due to the draught of any given chimney, also occurs when R equals 2.

7. The energy expended on the draught alone, and thus expended outside the flues, varies as the product of the total energy and the difference of the temperatures.

8. The maximum of energy available for expenditure within the flues and in proximity to the heating-surface of the boilers is, in the average case, obtained, not when R equals 2, nor when F_o equals 600°, but when R equals 1·57, or thereabouts. And no possible combination of circumstances would enable such maximum to occur at a value of R even so high as 1·7.

9. Where steam of any pressure higher than the very lowest

is required to be raised, economical draught can only be secured by making feed-heaters or economizers an essential detail of the plant.

———

The PRESIDENT (Sir Lindsay Wood, Bart.) moved a vote of thanks to Mr. Halbaum for his valuable paper.

Mr. M. WALTON BROWN seconded the resolution, which was cordially approved.

NOTES ON THE OCCURRENCE OF NATURAL GAS AT HEATHFIELD, SUSSEX.*

By H. B. WOODWARD, F.R.S.

In a paper read before the Geological Society of London in 1898, Mr. Charles Dawson directed attention to the discovery of natural gas in East Sussex.†

Referring first to the reported issue of inflammable gas in the deep boring made at Netherfield near Mountfield for the Sub-Wealden exploration, and which was (1875) attributed to the petroleum-bearing Kimeridge Clay, Mr. Dawson then gave particulars of the discovery of natural gas at Heathfield.

In 1895, when a boring was made in search of water at Heathfield Hotel, the foreman of the work found that inflammable gas was being given off at a depth of 228 feet. No special interest appears at the time to have been taken in this discovery, and as no supply of water was obtained the boring was abandoned.

In the following year, another boring for water was made by the London, Brighton and South Coast Railway Company, by the side of the railway between Heathfield Station and the tunnel to the north-west. This boring was abandoned at the depth of 377 feet, as "no useful amount of water" was obtained. At the depth of 312 feet, there was, however, a rush of gas so pronounced that it was tested, and a flame of about 16 feet in height was the result. A strong odour of gas had been noticed at lesser depths, and below 312 feet it "continued to increase during the remainder of the depth bored." No immediate steps were taken to utilize the gas, indeed the wrought-iron tubes were withdrawn from the bore-hole, with the exception of about one length which was left in the ground.

Experiments made in 1898 having proved that the quality of the gas was suitable, and that the pressure and supply were

* Reproduced from the *Memoirs of the Geological Survey: Summary of Progress of the Geological Survey of the United Kingdom for 1902,* pages 195-199, by permission of the Controller of H.M. Stationery Office.

† "The Discovery of Natural Gas in East Sussex," *The Quarterly Journal of the Geological Society of London,* 1898, vol. liv., page 564 ; and *Proceedings of the Geologists' Association,* 1901, vol. xvii., page 171.

maintained, the gas was in 1899 utilized in lighting up the railway-station, and subsequently for other purposes.

The possible importance of the gas-field did not, however, arouse public interest until the present year when the company entitled " The Natural Gas-fields of England, Limited," commenced a series of borings at Heathfield and at Mayfield. As a result, the village of Heathfield has been lighted from one of the bore-holes, and pipes have been laid down for the supply of Polegate and Eastbourne.

Five borings have now been made, or are in progress, between Heathfield Station and Britannia Mill, and a sixth is in progress by the eastern side of Waldron Gill, rather less than ½ mile south of the station. Through the courtesy of Mr. Inverness Watts, engineer to the company, the writer was enabled to visit the sites. Gas had been found in all but one (No. 5) of the bore-holes.[*] Unfortunately, the borings are made by means of a jumper, and the material is brought up as sludge, mostly a silty clay, so that details of the strata cannot be ascertained and recorded with anything like the precision that was possible with the cores from the two earlier borings, made by Messrs. Le Grand and Sutcliff. On those borings we must rely chiefly for our knowledge of the strata.

Summarized, the particulars are as follows[†] : —

I.—HEATHFIELD HOTEL, 493 FEET ABOVE ORDNANCE DATUM-LEVEL.

No.	Description of Strata.		Thickness of Strata. Feet.	Depth from Surface. Feet.
	Ashdown Sand—			
1	Sand, with *lignite* and marl (dug well)	...	21	21
2	Sandstone and marl	... ...	11	32
3	Sandstone	... ...	18	50
4	Sandstone and marl	... ...	21	71
5	Sandstone	... ...	5	76
6	Sandstone and marl	... ...	18	94
7	Marl	... ...	57	151
8	Sand-rock	... ...	4½	155½
9	Marl, with stone-band	... ...	60½	216
10	Sand-rock	.. ..	3	219
11	Marl	... .	30	249*

*Gas noticed at 228 feet.

* [Subsequently, gas was struck in the No. 5 bore-hole, the pressure being as high as at the other bore-holes.]

† " The Discovery of Natural Gas in East Sussex," by Mr. Charles Dawson, *The Quarterly Journal of the Geological Society of London,* 1898, vol. liv., pages 569-571 ; and *Memoirs of the Geological Survey of England and Wales: The Water-supply of Sussex from Underground Sources,* by Messrs W. Whitaker and Clement Reid, 1899, pages 87-89. Mr. W. Whitaker has grouped all the beds at Heathfield Station, down to 339 feet, with the Fairlight Clays.

II.—HEATHFIELD STATION, 450 FEET ABOVE ORDNANCE DATUM-LEVEL.

No.	Description of Strata.	Thickness of Strata. Feet.	Depth from Surface. Feet.
	Fairlight Clays and Ashdown Sand—		
1	Dug well	73	73
2	Sandy marl-rock, *lignite* and ironstone ..	67	140
3	Hard sandstone	1¾	141¾
4	Sandy marl-rock and ironstone...	19¾	161½
5	Sandstone	1¼	162¾
6	Marl-rock, sandstone and ironstone	9¼	172
7	Sandy marl-rock and ironstone-bands ...	24¼	196¼
8	Marl and sandstone	22¾	219
9	Sandy marl-rock and sandstone	45½	264½
	Purbeck Beds—		
10	Marl, with bands of marble	2½	267
11	Marl, shale and marl-rock (*Paludina*) ...	72	339*
12	Marl-rock, bituminous shale, shells ...	14	353
13	Marl, shale and shell-rock	24	377

*Gas first lighted at 312 feet.

Mr. I. Watts informed the writer that, in the new borings, gas is found to issue at about 280 feet, sometimes when hard ironstone-bands are pierced. The pressure at Heathfield Station was stated to be equal to 200 pounds to the square inch, and this had not been modified by the subsequent borings. The quality is so good that no refining is necessary.

Heathfield Station is in Waldron Parish, and the newer part of the village of Heathfield extends to the north-east and south-east of the station, at the head of a fine valley which opens out to the south-west and is drained by Waldron Gill. The ground rises on the north to 600 feet near Mutton Hall, and to the south to 598 feet at Gibraltar Tower in Heathfield Park. These higher and well-wooded grounds are formed mainly of sandstones which dip northwards (as shown on the Geological Survey map) on the north, while in Heathfield Park the beds are inclined to the south-east. Thus the anticline, which brings in the Purbeck Beds farther east along the course of the river Dudwell, probably extends in a north-westerly direction through Tilsmore Wood; and Heathfield Station would be on the southern side of the anticline, with perhaps an inlier of Lower Wealden (Fairlight) clays and sands in the Waldron-gill valley to the south-east.

The district may be affected by more faults than are shown on the Geological Survey map, but if the dip, as above stated, is maintained, it is probable that all the six borings have been commenced in approximately the same strata—perhaps No. 5 in slightly higher and No. 6 in slightly lower strata than those of the other four borings.

The strata exposed in the banks of Waldron Gill, below the site of No. 6 boring, comprise thinly-bedded sandstones with harder bands of ferruginous sandstone, and these are bent in one spot into a sharp fold. Lignite has been dug in the bed of this brook.[*]

In general the beds here comprise sandstone, fine sand, lignitic sands, white sandy clay and purplish clay. The brickyard, south of Heathfield Station, exposed whitish clay and sands in thin alternating beds.

In the two quarries in the north-west of Heathfield Park, 12 to 15 feet of thin fissile sandstones, with bands of pale purplish and whitish clay, overlie thickly-bedded sandstone much jointed and also false-bedded. The soil made up of the alternating fine sands and clays is impermeable in places, so that water stands here and there in pools in wet weather, and the roads are wet and silty. The thickly-bedded sandstones above noted would be well adapted to store gas under favourable circumstances. They are, however, above the horizon of the strata penetrated in the Heathfield borings.

Beds on the horizon of those bored into at Heathfield are exposed on the borders of the deep valley east of Heathfield Park, where also the Purbeck Beds come to the surface, at a distance of rather less than 2 miles from Heathfield Station. North of Punnett's Town, to the east of Cade Street, ferruginous sandstone and loam and fine-grained sands and sandstones are observable on the way from Lucks' Farm to Little Greenwood.

Farther on to the east of Greenwood Farm, purple and mottled clays are exposed, while south of the Ford on the river Dudwell, on the trackway from Three Cups Corner to Pounceford or Poundsford, there may be noted in descending order:—

> Wealden : Fairlight Clays—
>> Yellow sandstones and clayey beds.
>> Grey and purplish clay.
> Purbeck—
>> Leathery shales.

---

[*] *Memoirs of the Geological Survey of England and Wales: The Geology of the Weald (Parts of the Counties of Kent, Surrey, Sussex and Hants)*, by Mr. William Topley, 1875, pages 347 and 348.

On the north side of the Ford, Purbeck Beds were shown having a high northerly dip, but there is little doubt that the beds are disturbed by landslipping. Shelly limestone and shale and compact limestone occur here, and higher up leathery shales like those seen south of the river. The intermediate ground has cappings of grey and ferruginous clay with blocks of Wealden sandstone, suggestive of a southerly dip, and of the crest of the anticline being a little north of the Dudwell, an inference supported by dips marked on the Geological Survey map.

On the hill above Poundsford, the beds have a northerly dip, and the section is like that of the quarries in Heathfield Park. Fine white sand, sandy clay and ferruginous sandstone, in thin layers, overlie thick beds (3 or 4 feet) of yellow sandstone, much jointed. These beds appear to be slightly faulted, but are more probably displaced by landslip on the brow of the hill. The bold features, north of Poundsford, are formed by the strong beds of sandstone, and had there not been considerable faulting and downthrow of Wadhurst Clay and Tunbridge-Wells Sands, the Purbeck Beds might again have appeared farther north in the valley of the Rother. As it is, Mayfield, where another boring is being made, is on Tunbridge-Wells Sand and Sandstone. These are much more evenly bedded than the Ashdown Sands of Heathfield, but they also have sandy clay-partings.

The leathery shales which occur in the upper part of the Purbeck Beds at Poundsford are distinctly bituminous. Specimens have been tested by Dr. W. Pollard.

Attention however was directed to this fact by the late Mr. W. Topley, who stated in 1875 that Mr. R. Hallett, then owner of Swife's Farm, near Burwash "had several tons operated upon, from which was extracted in the first instance tar, and by subsequent processes a variety of products such as pitch, grease, oil, naphtha, and paraffin."[*]

Mr. W. Topley also called the writer's attention to the fact that the gypsum at the works at Netherfield becomes bituminous along a line of fault.[†]

---

[*] *Memoirs of the Geological Survey of England and Wales : The Geology of the Weald* (*Parts of the Counties of Kent, Surrey, Sussex and Hants*), by Mr. William Topley, 1875, page 38.

[†] *Memoirs of the Geological Survey of the United Kingdom: The Jurassic Rocks of Britain.*—Vol. v., *The Middle and Upper Oolitic Rocks of England* (*Yorkshire excepted*), by Mr. Horace B. Woodward, 1895, page 331.

Again, in the Kimeridge Clay at the sub-Wealden boring, a "hard light-coloured bed, very rich in petroleum" was passed through at depths of 600 to 602 feet,[*] and bituminous shales again occurred from 960 to 1,000 feet.

III.—GENERAL SECTION OF THE STRATA IN THE HEATHFIELD DISTRICT.

	Feet.	Feet.
Ashdown Sand, with Fairlight Clays—		
Sands, sandstones, loamy sands, shales and clays—on the whole more clayey towards the base (Fairlight Clays) 350 feet to	400	
		400
Purbeck Beds—		
Shales, etc.	30	
Limestones, calcareous sandstones and shales ...	50	
Shales and clays, with occasional stone-beds ...	130	
Limestones and shales	75	
Shales, with occasional stone-beds	35	
Shales, with gypsum	80	
		400
Portland Beds—		
Sandy shale	23	
Soft whitish sandstone	52	
Darker sandstone	5	
Sandy shale	35	
		115
Kimeridge Clay—		
Shales, with septaria, etc....	668	
Bituminous shales	40	
Clays and sandy beds	4	
Sandstone	41	
Sandy shale, etc.	47	
Shaly sandstone	45	
Limestone ... ... ` .	45	
Shale	94	
Limestone	29	
Shale	20	
Grit and limestone	45	
Limestone and shale	27	
Shale	23	
Limestone	10	
Shale	135	
		1,273

The information which is before us indicates that a certain amount of natural gas is pent up in the Lower Wealden Beds and in the Purbeck Beds. The bore-holes, however, have not penetrated any thick masses of porous strata such as we might

[*] *Memoirs of the Geological Survey of England and Wales: The Geology of the Weald (Parts of the Counties of Kent, Surrey, Sussex and Hants),* by Mr. William Topley, 1875, page 43.

judge capable of storing a vast accumulation of gas.   What has been obtained must have been stored in comparatively thin layers of sandstone and sand, interlaminated with the Wealden clays, and in the harder bands of the Purbeck.

It appears, therefore, that gas is derived from Purbeck shales and that it may be stored over a considerable area in the thin porous beds. The Portland Beds identified in the sub-Wealden Boring have been somewhat differently described in the several records which have been published. The record given by Mr. W. Topley* is noted in the accompanying general table.

Here, below the Purbeck Beds, there is evidence of important beds of sand and sandstone, 50 to 60 feet thick, and capable of storing a great quantity of gas.

At present, these strata have not been penetrated in any of the bore-holes put down in search of natural gas, and it would seem necessary to bore from 300 to 320 feet deeper, at Heathfield, to reach this horizon.

It is possible, however, that gas from these Portland Beds may have been indirectly tapped through crevices and fault-planes, but on this point nothing definite can be said.

The absence of water from the strata may be due to the presence of the gas, and water may replace it as it escapes.

---

* The Geology and Fossils of the Tertiary and Cretaceous Formations of Sussex, by Mr. Frederick Dixon, second edition, 1878, pages xxiii., 6 and 154 ; Memoirs of the Geological Survey of the United Kingdom : The Jurassic Rocks of Britain.—Vol. v., The Middle and Upper Oolitic Rocks of England (Yorkshire excepted), by Mr. Horace B. Woodward, 1895, page 229 ; and Memoirs of the Geological Survey of England and Wales : The Water-supply of Sussex from Underground Sources, by Messrs. W. Whitaker and Clement Reid, 1899, pages 65.70. The boring is at Mountfield.

UNDERGROUND FIRES.*

By F. W. HARDWICK.

Introduction.—An underground fire is one of the most serious disasters which may happen to endanger the working of a mine. Its importance will depend to a certain extent on its position relatively to the workings, but in any case its existence must cause grave anxiety to the management. If steps are not taken at once to isolate or to extinguish it, it will, especially in a colliery, spread with rapidity. On the other hand, the difficulty, in most cases, of access, rendered greater by the poisonous and, in some instances, explosive gases generated by the fire, coupled with the intense heat, makes it awkward for the necessary work to be carried on with speed and without undue risk to the workmen. The average mortality due to this cause has not been, comparatively speaking, great during the past 12 years; this, however, is due in all probability not so much to a scarcity of fires as to the careful arrangements made by mine-managers in dealing with them; but from time to time a few accidents from this cause have occasioned serious loss of life. The loss to property may, in some cases, be considerable, arising not merely from the destruction caused by the fire, but from the fact that its extinction may demand the isolation of a large area of workings, which, in most cases, implies a diminution of output; and, in any case, great anxiety is thrown on the management by the outbreak of such a fire.

It is quite impossible to estimate the number of fires which occur annually in and about mines in the United Kingdom. The annual reports of H.M. inspectors of mines deal, as a rule, only with accidents which are fatal or which cause injury to persons: hence, in these reports, the number of fires mentioned is extremely limited; although in some cases mention is made of fires

* Prepared for the International Fire-prevention Congress, held in London, July 6th to 11th, 1903.

which have not caused injury to persons, but which offer points of interest. For reference, a list is given, in Appendices I. and II., of the fires noticed in the reports of H.M. inspectors of mines during the period 1890 to 1901; but it must be remembered that the actual number of fires which have occurred is considerably greater than that set forth in the appendices.

The writer regrets that lack of time has prevented him from including in the appendices, fires and accidents due to fire previous to 1890. Great care has been taken to secure accuracy in compiling the appendices, but it is possible that some fires, or accidents due to them, may have been omitted. It will be noted that, under the heading of "Deaths" are included fatal accidents due both directly and indirectly to fire in and about mines.

As regards the cause, underground fires may be arranged in two classes:—(1) Fires due to various causes—chiefly ignition of coal, timber or other combustible material in the mine; and (2) fires due to spontaneous combustion of coal or bituminous matter.

CLASS I.—*Fires due to Various Causes.*—Ignition may be brought about, either directly or indirectly, in a variety of ways. It may happen in the ordinary course of working, or as the result of an accident. Some of the commoner causes under this head are:—

(*a*) Ignition of the coal, of adjacent carbonaceous strata, or of timber, by underground furnaces used for the purpose of ventilation,* or for raising steam.

(*b*) Ignition of timber. This is generally due to the use of naked lights, either candles or open lamps. The timber may be either timber used for supporting the roof and sides,† or timber used in connection with underground engine-houses or cabins.

---

* *Report to the Right Honourable the Secretary of State for the Home Department, on the Accident at Kinneddar Colliery, Fifeshire, May 31st, 1895, by Sheriff John Comrie Thomson and Mr. J. B. Atkinson, London, 1896.*

† *Reports of H.M. Inspector of Mines for the East Scotland District (No. 1), to H.M. Secretary of State for the Home Department, under the Coal-mines Regulation Act, 1887, the Metalliferous Mines Regulation Act, 1872, and the Quarries Act, 1894; for the Year 1895, by Mr. J. B. Atkinson, London, 1896, page 12; and Reports to the Right Honourable the Secretary of State for the Home Department, on the Circumstances attending an Underground Fire which occurred at the Snaefell Lead-mine, Isle of Man, in the Month of May, 1897, by Dr. C. Le Neve Foster and Dr. A. E. Miller, London, 1898.*

It is noticeable that two of the accidents enumerated in Appendix I. happened in connection with compressed-air engines used underground, where open lamps burning mineral-oil were employed to prevent the formation of ice in the exhaust-pipe.*

(c) Ignition of carburetted hydrogen (fire-damp) by a naked light or a defective safety-lamp. This may also happen after an explosion of fire-damp or coal-dust in a coal-mine.†

(d) Use of open lamps burning mineral-oil—either from defective construction or from carelessness in filling and handling.

(e) Heat generated by the action of steam or steam-pipes on the coal,‡ strata or timber, or possibly by friction.§

(f) Communication of fire on the surface to the underground workings.‖ Even if the fire be not communicated to the workings the products of combustion, by mixing with the intake-air, may produce a dangerous atmosphere in the mine.

(g) By the spontaneous combustion of material stored in the mine.

(h) By electricity. Mr. Robert Lamprecht quotes the case of a fire that occurred at the Herminegild shaft (Polnisch-Ostrau) in 1896,¶ and caused the loss of 16 lives. The cause

---

* *Reports of H.M. Inspector of Mines for the Liverpool District (No. 7), to H.M. Secretary of State, under the Coal-mines Regulation Act, 1887, the Metalliferous Mines Regulation Acts, 1872 and 1875, and the Slate-mines (Gunpowder) Act, 1882; for the Year 1892*, by Mr. Henry Hall, London, 1893; and *Reports of H.M. Inspector of Mines for the Newcastle District (No. 3), to H.M. Secretary of State for the Home Department, under the Coal-mines Regulation Acts, 1887 to 1896, the Metalliferous Mines Regulation Acts, 1872 and 1875, and the Quarries Act, 1894; for the Year 1899*, by Mr. J. R. Hedley, London, 1900.

† "Report on the Working of the Mines Regulation Acts in the Yorkshire and Lincolnshire District during the Year ended 31st December, 1886," by Mr. Frank N. Wardell, *Reports of the Inspectors of Mines to H.M. Secretary of State, under the Coal-mines Regulation Act, 1872, with the Stratified Ironstone-mines (Gunpowder) Act, 1881; and the Metalliferous Mines Regulation Act, 1882; for the Year 1886*, London, 1887, page 95; and "Suggested Rules for the Recovery of Coal-mines after Explosions," by Mr. W. E. Garforth, *Trans. Inst. M.E.*, 1897, vol. xiv., page 495.

‡ "The Application of Liquefied Carbonic Acid Gas to Underground Fires," by Mr. George Spencer, *Trans. Inst. M.E.*, 1899, vol. xvii., page 181.

§ *Report of H.M. Inspector of Mines for the West Scotland District (No. 2), to H.M. Secretary of State, under the Coal-mines Regulation Act, 1887, the Metalliferous Mines Regulation Acts, 1872 and 1875, and the Slate-mines (Gunpowder) Act, 1882; for the Year 1893*, by Mr. J. M. Ronaldson, London, 1894.

‖ *Reports of H.M. Inspector of Mines for the Manchester and Ireland District (No. 6), to H.M. Secretary of State for the Home Department, under the Coal-mines Regulation Acts, 1887 to 1896, the Metalliferous Mines Regulation Acts, 1872 and 1875, and the Quarries Act, 1894; for the Year 1901*, by Mr. John Gerrard, London, 1902; and "An Underground Fire at Bridgewater Colliery," by Mr. A. Dury Mitton, *Trans. Inst. M.E.*, 1897, vol. xiii., page 466.

¶ *Recovery-work after Pit-fires*, by Mr. Robert Lamprecht, London, 1901, page 8.

was a breakage of wire in an electric cable in the shaft. Mr. A. H. Stokes mentions a fire which was caused by sparking or short-circuiting of the electric current of a rheostat or starting-switch, in connection with an electric underground pump.* The rheostat was fixed on a bedplate in a sheet-iron frame, attached to a wooden base, and fastened to props supporting the roof of the engine-house. The switch being much worn, an electric arc was formed which fused the metal of the starter, and the molten metal falling on to the floor set fire to dust and grease, as well as to the bottom of the wooden prop to which the starter was fixed.

It will be seen that Class I. includes fires which originate from various causes. It is, however, unnecessary to make too minute a division of causes, and the distinction between fires which arise from ignition due to certain working conditions, and fires which owe their origin to spontaneous combustion of coal or bituminous matter, is very marked.

CLASS II.—*Fires due to Spontaneous Combustion of Coal or Bituminous Matter.*—Fires under Class I. may occur in any description of mine; those under Class II. are confined to coal-mines or mines working on or near to deposits containing bituminous matter. Spontaneous combustion in substances which are stored in the mine, and need not necessarily be present in the mine, is excluded from this class and placed in Class I (*g*). Although the mortality from this cause has not been as great during the past 12 years as that from fires under Class I., yet it is probable that the majority of underground fires in collieries are due to spontaneous combustion.

All coal-seams are not liable to the action of spontaneous combustion. In some seams, fires from this cause are common, in others they are rare, while in yet others they are unknown. It follows, therefore, that either the composition of a seam of coal, its surrounding conditions, or its depth from the surface, must have some effect on its liability to spontaneous combustion. These are matters which may be briefly referred to later. It

* *Reports of H.M. Inspector of Mines for the Midland District (No. 8), to H.M. Secretary of State for the Home Department, under the Coal-mines Regulation Acts, 1887 to 1896, the Metalliferous Mines Regulation Acts, 1872 and 1875, and the Quarries Act, 1894 : for the Year 1900,* by Mr. Arthur H. Stokes, London, 1901, page 20.

is advisable, first of all, to consider the cause to which such fires are due.

The cause of spontaneous combustion in coal-seams, or in the goafs or wastes from which the coal has been removed (gob-fires) is a matter upon which neither mining-engineers nor chemists are absolutely agreed. A similar difference of opinion is shown in discussions upon the best means of prevention and extinction. When, however, the great diversity in the physical conditions of various seams is considered, the different methods of working employed, and the different positions in the workings in which fires may break out, it will be seen that these apparently-conflicting opinions may often be reconciled one with the other.

Spontaneous combustion is attributed to three causes:—

(1) *Oxidation of Coal.*—Dr. Percy pointed out that coal absorbs oxygen from the air, "of which one portion combines with part of the carbon and part of the hydrogen of the coal, forming carbonic acid and water respectively; while another portion enters into an unknown state of combination with the organic substance of the coal, and the remainder is consumed in oxidizing the iron-pyrites which invariably exists in coal."[*] This chemical action produces heat, and according to the experiments of Prof. Richters at Waldenburg,[†] the absorption of oxygen by coal is promoted by a rise of temperature. The correctness of this inference has recently been shown by Dr. J. S. Haldane and Mr. F. G. Meachem.[‡] Consequently, as the coal absorbs oxygen, it heats, and this rise of temperature increases its capacity for taking up oxygen. Prof. Richters has stated also that "sunlight lessens or retards the absorption of oxygen by coal,"[§] consequently coal underground is favourably situated for absorption.

It must be borne in mind that coal in a fine state of division,

---

[*] *Metallurgy: The Art of Extracting Metals from their Ores: Introduction, Refractory Materials and Fuel*, by Dr. John Percy, London, 1875, page 289.

[†] *Ibid.*, pages 293 and 294.

[‡] "Observations on the Relation of Underground Temperature and Spontaneous Fires in the Coal to Oxidation and to the Causes which Favour it," by Dr. John S. Haldane and Mr. F. G. Meachem, *Trans. Inst. M.E.*, 1898, vol. xvi., page 474.

[§] *Metallurgy: The Art of Extracting Metals from their Ores: Introduction, Refractory Materials and Fuel*, by Dr. John Percy, London, 1875, page 296.

-such as slack or small coal, or a coal-pillar permeated by cracks, will offer a larger surface to the oxidizing action of the air-current and will, therefore, heat more rapidly than an uncrushed face of solid coal; consequently, though the surface of the coal in such a face may be open to this oxidizing action, yet the heat generated will not be so great as that generated in a crushed pillar or a heap of small coal.

Dr. Percy considered the oxidation of the coal as being the chief factor in producing spontaneous combustion, and most mining-engineers and chemists of late years have supported his opinion.

(2) *Oxidation of Iron-pyrites.*—Iron-pyrites, disulphide of iron (FeS_2), is one of the commonest of minerals. It is found in most coal-seams in varying quantity, although some seams are comparatively free from it. It occurs in balls, nodules, fibres, thin cakes and in fine particles in close union with the coal. On exposure to air, especially where moist, the pyrites in the coal becomes oxidized, and forms ferrous and ferric sulphates: sulphuric acid is also said to be produced.[*] The older theory attributed spontaneous combustion in coal to the oxidation of iron-pyrites, the rise in temperature set up by this reaction being supposed to cause the heating in the coal. Dr. Percy[†] threw doubt upon the universal application of this theory stating that "atmospheric oxidation of iron-pyrites is always a comparatively slow process, and, consequently, there must be much loss of heat." He further pointed out, on the authority of Prof. Richters, that coal most liable to spontaneous combustion is not always that which contains most iron-pyrites. Prof. Richters arranged eleven varieties of coal from the Carboniferous system in three classes, and the table[‡] shows that the increase of self-inflammability of these coals was accompanied by a decrease in the percentage of iron-pyrites present in the coal, and by an increase in the percentage of water.

[*] *Ibid.*, page 297; and "Observations on the Relation of Underground Temperature and Spontaneous Fires in the Coal to Oxidation and to the Causes which Favour it," by Dr. John S. Haldane and Mr. F. G. Meachem, *Trans. Inst. M.E.*, 1898, vol. xvi., page 478.

[†] *Metallurgy: The Art of Extracting Metals from their Ores: Introduction, Refractory Materials and Fuel*, by Dr. John Percy, London, 1875, page 299.

[‡] *Ibid.*; and "The Spontaneous Combustion of Coal," by Mr. Herbert W. Hughes, *Trans. Inst. M.E.*, 1893, vol. v., page 399.

Mr. H. W. Hughes[*] mentions that the Thick coal-seam of South Staffordshire (Dudley Ten-yards seam) contains not more than 0·6 to 1 per cent. of pyrites, and yet is very liable to spontaneous combustion. Of late years, this theory has been discredited, but it is only fair to point out that Dr. Percy[†] expressly stated that iron-pyrites, when present in coal in considerable quantity, might develop sufficient heat during its oxidation by atmospheric air to set the coal on fire, and that moisture favoured the oxidation of iron-pyrites, while it impeded rather than promoted the oxidation of coal. Dr. J. S. Haldane and Mr. F. G. Meachem[‡] have recently brought forward facts in support of the pyrites theory, and in the case of the fires which occurred in the Fanny-Chassée colliery in Silesia,[§] it was shown that they were caused by the decomposition of pyrites contained in layers of bituminous shale situated between the two seams worked.

In any case, it is not denied that the decomposition of pyrites may assist in initiating a fire, owing to the expansion in bulk which this substance undergoes when it oxidizes. This will tend to split up and fissure the coal, and so expose more surfaces to oxidation. Before the coal actually fires, sulphuretted hydrogen will be given off, and, being a gas with a very low temperature of ignition, it will materially aid the actual ignition of the coal.

(3) *Friction from Pressure.*—In many systems of working coal, the area of coal to be worked is laid out in pillars or ribs; even where this is not the case, as in some longwall systems, pillars of coal are often left to support the main haulage and

---

[*] "The Spontaneous Combustion of Coal," by Mr. Herbert W. Hughes, *Trans. Inst. M.E.*, 1893, vol. v., page 399.

[†] *Metallurgy: The Art of Extracting Metals from their Ores: Introduction, Refractory Materials and Fuel,* by Dr. John Percy, London, 1875, page 299.

[‡] "Observations on the Relation of Underground Temperature and Spontaneous Fires in the Coal to Oxidation and to the Causes which Favour it," by Dr. John S. Haldane and Mr. F. G. Meachem, *Trans. Inst. M.E.*, 1898, vol. xvi., page 478.

[§] "Der Grubenbrand auf der Steinkohlengrube Fanny-Chassée bei Laurahütte, O.-S.," by Mr. — Fiebig, *Zeitschrift für das Berg-, Hütten- und Salinen-wesen im Preussischen Staate,* 1890, vol. xxxviii., page 291; abs., *Trans. Inst. M.E.,* 1892, vol. iv., page 616; *Recovery-work after Pit-fires,* by Mr. Robert Lamprecht, London, 1901, page 25; and "Note sur les Incendies dans les Houillères," by Mr. — Durand, *Bulletin de la Société de l'Industrie Minérale,* 1883, series 2, vol. xii., page 43; abs., *Minutes of Proceedings of the Institution of Civil Engineers,* 1884, vol. lxxv., page 230.

ventilating roads. If these pillars are not left large, and they are subjected to considerable pressure, the coal in them will become ground by the weight, and the pillars will become cracked and fissured. Some engineers are of opinion that the heat so generated is the cause of some underground fires.[*] There is, however, a difficulty in deciding how much of the increased temperature may be due to the actual pressure, and how much may be due to the fissures started in the pillars, which must necessarily lay open greater areas of coal to the oxidizing action of the atmosphere.

It is quite possible that, in some cases, all these causes may combine to bring about combustion, while, in others, any one of them alone may be sufficient for this purpose. Opinions which have been formed from experience, or from experiment, are not to be lightly disregarded, and it is certain that a considerable difference of opinion exists on this subject.

The coals which appear to be most liable to the action of spontaneous combustion are lignites, and free-burning, bituminous coals, which contain a high percentage of oxygen and volatile constituents. In England, certain seams in the districts of South Staffordshire, North Staffordshire, Warwickshire and Leicestershire appear to be especially liable to its action. Steam-coals and anthracites seem to be exempt: at least, the writer has not been able to find any instance of gob-fires occurring in such coals, while the closer and harder nature of the coal and the absence of any considerable quantity of oxygen and volatile constituents in such coals, would *primâ facie* point to the same conclusion. It does not, however, follow that seams producing steam-coal are free from this danger; several instances to the contrary might be quoted, which have occurred in seams containing both hard and soft coal. The fires in these cases may, however, have been due to the more easily-combustible nature of the soft coal.

It is not necessary that the substance which causes heating should be pure coal. In some cases, the dirty inferior coal which sometimes occurs near faults is found to heat, if left in the goaf. Mr. H. Johnson, of South Staffordshire, in his evidence before

---

[*] *Recovery-work after Pit-fires*, by Mr. Robert Lamprecht, London, 1901, page 4.

the Royal Commission of 1881,[*] stated, as regards the Thick coal-seam that—" The more impure the coal, the more rapid the spontaneous combustion. A hard bright coal will produce less gob-fire than soft impure rubbishy coal." Mr. H. W. Hughes[†] gives a similar opinion.

In some cases, fires appear to be due to bituminous shale or similar substances containing coaly matter, which are not coal at all in a commercial sense; the fires in the Fanny-Chassée colliery form an example of this.

The presence of moisture in most cases induces a tendency to heat, in coals liable to spontaneous combustion. Dr. Percy, from experiments made by Prof. Richters, stated[‡] that the oxidation of coal is hindered by moisture, but that the oxidation of iron-pyrites is favoured by it. Whatever may be the cause, the effect of moisture is often to assist heating in coal. In one instance, in a seam in which gob-fires had seldom or never occurred previously, a fire, which broke out in a band of soft coal overlying the seam, was attributed to the presence of a little water.

In his evidence before the Royal Commission of 1881, Mr. W. Eley, speaking of a fire which occured in some old workings in a seam in South Derbyshire, from which water had been drawn off, stated[§] that, in his opinion, the moisture left in the coal had contributed considerably to the heating of the coal.

Some mining engineers think that the character of the strata in contact with the seam influences the liability of the coal to fire. Under a sandstone-roof, since sandstone is a good conductor of heat and porous, fires will be less likely to occur than under a shale-roof, shale being a bad conductor of heat.[‖] It would appear, however, from experience in some cases[¶] that this idea is not of universal application.

[*] *Preliminary Report of H.M. Commissioners appointed to inquire into Accidents in Mines, and the Possible Means of Preventing their Occurrence or Limiting their Disastrous Consequences, together with the Evidence and an Index.* London, 1881, page 378, No. 11,602.

[†] " The Spontaneous Combustion of Coal," by Mr. Herbert W. Hughes. *Trans. Inst. M.E.*, 1893, vol. v., page 402.

[‡] *Metallurgy : The Art of Extracting Metals from their Ores : Introduction, Refractory Materials and Fuel*, by Dr. John Percy, London, 1875, page 290.

[§] *Preliminary Report of H.M. Commissioners appointed to inquire into Accidents in Mines, and the Possible Means of Preventing their Occurrence or Limiting their Disastrous Consequences, together with the Evidence and an Index*, London, 1881, page 317, No. 9,636.

[‖] *Ibid.*, Mr. Alan Bagot's evidence, page 139, No. 4,332.

[¶] " The Spontaneous Combustion of Coal," by Mr. Herbert W. Hughes, *Trans. Inst. M.E.*, 1893, vol. v., page 404.

The depth of a seam from the surface will probably be found to influence the liability of the coal in it to spontaneous combustion: the greater heat, due to increased depth, should, if what has been previously stated is correct, induce a tendency to rapid absorption of oxygen, and consequently to spontaneous combustion. The extent, however, to which this will take place, must depend also on the composition and character of the coal.

The temperature at which coal will fire depends upon its composition, state of division, and upon the length of time during which it is exposed to air. Mr. Henri Fayol[*] experimented with a heap of small coal, which he alternately exposed to air and closed from access of air. Air being admitted, on the 27th day the temperature rose to 349° Fahr. (176° Cent.), at which temperature the coal took fire. Dr. P. P. Bedson[†] stated that he had ignited coal-dust experimentally at lower temperatures, 284° to 302° Fahr. (140° to 150° Cent.). Mr. W. Carrick Anderson[‡] found that volatilization of certain constituents of coal commenced at 374° Fahr. (190° Cent.). The heating of the small coal exposed to the oxidizing action of the atmosphere proceeds until distillation of inflammable gases takes place, and these, coming into contact with the oxygen of the air, cause combustion. The fire having once broken out, the presence of coal and coaly matter, possibly also of timber, favours its spread: while the combustible gases, carburetted hydrogen, sulphuretted hydrogen and carbon monoxide, help to feed it. In seams, in which no explosive gas has previously been found, or where it has occurred in very small quantities, large amounts of fire-damp have made their appearance as the result of a fire.

The outbreak of a fire is often preceded by a kind of musty smell, and a local rise of temperature; if the heating is taking place in or near coal, a sweating of the coal is observable. The next sign is the emission of fire-stink, and, as this is probably due to gases formed by the destructive distillation of the coal, its

---

[*] "Études sur l'Altération et la Combustion Spontanée de la Houille exposée à l'Air," by Mr. Henri Fayol, *Bulletin de la Société de l'Industrie Minérale*, 1879, series 2, vol. viii., page 639.

[†] Discussion of Mr. Arnold Lupton's paper upon "Spontaneous Combustion in Coal-mines," *Trans. Inst. M.E.*, 1892, vol. iv., page 488.

[‡] "A Contribution to the Chemistry of Coal, with Special Reference to the Coals of the Clyde Basin," by Mr. W. Carrick Anderson, *Trans. Inst. M.E.*, 1898, vol. xvi., page 347.

appearance shows that a fire is either imminent or has commenced. A rise of temperature in the return-airways may also be noticed, and in some cases when the fire has actually commenced, explosions of inflammable gas take place.

Prevention.—The prevention of fires under Class I. is a matter of care and foresight, and in several cases where such fires have occurred, rules have been enforced in order to prevent their recurrence.

The prevention of fires under Class II. is generally attempted by the adoption of some special system of working, or by alterations in the ordinary systems to meet the particular case. Very often the utmost that can be done is to minimize as far as possible the risk :—

(1) By removing from the working-places all slack or any material which may be liable to spontaneous combustion, and sending it out to the surface. In many seams, small coal below a certain size is not sent out, or only a proportion of it is sent to the surface; the small coal which is not sent out is thrown into the goaf—the open space from which the coal has already been extracted. If this slack, when left in the mine, causes gob-fires, it is better to incur the expense of sending it out of the mine than to leave it to fire. The same recommendation applies to bands of dirt or to impure coal which may occur in connection with the seam, or on the sides of faults.

(2) By excluding the air from the goaf, or from any place in which heating is liable to take place.

In longwall workings, the gateways (that is, the roads along which the tubs for carrying coal are taken into and out of the working-places, and which serve as roads along which the men and the ventilating current can travel to the coal-face) follow the advance of the coal-face. Consequently, they have to be maintained through the goaf or gob (the space from which the coal has already been removed by the advancing coal-face) by means of pack-walls, built of stones laid one on the other, without mortar. These pack-walls, though they become compressed by the weight of the strata overlying the seam, are seldom airtight: hence, if heating occurs in the goaf, the air is liable to scale through the pack-wall into the goaf, and oxygen is supplied to increase the heating. In order to prevent this, the gateway pack-walls are sometimes lined with clay-lumps, to prevent leakage of air; in other cases, sand, or finely-ground ashes, are employed for the purpose of making the pack-walls air-tight. If the air-current in passing along the coal-face scales into the goaf and produces heating, sand or clay-lumps are somewhat similarly applied to prevent infiltration of air. In some cases, the waste pack-walls (the pack-walls put up at the coal-face between two gateways) are placed chequer-wise, like the squares on a chessboard, to block the air out of the goaf; in other cases, the whole goaf is stowed as tightly as possible. This, however, is rather a Continental than an English method, close stowing of the goaf being far commoner abroad than in England. The material used for stowing must, of course, be non-inflammable.

(3) By adopting a special system of working. The method of working adopted in South Staffordshire for the Thick coal-seam may be instanced. The coal is won by driving a pair of roads to the boundary in the solid coal, and opening chambers or sides of work out of these roads. Only one side of work is opened at a time, and as it has only two entrances, these can be stopped off so soon as the coal shows signs of firing; when stopped off, it does not interfere with the working of the rest of the coal as it lies on the inbye-side (that is, farther away from the shaft) of the next side of work opened out, and is separated from it and others which may be opened near it by ribs of solid coal. A full description of this method of working will be found in the works mentioned in the footnote.*

In working a coal-seam liable to fires on the pillar-and-stall system, the area of coal is divided into small panels or districts separated one from the other by ribs of solid coal. The coal in such districts can be worked out more quickly if the districts are small; the number of openings into the districts being few, they can be stopped off more quickly, and the diminution of pit-room (that is, coal laid out ready for working) is less seriously felt than if the districts are large.

* A Text-book of Coal-mining, by Mr. Herbert W. Hughes, London, 1901 ; "A General Description of the South Staffordshire Coal-field, South of the Bentley Fault, and the Methods of Working the Ten-Yard or Thick Coal," by Mr. W. F. Clark and Mr. Herbert W. Hughes, Trans. Inst. M.E., 1891, vol. iii., page 25 ; "Observations on the Relation of Underground Temperature and Spontaneous Fires in the Coal to Oxidation and to the Causes which Favour it," by Dr. John S. Haldane and Mr. F. G. Meachem, Trans. Inst. M.E., 1898, vol. xvi., page 457 ; and "The Thick Coal of South Staffordshire," by Mr. Herbert W. Hughes, The Journal of the British Society of Mining Students, 1885, vol. ix., page 4.

(4) By working home. In the majority of cases, coal is worked outwards away from the pillar left to protect the shafts, consequently in a colliery, which has been working for some years, a large area of goaf intervenes between the shafts and the working-faces; if fires are liable to occur in any part of such a goaf, the danger and inconvenience can be easily imagined. In working home, the operation is reversed, the roads are driven in the solid coal to the boundary, and the coal-face retreats daily towards the shafts; consequently the goaf and its dangers are left behind. This system has the disadvantage of determining from the commencement the area of coal which can be worked from the shafts, and also of heavy initial expense in opening out, when the output of coal coming out can only be comparatively small.

(5) By rapid working. If the coal-face advances slowly, the air-current, which ventilates it, has time to make its way into the adjacent goaf and cause heating; the more quickly the face advances, the less likelihood is there of this action taking place.

(6) By passing a strong current of cool air through the place which is heating, in order to cool it and carry off the heat as fast as it is generated. If this is not possible, then the access of air to the place should be cut off altogether, so as to deprive it of oxygen.

Extinction.—The means mentioned above for prevention of fires are not always successful. Fires under Class I. may happen from some unforeseen cause or accident, while in the case of fires under Class II., although the means adopted for preventing fires may lessen their frequency and extent, yet it is not possible in some cases to prevent them. A fire having broken out, the next thing is to attempt to extinguish it, and, in this case, time is a most important factor, since if the fire is given time to gain ground its extinction will become more difficult, and most drastic remedies will have to be applied:—

(1) By filling out the fire. A road is driven into the seat of the fire, water is applied to quench the burning materials, and they are filled into tubs and removed to the surface. The difficulties, in doing this, are often very considerable. While air is especially necessary for the workmen, on account of the heat and the poisonous gases given off, yet such air, if it reaches

the fire, will help to increase its combustion. The use of water,. necessary for quenching the burning material, will, in some cases, weaken the roof, and so cause additional danger to the men at work. In some cases, instead of attacking the fire on the plane of the seam, a road has been driven in the strata above the seam and water poured on to the fire through boreholes,[*] but generally such a course is not practicable. When the heated material has been removed, the place is cooled down, and stowed up with non-inflammable material.

Many mining-engineers consider that this is the only really effective remedy; when, however, the fire has got firm hold over a large area, or is so situated that it cannot be attacked in this way, other methods must be adopted.

(2) By stopping off the fire. The object of this is, of course, to prevent access of air to the fire; but the situation and extent of the fire must determine the plan adopted in each particular case. In many cases, stoppings[†] built in roads leading to the seat of the fire are found sufficient; these stoppings are generally constructed in brickwork, well let into the strata at the sides, roof and floor, and are built in pairs, in threes, or as many more as may be requisite, the space between being stowed with sand or other close-setting material; but sometimes it has been found necessary to close the shafts by building scaffolds across them, and in this way rendering them air-tight. It is, of course, most important that any defect in the stoppings, due to construction, or to subsequent movements of the strata, should be detected at once and remedied, also that air shall not scale to the fire at other points; if this happens, the fire will only be subdued, not extinguished.

(3) By drowning out the fire. This very drastic remedy can be applied in districts of mines where solid coal or other material exists to prevent the water from flowing away, or where the whole mine is flooded under similar conditions. Its drawbacks are evident, and the expense of unwatering the mine, the deteriorated condition of the coal and strata from having water standing in them, when the mine is re-opened, make its adoption a

---

[*] "The Spontaneous Combustion of Coal," by Mr. Herbert W. Hughes, *Trans. Inst. M.E.*, 1893, vol. v., page 406.

[†] "Stoppings on Underground Roads,' by Mr. E. B. Wain, *Trans. Inst. M.E.*, 1893, vol. vi., page 572.

last resource. In some cases, a liability to spontaneous combustion is induced in the coal after the water has been drawn off, owing to the moisture left in the coal.

(4) By the use of carbonic-acid gas. At first sight the employment of this gas for the extinction of underground fires would appear advantageous, but its application seems to be attended with difficulty. A historic example of its use is the extinction of the fire at Wynnstay colliery, Ruabon, of which a description was given by Mr. George Thomson,[*] which is quoted by Mr. George G. André.[†] Mr. George Spencer[‡] gives an instance in which he successfully dealt with a fire by the use of liquefied carbon dioxide. On the other hand, Mr. W. H. Chambers[§] states that, in the case of a colliery near Rotherham where the shafts were sealed up, and carbonic-acid gas was passed through the stoppings, it was found that the gas merely suspended the burning, which broke out as badly as before when air was admitted. The first-mentioned method of extinction is of course attended with the danger of employing this suffocating gas where men are working.

Dangers.—As already mentioned, it is quite impossible to estimate the damage to property and the pecuniary loss caused by underground fires. The particulars set forth in Appendices I. and II. will, however, give some idea of the loss of life occasioned by them, and the actual cause of death. Table I. gives a summary of the facts stated in the appendices.

It is evident that the majority of the deaths are due to suffocation, or poisoning, by the gases produced by fires. The table also shows that during the years 1890 to 1901, underground fires, due to various causes, brought about double the number of deaths due to spontaneous combustion. Incidentally, it may be noted that, during the period 1890 to 1901, the mortality

---

[*] "A Brief Account of an Underground Fire in the Wynnstay Colliery, Ruabon, and the Measures adopted to Extinguish it, and to Re-enter the Workings," by Mr. George Thomson, *The Journal of the Iron and Steel Institute,* 1875, page 172.

[†] *A Practical Treatise on Coal-mining,* by Mr. George G. André, London, 1876, vol. ii., page 512.

[‡] "The Application of Liquefied Carbonic-acid Gas to Underground Fires," by Mr. George Spencer, *Trans. Inst. M.E.,* 1899, vol. xvii., page 181.

[§] "Notes on Gob-fires," by Mr. W. H. Chambers, *Trans. Inst. M.E.,* 1899, vol. xviii., page 154.

due to accidents in and about all mines in the United Kingdom was 12,896. Out of this number, only 188 deaths were due to fire, that is, 1·4 per cent.; this figure also includes deaths from fire on the surface of mines.

TABLE I.—CAUSES OF DEATHS FROM FIRE IN AND ABOUT MINES, DURING THE PERIOD FROM 1890 TO 1901.

Causes.	Class I.		Class II.		Both Classes.	
	No. of Accidents.	Deaths.	No. of Accidents.	Deaths.	No. of Accidents.	Deaths.
Suffocation	9	118	8	53	17	171
Burning	4	10	1	1	5	11
Falling down shaft ...	*	3	—	—	*	3
Fall of roof	*	1	1	1	1	2
Fracture of skull ...	—	·—	1	1	1	1
Totals	13	132	11	56	24	188

* Occurred in connection with accidents, causing other deaths by suffocation.

It must, however, be remembered that the effect of a fire in a coal-mine is to generate explosive gases, consequently the management of a colliery in which a fire occurs has to be on its guard against the contingency of an explosion.

It does not appear to have been accurately ascertained what reactions actually take place in the course of an underground fire, and the nature of the gases given off must, therefore, be a matter of conjecture. It would appear, however, that a fire in any mine must generate carbonic-acid gas and carbon monoxide, while a fire in a colliery will generate, in addition, carburetted hydrogen and sulphuretted hydrogen. The principal properties of these gases are as follows:—

Carbonic-acid gas or carbon dioxide (CO_2), which forms an important ingredient in the black-damp of mines, and, in some cases is probably identical with it, has a specific gravity of 1·52 and is a product of combustion.* Over 12 per cent. in air is said to be dangerous to life, while 15 per cent. will extinguish lights. It is ordinarily detected by the fact of its extinguishing a light.

Carbon monoxide (CO) is the result of incomplete combustion. Its specific gravity is 0·97. It is a peculiarly deadly gas, entering into combination with the blood, and depriving it of the

* Report to the Secretary of State for the Home Department on the Causes of Death in Colliery-explosions and Underground Fires, with Special Reference to the Explosions at Tylorstown, Brancepeth and Micklefield, by Dr. John S. Haldane, London, 1896.

power of taking up oxygen. Its effect is due partly to the amount of the gas present, and partly to the length of time of exposure to it. Dr. J. S. Haldane states[*] that 1 per cent. of this gas will cause loss of consciousness after a few minutes, and final death, while as little as 0·1 per cent., after a long exposure, will produce injurious symptoms. The presence of this gas cannot be satisfactorily ascertained by the flame-test, and Dr. Haldane has suggested the use of small animals, such as mice, in order to indicate its presence. Mice have since been used for this purpose, notably at Snaefell mine[†] in 1898, and at Hill-of-Beath colliery[‡] in 1901; in the latter case a canary was also used. In connection with the former accident, Sir C. Le Neve Foster points out that the bigger the animal used, the less quickly were the symptoms observed. Carbon monoxide is an inflammable gas.

Sulphuretted hydrogen (H_2S) is produced by the action of heat on mixtures of coal and pyrites, or by the distillation of coal alone. Its specific gravity is 1·17. It is an inflammable gas, and is highly poisonous. It may be detected by its blackening silvered surfaces, and, when present in quantity by its characteristic and offensive smell.

Carburetted hydrogen, methane (CH_4), specific gravity, 0·55, is a gas which is frequently present, naturally, in coal-mines. Even, however, in coal-mines, in which this gas has been either unknown or present in very small quantities, the destructive distillation of the coal caused by a fire may produce it in abundance. It is in some cases identical with the fire-damp of the miner, or, in any case, forms a large proportion of it. When from 4 to 16 per cent. of it is mixed with air, it is explosive, and over 60 per cent. of it causes suffocation. It is detected by the elongation which it produces on the flame of a

[*] *Report to the Secretary of State for the Home Department on the Causes of Death in Colliery-explosions and Underground Fires, with Special Reference to the Explosions at Tylorstown, Brancepeth and Micklefield*, by Dr. John S. Haldane, London, 1896, page 21.

[†] *Reports to the Right Honourable the Secretary of State for the Home Department on the Circumstances attending an Underground Fire which occurred at the Snaefell Lead-mine, Isle of Man, in the month of May, 1897*, by Dr. C. Le Neve Foster and Dr. A. E. Miller, London, 1898.

[‡] *Reports of H.M. Inspector of Mines for the East Scotland District (No. 1), to H.M. Secretary of State for the Home Department, under the Coal-mines Regulation Acts, 1887 to 1896, the Metalliferous Mines Regulation Acts, 1872 and 1875, and the Quarries Act, 1894; for the Year 1901*, by Mr. J. B. Atkinson, London, 1902, page 13.

safety-lamp, and by the formation of a pale-blue halo on a reduced flame when over 2½ per cent. of it is present.

It is the presence of these gases which perhaps constitutes the chief difficulty in dealing with underground fires and in recovering a mine after an explosion. The work of reopening a mine or district that has been sealed off to extinguish a fire is attended with similar risks.

Safety-appliances.—In order to enable necessary work to be carried on in these irrespirable gases, attempts have been made to enable persons to enter mines in which such gases were present, by arrangements similar to those used by divers. Of late years, a contrivance, known as the pneumatophore, has been introduced into this country from Germany.[*] This may be described as a bag or knapsack, which can be worn on the chest or back, fitted with a small cylinder of oxygen, a glass-flask filled with a solution of caustic soda, and a coarse net or sponge to suck up the caustic-soda solution when it is set free. The bag is connected by means of a pipe with the wearer's mouth, and his nose is closed by means of a clip. In this way, the man wearing the pneumatophore is independent of the external atmosphere, and breathes only the atmosphere contained in the bag. The deficiency of oxygen caused by his breathing is made up from the oxygen-cylinder, while the carbonic-acid gas given off by the breath is absorbed by the caustic soda. Such appliances are used already in Austria and Germany; and in England, Mr. W. E. Garforth has fitted up at Altofts colliery an experimental gallery for training men to their use.[†] On the Continent, regular rescue-stations have been organized, and recently in South Yorkshire three collieries have united to form a joint rescue-station.[‡] Such a rescue-station is equipped with a number of pneumatophores and accessories, and with oxygen

[*] "The Walcher Pneumatophore, and the Employment of Oxygen for Life-saving Purposes," by Mr. Richard Cremer, *Trans. Inst. M.E.*, 1897, vol. xiv., page 575 ; "A Joint Colliery Rescue-station," by Mr. M. H. Habershon, *Trans. Inst. M.E.*, 1901, vol. xxi., page 100 ; and "Experimental Gallery for Testing Life-saving Apparatus," by Mr. W. E. Garforth, *Trans. Inst. M.E.*, 1901, vol. xxii., page 169.

[†] "Experimental Gallery for Testing Life-saving Apparatus," by Mr. W. E. Garforth, *Trans. Inst. M.E.*, 1901, vol. xxii., page 169.

[‡] "A Joint Colliery Rescue-station," by Mr. M. H. Habershon, *Trans. Inst. M.E.*, 1901, vol. xxi., page 100.

and everything needful for the recovery of persons who have
been overcome by noxious gases. In connection with it there
is a trained corps of officials and workmen, who can use the
pneumatophore and work with it, and arrangements are also
made for furnishing first aid to persons found injured in the
mine. At the mining school at Bochum, in Westphalia, a
number of the students are trained each year in the use of the
pneumatophore. Useful as the apparatus may be made by a
person used to wearing it, and working with it, it is evident that
its use by an unskilled person might only lead to danger to the
wearer.

Conclusion.—This paper has been written primarily with a
view of giving a general *résumé* of the subject, for those who
have hitherto not devoted much attention to it. For the benefit
of those who wish to pursue their investigations further, the list
of books and papers given in Appendix III. may be found useful; it cannot be claimed that the list is complete, but it may
at least form the groundwork for a complete index to the
literature of underground fires.

APPENDIX I.—LIST OF FIRES OR FATAL ACCIDENTS FROM BURNING, IN OR ABOUT
MINES IN THE UNITED KINGDOM, MENTIONED IN THE ANNUAL REPORTS
OF H.M. INSPECTORS OF MINES, 1890 TO 1901. CLASS I.—VARIOUS
CAUSES.

Dates.	Mines-inspection Districts.	Deaths.	Mines.	Remarks.
1890 Jan. 9	Southern	0	Dolcoath	Fire broke out below the 264 fathoms level, near the main engine-shaft.
Jan. 22	Cardiff	5	Glyn	Two timbermen ignited some fire-damp with naked lights; this set the timber on fire, and suffocated 5 men.
March 29	Staffordshire	1	Walsall Wood	Explosion of petroleum-vapour from an underground oil-engine. The man was burnt.
May 7	Scotland, East	2	Dalziel	Fire in an engine-house at the pit-bottom. No cause assigned. The men were suffocated.
1891 Jan. 5	Scotland, East	1	Hermand	Accident on the surface, due to the man filling a naphtha-lamp. Man burnt.

APPENDIX I.—*Continued.*

Dates.	Mines-inspection Districts.	Deaths.	Mines.	Remarks.
1891 Dec. 8	Yorkshire ...	5	Wheldale ...	Fire broke out in an underground cabin, where open lamps were used. The men were suffocated.
1892 June 7	Staffordshire ...	1	Netherton ...	Accident on the surface. The man, who was not employed at the colliery, was filling a petroleum-lamp with the wick lighted, and set the engine-house on fire. The man was burnt.
Dec. 14	Liverpool ...	16	Bamfurlong	Fire in an underground hauling-engine-house, caused by an open paraffin-torch lamp setting fire to the floor. The engine was worked by compressed air, and the lamp was used for warming the exhaust-air to prevent freezing. Men were suffocated.
1893 April 11	Cardiff ...	63	Great Western	Fire caused by sparks emitted from the brake of a hauling-engine, which probably set fire to some brattice-cloth. Men were suffocated. 1 man killed by falling down the shaft, and 1 man buried by a fall.
May 26	Scotland, West	3	Orbiston ..	Fire supposed to have been caused by the heat produced by underground pump-rods rubbing on coal. Men were suffocated.
1895 May 31	Scotland, East	9	Kinneddar ...	Wooden lining of shaft took fire. Men were employed in putting in a dam. Flames were blown on to the men and 7 of them were severely burned. Two of the men were killed by falling down a shaft in the confusion.
1897 May 10	Newcastle-upon-Tyne	20	Snaefell ...	Fire was probably due to ignition of dry timber by a candle. The men were suffocated.
1899 July 7	Newcastle-upon-Tyne.	4	Southam Iron-ore Mine	Fire broke out near a pump driven by compressed air, a paraffin-burner being used to prevent the formation of ice. Cause not known. The men were suffocated.

APPENDIX I.—*Continued.*

Dates.	Mines-inspection Districts.	Deaths.	Mines.	Remarks.
1899 Oct. 4	Durham ...	0	Binchester ...	Fire on the surface, around the downcast shaft.
Nov. 24	Newcastle-upon-Tyne	0	Broomhill ...	Fire on the surface.
Dec. 20	Durham	1	South Pelaw	Fire broke out in a cabin at the pit-bottom, where paraffin-lamps for lighting the pit-bottom were trimmed. Man was suffocated.
—	Manchester ...	0	Bradley Fold	Fire due to an underground furnace.
-	Manchester ...	0	Outwood ..	Fire due to an underground hauling - engine, t h e wooden blocks on the brake-strap having taken fire without being noticed by the attendant.
1900 Jan. 13	Staffordshire ...	0	Silverdale ...	Fire caused probably by ignition of fire-damp.
July 31	Cardiff... ...	0	Great Western	Gas from an old goaf ignited at a lamp-station, and set fire to some old timbers, etc.
Dec. 27	Midland ...	0	Wyken	Fire caused by sparking or short-circuiting of the electric current of a rheostat or starting-switch. Electric pump with current at a pressure of 450 volts. The rheostat was fixed on a bed-plate in a sheet-iron frame, attached to a wooden base. and fastened to props supporting the roof of the engine-house.
1901 May 19	Manchester ...	1	Ashton Moss	Fire in the winding engine-house at the surface. The winding-rope got red-hot, and when the foot and steam-brakes were disengaged by the fire, the drum was free, and the red-hot rope went down the shaft and set fire to the timber in the pit-bottom. One man was suffocated. The fire was caused by the fusing of the cable for electric lighting.

APPENDIX II.—LIST OF FIRES OR FATAL ACCIDENTS FROM BURNING, IN OR ABOUT MINES IN THE UNITED KINGDOM, MENTIONED IN THE ANNUAL REPORTS OF H.M. INSPECTORS OF MINES, 1890 TO 1901. CLASS II. - SPONTANEOUS COMBUSTION.

Dates.	Mines-Inspection Districts.	Deaths.	Mines.	Remarks.
1891 May 7	Staffordshire ...	1	Shelton, Rowhurst	Explosion of fire-damp, ignited by a gob-fire. The men were burned.
Nov. 11	Scotland, West	0	Ross	Man injured by gas, supposed to have been distilled from the coal by a fire.
Dec. 2	Manchester ...	4	Agecroft ...	Men were suffocated by gases, when shutting off a fire in the mine, supposed to be due to spontaneous combustion.
1892 April 6	Staffordshire ...	2	East Cannock	Two men suffocated by gas, given off from a gob-fire.
May 2	Southern ...	1	Coalbrookvale	Man killed by a fall of roof and sides, while putting out an underground fire.
1894 Oct. 15	Staffordshire ...	1	Harecastle, Moss Pits	Explosion was caused by a gob-fire. The cause of death was probably a fracture of the skull. Four men were slightly burnt.
1895 Oct. 29	Staffordshire ..	2	Oldfield ...	Men were engaged in stopping off a gob-fire. Men were suffocated.
1896 Sept. 10	Staffordshire ...	1	Shelton, No. 1 Rowhurst	Manager was suffocated by gases given off from a gob-fire.
1898 April 19	Midland ...	35	Whitwick ...	The fire was caused by spontaneous combustion in the main intake airway of the colliery. The men were suffocated.
Feb. 6	Staffordshire ...	0	Podmore Hall, Minnie	Explosion was caused by a gob-fire. The pit was completely wrecked, but no lives were lost. There were eight gob-fires during the year.
Nov. 11	Staffordshire ...	0	Hamstead ...	The pit was lost by a gob-fire.
1899 Oct. 20	Scotland, West	1	Dalquharran	Man suffocated in trying to put out an underground fire.
—	Manchester ...	0	Ladyshore ...	Spontaneous combustion.
—	Manchester ...	0	Darcy Lever ...	Spontaneous combustion.
1901 Jan. 23	Scotland, East	1	Shotts	Man suffocated.
Feb. 15	Scotland, East	7	Hill of Beath	Men suffocated.

APPENDIX III.—List of Papers and Books on the Subject
of Underground Fires.

André, George G., *A Practical Treatise on Coal-mining*, 1876, vol. ii., pages 510-517

Atkinson, J. B., and John Comrie Thomson, *Report to the Right Honourable the Secretary of State for the Home Department on the Accident at Kinneddar Colliery, Fifeshire, May 31st*, 1895.

Chambers, W. H., "Notes on Gob-fires," *Trans. Inst. M.E.*, 1899, vol. xviii., page 154.

Channing, J. Parke, "The Underground Fire at the Lake Superior Mine, Ishpeming, Michigan," *The Engineering and Mining Journal*, 1892, vol. liii., page 106; abs., *Trans. Inst. M.E.*, 1893, vol. v., page 563.

Clark, W. F., and Herbert W. Hughes, "A General Description of the South Staffordshire Coal-field, south of the Bentley Fault, and the Methods of Working the Ten-yard or Thick Coal," *Trans. Inst. M.E.*, 1891, vol. iii., page 25.

Cremer, Richard, "The Walcher Pneumatophore, and the Employment of Oxygen for Life-saving Purposes," *Trans. Inst. M.E.*, 1898, vol. xiv., page 575.

—, "Wagner Portable Pneumatic Safety-stopping for Mining Purposes," *Trans. Inst. M.E.*, 1898, vol. xv., page 219.

Dunn, Matthias, *A Treatise on the Winning and Working of Collieries; including numerous Statistics regarding Ventilation and the Prevention of Accidents in Mines*, second edition, 1852, pages 179-193.

Durand, —, "Note sur les Incendies dans les Houillères," *Bulletin de la Société de l'Industrie Minérale*, series 2, 1883, vol. xii., page 43; translation and abstract, by Alfred Bache, *Minutes of Proceedings of the Institution of Civil Engineers*, 1884, vol. lxxv., page 230.

Evans, S. T., and J. T. Robson, *Reports to the Right Honourable the Secretary of State for the Home Department on a Disaster in the Great Western Colliery, Rhondda Valley, on the 11th April*, 1893.

Fayol, Henri, "Études sur l'Altération et la Combustion Spontanée de la Houille exposée à l'Air," *Bulletin de la Société de l'Industrie Minérale*, series 2, 1879, vol. viii., page 487.

Fiebig, —, "Der Grubenbrand auf der Steinkohlengrube Fanny-Chassée bei Laurahütte, O.-S.," *Zeitschrift für das Berg- Hütten- und Salinen-wesen im Preussischen Staate*, 1890, vol. xxxviii., Abhandlungen, page 291; abs., *Trans. Inst. M.E.*, 1892, vol. iv., page 616.

Foster, C. Le Neve, and A. E. Miller, *Reports to the Right Honourable the Secretary of State for the Home Department on the Circumstances attending an Underground Fire which occurred at the Snaefell Lead-mine, Isle of Man, in the month of May*, 1897.

Garforth, W. E., "Experimental Gallery for testing Life-saving Apparatus," *Trans. Inst. M.E.*, 1901, vol. xxii., page 169.

Gresley, W. S., "Spontaneous Combustion in Coal-mines," *Trans. Inst. M.E.*, 1894, vol. vii., page 206.

Habershon, M. H., "A Joint Colliery Rescue-station," *Trans. Inst. M.E.*, 1901, vol. xxi., page 100.

Haldane, John S., and F. G. Meachem, "Observations on the Relation of Underground Temperature and Spontaneous Fires in the Coal to Oxidation and to the Causes which Favour it," *Trans. Inst. M.E.*, 1898, vol. xvi., page 457.

Hassam, Arthur, "Gob-fires in Longwall Workings, with Special Reference to the Yard Seam," *Trans. Inst. M.E.*, 1894, vol. viii., page 332.

Hill, Frank A., "The Hill-Farm-Parrish Mine-fire," *Transactions of the American Institute of Mining Engineers*, 1892, vol. xxi., page 632; abs., *Trans. Inst. M.E.*, 1894, vol. vii., page 730.

Honl, Ant, "Vorsichtsmassregeln gegen Grubenbrände," *Oesterreichische Zeitschrift für Berg- und Hüttenwesen*, 1888, vol. xxxvi., pages 639 and 646; abs., *Trans. Inst. M.E.*, 1892, vol. iii., page 1100.

Hughes, Herbert W., "The Spontaneous Combustion of Coal," *Trans. Inst. M.E.*, 1893, vol. v., page 392.

—. and W. F. Clark, "A General Description of the South Staffordshire Coal-field, south of the Bentley Fault, and the Methods of Working the Ten-yard or Thick Coal," *Trans. Inst. M.E.*, 1891, vol. iii., page 25.

Jones, Chester, and A. H. Stokes, *Reports to Her Majesty's Secretary of State for the Home Department, on the Circumstances attending an Underground Fire which occurred at the Whitwick, No. 5 Colliery, Leicestershire, in the month of April, 1898*.

Jones, M. Y., "The Causes of Spontaneous Combustion of Coal, and Prevention of Explosions on Shipboard," *Trans. Inst. M.E.*, 1892, vol. iii., page 789.

Lamprecht, Robert, *Recovery-work after Pit-fires*, translated from the German by Charles Salter, 1901.

Lewes, Vivian B., "The Spontaneous Ignition of Coal, and its Prevention," *Journal of the Society of Arts*, 1892, vol. xl., page 352.

Lupton, Arnold, "Spontaneous Combustion in Coal-mines." *Trans. Inst. M.E.*, 1892, vol. iv., page 481.

Meachem, F. G., and John S. Haldane, "Observations on the Relation of Underground Temperature and Spontaneous Fires in the Coal to Oxidation and to the Causes which Favour it," *Trans. Inst. M.E.*, 1898, vol. xvi., page 457.

Miller, A. E., and C. Le Neve Foster, *Reports to the Right Honourable the Secretary of State for the Home Department on the Circumstances attending an Underground Fire which occurred at the Snaefell Lead-mine, Isle of Man, in the month of May, 1897*.

Mills, E. J., and F. J. Rowan, *Chemical Technology, or Chemistry in its Applications to Arts and Manufactures*, edited by Charles Edward Groves and William Thorp: Vol. I.—*Fuel and its Applications*, edited by Charles Edward Groves, page 83.

Mitton, A. Dury, "An Underground Fire at Bridgewater Colliery," *Trans. Inst. M.E.*, 1897, vol. xiii., page 466.

Official, *Preliminary Report of Her Majesty's Commissioners appointed to inquire into Accidents in Mines, and the Possible Means of Preventing their Occurrence or Limiting their Disastrous Consequences, together with the Evidence and an Index*, 1881.

Percy, John, *Metallurgy: the Art of Extracting Metals from their Ores: Introduction, Refractory Materials and Fuel*, 1875, page 289.

Richters, E., "Ueber die Veränderungen, welche die Steinkohlen beim Lagern an der Luft erleiden," *Dingler's Polytechnisches Journal*, 1870, vol. cxcv., pages 315 and 449; and 1870, vol. cxcvi., page 317; abs., *Jahres-bericht über die Leistungen der Chemischen Technologie in 1870*, edited by Johannes Rudolf Wagner, 1871, vol. xvi., pages 758-778.

Robson, J. T., and S. T. Evans, *Reports to the Right Honourable the Secretary of State for the Home Department on a Disaster in the Great Western Colliery, Rhondda Valley, on the 11th April, 1893*.

Rowan, F. J., and E. J. Mills, *Chemical Technology or Chemistry in its Applications to Arts and Manufactures*, edited by Charles Edward Groves and

William Thorp: *Vol. I.—Fuel and its Applications*, edited by Charles Edward Groves, page 83.

Settle, Joel, "Spontaneous Combustion in Coal-mines," *Trans. Inst. M.E* 1893, vol. v., page 10.

Spencer, George, "The Application of Liquefied Carbonic-acid Gas to Underground Fires," *Trans. Inst. M.E.*, 1899, vol. xvii., page 181.

Stokes, A. H., and Chester Jones, *Reports to Her Majesty's Secretary of State for the Home Department, on the Circumstances attending an Underground Fire which occurred at the Whitwick, No. 5 Colliery, Leicestershire, in the month of April, 1898.*

Thomas, J. W., *A Treatise on Coal, Mine-gases and Ventilation*, 1878, pages 60 and 242.

Thomson, John Comrie, and J. B. Atkinson, *Report to the Right Honourable the Secretary of State for the Home Department on the Accident at Kinneddar Colliery, Fifeshire, May 31st, 1895.*

Wain, E. P., "Stoppings on Underground Roads," *Trans. Inst. M.E.*, 1893, vol. vi., page 572.

Wight, Edward S., "Queensland Coal-mining: and the Method adopted to Overcome an Underground Fire," *Trans. Inst. M.E.*, 1893, vol. iv., page 548.

———

DISCUSSION OF MR. F. W. HURD'S PAPER ON "ELECTRICAL COAL-CUTTING MACHINES."[*]

Mr. W. W. MILLINGTON (Hollinwood) wrote that he did not agree with Mr. Hurd that the most probable reason for delay in adopting machines in this country had been due to colliery-owners doing well, so far, without them. The owners most in want of machines have, as a rule, been doing badly, and could only hold their own in time of fair trade. He was of opinion that machine-makers are a good deal to blame. Owners cannot have had much experience in the past, and have had to accept the assurance of the makers, perhaps, wedded to one machine, and they have recommended it, regardless as to whether it was suitable for that particular mine or not, and perhaps, to tempt the owner to adopt machine-cutting, they have tendered for a cheap electrical installation, the result being, in a very short time, disastrous failure, and the disappointment of the owner. Perhaps a disc-machine has been adopted, which could never be a success in the mine, or it may have been a bar-machine, or any other. He contended that the conditions of every mine required careful consideration by experienced experts, who were not pledged to any machine. Mr. Hurd stated that the expenditure of a "few hundred pounds are grudged for the purchase of machines, which would save 1s. to 3s. per ton."[†]   He was of

* *Trans. Inst. M.E.*, 1903, vol. xxv., pages 108, 231 and 251.
† *Ibid.*, vol. xxv., page 109.

opinion that Mr. Hurd should have said thousands of pounds spent, and pence per ton saved: but 1s. per ton saved would induce owners to spend thousands of pounds, and further, there would be a saving in the production of small coal and the employment of a coal-getter that never had a grievance, and never wanted a prolonged strike.

He (Mr. Millington) agreed with Mr. Hurd as to the necessity for strict supervision, but he hardly liked the manner in which Mr. Hurd insinuated that a manager should be paid a premium to make anything a success at the colliery under his charge (although probably Mr. Hurd did not mean it to be taken in that light). All managers are engaged to do their best for their employers, and should do so at all times, and no doubt they do get a reward for their assiduity. To make machine-cutting a success entails extra work on the manager, and for a long time, at the commencement, nothing goes smoothly. As a rule, the men are awkward and require careful handling, and it takes some time to convince them that they will lose nothing by the adoption of machines. During this time the manger has really the hardest work and the greatest anxiety.

The method of working, in his (Mr. Millington's) opinion, was one of the problems that required the greatest consideration on the part of the manager, as so many points have to be considered, such as the nature of the holing, the dip of the mine, the good and bad roof, the débris for packing, the mode of haulage, etc. A mistake may easily be made and probably cause the condemnation of machine-holing without the cause of the failure being realized. Mr. Hurd had correctly stated his system, but he would like to add that he had adopted a heavy rail, 24 pounds per yard, in lengths of 15 feet, laid on steel sleepers for the whole length of the face. A gang of packers go down at night, at the same time as the machine-driver, and as he holes with the machine, they remove the dirt, about 20 inches thick, from the top of the coal, and leave all clear for the collier to commence work on the following morning, the rails also being left clear for the transit of his tubs.

He quite agreed with Mr. Hurd's remarks that a straight face was most essential.

Where possible it is advisable to lay the rails along the entire length of the face, time and trouble is saved, especially if the machine-gauge is made of the same width as the tub-

used, the only objection that he had heard raised as to the suitability of the bar-machine for any seam had been the hardness of the holing. On this score he could only repeat that, in his experience, the bar-machine would hole in as hard material as any other machine, and would do good work in any seam workable by machines with a minimum of power, and at a low cost for maintenance.

The sums of 1s. to 3s. per ton cover all the gains made, and it is not a case of 1s. to 3s. per ton *plus* the advantages gained by more round coal, less upkeep in roads, extra output for man employed, etc., as Mr. Millington would seem to infer.

Referring to the matter of premium or its equivalent, all must agree with Mr. Millington that a manager must certainly do his best to make everything a success at the colliery under his charge, but the introduction of coal-cutters was probably never thought of when his agreement was made, and he ventured to suggest a method of payment for the extra work entailed. Of course, it would be understood that such an arrangement particularly applied to those smaller collieries where a great amount of the actual detail of management fell directly on the manager.

The methods of working suggested could not of course be taken as applicable in all cases. Every seam certainly required its own particular method, but generally the main idea could be adopted, although the details might be varied.

He was very pleased to see that Mr. Millington endorsed his remarks as to the working-face being straight, and he was convinced that there were many machine-faces which could be greatly improved in this respect. To straighten them would mean probably a loss in output for a few days, but a wonderful difference would be very quickly shewn, both in the tonnage filled and in the increased yardage cut per shift by the machine. Unless very carefully watched, a machine-face would fall back at the ends and become concave. Getting out of a bad end was vexatious work for the machine-man, and a severe strain on the machine and the rails. The method of laying the rails along the full length of the face helped considerably to keep the face straight.

With reference to the generating plant, the high boiler-pressures mentioned by Mr. Millington were not, as yet, found at many collieries, so that for general use a belt-driven or direct-driven slow-running engine was still the best.

Mr. W. N. ATKINSON (H.M. Inspector of Mines, Stoke-upon-Trent) moved a vote of thanks to the Geological Society for the use of their rooms, to the owners of works to be visited by the members, and also to the governors of the National Physical Laboratory.

Mr. JAMES BARROWMAN (Hamilton) seconded the resolution, which was cordially adopted.

————

Mr. W. H. PICKERING (H.M. Inspector of Mines, Doncaster) moved a vote of thanks to Sir Lindsay Wood for his services in the chair.

Mr. G. C. ROBINSON (Brereton) seconded the resolution, which was cordially adopted.

————

The following notes record some of the features of interest seen by visitors to works, etc., which were, by kind permission of the owners, open for inspection during the course of the meeting on July 2nd, 3rd and 4th, 1903 : —

NATIONAL PHYSICAL LABORATORY.[*]

The laboratory is at Bushy House, Teddington (Fig. 1), and the old house is admirably suited for many purposes.

The engineering laboratory (Figs. 2, 3 and 4) is an almost entirely new building, 80 feet long by 50 feet wide. It is divided longitudinally into two bays, each of which is lighted from the north by a weaving-shed roof. A power-shaft, driven by an electric motor, runs along one bay, and in it are placed the machine-tools, comprizing four lathes, a universal grinder, a shaping-machine, drilling-machine, etc.

The second bay, used for experimental work, contains the testing and part of the pressure-gauge testing apparatus (Fig. 4). The testing-machine is placed at one end (Fig. 3). A wind-pressure apparatus is also set up. A traversing-crane runs along this bay, and has already proved of great assistance in erecting the machinery and apparatus.

———

[*] Abstracted by permission of the Council from a lecture on " The National Physical Laboratory and its Relation to Engineering," by Dr. R. T. Glazebrook, *Proceedings of the Institution of Mechanical Engineers*, 1903, pages 57-8 .

A small forge and smelting-shop are situated near the engineering building.

The engine-room contains a condensing Parsons turbine of 75 kilowatts. There is also a gas-engine of 18 horsepower, driving a dynamo. A very convenient booster-set is used for charging the storage-batteries. The boiler-house contains a boiler of 100 horsepower, serving for the heating of the engineering laboratory and of Bushy House.

FIG. 1.—SOUTH FRONT OF BUSHY HOUSE.

Although most of the work specially interesting to engineers will be conducted in the engineering laboratory, several of the rooms in Bushy House are fitted for experiments which have a direct bearing on engineering practice. Fig. 5 shows the metallurgical room, in which it is hoped to carry on the work begun at the Royal Mint by the late Sir William Roberts-Austen, whose labours have helped so greatly to elucidate some of the changes which go on in metals. On the extreme left are the gas-furnaces, and in the centre is the Roberts-Austen pyrometer as fitted at the Royal Mint. To the right are the coke-furnaces: the room was the old kitchen, and the chimney and fire-place have been thus utilized.

Fig. 6 shows the thermometric laboratory, especially arranged for the accurate measurement of high temperatures. The various baths and ovens can produce and measure steady temperatures up to 1,250° Cent. For temperatures up to 240° Cent., an oil-bath is employed; from 220° to 650° Cent., a bath

consisting of fused nitrates of sodium and potassium is most
convenient; and from 500° to 1,250° Cent., electric ovens
afford by far the best means of obtaining a high temperature.

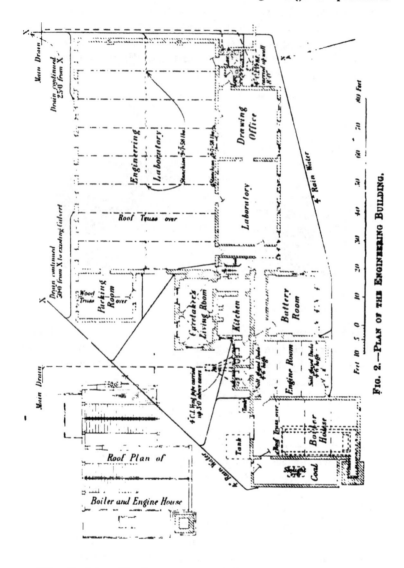

FIG. 2.—PLAN OF THE ENGINEERING BUILDING.

Fig. 7 shows the basement-room in which the measuring-
machine, dividing-engine and standards of length are housed.

work
of the
(3) e
perin
reque
other
fees a
(
is we
testii
zatio
gaug
colun
feet
been
Busb
side
per
of w
lowe
from
300]
alon;

pour
direc

Sir Benjamin Baker, the specific heat of superheated steam, the effects of rapidly alternating stress, and the temperature of the field - coils of dynamos.

(3) A list of the enquiries and experiments made at the request of engineers and others indicates

FIG. 7.—GAUGE ROOM IN BUSHY HOUSE.

that the results of experiments on small details may be of great service.

LONDON UNITED ELECTRIC TRAMWAYS, LIMITED: CENTRAL POWER-STATION, CHISWICK.

The boiler-house is a steel-frame structure, made especially strong so as to support a coal-store, containing 500 tons of coal, and the feed-water tanks, placed above the boilers. There are eleven boilers, of the water-tube type, each capable of evaporating 11,000 pounds of water per hour at a steam-pressure of 150 pounds on the square inch, with a draught, measured in the main flue, of 0·75 inch of watergauge. The heating surface of each boiler is 3,140 square feet. Each boiler is fired by a Vicars mechanical stoker, driven by shafting placed underneath the floor.

There are two coal-receiving hoppers, one at either end of the boiler-house, and the coal is lifted from these hoppers to the coal-store over the boilers by means of an endless chain of buckets, capable of handling 40 tons of coal per hour. The conveyor is driven by a 10 horsepower motor: it is also used for the removal of ashes, and the buckets run along the basement in front of the ash-pits, the ashes being dumped into an ash-bin which shoots them into a receiver outside the building.

There are two steam-driven feed-pumps, designed to deliver 6,000 gallons of water per hour against a boiler-pressure of 180 pounds per square inch, at a piston-speed of 50 feet per minute. A third steam-driven pump is capable of delivering 8,000 gallons per hour under similar conditions.

The flue is placed above the boilers, and the gases are taken into it by breeching-pieces. It is constructed of steel-plates, and is 150 feet long, $6\frac{1}{4}$ feet wide, and varies in height from 3 feet to $13\frac{1}{2}$ feet. Two brick-branches go from this flue to the chimney, in one of them being placed the economizer, which contains 360 tubes.

The chimney, 10 feet in diameter, is 240 feet high, and the upper length of 200 feet consists of steel-plates, $\frac{1}{2}$ and $\frac{9}{16}$ inch thick, lined with firebricks.

Between the flue and the coal-bunkers are placed two feed-water tanks, each having a capacity of 6,000 gallons. The water-softening plant is capable of treating 10,000 gallons of water per hour, reducing it to 5 per cent. of hardness on the Clarke scale. Water is obtained from three wells, each capable of yielding 3,600 gallons per hour, with a water-level of not more than 16 feet below the surface. An electrically-driven three-throw pump is capable of lifting 50,000 gallons of water per hour from the wells to the feed-water tanks.

The cooling-towers, of the twin-type, are arranged in three sections, cover an area of 40 feet by 15 feet, and are 38 feet high. They have a capacity for dealing with cooling water for 6,000 horsepower. The mats inside the towers are made of galvanized iron-wire netting. Fans, worked electrically, are provided for each section, and run at 250 revolutions per minute.

The engine-room is 60 feet wide and 154 feet long. An electric overhead travelling-crane runs its whole length. There are two direct-current and three three-phase generators: three of the units representing 500 kilowatts each, and two 1,000 kilowatts each. There are three vertical cross-compound Corliss condensing engines of 700 horsepower. In the case of the direct-current generators, each engine drives two compound-wound continuous-current generators of 250 kilowatts, with a flywheel between them. These generators are used for supplying current at 500 to 550 volts to the lines in the vicinity of the power-station. For the outlying districts, high-tension three-phase machines are used: there is one of 500 kilowatts at 5,000 volts, and 25 cycles per second. The three-phase generators, of 1,000 kilowatts, are direct-coupled to 1,400 horsepower engines. An engine of 100 horsepower is connected to a dynamo for lighting purposes.

There are two surface-condensers, each capable of dealing with the steam represented by an output of 1,000 kilowatts, and one for dealing with 2,000 kilowatts.

There is one rotary converter of 500 kilowatts, and two negative boosters, one being in circuit with the cable connected to the rails in Brentford, and the other with a similar cable to Acton.

The main switchboard consists of twenty-seven marble and angle-iron panels. A smaller switchboard deals with the lighting and the government of the various motors throughout the premises.

The station contains machines capable of developing 1,000 kilowatts direct current, and 2,500 kilowatts alternating current for conversion to direct current. Assuming that a car when fully loaded requires 8½ kilowatts to run it, and allowing for loss in transmission, etc., it is calculated that 375 cars can be worked from this station, or 315 cars, allowing 500 kilowatts to be held in reserve.

The car-sheds are capable of accommodating 100 cars; and repairing-shops, a machine-shop and two smiths' hearths are provided. The sheds contain eleven lines of rails, furnished with inspection-pits. The movement of the cars is brought about by overhead-trolley wires.

The buildings and plant were designed by Mr. J. Clifton Robinson, engineer to the London United Electric Tramways, Limited, and the construction and equipment were carried out by the staff of the company, under his supervision.

THE COUNTY OF LONDON ELECTRIC SUPPLY COMPANY, LIMITED: CITY-ROAD STATION.

The area supplied from the City-road station includes the St. Luke and Clerkenwell districts, the liberties of Saffron Hill, Glasshouse Yard and Charterhouse, together with the eastern portion of Holborn, parts of Holborn west and the parish of St. Giles. The districts were originally laid out for an alternating-current supply, but, subsequently, continuous-current plant was installed and a supply for power-purposes is now given at 530 volts in addition to the alternating-current supply at 104 volts for lighting.

The station was erected on a site having a frontage of 303½ feet along the canal-basin of the Regent Canal, near the spot where the latter is crossed by the City Road. Much of the subsoil proved to be of earth, though a considerable quantity of gravel and sand was encountered before reaching the blue clay at 30 feet below ground-level. It was decided to construct the buildings within a watertight tank, formed by a thick layer of concrete over the whole area, and having watertight walls on the canal-side. The buildings were constructed throughout of best selected stock-bricks, laid in Portland cement.

The boiler-room runs alongside the canal-front, and measures 198¼ feet in length by 41½ feet in breadth. There are 17 Babcock-and-Wilcox boilers, designed for a working pressure of 150 pounds per square inch, and fitted with superheaters, which raise the steam about 150° Fahr. above its temperature of saturation. Two Green economizers are provided at one end of the boiler-room, whilst on a short length of floor immediately above them are fixed two Weir steam-pumps, each capable of supplying 4,000 gallons of feed-water per hour. There are also two motor-driven pumps of similar capacity. The condensed steam, collected in a large hot-well, is passed through two Railton-and-Campbell oil-filters before being supplied to the feed-pump.

The flue-gases are analysed by an Ados carbonic-acid recorder.

The coal is brought up the canal in steel-barges, each holding 80 tons. A grab, holding 15 cwts., suspended from an hydraulic crane, picks up the coal from the barge and dumps it into the hopper of an automatic weighing-machine, through which the whole supply is weighed before being passed into the bunkers. The bunkers are situated on a floor over the boiler-room, on either side of a central tramway. After weighing, the coal is distributed to the bunkers by chain-conveyors running along the coal-store from end to end. The plant is capable of dealing with 30 to 40 tons of coal per hour, and is worked by 1 crane-man and 2 bargemen, who heap together the coal ready for the grab, with the occasional assistance of 1 trimmer, who closes the shoot-doors to each bunker as it is filled. The store is capable of holding about 1,000 tons.

A special weighing-car runs along the rails between the bunkers, and is filled by a trimmer, who notes every charge. This weighing car has inclined sides and an end-door, and every opening to discharge a load into the hoppers of the mechanical stokers is automatically recorded, thus maintaining a check on the work of the trimmer, whose list of weighings during his shift must tally with the record on the counter.

There are two Doulton water-softening plants, through which all the make-up feed-water, obtained from the New River main, is passed before use.

The engine-room, immediately behind the boiler-room, measures 60 feet in width by 180 feet in length. The machinery consists, in the first place, of two travelling-cranes used for erecting and overhauling the engines, and of auxiliary machinery in the shape of one motor-driven and two steam-driven centrifugal pumps, used for circulating the condensing water; and, in addition, there are three steam-driven exciters.

The generating-plant comprizes thirteen direct-coupled sets, as follows:—Six two-phase 50 periods per second alternators supplied by the Electric Construction Company, Limited, each having a normal output of 150 kilowatts, driven by Brush marine-type compound condensing engines, fitted with Raworth flywheel-governors. Three two-phase, 50 periods per second, Mordey inductors, each of 400 kilowatts output, driven by Brush universal compound condensing engines. The continuous-current machinery consists of two generators supplied by the British Thomson Houston Company, Limited, each of 525 kilowatts output, driven by Allis cross-compound condensing Corliss engines; and two generators each of 600 kilowatts output, also supplied by the British Thomson Houston Company, Limited. driven by Yates-and-Thom compound condensing horizontal engines. In addition, there are two Brown-Bouverie motor-generators, each of 250 kilowatts output; and a motor-generator of 420 kilowatts output, supplied by the Electrical Company, Limited.

The steam-pipes, supplying the engines, are arranged on the ring-plan. The plant is generally worked condensing, but can be worked non-condensing, the exhaust-steam being then turned up the chimney, 200 feet high. The condensing water is obtained from the canal, and is strained before being permitted to pass into the condensers.

The accumulator-plant consists of 266 Tudor cells.

APPENDICES.

I.—NOTES OF PAPERS ON THE WORKING OF MINES, METALLURGY, ETC., FROM THE TRANSACTIONS OF COLONIAL AND FOREIGN SOCIETIES AND COLONIAL AND FOREIGN PUBLICATIONS.

CONTACT-METAMORPHIC ORES OF BALIA MADEN, ASIA MINOR.

Beiträge zur Kenntniss der contactmetamorphen Lagerstätte von Balia-Maden. By G. BERG. *Zeitschrift für praktische Geologie,* 1901, *vol. ix., pages* 365-367.

The author describes the macroscopic and microscopic characters of some of the specimens from this ore-deposit, which are in the possession of the Freiberg Mining Academy.

The sulphidic ores are predominantly galena, iron-pyrites and blende. Copper-pyrites also occurs, and arsenic fahlore is associated with some of the coarsely-crystalline galena and iron-pyrites. Among the oxidic ores, that one to which the author draws special attention is smithsonite, occasionally encrusted with calamine, and betraying (by the blowpipe-test) the presence of a considerable proportion of cadmium.

The country-rock is represented by fossiliferous Carboniferous Limestone, granular limestone, and a recent tufaceous freshwater limestone. There is also a specially interesting (from the petrographic point of view) calcium-silicate rock which occurs in the contact-zone of the augite-andesite. It contains epidote, garnet, pyrites and galena, and shows that the pyrites was formed almost at the same time as the garnet, after the epidote had crystallized out. The galena was formed much at the same time as the anorthite-felspar, for both minerals make up the cement which fills the interstices between the older constituents. But, whereas the anorthite occurs in the more garnetiferous portions of the rock, the galena is almost invariably associated with the pyrites. This rock was doubtless originally a limestone.   L. L. B.

COPPER-ORES IN PORTUGUESE WEST AFRICA.

Das Kupfererzvorkommen bei Senze do Itombe in der Portugiesischen Provinz Angola, Westafrika. By F. W. VOIT. *Zeitschrift für praktische Geologie,* 1902, *vol. x., pages* 353-357, *with* 1 *figure in the text.*

Senze do Itombe is situated on the Loanda and Ambacca railway, 117 miles distant from the former town, which, besides being the capital of the province of Angola, ranks also as a seaport. One train runs daily thither, taking 12 hours to accomplish the journey. The ore-deposits lie 1½ miles west of Senze, but only ⅓ mile distant from the railway-line.

The author gives a brief description of the physiography and climatic conditions of the region, and then describes fully the geology of the more immediate neighbourhood of Senze. Under a thick mantle of alluvial and drift-deposits lie in descending order.—(1) White, fine-grained, calcareous sandstones, (2) iron-stained quartzose conglomerates, frequently interbedded with micaceous marls, and (3) fine-grained sandstones, partly calcareous and micaceous. All these beds contain fossils, by means of which the group of strata has been assigned to the Upper Cenomanian division of the Cretaceous series.

The copper-ore occurs mostly in connexion with the conglomerates, and cupric carbonates (associated with carbonate of lime and kaolin) form the cementing-material of the rock. In some exposures, the conglomerates are so completely impregnated with copper-ore that they may be regarded as ore-beds. Near the outcrop, the ores are chiefly malachite and azurite, but these are the secondary products of sulphidic ores; in fact, in places the original copper-glance is found, the proportion of it increasing as one goes deeper down. The chemical reactions set up by the percolating surface-waters, more particularly active under tropical conditions, account for the transformation of the sulphidic into oxidic ores nearer the outcrop (down to a depth of 36 feet or so). As the strata dip south-westward, the cover increases in thickness in that direction, and one may therefore expect to meet with a thinner but more richly-metalliferous bed along that line. So far as mining operations have gone at present, on the gossan, pockets of ore containing from 12 to 18 per cent. of metallic copper are found to alternate with sparsely-impregnated beds containing only 1 or 2 per cent. of the metal. The author speaks hopefully of the probable amount of sulphidic ores available, and he mentions an occurrence of limestone-gravels completely "saturated" with copper-ores, about ¼ mile north of the main deposit. This northern deposit appears to be an instance of metasomatic replacement of limestone by copper-compounds. The water-supply of the neighbourhood of the mines is amply sufficient for industrial purposes.   L. L. B.

THE AURIFEROUS DEPOSITS OF RAPOSOS, BRAZIL.

Beiträge zur Kenntniss der Goldlagerstätten von Raposos in Brasilien. By GEORG
   BERG. *Zeitschrift für praktische Geologie*, 1902, *vol. x., pages* 81-84, *with*
   5 *figures in the text.*

The deposits of Raposos, east of the Rio das Velhas, are perhaps among the least-known of the auriferous occurrences which enrich the province of Minas Geraes in Southern Brazil. It is true that the gold-output of Raposos can hardly be termed considerable, but the mode of occurrence of the auriferous pyrites is interesting: the ore is found in more or less cylindrical "pockets" which are traced obliquely downward through the rocks.

The predominant strata of the district are grey pre-Cambrian clay-slates and phyllites, sometimes represented by chlorite- and sericite-schists, with which are interbedded itabirites and various quartzites. The itabirites are so highly charged with magnetite that surveying-work with the compass-needle is practically impossible throughout the district. The quartzites vary in character from finely-schistose to coarsely-granular, the latter variety containing a remarkably large proportion of calcite. The slaty rocks are traversed by two diabase-dykes, and the slates continue with the same strike and petrographical character across country into the important gold-bearing rocks of Morro Velho.

Now, the slates of Raposos have been much folded, thrust and sheared-

and it is in such sheared portions that the pockets of ore occur, measuring generally between 10 and 20 feet, though in some cases they may measure as much as 40 feet, and in others as little as 8 inches. The "pockets" are arranged in rows parallel to the strike of the slates, and are infilled with quartz and pyrites, predominantly iron-pyrites. The quartz forms the nucleus of the "pocket," and is microbrecciated; the largest "pockets," however, instead of being directly enveloped by the slate, have a peripheral layer of fibrous quartz or quartzite. Arsenical pyrites is not of uncommon occurrence at Raposos, and its presence there is welcomed by the miner, as it contains a high percentage of gold. Of subsidiary importance only are the occurrences of chalcopyrite, magnetic pyrites and zinc-blende. The magnetic iron-ore, which is most probably the residuum of leached-out itabirites, is a frequent associate of the ore-pockets. Occasionally also geodes are found, containing true vein-quartz with auriferous pyrites, and very small nuggets of native gold.

There is no evidence to show that the "pockets" are mere widenings or bellyings of fissure-veins. On the other hand, quite independently of the lie and arrangement of these "pockets," small pyritiferous veins course through the country-rock and some of the unsheared slates are found to be impregnated with pyrites. It seems probable that the Raposos ores are the result of the percolation of metalliferous solutions (genetically connected with the diabase-eruptions) among the sheared schists and quartzites, and the author ingeniously explains how both "pockets" and small veins can thus be accounted for.

Down to depths of 10, 20 feet and more, the surface-rocks in this tropical area (19 degrees 58 minutes south latitude) are weathered and decomposed almost beyond recognition, and at one time the gold was got by simply washing these loose "lateritic" surface-deposits. L. L. B.

––––

THE CUPRIFEROUS DEPOSITS OF CHUQUICAMATA, CHILE.

Asiento Mineral de Chuquicamata. By CÁRLOS G. AVALOS. *Boletin de la Sociedad Nacional de Mineria, series 3, 1901, vol. xiii., pages 145-153.*

The mines of Chuquicamata have increased enormously their output within the last year or two, owing to the stimulus afforded by the rise in the price of copper. An output which averaged annually 5,000 tons (containing 18 per cent. of metallic copper) previous to 1899, reached in that year and in 1900 respectively 18,000 tons of about the same quality. The output for 1901 is estimated at 25,000 tons containing 17 to 18 per cent. of metallic copper. The slight increase recorded of late in the copper-output of Chile is due to these mines alone, the production of other and older mines in that country having diminished or remained stationary.

Chuquicamata lies 12½ miles north-west of Calama station, on the railway-line from Antofagasta into Bolivia. Calama marks the southern limit of the pluvial region of the Bolivian high plateau; southward and westward of it stretches the great desert of Atacama. The neighbourhood, of itself, offers no inducement to the agriculturist, but in consequence of the development of the mining industry strenuous efforts are being made to modify the natural conditions of the soil, so as to establish on the spot the necessary food-supply, etc. for a vast mining population. The Rio Loa, coming down from the snow-clad Andes, will furnish with its numerous water-falls a good source of motive power, and its flow is so regular and so abundant that it can be used for irrigation as well. The district, moreover, is rich in deposits of sulphur, hydroboracite, sulphates and nitrates of soda, whence the reagents necessary for treating the ores either in the dry or the wet way could be obtained. A small branch from the Antofagasta railway runs very nearly up to the mines, but as the intervening distance has to be traversed in carts, entailing increased cost of transport, many mine-owners prefer to send their output direct to Calama by road, instead of availing themselves of the small branch-railway. It would appear that the owners of the last-named at present control in some way the water-supply to the mines, and the author appeals to Government intervention as the remedy for the present state of matters.

The metalliferous deposits extend over an area of 2½ miles from north to south and 1 mile from east to west: but, in the latter direction especially, fresh discoveries are continually enlarging the boundaries of the mining-field. The district is one of low undulating hills with a general southerly slope, cut by gullies. The surface, down to a depth of 7 feet or so, is made up of very recent detrital material, the débris of the neighbouring rocks. The existence of this superficial mantle concealing the mineral outcrops explains how it is that their number and importance have been recognized only by slow degrees. Below it occur the granites, pegmatites and syenites, which are the true country-rock of the Chuquicamata ores.

These ores occur as true veins, the main strike of which is north and south, although the San Luis, one of the most important, runs east and west. In thickness they vary from 3 to several feet (at the intersections, etc.): near the surface they yield a natural sulphate of copper, lower down a sulphate of copper and soda (krönkite), deeper still and in far greater abundance atacamite, the oxychloride of copper. At a depth of about 200 feet, occur the black oxide and the sub-sulphide of copper with a ferruginous gangue; at 330 feet (the greatest depth so far attained at Chuquicamata) occur copper- and iron-pyrites.

The central portion of the field is taken up by what are known as *llamperas*, the rocks being traversed in every direction by a multitude of small fissures, filled with copper-ores exhibiting the same varietal succession in depth as was noted in the case of the true veins and having doubtless the same origin. In fact the *llamperas* are regarded by the author as the upward or lateral prolongation of the less numerous but far more considerable fissures constituting the true veins, and there is reason to believe that more of such veins will be ultimately struck by following the *llamperas* downward.

The principal mine is the Poderosa, with a monthly output of 300 tons containing 20 per cent. of metallic copper. In this and in the Zaragoza and Emilia mines, work is being actively pushed forward, with a view to an extension in depth when it is hoped to reach the sulphidic ores. Moreover, it is possible that in this way an artesian water-supply will be tapped.

The author thinks that a brilliant future lies before the copper-mining industry in this region, especially if what he considers the right treatment of the ores is adopted. He is in favour of the " wet way," using sulphurous acid as a solvent.　　　　　　　　　　　　　　　　　　　L. L. B.

THE COPPER-SULPHATE DEPOSIT OF COPAQUIRE, NORTHERN CHILE.

Eine Exkursion zur Kupfersulfat-Lagerstätte von Copaquire im nördlichen Chile. By
HANS OEHMICHEN. *Zeitschrift für praktische Geologie*, 1902, *vol. x., pages*
147-151.

The author describes in some detail the general features and geology
of the country as observed by him in a journey north-eastward from the
Challacollo silver-mines (78 miles south-east from Iquique, as the crow flies),
over the high dune-covered plateau of the Pampa Tamarugal, past Chelles,
Tiquima and Huatacondo, to Copaquire, remarkable for its hitherto unique
deposit of copper sulphate.

In the eruptive andesites and contiguous sedimentaries of Tiquima some
exploration-work has been done on oxidic copper-ores. This occurrence is,
as it were, the outpost, the first indication, of the cupriferous deposit farther
within the mountains, which is presently to be described. Huatacondo is
3 miles or so farther up the glen than Tiquima, and 12 miles beyond that
one comes to the granite-massif wherein the copper sulphate occurs.

This remarkable deposit had evidently been worked in olden times by the
native Peruvians, but it was re-discovered only 4 years ago. Copaquire is
about 10,500 feet above sea-level, and 44 miles east of Challacollo. The
deposit is associated with the contact-zone between the granite and certain
sedimentary rocks. Practically, all the rocks are impregnated with particles
of copper-pyrites, but the eastern ridge of the Pastillo ravine (granite proper)
is the most highly metalliferous. It is in decomposed belts of the granite
too that the copper sulphate mostly occurs in magnificent dark-blue
"stringers." Brown hæmatite and gypsum are commonly associated with
the mineral, and less commonly malachite, azurite and chrysocolla. The
talus below the ridges, consisting of débris largely impregnated or brecciated
with copper sulphate, would of itself afford scope for profitable mining
operations. Although rain very seldom falls at Copaquire, it does so sometimes, and thereby the uppermost layer of the débris is leached out and
compacted into a sort of crust. Where the copper sulphate is in direct
contact with the dry air, it loses, occasionally, its water of crystallization
and crumbles into a white powder. The percentage of metallic copper
varies between 2·5 and 3.

It seems that plans are being developed for working the Copaquire deposit
on a large scale. As to the primary origin of the copper-ores, it is noticeable throughout this portion of the Province of Tarapaca that they are
associated with acidic eruptive rocks. A more detailed study is promised
for a subsequent occasion. L. L. B.

THE COPPER-ORES OF AMOLANAS, CHILE.

Das Kupfererzlager von Amolanas im Departement Copiapó (Chile). By A. ENDTER.
Zeitschrift für praktische Geologie, 1902, *vol. x., pages* 293-297.

The mines of Amolanas rank among the most important in the department of Copiapó, and are situated about 15 miles south-east of the San
Antonio railway-station (on the line from Caldera to the Pacific seaboard),
at the upper end of a gorge, at a height of 7,216 feet above the ocean.
The district is comprised within the southernmost portion of the great
desert of Atacama, and 14 mining concessions have been pegged out in it,
in strips 2½ miles long and 820 feet broad.

The dark shales and quartzitic sandstones of the Lower Oolites and the Lias, astonishingly like the typical Jurassic of Southern Germany, are torn through, overlain by, or partly interbedded with dykes, sills, sheets and bosses of eruptive rock. Forming the ramparts of the Amolanas valley, two well-marked series of dykes strike east and west, parallel one with the other, about 3,000 feet apart: in the intervening space a complex of smaller dykes cuts through the Jurassic sedimentaries in various directions. The northernmost "rampart" consists mainly of a whitish liparite, which passes into a more distinctly porphyritic rock in depth. The liparite is full of irregular cavities which are in part infilled with copper-ore. This "rampart" of acidic rock is cut by sills of melaphyre and diabase-porphyrite, and associated with these, there are extremely coarse tuffs containing slaggy volcanic bombs.

The copper-ores occur chiefly in the form of impregnations, the general "habit" of which indicates that they crystallized out from mineral solutions circulating through the northern belt of quartz-porphyry, near the junction with the later basic intrusions. The last-named may therefore be regarded as the true ore-carriers, although on account of their own barrenness they are stigmatized as *caballos de piedra* (stone horses) by the miners. That the ores should occur over a length of only 1 mile, while the eruptive belt can be traced for several leagues, is attributable to a transverse disturbance and overthrusting, from north to south, fracturing the entire mass. The fissures thus formed were filled with metalliferous gangue, and caused an enrichment of the ores at those points where they traversed the ore-body.

The metalliferous particles were precipitated from the circulating solutions in the clefts and fissures and cavities, by the alkalies or alkaline earths contained in the rocks. These, with their alkalies thus leached out of them, became proportionately richer in silica and alumina, forming what is known in Chile as the *manto* (capel). The principal ores at Amolanas are dark leaden-grey copper-glance, red copper-ore (chiefly in compact or earthy masses at the outcrop), and blackish-green atacamite (in reniform, fibrous and compact masses). Of less frequent occurrence are malachite, chalcopyrite, bournonite, etc. In view of the fact that the ore as a whole is mainly an impregnation, the percentage of metallic copper in the deposit is low, ranging from 8 down to 5. Nevertheless the amount of copper in sight is reckoned at 1,215,000 tons. Working has been carried on, so far, in a terribly wasteful and unsystematic fashion.   L. L. B.

THE AURIFEROUS DEPOSITS OF SURINAM OR DUTCH GUIANA.

Geologisch-bergmännische Skizzen aus Surinam. By G. C. Du Bois. 1901 [*published at Freiberg for the Maatschappij Suriname, of the Hague*], *pages* i.-iv., 1-108, *with* 13 *figures in the text,* 2 *plates of microscope-sections and a geological map.*

The geography, climate and geological structure of the colony are dealt with, and a short history is given of the comparatively recent development of gold-mining there. It would appear that the most important auriferous deposits are those of secondary origin, that is, the alluvial and other placers. "Reefs" have been proved at some places, but it is too early yet to say whether they would repay working: the true reefs or veins will be referred to later on.

Iron-ores, containing as much as 52·6 per cent. of metallic iron, are associated with the laterites (rocks which are the result of the decomposition, under tropical climatic conditions, of diabases, diorites, etc.), but the cost of transport to the nearest available harbour is so enormous, that mining operations are at present out of the question. Another face would be put on the matter if railway-communication were provided. Copper-ores are said to occur, but the author has not seen them himself. The galena of the Carolina district is of no industrial importance. Concerning precious stones, only hearsay evidence as to diamonds is forthcoming, and the sapphires, zircons, garnets, etc., which the author has seen *in situ* do not call for special remark. There are widespread deposits of clays suitable for pottery purposes and of loams suitable for brickmaking. From this brief digression we return to the main subject of mining enterprise in Surinam, that is, gold.

The search for veins or reefs in the bush of Dutch Guiana is attended with unusual difficulty, as in most cases no outcrop or gossan or indicator of any kind is to be seen at the surface, and the prospector has usually to make trial-diggings in a happy-go-lucky fashion. The different complexes of "veins" are manifestly associated with the ancient eruptions of granites and diorites, and the richest of all occur in the old crystalline schists. All these rocks are much weathered, and both walls of the veins consist of an iron-stained, kaolinic, decomposed material. The older veins strike parallel to the schists and pitch steeply northward, while the newer strike north and south across the older, and pitch steeply eastward. The vein-stuff is quartz; this where richest in gold, assumes a reddish or a bluish-grey tinge, milky quartz being here a bad indicator. Iron-pyrites, tourmaline and iron-glance are frequent associates of the precious metal. The older veins average 3½ feet in thickness, while the younger are only 10 to 12 inches thick. The native gold occurs in a very fine state of division, and also in large flakes among the cracks and crevices of the quartz. The pyrites, too, is auriferous, and the immediately neighbouring country-rock is often so richly impregnated with gold as to be worth working.

At depths of 100 and 165 feet below the surface, the greatest yet reached, pyritic ores seem to become more abundant; but the author is unable to say whether the conditions here are similar to those in British Guiana—that is, whether the auriferous veins have been more or less leached out near the outcrop, and become richer lower down.

Auriferous placers are distributed broadcast through Dutch Guiana: indeed, there is scarcely a single valley where gold cannot be said to occur, not only on the valley-bottom but on the hill-slopes also. The author shows how these placers originate primarily from the weathering, most often of "greenstones," less often of crystalline schists, granitic contact-rocks, etc., into laterite; and how the lateritic material deposited in the shape of clays, loams and gravels, forms the "eluvium" of the benches or deep placers, and the "alluvium" of the creek-bottoms or shallow placers. He does not consider, therefore, that the gold has travelled far from its original matrix: this is opposed to the popular local traditions in Surinam, which point to the still unexplored mountain-ranges, where Dutch territory marches with that of Brazil, as the *locus* of the auriferous rock. As a general rule, the alluvial placers are found to become richer as they are worked deeper down to the bed-rock, but in the eluvial placers or benches the deposit is richest some little distance above the bed-rock. The gold is of very fine quality, containing in 1,000 parts 926 of gold, about 50 of silver, and traces of copper; it is of a deep yellow colour. There is a so-called "black gold," which will not amalgamate: the precious metal is covered by a deep black to brownish-red crust of iron oxide and organic matter. Ferruginous "greenstones" invariably occur in the immediate neighbourhood of the localities where this variety is found.

The author devotes a short chapter to hints on prospecting, and then gives a lengthy and well-illustrated description of the methods of working the placers. Great progress, in this matter, he says, has been achieved in Surinam within the last 15 years, although the greater number of placers are still worked on the primitive "long-tom" and wooden-sluice systems. But hydraulicking on a large scale with modern American plant has been started on several properties, and Government concessions have been obtained for the use of steam-dredgers in certain creeks.

The climate does not permit of the use of other manual labour than that of negroes or Indian coolies: the European soon breaks up under the strain, if he attempts to do hard physical work. In 1900, the gold-production of Dutch Guiana reached a total of 28,041 ounces troy. A bibliography of the colony appropriately terminates this useful little work.

L. L. B.

BITUMEN IN CUBA.

Bitumen in Cuba. By T. WAYLAND VAUGHAN. *The Engineering and Mining Journal* [*New York*], 1902, *vol. lxxiii.*, *pages* 344-347, *with 2 illustrations in the text.*

The writer was one of the three geologists appointed at the request of Governor-General Wood to make a reconnaissance of the island, and the data included in the account were obtained partly at first and partly at second-hand. The survey dealt with deposits in the provinces of Pinar del Rio, Habana, Matanzas, Santa Clara and Puerto Principe. Numerous deposits of greater or lesser value were examined.

From the descriptions it appears that all the deposits of asphalt about which specific information was obtained occur as veins, pockets, or exuding springs, usually in serpentine-rock, but occasionally in limestone. None of the material is now found in its original matrix, it having come into its present position from elsewhere, but it is not at present possible to state the original place of its occurrence.

As for the commercial value of the deposits, the bituminous substances, asphalt and mineral tar, occur from place to place in every province of Cuba. In some localities there is promise of fairly large quantities. Much of the material can be used in the manufacture of varnish and for insulating purposes, but the practicability of its use for other purposes is at present an undecided question. Apparently in the glance-pitch the proportion of asphaltene is too high to allow of those cementing qualities required for paving, roofing, or asphalt-cement. Some of the material possesses large amounts of petrolene, but the stability of its chemical combination is yet to be determined. It is suggested that the glance-pitches containing much asphaltene might be mixed with maltha containing much petrolene. In undertaking any commercial handling of these Cuban asphalts and maltha, there should first be a careful examination to determine quantity, and if proved that a property possesses sufficient quantity, it should be chemically studied in order to determine the precise uses to which it can be put. These bituminous substances offer an inviting field to the capitalist, but it is suggested that anyone contemplating such an investment should proceed with the greatest care.

X. Y. Z.

AURIFEROUS DEPOSITS IN SOUTHERN HAYTI.

Beiträge zur Geologie Haïtis: V. Das La Selle Gebirge, die Cul-de-Sac Ebene und das Salzseengebiet. By L. GENTIL TIPPENHAUER. *Petermanns Mittheilungen, 1901, vol. xlvii., pages 169-178, and 2 plates.*

This paper, on the whole, deals with a district which is interesting from every other point of view but that of the metalliferous miner. It contains, among other information, a catalogue of the 42 earthquakes recorded at Port au Prince, the capital of the Haytian Republic, between September, 1890, and November 25th, 1900.

In the latter portion of the paper, however, we learn that the Gosseline river, which runs down to the Caribbean sea at Jacmel, is, for a stretch of about 4 miles below its confluence with the Mabial, full of auriferous sands. These are estimated, from an assay made in New York, to contain 3 parts of gold per million. It is said, moreover, that gold has been found in the river-bed lower down its course, below the point where it joins the Grande Rivière, and also in alluvium near Jacmel.

The author found that the doleritic basalt through which the Gosseline cuts its way itself contains 3 parts of gold per million. It is evidently distributed throughout the rock in a very finely divided condition, so finely divided that the particles are invisible to the unaided eye. On the other hand, as gold has been found in flakes of about ⅛ inch in diameter, it would seem that there must be an outcrop somewhere in the district, of rocks still richer in gold than these basalts. Now on examining the auriferous river-sand under the microscope, the author discovered not only that each particle of gold was adherent to a fragment of reddish-violet quartz, but that wherever quartz of that colour came into view, so surely was gold to be found with it. The district where the violet quartz-reef must occur, if its outcrop is to be found at all, is a triangle of about 324 square miles, bounded on the south by the junction of the basalts with the Tertiary limestones, on the north by the watershed of the Selle mountain-range and on the west by the watershed between the Gosseline and the Grande Rivière. The basalt-dykes, with which quartzose rocks are apparently associated, will be easily distinguished from the superficial basalt-flows. The neighbourhood of the dykes is generally indicated by outcrops of nearly vertical shale, torn asunder and thrust aside by the eruptive magma. These shales evidently underlie the newer (?) Tertiary limestones and sandstones. The facilities of communication with the district, and the abundance of water-power available, would contribute greatly to the success of gold-mining enterprise, if it should turn out that a workable reef occurs there.   L. L. B.

— —

THE BROWN-COAL DEPOSITS OF MAÏSSADE, HAYTI.

Das Lignitlager von Maïssade und der Aufstieg zum Zentralplateau von Gonaïves und vom Norden aus. By L. GENTIL TIPPENHAUER. *Petermanns Mittheilungen, 1901, vol. xlvii., pages 193-199, with 3 figures in the text, and 2 plates.*

In the centre of Hayti, surrounded on every hand by limestone-mountains ranging from 3,500 to 5,000 feet in altitude, is a broad flattish tableland, some 1,500 feet or less above sea-level. The geological evidence shows that this area is a former bay of the older Tertiary sea, and that it rose by a slow process of elevation during Pliocene times, so slow that the beds in the centre of the basin have neither been fault~~ ~~ ~~ ~~~bed. It seems likely enough that artesian borings will tap a reservoir of petroleum—the product of the dry distillation of the long-buried marine fauna of the ancient gulf. As it is, petroleum-springs are known to exist some distance to the south, near Azua, in the same geological formation.

It is in this plateau-region, 6¼ miles northwest of Maïssade, that the brown-coal belt occurs. From north-west to south-east the belt extends for more than 7 miles, with an average breadth of 3¼ miles; but, as new outcrops of coal are being continually discovered, its real extent is probably much greater than this. The seams are almost horizontal, dipping at an angle of about 5 degrees at the edges of the basin. The best seam, about 5 feet thick, occurs in no case at a greater depth than 30 feet below the surface. It well repays working, and the quantity of coal available from it is reckoned at 50,000,000 tons. The analyses of this coal gave the following results:—Moisture, 16·05 per cent.; volatile matter, 41·30 per cent.; fixed carbon, 22·65 per cent.; and ash, 20 per cent. Moreover, it was shown to contain 3·19 per cent. of sulphur. Its calorific power is reckoned at 3,829 calories; and 1 pound of coal is required to evaporate 7·128 pounds of water. In some localities this coal could be worked by opencast. There are other seams of varying quality and thickness (1½ to 10 feet): generally speaking, the reddish-brown inferior lignites occur in the upper portion of the group, and the black, lustrous, more compact coals in the lower. The strata among which these lie are highly fossiliferous sandstones and clays, and the author holds that the coal- and lignite-seams are the remains of the mangrove-clumps which covered the swamps bordering the old Tertiary sea. Near the Canots hills, the lignites are suddenly thrust up into a nearly vertical position, conformably with the rest of the strata thereabouts.

There is plenty of timber available for mining purposes, and it is calculated that patent fuel or briquettes made from the Maïssade coal could be delivered at the Haytian seaports at exactly half the price now paid there for American coal (that is, at $3·50 per ton instead of $7). The actual cost of production or output [at the pit-mouth] is reckoned at 3s. 9d. per ton.

L. L. B.

IRON-ORES IN MEXICO.

Los Criaderos de Fierro del Cerro de Mercado en Durango, y de la Hacienda de Vaquerías, Estado de Hidalgo. By M. F. RANGEL, J. D. VILLARELLO and E. BÖSE. *Boletin del Instituto geológico de México*, 1902, No. 16, pages 1-44, with 5 figures in the text, and 6 plates.

The first portion of this memoir deals with the extremely rich ore-deposits situated among the mountains which cover so large a portion of the State of Durango. The Cerro Mercado, a ridge about 1 mile long, and some 2 miles north of the city of Durango, is covered with blocks of hæmatite, leading at first sight to the supposition that the entire range is made up of the ore. In reality, the ore forms a sort of dyke in a great mass of rhyolite, but it is believed to extend to a considerable depth below the surface. Besides hæmatite, martite, specular iron-ore, and red ochre also occur here, but no magnetites properly so-called. Among the associated minerals are quartz, apatite (in great quantity) and topaz. An analysis of the best hæmatite yielded the following results:—Iron peroxide, 91·5 per cent.; silica, 2·5 per cent.; alumina, 0·6 per cent.; and moisture, 3 per cent. Lime and phosphorus were determined as present in the residues, but no quantitative result is recorded as to these. More complete analyses, made in the United States and based on 27 different samples, show that the average yield of the Cerro de Mercado ore is 55·8 per cent. of metallic iron, and the average amount of phosphorus contained in it, 1·328 per cent. (or 2·379 parts in every 100 of metallic iron). Sulphur averages only 0·085 per cent. of the ore, and it is plain that the deposit does not represent an enormous mass of pyrites altered by subsequent oxidation into hæmatite, etc.

The second, and larger portion of the memoir deals with the iron-ores of Pleistocene age which crop out in certain *barrancas* or ravines in the Tulancingo district of the State of Hidalgo. The stratigraphy of this area comprises Lower Cretaceous shales and limestones (often much folded), Cenomanian limestones (also highly folded) much poorer in fossils than the underlying group, Tertiary eruptive rocks (rhyolites, labradorites and basalts) and breccias of volcanic materials, with rolled and waterworn fossil-remains (*Mastodon, Equus*, etc.). The rhyolite was erupted in late Miocene or early Pliocene times, and the breccias belong to two horizons, separated probably by a considerable interval of time. Quaternary deposits and recent alluvia overlie the basalt, etc. A so-called " lacustrine " deposit of lignite, overlying tuffs with badly-preserved plant-remains, occurs in the Arroyo de las Flores. The lignite is of no commercial value, either as to quality or quantity. The now moribund vulcanicity of the region is represented by the thermal springs of Arroyo Seco (depositing ferruginous concretions) and Acaseca.

The great upheaval of this part of Mexico must have taken place near the beginning of the Tertiary era, as no folding or faulting of any consequence is observable in rocks of later date than the Cretaceous.

The iron-ore deposits are of excellent quality, and the local economic conditions of the Tulancingo district are favourable; yet for 40 years past the mineral industry has been conducted there on a very small scale, and only the actual outcrops have been worked. The ore-bodies are intercalated between the older labradorite (foot-wall) and the later basalt (hanging-wall), and vary in thickness from 4 to 32 inches. On the whole, they may be described as great flattened lenticles, thinning out into almost imperceptibly tenuous layers. They occur at an altitude of about 6,000 feet above sea-level, dipping slightly westward, and occupy a belt which is at least 19 miles long and 1¼ miles broad. The labradorite-footwall is of a " spongy " texture, impregnated with limonite; between it and the ore-body intervenes a thin band of rhyolitic breccia, similarly impregnated with limonite and with semi-opal. The ore-deposit itself consists in its lower part of limonite, and in its upper of red hæmatite. Between this and the basaltic hanging-wall, there is a thin band of ferruginous semi-opal of various colours. The ore-deposit is very crumbly and therefore easily worked, the masses of hæmatite being bound together merely by a sort of argillaceous cement. A sample of limonite from the El Sabinal mine yielded the following analysis:—Iron peroxide, 75·72 per cent.; silica, 8·37 per cent.; combined water, 16 per cent.; and traces of phosphorus and lime.

The genesis of the deposits is discussed at considerable length, and the conclusion is arrived at that the iron was originally leached out of deep-lying rocks by carbonated thermal solutions at high pressure, and thus it came about that silicates of iron passed into soluble bicarbonates. These were precipitated (and subsequently oxidized), as the thermal waters lost some of their heat in their upward journey. L. L. B.

THE ORE-DEPOSITS OF TAXCO DE ALARCÓN, MEXICO.

Apuntes relativos al Mineral de Taxco de Alarcón, Estado de Guerrero. By
LEOPOLDO SALAZAR. *Memorias de la Sociedad científica " Antonio Alzate,"*
1901, *vol. xvi., pages* 167-177, *and* 1 *plate.*

As early as the middle of the sixteenth century the Spanish *conquistadores*
obtained from the neighbourhood of Taxco copper and tin, which they used
for casting cannon. But the real wealth of the district was only revealed
two centuries later by the enterprise of a French miner. He made money
enough to build a magnificent church, and appears to have retired to enjoy
his fortune in 1763. From that time onward until pretty late in the
nineteenth century, various mines were opened up haphazard, and worked
irregularly, the salient feature of the industry (such as it was) being the
absence of scientific method.

Certain mining engineers, who examined the district, more or less
officially, previous to 1884, conceived favourable views of its possibilities,
pointing out that none of the workings had been carried to a depth exceeding
330 feet below the surface. The statistics published by the Mexican Government in 1892 and 1898 showed, however, that the attempts to revive the
mineral industry at Taxco had met with but scant success. The present
author is now engaged in exploration-work there, on behalf of a mining
company, but he does not look for immediate results. So far, he has devoted
special attention to the area south-east of the town of Taxco. He describes
the andesitic eruptive rocks which he has found in contact with the metalliferous veins at a depth of 100 feet or so, and regards their presence as a
favourable sign. It is supposed that the eruption of the andesites fissured
the rocks of the district, and that the metalliferous ores were subsequently
deposited in the fissures by hydrothermal action. As in many other of the
mining districts of Mexico, a later period of eruption is denoted by the
abundant occurrence of rhyolitic lavas which cap the summits of the
mountains.

In the central portion of the district, that is, at Taxco and south of it,
mining operations will have to be carried pretty deep down, in order to
reach payable ore. The conditions for exploration-work are favourable,
labour, pit-timber and fuel, being all within reach; but a fair amount of
capital and patient plodding work will be necessary to make the mines pay.
The author omits to say clearly what the ores are, apart from the copper
and tin already mentioned, but one may infer that they are silver-ores.

L. L. B.

PEAT IN PERU.

La Turba en el Perú. By JOSÉ J. BRAVO. *Boletin del Ministerio de Fomento,*
1903, *No.* 1, *pages* 20-22.

Peat occupies the bottom of nearly all the valleys of glacial erosion in
the high Cordilleras, in the centre and south of the country. It is a spongy
mass of a yellowish colour, darkening to black deeper down. For local use,
the peat is generally not cut at a greater depth than 8 or 12 inches, although
the author has seen cuttings of 6½ and 10 feet. The peat is cut out in
rectangular blocks and dried in the sun. In the pampas of Callan, in the
province of Jauja, 12,600 feet and more above sea-level, there is an area
exceeding 750,000 square metres (200 acres) covered by peat 10 feet or so
thick, the analysis of which is as follows:—Fixed carbon, 27·5 per cent.;

volatile combustible matter, 36·7 per cent.; ash, 21·2 per cent.; and moisture, 14·5 per cent. The calorific power has been determined as equivalent to 5,021 calories, and this peat has been successfully used (when mixed with a little coal) for roasting ores in reverberatory furnaces.

A recapitulation is given of the turbaries and analysis of peat therefrom, mentioned by Dr. Raimondi in vol. iv. of his work on Peru; and the author then goes on to point out that, although the climatic conditions of the Pacific seaboard are not on the whole favourable to the formation of peat, some turbaries are to be found in the coastal region of Peru, such as those of Vientelo in the neighbourhood of Lima. Here the peat is of very recent formation; its analysis is as follows:—Fixed carbon, 11 per cent.; volatile combustible matter, 34 per cent.; ash, 29 per cent.; and hygroscopic water, 26 per cent. It is compressed into a kind of patent fuel, after preliminary dessication (to ¾ of its weight) and impregnation with tar.

According to the mining laws of the country, peat-deposits belong to the owner of the soil. These deposits generally occur on pastoral land of small value, and in districts where labour is cheap. There appears to be an opening for peat-cutting on a grand scale, under advantageous conditions.

L. L. B.

***

THE MINERAL RESOURCES OF THE PROVINCE OF CAJATAMBO, PERU.

Memoria acerca de las Riquezas Minerales de la Provincia de Cajatambo, y especial de los Cerros de Chanca. By JUAN TORRICO Y MESA. Anales de Construcciones Civiles, Minas e Industrias del Peru, 1901, series 2, vol. i., No. 2, pages iii. and 70.

The province here described covers an area of ·5,838 square miles, and the population (largely half-breeds and Indians) probably does not exceed 40,000. On the east and north Cajatambo is bounded by the snow-clad summits of the Cordillera Blanca, etc., and the spurs coming down from these high Andes divide the region into innumerable glens and valleys. Despite the rugged nature of the country the roads are not so steeply graded as might be imagined; those, however, which run from the chief city to the Pacific seaboard are not merely bad, but positively dangerous.

Proceeding eastward from the harbours of Huacho and Supe towards the interior, one passes over great eruptive masses of syenite, granite and diorite, which farther inland are mantled by alluvial deposits and clays. Then come the younger eruptive or porphyrites, occupying a vast area: with them are genetically associated the silver-bearing veins. The porphyrites are followed by a great belt of sedimentaries, and the summits of the Cordillera Blanca are well nigh reached before another series of eruptives comes in (syenites). There are no recent volcanoes or any eruptive phenomena of modern date in Cajatambo, except indeed the sulphureous and ferruginous thermal springs of Churin, Viroc and Laclla.

The author's brief reference to the fossils found among the sedimentaries leads one to infer that these are largely of Jurassic age. He certainly thinks that the coal-bearing sandstones belong to that period, but he suggests that the marls (with brine-springs) of Oyón and Colpa are remnants of the Triassic formation. The two most abundant useful minerals in the province are silver-ores and coal: a rapid sketch of its other mineral resources will not, however, be deemed out of place.

Gold is known to have been mined both by the Spanish *conquistadores*. and before them by the natives. There are, however, no gold-workings in activity at the present day. The precious metal is associated, sometimes in fairly considerable proportions, with the silver of the argentiferous ores.

Silver, invariably associated with antimony, occurs abundantly all over the province, from the Pacific seaboard to the summits of the Cordillera Blanca; but it is noticeable that the ores increase in value and quantity as one approaches the main chain. In many instances, iron-pyrites is associated with these, and then indeed forms the principal infilling of the fissure-veins. In the veins the silver seems to be concentrated in rich "pockets," between which completely barren "ground" intervenes. The most common mineral species are tetrahedrite, bournonite and panabase, then less commonly occur pyrargyrite, stephanite and dürfeldtite (a comparatively new silver-antimony ore discovered for the first time in the Irismachay mine in this province). Frequent secondary associates are galena, blende, iron-pyrites, stibine, chalcopyrite, and in some cases azurite, malachite and chrysocolla. The gangue is most usually quartz or calcite, more rarely pyrolusite, dialogite and barytine. As in nearly all Peruvian ore-deposits, the minerals are very seldom in the crystalline form.

Going from south to north, the author describes in succession the silver-mines of Mesías, Mercedes (Uticto district), Nazareno, Anamaray (Quichas district), Raurac, Gasuna, Tarasca, Tinta, Auquimarca, Tallenga, etc. In some cases, calcination, lixiviation and amalgamation are carried out *in situ*. the furnaces being fired with Conocpata coal.

Copper is almost as frequent an associate of silver in Cajatambo as antimony, but cupriferous deposits properly so called are of comparatively sparse occurrence, so far as is at present known. The recent rise in price of the metal has, however, stimulated prospecting work in this direction.

Galena occurs in many mines, but in small quantity and irregularly distributed. It almost always contains some antimony. The only mine where the lead-ore may be regarded as of industrial importance is that of Niñoperdido, in the Raurac district: the deposit here contains as much as 70 per cent. of lead, and never as little as 40 per cent.

Blende, mostly associated with tetrahedrites, bournonites, etc., can hardly be said to form independent deposits.

Antimony-ores and iron-pyrites have already been mentioned.

Coal occurs all over the province in a great series of sandstones interstratified with thin shales, in seams which vary in thickness from 2 to more than 60 inches. The character and quality vary from anthracite to long-flame bituminous coal. The most important deposit is that of Conocpata. An Italian company, as many as 15 years ago, drove an adit here, and experimented on the coking quality of the coal with extremely satisfactory results. Both roof and floor of the seam consist of a lustrous black shale. difficult to distinguish at first sight from the coal itself. The thickness of the seam varies from 16 inches to 5 feet. The dip, nearly horizontal at the outcrop, increases gradually to 60 degrees. The coal is excellent for blast-furnace and reverberatory-furnace purposes, it leaves only about 4 per cent. of very white ash, and contains scarcely any earthy matter or pyrites. The Conocpata coal is of unique quality so far as the province of Cajatambo is concerned, no other exactly like it being found there.

In the Gasuna district, three seams are worked, with a sandstone-roof and a shaly floor. The thickness of the seams is fairly uniform, averaging

20 inches. The coal is crumbly, dusty, burns with a short non-crepitating flame, leaves extremely little ash, and is unsuitable for coking purposes. The author classifies it as an anthracite.

Around Oyón, there are many seams of coal of intermediate character, almost all the varieties which range between anthracite and highly bituminous coal occurring there.

Very pure graphite is got from Antisicuar in the district of Auquimarca. Deposits of gypsum and native sulphur have been found recently near Cajatambo City. The occurrence of brine-springs has already been mentioned.

Reviewing generally the prospects of the mineral industry in Cajatambo the author takes a distinctly hopeful view, but points out that the great difficulty which demands the serious attention of the Government is the scarcity of labour. This applies not only to one province, but to the whole of Peru. **L. L. B.**

THE METALLIFEROUS DEPOSITS AND COAL OF THE CERRO DE PASCO, PERU.

Asiento Mineral del Cerro de Pasco. By MICHEL FORT. *Anales de Construcciones Civiles, Minas e Industrias del Peru,* 1901, *series* 2, *vol. i., No.* 1, *pages* 1-164.

This elaborate memoir, which may be regarded as embodying all the latest information in the possession of the Peruvian Government as to the important mineral region of the Cerro de Pasco, begins with the more or less obligatory and customary historical summary. Mining began there as far back as 1630, but one gathers that of late years the normal development of the industry has been checked by various causes, among which may be mentioned the heavy cost of unwatering the mines, and extravagant speculation in regard to concessions. A fairly detailed bibliography is given, and the author then deals with the topography, physiography, climate, flora and population of the district. He devotes very rightly considerable space to the question of means of communication and transport: this will be solved ere long by the projected railway from La Oroya, which appears to offer no insurmountable engineering difficulties.

The fourth chapter, wherein the author makes extensive use of the works and conclusions of previous writers on the subject, deals with the geology of the area. The rocks have been so greatly disturbed by the agencies which determined the " uprise of the metalliferous mass," that the geological structure is very complicated. Messrs. Raimondi and Hodges, however, have done much to unravel the tangled skein. Two groups of rocks are recognizable:—(1) The metamorphosed sedimentaries, consisting of calcareous conglomerates, limestones (of Cretaceous age), older grits and barren slates; and (2) eruptives—quartz-biotite-andesites, trachytes (Raimondi), and possibly diorites.

The metalliferous deposits are divided into:—(1) Superficial, that is, largely altered and decomposed; (2) intermediary deposits; and (3) deep-seated deposits. The first may be regarded as a vast ferruginous gossan, with locally respectable proportions of lead and copper, and traces of silver, zinc, etc. In the second are found sulphides of iron, copper and lead, occasionally argentiferous. Rich pockets are of sporadic occurrence, and the ores are occasionally oxidized or even reduced to native copper. Blende is not very common in this group. The third or deep-lying group comprises the unaltered sulphides, arsenides and antimonides of silver, bismuth, copper, lead, iron, etc., the mere fringe of which has been touched by mining operations thus far. The genesis of these great metalliferous deposits is evidently connected with those volcanic phenomena, one manifestation of which was the eruption of the andesites. Vast fissures were torn in the sedimentary rocks, and were ultimately infilled with metalliferous precipitates from thermal (siliceous and other) waters welling up from below.

The fifth chapter deals with the main difficulty, which it is essential for mining enterprise in the Cerro de Pasco to overcome—the elimination of the underground waters. If deep-level working is to continue and extend. the construction of great drainage-galleries is imperative.

From the sixth chapter, we learn that the methods of working are of the most imperfect and primitive description, and no progress is being made towards a better system.   The miners are exclusively natives (so-called Indians); the total number of claims in 1899 amounted to 546, of which 449 were nominally for silver-ores (but including copper), 12 for gold, 77 for coal, 3 for rock-salt, etc., and nominally only 5 for copper (that is. deep-level). No less than 117 *haciendas* for the treatment of the ores have been established in the various ravines of the Cerro. The author describes briefly the various methods of treatment, which appear to be of the old-fashioned order, and offer, as he says, no feature calling for especial notice.

With regard to coal, the number of concessions in 1900 had reached 174, covering an area of 2,500 square miles. The mineral varies in character from anthracite to highly bituminous coal, and some of the seams are more than 30 feet thick. A railway to convey the coal to a central smelting-works, which would treat all the ores got from the Cerro de Pasco metalliferous mines, is badly needed. For this, as for other things, the author believes that the necessary capital will be forthcoming, so soon as the Peruvian Government grants the necessary legal securities and facilities. Cerro de Pasco appears to have been too long the battle-ground of contending syndicates.

Appendices giving the results of silver-assays, copper-assays and coal-analyses, accompany the memoir. We may cite such coals as that from the Carmen mine, Yanahuanca, containing 45 per cent. of volatile matter. 35·4 per cent. of fixed carbon and 19·6 per cent. of ash; that from the Tucsi mine with only 22 per cent. of volatile matter, but 54 per cent. of fixed carbon and 24 per cent. of ash; or the long-flame coal of Yanacachi. with 49·5 per cent. of volatile matter, 33·3 per cent. of fixed carbon and 17·2 per cent. of ash, and a calorific power of 6,290 calories.   L. L. B.

THE GOLOVIN BAY REGION OF NORTH-WESTERN ALASKA.

The Golovin Bay Region of North-western Alaska. By J. D. LOWRY. *The Engineering and Mining Journal [New York],* 1901, *vol. lxxi., pages* 751-752.

While much has been written about Nome, very little has been said about the Golovin Bay region, lying from 50 to 75 miles northeast of the former, although mining operations have been carried on principally around Council City, now called Golovin City, for the last three or four seasons. The characteristics of the region as regards formation, topography, etc., are similar to those of the Nome country, except that the general elevation is much higher.   There is the same surface-covering of tundra, with the underlying strata of gravel, boulders, etc. Bed-rock, principally mica-schist and shales, is variously found from 1½ feet down. This, however, is held by many to be a false bedding, and the theory is advanced that true bedrock will be found at greater depth. Gold-dust from this region assays from £3 11s. to £3 19s. (17 to 19 dollars) or more per ounce. A quantity brought down from Ophir Creek last autumn by the writer ran nearly £3 18s. (18·62 dollars) to the ounce at the United States Assay Office at Seattle, Washington.

X. Y. Z.

THE NITRATE-DEPOSITS OF CALIFORNIA.

Natronsalpeter in Californien. By Dr. Carl Ochsenius. *Zeitschrift für praktische Geologie,* 1902, *vol. x., pages* 337-339.

The author points out that the increasingly widespread use of smokeless powder is causing the demand for Chile saltpetre to develop enormously. Mining-engineers reckon that the Chilian deposits, supposing that the consumption increases at the same rate as it has within the last ten years, will be able to satisfy the demand for good nitrate for another quarter of a century. Then, however, it will become necessary to work the poorer deposits, nowadays contemned, containing only from 5 to 20 per cent. of nitrate.

In the northern part of San Bernardino County, California, the newly discovered nitrate-fields are reckoned to contain 22,000,000 tons of workable mineral, lying in places 5 feet thick, of a quality equal to that of the Chilian saltpetre. The geological and physiographic conditions, in fact the whole natural environment of the Californian occurrence, are almost exactly parallel with those of Chile. The author draws attention, however, to the fact that California, although dry, is not so absolutely rainless a region as Northern Chile; and he has consequently some doubts as to the real wealth of the Californian nitrate beds, for sodium nitrate (as he remarks) is an extremely hygroscopic substance. Just now, it would hardly pay to work beds containing less than 30 per cent. of pure nitrate.   L. L. B.

THUNDER MOUNTAIN, IDAHO.

Thunder Mountain, Idaho. By Wm. E. L'Hame. *Mines and Minerals,* 1901, *vol. xxi., page* 558.

Although much is known concerning the mineral wealth of Idaho, there are portions practically unknown to the outside world. These lie almost in the central section of the state, and their retarded development is principally due to their inaccessibility and remoteness from transportation facilities. For 15 miles above its junction with the main river, the Middle Fork of Salmon river has carved its bed through solid rock, the work of ages; sheer drops of 1,000 feet are frequent, and the confining slopes rise from 2,000 to 3,000 feet higher. A few daring adventurers have braved the dangers of this rapid stream, and the fact that a pan of gravel can scarcely be picked along the stream without getting the "colour" suggests extensive belts of auriferous material above.

One of these deposits was discovered about 3 years ago: it is a veritable mountain of ore, whose estimated wealth throws the treasures of the Incas into the shade. The locality is now known as Thunder Mountain, and its elevation is 8,000 feet, occupying a central position among the headwaters of Marble, Holy Terror and Monumental creeks.

Thunder Mountain consists entirely of volcanic material. Mineralogically and structurally, it is strongly suggestive of the Cripple Creek region. The predominating rock is rhyolite of a whitish colour, with porphyritic development of sand in felspar and quartz-crystals, giving it the characteristics of liparite. The gold occurs in this material in native form, auriferous pyrite being found only sparingly, whilst magnetic iron, a constituent of most volcanic rocks, is almost entirely absent.   X. Y. Z.

PETROLEUM AND NATURAL GAS IN KANSAS.

Petroleum and Natural Gas in Kansas. By ERASMUS HAWORTH. *The Engineering and Mining Journal* [*New York*], 1901, *vol. lxxii., pages* 327 *and* 397.

The area in Kansas, throughout which oil and gas have been found in greater or lesser quantities, covers about 8,500 square miles in the southeastern part of the state. In the extreme southeastern corner, the sub-Carboniferous limestone comes to the surface. Along the southern line of the state, the limestone dips to the west about 20 feet and to the northwest from 4 to 8 feet to the mile, and constitutes a floor upon which the Coal-measure shales and limestones rest. The base of the Coal-measures is occupied with the Cherokee shales with interbedded sandstones, the whole aggregating a thickness of 450 to 500 feet. These shales are known to extend west and north as far as drilling has been done, and they probably reach much farther. The Cherokee shales are the principal producers, as all of the oil and the greater part of the gas comes from them. All the strong gas-wells obtain gas from a sandstone lying within the lower 200 feet of these shales. There is some doubt about the continuity of the sandstone from place to place, but present knowledge favours the opinion that the productive sandstone of any one locality is limited in extent, and that it is in reality a lenticular mass of sand which changes to shale in every direction, while some distance away a corresponding bed of sandstone is found at or near the same horizon.

The drilling of the gas or oil-wells is done by machinery similar to that used in Indiana, Ohio, Pennsylvania and West Virginia. But the general condition of the ground is quite different from that of Indiana and other fields. In a number of instances, drillers have come to Kansas from the Indiana fields assuming that all the casing required was a short length at the surface, and have got into difficulties by the wells caving in at 400 or 600 feet below the surface. Owing to the many soft shale-beds it is generally necessary to case the wells for more than half their depth, and sometimes for almost their entire depth. In this one respect, well-drilling in Kansas is more expensive than in Indiana or Ohio, but otherwise it is as cheap as anywhere in America.

Gas is used as the principal fuel and light-producer in 12 or 15 different cities and villages, besides being piped to many farm-houses throughout the gas-region. And, on account of the cheap and abundant fuel supplied by natural gas, a number of manufacturing enterprises have already been established in Kansas. The more extensive of these are zinc-smelters, brick-plants, and Portland-cement factories, and present indications are that at no distant date coal as a fuel will be entirely superseded by natural gas throughout southeastern Kansas.   X. Y. Z.

ROCK-SALT IN LOUISIANA.

Louisiana Rock-salt—Avery's Island. By HAROLD A. TITCOMB. *The Engineering and Mining Journal [New York],* 1901, *vol. lxxii., page* 780, *with plan and section in the text.*

Ten miles southeast of New Iberia, Louisiana, rising abruptly from a flat expanse of marshland, is a remarkable hill covering an area of about 5 square miles and from 150 to 200 feet high. It is called Avery's Island or Petite Anse. The entire island is underlain by a deposit of rock-salt, pure and absolutely free from foreign matter, and of unknown depth. A bore-hole nearly 2,000 feet deep failed to 'pass through the salt. Mining was carried on for many years, but the workings got flooded through being operated too near the surface.

A new shaft, however, has now been sunk; for 40 feet down from the surface it has been close-cribbed and cemented, so as to ensure the exemption of the mine from water. The cage thereafter descends through 470 feet of solid salt, and no pumping whatever is required from the mine. The salt is mined by a simple room-and-pillar system, 50 per cent. being left in pillars to support the roof. Galleries, 30 feet high and 30 feet wide, are driven; thirty holes, each 9 feet deep, are bored per round; each 9 feet hole is drilled in from 4 to 6 minutes by a compressed-air drill; the thirty holes take an aggregate charge of 100 pounds of 40 per cent. dynamite; and 65 tons of salt are blasted at each round. The salt is then loaded into cars, each carrying 2½ tons; hauled to the shaft by mules, then hoisted, crushed, sized, sacked, and loaded on cars for shipment. The total cost of mining and crushing is about 1s. 8d. (40 cents) per ton. X. Y. Z.

THE TUNGSTEN-DEPOSITS OF OSCEOLA, NEVADA.

The Osceola, Nevada, Tungsten-deposits. By FRED D. SMITH. *The Engineering and Mining Journal [New York],* 1902, *vol. lxxiii., pages* 304-305, *with 1 illustration in the text.*

In December, 1901, the writer made a careful examination of the deposit and believes it to be worthy of a more detailed description. Compared with other visible supplies of tungsten-ore in the world, these mines are of extraordinary economic importance, as they are apparently capable of producing more tungsten mineral than any other mine known and more, perhaps, than all other mines combined. The prominent outcrops of veins of white quartz in the brownish-grey granite were noticed by the earliest prospectors, but the mineral was for long supposed to be "specular hæmatite." More observant prospectors did not determine the true character of the ore until 1889, and it was not until 1899-1900 that any systematic attempt was made to really explore the deposits.

The veins are in the foothills and lowest slope of the Snake mountains; for several thousand feet up the mountain-side the formation is granite, which is overlain by Cambrian quartzite. The latter rock forms the main ridge of the mountains to the top of Mount Wheeler, or of Davis Peak, which is directly above the deposits in question. The veins, of which there are five prominent ones, all occur in the granite, and plainly cut across the bedding of the granite. Talus has covered the veins in many places, but they are discernible for a total distance of 1,800 feet up the mountain side, which has a general slope of about 18 degrees. The hübnerite-vein, shown in Mr. J. H. Marriott's tunnel, has a strike of north 70 degrees east, and a dip of 65 degrees north-westward. The walls are well defined, and part easily from the vein, the width of which varies from 18 to 36 inches, and averages 26 inches throughout the whole length of the tunnel.

The hübnerite occurs in the white quartz in variously sized crystals, many of which are 3 inches long, and plainly show the crystalline character. Massive specimens when broken show cleavage-planes 2 to 4 inches long and 1 to 3 inches wide. Much of the mineral, however, is in fine grains and in irregular bodies. The quartz is found in some cases entirely enclosing the hübnerite, whilst in other cases, apparently solid mineral specimens are found to enclose the quartz. This plainly shows that the two minerals were deposited simultaneously. The quartz is very solid, with practically no evidence of weathering. No oxidation-products are found. In one locality, a considerable amount of scheelite was found mixed with the hübnerite.

As far as studied, it appears that the mineralization of the vein has taken place across its whole width, although it often occurs in larger quantities on one wall than on the other. This concentration on the walls was found to change from side to side in short distances. It was also found that a concentration of the mineralization into so-called shoots had taken place. At the same time some mineral was found throughout the vein. These shoots are very prominent on the surface, as each shows an abundance of the hübnerite in massive form, owing to the high contrast in the colours of the hübnerite and the gangue. At a depth of 80 feet, at the face of the tunnel, it is reported that the mineral was found in as good a proportion as the surface-indications would suggest.

As far as examined, the ore is remarkably free from other minerals. All of the ore shipped has been either hand-sorted and cobbed, or concentrated by the crude methods outlined in the text. These ores carried an average of 68 per cent. of tungstic acid. From the ore as exposed in croppings and workings, it is the opinion of the writer that the whole vein matter, if carefully concentrated so as to save both the hübnerite and the scheelite, would produce 10 per cent. of mineral carrying 70 per cent. WO_3.

The mine appears to be singularly favoured with regard to accessibility, proximity to water for milling and power, and large quantities of wood for fuel and for timber. The nearest railroad point is Frisco, Utah, on the Oregon Short Line, a distance of 85 miles, over fairly good desert-roads.

X. Y. Z.

DISCOVERY OF PLATINUM IN WYOMING.

The Discovery of Platinum in Wyoming. By WILBUR C. KNIGHT. *The Engineering and Mining Journal* [New York], 1901, vol. lxxii., page 845.

The Rambler mine is located about 50 miles southwest of Laramie, in the Medicine Bow Mountains, Wyoming. At a depth of 70 feet, a huge vein of oxidized copper-ores was found, which made it possible to commence shipping ore immediately the strike was made. Below the oxidized ores, chalcosite and bornite were found, and, still deeper, covellite. The ores have varied considerably in value, but the blue ore has averaged about 30 per cent. of copper by the car-load. The vein occurs in a schistose formation, that is only a short distance from the Archæan granite.

A short time ago, the Denver assayers, while attempting to account for the variation in the assay-value of the ores, discovered that platinum was present, and usually in paying quantities, more especially in the blue copper-ore. The discovery of platinum in a vein is a matter of extraordinary scientific interest, especially so when it appears to have a commercial aspect, and it is probable that this region may prove to be of great import-ance as a platinum-producer. The average of ten samples showed 0·259 ounce of platinum per ton of ore. The second lot of samples were car-samples of from 20 to 45 tons each, and the platinum averaged 0·8 ounce per ton. Another test was made on a sample of matte carrying 1·92 ounces of platinum per ton.   X. Y. Z.

THE MINERAL RESOURCES OF VENEZUELA.

Escursioni geologiche al Venezuela. By E. CORTESE.   *Bollettino della Società geologica Italiana*, 1901, *vol. x.c., pages* 447-469.

This paper embodies the results of the observations made by the author during a recent stay, extending over two months, in the northern part of Venezuela. His collection of rock- and mineral-specimens has been handed over to the care of the Royal Geological Office at Rome.

Metalliferous Veins, etc.—A great many mining concessions have been granted by the Government, in the neighbourhood of Caracas and up to near La Guayra, for the purpose of working the metalliferous veins that seam the mica-schists. It is said that near Caracas copper-ores can be traced over an extent of 7 leagues, but the author saw nothing of this. Around La Guayra the concessions were obtained for working pyrites, galena, graphite, etc. The author only visited one of these, and found a deposit of iron-pyrites which he does not believe to be auriferous. As to the graphite which does really occur in the mica-schists near La Guayra, it is not in sufficient quantity to justify mining-operations. South of Carúpano, on the other hand, and in the west, are metalliferous veins which have been worked in the past, and may yet be actively worked again in the future. Thus, in the Ensenada Esmeralda, west of Carúpano, is an outcrop of malachite and azurite, which in depth passes into chalcopyrite: it occurs among diabasic greenstone-slates. In similar rocks, nearer Carúpano, on Guiria Bay, is a vein of blackish argillaceous matter which was analysed in New York with the following result:—Copper, 67 per cent. [?], silver 3·5 per cent., gold 0·05 per cent., and vanadium 0·008 per cent. More to the south-west, at Revilla, Juan Burro, etc., above Cariaquito and Muco, is an extremely rich hæmatite, more especially associated with crystalline limestones. It is fine-grained, occasionally very siliceous, elsewhere very pure and heavy, and passing into magnetite. Though the ore contains perhaps 65 per cent. or more of iron, the author does not consider that it would repay working, on account of the heavy cost of getting it down to the coast and put on board ship.

In the localities known as Carmen and El Encanto, above Cariaquito, galena occurs in compact masses and in stringers among crystalline lime-stones. The ore will prove valuable for its lead, rather than for silver, of which it contains only 0·4 per cent. The old mine of Gran Pobre seems to be still very rich in grey copper-ore. The author mentions some other occurrences of chalcopyrite and galena, to which, however, he is unable, at present, to attach any very great importance.

Coal-bearing Formations.—It is said that nearly 480 outcrops of coal occur among the Oligocene (or Miocene?) sandstones of the Barcelona-Naricual district, but the author confined his attention to such seams as are sufficiently thick to be workable at a profit. Of these he saw six in the Naricual district, two in that of Capiricual, and one in that of Araquita. The strike of the beds is generally east-south-east and west-north-west or south-east and north-west, but the dip varies greatly. The outcrops of coal are visible only in the valleys or ravines, being masked on the hills, either by the luxuriant tropical vegetation or by more recent fluviatile deposits. The coal is hard, compact, with a bright conchoidal fracture, and analysis has yielded the following results:—Moisture, 1·25 per cent.; ash, 2·83 per cent.; volatile substances, 38·63 per cent.; and coke, 58·49 per cent. The calorific power is tabulated as 9,052 [calories], and, everything considered, we have here an excellent fuel.

At the Naricual mine, 130 feet above sea-level, No. 5 seam is being won: it is more than 6½ feet thick. Exploration-work is being done on No. 3 seam, which is not far from 10 feet thick; and at the time of the author's stay in that district another seam was struck, still thicker than the last-named. There are many other seams at Naricual, ranging between 1 and 6 feet in thickness. Dips vary between 35 degrees and 55 degrees south-westward. Still another seam, the Vena Simplicio, 8½ feet thick, dips 60 degrees south-south-westward.

In the neighbourhood of Capiricual, in various *quebradas* or gullies, seams from 3 to 7 feet thick crop out, some 250 feet above sea-level. The estimated area of the coal-field is 309 square miles. It is worked exclusively by an Italian company, which also controls the railway from the coal-field through Barcelona to Guanta, and the breakwater at the last-named port.

Coal-outcrops have also been traced in sandstones of the same age as those of Naricual, etc., near Puerto Cabello and south of Carúpano.

Solfataras and Thermal Springs.—These occur more especially south of Carúpano in the Chaguarama Valley; they well out from among Miocene sandstones, making great bare white scars amid the tropical greenery, and depositing very pure native sulphur. A concession has been obtained by a German company: the concessionaires have put down a magnificent plant, and constructed a cable-railway, 11 miles long, to the seaboard at Carúpano. However, careful and repeated study of the locality seems to have convinced the author that sulphur will only be obtained thence in comparatively small quantity; on the other hand, he has ascertained the occurrence of cinnabar, and it is possible that this may prove to be of industrial importance.

Asphalt and Petroleum.—Rich asphalt-lakes occur at Guanoco and Guariquèn, among the lagoons which end off the Gulf of Paria on the west, also near Putucual, etc.

At Manicuare, opposite Cumanà, and near La Salina de Araya, the author observed indications of natural gas and oil. In fact he is persuaded that petroleum occurs along a vast extent of the coast-line of the Gulf of Cáriaco, and that there are the makings of a great industry, once the strike of the petroleum-deposits has been followed up and mapped out.   L. L. B.

THE IRON-ORE DEPOSITS OF NEW SOUTH WALES.

The Iron-ore Deposits of New South Wales. By J. B. JAQUET. *Memoirs of the Geological Survey of New South Wales*, 1901: *Geology, No. 2, pages* 1-186, *with numerous maps, sections and plates.*

The writer has made a careful survey of all the deposits of iron-ore within a reasonable distance of those centres which, by reason of their proximity to supplies of coal and limestone, may be regarded as suitable localities for the establishment of smelting-works.

The writer examined nearly 200 deposits of iron-ore. Many of these were of dimensions too small to be of value to the ironmaster, but the most important, to the number of sixteen, are tabulated below, shewing the districts in which they occur, the description of ore that they contain, and the estimated quantities that they will yield.

District.	Description of Iron-ore.	Estimated Minimum Quantity of Ore.
		Tons.
Breadalbane	Brown ore and hæmatite ...	700,000
Cadia	Specular hæmatite, magnetite, and carbonate ore	39,000,000
Carcoar	Brown ore and hæmatite ...	3,000,000
Chalybeate Spring	Brown ore	1,510,000
Cowra	Magnetic	100,000
Goulburn	Brown	1,022,000
Gulgong	Magnetic	120,000
Mandurama and Woodstock ..	Brown	609,000
Marulan	Brown ore and hæmatite ...	40,000
Mudgee	Brown ore with manganese ...	150,000
Newbridge, Blayney and Orange	Brown and magnetic ...	150,000
Queanbeyan	Magnetic	1,000,000
Rylstone and Cudgegong ...	Brown	443,000
Wallerawang and Piper's Flat ..	Brown	200,000
Williams and Waruah River District	Titaniferous magnetic ore ...	1,973,000
Wingello	Aluminous	3,000,000
	Total	53,017,000

X. Y. Z.

THE METALLIFEROUS ORES OF THE MENADO DISTRICT, DUTCH EAST INDIES.

(1) *Verslag omtrent de gedurende het jaar 1899 in de Residentie Menado gedreven geologisch-mijnbouwkundige Onderzoekingen en Mijnontginning.* By M. KOPERBERG. *Jaarboek van het Mijnwezen in Nederlandsch Oost-Indië,* 1900, *vol. xxix., pages* 30-50, *and* 4 *plates.*

(2) *Verslag van het geologisch en mijnbouwkundig Onderzoek in de Residentie Menado over het jaar* 1900. *By* M. KOPERBERG. *Jaarboek van het Mijnwezen in Nederlandsch Oost-Indië,* 1901, *vol. xxx., pages* 115-121.

In the Totok river-basin, andesites and andesite-conglomerates are overlain by the Miocene-Pliocene marine limestones, and at the point where the river saws through a gorge in the limestone, irregular " quartz "-masses are found with visible native gold. The productive workings are all above the 1,000 feet contour-line, and no " quartz " is found at the base of the limestone-wall. One is rather inclined to infer that the " quartz " is really a chert, formed by silicification of the limestone in past ages, and that the gold, originally associated with sulphidic ores, has been set free by the processes of weathering.

Reefs or veins of quartz with sulphidic ores occur in the andesites and andesite-conglomerates of several localities (Bolang Mongondo, Ranoiapo river, etc.) : these ores are chiefly pyrites, more or less auriferous, zinc-blende, and galena (generally containing some gold and silver). The two principal veins of Soemalata occur in an eruptive conglomerate, at first thought to be porphyrite, with a steep north-easterly dip. The gangue is quartz, and the ores are pyrites associated with magnetic and arsenical pyrites, and now and then blende. So too the principal vein of Dopallak courses through eruptive rock: it consists chiefly of iron and copper-pyrites. with galena and a little zinc-blende, in a gangue of brecciated quartz and " blanched " andesite.

Details are given of the prospecting and preliminary work done by various private companies, and the Soemalata, Paleleh and Totok mines are briefly described. From the Paleléh mines all ore containing more than 150 parts of gold per million is shipped to Europe, the remainder being treated in the colony. The Totok mines had 20 stamps at work at the time of writing.   L. L. B.

————

THE DIAMANTIFEROUS DEPOSITS OF SOUTH-EASTERN (DUTCH) BORNEO.

Les Gisements diamantifères de la Région sud-est de l'Ile de Bornéo (Possessions Hollandaises). By — GASCUEL. Annales des Mines, 1901, series 9, vol. xx., pages 5-23, and 1 plate.

Diamantiferous deposits have been worked by the Malays for many centuries, at the two extremities of the southern coast of Borneo. Those around Landak, near Pontianak, the capital of the western district, were not visited by the author, so his paper deals exclusively with those of Martapoera, near Banjermasin, capital of the southern and eastern districts. This town is built on piles, amid the marshes which extend inland from the swampy coast, and it, as well as the country right up to Martapoera (19 miles farther inland) is often under water during the rainy season.

Martapoera is situated a little below the confluence of two small rivers, the valleys of which, as well as their affluents, are diamond-bearing. From the miner's point of view, the most important area is covered by the 49,000 acres stretching north and south of the Banjoe-Irang river. On the east the diamantiferous gravels abut on the flanks of coal-bearing Eocene hills. while on the west they dip under the swamps that extend to the sea-coast, and no man can say where they end. Of all the prosperous and populous villages that formerly dotted this area, only one, Tjampaka (6 miles from Martapoera) has retained some sort of vitality.

The " pay-gravels " are only a few inches below the surface in the upper parts of the valleys, lower down they are 23 to 26 feet below, and on the borders of the swamps they have to be reached by pits as much as 82 feet deep. Moreover, the thickness of the gravels varies greatly, more especially near the upper ends of the valleys: in the lower, broader portions the average thickness is about 3 feet, and there too the continuity of the deposit is more regular. Perhaps the most variable factor of all is the tenour in diamonds, not only in regard to the number of precious stones per cubic yard, but also as to dimensions, lustre, colour, etc. The chief constituents of the gravel are white quartz-pebbles, often iron-stained: with these are a few porphyritic rocks, micaceous quartzites, shelly limestones, rolled flints, and semidecomposed granulites and pegmatites, but no basic igneous rocks. The whole is more or less bound together by stiff white clay (evidently the decomposition-product of felspars) and a less compact yellow clay. The overlying deposits are clays which pass downward into sands, and these in turn pass into gravels which become coarser and coarser until they pass into the "pay-gravels," the quartz-pebbles of which vary in size from that of a billiard-ball to that of a man's head. The bed-rock is either a compact red clay or dark bluish-grey tough shales. In the upper parts of the valleys a conglomerate of highly-decomposed greenstone-pebbles sometimes comes next below the diamantiferous gravels, and is underlain by the red clay. Both this clay and the shales are of Eocene age, and form the main mass of the hills. Among the shales are intercalated bands of grit and seams of bituminous coal, some of which are workable. There has been no change in the general physiography of the district since the deposition of the diamantiferous gravels, and we may therefore infer that they are of Quaternary age. The author, in contradistinction to the late Mr. J. A. Hooze's view* that the matrix of the diamonds of Martapoera is to be sought among the crystalline schists at the heads of the valleys, holds that the mother-rock of these diamonds is a granulite or pegmatite.

The diamonds are very little rolled, and have generally a distinct crystalline form, the most frequent being the octahedron. They are usually tinged with yellow, but are sometimes water-clear, very rarely red or blue, and still more rarely black. The last-named are highly valued by the natives as amulets against bad luck. The stones are on the whole small in size, renowned for their lustre and purity, and generally considered harder than the Cape diamonds. It is, however, the Cape diamond that is mostly hawked about by the Arab traders in Borneo as being the Martapoera diamond. Among the associates of the diamond in the pay-gravels are native gold, platinum, corundum (generally blue) and rutile. The black sand which forms the abundant residue of the washings contains much magnetite, titanite and chrome-iron-ore.

The deposits have never been worked on what Europeans would consider a methodical and regular system. A description is given of the primitive manner in which the natives set to work—but, however primitive, it implies a considerable amount of industry, patience and skill. In the later seventies, the South African diamonds made their appearance on the local markets, and it may be said that the diamond-industry of Martapoera, moribund since 1886, is now a thing of the past: the annual output is barely 1,000 carats.   L. L. B.

AURIFEROUS ORES OF NORTHERN CELEBES.

Ueber die Geologie der Umgegend von Sumalatta auf Nord-Celebes und über die dort vorkommenden goldführenden Erzgänge. By G. A. F. MOLENGRAAFF. Zeitschrift für praktische Geologie, 1902, vol. x., pages 249-257, with 4 figures in the text.

The gold-bearing coastal area of Northern Celebes is of a very mountainous character: along the whole length of it runs a mountain-range, the summits of which vary between 5,000 and 7,000 feet above sea-level, forming the main watershed of that region. In the neighbourhood of Soemalata, the district which forms the principal object of the present memoir, this mountain-range is about 6½ miles distant from the northern

* *Jaarboek van het Mijnwezen in Nederlandsch Oost-Indië, 1893, vol. xxii., pages 1-431, and 14 plates.*

coast and attains its highest elevation, that portion of it being known as the Bolio-Hutu range. Parallel to it, and halfway between it and the sea, runs another mountainous ridge (which the author calls the Soemalata coastal range) with summits varying from 3,300 to 4,500 feet in altitude. The Bolio-Hutu mountains consist of a granitic core, in part mantled by highly metamorphosed bedded rocks. To these latter, the author has applied the name of the Dolokapa series, as their best exposures are seen about the headwaters of the Dolokapa river. They consist largely of crystalline schists and contact-rocks, thrust up on end and showing the fan-like disposition familiar in the Alps, the Himalaya, etc. Overlying them unconformably is the Wubudu eruptive breccia, which constitutes almost the entire coastal range, from Matinang to Kwandang bay. The Wubudu river, from its sources to the sea, flows exclusively through this formation, and the pebbles rolled along its bed are made up of the eruptive breccia alone.   In places tuffs are intercalated among the breccias, frequently showing traces of subsequent silicification.

The author also describes a purely sedimentary conglomerate. the Obapi conglomerate, best seen on the farther side of the ridge, behind Bolontio, 6 miles or so west of Soemalata. There is reason to believe that this conglomerate is older than the eruptive breccias above mentioned, being possibly of early Tertiary or Upper Cretaceous age.

The Wubudu breccias are traversed by two systems of fissures, the older striking generally west 35 degrees north and east 35 degrees south, while the younger strikes due north and south. Nearly all the metalliferous veins so far discovered along the "Gold Coast" of Northern Celebes, are associated with the older system of fissures. It is curious that faults are extremely rare in them, and they are largely infilled with eruptive material. In the case of the best mineralized veins the gangue is a peculiar, very felspathic, easily-weathered diorite-porphyrite of andesitic character: the author terms it the Dopalak porphyrite, as having been first noted in the Dopalak vein of the Paleleh mine.   Mineralization seems to have taken place in several ways: sometimes the ore-deposit is a true contact-formation between the Wubudu breccias or country-rock and the Dopalak porpnyrite; sometimes the actual metalliferous vein is more or less split up in the gangue-rock, in which case the latter is always impregnated with ore. A gradual passage between the two modes of occurrence is also seen. There does not seem to be much doubt that the genesis of the ores is associated with the volcanic activity which caused the injection of the fissures with andesitic and porphyritic magmas.   This injection was probably accompanied and followed by the upwelling of thermal waters carrying the metalliferous particles in solution, and these were ultimately precipitated in the fissures and combined into various compounds.

The conditions of weathering are such that, inland the metalliferous veins do not crop out at the surface, whereas along the coast the eruptive dykes which contain them stand out like walls, the salbands especially offering considerable resistance to the continual beating of the surf.   The remnants of these dykes form nearly all the islets, reefs and rocky head-lands which characterize the northern coast of Celebes.   The Dopalak porphyrite, in consequence of its high percentage of felspar, weathers (even down to a considerable depth) to an easily-recognizable clay, of a pale bluish-grey, flecked with numerous white specks, and assuming a pink tinge only near the surface. This clay serves as a good "indicator" for mining prospectors.

One metalliferous vein was observed by the author in the Dolokapa series. It was 4 inches thick, and was infilled with beautifully transparent, very pure zinc-blende, associated with some pyrites and quartz. The Obapi conglomerate is traversed by numerous fissures which are in part mineralized and gold-bearing. In their general strike they coincide with the younger system of fissures previously mentioned: these sometimes fault the older system, and the author does not positively know of any case where they are metalliferous, although he thinks it possible that the Dunuki vein is to be reckoned among these later fissures.

The chief ore-bearing veins in the neighbourhood of Soemalata are then more particularly described: these are:—(1) The Parajava, some 5 inches thick, infilled with auriferous magnetic pyrites, and a very little quartz and calcite; (2) the Northern, of a maximum thickness of 12 inches, only the western portion so far proving profitable, infilled with auriferous sulphides: up to January 1st, 1902, 660 tons of ore had been got from this vein, containing 30 parts of gold per million; (3) the Southern, the most important of all, and the only one that has so far been worked to any considerable depth. It is characteristically lenticular, varying in thickness from 4 to 50 inches. The Wubudu breccia forms the hanging-wall, and the Dopalak porphyrite the foot-wall. The ore consists of the following minerals, in decreasing order of abundance: arsenical pyrites, iron-pyrites, magnetic pyrites, zinc-blende, galena and rarely copper-pyrites. All are more or less auriferous and argentiferous. There is very little quartz, and still less calcite, associated with the metallic sulphides. The average gold-content is 40 parts per million, and in the richer portions of the vein it attains a maximum of 400 parts per million, although the precious metal is invisible to the unaided eye. From the beginning of 1900 up to August 1st, 1901, some 5,500 tons of pure ore were got from this vein, yielding about 8,270 ounces of gold. The deepest level in the mine (recently driven) is 65 feet below the adit. (4) The Veta Nueva, discovered as lately as 1901, was proved for a length of little less than 1 mile. So far, only the ferruginous gossan has been touched, but iron-pyrites and arsenical pyrites are to be met with below, and the vein is known to consist of a series of lenticles intimately intermixed with the Dopalak porphyrite.

The ore from the Soemalata mine is roasted on the spot with a mixture of coke and coral-limestone, and a matte consisting of 74 per cent. of iron, 1 per cent. of copper, and 25 per cent. of sulphur is thus obtained. Its gold-content is 200 parts or more per million. If it contains less, it is broken up small and re-roasted. The slag that runs off from the smelting-furnace is found to be free from gold, and hence it is assumed that the process involves practically no waste. The good matte undergoes a second smelting with an additional amount of lead. The gold is thus concentrated into the lead up to about 1,400 parts per million, and this auriferous lead is thereupon shipped to Europe. · L. L. B.

MAGNETITE-DEPOSITS IN LUZON, PHILIPPINE ISLANDS.

Ueber eine Magneteisenerzlagerstätte bei Paracale in Nord-Camarines auf Luzon. By F. RINNE. *Zeitschrift für praktische Geologie, 1902, vol. x., pages 115-117, with 1 figure in the text.*

In the course of a visit to the Philippines in 1900, the author was enabled to examine an iron-ore deposit in the province of Northern

Camarines, on the eastern coast of the island of Luzon. The locality is a forest-clad hill rising some 130 feet above the surrounding district of Paracale, a small Tagala township, very near the embouchure of the Malaguit in the Pacific Ocean. It may be noted in passing that along the banks of the Malaguit are frequent outcrops of white and yellowish quartz, with which are associated pyrites and gold, evidently the débris of considerable reefs.

The hill of Bato-balani, or the Living Stone, is littered with blocks, great and small, of magnetite, and at first sight impresses the observer with the erroneous idea that the entire hill-mass is magnetite. Exploration-work, however, has shown that the ore occurs here only as loose blocks strewn on the surface, or accumulated in heaps, but not forming solid rock. Diggings prove that the weathered laterite on which these blocks rest, passes deeper down into a dioritic eruptive rock, and it may be plausibly assumed that the magnetite, at one time or another, has been chemically separated out of the diorite-magma. The occurrence reminds the author of the genesis of similar deposits in the Urals, such as Gora Blagodat and Wissókaya Sopka. In the case of Bato-balani, the former existence of a limestone overlying the diorite was doubtless an important factor in the contact-metamorphic phenomena which resulted in the setting-free of magnetite. Similar blocks of iron-ore are numerous near Dagang, and on the way up the Taguntung Hill, in the aforesaid district of Paracale.   L. L. B.

CUPRIFEROUS SANDS OF LUZON, PHILIPPINE ISLANDS.

Kupferreiche Sande im Malaguitgebiet bei Paracale, Luzon. By F. RINNE. *Zeitschrift für praktische Geologie*, 1901, vol. ix., pages 387-389.

Two deposits of cupriferous sand are worked by the Tagalas in the Malaguit river-basin, one of which occurs in the neighbourhood of the Bato-balani ("living stone") magnetite in the Submaquin Valley, and the other in the Calaburnay Valley.

The author washed samples of sand which he obtained from these deposits, and found native gold, more abundantly in the Submaquin than in the Calaburnay sand. As a matter of fact, gold occurs in the alluvia of all the affluents of the Malaguit. The copper occurs abundantly in grains not bigger than the head of a pin, and is very pure. Magnetite and iron-glance form the main mass of the heavy residue, zircon is found in numerous minute crystals, and iron-pyrites only occasionally.

The condition of the constituents of these cupriferous sands shows that they cannot have travelled very far from their original *locus*.

The Tagalas smelt the copper in shells, and it is sold in that form to Chinese and Tagala traders.   L. L. B.

UPRIGHT FOSSIL TREES IN THE FALISOLLE COLLIERY, BELGIUM.

Un Gisement de Troncs d'Arbres-debout au Charbonnage de Falisolle. By X. STAINIER. *Bulletin de la Société Belge de Géologie, de Paléontologie et d'Hydrologie*, 1902, vol. xvi., *Mémoires*, pages 69-76, and 2 *plates.*

The occurrence of upright boles or trunks of fossil trees is by no means common in the Belgian Coal-measures. Those just laid bare at Falisolle happen to occupy a position in a cross-cut that affords unusually favourable

-opportunities for close investigation. At this colliery, the Lambiotte coal-seam, the lowest worked in the principal pits of the Namur coal-field, over-lies a bed of extremely hard, highly quartzose, white or grey grit, some 22 or 23 feet thick. Sometimes there intervenes a clayey "floor" (at most 30 inches thick), with rootlets of *Stigmaria*, between the grit-bank and the ·coal-seam. Eastward the grit thins out, southward and westward it passes into an excessively hard quartzite. At Falisolle, the Lambiotte seam is contorted into no less than six folds, between the Gouffre and Carabinier faults. Above the seam, at an average distance of 13 feet, is a seamlet of coal. This at Arsimont and Ham-sur-Sambre, comes near enough the Lambiotte ·seam to be worked with it, and so too in the southern district of the Falisolle colliery. Going westward, the seamlet suddenly diverges widely from the main seam, and it was hereabouts that the upright tree-trunks were discovered, in the thickness of a bed of 28 feet. They are three in number, two of them quite destitute of the usual external carbonaceous ·crust, and apparently are now mere casts of hard, compact, grey shale.

The author concludes that they did not grow in place, but were drifted into their present position and silted up there. He points out that no great mass of roots extends from their bases into the rock-floor: there are merely a few isolated fragments of roots, evidently brought from afar. At the roof of the seamlet above described, not only in the Falisolle colliery, but in all those of the neighbourhood, there is an abundant fauna of marine fishes. The practically simultaneous existence of these and terrestrial vegetation in the same area would presuppose (in the author's opinion) changes of cataclysmic violence, for which there is not the slightest evidence. The upright position of the fossil trees does not affect the question. It is well known that drifted trees float with their roots downward, as the centre of gravity lies somewhere near the base, and this would be especially true of the ·conical, all but hollow, trunks of the Carboniferous forest-trees.　　L. L. B.

BAUMHAUERITE, A NEW SULPHARSENITE OF LEAD.

Bleisulfarsenite aus dem Binnenthal. Theil III. Baumhauerit, ein neues Mineral, und Dufrenoysit. By R. H. SOLLY. *Zeitschrift für Krystallographie und Mineralogie,* 1903, vol. xxxvii., pages 321-331, and 1 plate.

This new mineral has been found in the Binnenthal, Switzerland, in isolated monoclinic crystals, associated with other plumbosulpharsenites, in the white dolomite of the stream-bed of the Lengenbach. It is leaden-grey to steel-grey in colour, possesses a metallic lustre, and yields a chocolate-brown streak. The mineral exhibits conchoidal fracture, its hardness is 3, .and specific gravity 5·33; and the chemical formula is given as $4PbS,3As_2S_3$. Sometimes the crystals are of tabular form, deceptively like those of jordanite; sometimes they are rhombic prisms like dufrenoysite; and some apparently simple crystals are really built up of sensibly parallel growths of small crystals, often twinned. The paper contains an elaborate description and crystallographic measurements of the various forms, and also the ·chemical analyses by means of which the above-cited formula was arrived at.

The mineral is named after Prof. Baumhauer, of Fribourg.　　L. L. B.

THE FLORA OF THE ALPINE UPPER CARBONIFEROUS.

Begleitworte zur Demonstration eines Florenbildes des Alpinen Obercarbon. By F. VON KERNER. *Verhandlungen der kaiserlich-königlichen geologischen Reichsanstalt,* 1902, *pages* 125-127.

The vegetation generally, of which the remains are preserved in the Carboniferous strata of the Alpine region, was one that grew along the banks of rivers which brought down in their waters mud, sand and gravel. At the Steinach Pass, in the Tyrol, remains of tree-ferns are predominant (*Pecopteris, Alethopteris, Scolecopteris,* etc.) and the creeping fern, *Lygodium Stachei.* The Sphenopterids, so characteristic of the Middle Carboniferous, are not to be found in the Upper Carboniferous of the Central Tyrol. Other well-known Coal-measure plants which occur in the fossil flora of the Steinach Pass are species of *Lepidodendron, Lepidophyllum,* and *Calamites.*　　　　　　　　　　　　　　　　　　　　　　　　　　　L. L. B.

CALEDONITE FROM THE CHALLACALLO MINES, CHILE.

Ueber einen neuen Fundort des Caledonits in Chile. By GEORG BERG. *Tschermak's Mineralogische und Petrographische Mittheilungen,* 1901, *vol.* xx., *pages* 390-398.

This mineral is a sulphate and carbonate of lead and copper, in small blue crystals, and was found among the ore-specimens from the upper levels of the Challacallo silver-mines, in the Atacama desert. It forms part of a collection recently forwarded to the Freiberg Mining Academy.

The intense sky-blue coloration of the mineral differentiates it from nearly all previously-described occurrences of caledonite. The locality, moreover, is a new one for that mineral species. A chemical analysis of the Challacallo caledonite is given, showing its high percentage of carbonate; and, after an elaborate discussion of the analyses of caledonite from other localities, the author rejects the view (advanced by Prof. Story-Maskelyne and Mr. Flight) that carbonate occurs in it merely in the form of a mechanical admixture of cerussite. Still, what with differences of coloration and chemical analyses, it would appear that caledonite can hardly be regarded as a mineral of fixed composition, and it is here defined as an "isomorphic mixture."

The Challacallo ores occur as parallel veins, sometimes coalescing into one great vein, coursing through a quartz-trachyte cone which rises up from the high plateau of Atacama. The strike is north, nothing to 10 degrees west, and the dip westerly, very steep. The filling of the veins appears to be chiefly quartz, but it is a question whether deeper down the quartz is not in part replaced by other minerals. There are cross-veins, dipping steeply too, but they appear to have little effect on the principal vein. On the other hand there is a great crush-zone, 6 to 13 feet thick, to which the rather vaguely-used term *manto* has been applied, which has had an impoverishing influence on the principal vein—owing perhaps to the circulation of subterranean waters, leaching out and carrying off the ore-particles.

Many rare minerals have been got from the outcrop, but the only immediate associate of caledonite here is azurite. The principal silver-content is connected with the so-called *lechedor,* a decomposed mass of vein-stuff and country-rock, highly impregnated with chlorides, iodides and other salts of silver.　　　　　　　　　　　　　　　　　　　　　　　　　　　L. L. B.

CHALMERSITE, A NEW COPPER-SULPHIDE FROM BRAZIL.

Ueber Chalmersit, ein neues Sulfid der Kupferglanzgruppe von der Gold Mine " Morro Velho " in Minas Geraes, Brasilien. By E. HUSSAK. *Centralblatt für Mineralogie, Geologie und Palæontologie,* 1902, *vol.* ii., *pages* 69-72.

The gold-mine of Morro Velho, one of the oldest and most successful in Brazil, lies at the foot of the Serra do Curral, near the capital of the State of Minas Geraes. The ore-body, interbedded conformably with chloritic calc-schists, consists mainly of magnetic pyrites, carbonates, and arsenical pyrites, together with quartz. The strike is easterly and westerly, the dip about 45 degrees, and the depth to which workings have been carried well nigh 3,300 feet. There is no visible gold present in the free state: nevertheless the gold-content of the pyrites averages 18 parts per million.

After a reference to a detailed description of the mine,[*] the author draws attention to the druses lined with magnificent crystals of various minerals, which occur chiefly at the junction of the ore-body and the calc-schist.

The new mineral occurs invariably in such druses in association with the magnetic pyrites and the copper-pyrites, and has been described by the Prince Dom Pedro Augusto de Saxe-Coburg as millerite (which it is not).

It is strongly magnetic, has a specific gravity of 4·68, and corresponds to 3·5 in the scale of hardness. Called chalmersite after Mr. G. Chalmers, the superintendent of the mine, this mineral is found in opaque yellow acicular crystals, possessing a metallic lustre, often twinned. Chemical analysis yielded the following percentages:—Iron, 46·95; copper, 17·04; sulphur, 35·30; whence the formula $CuFe_3S_4$ (or $Cu_2SFe_6S_7$) is deduced. The author regards chalmersite as in all respects isomorphic with copper-glance; and a study of the order of crystallization of the minerals at Morro Velho shows that the new mineral was the last but one to separate out, being preceded by magnetic pyrites and followed by copper-pyrites.   L. L. B.

RICKARDITE, A NEW TELLURIDE-MINERAL.

Rickardit, ein neues Mineral. By W. E. FORD. *Zeitschrift für Krystallographie,* 1903, *vol. xxxvii., pages* 609-610.

This mineral was found at the Good Hope mine, Vulcan, Colorado. The chief ore worked there is pyrites, with which metallic tellurium is associated in unusually large masses. The new mineral occurs in lenticles intimately intergrown with the native tellurium. The mean of two analyses shows its percentage-composition to be: Copper, 40·51; and tellurium, 59·49. This corresponds to the chemical formula Cu_4Te_3; it is a pure telluride of copper, without any traces of sulphur, selenium, arsenic, antimony, lead, silver or gold. As it constitutes a new type of telluride, the suggestion is put forth that each molecule of the substance is really a combination of one molecule of cuprous with two of cupric telluride, or $Cu_2Te + 2CuTe$. It has a magnificent purple colour, is tough, and of irregular fracture, of hardness 3·5, and specific gravity 7·54. When molten it burns with a pale azure-blue flame, green-tinged at the edges.

The mineral has been named after Mr. T. A. Rickard, of New York.

L. L. B.

---

[*] *Engineering and Mining Journal* [*New York*], 1901, *vol. lxxii., pages* 485-489, *with 5 illustrations in the text.*

COLOURED TOURMALINES OF CALIFORNIA.

Precious Stones in the United States in 1901. By GEORGE F. KUNZ. *The Engineering and Mining Journal* [New York], 1902, *vol. lxxiii., page* 38.

Recent discoveries in Southern California have revealed a locality of coloured tourmalines, in a ledge of quartz and lepidolite, at an altitude of nearly 1 mile, on Mesa Grande mountain, San Diego County. As compared with the well-known locality of pink tourmalines in lepidolite at Palo, California, this new occurrence differs in presenting large and separate crystals in both lepidolite and quartz, many being translucent or even transparent, and with perfect terminations. Rubellite is the prevailing variety; but all the colours occur, sometimes several in one crystal—both in transverse sections, as at the Maine and Connecticut localities, and in concentric zones, as often in Brazil. Besides, there are some large and perfectly colourless achroites, and some yellow specimens. Frequently rubellite or other coloured crystals have a thin dark green, or nearly black, outer shell or coating, characteristic of this locality. Many gems have already been cut; and as specimens, the crystals are magnificent. A number of single rubellites, only partially perfect, weighing up to 70 carats each, and occasional yellow, green, and white gems, have also been found here.

X. Y. Z.

A NEW MANGANESE-IRON MINERAL, MANGANOSPHÆRITE.

Mittheilungen über Manganosphärit, Schwefel, Brookit, Augit und Pyrit. By K. BUSZ. *Neues Jahrbuch für Mineralogie, Geologie und Palæontologie,* 1901, *vol. ii., pages* 129-140, *with 6 figures in the text and 1 plate.*

At Horhausen in the Westerwald, Germany, the Luise mine is opened up in a spathic iron-ore deposit, which is associated with a basalt-dyke of a thickness varying from 2½ to 16 feet, cutting through the Devonian grauwacke-slates. At the 275 feet level the basalt is seen to be intrusive into the ore-body itself, inducing therein contact-metamorphic phenomena. Some of the cavities in the basalt are filled with reniform or grape-like aggregations of a mineral, the rhombohedral surfaces of which have a shimmering lustre. The colour is reddish-brown, much lighter than that of sphærosiderite. Careful microscopic examination and chemical analysis have proved that this is a new mineral, quite distinct from oligonite; and, on account of its high tenour in manganese, the author names it " manganosphærite." It stands in exactly the same relation to oligonite, as sphærosiderite does to spathic iron-ore. In the scale of hardness it stands a little below apatite, being represented by the number 4½ to 5; its specific gravity is 3·63; and its percentage-composition is as follows:—Ferrous oxide, 36·72; manganous oxide, 24·76; and carbon dioxide, 38·34. In other words, it is made up of three parts of iron carbonate to 2 parts of manganese carbonate.

L. L. B.

THE ARTIFICIAL REPRODUCTION OF MINERALS BY SUBLIMATION.

Ueber künstliche Darstellung von Mineralien durch Sublimation. By HERMANN TRAUBE. *Centralblatt für Mineralogie, Geologie und Palæontologie,* 1901, *vol. i., pages* 679-683.

It is pointed out that the reason why, up till recently, the method of sublimation in the experimental synthesis of minerals has not met with much success, is that the boiling-points of most inorganic compounds are very high, and thus at the high temperatures at which the experiment must be conducted there is enormous difficulty in excluding either the oxidizing action of the atmosphere or the reducing action of flame-gases. Even the use of the vacuum-pump, though it allows of the lowering of the boiling-points, does not do away with the technical difficulties of such experiments. But the boiling-point of a given body, A, may be lowered by heating it with a substance, B, which sublimes at a lower temperature. It is thus, no doubt, that certain inorganic crystalline residues found in blast-furnaces originate—sublimation being facilitated by the presence of other vapours and gases.

Dr. Manross, in fusing amorphous [cerium] wolframate with sodium and potassium chlorides in a porcelain crucible at a temperature of 1,400° Cent. (2,552° Fahr.) had noticed that a portion of the "cerium wolframate" thus artificially reproduced had sublimed in beautiful crystals on the crucible-cover, and also on the outside at those points where the chloridic vapours had escaped. This led to further experiments on the volatilization of substances in the presence of sodium- and potassium-chloride vapour, for which purpose the author used an electric furnace patented by Dr. C. A. Timme. By measurement with a Le Chatelier pyrometer it was ascertained that the highest temperature attainable in this furnace is 1,500° Cent. (2,732° Fahr.): its principal advantage is that the reducing-action of flame-gases, etc., is completely excluded.

The following were the substances experimented with:—Cerium wolframate and molybdenate, didymium, lanthanum, calcium and lead wolframate and molybdenate, barium sulphate, and mixtures of the foregoing salts. Platinum-crucibles were used in some cases, porcelain-crucibles in others, and tubes closed at one end in yet others. At a temperature of about 2,552° Fahr., the potassium and sodium chlorides with which the wolframate or molybdenate was mixed vaporized abundantly, and sparkling crystals of the wolframate, etc., sublimed on the cover of the crucible or on the walls of the tube. Thus too were obtained crystals of barium sulphate and cerium wolframate combined, and also of pure barium sulphate. Without the presence of the chloridic vapours, it would be impossible to obtain crystals of these salts by sublimation.

The door is thus opened to further experimentation with many other mineral substances, and in this way light may be thrown on the genesis of many ore-deposits. Thus, for instance, the fact that scheelite may originate by sublimation tends to support Prof. Daubrée's hypothesis as to the genesis of zinc-ore deposits. L. L. B.

SECULAR VARIATION OF THE EARTH'S MAGNETISM.

Sur les Variations séculaires du Magnétisme Terrestre. By V. RAULIN. *Comptes-rendus hebdomadaires des Séances de l'Académie des Sciences*, 1901, *vol. cxxxiii., pages* 760-763.

The author refers to a former paper of his, published by the Paris Academy in 1866, wherein he concluded that the hypothesis of a rotation of the magnetic north pole around the earth's pole in latitude 70 degrees, explained and co-ordinated all the facts so far observed in this connexion, in Europe and in the Atlantic basin. The rotation would last 600 years. He points out that the observations which have been accumulated, during the 35 years that have elapsed since his former paper, have fortified and confirmed his conclusions.

Declination at Paris increased from 1664 to 1814, decreased from 1814 to 1866, and will go on decreasing (*ex hypothesi*) till 1964. Then it would begin increasing again until 2114, decreasing thereafter till it returned to 0° in 2264. The decrease of declination has been just as regular since 1814 as was the increase in the 78 years preceding; and there is every reason to believe that the phenomenon will continue with the same regularity till the zero-point is reached in 1964. The present position of the magnetic north pole should be, according to the author, in longitude 141 degrees 7 minutes west [of Paris], somewhere beyond the mouth of the Mackenzie river.

Inclination has been decreasing for 230 years (since 1671) and it appears reasonable to assume that this will continue at least till 1964, thus completing a single decrescent phase of 300 years, parallel and simultaneous with the two phases (crescent and decrescent) of declination.

These phenomena are explicable, in the author's opinion, on the hypothesis of the existence of a fusiform magnetic body, probably ferruginous, the two extremities of which coincide with the earth's magnetic poles. The westerly displacement of the whole superficial magnetic system of the globe is explained, if we admit that this fusiform body and the liquid kernel of the earth travel less rapidly than the solid outer crust, in the diurnal east-to-west rotation, by an amount equal to $\frac{1}{100}$ of the velocity of that crust at all latitudes.

L. L. B.

TERRESTRIAL MAGNETIC PULSATIONS.

Pulsations de la Force Magnétique Terrestre. By W. van BEMMELEN. *Archives Néerlandaises des Sciences exactes et naturelles*, 1901, series 2, vol. vi., pages 382-384.

In studying the magnetograms of Batavia the author noticed series of pulsations of the horizontal component, evidently not due to solar disturbances, since they occurred for the most part during periods of magnetic calm. They took place during the night, and their amplitude was about $\frac{1}{10,000}$ of the horizontal component. The result of three years' study demonstrates that the mean period of these pulsations is subject to annual variations, from a minimum at the beginning of a year to a maximum in the middle of a year. Observations simultaneously made at Zikawei, near Shanghai, revealed series of magnetic pulsations, corresponding in amplitude and time with those recorded at Batavia, though on the whole the phenomenon is of less frequent occurrence at the former place. Other observations, carried out at Karang Sago, Sumatra, show a similar parallelism.

As to the immediate cause of such pulsations, it is pointed out that, during solar perturbations, circular electric currents are developed in great number in the earth's atmosphere, inducing polar auroras. The nearer one approaches the zone of greatest auroral frequency, the more irregular are these currents; but in China and the Malay Archipelago they are more regular. In studying what he has termed "magnetic post-turbation" the author has satisfied himself that there is a system of electric currents travelling round the globe, parallel with the "isochasms" or curves of equal auroral frequency. It does not seem unlikely that the magnetic pulsations are caused by small and rapid periodic fluctuations in the aforesaid currents, and that such fluctuations are of terrestrial origin. They are perhaps comparable with the undulations of a fluid surface, which is being subjected to continuous disturbance at a particular point.

L. L. B.

MAGNETIC ANOMALIES OBSERVED IN FRANCE.

Sur la Distribution Régulière de la Déclinaison et de l'Inclinaison magnétiques en France au 1er Janvier, 1896. By E. MATHIAS. Archives Néerlandaises des Sciences exactes et naturelles, 1901, series 2, vol. vi., pages 412-429; and Comptes-rendus hebdomadaires des Séances de l'Académie des Sciences, 1901, vol. cxxciii., pages 864-867.

The author had previously shown that the linear formula which gives, in functions of the geographical latitude and longitude, the law of regular distribution of the horizontal component for the region of Toulouse, is applicable to the whole of France.

He tabulates the results obtained (for declination) by means of a similar formula in 11 departments showing perfect regularity, and in 17 other departments showing on an average only one anomaly (with 5 or 6 observing-stations in each department). All the other departments (say 61) show variable proportions of regular and anomalous stations. On the whole, the measurements of magnetic declination at more than 300 stations in France and Corsica yield differences of less than 3 minutes 5 seconds, and it is evident that declination, although strongly influenced by local causes, even where the superficial strata of the earth's crust are not at all magnetic, conforms to a law of regular distribution just as much as the horizontal component does.

The author then attempted to trace the law of distribution of magnetic inclination. Here, by means of formulæ of the same order as that above mentioned, he found 462 stations yielding differences of less than 4 minutes. (In 16 departments there was no anomaly, in 17 only one.)

The final conclusion is that, for all the magnetic elements, only two departments are perfectly regular, namely, Ain and Hautes Alpes. The Loir-et-Cher, Mayenne and Sarthe, and also the Isère, are perfectly regular as regards the horizontal component and inclination, but anomalous as regards declination. In these cases it is argued that the irregularities of declination depend far more on the structure, than on the nature, of the underlying rocks. The structure indeed becomes an important factor, if we suppose that a part of the earth's magnetic field is due to telluric currents flowing in the neighbourhood of the earth's surface. It is then easy to account for the perfect regularity of the magnetic vector as contrasted with the anomalies of declination.

L. L. B.

MAGNETIC DISTURBANCE OBSERVED AT ATHENS.

Sur une Perturbation magnétique, observée à Athènes le 8 Mai, 1902. By D. EGINITIS. Comptes-rendus hebdomadaires des Séances de l'Académie des Sciences, 1902, vol. cxxxiv., pges 1425-1426.

On May 8th, 1902, from 1·35 a.m. until 9·30 a.m., at the very time of the destruction of St. Pierre in Martinique, a magnetic disturbance was recorded by the instruments set up in the Athens observatory.

It affected chiefly the horizontal component, the declination rather less, and the vertical component scarcely at all. The same phenomenon was observed simultaneously in Paris.

The author argues that if this perturbation was in fact one of the effects of the eruption of the Montagne Pelée, the originating cause must have been exclusively magnetic or electric, and did not consist in the mechanical transmission of the earth-shocks which doubtless accompanied that great eruption. For the Agamennone seismograph at the Athens observatory remained absolutely motionless at that time, and yet it is an instrument which may be relied upon to record clearly even very feeble shocks.   L. L. B.

PLUMBING DEEP SHAFTS.

(1) *Plumbing Deep Shafts of the Tamarack Mine. By* H. M. LANE. *Mines and Minerals*, 1902, *vol. xxii., pages* 247-248, *with 3 illustrations in the text.*

The No. 5 shaft of the Tamarack mine is the deepest shaft in the world, its depth being 4,938 feet. It was desired to connect this shaft with the workings from the No. 2 shaft, and a new survey was taken down both shafts. The shafts are 3,200 feet apart, and the proposed connecting tunnel starts at a point in No. 5 shaft situated 4,250 feet below the surface. The plumbing of No. 5 shaft was proceeded with, and No. 24 piano wire was used for the plumb-lines. The wires were lowered by a small engine, being weighted at their lower ends by wooden frames, or balloons, each 10 feet long, $2\frac{1}{4}$ feet in diameter, and 20 pounds in weight. In the centre of each balloon a lantern was hung, so that the progress of the balloon could be watched from above and below. For obvious reasons, the balloons were lowered as nearly as possible in the centres of the shaft-compartments; but once lowered, it was desirable to place the plumb-lines as far apart as possible; 8 pounds plummets were substituted for the balloons, and the distance between the lines at the surface was increased to 17·58 feet. The 8 pounds plummets were then replaced by plummets of 50 pounds, when to everybody's surprise, the lines stretched a distance of 15 feet. The wires were then cut to the proper length, and the plummets were immersed in pails of engine-oil in order to minimize vibration. Here, again the unexpected happened, for the wires immediately shortened up 25 inches, owing to the buoyancy of the oil diminishing the effective weight of the plummets. But the greatest surprise occurred when it was found that the spacing of the plumb-lines was 0·07 foot or 0·84 inch greater at their lower than at their upper ends. Measurements above and below were re-made and verified, and the wires were examined throughout the whole length of the shaft, but nothing was discovered that could in any way account for the divergence. At No. 2 shaft, practically the same conditions were found, but the divergence of the lines was even greater, the spacing at the lower ends in this case being 0·1 foot or 1·2 inches greater than at the upper ends. The substitution of leaden for cast-iron plummets had no effect on the divergence, showing that magnetic attraction was not the cause of it. No satisfactory explanation of the occurrence appears to be as yet forthcoming.

(2) *The Divergence of long Plumb-lines at the Tamarack Mine. By* PROF. F. W. MCNAIR. *The Engineering and Mining Journal* [*New York*], 1902, *vol. lxxiii., pages* 578-580, *with 2 illustrations in the text.*

Possible causes of the divergences have been offered and freely discussed. It was contended that the divergence was due to the greater attraction of the material at the end of the shaft for the plummet hanging nearest it. There seems to be a general lack of appreciation of the forces of gravitation. It is, of course, true that the attractions on each plummet towards the ends of the shaft differ, the stronger being toward the end nearest to which the plummet hangs. It is also true that these differences tend to diverge the wires, but the amounts are so insignificant as to put them quite out of consideration. Another explanation worth notice was offered by Prof. Hallock, of Columbia University. This was that of mutual repulsion between like poles at the lower extremities of the wires. It was subsequently modified, so as to include also repulsion between like consequent poles distributed along the wires. However, the management of the Tamarack mine, with its characteristic interest in scientific matters, granted the writer permission to carry on further experiments in the shafts. These experiments are fully described by the writer, and the results seem to afford ample proof that neither gravitation nor magnetism could account for the divergence of the lines.

On the contrary, the writer became convinced that a very simple cause lay at the bottom of the divergence which attracted so much attention. One or both wires were deflected by the air-currents circulating in the shaft. This suggestion of air-currents was early made, but was at first treated with scant courtesy because it did not seem probable that such currents could be steady enough in speed and direction to permit the constancy of mean position which had been observed. But the hypothesis once admitted, it accounted for all the phenomena observed. Experiments showed that the divergence increased when the ventilation was augmented. The divergence diminished when wires of smaller section were used, that is to say, when the area of impact was diminished, and a thinner pencil of air collided with the wire. In No. 5 shaft, under normal conditions of ventilation, the divergence varied from 0·07 to 0·11 foot. The top of the shaft was then enclosed, but it was found impracticable to stop entirely the circulation of the air. But the reduced circulation of the air reduced the divergence to 0·018 foot. Again, in No. 2 shaft, which is a downcast, the plumb-lines diverged as in No. 5 shaft, while plumb-lines hung down No. 4 shaft, which is an upcast, converged in four experiments, the fifth showing a slight divergence of 0·004 foot.

The writer concludes by saying that it is remarkable that the air-currents should be so constant in their action. But the experiments have shown that the engineer who has to plumb a deep shaft has great difficulties to overcome, with regard to the influence of the air-currents on his lines. Before deciding where to hang his lines, and in order to estimate the dependence to be placed on them when hung, he should study the shaft air-currents very intently. X. Y. Z.

GOLD-MINING IN HUNGARY.

Die montangeologischen Verhältnisse des Kornaer und Bucsumer Thales, sowie des Goldbergbaues um die Berge Botes, Korabia und Vulkoj herum. By ALEXANDER GESELL. *Jahresbericht der königlich Ungarischen Geologischen Anstalt, 1899, pages 97-103, with 1 figure in the text.*

In the Korna valley, the Carpathian Sandstones and the underlying Jurassic limestones are the predominant rocks, but at some points they are interrupted by andesites, dacites and volcanic breccias. Mining industry flourished formerly in the upper part of the valley, but at present there is only one mine at work, the Valca Verde. The nature of the rocks does not appear to affect the metalliferous occurrences, as mines have been opened up indifferently in the igneous masses and in the sedimentary strata.

In the Bucsum valley, the Concordia mine works pillars of ore, associated with a conglomerate which is interbedded with sandstone. Workings are pushed down to a depth of 280 feet from the surface, and the richest points are at the intersection of vertical with flat fissures. Native gold, in the free state, occurs in considerable quantity at such intersections. The infilling of the fissures here is calcspar and the country-rock is a siliceous breccia, believed to be of contact-metamorphic origin, resting upon slates: it too appears to be impregnated with gold. The author believes that gold-bearing rocks extend over a much larger area than that at present worked by the Concordia Company, and points to the fact that neighbouring mines are now also at work on flat auriferous fissures.

The core of the Korabia and Konczu range is made up of dacites and andesites, which in former ages burst through the Carpathian Sandstone Series. The fissure-veins in the eruptive rocks strike parallel with the mountain-range, and are equally productive in the (older) dacite and the (younger) andesite. Taken as a whole, the Botes-Vulkoi-Korabia ore-deposits are without doubt genetically connected with the famous Veres-paták ore-deposit which seems to be their northward prolongation.

Cursorily glancing at the Aráma mine, the author describes more at length the Botes mine, worked by an adit in the hill of the same name. Here the auriferous belt, dipping northward some 20 to 30 degrees, is over-lain by Carpathian Sandstone and underlain by shales. Instead of the pillar-like occurrences previously described, the ore-body here is more of the nature of a bedded mass 300 feet thick. The flat fissures, containing native gold in the free state, course parallel to the bedding-planes. The shales are absolutely barren, here as well as in the Korabia and Vulkoi mines. The Bucsum-Pojen and Valca-Csurtului mines are opened up in the eruptive dacites, and what was said of the Korna valley seems to hold good of the whole region traversed by the author, namely, the auriferous fissures are characteristic alike of the eruptive rocks and the sedimentary sandstones.

L. L. B.

THE ALSÓ-GALLA AND BÁNHIDA BROWN COAL-MINES, HUNGARY.

Der Alsó-Galla-Bánhidaer Braunkohlenbergbau der Ungarischen Allgemeinen Stein-kohlenbergbau-gesellschaft. By LUDWIG LITSCHAUER. Berg- und Hütten-männisches Jahrbuch der k. k. Bergakademien zu Leoben und Přibram und der königlich Ungarischen Bergakademie zu Schemnitz, 1902, vol. l., pages 351-418, and 3 plates.

These mines belong, as indeed do many others in Hungary, to the Hungarian Universal Coal-mining Company, a powerful organization which, founded as recently as 1889, has extended of late years enormously its sphere of operations.

The deposits lie in a basin, bounded on the south-east, east and north-east by the Vertes hills (1,400 feet above sea-level), flattening westward down towards the Danube, and narrowing south-westward into the Mély-árok or deep gorge. The Danube hereabouts is 357 feet above sea-level, while Bánhida lies 145 feet higher, and Alsó-Galla and Felsö-Galla are situated at altitudes of respectively 522 and 662 feet.

The Tata Vidék (Alsó-Galla, etc.) concession stretches alongside the state railway-line, some 45 miles distant from Budapest and 119 miles distant from Vienna, a very advantageous position in regard to the Austro-Hungarian coal-markets. It covers an area of about 935 acres.

Proceeding to consider the geology of the so-called " Tata coal-field."

which includes the three principal localities above mentioned, the author points out that the literature of the subject is very meagre. It appears probable that this coal-field is a continuation of that of Gran (Esztergom), and is possibly also to be correlated with the coal-basin of Pilis-St. Iván. The basement-rocks of the whole area are Triassic dolomites and Rhætic and Liassic limestones. Overlying these are Lower Eocene freshwater beds with coal-seams, succeeded by a brackish-water, and that in turn by a marine facies; next come Oligocene beds, in part also coal-bearing, and the whole complex is mantled over by æolian drift or lœss.

The coal is of excellent quality for a brown coal, and its calorific value nearly approaches that of ordinary (black) coal. It is pitch-black, with a conchoidal fracture and a greasy lustre, but crumbles too easily to furnish much round coal. A series of analyses, made during the years 1897, 1899 and 1900, yielded the following averages:—Combined water, 12 per cent.; ash, 7·5 per cent.; sulphur, 1·3 per cent. The heating power was 5,000 to 6,000 calories. In making briquette-fuel from this, petroleum-residues are used as a binding-material.

Coal is wound from three shafts, a fourth is in process of sinking, and there are also opencast workings. The average thickness of the seam is 33 feet, but this apparently includes some shaly partings. It is estimated that there are at least 72,000,000 tons of coal in sight, though it is expected that further exploration-work will show the true coal-reserve to be far greater than this. Previous to 1897, only one shaft had been sunk in this coal-field. Now, there is a central station furnishing electricity for transmission of power and for lighting to the three principal pits, endless-rope haulage is used on the inclined planes, etc., and the coal is loaded direct from the central picking-screens into railway-wagons by which it is conveyed to the Bánhida station of the Hungarian State Railway. Opencast workings were only seriously started in 1901: here, traces of Roman buildings and graves were discovered, but no signs whatever of ancient mining operations of any kind.

For shot-firing, progressite is the explosive used, with electric detonators. For ventilation of the pits Pelzer fans have been chosen, which pass 88,275 cubic feet of air per minute into the workings. Moreover, Munk and Capell reserve-fans have been erected. As the mines are fiery, Wolf benzine safety-lamps are used at the working-face, but the winding-shafts, inclined planes, etc., are lighted (as above mentioned) by electricity. The author enters into great detail regarding workmen's dwellings, clubs, schools and wages, and describes very fully the methods of working and the machinery in use at these collieries.   L. L. B.

―――――

THE PETROLEUM-INDUSTRY OF AUSTRIA-HUNGARY.

Die Erdöl-industrie in Oesterreich-Ungarn. By EDUARD WINDAKIEWICZ. *Bergund Hüttenmännisches Jahrbuch der k. k. Bergakademien zu Leoben und Pŕibram und der königlich Ungarischen Bergakademie zu Schemnitz,* 1901, *vol. xlix., pages 17-104, with 8 figures in the text, and 2 plates.*

This elaborate memoir is divided into thirteen chapters, the first of which is devoted to a brief history of the rise and progress of the oil-industry in the various territories of the Austrian Empire. Galicia saw its birth between 1810 and 1817, but the true beginning of it on a scale of any commercial importance dates from about 1859. It has been revived recently in the

Bukowina, and is making steady progress in Hungary. Despite the severity of American and Russian competition, Galician petroleum still holds its own, in the home-market at least.

Turning then to the geology of the subject, the author remarks that the structure of the Galician Carpathians constitutes a very thorny and complicated problem, which has been studied by a long succession of eminent scientific men, but no solution as yet proposed has been accepted unanimously as satisfactory. The Government geological-survey map of the petroleum-region of the Carpathians is now out of date; the geological atlas of Galicia begun in 1888 by the Provincial Government and the Cracow Academy of Sciences is not yet completed, and is moreover too full of local detail to facilitate a general view of the subject. Prof. Zuber, of Lemberg, however, has recently issued a most valuable map of the petroleum-region of Galicia, with explanations.

The petroleum of that province is intimately associated with the Carpathian Sandstone Series, a succession of alternating sandstones, conglomerates, shales, clays and marls. The want of characteristic fossils has long been a stumbling-block in the path of classification, but it is now pretty well ascertained that this series belongs in part to the Cretaceous, and in part to the Tertiary system. The various strata among which petroleum is most frequently found in Galicia are described, and the structural difference between the Carpathians of the western and those of the eastern district is emphasized. Thus, west of Stwiaz̧ and San, the mountain-ranges are comparatively short, and consist mainly of uplifted and irregularly-contorted Oligocene and Eocene beds, among which now and then crop up (unconformably or pseudoconformably) the broken and disturbed Ropianka beds, of Cretaceous age. Eastward this unconformity dies out, and all the strata from the Ropianka beds up to the Oligocene, follow each other in conformable and parallel succession, forming a series of regular folds often interrupted by longitudinal faults. Practical experience has proved that petroleum occurs in greatest abundance where the strata are folded into saddles, and further that it occurs only in certain definite groups of beds.

The reservoirs of natural oil in the Cretaceous Ropianka beds are not of very considerable extent: in Western Galicia the best localities are those where there is a cover of impervious Eocene clays and shales. The limits of the rich Eocene oil-belt in the Eastern Carpathians are determined by the outcrop of the Jamna Sandstone (Cretaceous). In the Oligocene beds only those occurrences of petroleum are of industrial value, where the strata form broad, flattish, regular saddles. Where traces of oil are discovered in saddles of menilite-shale, there is always a prospect of striking the richer oil-reservoirs of the underlying Eocene by means of deep borings.

There appears also to be a great accumulation of petroleum and ozokerite among the saddles of the Miocene saliferous clays. A general rule is stated for all petroleum-deposits in the Carpathians, namely, that the south-western side of the outcrops is more regularly bedded, and therefore easier to work, than the usually dislocated and sunken north-western side.

In the Bukowina, the oil-belt ranges north-west and south-east in prolongation of the Galician oil-region, and is traced along three parallel lines. The central or second line is the richest, it yields oil (comparable with that of Galicia) occurring in the Oligocene portion of the Carpathian Sandstone Series. Not much has been done, so far, in the way of putting down bore-holes or trial-shafts in the Bukowina, but a French company is said to have started operations recently.

In Moravia, petroleum has been found at a few localities, most recently of all at Göding. It occurs in beds which are correlated with the Miocene saliferous clays of Boryslaw.

Although petroleum has been found at several localities in Northern and Southern Hungary and Croatia, an oil-field comparable with the Galician is not to be looked for, so far as extent is concerned; but in regard to quality, the petroleum seems likely to prove equal to the best that is got in Galicia. As in that province, it occurs alike in strata of Cretaceous, Eocene, Oligocene and Miocene age.

Natural gas is of frequent occurrence in the strata overlying the petroleum-deposits, and is used at many localities for purposes of illumination. The author publishes the following percentage-analysis of the gas which burst forth at a pressure of 20 atmospheres from the Turoszówka boring, Krosno district:—Carbon dioxide 0·4, oxygen 0·6, ethylene 1·1, methane 82·3 and nitrogen 14·4.

Brine occurs in association with the petroleum in several localities, but as a rule in very small proportions.

The sixth chapter describes the physical and chemical characters of petroleum, and contains a very full comparative table giving the results of the distillation and fractionation of the crude oils of Galicia, Sumatra, Baku, Alsace and Pennsylvania.

The seventh chapter deals with the methods of winning, which have undergone a great change within the last 15 years, owing to the introduction of Canadian derricks and simpler pumping-plant. The various derricks and boring apparatus now in use are described and figured. In 1897, there were 571 shafts and 2,223 boreholes in the Galician oil-field; of these, 26 shafts and 1,263 bore-holes were yielding oil, and 21 shafts and 264 bore-holes were in process of sinking. There were 65 mineral-oil refineries in Austria and 16 in Hungary.

According to the Galician Naphtha Act of 1884, petroleum, ozokerite and bitumen are the property of the landowner on whose estate they occur. He may divest himself [in favour of others] of the right to win these products, a declaration being made before the mining authority, and the property being then entered as a naphtha-field in the official register. If the workings are continuously neglected in such wise as to be prejudicial to the common weal and common safety, the authorities may lay an embargo on the field, estimate its value, and sell it by auction. Such works as are necessary for the safety of mining operations may be extended even into neighbouring properties, with the obligation to make good any damage caused thereby. The authorities must receive four weeks' notice of the commencement of operations, and these can only be conducted by persons of certified capacity, duly confirmed by the authorities.

Statistics of output are given, whereby it is seen that Austria-Hungary in 1897 produced about 2 per cent. of the total amount of petroleum got from all countries, Russia and the United States accounting each for about 47 per cent. As a consumer of petroleum, Austria ranks next to Great Britain. Statistics of import and export are also given, and bear out the statement previously made that the Galician petroleum retains well its hold on the home markets. The memoir terminates with a list of all the petroleum-wells in Galicia and the Bukowina.　　　　L. L. B.

IRON-ORE MINES IN MORAVIA.

Die Eisenerzvorkommen und die ehemalige Eisenerzeugung bei Römerstadt in Mähren.
By JOSEF LOWAG. *Oesterreichische Zeitschrift für Berg- und Hüttenwesen,*
1901, vol. xlix., pages 129-133.

Little is known of the original working of the iron-ore mines at Römerstadt
in Moravia. The earliest name of the town was Raymerstadt, and the Romans,
probably, never penetrated to this district. The mines are, however,
very old, and seem to have been worked by the Celts long before the
Romans, for there are remains of open diggings and trenches far below the
present surface. The iron was extracted from the ore in charcoal-furnaces,
and this method was practised till quite recently. Mines of gold, silver,
copper, lead and iron are mentioned in 1535. The mines were repeatedly
pledged by the Crown to various people; and they are now the property of
Count Harrach. The introduction of railways gave the death-blow to the char-
coal-iron industry and although there were still stores of iron-ore in hand,
the charcoal-furnaces were blown out everywhere. Even at the present day,
immense quantities of good iron-ore exist in this formerly busy mining
district, and await the enterprise of some ironmaster, who will be able
to utilize them.

The ore is composed of brown, red, and magnetic ironstone. It occurs
in several parallel beds in Lower Devonian chlorite-schists, and was chiefly
dug from two places about 650 feet apart. The zone in which the ore is
found is about 3,600 feet wide, and about 7 miles in length. The seams
are from 3 to 18 feet thick and the percentage of iron varies from 30 to 60.
If magnetic and red iron-ore occur in the same seam, the former is usually
found on the floor, and the latter on the roof. The magnetic iron-ore is
black and coarse-grained, generally rich in iron and contains much lime; the
red iron-ore is usually iron-grey, siliceous, and contains red hæmatite.
Where the three ores are found mixed, the percentage of iron in them is
sometimes as high as 62. In some places the ore crops out at the surface,
and fine crystals of iron-pyrites are often found in it. The seams are sharply
defined, and are supposed to have been formed at the same period as the
chlorite-schist. Three mines showed seams of magnetic iron-ore from 3 to
6 feet thick, and containing from 30 to 50 per cent. iron, reached by
shafts 50 to 60 feet deep. In two others, the seams were about 80 feet
below the surface, 9 to 18 feet thick, and the mixture of magnetic and red
iron-ore contained 45 to 62 per cent. of iron. Another important mine had
three shafts and one gallery at a depth of 328 feet; the seam was 19 feet
thick, and the ore contained from 45·to 60 per cent. of iron. Many of the
outcrops are untouched. The mines have at present been so poorly worked,
and carried so little below the surface, that they might probably be driven
much deeper, with good results.   E. M. D.

GOLD-MINES IN THE HOHENBERG AND OELBERG HILLS. AUSTRIAN SILESIA.

Die Goldvorkommen am Hohenberg und Oelberg bei Würbenthal und Engelsberg in
Oesterreichisch-Schlesien. By JOSEF LOWAG. *Oesterreichische Zeitschrift*
für Berg- und Hüttenwesen, 1901, vol. xlix., pages 415-417.

Nothing is known of mining for gold in the Hohenberg and Oelberg
hills, near Würbenthal and Engelsberg in Austrian Silesia, until 1556. In

1828 an old shaft was re-opened and a quartz-vein containing gold and galena was found. In 1884, open cuttings were made in the Hohenberg and the Oelberg, and various further attempts to find gold proved fruitless. In 1893, the writer succeeded in discovering a vein of auriferous quartz, 6½ feet thick. A shaft was sunk and galleries driven, and 9 tons of quartz taken from it, yielded 17 dwts. of gold to the ton of ore. Later finds have given as much as 23 dwts. and 1 oz. 16 dwts. respectively.

The Hohenberg and the Oelberg, which are separated by a slight depression, form a contorted fold of Lower Devonian strata. The ridge is about 3 miles long, with a steep fall to the northwest. It consists of black or grey clay-slate, interspersed with quartz and mica, which impart a metallic glitter to the surface. The lode of auriferous quartz is embedded in the clay-slate, which also contains dark green chlorite-schist with magnetic iron-ore, quartzose schist with copper- and iron-pyrites, and masses of diorite. The Lower Devonian rests upon Laurentian gneiss. The fissures filled with auriferous quartz were probably formed during the folding of the strata and run parallel to the longitudinal axis of the folds. Long parallel depressions, eleven of which may be distinguished, denote the presence and the direction of the veins, which are thickest at the surface, and thin out as they descend to a depth of several hundred feet. The vein-quartz is coloured brownish-yellow with oxide of iron, it contains pockets filled with brown iron-ore, pyrites and galena, and is generally gold bearing. The more fissured the lode, the richer is it in gold; and the firmer and more compact the rock, the less gold does it appear to contain. To follow the vein, a perpendicular shaft was sunk for 40 feet, the quartz then again became solid, and the proportion of gold fell to 3 to 7 dwts. per ton.

Only these rich lodes were formerly worked, the poorer parts of the vein being left as supports. They were reached by sinking shafts, from 60 to 80 feet deep, below which the old miners could not go. At least 400 depressions have been counted, due to the filling up of these old shafts.

The auriferous zone lies in a valley, bounded by a copper-lode apparently of older formation, containing pyrites, malachite, a little gold, and the remains of old workings. The works at present consist of a shaft 46 feet deep; a gallery 984 feet long, at a depth of about 200 feet; a second gallery of the same length, and a shaft 65 feet deep. All the streams running from the Hohenberg and the Oelberg contain gold, which was formerly extracted by washing, and it is also found in the alluvial deposits forming the ancient bed of the river Oppa.   E. M. D.

COPPER-MINING NEAR KNITTELFELD IN STYRIA.

Der Kiesbergbau der Flatschach und des Feistritzgrabens bei Knittelfeld in Steiermark. By DR. KARL A. REDLICH. Oesterreichische Zeitschrift für Berg-und Hüttenwesen, 1901, vol. xlix., pages 639-643, and 1 plate.

From the beginning of the eighteenth century copper has been mined in the Flatschach, a district north of Knittelfeld in Styria, by the monastic order at Sekkau. From 1718 to 1741, 275 tons of copper were obtained from 3,850 tons of ore. In 1778, samples contained 6 per cent. of copper, 50 dwts. of silver and 18 dwts. of gold to the ton, but the average became so low that the mines were closed in 1789.

The veins are, as a rule, 5 feet thick, with a core of quartz and limestone. In some places copper and arsenical pyrites as large as a walnut have been found, and also tennantite. Another mine, at some distance, was also formerly worked, but the ore now contains little copper, and the percentage of gold was always small. The mines cover an area of 12¼ miles, but unfortunately the vein of ore is intermittent, and its course cannot be traced with precision.   E.M.D.

PYRITES-MINING IN UPPER STYRIA.

Die Walchen bei Oeblarn: ein Kiesberghau im Ennsthal. By Dr. Karl A. Redlich. Berg- und Hüttenmännisches Jahrbuch der k. k. Bergakademien zu Leoben und Příbram und der königlich Ungarischen Bergakademie zu Schemnitz, 1903, vol. li., pages 1-62, and 2 plates.

The ancient mining industry at Oeblarn has been recently resuscitated after a slumber of well nigh half a century, and the author gives in this memoir the results of investigations made among the newly opened-up ore-deposits, as well as the gist of the literature devoted to the district ever since Hans Adam Stampfer first described the mines in 1683. A complete bibliography (including manuscripts as well as printed works and maps), is followed by a detailed historical summary.

The rocks of the neighbourhood consist chiefly of garnet-mica-schists, among which are interbedded pure white crystalline limestones, folded into synclines and anticlines; and of quartz-phyllites (intermediate in age between the Archæan and the Silurian) in which occur the ore-deposits. The general strike is east-north-east and west-south-west, and the dip is 33 or 34 degrees to the north-north-east. There is plenty of evidence of disturbance and faulting.

Three main deposits, termed "veins," by the older authors, are recognizable. The highest, the Walchen ore-body, is really a partly-broken lenticle. 3¼ feet thick at the outcrop, double that thickness at a depth of 200 feet, and thinning out to nothing towards the dip, at a depth of 650 feet. The Gottesgabe or God's Gift " reef," 590 feet lower down, separated at first from the Dreifaltigkeit or Trinity "reef" by 45 feet of barren ground, unites with this to form one deposit, varying in thickness from 4 inches to 3¼ feet, but occasionally attaining a maximum of 10 feet. On the whole, this deposit may be regarded as originally consisting of two lenticles, the uppermost of which has been worked out. Striking parallel with the main deposits are innumerable thin seamlets of pyrites, so thin indeed as to be unworkable. At the outcrop the ores are recognizable by their brown weathered gossans, really masses of ochre, which, in conjunction with the sericite-schists, serve as good indicators.

The Thaddeus adit, dating from the early years of the 18th century, over 3,400 feet in length, was reopened and somewhat lengthened in the autumn of 1897. About 40 waggon-loads are got from it monthly and are used up in a celluloid factory. Ordinary pyrites, indeed, forms the mass of the ore, in association with copper-pyrites. Occurrences of secondary importance are arsenical and magnetic pyrites, galena, blende, argentiferous fahlore, antimonite, etc. Small proportions of gold and silver are found in the iron- and copper-pyrites: several recent analyses of these are tabulated by the author.

With regard to the genesis of the deposits, he enunciates the following theory. Suppose a rock (unlike the surrounding country-rock) to have occupied their present position in the form of a bedded mass. Mountain-building movements take place, fissures are torn in the rocks and are infilled with crush-material (?). Subsequent metamorphism occurs, obliterating nearly all traces of the fissures, and the whole appears to be one compact mass. Then fresh disturbances take place, fissures are once more opened up in it, and this time the presence of pyritous elements results in pyritic infilling. The substance which originally occupied the place now taken by the ore-deposits was probably limestone.

Statistics of the output of the mines, from 1570 to 1857, are followed by a detailed description of the old method of treatment of the ores. From this one gathers that the operation must, in later years at any rate, have been carried on at a loss. As a matter of fact, sulphur, under the new conditions, will be the chief product obtained from these mines; and then, if the annual output is much increased, a payable quantity of metal may be got from the roasted pyrites. L. L. B.

MINERALS OF THE TYROL.

Beitrag zur Geschichte der Baue des Berggerichtes an der Etsch, 1472-1659. By MAX REICHSRITTER VON WOLFSKRON. *Oesterreichische Zeitschrift für Berg- und Hüttenwesen*, 1901, *vol. xlix., pages* 91-94.

In the Tyrolese valleys on the banks of the Adige extensive mining operations have been carried on for centuries for silver-bearing lead-ore, the existence of gold being little more than a tradition. Works were carried on throughout the sixteenth century at Nals, and in 1525 at Terlan; from the latter place the lead-ore was exported to Spain, while the ore dug at Nals was sent to Rattenberg. The works at Nals and Terlan employed at that time 163 men, but the mines were gradually worked out, as the ore became poor, and ceased to yield either silver or lead.

In Gsteyr, near Meran, a vein of lead-ore was discovered in 1548, and thirteen mines were opened, some being at a depth of only 4 or 5 feet, and therefore worked at small cost. Like the mines in Nals and Terlan, however, these mines gradually decayed, in spite of an effort made towards the end of the century to revive the industry. One cause for their failure was that the working of the mines interfered with the cultivation of the rich land in the Adige valley. The miners being mostly men foreign to the district, there was little desire to continue the work after the rich yield of the mines had begun to diminish, because the country-folk were able to earn a more comfortable living by agriculture. E. M. D.

THE LAURIUM MINES, GREECE.

Ueber den Bergbau im Laurion. By C. VON ERNST. *Berg- und Hüttenmännisches Jahrbuch der k. k. Bergakademien zu Leoben und Přibram und der königlich Ungarischen Bergakademie zu Schemnitz*, 1902, *vol. l., pages* 447-501, *with* 8 *figures in the text*.

The mining district of Laurium in Southern Attica covers an area of about 50,000 acres. The various centres are connected up with each other and with the harbours of Theriko and Ergastiria by tram-lines and narrow-gauge railways, and with Athens by a broad-gauge railway.

Speaking broadly, the rocks of the district consist of an alternation of marbles, dolomitic limestones and mica-schists, which overlie unconformably the Cretaceous limestones of Athens and Lykabettos. The whole series has undergone folding and dislocation, with the result that the massif is traversed by numerous fissures, partly infilled with apophyses of the fine-grained granite or eurite of Plaka and partly with sulphidic ores. The rock-succession is as follows, in descending order:—(1) Upper ferruginous limestone, which forms the capping of many hills. (2) Upper micaceous slate or schist, partly weathered, containing white mica. (3) Middle lime-stone (marble): one variety is pink and schistose, and the other is bluish-grey and fine-grained. This forms the foot-wall, and No. 2 the hanging-wall, of an ore-body which consists chiefly of iron-ores, impregnated with galena, etc. (4) Lower micaceous slate or schist, forming the foot-wall (and No. 3 the hanging-wall) of a second ore-body, consisting in the northern area of manganiferous iron-ores, and in the southern of finely-granular argentiferous galena associated with chalybite, fluorspar and calcspar. (5) Lower marble, with impregnations of iron-pyrites and galena. (6) Mica-ceous slate or schist, between which and the underlying saccharoidal marble (of unknown thickness) lies the third ore-body, varying from 20 inches to 40 feet in thickness, and consisting of galena and cerussite. This deposit was largely worked by the ancients: in the foot-wall occurs a good deal of zinc-ore, in the form of nests, etc. Nearly all the above-described rocks are seamed by a network of fissure-veins infilled with calamine and galena. Besides the granite, already mentioned, various greenstone-dykes break through the massif. The ores are richest where schist forms the hanging-wall and limestone the foot-wall, and the proportion of zinc-ore tends to increase the deeper down one goes.

The explanation appears to be that the ore-bearing fluids, coming up from below reacted with greatest effect on the lowermost marble, leaching most of it away and depositing in its place zinc oxides. Thus impoverished, when they reached the higher limestone-beds the ore-bearing fluids proved less potent in inducing double chemical decomposition, etc. On the other hand, as the schists by their very texture would tend to check the upflow of the fluids, the limestone with a schistose roof would there be richest in ore.

The usually very compact lead-ore contains indeed only from 9 to 10 per cent. of metallic lead, but it is rich in silver, yielding 4½ pounds per ton of lead. The quality of the calamine varies: from the lowermost deposits 2,000 to 3,000 tons are got monthly, and have yielded as much as 65 per cent. of metallic zinc, but ores as rich as this are rarely obtained nowadays.

The author gives an elaborate and interesting description of the mining industry carried on in the Laurium district by the ancient Greeks. The slag-heaps left by them still contain from 7 to 14½ per cent. of metallic lead, but they appear to have been very successful in extracting most of the silver. The more recent history of the industry is fairly well known from other sources. It will be remembered that after a lapse of 2,000 years working was started on the slag-heaps in 1864. Since then enormous masses of slag have been found below water-level, where they were tipped over by the ancients into the sea, and the total amount of slag was still estimated in 1893 at 2½ millions of tons. The richest portion has now been almost all worked over. In addition to this there are the waste-heaps of the ancient mines, or *ekvolades*, and the remains of the ore-washings or *plinites*, from which a good deal of lead and silver has been got. The Greek metallurgical com-pany founded in 1873 has not, however, confined itself to working over the old heapsteads, etc., but has opened up fresh mines, and in 1901, the output from these for the year amounted to 21,346 tons of manganiferous iron-ore, 12,679 tons of plumbiferous iron-ore, and 339 tons of zinc-ore.

The French Laurium company, founded in 1875, works some of the richest mines in the district, at Kameresa (where the Serpèri and Hilarion shafts have been sunk to a depth of 590 feet or so), Plaka, Suresa and Despilesa. The ores are sorted and smelted at Kypriano, where works on a magnificent scale have been laid out. The output for 1901 amounted to 10,730 tons of roasted calamine, 803 tons of plumbiferous calamine, 3,942 tons of ferruginous calamine, 454 tons of zinc-blende, 11,500 tons of red hæmatite, and 84,973 tons of manganiferous iron-ore. In addition to the two great companies, two or three other smaller companies are at work in the district, their mines being situated at Dardesa, Sunion, Daskalio, Seriphos, etc.

The total number of work-people employed in 1899 was 9,346, but the industry has since then fallen on evil days, and in the first half year of 1901 more than 2,000 workmen were perforce dismissed, as the prices of lead and zinc were so low that it did not pay to work any except the richest deposits. The constant inflow of water into the mines is one of the great difficulties with which the Laurium engineers have to battle.

L. L. B.

COAL-MINING IN DUTCH LIMBURG.

Steinkohlen in Niederländisch-Limburg. By HENRI ZONDERVAN. *Petermanns Mittheilungen*, 1901, *vol. xlvii.*, *pages* 187-190.

The Netherlands being mainly an accumulation of drift-deposits, and outcrops of older rocks occurring only in the east of the provinces of Gelderland and Overijssel and in the south of Limburg, mining enterprise in that country is restricted to a very small area. In fact, mining operations, in the true sense of the word, are in progress in Southern Limburg alone, and although at present conducted on a very modest scale they are destined to undergo considerable development in the near future. This hopeful view is based on the comparatively-recent discovery of a great extension of the coal-field, the seams in the new area being well worth working.

In part, the Limburg coal-field belongs to the Wurm basin (which is itself a small portion of the great belt of Coal-measures that stretch from Stadtberge in Westphalia to Valenciennes in French Flanders). But a number of new borings have revealed the presence in Limburg, north of the Wurm Coal-measures and separated from them by the saddleback ridge of Kerkrade, of another group of coal-bearing strata. This coal-field strikes south-west and north-east, and is cut by two main faults running south-east and north-west: the throw of one of these is 1,000 feet or so, and the maximum downthrow of the other is not yet known, as a bore-hole near Gangelt went down 1,640 feet without reaching the Coal-measures. The strata south-west of the last-named fault (known as the Sandwall or northern fault) can alone be regarded as workable, for north-east of it the thickness of the Tertiary deposits is so enormous, that shaft-sinking would prove too costly. East of Valkenburg, another bore-hole has struck the Coal-measures, but they are here unproductive; and, in view of this, the actual extension of the workable measures westward from the Sandwall fault may be reckoned at 6 or 7 miles. The probable boundaries of the new Limburg coal-field are then: east and north, the Sandwall; west, the barren strata of Valkenburg; and south, the Kerkrade ridge. Its area measures about 81 square miles: the overlying rocks are mainly Tertiary sandstones, Chalk and Greensand, none of which appear likely to enhance the difficulties incident to mining operations. On the other hand, the frequent occurrence of heavily-watered sand-beds will complicate somewhat the labour of shaft-sinking, but will hardly prove a serious obstacle, unless the shafts have to be sunk to a very great depth.

The two Dutch localities where coal is now being mined are Kerkrade (for the Wurm basin) and Heerlen (for the Limburg basin). It is expected that more new pits will be started at the last-named place before long.

At Kerkrade, four seams are worked, down to a depth of 1,100 feet from the surface (or 600 feet below sea-level); the output in 1899 amounted to 122,796 tons, and in 1900 to 124,538 tons. The number of workpeople employed belowground was 291 and abovebank 80. Not far off from here is the Neu Prick colliery, which produced in 1899, from three seams, 46,040 tons of coal, the total number of workpeople employed being 145.

About 1870, boreholes were put down at thirty different places in what is now the new coal-field, and in nearly every case they struck coal. Mining concessions were obtained from the Dutch Government; but practically no use was made of them, partly in consequence of industrial depression, and partly because of the absence of means of transport. The last named desideratum, however, was attained by the building of the local railway-line which runs from Sittard to Heerlen and Herzogenrath. Thereafter, a serious beginning was made with mining work, and from 1893 onwards, company after company opened up collieries over a total area of 14,656 acres.

The seams west and south of the Feldbiss fault are chiefly " meagre " or " sand-coals "; at Heerlen, however, a " flaming " coal is found. North of the Feldbiss, both " flaming " and bituminous coals occur. The " meagre " coals in the government collieries and at Neu Prick-Bleyerheide, are anthracitic in character. They are tough and clean, burn slowly, almost without flame, and give off very little soot or dust. The only colliery at work at present near Heerlen is the Oranje-Nassau pit. Working four seams, it produced in 1899, 44,136 tons of coal, and employed 271 workpeople (192 belowground and 79 aboveground). It is estimated that when all the concessions recently taken up are being worked, the total production of the new Limburg field will amount to 10,000 tons *per diem*. If the whole of the available coal-area were worked, this figure could be trebled or quadrupled, and reckoning 290 working days in the year, the annual output could be increased to 11,500,000 tons. If this were so, the Netherlands, which imported in 1898 a little over 5,000,000 tons of coal, could in future be entirely self-supporting in that respect; the supply from their own collieries would be amply sufficient to cover all the requirements of the consumers, and not a single ton of coal need be got from abroad. On the contrary, Holland could enter the lists with other nations as a coal-exporter. The author confesses, however, that more than one decade is likely to elapse before this consummation is reached. Meanwhile, the Government have brought forward a Bill in the Netherlands Legislature, for the purpose of enacting that no further mining concessions shall be granted, but that the remaining area of the coal-field (two-thirds of the whole) shall be held and worked by the State. The committee of experts appointed by the Government to examine into the question had, however, previously reported in favour of conceding four-fifths of the coal-field to private enterprise and reserving only the remaining fifth (the choicest part, calculated to contain 800,000,000 tons of coal) for the State. This question is not yet settled.

L. L. B.

COAL-SEAMS AT TKVARTSHALI IN TRANS-CAUCASIA.

Die neu entdeckten Kohlenflötze von Tkvartschali im Kaukasus. By EM. LADOFF. *Oesterreichische Zeitschrift für Berg- und Hüttenwesen, 1901, vol. xlix., pages 173-174.*

Tkvartshali, in the province of Kutais, is about 34 miles from the coast, whence a good road leads to it. It is a wild and mountainous district traversed by numerous deep rivers, with vast primeval forests. Only the semi-civilized Abbasiers wander through it. These wild inhabitants have long known of the existence of coal, which they call "devils' blood," and hold in abhorrence. It was from their description that the first seam, 2½ feet thick, was discovered. It contains more ash than coal, but two other seams of good coal, 7 and 14 feet thick, were soon afterwards discovered. The coal-district covers an area of about 77 square miles, but has been little explored. The thill of the seams is sandstone, containing clay, and sometimes passing into shale, and the roof is conglomerate. In several places there are hot springs of sulphuretted hydrogen. There are many faults, the strata are much contorted, and in one place a vein of grey porphyrite occurs. A few fossils have been found. Variations in the thickness and colour of the seams are characteristic of the district. The coal-measures are cut through by rivers, which in their upper reaches run at a steep angle, and the coal often crops out at the surface.

The method of working the seams is primitive. In the deep river-ravines, the coal at the surface is simply dug out with pickaxe and shovel. Elsewhere, the strata are inclined at the same angle as the bed of the rivers, and lie far below them. Here, three holes, 300 feet apart, were bored by hand. One of them, sunk to a depth of 250 feet, cut through three seams of coal, 7, 11 and 18 feet thick respectively; and at a depth of 180 feet, a hot spring impregnated with sulphuretted hydrogen was tapped. The coal contained about 34 per cent. of ash. There are seven seams at Tkvartshali, one of which is 40 feet thick, but this is chiefly shale with thin beds of coal, and only three of the other seams are valuable. The ash usually exceeds 7 per cent. The coal can be coked and in this respect, Caucasian coal is remarkable, as even with 33 per cent. of ash it cakes into a solid mass. The coke is light, not brittle, and burns well in furnaces.

Unfortunately the development of the industry is hindered by the difficulty and expense of connecting the district with the sea. An ordinary railway would be too costly, but an electric or funicular railway might be constructed, if the natural features of the country were utilized. There are several waterfalls, from which power might be obtained.   E. M. D.

COPPER-ORES OF SÜNIK, TRANSCAUCASIA.

Die Kupfererze der Sünikgruben im Gouvernement Elisabetpol, Transkaukasien. By — ERMISCH. *Zeitschrift für praktische Geologie, 1902, vol. x., pages 88-89.*

Transcaucasia is extraordinarily rich in metalliferous ores, and among these copper seems to play an important part. The Sünik mines, district of Sangesur, in the government of Elisabetpol, are one of the three most considerable copper-mining enterprises in that part of the Russian Empire: they are near the frontiers of Persia and Asiatic Turkey, and their situation therefore is somewhat remote. The nearest railway-station, Evlach, is 125 miles distant, and it may be for this reason that, from the technical point of view, the Sünik mines have much leeway to make up. In the old days, the softer ore (bornite) used to be worked out by means of narrow, deep and sloping shafts, while the somewhat harder ore (such as chalcopyrite) was left behind. Within the last five years, however, mining has been actively carried on, more in accordance with modern methods, and, recently, smelting-works have been built at Sünik, on the Manhes system.

The ore-bodies are undoubted vein-deposits, and the gangue is almost exclusively quartz. The chief ore is chalcopyrite, containing from 14 to 24 per cent. of metallic copper, seldom more. With it are associated, mixed together in one mass, iron-pyrites, bornite and antimony fahlore, and (very rarely) some native copper. At present no economic importance is attached to the gold-and-silver content of the copper-ores, which is said not to be inconsiderable. Precise data, however, are not forthcoming. So far, twenty veins are known, varying in thickness from 7 to 45 inches: taken altogether they form a channel striking north-westward through a massif of dark-green to blackish diabase. Fault-fissures in this diabase are numerous; they cut through the cupriferous veins, and are very unwelcome to the miner.

The workings are irregular narrow adits driven some 100 to 300 feet or so horizontally into the hillside: no true shaft, in the modern sense, has yet been sunk. In regard to water, all the conditions are very favourable for the mining industry. The use of dynamite has only just begun at Sünik, ordinary blasting-powder having been employed there till quite recently. Haulage is in a primitive stage indeed, for the miners bring to the surface, both the mine-water and the ores that they have won, in great leathern sacks. Yet the annual production has risen from 230 tons of metallic copper in 1890 to 806 tons in 1900.   L. L. B. .

THE PRESENT POSITION OF COPPER-MINING AND WORKS IN THE ZANGAZURSKY DISTRICT, ELISABETHPOL, TRANS-CAUCASIA.

The Present Position of Copper-mining and Works in the Zangazursky District, Elisabethpol, Trans-Caucasia. By N. SHOSTAK. 1901, 16 *pages.*

The copper-deposits of Zangazursky are situated near Evlach station on the Trans-Caspian railway. The absence of good communication with the railway has prevented these deposits from being opened up on a large scale.

The country-rock of the district is chiefly andesite. The ore is a copper-glance, together with lead-glance and zinc-blende in quartz-veins. The direction of the veins is north-west to south-east without any appreciable divergence or faulting. The veins do not narrow as they attain depth.

The development-work shows that the veins extend to a length of 7,000 feet with a width varying from 6 inches to $3\frac{1}{4}$ feet, and the workings have extended to a depth of 700 feet. From 1858 to 1900, 9,643 tons of copper have been produced. In 1900, the output of copper was 800 tons. All the works are run with water-power, as coal costs £2 1s. per ton, wood, 10s. to 12s. per ton, and coke £7 per ton. The mine is robbed, only the rich parts of the veins being worked, without systematic development.

The cost of mining amounts to £2 10s. per ton of ore, and allowing for capital-expenditure the cost will be about £3 10s. per ton. The smelting of the ore, in old-fashioned furnaces, costs £3 12s. per ton, thus making a total cost of about £7 2s. per ton. In order to work at a profit under these conditions, 8 per cent. of copper is the minimum, and ore containing less than 7 per cent. is thrown on to the waste-heap. At the new works, erected at Sounthsky in 1897, the ore is now being smelted for £1 8s. per ton. The ore, in the first instance, is smelted to copper matte, and refined in converters to black copper and to pure copper in the *schpleizoffen.*   R. W. D.

THE DONETZ COAL-FIELD, SOUTHERN RUSSIA.

Die Kohlen-industrie im Donetzbecken. By — TITTLER. *Zeitschrift für das Berg-, Hütten- und Salinen-wesen im Preussischen Staate,* 1901, *vol. xlix., Abhandlungen, pages* 477-480.

The Donetz basin is really a typical "plateau of erosion," the highest summits of which do not exceed an altitude of 1,130 feet. The elevations and depressions of the surface coincide here exactly with the geological structure of the Coal-measures. The Carboniferous deposits are of an unmistakably littoral type, and cover an area which extends over a length of 233 miles from east to west with a maximum breadth of 100 miles: they are in part overlain by saliferous red marls of Permian age, but they crop out at the surface over an extent of no less than 7,720 square miles.

The seams, as a rule rather thin, number more than 200: the workable coal mostly occurs in the Middle Group (horizon of *Spirifer mosquensis*) of the Upper Carboniferous system—as defined in that region. The seams usually dip between 15 and 20 degrees, high dips of 50 to 70 degrees being uncommon; in thickness they range from 2 to 5 feet, rarely attaining 6 feet. The coal varies in character from a good gas-coal to an excellent anthracite; the upper seams are generally of bituminous coal, and the lower of semi- or non-bituminous coal. One case is known of a seam which, on being followed along the strike, was found to contain all these varieties of coal passing one into the other. Sulphur is present in so large a proportion that the coal from the Ruchenko mine, containing from 1 to 1·3 per cent. of sulphur, is regarded as comparatively free from that undesirable ingredient. The percentage of ash varies from 5 to a maximum of 10. The coal-seams are interbedded with shales and sandstones, with occasional thin beds of *Fusulina-* and *Productus-*limestones.

It cannot be said that mining here is attended with any special difficulty. Pillar-work and overhand stoping are the methods usually employed. Some mines are very fiery, and all of these are provided with ventilators (30 ventilators on 17 mines, with a total capacity of 1,220,000 cubic feet per minute). Altogether at the present time there are 135 collieries at work in the field, with a combined annual output of about 7,500,000 tons. The depth of the shafts ranges from 100 to 1,250 feet. The 205 haulage-engines represent a total of 16,820 horsepower and the 212 pumping-engines can raise between them 3,530,000 cubic feet of water to the surface in 24 hours.

The high protective tariff alone enables the Donetz coal to compete with British coal in the harbours of the Sea of Azov and the Black Sea. On the other hand, the Donetz coal dominates completely the markets of Southern and Central Russia, right up to Moscow. The coke produced in the district by 2,860 ovens (mostly Coppé ovens), averaging about 1,000,000 tons *per annum*, is all consumed locally.

The number of workpeople employed in 1899 belowground was 40,000 and aboveground 10,000. Hewers earn from £3 5s. to £5 10s. a month, trammers from £2 12s. to £3 5s., surface-labourers the same, and the wages of the other workpeople are on a similar scale. This is in part due to the great number of religious festivals and saints' days which are kept as holidays, so that there are but 240 working-days in the year or 20 in the month, and the annual output per head is only 146 tons.   L. L. B.

PLATINUM-MINING IN THE URAL.

Vorkommen und Gewinnung des Platins im Ural. By L. St. Rainer. *Berg. und Hüttenmännisches Jahrbuch der k. k. Bergakademien zu Leoben und Příbram und der königlich Ungarischen Bergakademie zu Schemnitz, 1902, vol. I., pages 255-298, with 2 figures in the text, and 3 plates.*

Premising that 95 per cent. of the world's output of platinum comes from the Ural, the author gives a brief history of platinum-mining in that region, supplemented by statistics of annual production from 1825 to 1900.

Scarcely a single gold-placer is worked along the whole chain of the Ural that does not contain some traces of the rarer metal platinum. The platinum-workings properly so-called are, however, practically confined to two districts:—that of Tagil and that of Goroblagodat, 74 miles farther north, their combined productive area being reckoned at 64 square miles or so. Economically the Goroblagodat field is by far the most important of the two; and the Tagil workings show signs of exhaustion, as they now yield only 17 per cent. of the total platinum-output of the Ural.

A brief sketch of the geology of the region leads to the pregnant conclusion that the occurrence of the platinum is genetically associated with that of the peridotite-rocks. Some Russian observers assert that the precious metal, in the Goroblagodat district, is also to be found in porphyrite-sills, in gabbro-diorites, syenite-gneisses and syenites; but the evidence for this is not yet regarded as altogether beyond doubt. Analyses of the crude ore are tabulated, showing that it always contains more or less iron and copper, besides osmium, iridium, ruthenium and palladium. The miners divide the ore into three grades: No. 1, containing 85 per cent. of pure platinum, from the Lower Iss and Tura river-valleys; No. 2, containing 82 per cent. of the metal, from the Upper Iss valley; and No. 3, containing 75 per cent. from Nizhni Tagilsk, where the ore is got out of the original matrix (chromite-bearing serpentine).

The so-called "eluvial" placers, consisting of weathered detritic material which has not been carried by water to any distance from its original *locus*, are generally poor in platinum, precisely because they have undergone so little natural concentration. The true alluvial deposits along the valleys are mantled by barren layers of varying thickness (from 10 to 130 feet or more), but the barren layers are seldom intercalated with platiniferous bands. In many places the bottom portion of the placer is indistinguishable from the highly-decomposed bed-rock. The rich paying alluvia consist of sand, loams and gravels, sometimes containing tusks and bones of mammoth, and vary in thickness from a minimum of 5 inches to a maximum of 10 feet.

The method of working the placers is described, and it is notable that in cases where an attempt has been made to economize time and labour by means of British machinery, the stupidity and want of discipline of the Tartar miners have in part nullified the advantages derivable from modern apparatus, with the result that slow manual labour has retained its primacy. The method of washing the alluvia is also described, and the author appears to think that a considerable amount of platinum is lost in the tailings, and that not a little is stolen by the miners (despite the vigilance of the authorities).

Within the past two years, dredging has been introduced in workings in the Iss valley, and the resulting economy in cost of output is estimated at 51 per cent. or even 57 per cent. The author points out, however, that the dredger often fails to scour away thoroughly the lowest (not seldom the-richest) layer of the placer immediately above the bed-rock; it fails too to scoop out the heavy metallic grains lying in the inequalities and hollows of the surface of the bed-rock itself. He remarks that no one yet seems to have thought of attaching a *caisson* to the dredger, capable of accommodating a certain number of workmen whose task it would be to scour completely the surface of the bed-rock.

In conclusion, the author draws attention to the fact that the richest platinum-deposits are practically exhausted, and that in many cases the miners are reduced to washing over and over again the old tailings. A marked diminution of output within the next 10 years is inevitable, unless new sources of supply be found, along some of the unexplored river-valleys on Count Shuvaloff's property and north of it (that is, beyond 61 degrees north latitude). It is curious to note that 44 per cent. of the platinum consumed in the United States is used up by the makers of artificial teeth, and equally remarkable statistics are cited from Austria-Hungary.

L. L. B.

LÅNGFALL ZINC-BLENDE MINE, SWEDEN.

Ueber die Gesteine der Zinkblendelagerstätte Långfallsgrube, bei Räfråla in Schweden. By R. BECK. *Tschermak's Mineralogische und Petrographische Mittheilungen,* 1901, *vol. xx., pages 382-389, with 4 figures in the text.*

This mine is situated in the province of Dalecarlia, south of Lake Vessman, west of Ludvika, and belongs to the Saxberget Company. The first discovery of the ore-deposit was as recent as 1881 or 1882, and it appeared in the statistics of ore-production for the first time in 1891. The author has given a short description of the mine in his recently-published text-book,[*] but in the present paper he gives fuller particulars in regard to the country-rocks.

The ore occurs in the form of a bed 10 to 16 feet thick, underlying drift, and intercalated conformably with crystalline schists, which dip steeply southward. The predominant rock is a fine-grained flaky biotite-gneiss (of widespread occurrence in Central Sweden, and well known as the *locus* of many ore-deposits). It forms the floor of the ore-bed here, but passes into a coarsely slaty biotite-quartz rock, containing much sillimanite and extremely little felspar. The immediate roof of the ore-bed is formed by actinolite-rocks of varying mineralogical composition. Nevertheless, they are conspicuous for the occurrence of amphibole-anthophyllite, a type of actinolitic hornblende apparently new to Sweden. With the fibrous aggregates of this mineral, brown zinc-blende, copper-pyrites, and some galena are intimately intergrown. Microscopic examination shows the blende in extremely thin lamellæ thrust in between the cleavage-folia of the amphibole-anthophyllite, or matted together with it in an inextricable felt-like plexus.

The ore, then, is mainly zinc-blende; magnetic-pyrites, copper- and iron-pyrites, and argentiferous galena playing a comparatively subordinate part. Reasons are assigned for explaining the genesis of the deposit, as due to a secondary mineralization of a hornblendic rock originally rich in cordierite and spinel.

In many respects the Långfall ore-deposit is similar to the pyrites-deposits of Silberberg near Bodenmais (Bavaria), recently described by Dr. E. Wein-

* *Lehre von den Erzlagerstätten,* 1901, pages 476.477.

schenk.† With the latter's genetic hypothesis, however, the present author does not agree. He thinks that all the phenomena of corrosion and impregnation characteristic of both deposits are ultimately due to the intervention of thermal solutions.   L. L. B.

_____

IRON-MINES IN CENTRAL AND SOUTHERN SPAIN.

Datos geológico-mineros de varios Criaderos de Hierro de España. By L. MALLADA. *Boletín de la Comisión del Mapa geológico de España,* 1902, *series* 2, *vol. vi., pages* 153-203, *with* 5 *figures in the text.*

The author deals first of all with the mines in the Fuente del Arco and Guadalcanal districts, in the provinces of Badajoz and Seville. The Sierra de la Jayona forms the northern extremity of a ferruginous belt, some 38 miles in length from north-west to south-east, and varying in breadth from 600 to 5,000 feet. The deposits, miscalled "veins," are really long strings of "pockets," of irregular dimensions, the chief ore being a micaceous specular iron-ore, in places entirely pure, and containing 69 per cent. of the metal. Sometimes it is mixed with varying proportions of calcium, magnesium and iron carbonates. The genesis of these deposits is undoubtedly associated with the eruption of hornblendic porphyries, which invaded the surrounding Cambrian slates, carrying with them thermal waters or fumes charged with iron, which latter was ultimately precipitated in the cavities of the limestone, also of Cambrian age. In certain localities the ancients exploited the richer and more easily-workable portions of the deposits, but what they have left would well repay working nowadays. Great confidence is expressed in the still untouched mineral wealth of the Sierra de la Jayona, and the ores could be exported through Seville, by the suggested light railway (12¼ miles long) which would join up with the line from Mérida to that city. On the other hand, the ores could be brought into the coal-field of Belmez for metallurgical purposes, by means of a still shorter light railway (5 miles) to the station of Fuente del Arco, which is joined by a line 43 miles long to Peñarroya, within the aforesaid coal-basin. The six mines of Guadalcanal proper are not working such rich deposits as those just described.

Proceeding southward, the author deals with the San Nicolas and Constantina districts in the province of Seville, and the Hornachuelos district in the province of Córdoba. The geological conditions are very similar to those described in the Sierra de Jayona. The same statement applies to the Feria deposits in the province of Badajoz, with the exception that here the intrusive igneous rocks are largely impregnated with ore. So too the extensive Luisa and Los Pedros mines in the Sierra de Córdoba, which forms the northern wall of the Guadalquivir valley, work ore-deposits varying in thickness from 40 inches to 6¼ feet, the foot-wall being Cambrian slates, and the hanging-wall limestones, sometimes faulted or eroded away. The strings of pockets contain sometimes hæmatites and sometimes specular iron-ore.

In the Sierra del Madroño, near Hellín, province of Albacete, a vein of very pure hæmatite, averaging from 1 to rather more than 10 feet in thickness, is worked amid ashen-grey or white marls and limestones of Lower Cretaceous age. At the contact with the vein, the limestones are con-

_____

† *Trans. Inst. M.E.*, 1901, vol. xxii., page 693.

verted into brecciated pinkish dolomites associated with carbonate of iron. The ore is not very siliceous, and analysis shows it to contain between 54 and 57 per cent. of metallic iron.

The author then turns to the Sierra Alhamilla in the province of Almeria. This mountain-range is mainly built of pale chlorite- and mica-schists of pre-Cambrian age, overlain uncomformably by yellowish and pinkish, more or less magnesian, limestones of Triassic age. The ore-deposits occur partly in pyritiferous quartz-veins which traverse the schists, and partly in a more bedded form, intercalated between these schists and the limestones. Here also the genesis of the ores is traced back to thermal waters, the upflow of which formed the moribund stage of the volcanic phenomena that took place in the district in late Tertiary (?) times. The hæmatites occur in irregular masses, varying according to the cavities eaten away in the dolomite by the lava-flows: sometimes they measure as much as 100 feet in thickness. The ore is strongly impregnated with manganese, containing from 3 to 5 per cent. of that metal. Phosphorus and sulphur are absent, and silica almost entirely so. The percentage of iron oscillates between 50 and 60, some samples frequently yielding even more. The ore is tough, and the fact that it does not tend to make much "small" or "dust" is considered an additional advantage.

The Sierra de Almagro, in the Cuevas de Vera district, in the same province, presents much the same geological conditions as those just described, but great masses of gypsum are interbedded with the limestones. And the constant association of the gypsum with the iron-ore deposits as well, leads to the inference that they are of contemporaneous origin. Wherever these minerals crop out, hard green diabases are found with them. The author gives a detailed description of the mines, the most important of which is that of Los Tres Pacos. The ore is an excellent dark-red hæmatite, never containing less than 60 per cent. of metallic iron. The distance to Villaricos harbour is only 8¼ miles.

The mines of Atienza, in the province of Guadalajara, work masses of hæmatite interbedded with Silurian slates and sandstones. The author terminates with the description of a few deposits in the extreme north of Spain—those of Begonte in the province of Lugo (of no industrial importance) and those of Berástegui, in the province of Guipuzcoa. The latter occur in the form of bedded veins in the Cambrian slates. Here various ores of iron are associated with galena, blende, chalcopyrite, etc.   L. L. B.

THE AURIFEROUS QUARTZ- AND OTHER ORE-DEPOSITS OF SIBERIA.

Les Gisements de Quartz aurifère en Sibérie. By A. BORDEAUX. *Annales des Mines*, 1902, series 10, vol. ii., pages 499-549, and 1 plate.

The author begins by stating that, on the one hand he does not propose to deal with the Altai district, and that, on the other he will not confine himself to the gold-bearing quartz-reefs, but will also describe the iron-, lead- and copper-ore deposits as opportunity arises.

The Achinsk district in Western Siberia forms part of the Obi river-basin, and here the crystalline schists are overlain by some 5,000 feet or more of Silurian, Devonian and Mesozoic sedimentaries with Tertiary and Glacial drifts. In the sedimentaries are numerous thermal springs containing iron, manganese, magnesia and sulphur, and some beds are impregnated with cupriferous ores just as at Mansfeld. There are considerable iron-ore deposits at Abakansk, Irba and Kuznetzk. At the last-named locality coal of Permian age occurs, and Jurassic (?) coal-seams are of frequent occurrence elsewhere. The strata are nearly horizontal, the mountains take on tabular forms, and the scarps formed by the watercourses are often vertical.

From the station of Itate on the Trans-Siberian railway, a journey of 100 miles or so by road brings the traveller to Chebaki, which is a sort of central depôt whence all the mines of the district draw their supplies, stores, etc. The first mine described is that of Yoannovski or Podvintzeff, where gold was first discovered in September, 1899, and first worked about a twelvemonth later, yielding 35 to 36 parts of gold per million. The reef, some 15 or 20 feet thick, occurs as a dyke in diorite. At the foot of the mountain are rich placers (worked since 1853), containing gold in rounded nuggets. At the Tumani mine (where work had not yet begun in May, 1902), is another quartz-reef, with gold visible to the unaided eye, in diorite. Assays show it to yield 1 ounce per ton.

At the Bogom-Darovanny (or Ivanitzki) mine, gold was first found in 1896, but working was not begun till 1898. Its output has already attained considerable dimensions. The country-rock, here, is syenite: the quartz is milky, the gold very finely divided but very pure, and pyrites is of rare occurrence. The total cost of the mining plant: turbine, stamp-batteries, etc., is about £16,000. About 32 tons of veinstuff are worked through per day; wood-fuel exclusively is used. The prime cost is very irregular, but including taxes, wages, salaries, fuel, sinking-fund, etc., it amounts to exactly £2 per ton, and with a yield of 24 parts of gold per million, a very fair profit can be made. The yield, at the time of the author's visit in 1902, averaged 30 parts per million.

At present, the industrial development of the Achinsk district is bound up with its immense forests, but later on the Permian coal of Kuznetzk, yielding only 3 per cent. of ash, and said to be equal in quality to the best English coal, will come into play. The seams are numerous, and extend over a vast area. At Kuznetzk also, no less than 80,000,000 tons of excellent iron-ore have been proved.

In the north-eastern portion of the Minussinsk district are numerous auriferous placers, and there are some gold-quartz reefs. Hæmatite-deposits occur in the Suida valley, and yellow and red ochre in that of Karaskuir, while galena and other lead-ores are recorded from the Seiba, Pieshka and Izuirtak river-basins. The cupriferous veins observed in the porphyries of this region do not seem to be of industrial importance.

In the east, the important mining concession of the Irba basin includes splendid magnetite-deposits, yielding from 64½ to 67 per cent. of metallic iron, galena (not in sufficient quantity to repay working), and copper-ores (equally unsatisfactory from the industrial point of view).

In the south, copper-ores were worked and smelted for several years (1837-1847, and again after 1874), but the industry is now extinct. At Abakansk, the magnetite-mines, worked since 1869 or so, are increasing considerably their output. The ore yields from 61½ to 63½ per cent. of metallic iron.

Only 25 miles away from the town of Minussinsk, at Zhazikh, are coal-seams several feet thick. The coal yields the following results to analysis:— Carbon, 62·22 per cent.; combined water, 6·42; oxygen and nitrogen, 8·48;

hygroscopic water, 14·60; and ash, 8·28. The heating power is 7,239 calories. The auriferous quartz-reefs along the Chinese frontier near Abakansk are rich, but have not been developed, because of the general insecurity of the border-districts.

Splendid deposits of white marble occur over an immense area, and many other occurrences of lead and copper-ores (besides those already mentioned) were reported to the author in Western Siberia.

The remainder of the paper is devoted to a brief account of the mineral wealth of Eastern Siberia. The region of the Middle Vitim river, difficult of access, has aroused much attention of late years, because of the recent discovery there of rich placers and rich auriferous quartz-reefs. Copper-ore deposits also occur. As to the thin and not very promising quartz-veins of the Upper Vitim, the author declines to pronounce definitely an opinion.

Two syndicates are at work on the gold-quartz reefs of the Onone district, and the annual output of the precious metal at the Bielogolovy mine there (somewhat fluctuating) amounted in 1901 to 93 pounds (42 kilogrammes). Forewinning is not always pushed with sufficient rapidity to keep the stamp-batteries continuously at work, and this partly accounts for the disappointing results often obtained from Siberian gold-mines. The district of Nerchinsk is, or was, famous for its silver-mines, but it also contains gold-bearing veins. The mineral industry conducted there for well nigh two centuries with convict-labour (a very costly form of labour because of its bad quality), is now, however, practically extinct. In the Shilka district, the author found gold associated in a quartz-reef with pyrites and galena. Nearer the surface, the pyrites is oxidized to iron oxide. The reef, 6$\frac{1}{2}$ feet broad at the outcrop, widens to 26 feet at the bottom of a shaft some 50 feet deep. Assays give very promising results. Other quartz-reefs are known to occur, also antimony-ores, phosphorites, asphalt, and (it is whispered) petroleum.

On the island of Askold, 37 miles east of Vladivostok, the plant at a gold-mine crushes about 120 tons of quartz per month during the nine working months of the year (so long as the sea-coast is not icebound). About 130 workpeople (Koreans and Russians) are employed at this mine. Other quartz-veins have been proved on the neighbouring mainland, and in the Tuteka district are argentiferous lead-ore mines, the output of which goes mainly to England. The Tuteka ore yields 62 per cent. of metallic lead, and about 1 part per thousand of silver. In Nahodka Bay, which is government property, three thick seams of semi-anthracite have been proved, and at least 10,000,000 tons of coal are in sight, but the deposit is probably much more extensive.   L. L. B.

THE MINERAL RESOURCES OF NORTHERN CHI-LI, CHINA.

Reise durch den nördlichen Theil der Provinz Chi-li. By KARL VOGELSANG. *Petermanns Mittheilungen*, 1901, *vol.* xlvii., *pages* 243-250 and 278-284, *and* 1 *plate.*

In September, 1899, the author made an excursion from Peking to the western hills, and visited the small coal-mines in the neighbourhood of the village of Wan-ya-tsen. These appear to be worked by the Chinese in a most primitive and unsystematic fashion. The seams are thin and much broken, the coal is of bad quality and is chiefly used for lime-burning. Nevertheless, from the slack, small spherical " briquettes " are made, and find a ready sale in Peking. The selling price (quoted by the author) at the mine appears incredibly low: it works out at barely 1½d. per cwt. (?) The miner's daily wage, for 12 hours' work, is about 6d.

In the valley of Montakou, which opens out eastward, and narrows westward into a steep gorge, coal has been mined for untold ages, but in a very wasteful manner. Here also the seams are rather thin, but the quality of the coal is good: it is carried to Peking on camels, and sells there for about 28s. a ton, costing at the pit-mouth just half of what it costs in the capital.

The first auriferous occurrence described by the author consisted of narrow gold-quartz veins coursing through the hornblende-rock of the Sanyuen-pei-kou valley. Here, the Chinese had driven an adit about 100 feet long, and brought the quartz to-day in baskets. There were evidences of several unsuccessful diggings on the opposite (eastern) slope of the valley. Here, again the author was struck by the absence of method and wastefulness characteristic of Chinese mining. The gold-output in that neighbourhood is too insignificant to be of any industrial importance.

Wonderful stories were current concerning a new Eldorado, Che-lin on the banks of the Chao-ho, and wealthy Chinese had spent large sums on prospecting; but, so far as the author could ascertain, only one unimportant quartz-vein occurs thereabouts, in hornblende-rock. It is said that placers were successfully worked in the district in bygone years, but of these not a vestige is left. Nor does the immediate neighbourhood of Jehol (Cheng-te) boast any auriferous deposit worthy of mention. At Chun-ku-lo-kow, some distance north of the above-named town, workings had been begun, but were subsequently prohibited by the authorities and abandoned, on a quartz-reef which cropped out in places to a thickness of 10 feet. Some specimens assayed in Shang-hai yielded a very moderate amount of gold, 4 to 6 parts per million.

In the barren valley of Hung-hua-kow, in the neighbourhood of Chih-feng, are abandoned workings on a quartz-reef (in diorite-gabbro), specimens from which yielded to the author very unsatisfactory assays. Nevertheless, he is of opinion, from the general appearance of the country, that it would repay systematic prospecting.

North-east of Chih-feng, a town of 100,000 inhabitants, the author visited the coal-deposits of Yuen-pao-shan, in the Lao-ho valley. The coal is hard, and very like the brown coal of Bohemia. The seam, now being worked, varies in thickness from 5 to 10 feet, and underlies about 150 feet of gravel and sand. It is worked by means of long sloping galleries rather than shafts, in which steps are cut, one only being provided with a windlass. The coal is sold at about 14s. a ton, and is nearly all sent by pack-mules to Chih-feng. About 200 workpeople are employed: they are all housed in one huge shelter, and in addition to food, receive piece-wages (which work out at about 6d. per man per ton).

The author further describes his visits to the comparatively insignificant gold-mines of Chuan-shan-tse, and to the mines of Chin-chang-kow-liang (where both auriferous quartz-reef in hornblende-rock, and alluvial gold-deposits are worked). He concludes that, although gold-quartz reefs are of widespread occurrence in Northern Chi-li, they are mostly so narrow as to hardly repay working; new finds of importance are not to be expected, as the province has been pretty minutely prospected by the Chinese themselves, and he does not recommend the investment of foreign capital in that quarter.

L. L. B.

THE SHANSI COAL-FIELDS, CHINA.

The Coal-fields around Tsê Chou, Shansi, China. By NOAH FIELDS DRAKE. *Transactions of the American Institute of Mining Engineers,* 1900, *vol. xxx., pages* 261-277, *with 8 figures in the text.*

Tsê-chou is the principal city in that part of the Shansi coal-field here described. It lies some 300 miles south-west of Tientsin and 500 miles north-west by west of Shanghai. All the coal of the Tsê-chou region is anthracite; its average specific gravity is 1·5; and it is hard enough to support any weight that might be put on it in the blast-furnace. The percentage of sulphur is uniformly low, and that of ash—averaging 10 per cent.—is comparatively so for anthracite.

The rock-formations may be divided into three groups:—a lower group of limestone about 2,000 feet thick; a central group of shales, coal, sandstone, and a flint-bearing limestone aggregating about 300 feet thick; and an upper group of shales, clays and sandstones, exposing about 1,000 feet of strata. Over nearly all the valley and rolling lands there is a covering of loess, which reaches in places 100 feet in thickness. With the exception of this covering, the rock-beds are apparently conformable throughout and, as a whole, lie comparatively level; but there is some conspicuous folding and faulting.

The workable coal lies in one bed about 250 feet above the flint-bearing limestone; there may, however, be a little coal below the limestone, since the principal coal-bed in Western Shansi lies underneath it. This workable coal-seam is probably not less than 22 to 23 feet thick, with a few thin shaly partings. If 22 feet be taken as the average thickness and 1·5 as the average specific gravity of the coal, there are about 3,000,000,000 tons of coal within the 150 square miles of coal-area in the Tsê-chou region, which is only a little of the ragged edge of the great coal-fields of Shansi.

There is a narrow strip in the northern part of the district, where iron-ore (hæmatite) is being mined by open pits from a stratum 2 to 3 feet thick. Fire-clays of good quality are abundant, and considerable quantities are now used by the Chinese for making cheap pottery. Sandstones occur in abundance, and, though too friable to be durable, are used by the Chinese for bridges, houses and road-paving. Massive limestones abound of good quality for building-stone, fluxing material, the manufacture of lime and probably of cement. Reddish-grey marble of fine structure, and black limestone suitable for monuments, are also found.

Practically all of the coal is mined through shafts varying from 50 to 300 feet in depth. The mining is done with pick and gad, no explosives being employed. The coal is hoisted in baskets containing about 300 pounds by windlasses, which are run by men, no steam being used. The coal is mined by driving tunnels through the coal-bed from the bottom of the shaft, and at intervals along the tunnel-ways large quantities of coal are removed, leaving more or less circular rooms, from 40 to 50 feet in diameter.

The output of the district is probably not less than 50,000 tons per annum. For local use, the coal is carried away in little carts drawn by oxen; but the greater part is taken down the mountains by pack-animals, along rough and steep trails paved with stone and 12 to 14 feet wide.

The Pekin Syndicate, a joint English-Italian company, has concessions from the Chinese government to work the coal of the Tsê-chou region as well as most of the Shansi coal-fields. Some proposed railways leading from the Tsê-chou region to outlets on the plains and to shipping ports are described and indicated on a map.   A. W. G.

THE MINERAL RESOURCES OF SOUTH-WESTERN CHINA AND TONGKING.

Étude géologique et minière des Provinces Chinoises voisines du Toukin. By A. LECLÈRE. *Annales des Mines*, 1901, series 9, vol. xx., pages 287-40: and 405.492, and 12 plates.

This exhaustive memoir gives a very full account of the observations made by the author, in the course of a journey which he undertook on behalf of the French Ministry of the Colonies, through those districts which are to be opened up by the prolongation of the (French) Indo-Chinese railways. He started on December 5th, 1897, and returned on July 15th, 1899, bringing back with him more than a ton of rock- and mineral-specimens and fossils. With the exception of the Tongking delta, the whole of the region traversed is extremely rugged and mountainous, attaining altitudes of 10,000 feet or so near the Blue river. The rock-succession, described in ascending order, is:—(1) Crystalline schists; (2) pre-Cambrian or Sinian quartzites, etc., presenting features very similar to those of Britanny and the Ardennes; (3) Devonian limestones and shales; (4) Lower Carboniferous shales, grits, and limestones with coal-seams, with which are associated contemporaneous eruptive rocks (sills, dykes, etc.); (5) Upper Carboniferous limestones, the basal beds being "fetid limestone"; (6) Lower Permian limestones and shales; (7) Middle Permian limestones, etc., with coal-seams; (8) Upper Permian clays, sandstones and conglomerates, in part saliferous and gypsiferous; (9) Lower and Middle Triassic limestones and shales; (10) Upper Triassic variegated marls, overlain by Rhætic shales and grits, with coal-seams; (11) Liassic cavernous dolomite and grits, the latter sometimes containing angular bits of coal; (12) Tertiary deposits, characterized in Tongking by a string of small coal-basins, forming mostly inliers in the gneiss along the Red river. (The best-known of these basins is that of Yen Bay.) These Tertiary deposits exhibit all the features of torrential sedimentation; (13) Quaternary lacustrine formations, calcareous tufas, and peatbeds often with seams of lignite. The Silurian, Oolitic and Cretaceous systems appear to be entirely wanting in the region here described.

About 100 pages are occupied with a record of the facts noted by the author along fifteen different lines of route. His first itinerary was along the coast from Haiphong north-eastward to Kebao. Here in the Konga district, about 260,000 tons of coal are being wrought annually, the greater part of the opencast working. The coal-bearing formation (of Rhætic age) dips very slightly south-eastward, and the seams near the outcrop are sufficiently rich to have precluded for the present any endeavour to carry the workings very deep down. In the island of Kebao, the seams striking north-north-eastward assume an increasingly steep dip until they pitch almost vertically, and often pinch out altogether deeper down. The Lanessan shaft has been sunk to a depth of 1,000 feet, but further exploration-work has been temporarily suspended. What coal has been got shows a high percentage of ash, whereas the Hongay coal is very pure, though friable and "meagre." To make patent fuel out of it, Japanese coal has to be added, as well as a large proportion of pitch. Other Rhætic coal-deposits are worked on a small scale in the north of Haiphong.

A large number of mining-concessions have been obtained for working copper-, antimony-, lead- and silver-ores, the district of Monkay being especially rich in these. Moreover, cobaltiferous manganese is said to occur in the lagoon east of Kwangyen, and nickel north-east of Hongay.

The second itinerary was along the Red river, from Haiphong to Laokay. Navigation is increasingly obstructed by moving sandbanks, the origin of which the author traces to the comparatively recent, extravagant deforestation of Yunnan. Small Tertiary coal-basins occur at several localities along the river, and exploration work has been carried on at the Haidzuong pit (drowned at the time of the author's visit), at a depth of about 130 feet below the surface. The coal yields on analysis 35 per cent. of volatile matter and a considerable quantity of ash. A graphite-deposit has been discovered in the Lower Carboniferous rocks 6¼ miles east of Yenbay, and traces of petroleum are observed on the right bank of the river, 2¼ miles west of that town. Mining concessions have been granted for certain important coal-outcrops in the Lower Carboniferous near Pho-lu.

The third itinerary consisted in a series of prospecting traverses around Laokay. At Langhang, 2¼ miles up stream, seamlets of lustrous coal occur in a torrential deposit of sands, clays and conglomerates. This coal yields 35 per cent. of volatile matter and 4 per cent. of ferruginous ash. Outcrops of Lower Carboniferous coal are also described. Near Baxhot, Banmac, etc., and other localities, both on the French and on the Chinese bank of the river, excellent magnetite-deposits occur in association with milky quartz among strata of Devonian age. Mention is made of the old copper-mine of Mot-dong, 5 miles south-east of Longpo, worked 50 years ago by the Annamites, but no opinion is expressed by the author as to the present value of the deposit.

The fourth itinerary was devoted to a series of similar traverses around Mongtse. Near Manhao are considerable outcrops of a rather earthy coal, in yellowish grit of Tertiary age. This miserable village marks the uttermost limit of navigation at low water (that is, in the dry season); junks of not more than 3 tons' burthen may sail that far up the river. A sometimes very shaly coal occurs in the *Productus*-limestone (Permian?) of Sin-chiem, Pa-che-kay, etc. It is largely used in the lime-kilns of the district, but the coal needs first to be moulded into briquettes and dried, and these are "coked" in large heaps. There is a residue of 7 per cent. of ash.

The author visited the tin-mines of Tomuko, but circumstances (in the shape of a raid by 250 marauding miners on his small caravan) prevented him from seeing the much more important stanniferous deposits of Tsc-mentong. These miners, assembled in bands, periodically pillage the surrounding country, the desolate aspect of which is enhanced by the complete absence of trees (all destroyed for wood-fuel and charcoal) and the great scarcity of water. Some of the tinstone occurs in veins in the Triassic limestones, but there are also what may be termed "secondary ore-bodies," masses resulting from the decomposition of the limestones and the veins. These masses can be generally worked opencast. The tinstone, after washing, calcining, etc., yields 75 per cent. of raw tin.

After providing himself with a larger escort of soldiery, the author proceeded to Kotin, the great centre of the tin-mining industry in that region. At Malaken, the red clay-veins worked for tinstone are proving more and more barren in that ore, while the proportion of copper in the infilling increases. The stuff thrown on the waste-heaps by the Chinese miners is really a rich copper-oxide ore. In the Permo-Triassic grits of Veitaochan, 7¼ miles south of Mientien, is a copper-mine employing a score of workpeople. The ore, after being hand-picked, is sent by pack-horses to Yunnan-sen. It contains from 30 to 40 per cent. of metallic copper, but the percentage of silica is sometimes as high as 40.

The fifth and sixth itineraries consisted in a journey from Mongtse to Yunnan-sen, and in excursions around the last-named city. About 40 miles from Mongtse, at Puchao-pa, a coal-outcrop, up to 65 feet thick, is worked opencast. It is overlain by variegated grits, which are themselves capped by remnants of infra-Liassic limestone. The mineral yielded the following analysis:—Hygroscopic water, 6 per cent.; volatile matter, 38 per cent.; fixed carbon, 51 per cent.; and ash, 5 per cent. The calorific power is estimated as equivalent to 7,060 calories. At present, the annual output hardly exceeds 3,000 tons.

The coal-workings of Niuké are only 22 miles distant from Mongtse. They are opened up in a seam about 4 feet thick, dipping eastward below the limestones. Sloping galleries have been driven downward for a length of about 1,000 feet. The mineral contains from 30 to 35 per cent. of volatile substances, while the ash does not average more than 6 per cent. East and north of Mi Leu, a seam about 4 feet thick is worked by comparatively shallow pits (35 to 50 feet deep) in the Rhætic shales. An average sample yielded the following analysis:—Ash, 9·54 per cent.; volatile matter, 36·8 per cent.; and coke, 63·2 per cent. The calorific power is 7,081 calories. The coal is conveyed in wains, holding not less than ½ ton, drawn by oxen. down to the city of Mi-leu.

Near Tutsa, three or four coal-seams, varying in thickness from 5 to 6½ feet, occur in a group of grits and shales, among which are intercalated sills of eruptive rock (porphyrite), and underlain by a great mass of the same porphyrite. The seams are actively worked by means of several well-timbered drifts. A very lustrous coal is got, having the following composition:—Hygroscopic water, 1·3 per cent.; volatile matter, 15 per cent.; fixed carbon, 73·7 per cent.; and ash, 10 per cent. The calorific power is 7,500 calories. The coal is so friable, that it is found necessary to convert it immediately into coke at the very pit-mouth. The coke sells at about 14 shillings per ton (500 sapeks per horse-load). Some hundred miners are employed at a daily wage ranging from 2½d. to 5d., and the annual output does not exceed 2,500 tons.

A very similar coal is got at Yunnan-sen, where it has been largely used for lime-burning, and from the neighbourhood of Ilyang.

In the Mitsao and Imen district are important copper-ore and iron-ore deposits, and brine-springs. The author describes and figures the primitive plant with which the iron-foundries there are provided.

The seventh itinerary was from Yunnan-sen to Tungchuan. In the province of Yunnan, cupriferous deposits are numerous, but the bad condition of the wheel-tracks and the growing scarcity of timber combine to reduce the number of possible workings, as only those deposits can now be mined which are close to smelting-furnaces, and only those smelting-furnaces can be kept up which are close to a forest. The mandarins have so worked the law that the landowners have no interest in the preservation of the forests, and these are being laid waste beyond hope of recovery. At present, Tung-chuan and Wisi are the principal centres of the copper-industry in Yunnan: moreover a good deal of coal is worked in the Carboniferous strata around Tungchuan. At the Lupu mine, the copper-ores occur actually in a great mass of porphyrite, in a sort of horizontal or bedded vein: the infilling consists of barium sulphate, with which are associated copper oxide and native copper. The Tangtan mines are, however, of still greater importance: the ores (in the Laochang working) occur in a sort of stockwork, and were originally pyritous, but are almost entirely decomposed into carbonate, and these are mixed with the Laosinchang copper-pyrites, the whole being calcined in furnaces fed with wood-fuel. The annual output from Tangtan averages 500 tons of raw copper: this undergoes further treatment at Tungchuan. Deposits of argentiferous galena and manganiferous cobalt-ore are mentioned as occurring in the neighbourhood of Lupu.

The eighth itinerary was westward from Tungchuan to the mouth of the Kinho, a tributary of the Kinshakiang or Blue river. Round the city of Wei-li, much coal is worked in the Coal-measures. The brass-foundries and such like industries are especially active. The auriferous sands of the Kinho are extensively washed by the Chinese, in a patriarchal fashion: they afford a bare living to the easily contented natives, but appear to be of no industrial importance.

The ninth itinerary was south-westward, from the Gold river or Kinho-mouth, to Tali on the headwaters of the Mekong. The author records repeatedly outcrops of Rhætic coal on this traverse. The mineral is of excellent quality, and is worked in various localities by the Chinese. The average of five analyses shows the following percentage composition:—Volatile matter, 31·5; fixed carbon, 54·7; hygroscopic water, 5; and ash, 5·35. The calorific power is 7,965 calories. The land-sale price (in the neighbourhood of Mashang) is stated to average barely 4s. per ton. Tali is the great market for the produce of the copper, lead and zinc-veins of the Likiang and Wisi district. Moreover, rock-salt and gypsum are worked in the Upper Permian formation north of that city.

The author went back south-eastward from Tali to Yunnan-sen (tenth itinerary), passing on the way the great salt-mines of He-tsin. At Chao-chiu, there is a big trade in orpiment, brought from Monghoa, some 25 miles south. About 200 tons of this ore (finest quality) are marketed there annually. The "salt-mines" of He-tsin and Langtsin should perhaps be properly called "brine-shafts." The brine is pumped up by hand-pumps from the Triassic strata, some of the shafts going down as much as 300 feet from the surface, and delivered into batteries of evaporating-pans. The industry constitutes an Imperial monopoly, but is leased out to syndicates which receive a fixed price for the salt according to quality; they are bound to deliver the whole output to the Imperial authorities, who resell the salt to the consumer at a high profit (150 per cent. or so), while the actual miners are assumed to make a profit of about 10 per cent.

The eleventh and twelfth itineraries were from Yunnan-sen to Kweiyang-sen by Hing-gni. At Lanmuchang rather poor coal is worked in the Middle Permian strata. It was formerly used for the distillation of mercury, obtained from cinnabar-veins in the grey *Productus*-limestone. Cinnabar-deposits are numerous in this region, but they remain unworked, as the Chinese are hostile to any re-invasion of the province by Musulman miners. Other outcrops of Permian coal are described, but these are of less importance than the Rhætic coal worked around Ngan-shuen. The mineral is very lustrous, hard, generally "meagre," and does not make good coke. It contains little water, but yields as much as 20 per cent. of ash. Good Rhætic coal crops out in many localities north-west of Kweiyangsen, and its land-sale price varies from 8s. 4d. to 10s. per ton. Moreover, outcrops of bituminous coking coal are described in the shales underlying the Permo-Carboniferous limestones of Wei-ning, etc.

Fron Kweiyangsen the author proceeded to Kweilin-sen, the capital city of Kwang-si. He saw workings of Rhætic coal and Permo-Triassic gypsum on this thirteenth itinerary. He returned then southward by Nan-ning to Hanoi.

The third part of the memoir deals largely with the future prospects of mining industry in the region described. With regard to Palæozoic coal (Carboniferous to Middle Permian), the most favourable field of operations is the neighbourhood of Yunnan-sen. As we have seen, however, it is the Rhætic coal-measures which in this portion of the Chinese Empire assume an importance and an extension far beyond anything [of similar age?] in Europe. The mineral is generally of good quality, and the amount available is estimated at something like 20,000,000,000 tons. From the French point of view (facility of communications with Tongking) the best area for starting operations would be the western border of the Mongtse massif.

The iron-ore deposits, especially those near Yunnan-sen and Trinh-thuong in Tongking, may possibly give rise to an active metallurgical industry in combination with the coal-mines. Of the stanniferous deposits, the author speaks less hopefully, but the reserves of copper-ore in Yunnan are considerable and would probably repay European methods of working.

A translation in full is given of the mining regulations drawn up by the Chinese Government in fulfilment of the agreement of June 12th, 1897, whereby it was stipulated that French engineers and industrialists should work the mines in the three provinces of Kwangtong, Kwangsi and Yunnan.

The general tone of the memoir indicates clearly that those three provinces are regarded by the French government officials as an exclusively French "sphere of influence," a belief which such journeys as that of the author help, more or less, to translate into fact.　　　L. L. B.

STANNIFEROUS DEPOSITS OF HIN-BUN, TONGKING.

Sur les Gisements stannifères de Hin-Boun (Laos). By A. LACROIX. *Bulletin de la Société française de Minéralogie,* 1901, *vol. xxiv.,* pages 422-425.

A French company is engaged in exploration-work on certain tin-ore deposits in the province of Kammun (Laos). These have long been worked, in a very primitive fashion indeed, by the natives. The mine is called Hin-bun, because it is situated in a small tributary valley of that river, which is itself a left-bank tributary of the Me-kong. The ore is dug out of a hillside in a limestone-district where there are no eruptive rocks. It does not appear to be in place, but it is so little water-worn that it cannot have travelled far, and there is every expectation of finding shortly the original ore-body. The poor ore consists of limonite, varying in structure from cavernous to compact or concretionary, in which are disseminated prismatic quartz-crystals. The rich ore is a very compact crystalline yellow mass, made up of cassiterite-needles and quartz-prisms, impregnated with limonite. The cassiterite in many instances resembles very closely in appearance the "needle tin-ore" of Cornwall. None of the usual associates of cassiterite (topaz, mica, apatite, wolfram, etc.), occur in the Hin-Bun deposit. The limonite is possibly a pseudomorph after pyrites, although no absolute evidence of this is forthcoming. On the whole the author concludes that the cassiterite here occurs as a primary mineral in the ferruginous gossan of a sulphide-deposit, analogous to that of Campiglia in Tuscany. Comparison may also be made with the sulphidic antimony-bearing veins of Potosi (Bolivia), as analysis shows that the Hin-bun ore contains about 4 per cent. of antimony and a little lead.　　　L. L. B.

THE PULACAYO SILVER-MINES, BOLIVIA.

Mina Pulacayo. By CÁRLOS G. AVALOS. *Boletin de la Sociedad Nacional de Mineria*, 1901, *series* 3, *vol. xiii., pages* 171-184, *with* 1 *figure in the text.*

The output of these mines in 1899 amounted to about 230 tons of silver, say between 4 and 5 per cent. of the total output from all countries in the world. The Pulacayo deposit occurs in one of the mountain-ranges which branch off from the main cordillera of the Andes; the mountains are made up of trachytic rocks which bear testimony to the volcanic eruptions that at one time burst through the conglomerates, sands and clays of Tertiary age forming the southern portion of the Bolivian high plateau. It is with these trachytes that the great argentiferous vein is intimately associated: where the volcanic rocks give place to sedimentaries, so too does the ore disappear. The workings extend over a length of 7,700 feet and a depth of 2,500 feet (reckoning from the highest point of the outcrop to the lowest or sump-level). The great vein strikes east and west, pitches northward, possesses a maximum thickness of 13 feet and tends to narrow as it goes deeper down. The principal ores found in it are tetrahedrite (containing up to 10 per cent. of silver), galena (with 2 per cent.), blende, copper-pyrites and iron-pyrites (the three last-named contain still smaller percentages of silver than the galena). Towards the surface all these metallic sulphides give place to their respective oxides and sulphates. A frequent non-metallic associate is barium sulphate, which diminishes in quantity concurrently with the increase of depth from the surface, and finally disappears altogether at the lower levels. Galena is most abundant at the intermediate levels; lower down its place is taken by blende and pyrites, and, at a point very nearly reached by the present workings, the vein is entirely made up of the two last-named sulphides and therewith ceases to be of industrial importance. Not much water was met with at first; but, of late years, as the workings were pushed deeper down, thermal springs of which there was no trace at the surface, were tapped, and they flooded the lower levels in continually increasing volume; in fact, the lowest levels remain flooded to the present day. A curious circumstance is that the superficial, pluvial waters which percolate downward, decomposing on their way the sulphides left in the old workings and becoming thereby strongly acid, are neutralized by the thermal waters; and thus the water pumped from the mine does not corrode the pumping-plant. The thermal springs met with at the lower levels indicate temperatures of 108° and 136° Fahr.; they hold in solution considerable quantities of carbon dioxide and small proportions of sulphuretted hydrogen. Fatal accidents have occurred in the mine, due to the evolution of the carbonic-acid gas from these waters: in one case, more than 20 lives were lost. Only one mine, the Pique Central, possesses pumping machinery of any importance, consisting of a Bullock compound engine (American) of 300 horsepower, and a Cornish duplex pump. Steam is provided by seven Lancashire boilers (Galloway) capable of developing 200 horsepower each, and from them the San Leon, Rothschild and Ramérez mines derive their motive power, as well as the Pique Central. The author shows in detail that the present plant is inadequate to meet the demands made upon it. British coal appears to be the sole fuel used.

The remedy for the difficulties hinted at appears to be the construction of a tunnel some 28,000 feet in length, which would reach the sump-level of the workings, and would be available both for drainage and ventilation.

L. L. B.

THE PARRY SOUND DISTRICT, EASTERN ONTARIO, CANADA.

Mining Developments in Eastern Ontario. By KIRBY THOMAS. *The Engineering and Mining Journal* [*New York*], 1902, *vol. lxxiv.*, *pages* 186-187.

The Parry Sound district is located on the eastern centre of Georgian Bay and along Parry Sound. The city of Parry Sound is reached by deepwater navigation from the Great Lakes by Georgian Bay, and it is the terminus of the Canada-Atlantic railroad.

The rocks of the Parry Sound district are of the older geological series, although their exact horizon is not yet fully determined. The prevailing rocks are gneisses and schists. A peculiar and prevailing rock about Parry Sound is a garnetiferous schist. The rocks are all very much disturbed and contorted, and, at Parry Sound particularly, show extreme flexions. There are numerous small quartz-veins and several large pegmatitic dykes. In some cases these dykes may be traced for miles, and the prevailing crystals of felspar and mica, with some uncrystallized quartz and hornblende, are massive. In the Valentyne location, about 12 miles north of Parry Sound, one of these pegmatitic veins is mined for mica and felspar. The country is very rough and much glaciated. Probably 50 per cent. of the surface is glistening with uncovered rock.

The principal mineral channel appears to lie parallel to Parry Sound and extends for many miles either way. In general, it is from 1,000 to 1,400 feet wide, and the mineral veins or beds occur as extended and more or less parallel lenses, having the general direction of the mineral channel (north-east and south-west), and generally dipping at a high angle or almost vertical. At several points, the beds aggregate from 300 to 400 feet wide. They are from a few feet to 30 or 40 feet thick, and are generally indicated on the surface by an iron-stained capping. The ore appears to be in the zones or fractures along what is probably the line of an uprise, and there is every geological indication of going to depth.

The ore varies at different points, but generally contains iron-pyrites and chalcopyrite. At some points, the copper occurs as chalcocite and bornite. In some locations, zinc occurs in the form of spalerite, sometimes running as high as 15 per cent. Some gold is found with nearly all the ores, the values running from 12s. 6d. to £2 (3 to 10 dollars) per ton. In the Lafex mine, an assay of 2½ per cent. nickel was shown. The copper-values vary:—The Wilcot, at a depth of 80 feet, gave a value of 9½ per cent. across a 30 feet vein. At the McGowan location, masses of bornite weighing several tons, running as high as 54 per cent. in copper, were taken out at or near the surface in open-cuts. Seven car-loads of this ore shipped to New Jersey gave an average value of 16 per cent. in copper. It is claimed that the McGowan ore has been thrown up from below, and is an indication of what will be found at depth in other parts of the zone.

The Parry Sound district promises to become primarily a producer of copper, and incidentally a producer of nickel, zinc and gold.

X. Y. Z.

ASBESTOS IN CANADA.

Vorkommen und Gewinnung von Asbest in Canada. By FRITZ CIRKEL. *Zeitschrift für praktische Geologie*, 1903, *vol. xi.*, *pages* 123-131, *with 3 figures in the text.*

In 1901, the output of asbestos throughout the world amounted to 41,800 tons, 38,500 of which were obtained from Canada. About 1870, asbestos-deposits were discovered in the serpentine-hills of Thetford, district of Coleraine, in the east of the province of Quebec, but it was not until many years later that the discovery was turned to practical account. From 1879 onwards, the development of the asbestos-mining industry in the dominion has been extraordinarily rapid, and 12 companies are now at work, with a total capital verging on £1,000,000.

The author devotes his attention mainly to the serpentine- or chrysotile-variety of asbestos (hydrated silicate of magnesia) as being of far greater industrial importance than the brittle amphibole- or hornblende-asbestos (a silicate of lime and magnesia). The colour varies from olive-green to whitish, and there is one rare occurrence known to the author in the Ottawa district, which presents an extraordinarily fine fibrous texture with a dark-blue colour. He points out that a fine elastic quality of asbestos always contains more water than a poorer brittle variety. Add to this the fact that, if an asbestos-fibre be subjected for some time to a very high temperature, it becomes brittle. Evidently, then, there is an intimate re-lationship between the quantity of combined water in the mineral and its fibrous structure.

The Templeton variety was worked about 1890 in the crystalline lime-stones of the Ottawa district, but its occurrence was found to be of so restricted a nature that the workings were given up after two years. Nevertheless the author considers the Templeton asbestos of such extreme geological interest that he gives a lengthy description of the occurrence.

He then proceeds to describe the (commercially) far more important Thetford and Black Lake deposits, situated close to the railway between Sherbrook and Quebec, about 150 miles west of Montreal. He gives a map showing the distribution of the serpentines in the province of Quebec: these occur in three distinct areas. The easternmost, that of the Gaspé Peninsula, yields small veins of asbestos in the serpentine, but not in sufficient number to repay working. Of the Mount Albert area too, the same statement holds good, but with the reservation that, owing to its inaccessibility, that region has not been sufficiently investigated. In the Thetford and Black Lake district, the serpentine forms entire ranges of hills from 650 to 1,000 feet high. This part alone of the central area yields asbestos in sufficient quantity to repay working.

The third or south-western district stretches from Danville into the State of Vermont. The serpentine occurs often in the form of isolated hills, but hardly anywhere has asbestos been found in large quantities so far. How-ever, that area may repay more thorough prospecting.

The Thetford type of asbestos occurs in serpentine associated with black, grey or green schists, also subsidiary quartzitic conglomerates and sand-stones, all of Cambrian age. The association of diorite with these is noticeable, and in places the actual passage of diorite into the serpentine can be seen. Granulite, which is often found between the diorite and the serpentine, is generally regarded in the district as a good indicator of asbestos. This mineral occurs in irregular narrow veins (4 to 5 inches thick at most) in the serpentine, is very elastic and silky, and of a dark colour. Black Lake, only 4 miles from Thetford, produces a somewhat more vitreous and brittle asbestos, the mineral having been probably deprived of part of its combined water by the intrusion into the neighbouring rock of masses of eruptive granulite.

The author mentions, but does not decide as to, the relative merits of the two theories in vogue concerning the genesis of the asbestos-deposits. The lateral-secretion theory assumes that fissures were formed in the serpentine-magma as the result of cooling and contraction, which fissures were later slowly infilled with magnesia-bearing solutions. The other theory does not postulate the existence of fissures, but derives the asbestos from the alteration of olivine and serpentine at a very high temperature. The workings are mainly opencast, and the better mineral is (or was until lately) graded or sorted and picked by hand, as it was found that machine-" picking " damaged the fine fibre.   Details are given as to the cost of production and relative amount of output of the different qualities, from " first crude " down to " asbestic."   L. L. B.

THE AURIFEROUS PLACERS OF CHILE.

Los Lavaderos auriferos de Chile. By A. ORREGO CORTES. Boletin de la Sociedad Nacional de Mineria, 1903, series 3, vol. xv., pages 49-64.

In the Chilian placers, the gold occurs as a rule in the form of dust or very small grains.   Occasional finds of fairly large nuggets are, however, recorded, such as those seen by the author at the Nahuelbuta diggings in 1895, which varied in weight from 6 to 22 ounces troy.   According to the analyses tabulated by him, Chilian native gold always contains, in addition to varying percentages of silver, small amounts of copper and iron.   He estimates the value of the gold obtained from Chilian territory, from the time of the Spanish *conquistadores* down to our own day, at about £100,000,000.

The coastal cordillera, which ranges from north to south along the Pacific seaboard, is chiefly made up of granitic rocks, which here and there throw off transverse spurs penetrating some 15 or 20 leagues inland.   The formation is seamed by innumerable gold-bearing and copper-bearing quartz-veins of extremely variable thickness (from several feet down to fractions of an inch).   The gold in many of the placers is evidently derived from these rocks, the result of weathering by various agencies and decomposition.   In the fine sands, the precious metal is distributed throughout the deposit from top to bottom, but in the so-called *cascajos* or coarse gravels the gold is concentrated near the floor of the placer.   Some of the coast-district placers lying at the foot of the granite-ridges, are covered, and have indeed been preserved by, a deposit many feet thick, of a clay which is a decomposition-residue of the granite.   Accordingly, in order to work the placers, shafts have to be driven through this clay to various depths, until the rich *circa* or bed-rock is reached, as at Loica, Melipilla, etc.

Entering upon the detailed description of the placers, the author deals first of all with the fine sands of Carelmapu, on the extreme southern coast of Chile.   Certain assays carried out by him showed the proportion of gold here to vary from 0·09 to 0·70 per million, but down on the sea-beach, where the wash of the waves has effected a sort of natural concentration, much richer sands occur, containing from 10 to 35 parts per million.   These auriferous sands have been proved to a depth exceeding 26 feet.   There are similar placers at Cucao, and at Chacao Channel in the island of Chiloé.

In the placers of Quilacoya, 8 leagues north-east of Concepcion, the sands are coarser, and consist of quartz-grains (in part crystalline), mica, and the other constituents of granite.   The native gold occurs in the form of flakes, and was evidently derived from quartz-veins in the neighbourhood.

The coarse auriferous gravels of Marga Marga rest upon granitic rocks and those of Catapilco on syenites. An American syndicate was organized several years ago to work the last-named placers by hydraulicking, and expended nearly the whole of its capital in cutting canals and other preparatory works, only to find that the water-supply was inadequate. Nothing further appears to have been done in the matter since then.

After referring to the placers of Loica, Yalé, Andacollo (the coarse gravels of which have been profitably worked for centuries), Casuto, etc., the author points out that placer-deposits all over Chile are still awaiting development by modern methods. Those concerned appear to be hesitating as to the best method of working them. For the instruction of his countrymen, the author proceeds to give a short description of the principal methods in use in the United States and New Zealand, such as hydraulicking, dredging, etc.

L. L. B.

THE COPPER-MINES OF TAMAYA, CHILE.

El Mineral de Tamaya. By ISAAC VARAS C. *Actes de la Société Scientifique du Chili*, 1901, *vol. xi., pages* 30.41.

This mining field, which in its day ranked among the first of those where cupriferous veins were worked, is at present in a state of decadence, chiefly for two reasons:—(1) Many of the best veins have been worked out, and (2) the levels in several of the mines have been invaded by water and submerged. On the other hand, it is yet possible that in the still unexplored portions of the field, new metalliferous veins may be discovered which will serve to keep the industry alive for a longer period than can at present be foreseen. Its forlorn condition is, in the author's opinion, mainly due to the drowning of the workings. This began with the rains of 1888; and then, the exceptionally wet winter of 1891 caused floods in the mines, the effects of which it was the less possible to stave off that there was a concurrent scarcity of labour due to the unsettled condition of politics. From that time, a sort of paralysis seems to have overtaken Tamaya, but the recent rise in the price of copper, though it has hardly made itself felt in its full force there as yet, will undoubtedly prove beneficial—from the miners' point of view. A few fresh workings have been opened on the secondary veins, and renewed activity has been lately evident in old mines that were formerly of considerable importance. Among these may be mentioned the Llano Blanco and the Patos mines.

The author describes briefly the mines working the principal vein. The deepest is the Rosario mine, which goes down to 1,935 feet vertical below the outcrop. At this depth, the vein still yields good ore, containing 28 per cent. of metallic copper. The Lecaros gallery, the greatest work of its kind in South America, after a course of upwards of 1½ miles, debouches into this mine at the depth of 920 feet; and, forming a natural drainage-level, it has in part preserved the several mines with which it communicates from being flooded out. The Guias mine, not reaching a depth of more than 623 feet below the outcrop, comprises an immense extent of unexplored ground, and there is reason to believe that it will prove of great importance in the future. Innumerable mines have been opened up on the secondary veins: among these, the Patos mine, already mentioned, has yielded good masses of ore, containing from 28 to 30 per cent. of metallic copper.

The revival of the mining industry of Tamaya depends, in the author's opinion, firstly, on a systematic exploration of the still untouched areas; secondly, on a combined drainage of the mines by means of an electric plant, and by lengthening the Urmeneta gallery so as to communicate with and drain the mines of the northern portion of the field; and thirdly, on the unification of the mines, as far as possible, under one management. The railway-freights are very high, and the remedy would seem to be the purchase by the Chilian Government of the line which runs to Tongoi.

L. L. B.

THE MINING-FIELD OF CARACOLES, CHILE.

Caracoles: Revista Minera. By FELIPE LABASTIE. *Boletin de la Sociedad Nacional de Mineria,* 1903, series 3, vol. xv., pages 113-125, and 1 plate.

For the past fifteen years the four silver-producing provinces of Northern Chile have suffered much from the long-continued crisis which has befallen the world's trade in that metal, and the annual output is now barely 10 per cent. of what it was in the years between 1840 and 1894. But what affects the district of Caracoles more especially nowadays is the want of enterprising capitalists, and the need for more extensive exploration-work. In former times, mining was carried on there in an unmethodical, wasteful, reckless fashion, but the author appears to think that capital and knowledge prudently and methodically applied will assuredly bring about a revival of the mineral industry on a grand scale. Attention is directed to the three principal mines still at work in the district.

The Valencianita mine works the great reef known as the Corrida de Senda, which has been proved over a length of 2 miles; parallel to this (3,300 feet to the west) runs the reef known as the Gran Corrida de Caracoles, which, with its bifurcations extends over a length of 5¼ miles. Both are extremely rich. There are innumerable veins and veinlets transverse to these, varying in thickness from 2 inches to 6½ feet. The ore is got in the Valencianita mine between the depths of 385 and 590 feet; the Recuerdo mine did not strike really payable ore till a depth of 846 feet had been reached, and the author makes a point of the fact that much ore in Caracoles remains to be worked at fairly considerable depths. The Sud America mine also works the Gran Corrida de Caracoles.

A brief description is given of the geology, from which it may be gathered that the Jurassic rocks are overlain, as well as intruded into and much crushed, disturbed and otherwise altered by later eruptive porphyries and trachytes. Especially is it at the contact with the porphyries that the metalliferous veins are observed. To the eruptive phenomena also does the author attribute the rugged contours of the district.

Stress is laid on the splendid results attained in the Recuerdo mine, though inflows of water at first, and then difficulties of ventilation, had to be overcome. The infilling of the vein is very complex—there are both carbonated and sulphuretted ores of lead, silver and iron, as well as native silver.

Although not explicitly stated to be so, this paper appears to constitute the first of a series on the Caracoles district.

L. L. B.

THE MINERAL RESOURCES OF ECUADOR.

Les Gîtes minéraux de l'Équateur. By C. VAN ISSCHOT. *Annales des Mines*, 1901, series 9, *vol. xx., pages* 97-102.

At present, in the statistics of production of precious metals, Ecuador occupies a lower place than her neighbours, Peru on the south and Colombia on the north, and yet in the old days of the Spanish domination the territory of Ecuador was one of the chief areas of gold-production on the American continent. The output of silver on the other hand was never considerable, and by no means corresponded to the natural resources which Ecuador possesses in the shape of argentiferous veins. Mining enterprise, checked by the war of independence against Spain, remained in abeyance until 1876 when the Ecuadorian government initiated some prospecting work, especially in the Zaruma district. Thereupon several companies were formed to work the gold-quartz reefs, but what with insufficient capital, and what with bad management, they all came to grief within a short time.

At present, only two mining companies are at work in the Republic of Ecuador. One works the auriferous quartz-reefs which cut through the porphyries of the Zaruma district, some 50 miles distant from the Gulf of Guayaquil. The "reefs" average 40 inches in thickness, and contain gold in the free state down to depths of 80 and 100 feet: below this the gold is generally found in combination with sulphides. In the Mina Grande, where the "reef" attains a maximum thickness of 20 feet, the average yield of gold is 20 parts per million, of which only one third can be extracted by amalgamation. So far it has not been found possible to treat the remainder with economically satisfactory results, even by the cyanide process. The district, which covers an area of about 40 square miles, is traversed by numerous streams and torrents capable of furnishing abundant motive power. Timber, for structural purposes, is very costly on account of the difficulties of transport, and there is no fuel naturally available on the spot. The free gold is really an alloy, of 734 parts of gold with 266 parts of silver per 1,000.

In the province of Esmeraldas, huge auriferous placers are being worked by a company on the hydraulicking system along the banks of the Rio Santiago. The average yield is estimated at 4d. or 5d. per cubic yard. The author has prospected other placers near Angostura, Uimbi, Cachavi and Bogota, and estimates that a similar output could be obtained from them, but these at present being worked by the blacks, who are content to extract just enough gold to supply their very primitive wants. All these gold-bearing alluvia contain platinum.

No serious work is being done on silver-mines at present. Some ores were exported to Germany in 1893-1894, and one lot of 77 tons sold at Freiberg yielded 1 per cent. of silver, and a small proportion of gold. The mine whence these were got is now abandoned, for want of capital. Water-power (in abundance) and timber (both for structural purposes and for fuel) are available on the spot. The ore, chiefly iron-pyrites and copper-pyrites, with argentiferous grey copper-ore, could be very economically treated by lixiviation. Abandoned silver-mines are also to be found in the neighbourhood of Canar.

Argentiferous and auriferous pyrites, mixed with galena, occurs in a thick vein, off the Cuenca road, some 34 miles from Guayaquil. The ore yields on assay 15 to 22 per cent. of copper, 10 to 40 per cent. of lead, and as much as 0·6 per cent. of silver. Moreover argentiferous galena, containing up to 0·25 per cent. of silver, has been found at several other localities, especially at Malacatos near Loja.

Mercury does occur in the metallic state in various alluvial deposits, but nowhere in such quantity as to repay working.

Petroleum is found along the Pacific seaboard, in strata belonging to the same geological formation as those whence the oil is obtained farther south, in Peru.   It impregnates sandy beds close to the surface, and the inhabitants dig shallow wells, 10 to 12 feet deep, where the oil, dripping out from the sand, collects, together with brine.   The author thinks that if adequate search were made, by means of deep borings, industrially important natural reservoirs of oil would be struck.

Anthracite is recorded from Riobamba, and exploration work is proceeding on a lignite-deposit near Loja. The coal-outcrops of Azogues, Malacatos and Penipo, even if they prove workable, are so far in the interior of the country that the difficulties connected with transport are likely to constitute an insuperable obstacle to successful enterprise.

The mining legislation of Ecuador is conceived in a very liberal spirit. An annual tax of £4 on mines in full work is payable to the State, and in cases of suspension of work the tax is reduced to 17s. The necessary plant, tools, explosives, chemical reagents, etc., can be imported into the country free of duty.   In fact, the sole obstacles to mining enterprise are the absence of roads and the want of capital.   L. L. B.

THE PETROLEUM-INDUSTRY OF PERU.

La Industria del Petróleo en el Perú en 1901. *By* ALEJANDRO GARLAND. *Boletín del Cuerpo de Ingenieros de Minas del Perú*, 1902, No. 2, 14 pages and 4 plates.

At the end of 1901, no less than 269 concessions for working petroleum had been taken out, most of them being situated in the districts of Tumbes and Payta. In the former district, a French syndicate and an Italian firm appear to be the chief concessionaries, while in the latter district a British company and an American syndicate take the front rank.   About a fifth of the total number of concessions, in the hands of various individuals and companies, remain so far unworked.

The total sales of Peruvian crude oil in the year 1901 amounted to 6,400,000 gallons, those of kerosene to 516,920 gallons, benzine 667.412 gallons, and residues 1,983,500 gallons. Of this output by far the greater portion is credited to the London Pacific Petroleum Company of Talara, and nearly the whole of the remainder to the Piaggio establishment at Zorritos.

The French syndicate, after a lapse of two years, resumed boring operations, putting down 5 bore-holes, in some cases to a depth of 1,800 feet, without striking oil (except an insignificant quantity at about 500 feet belowground).   The American syndicate, whose property lies about 15 miles farther north of Talara, at Amarillos, had so far put down only one boring, to a depth of 1,200 feet, and this without striking oil.

In round numbers, the oil-field employs at present about 500 persons, and the concessions cover an area of 3,700 acres or so.

No very definite opinion can be hazarded at present (according to the author) as regards the future of the industry. The result of the borings carried out so far is to show that the nearer the sea-shore they are put down, the greater are the probabilities of striking oil. Natural reservoirs of any importance have only been found on the coast, at Zorritos, Punta Pariñas and Punta Agujas. It looks as if the great bulk of the petroleum-deposits lay out to sea, beneath the waters of the Pacific, and as if a portion of the Peruvian coast-line merely marked the eastern fringe of the field. Evidence is cited in confirmation of this hypothesis, and attention is drawn to the fact that the newly-discovered petroleum-deposits near Los Angeles in California are largely worked beneath the sea.* Their output is more than twelve times as great as the Peruvian.   L. L. B.

THE CROWN MOUNTAIN GOLD-MINE AND MILL, GEORGIA.

The Crown Mountain Gold-mine and Mill, Georgia. By HENRY V. MAXWELL.
  The Engineering and Mining Journal [New York], 1901, vol. lxxii., pages 355-356, with 4 illustrations in the text.

The Crown Mountain mine, Dahlonega, Georgia, possesses the first thoroughly equipped, combined mining, milling and sluicing gold-plant in the south, and is the first to utilize water-power in generating, and transmitting electrically, power for gold-mining purposes.

Hydraulic mining being the basis of this enterprise, water-rights were secured at a point 12 miles from Dahlonega, where, by making 2 miles of canal, the union of three streams was effected without dams, and the head-waters of the Chestatee river dropped 97 feet upon a wheel of 800 horsepower capacity.   Directly connected on the water-wheel shaft is a two-phase generator of 500 kilowatts.   The wheel and dynamo run at 514 revolutions per minute, and deliver 568 ampères per phase. The current is generated at 440 volts, transformed to three-phase at 12,000 volts, then transmitted 12 miles to the mill, and 13 miles to a pumping station on the Chestatee river, at the foot of Crown Mountain, where, at both points it is again transformed to two-phase, at 400 volts, and used at that pressure on all the motors.

At the pumping station, a Dean triplex pump, operated electrically and geared to its motor in the ratio of 20 to 1, easily lifts 1,500 gallons of water per minute through a 12 inches pipe to a reservoir on Crown Mountain about 550 feet above the river. From the reservoir, the water is distributed to four monitors or giants, three operating under a pressure of some 200 feet near the base of the mountain, while the fourth is working upon the saprolite-bodies of the summit; this giant is acting under direct pressure from a force-pump driven by compressed air. The air-compressor also furnishes air for hoisting, drilling, etc., at two working shafts as well as for operating the force-pump.   From the points of operation of the giants over 5,000 feet of flumes, supplied with riffles throughout their length, have been constructed, and through these flumes the entire product of the mine is sluiced to the mill.

The heavy ore is broken with hammers into sizes admitting of transportation through the flumes, and an arrangement of grizzlies within the flumes separates the finer from the coarser sizes. The ores are dealt with in the usual way, the heavier ore first passing through a Dodge crusher, and thence into the feed-bins.   Here are placed 50 stamps of 950 pounds

---

* *Trans. Inst. M.E.,* 1901, vol. xxii., page 729.

in 10 batteries of 5 each.   Through these the ore passes on to 10 Wilfley tables.   The finer ore passing through the grizzlies is treated on two Huntingdon mills, below which are also Wilfley tables.   In both mills, electro-plated copper-plates are used for the amalgamation of the free gold.

X. Y. Z.

GOLD-MINING IN THE SOUTHERN APPALACHIANS.

Gold-mining in the Southern Appalachians.   By JOSEPH HYDE PRATT.   *The Engineering and Mining Journal [New York], 1902, vol. lxxiv., pages 241-242.*

The area embracing the gold-fields of the Southern Appalachians extends from near Baltimore in a south-westerly direction across Maryland, Virginia, North and South Carolina, Georgia and Alabama.   The writer gives a synopsis of the mining being done in the south, with the exception of Georgia, which is not considered in the paper, either statistically or otherwise.

Gold-mining in the region has been prosecuted for over a century, but systematic mining has not been undertaken until within the last decade of the nineteenth century.   With the exhaustion of the placer-deposits, attention was turned to the veins from which they had been derived.   As a rule these veins contain low-grade ores in quantity, and it is the method of treatment rather than the amount of ore that requires consideration. Many of the ore-bodies consist of beds in the schists and slates impregnated with auriferous pyrites and of imperfectly conformable lenticular veins and stringers of quartz.   Where the upper portion of these have been thoroughly altered and decomposed, they were formerly worked for their free-gold contents, but as sulphides were encountered they were abandoned.   These large deposits of low-grade ores are capable of being mined in enormous quantities, and offer a promising field for a cheap process of recovering the values.   While the chlorination process has been worked successfully, it is highly probable that the cyanide process can be adapted to these ores.

During the past ten years, the production (apart from Georgia) has averaged about £40,000 (200,000 dollars) per annum, the larger proportion of which has come from one or two mines, showing clearly that when these low-grade ores are worked on a large scale they they can be treated profitably.

Many natural conditions favour mining in this region : forests for lumber, mine-timber and cord-wood are near at hand ; cheap labour is abundant, and skilled labour is also cheaper than in the western mining camps ; climatic conditions permit nearly continuous mining all the year round ; and at many of the mining localities there is abundance of water-power available for the installation of electric-power plants.   X. Y. Z.

COAL-MINING IN KANSAS.

Kansas Coal-mining.   By W. R. CRANE.   *The Engineering and Mining Journal [New York], 1901, vol. lxxii., pages 748-752, with 7 illustrations in the text.*

The coal-area of Kansas comprises about 20,000 square miles, within the western Central area of Carboniferous strata.   About three-fourths of this field is productive, the western portion being barren.   Mining is most extensive in the south-eastern part of the State, principally in Cherokee and Crawford counties. Taken as a whole, the coal is uniform in quality, and the bulk of it is bituminous coal. At Cherokee, Cherokee county, the coal was found by test to have the highest calorific value of any in the State, 15 pounds of water being evaporated for every pound of coal burned (sic). At Osage City, the calorific power drops to about 10 pounds, and continues to decrease towards the higher geological horizons on the north and west.

The Kansas coals occur in shale-beds, alternating with limestone-strata of varying thickness and an occasional unconformable sandstone-bar, usually lenticular in form, which partially or entirely cuts out the coal. These beds of sandstone are more abundant in the southern part of the State. The Weir-Pittsburg lower seam averages about 40 and the upper about 30 inches in thickness. The lower stratum dips westward at about 16 or 20 feet to the mile, while the lower stratum dips with an inclination of from 6 to 8 feet per mile less, indicating that during the interval between the deposition of the two beds, there was an excessive westerly thickening of the shale-deposits. No serious faulting exists, although it is true that faults with displacements varying from a few inches to 8 feet have been met with.

A peculiar feature, especially in Cherokee and Crawford counties, is the common occurrence of what are known as " horse-backs " or " clay-veins." The source of these disturbances was probably the Ozark uplift, the centre of which upheaval is located a hundred miles or so to the eastward. The Kansas coal-beds, lying as they do at the outer limit of the disturbed area, were only slightly affected, and as a rule, only fissures were formed. Afterward, the pressure of the superincumbent strata, acting on the clay-beds beneath the coal, forced the clays upward into the fissured strata, thus forming the clay-veins known as " horse-backs."

Besides horse-backs, there are other formations known as " pots," " kettles " or " bells," which seem to be exaggerated forms of unconformable strata commonly known as " rolls " and " slips." The origin of the horse-backs and bells is quite different; the horse-backs are younger than the coal-seams that they intersect, while the bells and the coal-strata were formed contemporaneously. As for the rest, the bells were probably the result of cessations in the deposition of the formative materials produced by external conditions, such as exist to-day in swamps and sluggish streams, where eddy-currents and sudden inrushes of sediment and organic matter are of common occurrence.

The existence of these numerous horse-backs and bells exerts its effect both on methods of prospecting and actual working. Test-holes must be drilled at any point where a shaft is to be sunk. Serious loss has resulted from the ignoring of this rule. In one case, a shaft was sunk ¼ mile distant from a test-hole in all confidence that the same conditions would be found, but the completion of the shaft discovered the fact that the coal had been cut out or rendered useless by the great number of horse-backs that crossed and re-crossed each other.

The most prevalent system of working is room-and-pillar. There are two reasons for this:—(1) The mines are comparatively shallow and the surface-lands valuable; and (2) the coal being so much cut up by horse-backs, these can without difficulty be utilized as supports for the roof. The room-and-pillar system is most practised in the counties of Cherokee, Crawford and Bourbon, in which counties the horse-backs are most abundant. In the mines farther north and north-west, the seams are thinner, the horse-backs are less numerous, and more longwall working is found. But by far the greater part of the coal is mined by room-and-pillar, though nearly as many mines are worked by one system as the other. Coal-mining machines have been used quite extensively in some mines, and are still employed in longwall work, although not so much as formerly.   X. Y. Z.

THE QUICKSILVER-MINES OF BREWSTER COUNTY, TEXAS.

The Quicksilver-mines of Brewster County, Texas. By E. P. SPALDING. *The Engineering and Mining Journal* [*New York*], 1901, *vol. lxxi., pages* 749-750, *with 8 illustrations in the text.*

In the south-western part of Brewster county, Texas, present indications point to one of the most important and productive deposits of cinnabar in the United States. The locality is within what is known as the " Big Bend County," and lies about 10 miles north-east of the entrance to the Grand Cañon of the Rio Grande.

The country-rock is a Cretaceous limestone upheaved by igneous protrusions and cut by igneous dykes and faulting planes. Much evidence exists of the action of powerful currents of water. Stream-beds and dry washes have cut several hundred feet through the hard limestone, and the denudation has been enormous. A section of the strata obtained in the Grand Cañon is 1,700 feet thick. The main belt, in which the cinnabar occurs in payable quantities, is confined to certain well-defined limits, constituting a zone 6 miles long and 2 miles wide. Prospecting has proved the existence of large quantities of surface-ore ranging from ½ to 5 per cent. of quicksilver, and considerable quantities have been found ranging up to 18 per cent. Very high-grade ore occurs in bunches and pockets, assaying from 40 to 78 per cent., and these pockets have been found with sufficient frequency to raise materially the average tenour of the large quantity of low-grade surface-material now being furnaced.

The cinnabar found here, as a rule, is of the crystalline variety, having the characteristic colour of pure cinnabar. Quicksilver is also found in the native state filling cavities in the limestone, and as globules in crevices of rocks impregnated with cinnabar. As a rule, the cinnabar is found impregnating a clay-like rock or marl, and the softer portions of the limestone and breccia along faulting lines and in calcite-veins and stringers. But the writer has noted it in every class of rock found in the district, even as a coating on an eruptive basaltic rock. This points to a very general distribution of the metal throughout the zone.

The rich pockets occur as replacements of the country-rock, and in places there is evidence that the mineral-bearing solutions have come up through fissures and deposited their metallic contents at points where conditions favoured concentration and deposition. This would point to a surface rather than to a deep-seated deposit, and while there are reasons for supposing that workable values will be found at considerable depth, much remains to be proved seeing that, as yet, the deepest working is only 65 feet from the surface. During the visit of the writer, several nuggets of nearly pure cinnabar, worn smooth and round, were found in a ravine. Investigation resulted in the discovery of a pocket from which about 20 tons of ore was taken, assaying from 40 to 75 per cent., or between £2,000 and £3,000 (12,000 to 15,000 dollars) in value.   X. Y. Z.

II.—BAROMETER, THERMOMETER, Etc., READINGS FOR THE YEAR 1902.

By M. WALTON BROWN.

The barometer, thermometer, etc., readings have been supplied by permission of the authorities of Glasgow and Kew Observatories, and give some idea of the variations of atmospheric temperature and pressure in the intervening districts in which mining operations are chiefly carried on in this country.

The barometer at Kew is 34 feet, and at Glasgow is 180 feet, above sea-level. The barometer readings at Glasgow have been reduced to 32 feet above sea-level, by the addition of 0·150 inch to each reading, and the barometrical readings at both observatories are reduced to 32° Fahr.

The statistics of fatal explosions in collieries are obtained from the annual reports of H.M. Inspectors of Mines, and are also printed upon the diagrams (Plates XXVI. and XXVII.) recording the meteorological observations.

The times recorded are Greenwich mean time, in which midnight equals 0 or 24 hours.

TABLE I.—SUMMARY OF EXPLOSIONS OF FIRE-DAMP OR COAL-DUST IN THE SEVERAL MINES-INSPECTION DISTRICTS DURING 1902.

Mines-inspection Districts.	Fatal Accidents.			Non-fatal Accidents.	
	No.	Deaths.	Injured.	No.	Injured.
Cardiff	2	9	1	5	6
Durham	1	1	1	3	5
Ireland	0	0	0	0	0
Liverpool	4	14	9	2	6
Manchester	0	0	0	0	0
Midland	0	0	0	12*	11
Newcastle-upon-Tyne ...	2	4	0	8	13
Scotland, East	4	9	1	18	22
Do. West	8	4	1	45	57
Southern	1	16	18	0	0
Staffordshire	2	2	1	11	14
Swansea	2	2	0	24	32
Yorkshire	2	2	1	5	6
Totals	23	63	33	133*	172

* Including an explosion, by which no persons were injured.

TABLE II.—LIST OF FATAL EXPLOSIONS OF FIRE-DAMP OR COAL-DUST IN
COLLIERIES IN THE SEVERAL MINES-INSPECTION DISTRICTS DURING 1902.

1902.		Colliery.	Mines-Inspection Districts.	Deaths	No. of Persons Injured.
Mar.	3, 18·20	Pemberton	Liverpool	2	0
April	2, 20·15	Garswood Hall (No. 9 Sinking Pit) ...	Do. ...	9	2
,,	3, 9·30	Glencraig...	Scotland, East ...	4	0
,,	19, 8·30	Benarty	Do. ...	1	1
May	14, 6·0	Todmorden Moor ...	Yorkshire	1	0
,,	21, 7·30	Dinnington	Newcastle-upon-Tyne	1	0
June	4, 3·0	Fochriw	Cardiff	8	0
July	4, 20·15	Woodhorn	Newcastle-upon-Tyne	3	0
,,	8, 13·30	Elba	Swansea	1	0
,,	9, 8·30	Wrexham and Acton ..	Liverpool	1	6
,,	19, 7·45	Collena	Cardiff	1	1
..	26, 6·40	Shut End (No. 13 Pit) ...	Staffordshire ..	1	0
.,	26, 7·0	Cleland	Scotland, East ...	1	0
,,	29, 20·30	Darran	Swansea	1	0
Aug.	1, 7·30	Pumpherston (Oil-shale)	Scotland, East ...	3	0
,,	15, 14·40	Skellyton (No. 3 Pit) ...	Scotland, West ..	2	0
,,	23, 2·30	Park Mill	Yorkshire	1	1
Sept.	3, 23·30	McLaren	Southern	16	18
Oct.	3, 9·30	Blair (No. 2 Pit)	Scotland, West ...	1	1
,,	15, 8·30	Holmside (Oswald Pit)...	Durham	1	1
Nov.	25, 7·0	Prior's Lee (Stafford No. 2 Pit)	Staffordshire ...	1	1
Dec.	1, 15·30	East Plean (No. 3 Pit)...	Scotland, West ...	1	0
,,	9, 7·0	Orrell	Liverpool	2	1
				63	33

TABLE III.—LIST OF NON-FATAL EXPLOSIONS OF FIRE-DAMP OR COAL-DUST IN
COLLIERIES IN THE SEVERAL MINES-INSPECTION DISTRICTS DURING 1902.

1902.		Colliery.	Mines-Inspection Districts.	No. of Persons Injured.
Jan.	1, 5·0 ...	Cowdenbeath	Scotland, East ...	1
,,	1, 19·30...	Gospel Oak	Staffordshire	1
,,	1, 20·30...	Netherton (Bridge Pit) ...	Do. ...	1
,,	2, 10·45...	Emley Moor	Yorkshire	1
,,	6, 7·30...	Gateside	Scotland, West ...	1
,,	8, 5·30...	East Plean	Do. ...	1
,,	10, 8·30...	Kelty	Scotland, East ...	1
,,	18. 18·15...	Ellistown	Midland	1
,,	23, 6·30...	Fairlie (No. 3 Pit) ...	Scotland, West ..	3
,,	24, 8·15...	Shakenhurst ...	Staffordshire	2
,,	24, 21·30...	Roseball (No. 5 Pit)	Scotland, West ...	2
Feb.	6, 10·0 ...	Eglinton (No. 1 Pit) ..	Do. ...	1
,,	7, 7·30...	Cymmer Glyncorrwg ...	Swansea	1
,,	9, 14·30...	Clydach Merthyr ...	Do. ...	1
,,	10, 11·30 ...	Cwm	Cardiff	1
,,	13, 16·30...	Glencraig	Scotland, East ...	1
,,	14, 16·0 ...	Ross (No. 1 Pit) ...	Scotland, West ...	1
,,	17, 7·30...	Horseley	Staffordshire ...	1
,,	18, 3·0 ...	Lochwood (No. 3 Pit) ...	Scotland, West ...	1
,,	19, 3·30...	Glangarnant ...	Swansea	3
,,	19, 19·30...	Wharncliffe Gannister ...	Yorkshire	1
,,	23, 23·45...	Canderigg (No. 1 Pit)	Scotland, West ...	1
,,	24, 0·30 ...	Garw Fechan (No. 2 Pit)	Swansea	1
,,	24, 22 30...	Tanfield Lea	Newcastle-upon-Tyne	1
Mar.	7, 16·0 ...	Pantmawr	Swansea	2

TABLE III.—*Continued.*

1902.		Colliery.	Mines-Inspection Districts.	No. of Persons Injured.
Mar.	8, 9·0 ...	Blairhall	Scotland, East ...	1
„	10, 5·45...	Greenhill	Do. ...	1
„	11, 12·30...	Dunnikier	Do. ...	2
„	17, 8·0 ...	Maypole	Liverpool ...	5
„	17, 9·30...	Noyadd	Swansea ...	1
„	20, 12·0 .	Woodhall	Scotland, West	1
„	21, 18·30...	Noyadd	Swansea	1
„	21, 22·0 ...	Allanton (No. 2 Pit) ...	Scotland, West	1
„	26, 10·0 ...	Allanton (No. 2 Pit) ...	Do.	1
„	29, 7·0 ...	Greenfield (No. 7 Pit) ...	Do. ...	2
April	14, 20·0 ...	Duffryn Rhondda	Swansea ...	1
„	14, 20·0 ...	Nethercroy (No. 3 Pit) ...	Scotland. West ..	1
„	15, 14·30...	Rigfoot	Do. ...	1
„	16, 13·0 ...	Lochwood (No. 3 Pit) ...	Do. ...	1
„	19, 20·0 ...	Clydach Merthyr	Swansea ...	2
„	21, 12·45..	Gartshore (No. 9 Pit) ...	Scotland, West ...	1
„	30, 8·30...	Blackwell	Midland ..	1
May	2, 23·30...	Crigglestone	Yorkshire ...	2
„	6, 6·30...	Mud Hall	Staffordshire	1
„	9, 6·30...	Greenfield (Threestonehill No. 7 Pit)	Scotland, West	1
„	9, 15·0 ...	Birchrock	Swansea ...	1 ·
„	14, 11·15..	Coneygre (Lye Cross Pit)	Staffordshire ...	1
„	16, 7·0 ...	Dillwyn	Swansea ...	1
„	17, 6·30...	Caenewydd	Do. ...	1
„	26, 14·0 ...	Monkland	Scotland, East ..	1
June	3, 10·0 ...	Donisthorpe	Midland ...	1
„	3, 18·0 ...	Braidhurst	Scotland, East ...	1
„	10, 4·0 ...	Houldsworth	Scotland, West ...	1
„	11, 16·30...	Stargate	Newcastle-upon-Tyne	1
„	21, 10·30...	Kiltou (ironstone-mine) ...	Durham ...	1
„	23, 13·30...	Diamond	Swansea ...	2
„	25, 21·0 ...	Skellyton (No. 3 Pit) ...	Scotland, West	2
„	28, 7·15...	Glencraig	Scotland, East ...	1
July	2, 14·30...	Morriston	Swansea ...	1
„	4, 7·20...	Avonhead	Scotland, East ...	1
„	5, 8·50...	Clifton	Midland ...	1
„	8, 6·30...	Swalwell	Newcastle-upon-Tyne	1
„	8, 8·30...	Pantycelyn	Swansea ...	1
„	8, 15·30..	Dunston	Newcastle-upon-Tyne	4
„	9, 6·15...	Machan	Scotland, West ...	1
„	11, 22·30...	Portland	Midland ...	1
„	14, 8·0 ...	Wester Gartshore (No. 1 Pit)	Scotland, West	1
„	21, 3·0 ...	Blaenclydach	Cardiff	2
„	23, 15·30..	Barglachan (No. 2 Pit) ...	Scotland, West	1
„	26, 6·15 ..	Grange	Staffordshire ...	1
„	29, 12·15...	South Hiendley	Yorkshire ...	1
„	30, 15·30...	Griff	Midland ...	1
Aug.	4, 11·0 ...	Cadder (No. 16 Pit) ...	Scotland, West ...	1
„	6, 14·45...	Cousland (Oil-shale) ...	Scotland, East ...	1
„	6, 23·30 ..	Hook	Swansea ...	1
„	8, 14·0 ...	Eston (ironstone-mine) ...	Durham ...	3
„	16, 13·0 ...	Bowhill (No. 2 Pit) ...	Scotland, West	1
„	17, 21·0 ...	Greenfield (Threestonehill No. 7 Pit)	Scotland, West ...	1
„	18, 15·30 .	Wood Farm (No. 4 Pit) ..	Staffordshire	2
„	22, 8·30...	Meiros	Cardiff	1
„	22, 11·30...	Portland (No. 4 Pit) ...	Scotland, West	2
„	22, 15·0 ...	Dalziel	Scotland, East ...	4

TABLE III.—*Continued.*

1902.		Colliery.	Mines-Inspection Districts.	No. of Persons Injured.
Aug.	23, 5·30...	Dalziel	Scotland, East ...	1
,,	23, 7·10...	Ibstock	Midland	1
,,	25, 1·30...	Ellismuir (No. 3 Pit)	Scotland, West ...	2
,,	26, 11·30...	Wharncliffe Gannister ..	Yorkshire	1
Sept.	3, 8·0 ...	Birkrigg (No. 2 Pit) ...	Scotland, West ...	1
,,	3, 9·0 ...	Rigfoot	Do. ..	1
,,	3, 12·0 ...	Thankerton (No. 4 Pit)	Do. ...	2
,,	6, 7·30...	Dumbreck (No. 2 Pit) ..	Do. ...	1
,,	8, 6·0 ...	Main Coal	Liverpool	1
,,	10, 14·30...	Noyadd	Swansea	1
,,	15, 20·0 ...	Onllwyn	Do. ...	1
,,	16, 9·30...	Pye Hill	Midland	1
,,	16, 19·30...	Birchrock	Swansea	2
,,	19, 7·30...	Law	Scotland, East ...	1
,,	22, 9·0 ...	Gartsherrie (Gartcloss Pit)	Scotland, West ...	1
,,	23, 14·30...	Shawsburn	Do. ...	1
,,	29, 6·30...	Arniston	Scotland, East ..	1
Oct.	6, 21·0 ...	Hazlerigg	Newcastle-upon-Tyne	1
,,	7, 16·0 ...	Onllwyn	Swansea	1
,,	8, 5·45...	Wallyford	Scotland, East ...	1
,,	9, 13·0 ...	Donisthorpe	Midland ...	1
,,	9, 16·15...	Tower	Cardiff	1
,,	11, 8·0 ...	Donisthorpe	Midland	1
,,	14, 9·30...	Ynyscedwyn	Swansea	2
,,	15, 23·30...	Backworth	Newcastle-upon-Tyne	3
,,	21, 10·0 ...	Exhall	Midland	0
,,	23, 3·0 ...	Duffryn Rhondda ...	Swansea	1
,,	24, 12·0 ...	Oldnall	Staffordshire	1
,,	26, 10·0 ...	Houldsworth	Scotland, West ...	2
,,	31, 7·30...	Tannochside (No. 3 Pit) .	Do.	2
Nov.	6, 2·0 ...	Bogleshole (No. 4 Pit) ...	Do.	1
,,	6, 9·0 ...	Ross (No. 1 Pit) ...	Do.	1
,,	8, 19·45...	Wallyford	Scotland, East ...	1
,,	18, 0·0 ...	Cadder (No. 17 Pit) ...	Scotland, West ...	1
,,	20, 12·45...	Swanwick	Midland	1
,,	20, 15·30...	Meiklehill (No. 5 Pit)	Scotland, West ...	2
,,	27, 11·0 ...	Wellshot (No. 2 Pit) ..	Do.	1
,,	28, 8·0 ...	Cwmavon	Swansea	2
,,	28, 9·30 .	Bourtreehill (Capringstone No. 7 Pit) ...	Scotland, West ...	1
,,	28, 21·0 ..	Pumpherston (Oil-shale) ..	Scotland, East ...	1
,,	29, 9·0 ..	Ellismuir (No. 3 Pit) ...	Scotland, West ...	2
Dec.	1, 22·0 ..	Goatfoot	Do.	1
,,	5, 7·30...	Maritime Level	Cardiff	1
,,	5, 21·0 ..	Seaton Delaval	Newcastle-upon-Tyne	1
,,	9, 7·15...	Waunycoed	Swansea	1
,,	9, 20·30...	Grange	Staffordshire	1
,,	12, 7·30...	Westburn·(No. 2 Pit) ...	Scotland, West ...	1
,,	16, 9·15...	North Elswick	Newcastle-upon-Tyne	1
,,	17, 14·0 ...	Bentilee	Staffordshire	2
,,	29, 18·0 ...	Lochwood (No. 3 Pit) ..	Scotland, West ...	1
,,	30, 17·30 .	Brusselton	Durham	1
				172

TABLE IV.—BAROMETER, THERMOMETER, ETC.. READINGS, 1902.

JANUARY, 1902.

	KEW								GLASGOW						
	BAROMETER				TEMPERATURE		Direction of wind at noon		BAROMETER				TEMPERATURE		Direction of wind at noon
Date	4 A.M.	10 A.M.	4 P.M.	10 P.M.	Max	Min		Date	4 A.M.	10 A.M.	4 P.M.	10 P.M.	Max	Min	
1	29·772	29·834	29·651	29·307	52·0	43·3	SW	1	29·491	29·462	29·211	28·765	44·9	39·8	SSW
2	29·189	29·329	29·448	29·663	52·3	45·1	W	2	28·779	28·916	29·192	29·444	47·0	40·9	W
3	29·841	29·986	29·983	30·001	52·3	39·0	SW	3	29·617	29·585	29·572	29·576	51·1	43·2	SW
4	29·963	29·913	29·819	29·942	52·2	43·0	W	4	29·537	29·503	29·454	29·426	48·6	40·8	WSW
5	30·006	30·117	30·281	30·437	47·3	38·6	W	5	29·615	29·995	30·109	30·060	46·4	40·1	WNW
6	30·421	30·451	30·456	30·550	50·1	39·2	W	6	30·028	30·147	30·233	30·270	49·9	45·4	WNW
7	30·583	30·629	30·618	30·602	46·6	39·4	W	7	30·272	30·217	30·283	30·291	49·4	45·5	W
8	30·565	30·536	30·442	30·352	44·3	41·6	W	8	30·256	30·259	30·138	30·006	46·9	42·9	SW
9	30·257	30·217	30·162	30·156	50·7	41·0	SW	9	29·813	29·666	29·6·5	29·758	48·1	43·8	WSW
10	30·121	30·069	30·043	30·058	52·8	47·7	SW	10	29·756	29·774	29·782	29·900	44·8	36·0	WNW
11	30·039	30·128	30·170	30·181	50·7	41·6	NE	11	29·985	30·102	30·135	30·125	37·0	32·3	N
12	30·125	30·099	30·048	30·097	49·0	41·0	S	12	30·006	29·967	29·971	30·157	39·4	29·2	W
13	30·179	30·298	30·352	30·440	43·1	33·7	NE	13	30·248	30·347	30·415	30·506	33·4	23·7	ESE
14	30·506	30·619	30·673	30·792	35·7	28·3	NE	14	30·565	30·652	30·647	30·675	28·9	20·8	SW
15	30·839	30·848	30·785	30·720	38·0	24·1	W	15	30·558	30·508	30·410	30·462	47·6	25·3	W
16	30·655	30·665	30·681	30·612	44·3	39·0	W	16	30·440	30·494	30·504	30·536	49·1	46·3	W
17	30·600	30·532	30·440	30·379	44·8	40·3	W	17	30·472	30·410	30·305	30·275	46·6	39·9	W
18	30·330	30·346	30·330	30·377	44·0	31·1	N	18	30·245	30·252	30·223	30·215	42·1	38·3	SW
19	30·398	30·438	30·370	30·341	41·8	27·0	N	19	30·146	30·063	29·973	29·833	44·6	39·4	W
20	30·269	30·264	30·231	30·279	49·7	41·6	WSW	20	30·735	29·794	29·892	29·966	49·3	43·7	W
21	30·292	30·324	30·323	30·372	51·1	45·2	W	21	29·980	30·035	30·065	30·068	48·6	43·1	WSW
22	30·342	30·330	30·267	30·268	51·9	47·5	W	22	30·060	30·102	30·085	30·023	49·1	42·8	W
23	30·230	30·203	30·093	29·959	48·2	42·2	WSW	23	29·973	29·904	29·729	29·550	48·5	45·1	SW
24	29·694	29·431	29·256	29·207	47·0	36·0	S	24	29·232	29·056	28·971	28·960	45·9	32·5	WSW
25	29·116	29·197	29·275	29·446	39·2	31·9	W	25	28·943	28·978	29·164	29·472	38·1	30·7	W
26	29·673	29·811	29·813	29·722	38·7	32·1	W	26	29·575	29·622	29·420	29·578	32·1	25·9	NW
27	29·547	29·443	29·505	29·486	47·6	35·0	W	27	29·463	29·454	29·339	29·150	30·3	25·8	E
28	29·274	29·198	29·214	29·562	44·5	35·5	SW	28	29·029	29·154	29·443	29·670	36·9	28·3	NNE
29	29·785	30·007	30·120	30·180	38·0	29·6	NNW	29	29·814	29·840	30·066	30·282	34·3	24·1	N
30	30·263	30·417	30·516	30·619	38·1	29·9	N	30	30·433	30·588	30·673	30··21	28·5	16·3	N
31	30·629	30·595	30·373	30·330	37·5	34·3	NE	31	30·915	30 969	30·958	30·982	34·7	14·8	ENE

FEBRUARY, 1902.

Date	4 A.M.	10 A.M.	4 P.M.	10 P.M.	Max	Min	Dir.	Date	4 A.M.	10 A.M.	4 P.M.	10 P.M.	Max	Min	Dir.
1	30·628	30·595	30·373	30·330	38·8	32·5	NE	1	30·952	30·990	30·811	30·732	37·9	28·4	E
2	30·190	30·162	30·072	30·065	34·9	31·5	NE	2	30·607	30·538	30·436	30·391	40·0	36·9	E
3	30·050	30·044	30·022	30·074	35·3	32·5	NE	3	30·303	30·277	30·198	30·229	41·3	37·1	E
4	30·073	30·111	30·094	30·096	35·6	32·8	NE	4	30·209	30·212	30·11²	30·061	38·9	34·1	ESE
5	30·043	30·020	29·889	29·797	36·4	32·2	ESE	5	29·967	29·870	29·72	29·596	35·3	28·9	SW
6	29·620	29·514	29·403	29·366	38·3	32·9	NNE	6	29·464	29·403	29·366	29·326	38·6	33·8	SW
7	29·348	29·413	29·441	29·455	38·8	31·7	N	7	29·356	29·419	29·417	29·418	36·1	29·5	ENE
8	29·380	29·328	29·352	29·338	37·3	32·0	ESE	8	29·398	29·435	29·420	29·380	33·3	24·2	ENE
9	29·363	29·478	29·475	29·523	38·9	30·1	W	9	29·401	29·329	29·369	29·485	38·1	24·8	W
10	29·531	29·547	29·520	29·634	34·4	27·3	NE	10	29·529	29·520	29·526	29·547	35·0	26·1	W
11	29·704	29·771	29·790	29·856	39·2	24·4	W	11	29·557	29·649	29·72²	29·782	33·2	26·4	W
12	29·884	29·940	29·911	29·921	37·0	24·0	N	12	29·907	29·973	29·84	29·892	34·7	26·2	NW
13	29·887	29·913	29·896	29·969	37·2	20·6	N	13	29·907	29·973	29·984	30·040	34·1	20·3	SSE
14	29·992	30·073	30·088	30·135	33·7	27·4	N	14	30·003	29·941	29·967	30·083	41·1	18·3	W
15	30·139	30·246	30·322	30·399	36·1	22·7	NNE	15	30·136	30·163	30·11²	30·078	39·5	26·9	SSW
16	30·377	30·296	30·160	30·061	36·2	17·7	ESE	16	29·991	29·952	29·87	29·865	40·1	35·1	SSW
17	29·927	29·929	29·900	29·976	40·0	21·0	NE	17	29·832	29·886	29·904	30·000	38·9	30·4	N
18	29·987	30·036	30·011	30·037	36·0	31·9	NE	18	30·010	30·032	29·974	29·982	37·1	28·6	ENE
19	30·029	30·059	30·049	30·087	36·4	33·7	ESE	19	29·934	30·010	30·02²	30·076	40·3	35·4	E
20	30·073	30·081	30·046	30·069	37·1	33·9	E	20	30·097	30·109	30·05	30·010	38·3	35·6	S
21	30·051	30·065	30·017	30·022	47·0	34·3	E	21	29·934	29·912	29·85	29·827	37·4	34·3	E
22	29·971	29·945	29·850	29·807	49·1	33·1	SSE	22	29·739	29·703	29·630	29·568	42·5	37·2	ESE
23	29·745	29·759	29·741	29·726	50·3	43·7	S	23	29·495	29·560	29·547	29·515	48·0	38·2	NNE
24	29·668	29·632	29·607	29·638	48·0	43·8	SE	24	29·497	29·509	29·51²	29·566	49·8	44·3	SE
25	29·627	29·628	29·613	29·634	47·7	38·9	E	25	29·588	29·629	29·605	29·617	46·9	40·6	ENE
26	29·594	29·536	29·433	29·275	47·1	37·8	E	26	29·580	29·562	29·436	29·341	42·1	33·3	E
27	29·226	29·248	29·252	29·318	49·8	42·0	SW	27	29·206	29·189	29·179	29·216	37·4	33·9	E
28	29·350	29·478	29·493	29·554	52·6	39·2	SW	28	29·214	29·233	29·29⁰	29·380	38·9	34·9	E

MARCH, 1902.

	KEW.							GLASGOW.						
	BAROMETER.				TEMPERATURE.		Direction of wind at noon.		BAROMETER.				TEMPERATURE.	Direction of wind at noon.
Date.	4 A.M.	10 A.M.	4 P.M.	10 P.M.	Max.	Min.		Date.	4 A.M.	10 A.M.	4 P.M.	10 P.M.	Max. Min.	
1	29·557	29·602	29·629	29·727	51·6	35·6	SW	1	29·416	29·484	29·506	29·555	47·2 36·3	SSE
2	29·771	29·862	29·881	29·967	53·1	34·0	SSE	2	29·582	29·660	29·736	29·780	47·4 41·9	SW
3	29·986	30·012	29·999	30·040	51·3	37·2	SW	3	·758	·714	29·655	·719	46·9 39·1	S
4	30·037	30·060	30·020	30·074	46·9	37·2	SE	4	·759	·825	29·857	·897	49·9 43·3	SW
5	30·063	30·095	30·043	30·060	47·0	35·2	E	5	9·870	·873	29·929	·931	47·6 42·2	WSW
6	30·024	30·038	29·934	29·937	54·9	31·7	NW	6	·876	·877	29·930	·833	46·5 42·9	WSW
7	29·904	29·928	29·903	29·947	51·9	30·0	W	7	·831	·879		·871	52·1 43·6	NNW
8	29·938	29·888	29·822	29·775	51·3	42·7	W	8	·768	·686		·459	48·1 43·3	W
9	29·702	29·768	29·839	29·906	55·1	46·9	NW	9	·545	·687	29·873	·732	48·4 42·7	W
10	29·895	29·960	29·963	30·000	49·7	43·7	E	10	·765	·840	29·947	·929	51·2 42·3	W
11	29·988	30·006	29·988	30·013	49·9	40·6	SSW	11	·948	·987		·942	46·4 39·9	E
12	30·020	30·039	29·991	30·035	53·7	40·7	S	12	·882	·834		·795	47·9 39·6	SSW
13	30·031	30·050	30·012	30·045	55·1	35·8	S	13	·738	9·645		29·780	48·1 36·9	SW
14	30·024	29·990	29·831	29·535	53·0	38·2	SSW	14	·742	29·709		29·512	48·4 39·0	SW
15	29·595	29·817	29·903	30·014	52·8	42·8	WNW	15	29·540	·641		·860	47·2 38·2	WNW
16	30·035	30·120	30·133	30·207	54·2	40·9	NW	16	29·940	·029		·012	50·5 37·3	W
17	30·195	30·218	30·170	30·172	59·2	39·9	W	18	30·007	·002		·891	51·3 45·8	SW
18	30·101	30·091	30·008	29·989	50·6	43·0	SW		29·769	·684		·694	49·1 42·9	SW
19	29·931	29·911	29·793	29·674	55·1	41·8	W		29·660		49	·905	44·9 38·4	WSW
20	29·473	29·363	29·284	29·305	49·7	40·0	WSW		28·916		28·905	·917	44·6 33·6	W
21	29·291	29·271	29·228	29·250	50·1	39·9	SW		28·903		28·952	·009	42·3 33·8	WSW
22	29·248	29·263	29·294	29·367	46·6	35·7	NW		24·988		29·047	·142	44·8 34·2	WSW
23	29·367	29·416	29·442	29·540	49·1	31·8	W		29·202		29·349	·441	44·1 34·6	N
24	29·578	29·551	29·288	29·082	46·6	32·5	S		29·397			·243	44·1 29·4	ESE
25	29·196	29·521	29·697	29·846	48·6	35·2	NW	25	29·606		2	·659	43·4 31·3	WNW
26	29·899	29·964	29·953	29·851	47·4	32·6	W	26	29·670		29·705	·643	41·8 31·5	W
27	29·663	29·723	29·736	29·854	57·7	39·8	W	28	29·575		29·643	·719	46·0 34·1	ENE
28	29·882	29·911	29·911	29·968	56·3	43·7	W		29·742		29·800	29·772	47·1 37·5	NNE
29	29·874	29·735	29·577	29·799	55·8	41·4	SW		29·581		29·575	29·796	49·5 38·8	N
30	29·917	29·946	29·893	29·925	45·0	36·5	SW	30	29·815		29·805	·767	43·4 33·2	W
31	29·878	29·882	29·796	29·737	60·0	43·2	W	31	29·633		29·525	·463	50·1 37·1	W

APRIL, 1902.

1	29·621	29·576	29·535	29·665	55·1	42·0	SW	1	29·363	29·338	29·418	29·531	45·9 37·4	W
2	29·727	29·817	29·815	29·849	50·1	38·6	NNE	2	29·592	29·680	29·663	29·621	48·1 34·9	WNW
3	29·813	29·795	29·689	29·680	52·7	36·9	SW	3	29·502	29·481	29·415	29·481	42·8	W
4	29·812	29·893	29·926	29·961	51·9	35·7	W	4	29·440	29·597	29·702	29·770	48·3	WNW
5	29·931	29·839	29·708	29·737	52·7	37·9	SW	5	29·777	29·812	29·873	29·974	46·4	ESE
6	29·855	30·016	30·115	30·230	47·7	35·7	N	6	30·071	30·151	30·190	30·252	46·8	ESE
7	30·275	30·317	30·285	30·285	45·1	32·1	NE	7	30·267	30·285	30·237	30·224	45·6	ESE
8	30·232	30·227	30·150	30·156	44·0	36·0	E	8	30·182	30·171	30·124	30·159	46·3 1·	NNE
9	30·129	30·139	30·091	30·114	47·4	35·3	ENE	9	30·167	30·185	30·145	30·161	52·3	ENE
10	30·074	30·058	29·995	29·996	46·2	36·3	NE	10	30·176	30·181	30·117	30·160	52·4	E
11	29·915	29·830	29·742	29·731	48·1	36·6	E	11	30·109	30·127	29·893	29·871	44·2	ESE
12	29·709	29·733	29·740	29·804	52·3	40·2	S	12	29·779	29·740	29·685	29·696	43·0 7·	ENE
13	29·850	29·931	29·920	29·979	56·3	35·8	NW	13	29·690	29·694	29·674	29·735	45·0	S
14	29·970	29·964	29·842	29·817	55·0	32·8	E	14	29·768	29·800	29·790	29·787	52·6	S
15	29·704	29·615	29·375	29·667	59·7	44·3	E	15	29·770	29·756	29·671	29·666	52·1	E
16	29·777	29·851	29·849	29·871	59·0	44·2	NW	16	29·706	29·799	29·817	29·848	55·3	NNW
17	29·873	29·913	29·915	29·982	60·3	40·9	W	17	29·826	29·836	29·833	29·893	54·8	NW
18	29·991	30·016	29·950	29·963	61·0	40·3	S	18	29·882	29·888	·845	29·836	56·7	SSE
19	29·889	29·848	29·804	29·870	65·6	42·3	SSE	19	29·760	29·716	·602	29·622	56·1	E
20	29·872	29·900	29·921	29·969	61·1	46·0	SW	20	29·661	29·636	·637	29·631	58·1 4·	SE
21	30·009	30·036	29·968	29·849	60·2	47·3	S	21	29·642	29·663	·680	29·601	56·0 4·	WSW
22	29·694	29·527	29·542	29·648	58·3	47·7	S	22	29·378	29·272	169	29·284	55·8	ESE
23	29·711	29·767	29·827	29·911	60·2	45·0	SSW	23	29·400	29·528	611	29·716	55·5	SSW
24	29·982	30·070	30·078	30·124	63·4	41·7	S	24	29·776	29·886	·951	30·011	55·7	SW
25	30·092	30·040	29·954	29·927	64·5	38·7	ESE	25	30·038	30·058	·992	30·042	59·6	NE
26	29·861	29·852	29·808	29·862	55·3	44·4	ENE	26	30·068	30·107	·106	30·160	51·8	E
27	29·860	29·909	29·926	30·017	55·1	43·6	ENE	27	30·176	30·222	·246	30·281	46·9	E
28	30·038	30·087	30·071	30·153	52·7	38·7	NE	28	30·302	30·298	·262	30·273	52·2	ENE
29	30·171	30·209	30·173	30·158	51·0	35·9	N	29	30·256	30·199	·086	29·991	53·7	WNW
30	30·085	29·990	29·884	29·815	55·2	40·0	WNW	30	29·934	29·845	·712	29·653	50·8	WSW

MAY, 1902.

	KEW.								GLASGOW.						
	BAROMETER.				TEMPERATURE.		Direction of wind at noon.		BAROMETER.				TEMPERATURE.		Direction of wind at noon.
Date.	4 A.M.	10 A.M.	4 P.M.	10 P.M.	Max	Min.		Date.	4 A.M.	10 A.M.	4 P.M.	10 P.M.	Max	Min.	
1	29·672	29·691	29·705	29·800	54·7	40·2	NW	1	29·627	29·678	29·676	29·666	53·1	39·9	NW
2	29·798	29·747	29·663	29·659	57·8	35·9	W	2	29·554	29·474	29·424	29·425	51·8	39·5	W
3	29·623	29·580	29·515	29·665	54·1	41·2	W	3	29·404	29·480	29·572	29·713	50·0	35·8	N
4	29·753	29·884	29·929	30·040	53·3	39·3	N	4	827	29·905	29·906	29·881	51·6	38·0	NW
5	30·036	29·999	30·045	30·163	52·8	38·0	WNW	5	952	30·056	30·064	30·174	48·9	36·4	WNW
6	30·158	30·200	30·159	30·170	49·3	35·3	NW	6	·202	30·203	30·161	30·248	48·6	33·1	NW
7	30·158	30·187	30·169	30·224	49·9	34·9	N	7	·283	30·287	30·271	30·316	55·1	38·0	NNE
8	30·190	30·205	30·188	30·247	51·1	38·9	N	8	·316	30·321	30·287	30·270	51·7	37·5	NE
9	30·210	30·175	30·095	30·137	50·9	36·4	N	9	29·207	30·190	30·175	30·244	51·4	41·1	E
10	30·131	30·125	30·063	30·059	51·0	34·8	N	10	30·231	30·160	30·055	30·010	54·7	32·8	W
11	30·014	29·991	29·921	29·898	53·3	40·0	NE	11	29·952	29·917	29·875	29·851	53·1	41·9	N
12	29·833	29·826	29·816	29·913	54·3	39·9	N	12	29·902	29·902	29·935	29·981	51·1	41·3	NE
13	29·922	29·919	29·896	29·918	48·4	35·9	N	13	29·947	29·933	29·873	29·856	50·3	36·3	N
14	29·571	25·881	29·857	29·874	52·1	32·3	WNW	14	29·888	29·873	29·823	29·758	48·4	35·8	NNW
15	29·801	29·744	29·631	29·592	55·3	34·2	S	15	29·641	29·571	29·524	29·575	45·1	39·8	ESE
16	29·571	29·640	29·536	29·500	55·4	48·4	WSW	16	29·571	29·496	29·363	29·287	46·4	37·7	SSW
17	29·467	29·452	29·295	29·253	57·2	44·2	WSW	17	29·106	29·161	29·126	29·275	54·2	41·2	WNW
18	29·291	29·518	29·558	29·598	54·6	43·1	W	18	29·408	29·495	29·604	29·761	49·5	37·6	ENE
19	29·680	29·774	29·821	29·881	53·0	41·3	NNW	19	29·825	29·863	29·890	29·948	51·1	39·8	NNW
20	29·903	29·961	30·033	30·124	49·9	41·0	NNW	20	29·996	30·044	30·091	30 179	58·2	35·2	NNE
21	30·186	30·261	30·293	30·335	55·6	43·1	N	21	30·243	30·280	30·227	30·223	53·4	35·0	W
22	30 322	30·313	30·247	30·177	55·9	42·9	W	22	30·126	30·053	30·070	30·075	52·3	43·6	W
23	30·170	30·234	30·293	3 ·363	60·3	51·1	NNE	23	30·137	30·198	30·261	30·322	59·4	50·2	W
24	30·365	30·388	30·364	30·397	70·0	49·4	NW	24	30·320	30·295	30·236	30·266	57·8	49·6	WSW
25	30·395	30·435	30·383	30·382	66·4	51·1	NNW	25	30·276	30·284	30·243	30·239	54·4	44·8	WNW
26	30·354	30·329	30·232	30·186	68·5	47·1	W	26	30·208	30·142	30·005	29·965	52·1	47·9	WSW
27	30·126	30·064	29·929	29·888	69·8	50·0	WSW	27	29·968	29·754	29·559	29·417	57·3	49·1	SW
28	29·769	29·720	29·695	29·717	65·1	48·9	WSW	28	29·333	29·343	29·410	29·511	52·8	43·9	W
29	29·739	29·817	29·800	29·792	62·6	49·9	W	29	29·568	29·702	29·772	29·826	55·7	38·9	WNW
30	29·684	29·652	29·594	29·574	63·3	48·1	NE	30	29·826	29·829	29·820	29·860	52·4	41·4	E
31	29·605	29·668	29·695	29·715	70·2	54·1	SE	31	29·825	29·835	29·920	30·020	46·8	42·8	E

JUNE, 1902.

Date	4 A.M.	10 A.M.	4 P.M.	10 P.M.	Max	Min.	Dir.	Date	4 A.M.	10 A.M.	4 P.M.	10 P.M.	Max	Min.	Dir.
1	29·686	29·683	29·691	29·873	73·6	54·0	E	1	30·009	30·010	29·981	29·933	·2	43·4	E
2	29·9 7	30·082	30·068	30·085	70·	51·6	SSE	2	29·909	30·012	30·042	30·110	·1	45·9	E
3	30·058	30·052	30·034	30· 72	75·	51·4	NW	3	30·121	30·075	29·979	30·023	·1	47·5	E
4	30·064	30·023	30·030	30·061	62·	54·9	NW	4	30·001	29·976	29·918	29·901	·1	47·0	N
5	30 052	30·061	30·025	30·019	66·	52·9	WSW	5	29·858	29·841	29·803	29·787	·4	50·7	SW
6	29·967	29·883	29·697	29·639	59·	50·7	SW	6	29·732	29·658	29·519	29·471	·1	47·9	WNW
7	29 566	29·510	29·411	29·547	59·	47·3	WSW	7	29·427	29·505	29·514	29·567	9	43·9	E
8	29 586	29·635	29·660	29·715	57·	46·7	W	8	29·632	29·719	29·749	29·788	·8	45·9	ESE
9	29 730	29·752	29·758	29·807	53·	45·0	N	9	29·792	29·791	29·791	29·834	·2	43·8	NE
10	29 812	29·820	29·800	29·817	57·	40·1	N	10	29·834	29·803	29·723	29·719	·9	40·5	NW
11	29 794	29·781	29·732	29·718	60·	46·2	SSW	11	29·677	29·646	29·596	29·594	·1	43·9	WNW
12	29 599	29·562	29·454	29·371	56·	48·7	SE	12	29·532	29·517	29·483	29·496	·2	44·9	W
13	29 323	29·407	29·484	29·596	54·	48·2	W	13	29·500	29·547	29·579	29 587	·3	45·9	ENE
14	29 645	29·724	29·692	29·694	55·	48·3	NW	14	29·608	29·618	29·587	29·609	·7	45·3	WNW
15	29 672	29·681	29·676	29·699	57·	47·4	WSW	15	603	29·651	29·705	29·765	·0	47·9	RNE
16	29 685	29·703	29·713	29·737	57·	48·0	S	16	785	29·823	29·861	29·920	·5	48·1	E
17	29 8·8	29·947	29·992	30·083	61·	45·0	N	17	·942	30·0 0	30·034	30·086	·3	44·5	ESE
18	30 102	30·122	30·062	30·025	64·	45·8	ESE	18	076	30·060	29·975	29·919	·6	45·0	E
19	29 925	29·820	29·665	29·590	70·	48·7	SE	19	841	29·799	29·668	29 ·663	·9	47·0	E
20	29·524	29·576	29·608	29·687	64·6	51·3	SW	20	617	29·641	29·682	·621	·4	48·9	ENE
21	29·753	29·849	29·925	30·047	66·	52·8	WNW	21	754	29·798	29·864	925	·3	47·1	ENE
22	30·101	30·135	30·131	30·139	68·	49·8	SW	22	29·941	29·955	29·927	922	·3	45·6	SW
23	30·140	30·166	30·160	30·205	74·	54·8	SW	23	29·963	30·041	30·077	1	·8	54·1	WNW
24	30·225	30·232	30·205	30·229	78·	54·9	SSE	24	30·139	30·178	30·171	·1	·2	53·3	WNW
25	30·231	30·208	30·149	30·146	70·	58·2	E	25	30·171	30·141	30·107	1	·8	55·9	SSW
26	30·134	30·125	30·089	30·124	73·	54·0	ENE	26	30·151	30·165	30·110	1	5	58·6	SE
27	30·140	30·171	30·157	30·171	77·	56·2	E	27	30·214	30·243	30·171	1	2	48·1	NE
28	30·131	30·088	30·014	29·992	80·	57·2	E	28	30·180	30·179	30·128	1	51·7	52·1	E
29	29·934	29·931	29·956	30·017	77·	57·9	S	29	0·139	30·118	30·041	29·968	·1	50·7	E
30	29·996	29·981	29·880	29·873	80·	54·1	SE	30	29·959	29·925	29·864	29·857	71·9	50·6	W

JULY, 1902.

| | KEW | | | | | | | | GLASGOW | | | | | | |
| | Barometer | | | | Tempera-ture | | Direction of wind at noon | | Barometer | | | | Tempera-ture | | Direction of wind at noon |
Date	4 A.M.	10 A.M.	4 P.M.	10 P.M.	Max	Min		Date	4 A.M.	10 A.M.	4 P.M.	10 P.M.	Max	Min	
1	29·846	29·882	29·881	30·004	72·9	51·9	SW	1	29·867	29·952	29·985	30·067	63·2	49·8	E
2	30·083	30·198	30·230	30·286	61	49·0	SE	2	30·110	30·165	30·149	30·154	64·7	49·0	W
3	30·272	30·270	30·203	30·183	67·	46·6	SW	3	30·091	30·062	29·996	29·935	60·3	52·6	SW
4	30·123	30·110	30·067	30·124	76·	49·0	SW	4	29·837	29·862	29·926	30·029	67·1	56·3	WSW
5	30·146	30·181	30·139	30·178	78·	59·5	W	5	30·105	30·142	30·133	30·117	71·6	49·1	W
6	30·184	30·166	30·096	30·094	79·	59·4	S	6	30·070	30·011	29·988	29·992	64·6	53·5	W
7	30·075	30·098	30·093	30·129	78·	60·0	NW	7	29·982	30·015	30·050	30·068	60·1	53·9	W
8	30·149	30·149	30·069	30·063	81·	56·6	SW	8	30·042	29·999	29·878	29·857	59·9	53·5	WSW
9	30·000	29·933	29·812	29·759	71·	55·7	S N	9	29·781	29·676	29·548	29·501	61·1	53·2	W
10	28·698	29·646	29·641	29·797	63·	53·2	WSW	10	29·451	29·519	29·673	29·836	59·1	50·0	N
11	29·941	30·066	30·123	30·195	63·	51·3	WNW	11	29·948	30·042	30·070	30·093	60·3	44·3	W
12	30·201	30·187	30·121	30·126	70·	45·6	W	12	30·049	29·959	29·861	29·843	57·3	44·1	SW
13	30·101	30·107	30·050	30·064	77·	52·9	SW	13	29·855	29·892	29·897	29·923	61·9	53·9	SW
14	30·032	30·038	29·988	30·000	83·	52·4	SE	14	29·911	29·935	29·943	29·927	63·3	56·9	SW
15	29·981	29·968	29·945	30·000	81·	56·5	WSW	15	29·877	29·893	29·934	29·991	65·4	55·9	W
16	30·029	30·003	29·954	29·980	73·	57·3	N	16	29·9n7	29·995	29·957	29·927	61·7	52·1	SW
17	29·979	30·007	29·987	30·019	7·	53·7	NW	17	29·860	29·877	29·901	29·943	59·8	50·9	W
18	30·025	30·052	30·016	30·075	65·	50·9	NW	18	29·966	30·015	30·027	30·049	59·8	44·8	NW
19	30·067	30·031	29·926	29·895	64·	46·6	NW	19	30·015	29·549	29·809	29·977	59·9	47·9	WSW
20	29·815	29·846	29·862	29·910	57·	50·4	NE	20	30·008	30·017	29·982	29·997	59·2	45·7	N
21	29·875	29·923	29·958	29·988	56	49·7	NNW	21	29·979	29·966	29·890	29·858	61·8	45·6	SW
22	29·941	29·945	29·905	29·871	58·	50·8	SW	22	29·807	29·816	29·818	29·846	60·3	48·1	NE
23	29·859	29·903	29·900	29·895	69·	52·3	N	23	29·845	29·834	29·799	29·770	64·5	48·1	SW
24	29·849	29·816	29·796	29·839	67·	53·8	SW	24	29·710	29·714	29·714	29·758	59·2	46·3	W
25	29·860	29·906	29·894	29·859	68·1	55·0	SW	25	29·779	29·792	29·807	29·798	59·3	49·1	WSW
26	29·650	29·550	29·478	29·459	71·1	55·8	SSW	26	29·611	29·387	29·259	29·215	54·1	47·2	E
27	29·493	29·590	29·784	29·968	64·	51·4	W	27	27·299	29·441	29·614	29·614	59·8	50·2	NW
28	30·054	30·124	30·129	30·174	68·	49·7	WSW	28	29·839	29·902	29·938	29·899	61·3	47·3	W
29	30·152	30·148	30·122	30·141	68·	51·0	W	29	29·855	29·898	29·928	29·963	59·6	50·9	WNW
30	30·126	30·119	30·079	30·088	65·	50·8	W	30	29·927	29·959	29·969	30·006	59·3	49·2	NW
31	30·050	30·072	30·068	30·109	64·	53·0	WNW	31	30·025	29·969	30·079	30·106	58·4	45·2	NE

AUGUST, 1902.

Date	4 A.M.	10 A.M.	4 P.M.	10 P.M.	Max	Min	Wind	Date	4 A.M.	10 A.M.	4 P.M.	10 P.M.	Max	Min	Wind
1	30·122	30·158	30·109	30·108	64·8	50·0	N	1	30·096	30·094	30·025	29·949	60·8	45·0	W
2	30·03	29·997	29·928	29·873	66·5	46·	W	2	29·786	29·721	29·667	29·664	59·	50·2	WSW
3	29·825	29·843	29·855	29·875	65·8	52·9	NW	3	29·660	29·675	29·697	29·735	58·	49·8	W
4	29·863	29·858	29·846	29·84	64·6	52·4	W	4	29·749	29·764	29·773	29·836	64·	49·5	NW
5	29·848	29·898	29·896	29·945	73·6	55·4	S	5	29·866	29·921	29·932	29·960	58·	51·4	NE
6	29·831	29·821	29·737	29·729	67·9	57·2	S	6	29·919	29·886	29·835	29·890	54·	47·1	S
7	29·701	29·722	29·736	29·788	63·8	57·0	SW	7	29·802	29·824	29·834	29·863	54·	49·9	ESE
8	29·781	29·787	29·867	30·040	65·5	52·8	N	8	29·872	29·911	29·907	29·949	60·	48·2	W
9	30·105	30·136	30·133	30·130	60·8	48·3	WSW	9	29·965	30·012	30·008	29·959	59·	49·4	W
10	30·159	30·004	29·959	30·000	63·8	49·0	W	10	29·899	29·948	29·930	29·971	57·	48·0	W
11	29·994	30·039	30·032	30·047	60·6	44·4	NW	11	29·952	29·950	29·949	29·943	59·	43·9	WNW
12	29·983	29·989	29·998	30·028	61·6	49·7	NW	12	29·932	29·964	29·947	29·955	65·	44·6	W
13	29·998	29·961	29·942	29·908	63·6	48·7	W	13	29·918	29·896	29·867	29·869	61·	48·8	NE
14	29·878	29·896	29·892	29·941	68·9	56·6	WNW	14	29·865	29·879	29·879	29·919	65	50·3	SW
15	29·954	29·988	29·832	29·943	71·3	55·9	S	15	29·905	29·912	29·856	29·846	69	49·7	WSW
16	29·88	29·832	29·714	29·656	76·5	51·3	S	16	29·798	29·783	29·770	29·785	66	52·8	W
17	29·64	29·706	29·715	29·765	67·5	59·0	S	17	29·717	29·711	29·640	29·558	56	51·0	E
18	29·780	29·776	29·676	29·625	65·3	57·7	S	18	29·506	29·487	29·404	29·458	63	51·9	NNW
19	29·621	29·686	29·683	29·753	70·1	58·6	SW	19	29·474	29·572	29·609	29·657	59	49·7	E
20	29·81	29·853	29·726	29·951	65·9	50·6	SW	20	29·603	29·634	29·726	29·841	60	48·2	WNW
21	29·99	30·083	30·113	30·165	66·0	51·6	N	21	29·914	29·987	30·024	30·028	57	44·5	SW
22	30·18	30·191	30·127	30·114	71·6	47·4	SSW	22	29·955	29·910	29·849	29·754	65	50·9	SW
23	30·023	29·977	29·943	29·897	66·8	56·4	SW	23	29·678	29·706	29·702	29·695	65	56·2	SW
24	29·816	29·813	29·791	29·860	69·6	56·9	W	24	29·675	29·685	29·683	29·746	62	53·6	WSW
25	29·91	29·970	29·969	30·024	67·2	51·6	NW	25	29·787	29·853	29·867	29·920	62	51·8	W
26	30·02	29·995	29·924	29·901	68·9	51·0	SE	26	29·920	29·916	29·860	29·844	64	49·4	SW
27	29·83	29·841	29·824	29·895	68·5	50·1	SW	27	29·803	29·809	29·802	29·851	60·	47·3	E
28	29·91	29·911	29·864	29·865	74·3	49·0	E	28	29·852	29·841	29·796	29·794	64·	51·3	SW
29	29·806	29·770	29·675	29·637	76·5	55·3	E	29	29·760	29·760	29·795	29·795	60·	49·5	W
30	29·598	29·642	29·751	29·830	62·8	56·3	NW	30	29·785	29·777	29·752	29·781	59·	30·0	NNW
31	29·804	29·782	29·756	29·827	66·3	56·9	E	31	29·778	29·782	29·736	29·775	60·	46·5	ENE

SEPTEMBER, 1902.

	KEW.								GLASGOW.						
	BAROMETER.				TEMPERATURE.		Direction of wind at noon		BAROMETER.				TEMPERATURE.		Direction of wind at noon
Date.	4 A.M.	10 A.M.	4 P.M.	10 P.M.	Max	Min.		Date.	4 A.M.	10 A.M.	4 P.M.	10 P.M.	Max	Min.	
1	29·836	29·851	29·825	29·867	71·2	56·0	S	1	29·761	29·742	29·658	29·612	62·6	47·3	ENE
2	29·840	29·840	29·748	29·664	67·8	54·6	S	2	29·556	29·391	29·561	29·532	61·8	56·4	WSW
3	29·544	29·584	29·744	29·828	68·7	58·4	SW	3	29·271	28·986	·212	365	58·6	53·6	
4	29·832	29·875	29·889	29·925	68·6	55·9	SW	4	29·394	29·473	·654	835	·1	51·8	W
5	29·948	30·018	30·013	30·050	66·9	53·9	W	5	29·881	29·929	2·930	927	·7	47·3	W
6	30·049	30·054	30·025	30·084	67·3	48·7	SW	6	29·902	29·913	29·946	·038	·6	44·3	
7	30·100	30·170	30·166	30·218	68·1	44·8	SW	7	30·076	30·139	30·154	193	·4	·9	
8	30·214	30·215	30·146	30·127	68·2	45·6	E	8	30·203	30·202	30·158	·135	·9	·5	W
9	30·031	30·001	29·92	29·953	68·0	53·9	E	9	30·089	30·080	30·033	30·089	·9	·4	ENE
10	29·925	29·958	29·91	29·926	60·2	56·3	E	10	30·090	30·109	30·071	·070	·3	·8	E
11	29·853	29·835	29·77	29·75²	64·9	55·7	E	11	30·007	29·996	29·926	876	·6	·9	E
12	29·690	29·688	29·648	29·780	58·9	48·0	N	12	29·713	29·626	29·757	917	5·9	·8	NNE
13	29·921	30·022	29·997	29·969	55·2	41·0	NW	13	29·968	29·952	29·818	·704	53·3	·0	NNW
14	29·85	29·821	29·819	29·91	50·2	46·3	W	14	29·341	29·681	29·700	·721	58·5	·8	WNW
15	29·92	29·956	29·890	29·81	61·1	52·9	WSW	15	29·652	29·569	29·485	·415	54·7	·8	SW
16	29·72	29·694	29·710	29·77	63·7	50·1	W	16	29·341	29·355	29·482	·802	56·5	·8	W
17	29·81	29·911	29·976	30·10	59·7	46·0	NW	17	29·681	29·792	29·876	·998	56·7	·5	WNW
18	30·14	30·218	30·238	30·299	58·8	41·7	NW	18	30·066	30·117	30·135	·202	57·1	·9	W
19	30·326	30·376	30·336	30·352	62·6	37·0	SE	19	30·245	30·305	30·278	·273	55·4	·3	E
20	30·305	30·286	30·206	30·225	63·3	41·1	SE	20	30·207	30·193	30·111	·111	56·1	48·1	SW
21	30·202	30·222	30·151	30·137	66·4	49·7	E	21	30·074	30·075	30·032	·024	55·5	50·8	E
22	30·063	30·068	30·017	30·009	73·7	53·8	E	22	29·932	29·881	29·874	904	61·9	58·1	E
23	29·91	29·864	29·805	29·877	69·3	57·0	SE	23	29·798	29·757	29·644	·628	60·7	·7	SK
24	29·95	30·068	30·098	30·197	63·5	46·7	WSW	24	29·774	29·947	30·067	·158	59·7	·9	WSW
25	30·27	30·348	30·359	30·420	64·0	42·7	NNE	25	30·199	30·252	30·333	·347	56·4	·6	SW
26	30·42	30·451	30·396	30·426	65·3	40·6	ENE	26	30·374	30·394	30·	·389	60·4	·6	SW
27	30·39	30·416	30·385	30·386	65·6	40·1	NNE	27	30·385	30·364	30·	·422	63·1	·8	WSW
28	30·37	30·383	30·352	30·352	58·1	47·4	N	28	30·444	30·4?7	30·43	·435	61·3	49·1	N
29	30·272	30·211	30·130	30·094	55·3	46·2	NE	29	30·397	30·345	30·	·236	59·3	52·6	NW
30	30·001	29·950	29·859	29·849	56·2	47·4	NE	30	30·182	30·147	30·088	·074	59·1	52·6	ENE

OCTOBER, 1902.

1	29·821	29·805	29·792	29·819	57·4	50·2	NE	1	30·063	30·100	30·108	30·148	54·4	49·5	E
2	29·846	29·959	30·021	30·061	53·0	42·8	NE	2	30·17	30·200	30·246	30·303	53·1	44·4	ENE
3	30·094	30·110	30·081	30·093	47·0	42·8	NE	3	30·3	30·307	30·255	0·277	49·5	41·9	EB
4	30·074	30·081	30·008	29·998	47·6	43·7	NE	4	30·25	30·2·9	30·206	·210	52·1	·0	E
5	29·949	29·923	29·887	29·872	52·3	45·9	NE	5	30·16	30·151	30·102	·100	53·1	·2	ESE
6	29·814	29·796	29·780	29·792	51·4	46·6	N	6	30·02	29·980	29·921	·915	50·6	·4	E
7	29·767	29·790	29·789	29·832	56·0	45·0	NW	7	29·883	29·894	29·884	901	50·4	·8	E
8	29·839	29·885	29·844	29·859	56·7	41·8	N	8	29·895	29·915	29·880	·914	51·2	·3	ESE
9	29·788	29·672	29·548	29·471	55·1	46·4	ENE	9	29·636	29·888	29·802	·768	50·3	·8	E
10	29·397	29·474	29·478	29·477	64·3	53·0	S	10	29·651	29·601	29·561	·609	50·8	·3	NE
11	29·457	29·536	29·662	29·842	57·0	46·0	N	11	29·698	29·	29·755	·881	52·4	·1	WNW
12	30·005	30·180	30·238	30·240	58·1	46·0	N	12	29·949	30·0	29·953	·846	55·0	·9	S
13	30·218	30·181	30·108	30·011	61·0	49·7	SSW	13	29·892	29·9	29·859	·907	54·8	·9	WSW
14	29·832	29·880	29·880	29·919	59·2	48·6	W	14	29·680	29·65	29·613	·559	54·3	·3	W
15	29·816	29·656	29·492	29·450	59·0	48·2	SSW	15	29·186	28·9	28·773	28·683	54·1	·0	SW
16	29·408	29·494	29·530	29·628	56·0	43·5	W	16	28·981	29·1	29·283	·377	51·4	4·4	WNW
17	29·644	29·717	29·761	29·775	53·1	42·1	WNW	17	29·482	29·5	29·592	·578	52·2	·2	WNW
18	29·593	29·567	29·734	29·890	51·5	42·0	N	18	29·615	29·7	29·787	·850	49·4	7·5	ENE
19	29·982	30·024	29·953	29·870	53·9	34·9	N	19	29·851	29·	29·736	·637	50·3	·3	SW
20	29·823	29·721	29·719	29·783	60·1	45·9	SSW	20	29·530	29·55	29·605	·654	51·9	·9	W
21	29·894	30·030	30·085	30·142	55·9	41·5	W	21	29·691	29·7·4	29·810	·768	49·8	4·5	WSW
22	30·065	29·994	30·032	30·287	55·0	41·1	WNW	22	29·636	29·	30·070	·118	52·9	·3	W
23	30·327	30·451	30·454	30·507	54·1	40·5	W	23	30·130	17	30·187	·196	55·8	·2	SW
24	30·501	30·545	30·484	30·475	57·0	45·3	W	24	30·161	·6	30·099	·254	55·2	4·1	SW
25	30·445	30·455	30·393	30·345	59·8	45·9	W	25	30·312	·3	30·226	·079	52·0	·2	E
26	30·235	30·114	29·976	30·007	55·9	44·3	SSW	26	29·749		29·808	·964	56·0	·5	NW
27	30·010	30·050	30·064	30·105	54·0	41·7	N	27	30·020	·0	29·966	·941	56·6		NNE
28	30·102	30·122	30·107	30·130	55·7	39·4	SW	28	29·881	·9·2	29·933	·933	·9	·3	NE
29	30·101	30·108	30·043	30·061	54·7	41·9	S	29	29·855	·83	29·814	·831	6·3	·3	WSW
30	30·025	30·050	30·064	30·150	55·6	40·5	WNW	30	29·856	·9	30·018	·079	51·1	·8	WNW
31	30·170	30·210	30·156	30·122	51·9	36·0	WNW	31	30·008	·8	29·851	·859	55·0	1·1	WSW

NOVEMBER, 1902.

	KEW								GLASGOW					
	BAROMETER.				TEMPERA-TURE.		Direction of wind at noon.		BAROMETER.				TEMPERA-TURE.	Direction of wind at noon.
Date.	4 A.M.	10 A.M.	4 P.M.	10 P.M.	Max	Min.		Date.	4 A.M.	10 A.M.	4 P.M.	10 P.M.	Max	Min.	
1	30·098	30·104	30·064	30·158	57·8	45·7	WSW	1	29·825	29·862	30·012	30·150	54·5	42·9	WNW
2	30·205	30·269	30·211	30·192	53·6	43·6	E	2	30·197	30·197	30·100	30·037	46·7	34·8	NNE
3	30·127	30·119	30·077	30·109	51·1	37·9	NNW	3	29·946	29·911	29·960	30·021	51·4	39·9	SW
4	30·085	30·069	29·945	29·862	52·3	41·0	ESE	4	29·995	29·993	29·913	29·864	52·1	43·7	E
5	29·933	29·659	29·560	29·612	56·0	43·9	E	5	29·757	29·971	29·503	29·431	51·6	47·8	ESE
6	29·40	29·604	29·463	29·369	58·1	50·7	E	6	29·416	29·475	29·452	29·351	56·2	49·3	SSE
7	29·339	29·536	29·613	29·671	58·2	48·0	SW	7	29·191	29·151	29·246	29·390	56·1	46·0	SE
8	29·651	29·486	29·421	29·474	53·7	44·9	SSE	8	29·360	29·215	28·926	28·910	49·1	43·2	ESE
9	29·473	29·541	29·590	29·752	53·5	45·4	SW	9	28·872	29·034	29·245	29·419	51·1	44·2	W
10	29·915	29·887	29·870	29·859	52·6	42·9	SW	10	29·500	29·609	29·637	29·648	50·1	43·2	SW
11	29·77	29·678	29·586	29·605	53·5	44·0	E	11	29·619	29·584	29·440	29·362	49·4	43·2	ESE
12	29·13	29·931	30·029	30·121	55·8	36·0	SW	12	29·439	29·537	29·701	29·756	50·2	45·2	SW
13	30·39	30·168	30·127	30·188	52·4	31·8	WSW	13	29·747	29·765	29·715	29·912	54·4	46·3	SW
14	30·29	30·318	30·314	30·360	53·8	35·6	SSW	14	30·087	30·217	30·208	30·194	49·7	41·5	NNE
15	30·21	30·347	30·252	30·227	48·5	42·3	E	15	30·198	30·233	30·209	30·185	53·7	45·4	ESE
16	30·10	30·248	30·237	30·274	44·5	39·5	E	16	30·157	30·216	30·250	30·285	49·9	43·1	ESE
17	30·279	30·297	30·285	30·306	43·2	36·3	ENE	17	30·291	30·326	30·323	30·349	46·3	38·1	ENE
18	30·294	30·290	30·213	30·177	40·1	34·0	ENE	18	30·388	30·388	30·366	30·374	43·6	35·8	ENE
19	30·119	30·084	30·037	30·094	34·8	32·9	NE	19	30·326	30·287	30·251	30·268	41·8	35·3	E
20	30·216	30·158	30·198	30·279	35·6	29·9	N	20	30·298	30·275	30·250	30·250	43·9	37·0	ESE
21	30·078	30·250	30·143	30·137	38·0	29·9	E	21	30·202	30·158	30·054	30·012	39·8	32·7	E
22	30·018	30·114	30·040	29·991	41·3	34·1	SSE	22	29·932	29·881	29·803	29·741	38·3	32·4	ENE
23	29·017	29·951	29·950	29·218	57·8	37·6	SW	23	29·693	29·723	29·703	29·721	46·3	36·1	WSW
24	29·856	29·694	29·446	29·357	51·6	34·9	SE	24	29·665	29·546	29·370	29·293	44·3	36·9	E
25	29·269	29·220	29·164	29·218	51·7	48·7	SE	25	29·250	29·284	29·297	29·325	49·1	43·6	ESE
26	29·281	29·366	29·408	29·506	50·6	47·3	E	26	29·330	29·378	29·420	29·475	47·9	45·1	ESE
27	29·556	29·619	29·596	29·530	48·1	42·0	N	27	29·389	29·465	29·389	29·265	47·9	43	E
28	29·357	29·219	29·207	29·341	49·6	40·2	N SSW	28	29·097	29·051	28·974	28·995	47·5	40	E
29	29·339	29·364	29·383	29·396	50·0	39·0	SE	29	29·162	29·308	29·387	29·489	45·1	38	ESE
30	29·365	29·459	29·497	29·549	47·5	45·1	E	30	29·596	29·695	29·699	29·675	45·5	38	ESE

DECEMBER, 1902.

Date.	4 A.M.	10 A.M.	4 P.M.	10 P.M.	Max	Min.	Dir.	Date.	4 A.M.	10 A.M.	4 P.M.	10 P.M.	Max	Min.	Dir.
1	29·582	29·611	29·475	29·288	49·6	44·3	W	1	29·553	29·451	29·319	29·208	43·6	38·9	SE
2	29·291	29·519	29·659	29·690	49·7	37·8	NE	2	30·348	30·489	29·655		41·9	38·5	ENE
3	29·742	29·935	30·128	30·313	38·7	29·1	N	3	29·859	30·056	30·241	30·390	40·3	34·5	E
4	30·411	30·488	30·497	30·537	34·9	27·3	K	4	30·470	30·537	30·547	30·574	35·7	29·4	SE
5	30·489	30·449	30·382	30·352	33·0	25·9	ENE	5	30·540	30·525	30·466	30·419	34·5	29·8	ESE
6	30·322	30·316	30·325	30·325	33·0	25·7	E	6	30·380	30·386	30·347	30·360	35·1	26·5	SE
7	30·320	30·351	30·271	30·230	35·7	25·7	NE	7	30·363	30·391	30·371	30·370	29·6	24·2	ENE
8	30·184	30·148	30·067	30·057	35·7	33·2	NE	8	30·342	30·314	30·270	30·265	35·4	22·0	NW
9	30·021	30·037	30·030	30·081	37·4	34·0	NE	9	30·257	30·287	30·307	30·351	39·8	34·5	E
10	30·086	30·109	30·098	30·166	38·7	35·1	ENE	10	30·375	30·410	30·388	30·349	39·3	36·3	ESE
11	30·152	30·194	3·138	30·124	36·7	33·8	ENE	11	30·240	30·161	30·109		39·2	35·2	ENE
12	30·077	30·075	30·017	29·997	44·2	32·4	R	12	30·044	29·980	29·901	29·810	36·8	34·1	ENE
13	29·994	30·043	30·063	30·012	49·5	44·4	R	13	29·932	29·636	29·652	29·773	47·3	34·1	SSW
14	30·210	30·212	30·128	30·004	52·1	44·1	SSW	14	29·785	29·511	29·450	29·489	52·7	41·5	SSW
15	29·884	29·911	30·100	30·227	50·7	38·3	WSW	15	29·362	29·305	29·292	29·792	46·1	38·3	WNW
16	30·111	29·920	29·838	29·865	55·7	40·6	SW	16	29·301	29·352	29·476	29·504	51·7	40·1	W
17	29·873	29·920	30·063	29·900	51·7	47·3	W	17	29·492	29·728	29·787	29·669	45·2	39·3	W
18	29·822	29·938	30·062	30·129	55·1	44·0	W	18	29·394	29·571	29·711	29·827	45·2	39·4	W
19	30·153	30·301	30·345	30·320	48·1	44·0	WNW	19	30·007	30·147	30·119	30·081	48·1	40·9	WSW
20	30·230	30·218	30·234	30·284	52·6	45·2	NNW	20	30·039	30·113	30·166	30·169	49·4	46·3	W
21	30·282	30·334	30·347	30·401	51·7	43·8	NW	21	30·210	30·281	30·319	30·350	51·0	46·8	W
22	30·437	30·503	30·499	30·527	49·3	46·2	NNW	22	30·367	30·386	30·381	30·383	47·1	42·3	WSW
23	30·533	30·580	30·542	30·534	46·3	39·3	WSW	23	30·354	30·392	30·334	30·327	46·0	41·8	SW
24	30·523	30·499	30·449	30·441	40·7	35·8	WSW	24	30·158	30·034	30·018		49·1	45·4	WSW
25	30·324	30·246	30·093	30·053	50·9	37·2	W	25	29·918	29·748	29·534	29·792	50·2	44·4	WSW
26	30·094	30·161	30·139	30·143	51·5	46·1	W	26	29·846	29·840	29·732	29·786	51·8	39·9	W
27	30·144	30·177	30·090	29·932	51·2	47·4	W	27	29·821	29·772	29·562	29·284	51·0	47·3	WSW
28	29·732	29·691	29·592	29·371	49·3	41·7	W	28	29·186	29·556	28·955	28·838	48·1	32·5	WSW
29	29·201	29·103	29·008	28·932	44·0	35·4	SW	29	28·656	28·552	28·417	28·543	41·3	31·9	SW
30	28·911	28·967	28·945	29·096	40·5	34·8	SW	30	28·626	28·767	28·824	28·967	41·1	37·9	WNW
31	29·081	29·206	29·337	29·474	42·0	32·9	WNW	31	29·107	29·225	29·306	29·368	41·3	32·3	NW

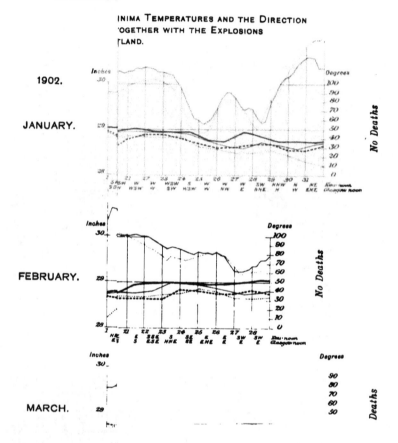

INIMA TEMPERATURES AND THE DIRECTION
'OGETHER WITH THE EXPLOSIONS
TLAND.

1902.

JANUARY.

FEBRUARY.

MARCH.

INDEX TO VOL. XXV.

A.

ABELL, W. PRICE, *alternating currents*, 172.

—, *application of coal-cutting machines to deep mining,* 167.

—, *sinking through heavily watered strata,* 146.

—, *sparkless electric plant,* 172.

Accidents, caused by high temperatures in mines, 273.

—, — — want of plans, Chinese mines, 139.

—, coal-mines, India, Giridih, 400.

—, — —, U.K. and U.S.A. compared, 546.

—, compressed-air underground locomotives, 544.

—, electricity, underground, 536, 541.

—, gas, Manchurian mines, 140.

—, hæmatite-mines, Cumberland, 293.

Accounts, S., 230.

Accumulator surface-condensers, 79.

Accumulators, electric, Max, 36.

—, —, winding-engines, 597, 598.

—, utilization of exhaust steam.—Discussion, 40, 407.

ACHESON, F., quoted, 487.

Acid-springs, U.S.A., 415.

ACKERMANN, A. S. E., pneumatic and electric locomotives in and about coal-mines, 529.—Discussion, 538.

Acorn-bank, Cumberland, gypsummine, 421.

ADAIR, JOSEPH, *hæmatite-mining in Cumberland,* 300.

ADAMSON, THOMAS, working a thick coal-seam in Bengal, India, 10.—Discussion, 13, 192, 396.

Africa, Portuguese west, copper-ores, 763.

Agglomerates, Lake district, 308.

AGRICOLA, GEORGIUS, quoted, 435.

AGUILLON, L., quoted, 52.

AINSWORTH, HERBERT, election, N.E., 1.

AINSWORTH, Messrs., quoted, 297.

Air, analyses, Cornwall, Dolcoath mine, 644.

—, —, mines, 657.

—, compressed, Cumberland, Threlkeld granite quarry, 343, 353.

—, —, objection to use for driving coal-cutting machines, 152.

—, —, pumping, 181, 182, 183.

—, —, storage cylinders, locomotives, 530.

—, liquid, deep mines, 275, 277, 285.

Air-blasts, fatal, caused by roof-falls, India, Bengal, 193.

Air-compression by water power, Belmont gold-mine, 206.

Air-compressors, British Columbia, Le Roi mine, 692.

— —, — —, War Eagle mine, electric, 691.

B.

D.

F.

G.

H.

L.

Looss, Prof. —, quoted, 651.
Loss in working coal, India, Bengal, 193.
Lostwithiel, Cornwall, gold, 439.
Louis, D. A., *gypsum of Eden valley,* 429, 430.
Louis, Henry, *granite-quarrying, sett-making, etc.,* 352.
—, *gypsum of Eden valley,* 430.
—, quoted, 176.
Louisiana, U.S.A., rock-salt, 781.
Low Moor jacket, ambulance, 360.
Low wood, Lake district, Ambleside, limestone, 309.
Lowag, Josef, gold-mines in Hohenberg and Oelberg hills, Austrian Silesia, 804.
—, iron-ore mines in Moravia, 804.
Lowry, J. D., Golovin Bay region of north-western Alaska, 778.
Lubrication, deficient, cause of electrical-plant failures, 566.

Lucas, Alfred, *alternating currents,* 170.
—, election, M., 90.
—, *sparkless electric plant,* 170, 173.
Luhrig tables, Greenside mines, 338.
Luny, Cornwall, St. Ewe, gold, 439.
Lupton, Arnold, *measuring-tape for mine-surveying,* 21.
—, *underlay-table,* 33.
Lupton, H., and A. H. Meysey-Thompson, some considerations affecting choice of pumping machinery.—Discussion, 175.
Luxemburg, mining legislation, 589.
— and its iron-ore deposits, 580.—Discussion, 589.
Luzon, Philippine islands, cupriferous sands, 790'
—, ——, magnetite-deposits, 789.
Lyburn, E. St. John, quoted, 497.
Lyell, Sir Charles, quoted, 415.
Lyra, Ireland, Wicklow, 491, 496, 498.

M.

McCreath, James, president, S., 226.
MacFarlane, Dr. —, quoted, 673.
MacGregor, Hugh Scott, election, S., 250.
McGregor, John Edward, election, N.E., 175.
Machine shops, equipment, Manchuria, Liao-yang coal-mine, 142.
— tools, National Physical Laboratory, 752.
Machinery, Cumberland, hæmatite-mines, 293.
Maclaren, J. Malcolm, occurrence of gold in Great Britain and Ireland, 435.
McLaren, Robert, vice-president, election, S., 231.
M'Phail, James, member of council, election, S., 231.
Macquet, Prof. A., *Guibal fan compared with dynamo,* 217.
—, quoted, 215, 216.
Macrone, Harry W., election, M.C., 165.
Maentwrog beds, Merionethshire, 448.
Magnesian limestone, Lake district, 314.
——, origin, 425.
Magnetic anomalies observed in France, 797.
— cut-out, electric coal-cutters, 172.
— disturbance observed at Athens, Greece, 797.
— pulsations, terrestrial, 796.
— sands, Sutherland, analyses, 477.
— unlocking safety-lamp, 285.
Magnetism, earth's, secular variation, 796.
Magnetite, Austria, Moravia, 804.
—, China, 823.

Magnetite-deposits, Philippine islands, Luzon, 789.
Maiden moor, Lake district, 303.
Maïssade, Hayti, brown-coal, 771.
Male-fern extract, ankylostomiasis, 654.
Mallada, L., iron-mines in central and southern Spain, 816.
Mallet, W., quoted, 494.
Managers, colliery, examinations, suggestions, 101.
—, —, premium for introduction of coal-cutting machines, 110.
Manchuria, mining in, 139.
Manganese-steel, jaws, stone-breakers, 348, 352.
——, offtake-socket for hauling-ropes, 2.
——, wheels, spoiled by red heat, 2.
Manganosphærite, Germany, 794.
Manica, Portuguese, gold-field, further remarks, 637.
Manross, Dr. —, quoted, 795.
Manto or capel, Chile, 768, 792.
Map and sheet memoir of north Staffordshire coal-field, 52.—Discussion, 60.
Markham, A. B., quoted, 541.
—, *Sherwood colliery,* 136.
—, *sinking through heavily watered strata,* 147.
Marles collieries, France, Pas-de-Calais, electric locomotives, 543.
Marr, John E., quoted, 303, 307, 328.
Marsaut, J. B., quoted, 70, 71.
Marsaut safety-lamp, 63.
Martin, C. W., quoted, 177.
Martin, J. S., quoted, 643.
Martin, Robert, sinking on the sea-shore at Musselburgh.—Discussion, 3, 95.

MARTIN, TOM PATTINSON, election, N.E., 174.

Maryport, Lake district, new red rocks, 313.

Maryville colliery, Newcastle, New South Wales, sinking shaft, 8.

MATHIAS, E., magnetic anomalies observed in France, 797.

MAURICE, WILLIAM, *alternating currents*, 170.

—, *electrical-plant failures*, 575.

—, *sparkless electric plant for use in mines and ironworks*, 170.

MAVOR, S., *electrical coal-cutting machines*, 235.

—, — *driving of winding gears*, 610.

—, *electrical-plant failures*, 574.

—, *pneumatic and electric locomotives*, 539.

Mawddach Afon, Merionethshire, gold, 444, 445.

MAX electric mining lamp, 36.—Discussion, 37, 219.

MAXWELL, HENRY V., Crown Mountain gold-mine and mill, Georgia, U.S.A., 835.

Mayola or Miola river, Ireland, Londonderry, gold, 484.

Maypole colliery, Wigan, sinking through heavily watered strata.—Discussion, 146.

MRACHEM, FREDERICK G., quoted, 728, 730.

—, underground temperatures, 267.—Discussion, 275, 283.

Measuring-tape, and its use in mine-surveying, 17.—Discussion, 21, 205.

Mellfell, Lake district, Skiddaw slates, 303.

Menevian beds, Merionethshire, 448.

MENZIES, JOHN, member of council, election, S., 231.

Mercury-ore, U.S.A., Texas, 838.

Merionethshire, gold-occurrences, 442.

—, slate mines, report of committee, quoted, 362.

MERRY, J. P., quoted, 442.

MERZ, CHARLES HESTERMAN, election, N.E., 291.

Metalliferous deposits, Peru, Cerro de Pasco, 777.

— ores of Dutch East Indies, 785.

Metallurgy, etc., notes of foreign and colonial papers, etc., 763.

Metamorphosis, Skiddaw slates, 306.

METCALFE, A. T., quoted, 425, 426.

Methane, properties, 740.

Mexico, iron-ores, 772.

—, Taxco de Alarcón, ore-deposits, 774.

MEYSEY-THOMPSON, A. H., AND H. LUPTON, some considerations affecting choice of pumping machinery.—Discussion, 175.

Mho, unit of electrical conductivity, 48.

Mica, Brazil, Rio Grande do Sul, with wolfram, 524.

Mice, indicators for carbon monoxide, 740.

Microbes, absent in coal, 367.

Microgranite, Lake district, St. John's, 316, 319.

MIDGLEY, JOHN, election, S., 95.

Midland railway, Eden valley gypsum-mines, 411.

Milan electrical-transmission line, 628.

MILLER, JOHN HENRY, election, N.E., 174.

MILLIGAN, PETER, member of council, election, S., 231.

MILLINGTON, W. W., *electrical coal-cutting machines*, 748.

—, quoted, 112.

Millom, Lake district, limestone, 309.

MILLOM AND ASKAM IRON COMPANY, Cumberland, hæmatite, 297.

MILLS, A., quoted, 485, 495.

MILLS, MANSFELDT HENRY, *application of coal-cutting machines to deep mining*, 152.

—, *mineral resources of Rio Grande do Sul*, 528.

—, *pneumatic and electric locomotives*, 540.

Millstone grit, north Staffordshire, 57.

MILLWARD, GEORGE ANTHONY, *condensing-plant for winding-engines*, 160.

MILNER, J. W., election, N.S., 220.

Milreis, Brazil, value, 510.

Mine-managers' examinations, suggestions, 101.

Mine royal, definition, 441.

Mine-surveying, measuring-tape and its use, 17.—Discussion, 21, 205.

Mineral colza oil, safety-lamps, 221.

Mineral-deposits, Lake district, 328.

Mineral resources, Brazil, Rio Grande do Sul, 510.—Discussion, 524.

— —, China, northern Chi-li, 819.

— —, —, south-western, 822.

— —, Ecuador, 833.

— —, Peru, Cajatambo, 775.

— —, Tongking, 822.

— —, Venezuela, 783.

Minerals, artificial reproduction by sublimation, 794.

—, Austria, Tyrol, 807.

—, Sutherland, accompanying gold, 477.

Miners, Spanish, mode of life, 294.

—, technical instruction, 101.

—, trained, scarcity, 108, 109.

Miners' anæmia, or ankylostomiasis, 643.—Discussion, 656.

— wounds, rapid healing, 367.

Mines, Austria, Moravia, iron-ore, 804.

P.

Q.

R.

ROBINSON, J. CLIFTON, *electrical-plant failures*, 576.
—, quoted, 759.
ROBINSON & COMPANY, JOHN, gypsum-mines, 411.
ROBINSON & COMPANY, JOSEPH, gypsum-mines, 412.
ROBSON, J. T., quoted, 226, 356.
ROCHE, EUSTACHIUS, quoted, 459.
Rock-drills, Cumberland, Threlkeld, granite-quarry, 342.
Rock-salt, U.S.A., Louisiana, 781.
Rodange, Luxemburg, iron-ore, 585.
ROGERS, DANIEL, JUN., election, S.S., 264.
Rolls, Cumberland, Threlkeld granite quarry, 347.
—, Westmorland, Greenside mines, 331.
Roman gold-workings, Wales, 436.
— slags, Spain, low copper-content, 204.
RONALDSON, J. H., *improved offtake socket for coupling hauling ropes*, 218.
RONALDSON, JOHN M., president, S., 226.
Roof, bad, preventing coal-cutting by machines, 152.
—, heavy fall, Serampore colliery, India, 12.

Rootlets, above coal-seams, north Staffordshire, 59.
Rope-pulley, Canada, Belmont, 209.
Ropes, hauling, improved offtake socket.—Discussion, 218.
Rosario mine, Chile, Tamaya, 831.
Rosebridge colliery, underground temperature, 276.
Ross, J. A. G., *granite-quarrying, sett-making, etc.*, 352.
Ross, MICHAEL, treasurer, S., 226.
Ross slate-quarries, Ireland, Waterford, 675.
Rossland, British Columbia, mines, 690.
ROUBIER, E., Max electric mining lamp, 36.—Discussion, 37, 219.
ROUTLEDGE, ALFRED JAMES, election, M.C., 165.
Royalty, gold, Wales, 443, 447.
Rubellite, U.S.A., California, 794.
Ruhr coal-field, Germany, ankylostomiasis, 665.
Rumelange, Luxemburg, iron-ore, 585.
RUSSELL, ARCHIBALD M'KERROW, election, S., 250.
RUSSELL, THOMAS, election, S., 250.
Russia, Donetz, coal-field, 813.
—, petroleum output, 803.
—, Urals, platinum-mining, 814.
RUTLEY, FRANK, quoted, 340.

S.

Safety-appliances, underground fires, 741.
Safety-lamps, failure, causes, 163.
— —, Gray type, 62.—Discussion, 72, 162, 220.
— —, locks, 222.
— —, magnetic unlocking and electrically ignited, 265.
St. Austell moor, Cornwall, gold, analyses, 440.
— Bees head, Cumberland, gypsum-mines, 412.
— — —, —, sandstones, 313.
ST. DAVID'S GOLD AND COPPER MINES, LIMITED, 447, 451.
St. David's lode, Merionethshire, Clogau, gold, 444, 447.
— Etienne collieries, France, ankylostomiasis, 646.
— Gothard tunnel, ankylostomiasis, 645, 665.
ST. JOHN AMBULANCE ASSOCIATION, 354.
ST. JOHN AMBULANCE BRIGADE, 359.
St. John's microgranite, Lake district, 316, 319.
ST. RAINER, L., platinum-mining in the Urals, 814.
SAISE, DR. W., quoted, 15, 397.
SALAZAR LEOPOLDO, ore-deposits of Taxco de Alarcón, Mexico, 774.
Sale Fell minette, Lake district, 317.
Salt, rock, U.S.A., Louisiana, 781.

Salt-pans, natural, 427.
Salt-water, action on iron and steel, 4.
Salt-working, China, He-tsin, 825.
Sampling, British Columbia, 681, 684, 687.
San Gabriel, Brazil, Rio Grande do Sul, 512.
Sand, sinking shaft, removal by endless bucket pump, 9.
—, used on underground rails, 533.
Sandstones, China, Shansi, 821.
—, Sussex, Heathfield, 720.
Sanitary precautions, mines, ankylostomiasis, 655, 656, 661.
Santa Maria, Brazil, Rio Grande do Sul, 512.
São Jeronymo coal-mine, Brazil, Rio Grande do Sul, 513, 521, 527.
— Sepé, Brazil, gold, 515.
Sapper stone or quartz, 461.
Satinspar, Eden valley gypsum-deposits, 423, 426.
Sawing granite setts, Threlkeld, 346.
SAWYER, A. R., further remarks on Portuguese Manica gold-field, 637.
—, quoted, 74, 162.
SAXTON, ISAAC H., *miners' anæmia*, 659.
Scawfell banded ashes and breccias, Lake district, 307, 308.
Schemnitz, Hungary, ankylostomiasis, 645.

T.

W.

CPSIA information can be obtained
at www.ICGtesting.com
Printed in the USA
BVHW08*1005031018
529155BV00010B/74/P